Climate Change Management

For further volumes:
http://www.springer.com/series/8740

Walter Leal Filho
Editor

Climate Change and the Sustainable Use of Water Resources

 Springer

Prof. Walter Leal Filho
Hochschule für Angewandte Wissenschaften
  Hamburg Forschungs- und Transferzentrum
Applications of Life Sciences
Lohbruegger Kirchstr. 65
21033 Hamburg
Germany
e-mail: walter.leal@haw-hamburg.de

ISSN 1610-2010   e-ISSN 1610-2002
ISBN 978-3-642-22265-8   e-ISBN 978-3-642-22266-5
DOI 10.1007/978-3-642-22266-5
Springer Heidelberg Dordrecht London New York

© Springer-Verlag Berlin Heidelberg 2012
This work is subject to copyright. All rights are reserved, whether the whole or part of the material is concerned, specifically the rights of translation, reprinting, reuse of illustrations, recitation, broadcasting, reproduction on microfilm or in any other way, and storage in data banks. Duplication of this publication or parts thereof is permitted only under the provisions of the German Copyright Law of September 9, 1965, in its current version, and permission for use must always be obtained from Springer. Violations are liable to prosecution under the German Copyright Law.
The use of general descriptive names, registered names, trademarks, etc. in this publication does not imply, even in the absence of a specific statement, that such names are exempt from the relevant protective laws and regulations and therefore free for general use.

Cover design: eStudio Calamar, Berlin/Figueres

Printed on acid-free paper

Springer is part of Springer Science+Business Media (www.springer.com)

Preface

Climate change is and will continue to be one of the central issues in the world's agenda. The 16th Conference of Parties (COP-15) of the UN Framework Convention on Climate Change held in Cancun, Mexico, in December 2010 has reiterated how much still needs to be done in order to tackle the various challenges climate change and its various ramifications pose to mankind.

This book, prepared as a follow-up to the third online climate conference CLIMATE 2010/KLIMA 2010, held on 1–7 November 2010, focuses on *Climate Change and the Sustainable Management of Water Resources*. There are two key arguments for the choice of this particular topic:

- First, it is widely believed that climate change has a serious impact on global water supplies and may worsen water scarcity–a problem which threatens a large part of the world already today. Under present conditions, approximately 1.2 billion people—especially in developing countries—have no access to drinking water. In order to address this problem, the United Nations has set the goal of increasing access to a further 600 million people by 2015, i.e. better access to drinking water for around 100 million people per year between 2010 and 2015.
- Second, there is a pressing need to use the presently available water resources, which are very scarce in some areas, more sustainably. Even though in parts of Africa, Latin America and the Middle East water resources are already scarce, the proportion of water wasted partly to leaks but also due to the lack of adequate systems to retain, recycle and reuse water is considerably high. Urgent action is needed to address this issue to keep up the UN targets.

The thematic focus of this book, which is also prepared in the context of the Interreg IVB (North Sea) project North Sea Skills Integration and New Technologies (SKINT), will allow in-depth discussions and support the search for global and regional solutions for the impacts climate change has on water supplies and will address the need to promote sustainable water use across the world.

Part I contains a set of papers on geochemical and physical impacts of climate change on water supplies, as well as on aspects of modelling, forecasting and software applications. Part II includes papers on the socioeconomic aspects of climate change in relation to water supplies and use, whereas Part III presents papers on the links between climate change, policy-making and sustainable water use. Part IV presents a number of projects and initiatives, which focus on addressing the links between climate change and sustainable water use, including educational and awareness-raising initiatives.

I want to thank all authors for sharing their knowledge and their experiences, as well as Mrs Marika Rudzite-Grike for the editorial support provided. Thanks are also due to the ICCIP Team (Franziska Mannke, Natalie Fischer, Kathrin Rath and Johanna Vogt) and Olaf Gramkow for contributing to Climate 2010 and to this book project.

It is hoped that this book, which is Volume 3 of the "Climate Change Management Series", initiated as part of the "International Climate Change Information Programme" (ICCIP), will be useful and allow a better understanding of the problems, barriers, challenges, opportunities and possibilities related to the promotion of the sustainable use of water resources worldwide.

Summer 2011   Walter Leal Filho

Contents

Part I

1 Climate Change Impacts on Green Water Fluxes in the Eastern Mediterranean 3
Ibrahim M. Oroud

2 Stormwater Reuse via Aquifer Storage and Recovery: Risk Assessment for Sandy Aquifers 17
Anke Steinel

3 Hydrologic Balance of Citarum Watershed under Current and Future Climate 43
Rizaldi Boer, Bambang Dwi Dasanto, Perdinan and Delon Marthinus

4 Towards a Database for an Information Management System on Climate Change: An Online Resource 61
V. Ramani Bai, S. Mohan and Reza Kabiri

5 Assessing the Impact of Climate Change on Evapotranspiration and Soil Salinization 69
Mohammad Hassan Rahimian and Samaneh Poormohammadi

Part II

6 The South African Water Sector: On its Way Towards Adaptive Water Governance? 79
Elke Herrfahrdt-Pähle

7	**Increase in Port Downtime and Damage in Vietnam Due To a Potential Increase in Tropical Cyclone Intensity** Miguel Esteban, Nguyen Danh Thao, Hiroshi Takagi and Tomoya Shibayama	101
8	**Hydro-Economic Analysis for Water Resources Management in a Changing Climate** David Corderi Novoa	127
9	**Climate Change and Water Resource Availability: What to Do?** Eugenia Aloj, Mariagrazia De Castro, Michela Totàro and Anna Zollo	143
10	**Adaptations to Climate Change and Variability Among Smallholder Farmers in Tanzania** Stephen Justice Nindi and David Gongwe Mhando	153
11	**"Greening" Integrated Water Resources Management Policies for Tackling Climate Change Impacts: A Call for Sustainable Development** Nazmul Huq and Jean Hugé	173
12	**Urbanization and Flood Vulnerability in a Peri-Urban Neighbourhood of Dakar, Senegal: How can Participatory GIS Contribute to Flood Management?** Audrey Maheu	185
13	**Vulnerability of Andean Communities to Climate Variability and Climate Change** Harry Diaz, Rosa Garay-Fluhmann, Julia McDowell, Elma Montaña, Bernardo Reyes and Sonia Salas	209
14	**Water Management Issues in Southern Tunisia Under a Climate Change Context** Nizar Omrani and Dieter Burger	225
15	**Climate Change and its Impacts on the Livelihoods of the Vulnerable People in the Southwestern Coastal Zone in Bangladesh** Md. Afjal Hossain, Md. Imran Reza, Sania Rahman and Imrul Kayes	237

16	Conservation and Management of Water Resources for Livelihood Sustainability in Mountainous Regions of Uttarakhand, India....................... Anil Kumar	261
17	Ports and Climate Change: Building Skills in Climate Change Adaptation, Australia.......................... Melissa Nursey-Bray and Tony Miller	273
18	Preliminary Report on the Socioecology of Rural Groundwater Markets in the Gingee Watershed........................ Susan Varughese and K. V. Devi Prasad	283
19	Impact of Climate Change on Water Supply and Food Security in Seraro District, Southern Ethiopia .... Fikadu Reta Alemayehu	297
20	Climate Change and Rural Water Supply Planning in Nigeria... Salisu Lawal Halliru and Da'u Abba Umar	305
21	Impact of Climate Change in Bangladesh: Water Logging at South-West Coast............................... Md. Moniruzzaman	317

Part III

22	Characterizing Adaptive Capacity in Water Governance Arrangements in the Context of Extreme Events............. Margot Hill	339
23	The Contribution of Rural Development Programmes in Mitigating Greenhouse Gas Emissions in Italy............. Rocío Dánica Cóndor, Marina Vitullo, Domenico Gaudioso and Marina Colaiezzi	367
24	"Climate Proofing" Water Resources Development Policy: The Evidence from Bangladesh......................... Nazmul Huq, Zakir Hossain, Rashedul Hasan and Al Mamun Azad	389
25	Developing an Environmental Sustainability Toolkit to Integrate Climate Change Issues in Development Cooperation... Tom Waas and Jean Hugé	401

26 **Spatio-Temporal Variation of Drought Severity in the Sudano-Sahelian Region of Nigeria: Implications for Policies on Water Management**........................... 415
Maruf Sanni, Theophilus Odeyemi Odekunle and Francis Adeyinka Adesina

27 **Climate Change and Sustainable Management of Water Resources** 431
Golda A. Edwin and G. Poyyamoli

28 **Problems of Water Resources Management in the Drainage Basin of Lake Balkhash with Respect to Political Development**............................. 449
Pavel Propastin

29 **Saline Irrigation Management for Sustainable Use** 463
Medhat Mekhail Tawfik, M. A. Ahmed, Amany A. Bahr, M. F. El Karamany and M. S. Zeidan

30 **Modelling of Water Cycle Processes** 473
Viktor Batyreu and Siarhei Zenchanka

31 **Bridging the Communication Gap: An Exploration of the Climate Science–Water Management Interface** 485
Katinka Lund Waagsaether and Gina Ziervogel

32 **Influences of Public Ecological Awareness and Price on Potable Water Consumption in Geneva** 499
Giuseppe Catenazzo, Jennifer D'Urso, Emmanuel Fragnière and Jean Tuberosa

33 **Sustainable Development Through Pathways of Mitigation and Adaptation to Offset Adverse Climate Change Impacts**..... 515
A. N. Sarkar

Part IV

34 **Activity-Based Water Resources and Climate Change Education Among School Students in Puducherry**............ 557
R. Alexandar and G. Poyyamoli

Contents

35   Water Resources Management in the Peruvian Andes: Participatory Adaptive Measures to Climate Change.......... 579
Rebecca Clements and Juan Torres

36   Community-Level Environmental and Climate Change Adaptation Initiatives in Nawalparasi, Nepal............... 591
Eloise M. Biggs, Gary R. Watmough and Craig W. Hutton

37   Multifunctional Land Use in Urban Spaces to Adapt Urban Infrastructure......................... 611
Marko Siekmann, Nina Vomberg, Michael Mirgartz, Johannes Pinnekamp and Sarah Mühle

38   Integrated Industrial Water Management as a Tool for Mitigating Climate Change...................... 627
Rasmi Patnaik and Gopalsamy Poyyamoli

39   Unjust Water: An Investigation of the Drinking Water Situation in Argentina 641
Sonja Maria Geiger and Natalia Salvático

40   Socio-Environmental Evaluation of Drip Irrigation System Implementation as a Climate Change Adaptation Measure Within the N'hambita Community Carbon Project Area, Mozambique......................... 663
Jelena Barbir and Walter Leal

41   Social Learning Through Local Water Governance Institutions 685
Margot Hurlbert

42   These are Our Water Pipes—Sand Dams, Women and Donkeys: Dealing with Water Scarcity in Kenya's Arid and Semi-Arid Lands............................ 701
Abigail Cruickshank and Velma I. Grover

43   Urban Rainwater Management Using the Aquifer for Storage as One Measure to Confront Changed Water Availability............................ 727
Reinhard Marth and Leonel González González

44	Educating for Sustainable Consumption as a Response to the Global Water Crisis: An Investigation of an Embedded Learning Approach. . . . . Daniel Fischer and Eva Freund	743
45	Achieving Sustainable Wastewater and Organic Solid Waste Management Taking Advantage of the Clean Development Mechanism . . . . Sören Rüd and Reinhard Marth	761
46	Local Wisdom of Danowudu Community in Preserving Forest as a Water Source for the City of Bitung . . . . Trina E. Tallei and Saroyo Sumarto	773
47	Education in Global Climate Change at a Botanical Garden: Students' Perceptions and Inquiry-Based Learning . . . . Daniela Sellmann and Franz X. Bogner	779
48	The Climate Change Challenge for the Urban Environment: The Use of an Integrated Management System . . . . Esther Kreutz and Kirsi-Marja Lonkila	787

About the Authors. 801

Part I

Chapter 1
Climate Change Impacts on Green Water Fluxes in the Eastern Mediterranean

Ibrahim M. Oroud

Abstract The present paper is part of the Glowa Jordan River project, which has been focusing on climate change impacts on environmental, economic and social issues within the lower Jordan river riparian states. The eastern Mediterranean is characterized by scarce and erratic precipitation with relatively cool, wet winters and dry hot summers. Water is the biggest growth-limiting factor. The present paper discusses the use of climate gradient as a tool to examine the impact of climate change on precipitation partitioning over field crops. The present experiment is carried out using a multi-layer, multi-year model with a daily time step. Six years of daily data for five locations, with average annual precipitation ranging from 170 to 580 mm, were used in this investigation. Results show that the ratio of soil evaporation (BE) to annual precipitation (P) during the growing season depends strongly on precipitation regime and amount, ranging from $\sim$15 to 20% when P > 600 mm to $\sim$60% when annual P < 200 mm. A decrease of 10% in precipitation along with a temperature rise of 2°C increases bare surface evaporation, on average, by $\sim$10% compared to average current conditions. The implications of this would be a tangible reduction in blue and green water fluxes, leading to compulsory land use shift and further water stress in the region.

Keywords Climate change · Mediterranean · Soil moisture partitioning · Rain-fed field crops

I. M. Oroud (✉)
Department of Geography, Mu'tah University,
Karak, 61710, Jordan
e-mail: ioroud@mutah.edu.jo

Introduction

Rain-fed field crops are widely grown in western Asia and North Africa. Being located in a transitional zone, this area experiences limited amounts of precipitation with substantial interannual and within season variability. For instance, a 30-year record of precipitation in a station located in this area shows that the average, lowest and highest annual precipitation there were 340, 123 and 639 mm, respectively, with a coefficient of variation of precipitation exceeding 30% in many locations. Precipitation in the eastern Mediterranean exhibits strong spatial gradients both latitudinally and across elevation contour lines. This is clearly depicted in the geographic distribution of water resources, floral composition and dominant agricultural practices.

General circulation models and long-term regional meteorological observations suggest strongly that the eastern Mediterranean will experience a warming trend along with a reduction in annual precipitation during the twenty-first century. This climate change is expected to adversely affect soil moisture availability at different depths, with subsequent impacts on the evapotranspiration regime and biomass production. One of the objectives of this paper is to demonstrate the impact of climate gradient on soil moisture partitioning when planted with a field crop, wheat for instance. The use of climate gradient, or space-for-time approach, provides an assessment of what to expect following a change in climate.

The objective of the present investigation is to examine how soil moisture regime within the active root zone of a wheat crop is influenced by the amount of precipitation and its temporal distribution, and how it is partitioned via direct evaporation and transpiration along a climate gradient. A multi-year, multi-layer simulation model was used. Six years of daily meteorological data for five stations representing the high precipitation zone in Jordan were used in this investigation.

Study Area and Data Quality

The study area represents a semi-dry Mediterranean climate regime with "average" Koppen climate classification of Csa and Csb. The study area is located in the mountainous areas of Jordan, with average annual precipitation ranging from 170 mm to about 550 mm (Fig. 1).

Precipitation falls in the cold season, October/November, and ceases around the end of March/early April. Figure 2 shows the annual course of precipitation in two locations. Annual potential evaporation (PE) in the study area is around 1,000 mm, with the index of aridity ranging from 1.5 in a small mountainous enclave to about 4 in the drier mountainous regions. Six years of continuous daily meteorological data (precipitation, maximum and minimum air temperatures, sunshine hours, cloud cover, wind speed, and ambient vapour pressure) covering the period 1996/1997–2001/2002 were obtained from the Department of Meteorology, Jordan.

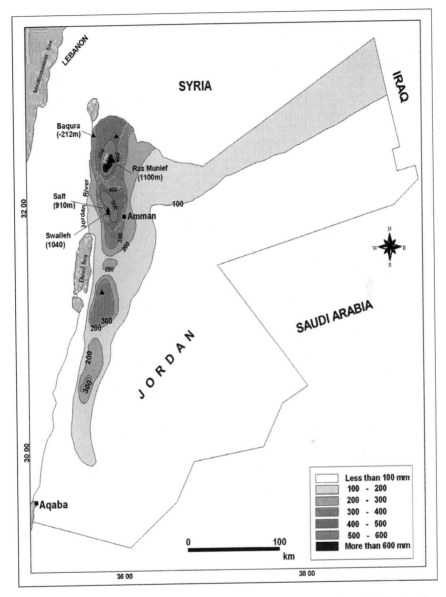

Fig. 1 Location of stations along with elevation (m) and average annual precipitation (mm)

This period covers several growing seasons with wet, average and dry years for the five stations used in the investigation. The selected stations provide good-quality meteorological data with elements being observed on an hourly or three-hourly basis. The data set was quite continuous with few gaps in certain elements which were estimated from neighbouring stations using linear regression, and were checked for consistency.

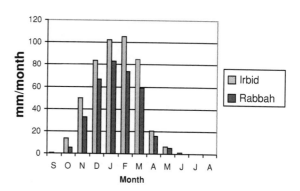

Fig. 2 Monthly distribution of precipitation in two locations, Irbid in northern Jordan and Rabbah in southern Jordan

Method of Investigation

The present investigation is carried out using a cascading water balance model. The water balance of a soil column may be expressed in the following form (e.g. Gleick 1987):

$$\frac{\delta S}{\delta t} = P - BE - A_T - R_O - D_p \qquad (1)$$

The first term represents soil moisture change with time; P, BE, A_T, R_O, and D_p are precipitation, bare surface evaporation, actual transpiration, surface runoff, and deep percolation, respectively. In this formulation the soil profile is divided into four equal layers, 0.25 m each, and thus it is assumed that maximum root extension is 1 m. Soil evaporation is determined by atmospheric demand, soil water content, and soil hydraulic properties (e.g. Ritchie 1972).

Atmospheric demand, or PE, is a thermal index which represents the amount of available energy, radiative and advective, that can be used to convert water from its liquid phase into vapour phase. A widely used expression to calculate PE is the Penman–Monteith expression, in which radiative and advective terms were combined to calculate PE (e.g. Dingman 2002).

Actual soil evaporation is either energy-limited or moisture-limited. Most of bare soil evaporation (BE) takes place from layers close to the surface-atmosphere interface. In this formulation, evaporation from the top layer is calculated using the concept of readily available water such that soil evaporation proceeds at its potential rate when skin layer moisture (ω) exceeds atmospheric demands:

$$BE = \tau P_E, \quad \tau P_E \leq \omega, \quad 0 \leq \omega \leq 4 \text{ mm} \qquad (2)$$

When skin moisture content does not meet evaporative demands, then direct evaporation is proportional to moisture content of the upper layer,

$$BE = \tau\omega + \tau(P_E - \omega)\left(\frac{\theta_i - \theta_h}{\theta_f - \theta_h}\right)^{1.8}, \quad \tau PE > \omega \qquad (3)$$

where BE, P_E, ω, θ_i, θ_h, θ_f, and τ are bare surface evaporation, PE, readily available water for evaporation, actual soil water content, residual water, field capacity, and solar radiation fraction reaching the soil surface, respectively.

The readily available soil water is assumed to be 4 mm. Following rainy events, bare surface evaporation proceeds at its potential rate until ω is less than the potential rate, with Eq. 3 being used instead.

Soil evaporation proceeds only from the top layer when its moisture content is above a threshold value, but when it drops below that, then evaporation proceeds from the deeper layers as well, but with a much reduced rate.

Evaporation/transpiration fraction distribution is assumed to have the following exponential form as suggested by Novak (1987) (see also Zhang et al. 2004),

$$k_\lambda = \frac{D \exp(-D(z/z_D))}{z_D(1 - \exp(-D))} \qquad (4)$$

where D is the water use distribution parameter, and Z_D is the maximum depth that responds to evaporation processes at the surface-atmosphere boundary. The water use distribution parameter for evaporation depends primarily on soil hydraulic properties, and its value may be in the range 6–10 (e.g. Zhang et al. 2004), taken in this paper as 8. When the top layer becomes desiccated, as is the case during the long summer days, evaporation proceeds from the deeper layers.

Surface albedo is parameterized as follows,

$$\alpha = 0.3 - (0.1\tau + 0.04(1 - \tau)), \quad 1 < \omega < 4\,\text{mm} \qquad (5)$$

$$\alpha = 0.3 - (0.1(\theta/\theta_f)^\beta \tau + 0.04(1 - \tau)), \quad \omega < 1\,\text{mm} \qquad (6)$$

where ω is skin layer soil moisture wetness factor (taken in this case to be 4 mm), θ and θ_f are the upper surface layer soil moisture and its field capacity, respectively, τ is the fraction of global radiation reaching the ground surface, calculated by (e.g. Oweis et al. 2000; Zhang et al. 2004),

$$\tau = \exp(-K \cdot LAI) \qquad (7)$$

where K is solar radiation extinction coefficient, taken in this paper as 0.6 (e.g. Zhang et al. 2004), and LAI is leaf area index. Because of rapid drying of the skin soil layer which determines surface albedo, the exponent β appearing in Eq. 6 is taken as 2.2. Equations 5 and 6 reflect adequately alterations to surface albedo due to upper soil wetting following rainy events and also due to development of foliage coverage.

Soil Moisture and Runoff

Runoff is a function of several parameters (e.g. soil texture and depth, topography, vegetation cover, antecedent soil moisture, precipitation intensity). Because of difficulty in obtaining precipitation intensity, runoff is assumed to be a function of total daily precipitation (p), soil moisture content of the upper layer (θ_1), and leaf area index (LAI),

$$R_O = f(P, \theta_1, LAI) \tag{8}$$

Runoff may be calculated using the SCS curve. In this formulation, precipitation threshold causing runoff increases as the upper soil layer moisture decreases and as LAI increases. Under such conditions, runoff may occur without the need for the entire soil profile being at its field capacity. The procedure adopted in this paper is similar, in principle, to that presented by Rushton et al. (2006).

Transpiration

The specification of transpiration is crucial for evaluating soil moisture dynamics, leaf area development, and dry matter accumulation, and thus its accurate determination is probably the most significant parameter in this type of study. Transpiration occurs as a result of water potential difference between the soil and plant roots; it is a function of climatic (mainly PE and physiologically active radiation), physiological (LAI, root development and distribution, plant health) and edaphic (soil texture, structure, depth and hydraulic properties) factors (e.g. Brisson 1998; Novak and Havrila 2006). For a well-managed soil, moisture extraction is determined by vertical root extension,

$$A_T = \int_z r(z,t) dz \tag{9}$$

where A_T is actual transpiration and r(z, t) is vertical root distribution. The plant moisture extraction parameter depends on vertical root distribution within the soil layer, and it ranges from 0.5 to about 5 (e.g. Novak 1987; Zhang et al. 2004). The D parameter, presented in Eq. 4, is assigned a value of 1.45 which gives water uptake fractions of 0.41, 0.27, 0.19, and 0.13 for the 0–0.25 m, 0.25–0.5 m, 0.75–0.75 m, and 0.75–1.0 m, respectively. These values are congruent with the widely reported fractions of 0.4, 0.3, 0.2 and 0.1 from the first, second, third and fourth soil layers, respectively (e.g. Gardner 1983). Moisture extraction from each soil layer depends on root extension in each layer and water availability.

Transpiration from a given soil layer is a dynamic parameter which is determined for a well-developed root system by critical leaf water potential and atmospheric forcings (e.g. Nishat et al. 2007). It is assumed that transpiration proceeds at its potential rate when the available moisture in a soil layer exceeds a given threshold level (θ_*), and then decreases linearly until reaching the wilting point (e.g. Hanks 1983; Dingman 2002; Kang et al. 2003),

$$A_T = P_E(1-\tau)\sum_i^n K_s \cdot y \cdot \gamma_i, \quad \theta_i > \theta_* \tag{10}$$

Fig. 3 Simulated transpiration coefficient of winter wheat as a function of plant stage (after Allen et al. 1998)

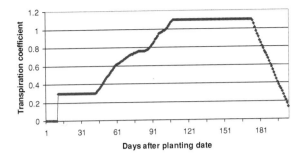

$$A_T = P_E(1-\tau)\sum_{i}^{n}\frac{\theta_i - \theta_{wp}}{\theta_* - \theta_{wp}} K_s \cdot y \cdot \gamma_i, \quad \theta_i < \theta_* \qquad (11)$$

where K_s is the plant transpiration coefficient, which depends on plant stage (Allen et al. 1998), y is the fraction of root extension in the ith soil layer, γ_i is the ith soil layer moisture contribution fraction, and θ_* is the critical soil moisture above which transpiration proceeds at its potential rate. This critical level is influenced primarily by meteorological factors, namely PE; it ranges from values close to 0.8 when atmospheric evaporation demands are high, to values close to the wilting point when evaporative demands are small. The transpiration coefficient was formulated after Allen et al. (1998) for winter wheat crop grown in a Mediterranean environment which ranges from less than 0.2 in late growth stages to 1.1 during the mid growth stage (Fig. 3).

Root vertical development of a cereal crop depends on edaphic elements (e.g. moisture, compaction), weather elements (e.g. PE, growing degree days) and growth stage (e.g. Izzi et al. 2008). Izzi et al. (2008) found that average vertical root growth of wheat in a typical Mediterranean soil in northern Syria is ~ 7 mm/day. In this paper, root growth from seeding to maturity is assumed to be a function of growing degree days. Root vertical growth within a given soil layer, however, depends also on soil water potential. It was indicated that root growth is severely restricted when soil water content is less than ~ -1.0 MPa (e.g. Taylor 1983). In this paper, root vertical extension within the second to fourth layers was not allowed when water content of a layer is less than ~ -1.0 MPa. This occurs during drought-stricken years when precipitation is not sufficient to penetrate to the deeper desiccated soil layers, thereby impeding root development. The germination period is assumed to be 10 days, during which roots are allowed to develop but no leaf development was allowed.

LAI was assumed to be a function of biomass accumulation and senescence. Sowing date was assumed to occur after 25 November, provided that accumulative antecedent precipitation for the period starting from 1 October was at least 40 mm. Farmers tend to plant their land around this time following sufficient amounts of precipitation to ensure germination and subsequent growth for a few weeks to come.

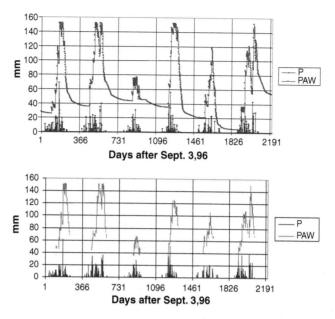

Fig. 4 Plant available water (mm) during the six-year simulation period under current conditions and assuming a 10% precipitation reduction along with a temperature increase of 2°C

Results

Soil Moisture Dynamics

Soil moisture reflects the overall budget of recharge and depletion via evaporation and transpiration due to root-soil moisture differential. Figure 4 shows the annual patterns of precipitation and plant available water (PAW) for the soil profile under current conditions and following a climate change in northern Jordan with average annual precipitation during the study period of 400 mm. There is an annual pattern of soil moisture availability, reaching its maximum during winter and declining during the growing season via both evaporation and root water uptake, and via evaporation during the long dry season. During years with low rainfall, deeper layers are not being replenished, with moisture there being further depleted by evaporation during the following long summer. Long- and short-term drought events are easily detected from moisture curves; poor years are also identified (2000/2001), as the presented moisture curves were narrow and never reached field capacity.

A climate change scenario assuming 2°C temperature increase along with a 10% reduction in precipitation shows that PAW is smaller than under current conditions in all seasons. Although annual patterns are somewhat similar, PAW under climate change scenarios is always less, reflecting the loss of water via direct

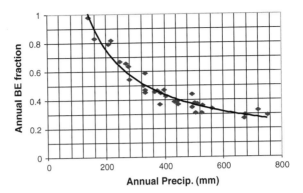

Fig. 5 Annual fraction of direct evaporation from soil for the five stations

evaporation because of the increased evaporative power of the contiguous atmosphere. With an increase of 2°C, annual PE increases by ~12%. The corresponding value for direct surface evaporation (actual) following a climate change is an increase of ~10%. This means that an increase of direct evaporation will come at the expense of blue and green water fluxes. This decrease, although relatively small, is quite significant in this marginal environment, given the very small amount of precipitation in this fragile ecosystem.

Direct Evaporation

Current seasonal direct evaporation ranges, on average, from ~80 mm for drier locations/years to ~120 mm for larger P values. The late response of vegetative growth in the mountainous areas, along with frequent surface wetting, intensify direct soil moisture depletion via direct evaporation early in the season.

To demonstrate the impact of climate change on soil moisture dynamics, we used data along a climate gradient. This data set is characterized by substantial interannual and spatial variations. Seasonal bare surface evaporation fraction (BE/P) ranges from ~60% in drier areas/years with annual precipitation <200 mm to ~15–20% when it exceeds ~600 mm. Corbeels et al. (1998) found in a similar environment that direct soil evaporation during the growing season was influenced strongly by precipitation amount, ranging from ~35% in a wet year (P ~400 mm) to ~80% during a dry year (P ~ 260 mm).

Annual soil evaporation fraction in drier realms/years accounts for >90% of total precipitation, whereas in areas/years with more precipitation this fraction is ~25–30% (P > 500 mm). Figure 5 shows the linkage between direct evaporation and precipitation amount for the stations used during the study period.

In drier Mediterranean regions/years, intermittent and scanty precipitation wets the upper soil layer, making it subject to continuous depletion by direct evaporation resulting from large insolation and dry winds following rainy events. Figure 5 demonstrates the steep response of BE to precipitation reduction,

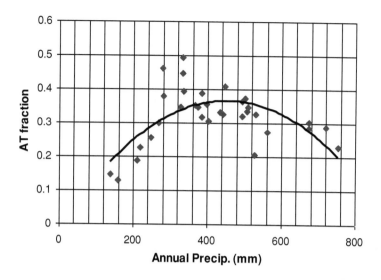

Fig. 6 Transpiration fraction as influenced by seasonal precipitation

particularly at low P values. For instance, a reduction of P from 400 to 360 mm leads to an increase in direct evaporation fraction by $\sim 5\%$, whereas a reduction in annual P from 240 to 200 leads to an increase of BE by $\sim 12\%$.

Transpiration

Transpiration is the most limiting factor in biomass and grain production in dryland farming (e.g. Gardner 1983; Tanner and Sinclair 1983). Wheat is usually sown following sufficient amounts of precipitation. This may occur from mid-November till the end of January. Surface frost occurs frequently in the mountainous areas during the period November to March (Oroud 2007) which, along with low temperatures, slows down vegetative growth. On average, more than 75% of precipitation falls before the end of February, whereas growth begins near the end of February. Figure 6 shows the ratio of wheat transpiration to total crop water demand (potential transpiration) (A_T/P_T) for the five stations during the simulation period. Transpiration fraction reaches its maximum when annual P is between 300 and 500 mm, and then decreases at both ends of precipitation. When P is small, soil moisture is lost via direct evaporation, and at large P values, it is lost as blue water.

Discussion

The concept of climate gradient as represented by interannual precipitation variability for a total of 30 years provides a measure of soil moisture dynamics and how precipitation is partitioned via various avenues. The present method provides realistic representations of precipitation partitioning via various avenues. Surface evaporation proceeds at its potential following rainy events until the wetness fraction drops below a critical level and then it continues at a reduced rate. Unlike other methods which deal with surface albedo as a constant, the present paper treats this element dynamically such that it changes following wetting/drying episodes and also in response to foliage development. This treatment is important given the fact that a change in surface albedo influences water loss appreciably following rainy events. Following rainy events, net radiation increases as a result of surface temperatures reduction and suppressed albedo due to strong spectral absorption ($\lambda > 0.75$ μm). The combined effects of increased global radiation along with a reduced albedo due to surface wetting enhance direct evaporation from soils. A 5% difference in surface albedo leads to ~12–15% difference in net radiation, and correspondingly a similar evapotranspiration difference. This is important following rainy events, as actual evapotranspiration is energy-limited following these events.

The meagre amount of precipitation, which usually wets the top soil layer, along with substantial evaporative demands, enhances direct evaporation in this environment. Results show that bare surface evaporation represents a significant fraction of precipitation for areas receiving less than 300 mm of annual precipitation, with a total annual fraction exceeding 90% in these areas. These results are congruent with those presented by Zhang et al. (1998), Oweis et al. (2000) in northern Syria, Corbeels et al. (1998) in Morocco, and Monzon et al. (2006) in southwestern Australia where Mediterranean-like conditions prevail. With a climate change, drought frequency and severity will be more intense in the near future compared to those experienced during the twentieth century. Under such conditions, traditional areas which sustained dry agriculture practices during the Holocene will be less suitable for rain-fed agriculture, with more frequent crop failure. Although people and decision-makers will continue denying these environmental changes, more intense drought episodes and frequent crop failures will force these new painful realities. Consequently, a shift in land use from agricultural towards grazing is an inevitable outcome of a climate change in this fragile habitat.

Conclusion

A simulation model with a daily time step is developed to examine the partitioning of precipitation along a climate gradient. Results show that direct evaporation represents the largest water loss component in areas receiving less than 300 mm of annual precipitation. The substantial evaporation fraction in dry areas indicates clearly that rainwater harvesting (deep trenches for underground recharge or small-scale collection systems directed to enhance green water fluxes) is the best available option to conserve water and to maximize its usefulness in these marginal areas.

Acknowledgments This paper is part of the GLOWA Jordan River Project Phase 3 which is funded by BMBF, Germany.

References

Allen RG, Pereira LS, Raes D, Smith M (1998) Crop evapotranspiration guidelines for computing crop water requirements. FAO irrigation and drainage paper 56, FAO, Rome
Brisson N (1998) An analytical solution for the estimation of the critical available soil water fraction for a single layer water balance model under growing crops. Hydrol Earth Sys Sci 2:221–231
Corbeels M, Hofmann G, van Cleemput O (1998) Analysis of water use by wheat grown on a cracking clay soil in a semi-arid Mediterranean environment: weather and nitrogen effects. Agric Water Manage 38:147–167
Dingman SL (2002) Physical Hydrology, 2nd edn. Prentice Hall, New Jersey
Gardner WR (1983) Soil properties and efficient water use: an overview. In: Taylor H, Jordan W, Sinclair TR (eds) Limitations to efficient water use. American Society of Agronomy, Inc., USA
Gleick PH (1987) The development and testing of a water balance model for climate impact assessment: modeling the Sacramento basin. Water Resour Res 23:1049–1061
Hanks RJ (1983) Yield and water-use relationship: an overview. In: Taylor H, Jordan W, Sinclair TR (eds) Limitations to efficient water use. American Society of Agronomy, Inc., USA
Izzi G, Farahani HJ, Bruggeman A, Oweis TY (2008) In-season wheat root growth and soil water extraction in the Mediterranean environment of northern Syria. Agric Water Manage 95:259–270
Kang S, Gu B, Du T, Zhang J (2003) Crop coefficient and ratio of transpiration to evapotranspiration of winter wheat and maize in a semi-*'humid'* region. Agric Water Manage 59:230–254
Monzon JP, Sadras VO, Andrade FH (2006) Fallow soil evaporation and water storage as affected by stubble in sub-humid (Argentina) and semi-arid (Australia) environments. Field Crops Res 98:83–90
Nishat S, Guo Y, Baetz BW (2007) Development of a simplified continuous simulation model for investigating long term soil moisture. Agric Water Manage 92:53–63
Novak V (1987) Estimation of soil water extraction patterns by roots. Agric Water Manage 12:271–278
Novak V, Havrila J (2006) Method to estimate the critical soil water content of limited availability for plants. Biol Bratisl 61(Suppl 19):S289–S293
Oroud IM (2007) Spatial and temporal distribution of frost in Jordan. Arab World Geogr 10:81–91
Oweis T, Zhang H, Pala M (2000) Water use efficiency of irrigated bread wheat in a Mediterranean environment. Agron J 92:231–238
Ritchie J (1972) Model for predicting evaporation from a row crop with incomplete cover. Water Resour Res 8:1204–1213
Rushton KR, Eiler VHM, Carter RC (2006) Improved soil moisture balance methodology for recharge estimation. J Hydrol 318:379–399
Tanner CB, Sinclair TR (1983) Efficient water use in crop production. In: Taylor H, Jordan W, Sinclair TR (eds) Limitations to efficient water use. American Society of Agronomy, Inc., USA
Taylor HM (1983) Managing root system for efficient water use: an overview. In: Taylor H, Jordan W, Sinclair TR (eds) Limitations to efficient water use. American Society of Agronomy, Inc., USA

Zhang H, Oweis TY, Garabet S, Mustafa P (1998) Water-use efficiency and transpiration efficiency of wheat under rain-fed conditions and supplemental irrigation in a Mediterranean-type environment. Plant Soil 201:295–305

Zhang Y, Yu Z, Liu C, Jiang L, Zhang X (2004) Estimation of winter wheat evapotranspiration under water stress with two semi-empirical approaches. Agron J 96:159–168

Chapter 2
Stormwater Reuse via Aquifer Storage and Recovery: Risk Assessment for Sandy Aquifers

Anke Steinel

Abstract In the light of increasing demand and diminishing supplies due to climate change and related alterations in rainfall pattern, a sustainable urban water management for Melbourne and other cities will need to include water recycling and reuse of reclaimed water and stormwater. One key issue in water reuse is the need for storage between times of collection and demand. Aquifer storage and recovery (ASR), which is defined as well injection of source water into a suitable aquifer for later recovery, would be a valuable option as it has limited space requirements, restricts loss from evaporation and limits the potential for contamination. However, stormwater commonly contains elevated levels of heavy metals, of which Zn and Cu are the most frequent and mobile. Stormwater and treated wastewater also contain suspended solids, organic carbon, oxygen and nutrients, which influence the behaviour of injected metals and induce geochemical changes in the aquifer. While ASR has been practised in limestone aquifers in South Australia, field data for sandy aquifers, which are more prevalent around Melbourne, are very limited. After giving an overview about current ASR practices, the general aim of the study was the risk assessment of injected metals derived from stormwater during ASR in sandy aquifers. As no field sites fulfilling these criteria are in operation in Melbourne or anywhere else in Australia, this study was limited largely to literature review and laboratory experiments.

Keywords Aquifer storage and recovery · Stormwater · Heavy metals · Risk assessment · Groundwater protection · Urban water management · Water reuse

---

A. Steinel (✉)
Federal Institute for Geosciences and Natural Resources (BGR),
Stilleweg 2, 30655 Hannover, Germany
e-mail: anke.steinel@bgr.de

Introduction

Water is vital to the survival and growth of all life and all economic and environmental processes, but the world's supplies of freshwater are under increasing threat from overexploitation, pollution and climate change. Among other countries, Australia has experienced a severe drought over the last 10 years and in Melbourne water supply reservoirs are at an all-time low with the lowest storage levels on record of below 30% at the end of June 2009 (Fig. 1).

A climate change study for Melbourne predicts higher temperatures and reduced rainfall for the future, which would increase the demand and decrease the supply (Howe et al. 2005). A sustainable water management is therefore vital for Melbourne and other cities around the world facing a similar challenge. The sustainable water management will also be an essential part of rural Australia, where the drought has resulted, for example, in a decline of more than 95% in rice production in the past 10 years (Barclay 2010).

Decision-makers have largely opted for desalination as the main solution to the problem. The high energy needs and the problems associated with disposal of highly saline brine make it an environmentally challenging option (Lattemann and Höpner 2008). On the other hand, urban areas produce stormwater and treated effluent streams, in volumes matching the demand (Mitchell et al. 2002) and dispose of them into urban waterways and adjacent coastal areas leading to degradation of these ecosystems (e.g. Walsh et al. 2005). It is therefore of twofold benefit to recycle stormwater and reclaimed water and this has been identified as an important strategy in maintaining a sustainable water supply for Melbourne.

Benefits and Constraints of ASR

The progress in this direction is often hindered, not by technical or economical reasons, but due to perceived health concerns by the public and decision-makers (Dillon et al. 2009; Po et al. 2003; Toze 2006) and by institutional structures that are not suitable for non-centralized systems. Water treatment is capable of rendering wastewater to a quality superior than that of potable water (e.g. Cheremisinoff 2002) and hence reuse as drinking water is possible. There is an even greater potential to use water treated to lower standards, hence using less energy, for non-potable uses, which constitutes more than 70% of urban water demand (Mitchell et al. 2002).

The benefits of ASR are manifold. Public confidence in recycled water is increased when the water has been put back into the natural system before reuse (Dillon et al. 2009; Po et al. 2003) as ASR increases water quality, e.g. through the removal of pathogens and nutrients (Toze 2006). ASR could provide capacity for inter-seasonal or inter-year storage, especially in aquifers of higher salinity that are of limited beneficial use (Pyne 1995; Maliva et al. 2006). Compared to surface storage, ASR has limited loss of water due to

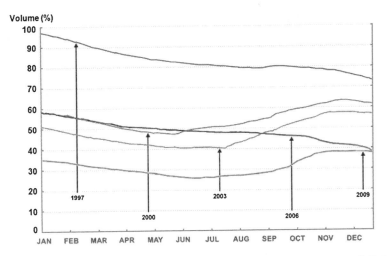

Fig. 1 Melbourne's total water system storage (10 reservoirs combined, total volume 1810.5 GL) development from 1997 to 2009, adapted from (Melbourne Water 2010)

evaporation and is also less affected by contamination or algal blooms and does not provide mosquito breeding grounds (Pyne 1995; Dillon et al. 2006). Space requirements are also limited, which is especially valuable in urban areas for projects larger than household scale (Hatt et al. 2006). And additionally, the costs of artificial groundwater recharge are less than investments necessary for large traditional dams (BGR 2008) or desalination (Dillon et al. 2009).

Despite these benefits, uptake of ASR in Melbourne has been limited by a number of restraints. Insufficient legislation and guidelines for water recycling including water ownership issues (Hatt et al. 2006), lack of experience and knowledge in the water industry and relevant authorities (Dillon et al. 2005; Hatt et al. 2006) and lack of detailed hydrogeological properties (Dudding et al. 2006) have restricted the application of ASR. In addition, more research on well-designed and monitored field projects with long-term performance data, as well as increased hydrogeological knowledge on contaminant attenuation processes and contaminant mobility (Dillon et al. 2005; Maliva et al. 2006) are needed to increase public acceptance (Toze 2006; Dillon et al. 2009).

Stormwater Versus Treated Effluent

Compared to treated effluent, stormwater has the advantage of lower concentrations in nutrients, organic matter and total dissolved solids (Mitchell et al. 2002) and is perceived as more acceptable for reuse by the public (Po et al. 2003). On the other hand, stormwater is generated in single events of different magnitudes and is often seasonally confined. Stormwater quality and quantity thus is rather variable and depends on a range of parameters, e.g. land use, rainfall intensity, frequency and volume (e.g. Wong 2006). Even within a catchment, pollutant concentration can change by more than one magnitude (Sansalone et al. 1995) during the hydrograph of a single event [e.g. first flush effect (Sansalone and Buchberger 1997; Barbosa and Hvitved-Jacobsen 1999)], from one event to the next and over the seasons [e.g. seasonal first flush (Lee et al. 2004)]. A vital part of stormwater reuse is therefore the collection and storage of the variable sources for later access.

Treated effluent, on the other hand, is a relatively constant and continuous source, but storage for summer months would commonly be needed nevertheless. It contains higher concentration of nutrients, salinity and organic matter. Its use for agricultural purposes, which commonly comprise about 70% of the total water demand, is therefore of twofold benefit, as the addition of fertilizers can be reduced due to the nutrients contained in the water. Another advantage of using reclaimed wastewater instead of stormwater is its lower impact on stream flows. Firstly, treated effluent would not be discarded into streams, deteriorating its quality and, secondly, stormwater flows would still be available as environmental flows. The reuse of reclaimed wastewater with and without aquifer recharge has been successfully implemented in many countries around the world and is being promoted in rural areas (e.g. Abu-Madi et al. 2008).

ASR Technique: Theory and Case Studies

The artificial recharge of groundwater can occur via infiltration or injection. The infiltration via infiltration basins, seepage trenches or vadose zone wells requires the availability or construction of an adequate permeable soil and unconfined aquifer (Bouwer 2002). Injection of water into the aquifer can involve one well for injection and recovery and is defined as ASR, while the use of one well for injection and another well for recovery is defined as aquifer storage, transport and recovery (ASTR). It requires a suitable aquifer and better infrastructure, but injection rates are usually higher than infiltration rates (Bouwer 2002).

In general, managed underground storage involves the following elements (Fig. 2):

1. Collection or harvesting: Water sources can include potable water, surface water, groundwater, treated effluent, and stormwater. This study will be mainly concerned with urban stormwater runoff.
2. Pretreatment: Pretreatment options can include standard stormwater pollution control design practices such as litter and sediment traps, wetlands and ponds, advanced treatment and disinfection (Victoria Stormwater Committee 1999; Ellis 2000; Hatt et al. 2006).
3. Injection: Suitable aquifers can be of different types: confined or unconfined, from unconsolidated sands and gravels to limestones and fractured sedimentary or volcanic rocks.

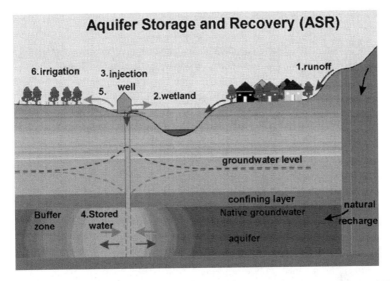

Fig. 2 Schematic aquifer storage and recovery system, modified after (CSIRO Land and Water graphics 2010)

4. Storage: A buffer zone separates the native groundwater from the injected water.
5. Recovery: Recovery is typically achieved in a number of recovery stages depending on demand.
6. End use: Recovered water can be used for drinking water, irrigation, firefighting, environmental flows, toilet flushing, industrial cooling and other purposes, and might be subjected to post treatment.

Site Selection Considerations

The site selection for an ASR scheme will depend on a range of factors. Firstly, a suitable water source and a local demand together with the availability of some space for stormwater collection and pretreatment are needed. Secondly, administrative, economic and operational constraints have to be met and finally, a suitable aquifer must be available (Pyne 1995; Dillon and Molloy 2006).

The suitability of the aquifer is mainly defined by its storage capacity and recovery efficiency (Pyne 1995; AGT et al. 2002). Both fractured and unconsolidated rock systems could be used for ASR systems. While storage capacity in fractured rock can be quite high, its recovery efficiency is often lower due to the usually not well-known heterogeneous characteristics of the fractures. Dual porosity limestone aquifers have been favoured in many regions due to their potential for increasing the storage volume by calcite dissolution and water quality improvements even with low recovery efficiency in karstic regions (Gerges et al. 2002) Unconsolidated porous aquifer systems have potentially the highest storage capacity. Predictions about the hydraulic performance of the system are much more reliable than for fractured rocks. In general, the aquifer should be sufficiently transmissive to avoid excessive build-up of groundwater mounds, preferably homogeneous and of sufficient extent for the scale of the ASR scheme.

Another consideration is the groundwater quality. It is commonly advisable to have a lower-quality groundwater, so beneficial use will not be impaired by the injection of non-potable water. Very high salinity levels, on the other hand, will increase the buffer zone of an ASR scheme substantially or will limit the recovery efficiency. Contaminated aquifers need to be avoided, as this would compromise the quality of the recovered water.

One problem not only experienced with infiltration but also associated with injection ASR is clogging. This reduction in porosity mainly around the injection well leads to a decrease in injection rate and an increase in hydraulic head. Clogging is a process that will usually occur, but in most cases can be managed with pretreatment and redevelopment (Pyne 1995; Brown et al. 2006). The pretreatment can include filtration/coagulation of suspended particle and microorganisms, addition of pH-elevating substances to reduce iron precipitation, and disinfection to avoid microbial growth (Pyne 1995). Nevertheless, periodical redevelopment by purging the well will be undertaken in regular intervals at all sites. The disposal of these backflushed waters that are usually of low quality have to be accounted for in the planning stage of ASR schemes. Limestone aquifers are likely to compensate for loss in porosity by dissolution of carbonic matrix. Addition of acidic agents has been used to enhance this process and decrease the effect of clogging (Gerges et al. 2002).

Water Quality Considerations

A number of issues such as the uncertainty regarding the possible impacts of recharge water on the beneficial use of the native groundwater resources, the potential for increased discharge into streams and wetlands as well as the quality of the recovered water have to be considered.

A better understanding of processes such as the fate of contaminants present in the recharge water and the interactions taking place in the disturbed groundwater system will also help with guideline development. These are of crucial importance for protecting environmental values and beneficial uses of the groundwater as required by statutory authorities such as the Environmental Protection Agency (EPA Victoria 2003).

Depending on the source water and the end use, different aspects of water quality will be of importance. A good compilation of water quality improvements that can be achieved during ASR has been presented by Dillon et al. (2005). When using lower quality water as recharge, the removal of pathogens, natural organic matter (NOM) and nutrients and associated redox reactions are of more importance. For all schemes, biogeochemical interactions with the aquifer matrix such as ion-exchange and dissolution/precipitation will have an impact on the recovered water quality, however.

Microbial Pathogens

Pathogens comprise a diverse group of microorganisms and have been found in high concentrations in wet weather urban runoff originating from animal faeces and sewer overflows (Wong 2006). The fraction of organisms associated with settleable particles varies by type of microbe, and the partitioning behaviour of each organism generally changes between dry weather and storm conditions (Characklis et al. 2005). Once injected, their inactivation in the subsurface can be achieved by adsorption, filtration, oxidation, degradation by native groundwater microorganisms and die-off (Pitt et al. 1999; Foppen and Schijven 2006). The main factors favouring long-term survival of pathogens in the ground are small size (reduced straining especially for viruses), soluble organics (reduced adsorption), low oxygen levels (reduced oxidation), low temperature, high pH and low salinity (prolonged life span) and the native microbiota (John and Rose 2005). Monitored ASR sites achieved complete removal of pathogens at storage times longer than 90 days (Dillon et al. 2005). It is therefore not necessary to disinfect water before injection. This will also avoid the formation of carcinogenic disinfection by-products.

Natural Organic Matter

Natural organic matter (NOM), usually measured as total organic carbon (TOC), consists of both particulate (POC) and dissolved (DOC) organic carbon and comprises a complex mixture of low to high molecular weight organic molecules. The bulk of NOM found in stormwater are fulvic and humic acids. The degradation rate of NOM varies widely. In general, smaller molecules with carboxylic and phenolic groups are favourable for biodegradation while large aromatic molecules are more stable (e.g. Howard 2000). Fulvic acids are soluble over the wide range of pH, while humic acids are insoluble at lower pH. Humic substances are most elongated and flexible at high pH, low salinity and low NOM concentration (e.g. stormwater), while they form spheres at higher salinity (>0.05 M), in the presence of trivalent cations and at high NOM concentrations (e.g. Ghosh and Schnitzer 1980). This behaviour plays an important role for cation and metal binding and pH buffering. Depending on the properties of the organic molecule, pH and surface properties of the clay or oxide, NOM can adsorb to clay minerals and oxides via a range of mechanisms (Stevenson 1982; Jardine et al. 1989). The usually positive surface charge of the underlying clay or oxide thus changes to negative surface charge and thereby influences the cation exchange capacity, sorption of other organic molecules and colloidal stability of the particles (e.g. Stevenson 1982; Kretzschmar and Sticher 1997). A complete reversal of surface charge and complete coating of matrix with NOM can be achieved with as little as 0.5 mg-C/L TOC (Day et al. 1994). Mono-layer coverage of organic matter may impede further sorption of NOM (Jardine et al. 1989; Day et al. 1994), but competitive exchange of more hydrophobic macromolecules with less hydrophobic molecules has been observed (Gu et al. 1996a, b), resulting in higher mobility of smaller organic molecules.

Redox Reactions

NOM in the injectant is the major driving force for microbial growth. Its oxidation acts as an energy source for microorganisms and controls the redox status especially near the injection well. If the electron acceptor is an aqueous species such as O_2, protons are consumed, while reactions involving reduced minerals (e.g. sulphides, ferrous oxides) can release protons (Appelo and Postma 1999). In the natural subsurface environment, these processes will be partly taking place simultaneously due to heterogeneities of the matrix, distribution of reactants and microorganisms, meaning that there is usually no equilibrium between all redox couples in the groundwater. Typically, groundwaters are low in both particulate and dissolved organic carbon (<0.5 mg-C/L). The recharge with organic, oxygen and nutrient-enriched injectant, therefore, is a major disturbance to the system (Bahr et al. 2002). Around the well (<10 m) where particulates are entrained, a so called "treatment" or "proximal" zone with highly elevated microbial activity develops. Here dissolved oxygen is eliminated in a few days, followed by nitrate reduction (Greskowiak et al. 2005; Vanderzalm et al. 2006) triggering further geochemical reactions such as degradation of organic contaminants, mineral dissolution/precipitation, sorption and pH changes that influence the quality of the recovered water (Dillon et al. 2005).

Ion Exchange

The aquifer matrix provides a number of cation exchange sites on clay minerals, oxides and organic matter (Appelo and Postma 1999). The injection of low salinity, mostly Ca-dominated recharge water, into brackish often Na–Mg-dominated groundwater, firstly decreases the salinity, which might cause swelling of clays and a decrease in permeability (e.g. Konikow et al. 2001). Secondly, it will lead to simultaneous homo- and heterovalent ternary cation exchange. The selectivity sequence would usually favour Ca, which would replace Mg, which in turn replaces Na, leading to a high loading of Ca on the exchanger and increased concentrations of Na and Mg in the solution (Appelo and Postma 1999). This usually helps to restore the permeability. During the establishment of an ASR scheme, these ion exchanges will be moved outwards into the buffer zone and not play a major role for later ASR cycles (Bahr et al. 2002).

Dissolution/Precipitation

The chemical stability of minerals depends on redox and pH conditions as well as the activity of ions in solution (Appelo and Postma 1999). As stormwater is usually of neutral pH, the change in redox state is the main driver for mineral dissolution and precipitation reactions, which are therefore mainly taking place near the injection well. Minerals of high reactivity and main importance in ASR are carbonates, hydr-/oxides and sulphides. Dissolution of minerals, especially carbonates, is induced by undersaturation with low ion activity in the recharge waters, especially by highly treated injectants (Johnson et al. 1999). The decrease of pH due to release of CO_2 during organic matter degradation also accelerates dissolution of carbonates (Herczeg et al. 2004). In limestone aquifers this relieves the effect of clogging but might lead to collapse of the aquifer matrix structure (Pavelic et al. 2006). The interplay between oxidative/reductive changes induced by ASR and dissolution/precipitation reactions with minerals can be complex. The injection of oxygen into an anoxic aquifer with dissolved Fe and Mn will lead to the precipitation of amorphous hydroxides, while reductive dissolution of Mn- and Fe-oxides starts after oxygen is completely depleted and nitrate is in the progress of reduction (Stumm and Morgan 1996). Both reactions occur in ASR operations close to the injection well, where the injected organic matter induces redox and pH changes (Greskowiak et al. 2005). While the precipitation of minerals will lead to co-precipitation and adsorption of trace metals, the dissolution will in turn mobilise previously incorporated or adsorbed trace metals (Bahr et al. 2002). Pyrite oxidation is of special importance as it releases mobile oxyanions such as arsenate and selenate and mobile cations such as nickel and zinc. It also significantly reduces the pH due to the formation of sulphuric acid in solution, which can mobilise other adsorbed metals. Multiple ASR sites have experienced a decrease in recovered water quality due to geogenic metal release related to pyrite oxidation (Brown et al. 2006). Over a number of ASR cycles, these effects seem to be decreasing and shifting towards the buffer zone of the ASR scheme.

ASR in Use

Research on potable injection and infiltration ASR has been going on for some decades most prominently in the USA (Pyne 1995; Brown et al. 2006), with most emphasis on quantitative aspects such as clogging and recovery efficiency mainly for long-term and emergency storage. It is becoming more popular in the Middle East together with desalination plants (e.g. Almulla et al. 2005). All schemes in the USA need to comply with the ASCE (2001) standards, which do not allow the injection of poor-quality water (Brown et al. 2006). Therefore the only water quality problem in these schemes is geogenic metal mobilization, mainly concerned with arsenic (Jones and Pichler 2007). Australia appears to be at the forefront of non-potable water recycling and as one of the driest countries in the world has focused on water reuse in urban areas. Adelaide, South Australia, has taken on a leading role in ASR research and has operational, trial and proposed sites, which are more or less well studied. Well-studied examples for stormwater ASR include Andrews Farm (Herczeg et al. 2004; Pavelic et al. 2006) and Parafield (Marks et al. 2005). For treated effluents, the ASR investigations at Bolivar (Dillon et al. 2005; Greskowiak et al. 2005; Vanderzalm et al. 2006) have to be mentioned. As they are all situated in limestone, these schemes cannot directly be compared to potential sites in Melbourne.

Generally, performance results are somewhat limited, as some projects are still relatively new and also because water quality monitoring has only been carried out or documented infrequently. This is partly due to the fact that in Australia no specific guidelines for stormwater recycling are currently in place and the existing (usually treated wastewater recycling) guidelines are not enforceable standards (Hatt et al. 2006).

First Conclusions

From the review so far it can be stated that ASR offers a valuable solution to the storage problem associated with stormwater and treated effluent reuse in urban areas, but implementation has been slow due to administrative hurdles and a lack of political will. In addition, the interactions between microbial, geochemical and hydraulic processes during ASR are complex. They are dependent on injectant, aquifer and groundwater site specifics and have not yet been fully understood. ASR scheme feasibility depends largely on the storage capacity of the aquifer and they are preferably implemented in transmissive sandy or limestone aquifers (Pyne 1995). However, ASR schemes have to take into account the protection of the aquifer and groundwater to reduce the impact on downstream environments and be sustainable in the long term. Limestone aquifers have been used successfully for stormwater ASR in South Australia and provide a high retention capacity for metals due to the alkaline pH. The hydrogeological setting of Melbourne is quite different and potential ASR sites are situated in sandy alluvial aquifers (AGT et al. 2002; Dudding et al. 2006), which have lower pH and might have limited sorption capacity for heavy metals. A clear gap in ASR experience in alluvial sands has been recognized (AGT et al. 2002).

Stormwater ASR in Sandy Aquifers

The risk associated with water recycling to augment drinking water supplies is currently under review and stormwater reuse and managed aquifer recharge are part of the Australian guidelines for water recycling (NWQMS 2007). As elevated levels of metals are ubiquitous in urban stormwater (Makepeace et al. 1995; Wong 2006) and metals cannot be degraded, their behaviour during stormwater ASR was investigated. In contrast to South Australia, where stormwater ASR has been practised for a number of years in limestone, Melbourne's potentially viable aquifers are mainly silicious (Dudding et al. 2006). As no field site was available for this study, laboratory experiments with local aquifer sediments, groundwater and stormwater were undertaken instead to evaluate the risk to the environment posed by injected metals during stormwater ASR. Common empirical methods to investigate metal solid-solution interactions are batch (OECD/OCDE 2000) and column tests (e.g. Schweich and Sardin 1981; Bürgisser et al. 1993) and were hence employed. Currently available hydrogeochemical models (Prommer et al. 2003) were adapted and enlarged to help interpret the empirical results. A detailed description of the experiments goes beyond the scope of this paper and can be found in Wendelborn 2008.

Stormwater Quality

Urban stormwater is a valuable water resource that needs to be integrated into an urban water conservation management to meet future water demands. Nevertheless, urban stormwater contains a range of pollutants that have negative impacts on receiving ecosystems and potentially need to be treated before reuse (Pitt et al. 1999). Stormwater quality and quantity is rather variable, however, and depends on a range of parameters, but commonly occurring priority pollutants have been identified (Eriksson et al. 2007).

Total Suspended Solids

Total suspended solids (TSS) comprise organic and inorganic particulates (>0.45 μm) deriving from natural and anthropogenic sources. TSS concentration in stormwater ranges typically from 50 to 450 mg/L with lower concentrations found in roof runoff and highest concentrations found in urban road runoff. Apart from their clogging potential during infiltration or injection, their importance arises from the attachment of other pollutants (e.g. nutrients, hydrophilic organics, heavy metals) mainly with the finer fraction of particulates (e.g. Dempsey et al. 1993; Mikkelsen et al. 1994; Deletic and Orr 2005). While the coarser particles are likely to be settling out before use in ASR schemes, the finer, more contaminated proportion is likely to be injected into the aquifer and not to be recovered, as observed at Andrews Farm (Pavelic et al. 2006). Particulate facilitated transport is therefore a potential pathway for contaminants to be distributed in the aquifer and should not be neglected (Dempsey et al. 1993; Liebens 2001) especially as particle size distributions (PSD) analysis of Australian surface runoff showed that higher fractions of smaller particles are generated compared to Europe or the USA with a mean particle size of ~ 50 μm (Wong 2006).

Nutrients and Oxygen

Nitrogen and phosphorus in stormwater derive from fertilizers, faecal matter, plant debris and combustion processes. While they are essential nutrients for plant growth, increased levels may lead to eutrophication and toxic algal blooms. While phosphorus attaches readily to particulates, nitrate is highly soluble and very likely to be injected during ASR. Dissolved oxygen and nitrate are the major electron donors injected into the aquifer stimulating microbial activity, significantly altering the aquatic ecosystem community (Datry et al. 2004).

Organic Carbon

Fulvic acids are the major colloidal and dissolved constituent in natural organic matter. Concentrations typically range from 8 to 35 mg/L. Interactions between DOC, dissolved constituents and solid matter are numerous. DOC stimulates the growth of biofilms, biological production and subterranean ecosystem metabolism and redox reactions (Stumm and Morgan 1996; Baker et al. 2000). Its presence influences the speciation of heavy metals and can increase their solubility through complexation. From the column experiments it is clear that the distinction between mobile and immobile organic carbon is crucial, as mobile organic carbon increases solubility of metals and facilitates their transport, while immobile organic carbon enhances the accumulation of metals in the subsurface. Organic carbon in stormwater consists of a conglomerate of organic molecules with different molecular weights and functional groups. Accordingly, a fraction will be mobile in the subsurface, while other fractions will either be strained, precipitate or sorb to the matrix, becoming immobile. In low ionic strength solution, remobilization of organic carbon can occur by biodegradation, increase in pH or dissolved small organic acids.

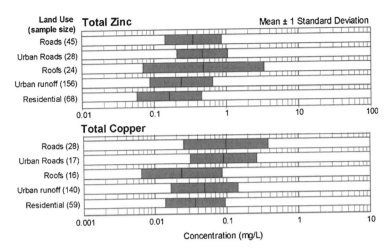

Fig. 3 Typical concentrations of worldwide stormwater samples for Zn and Cu in stormwater (modified after Wong 2006)

Heavy Metals

Heavy metals in stormwater derive from natural sources, as well as from different anthropogenic sources such as hydrocarbon combustion, vehicle exhaust, brake and tyre wear, roof materials, corrosion of metal objects, etc. (Makepeace et al. 1995; Wong 2006). They are often significantly toxic at low concentrations and are therefore considered priority pollutants in stormwater (Eriksson et al. 2007). Their importance stems from the fact that they (in contrast to organic contaminants) cannot be degraded, but are persistent and tend to bioaccumulate.

Metal concentrations in stormwater are highly variable depending on a number of factors such as land use, climate and rain event characteristics. According to their prevalence in stormwater and toxicity Pb, Zn, Cu and Cd (e.g. Barbosa and Hvitved-Jacobsen 1999; Legret and Pagotto 1999; Pitt et al. 1999) are the most important metals for study. Zn and Cu concentrations are highest from urban road runoff and Zn concentrations are increased through runoff from Zn roofs (Fig. 3).

This speciation of metals is vital to assess their potential for injection during stormwater ASR, as larger particulates will be settling out in the detention ponds before injection, while fine particulates and dissolved metals would not (Wong 2006). While Pb has been found to be nearly completely attached to particulates, Zn, Cu and Cd are more associated with dissolved solids and colloidal material (e.g. Sansalone and Buchberger 1997; Legret and Pagotto 1999). As Cd is frequently below detection limit in Melbourne's stormwater and non-settleable fractions of Pb are also low, the main focus of this study is Zn and Cu. Relative to overseas studies catchments in Melbourne showed elevated concentrations of Zn (~ 1.1 mg/L) and comparable concentrations of Cu (~ 50 μg/L). Compared to other metals Zn and Cu were predominantly found in the dissolved and fine particulate fraction. Hence their concentrations were still considerably above freshwater ecosystem guideline values after sedimentation. Without further pretreatment noteworthy amounts of Zn and Cu are therefore likely to be injected into the aquifer in dissolved and colloidal form (Wendelborn 2008).

Metal Behaviour During the ASR Cycle

Some of the processes described for water quality considerations are more or less important for stormwater ASR in general and come into play at different stages of the cycle. The main considerations will be highlighted below.

Pretreatment

In most cases stormwater will not be used for injection straight from the source, but would be collected and pretreated to a degree. Best management practices for urban stormwater (Victoria Stormwater Committee 1999; Hatt et al. 2006) offer a range of primary treatment methods to reduce contaminant loads, the most common being wetlands and detention ponds, which utilize the effect of sedimentation of suspended solids and attached contaminants (e.g. Braskerud 2000; Walker 2001). This helps to prevent clogging of bores and also decreases metal loads attached to settleable solids. However, sedimentation does not affect loads from dissolved metals or metals attached to particles about <20 μm (Walker and Hurl 2002), and the accumulated metals and organic matter in the sediment have the potential to be remobilized (e.g. Kalbitz and Wennrich 1998; Walker and Hurl 2002). Further improvements to the water quality has been achieved by using constructed wetlands with vegetation, which takes up nutrients and accumulates heavy metals (e.g. Zhao et al. 2007; Read et al. 2008). In return, a release of metals can be expected during decay of plant material, and removal rates in the long term are uncertain. Release of nutrients and dissolved organic matter from wetlands has also been reported and can be increased by wildlife, e.g. birds. Further treatment with filtration through gravel, sand or roughing filters has been shown to decrease particle concentration and associated metals further (e.g. Page et al. 2006; Hatt et al. 2007). Secondary treatment with filter material of different material (e.g. alumina, bark, Bauxsol-coated sand, fly ash, granulated activated carbon, iron oxide-coated sand, zeolite, etc.) have been tested with varying degrees of success (e.g. Sansalone 1999; Genç-Fuhrmann et al. 2007). Conventional secondary treatment (Kurniawan et al. 2006) with flocculation/coagulation, liming or dissolved air flotation are possible and effective, but require the right dosing of chemicals and the handling of sludge waste that are usually not practicable within the ASR settings.

Clearly, the pretreatment step is the most crucial step for metal behaviour as their fate might already be established if no metals are injected into the aquifer. This would be a favourable outcome. In this study metal concentrations dropped about 40–50% for Zn and 75% for Cu from mean stormwater concentrations through settlement (Wendelborn 2008).

Injection

The injection phase is the main disturbance of the groundwater system, as oxic, nutrient and organic matter enriched, low ionic strength water enters a commonly anaerobic, nutrient poor, brackish groundwater. The pH is commonly around neutral for stormwater and free dissolved metals are likely to adsorb to the immobile or mobile groundwater matrix near the injection well. Complexed metals and colloidal metals are more mobile and are likely to get transported until they reach areas of higher salinity (Ryan and Gschwend 1994; Bunn et al. 2002). Due to the increase in flow velocity (McCarty et al. 1993; Ryan and Gschwend 1994) and low salinity of the injected water, colloids near the well could also be mobilized (Johnson et al. 1999; Konikow et al. 2001) carrying attached metals with them. The increase in dissolved organic carbon is able to desorb previously adsorbed metals, if complexation is more favourable. If free Fe^{2+} are present in the groundwater, this would precipitate and scavenge dissolved metals. Other mineral phases are likely to dissolve due to undersaturation or change in Eh–pH stability conditions in the solution and release metals. Metals leached from the aquifer matrix near the well are then redeposited further away from the well, within the radius defined by the storage bubble.

Mixing along the flow path will diminish the difference between the injectant and the native groundwater, but due to heterogeneities in the aquifer, zones of fast preferential flow will develop next to diffusion-dominated areas and local disequilibrium might increase, meaning that the geochemical conditions and hence the dominating process can be quite different depending on the location.

During stormwater injection in the column experiments, Zn showed complete breakthrough for sediments with low sorptive capacity and starting breakthrough for sediment with organic coating. Solubility was slightly increased due to the addition of DOC. The breakthrough curve for the former sediments pointed to a high and a low affinity surface site, while the breakthrough for the latter sediment was much more gradual, indicating a more gradual change in surface site metal affinity exhibited by different organic functional groups (Wendelborn 2008). Cu was completely adsorbed during stormwater injection without added DOC, while solubility was significantly increased with the addition of DOC, with the organically coated sediment again showing the highest retardation of all sediments. The shape of the Cu breakthrough curve suggested transport as organic colloids and complexes and ligand influenced sorption for all three sediments (Wendelborn 2008). Even though clay was ≤0.2% for all sediments they showed measurable release of inorganic and organic colloids (~200 nm) when injected with neutral low ionic strength stormwater.

Storage

The injection of organic matter and nutrients will have stimulated microbial activity in the vicinity of the injection well. During storage, organic matter degradation and associated reduction in redox potential will be the main factor for metal behaviour. Metals that were attached to the organic matter might be released and reductive dissolution of minerals might also release metals. If redox potential drops low enough sulphide precipitation of metals might occur. Changes in pH during this stage might also influence the solubility of metals.

In general, disequilibrium will be lessened due to time for rate limited reactions and diffusive processes. After a number of ASR cycles, diffusive processes will have replaced the native groundwater with infiltration water completely.

In the column experiments, particle mobilization from the aquifer was observed during storages related to organic biodegradation and probably mineral dissolution. In conjunction, increased concentrations of Zn and Cu after storages were observed due to colloidal transport and release from matrix induced by changes in solution parameters during storage. Kinetic biodegradation resulted in sharp metal concentrations and pH increases after storage (Wendelborn 2008).

Recovery

The extent of geochemical changes during recovery is much lower than during injection and depends also on the change in volume and the extent of the buffer zone. Free metals in solution will be recovered and especially non-specifically adsorbed metals are likely to be desorbed with an increase in ionic strength. In the well vicinity, particles and all attached metals will also be recovered.

In the column experiments Zn was released continuously due to dilution and competition with major ions and protons, resulting in high recovery efficiency. Cu release was triggered by a decrease in pH. Both metals showed decreased desorption when DOC was added, resulting in very limited release of Cu. The retained fraction of added DOC had significantly increased specific sorption sites and anoxic conditions were less favourable for metal release (Wendelborn 2008). Hydraulic non-equilibrium conditions due to small scale heterogeneities in hydraulic properties and diffusion limited flow were observed. Chemical non-equilibrium was due to rate-limited sorption visible in extensive tailing (Wendelborn 2008).

Overall

Speciation of Cu and Zn depends on a range of different processes and factors. Cu speciation is mostly influenced by the presence of organic matter, in solution it is mainly found as organic complex and sorption is dominated by specific surface complexes, resulting in slow desorption. Zn speciation is mostly influenced by pH, in solution it is mainly found as free ion and sorption is dominated by non-specific surface complexes, resulting in fast desorption (Wendelborn 2008).

Overall, metals are persistent in the environment and mechanisms are reversible, meaning that metals retained in the aquifer could eventually become mobile again with time and change in geochemical conditions. In conclusion, the interactions of metals with solution and solid described in this section are complex and intertwined. At this stage, no single theoretical approach or model is capable of incorporating all processes and factors and empirical approaches are therefore still common. It is therefore critical to undertake site-specific studies and ongoing monitoring to assess key processes and outcomes.

In conclusion, injected amounts of Zn are mobile and will mainly be recovered. Cu can be mobile in the presence of mobile organic carbon, but will mainly accumulate in the aquifer until the sorption capacity is exceeded. The release of metals was triggered by reduction in pH, increase in ionic strength and particle mobilization. Metal concentrations were high after storage phases. Minor sediment constituents, especially organic matter, significantly reduce metal mobility.

Implications and Recommendations

Implications for Stormwater ASR in Sandy Aquifers

While the results obtained during the laboratory experiments are only representative for short-term behaviour, the lessons learned could be translated into long-term behaviour in sandy aquifers. It should be remembered that spiked metal concentrations during the experiments were considerably higher than expected injected stormwater concentrations, i.e. about four times higher for Zn and about 50 times higher for Cu. One also has to consider that other metals would be present in stormwater and especially Al and Fe would strongly be competing for sorption sites.

The prognosis for injected Zn would be that it would be relatively mobile within the aquifer and would be present throughout the injected freshwater zone. It would be partially adsorbed onto surface sites especially during the first number of ASR cycles, but largely be recovered in similar amounts as injected. If we assume a mean injected stormwater concentration of about 500 µg/L and assume that Zn is on average recovered in similar concentrations, then this concentration would be well below the drinking water guideline of 3 mg/L (NHMRC 2004) and below the irrigation water guideline of 2 mg/L (ANZECC and ARMCANZ 2000) and should be fit for these purposes. In contrast, these concentrations would be drastically exceeding freshwater ecosystem guidelines of 8 µg/L (ANZECC and ARMCANZ 2000) and should not be used for environmental flows.

The prognosis for injected Cu would need to be differentiated into different scenarios.

- Scenario A: Low concentrations of organic carbon

For scenario A, injected Cu would be largely adsorbed to the aquifer matrix slowly filling up sorption sites from the injection well outwards to the edge of the freshwater zone, expelling less strongly sorbed metals such as Zn in its wake. Desorption due to dilution would be highest close to the injection well, where high volumes of recovered water are flowing past. Unless pH values of recovered water are not significantly lower than pH 6, release of Cu would be limited though and Cu would accumulate in the aquifer over the long term. If the surrounding groundwater has lower pH values, the creation of a larger buffer zone and recovery volumes less than injected volumes would drastically reduce the impact of the original groundwater. Nevertheless, a breakthrough of low pH groundwater in zones of higher transmissivity could result in a surge of desorbed Cu in the recovered water, which could be many times higher than injected concentrations depending on the sorption capacity of the aquifer. Hence, special attention should be paid to recovered Cu values in low pH groundwater ASR settings. In these settings, continuous monitoring of pH would be recommendable, as pH would be a good surrogate indicator for elevated metal concentrations.

If we assume a mean injected stormwater concentration of about 15 µg/L and assume that Cu is largely retained in the sediment, Cu concentrations in recovered water would be well below drinking water guidelines of 2 mg/L (NHMRC 2004) and irrigation water values of 0.2 mg/L (ANZECC and ARMCANZ 2000). Use would only be compromised by release of previously accumulated Cu at low pH. As freshwater ecosystem guidelines are set as low as 1.4 µg/L (ANZECC and ARMCANZ 2000), recovered water would most likely be still not fit for environmental flows, especially in the later years of operation of the ASR scheme, when the sorption capacity has been exceeded.

- Scenario B: Elevated amounts of injected organic carbon

Injected organic carbon can be mobile or immobile in the subsurface. The formation of mobile Cu-ligand complexes would increase Cu solubility and transport, spreading Cu faster throughout the freshwater zone than in the previous scenario and would result in higher recovery of Cu. However, significant portions of the previously mobile organic complexes would presumably be deposited with an increase of ionic strength during recovery, and hence increase Cu accumulation throughout the subsurface. The immobile carbon would accumulate close to the injection well and fuel redox reactions. As not all organic carbon will be degraded quickly, retained amounts of organic carbon would significantly increase the sorption capacity of the aquifer allowing for a longer and higher accumulation of

Cu. On the other hand, release of Cu after storage phases would likely be increased due to the biochemical changes and dissolution and degradation of part of the immobile organic carbon during storage. Backflushed solutions and initially recovered volumes could therefore be unsuitable for reuse.

- Scenario C: Elevated amounts of colloids

Relevant concentrations of colloids could result from injected particles, may be due to mobilization of aquifer fines or be produced during mineral dissolution and precipitation. They have the potential to transport adsorbed metals that would usually be classified as immobile. In general, colloids will be transported along preferential flow path, e.g. zones of higher permeability, and will aggregate and be deposited once ionic strength reaches the critical limit at the outer edge of the freshwater zone (Ryan and Gschwend et al. 1994; Bunn et al. 2002). Over a number of ASR cycles, this could lead to a localized accumulation of fines, potentially lowering permeability and leading to a localized higher metal concentration at the fringe of the storage zone. Overall, this would lead to lower recovered metal concentrations.

To estimate the possible accumulated Cu concentrations a rough calculation could be as follows: With an injected Cu concentration of 15 µg/L it would add up to an injected amount of 15 kg of Cu per year and GL and amount to 300 kg/GL in a life span of 20 years. If we further assume a mean porosity of 0.33 an average Cu concentration of 67 mg/kg would result over 20 years. The experiments showed that even sediments consisting to >98 wt% of sand exhibited considerable sorption capacity with more than 200 mg/kg for Cu and hence Cu concentrations could easily be accommodated within the sediment. Comparing the potentially accumulated concentration with the sediment quality guideline target value of 34 mg/kg (ANZECC and ARMCANZ 2000), shows that the accumulation of Cu would be well above the threshold effect level. Similar calculations for an injected Zn concentration of 500 µg/L would result in 2,200 mg/kg Zn compared to a sediment quality guideline value of 150 mg/kg. This would mean that more than 90% of Zn would have to be recovered to stay below the sediment target value, which could be achievable in low sorption capacity sediments.

It is clear that ASR schemes would be decommissioned once they are no longer viable, e.g. due to clogging of the aquifer or decreasing volumes of stormwater available for injection. Once the injections have stopped, the previous groundwater flow and original groundwater conditions re-establish themselves over time. In the long term, this would result in desorption of accumulated metals due to dilution effects. Desorption would be accelerated if groundwater was of low pH, which could result in an increased release of metals into groundwater fed surface water ecosystems, which would be potentially problematic. Desorption would be limited if groundwater was highly reducing and metals would be precipitated as sulphides.

Aquifer Selection Recommendations

The choice of the aquifer would be restricted by the available aquifer in the location of source and demand of water, and the storage capacity needed for the ASR scheme. From the point of view of this study, the recommendations for an aquifer selection are divided by the question: (a) Do metal concentrations in the recovered water need to be very low (e.g. use for environmental flows) or (b) would long-term metal accumulation in the aquifer that might later pose a risk to downstream systems be of concern and hence recovery of injected metals would be preferred? In the first case high sorptive capacities are a bonus, while in the latter case low sorptive capacities are needed.

Accordingly, a local aquifer with limited recharge to surface water and adjacent aquifers would be the most suitable situation. In this case the aquifer could be used for water purification and high sorptive capacity would be appreciated to lower metal concentrations of the recovered water to a minimum. The accumulated metals in the aquifer would pose very limited risk to surrounding ecosystems. If, on the other hand, the aquifer is highly connected to other aquifers and discharges into adjacent surface waters, then low sorptive capacity and a high recovery efficiency of injected metals would be more appropriate. Given that irrigation and drinking water guidelines allow for much higher metal concentrations than are tolerated by freshwater organisms, high metal recovery would seem to be the most preferred option. It should be taken into account that the sorptive capacity of the aquifer will generally be increased with the injection of colloids and organic carbon as well.

Analysis of aquifer sediments before the implementation of an ASR scheme should include: fine clay, to assess potential mobilization of colloids from the matrix; organic carbon, which offers high amounts of specific sorption sites; pyrite, which is generally an undesirable mineral due to release of arsenic and other trace metals; and cation exchange capacity and possibly specific sorption capacity to assess the accumulation potential of metals. The evaluation of sediment analysis should take into account that surface coatings consisting of oxides and organic matter do not contribute a great deal of mass but offer a large amount of reactive surface area. Low values of Fe^{2+} in the native groundwater would also be favourable as this would prevent the precipitation of amorphous iron hydroxides, which would also increase the sorption potential.

Monitoring Recommendations

Since the accumulation of metals in the aquifers constitutes a concern, the ASR project should regularly be monitoring metal concentrations of injected and recovered (including backflushed) waters, to be able to calculate the retained amount of metals in the subsurface. Currently no ASR project has undertaken this mass balance for metals. As no stormwater ASR project in sandy aquifers has been established yet, field data to validate the current laboratory experiments is desirable. Additionally, the monitoring of backflushed waters is of special importance, as these would potentially contain the highest metal concentrations and should include analysis for As. These volumes should not be discharged into streams without the knowledge of metal concentrations and have to be disposed of correctly or treated before discharge. The continuous monitoring of pH would be highly recommended, as pH would be a good surrogate indicator for elevated metal concentrations. Low pH readings could then trigger additional sampling and analysis for metals before reuse.

Pretreatment Recommendations

The remediation of aquifers from heavy metal contamination is a difficult and expensive task and it would therefore be advisable to limit the potential for metal accumulation in the aquifer regardless of the specific situation. Recommendations for pretreatment would therefore include the lowering of injected metal concentrations (including Al and Fe) as well as organic carbon and colloid concentrations, as these increase the retention potential of the aquifer. This kind of pretreatment would also be beneficial in other respects as the potential for biological and mechanical clogging would be reduced.

Pretreatment for metals could be achieved with biofiltration, exchangers or reactive filters, while organic matter and colloids would best be reduced via coagulation and precipitation (Kurniawan et al. 2006). While this kind of pretreatment would be desirable, the benefits have to be weighed against the disadvantages of use of chemicals, disposal of waste sludge and associated carbon emissions. Pretreatment with natural or recycled substances and low carbon imprint should be investigated. For particulates and organic carbon, slow sand filtration or roughing filtration would be possible (Page et al. 2006). Metal removal could be achieved with oxide-coated sands (Sansalone 1999; Genç-Fuhrman et al. 2007), biosorbents such as tree fern (Ho 2003), agricultural waste products such as hulls (Marshall and Champagne 1995) or wetlands (Walker and Hurl 2002).

Conclusions

The final conclusion from this study seems to be that water recycling and reuse is a necessity in the current situation of many cities in the world with dwindling surface and groundwater resources. Stormwater ASR is a valuable tool in fulfilling this task. Additional pretreatment that might be necessary to bring recycled water up to acceptable standards for humans should in reality be undertaken anyway before discarding these waters into waterways.

It must be emphasized that there is no one solution to water shortage and a range of options should be implemented simultaneously. The trend is to decentralize supply according to localized demand and use holistic approaches. The most common form of small-scale stormwater reuse is the collection of roof runoff in rainwater tanks. It has been widely employed in rural areas and is increasingly implemented in urban households for domestic use (Mitchell 2004). Greywater tanks are also a favourable option at household scale. The local reuse also saves the cost of drainage, central treatment and redistribution. For larger catchments, detention ponds, retarding basins and reservoirs are needed (Hatt et al. 2006) requiring increasingly more space, which is a limited and costly asset in the urban environment and poses a major problem. Nevertheless, separation of stormwater and wastewater streams limits the treatment costs and energy needs. All schemes should be monitored for pathogens and heavy metals to limit potential health issues.

Apart from technical considerations (Dillon and Molloy 2006), increased efforts have to be made to educate and involve the population, which is accustomed to a centralized system. The community needs to be made aware of the problems and possible solutions. Information about actual dangers and realistic potentials needs to be given. To increase the acceptance for recycled water, the community should be incorporated in decision-making processes and the administrative and legislative frameworks need to be in place to allow innovative ways of water supply.

References

Abu-Madi M, Al-Sa'ed R, Braadbaart O, Alaerts G (2008) Viability of increasing the tariff of freshwater for irrigation as a tool to stimulate wastewater reuse in the MENA region. in Water Sci Technol 57(9):1475–1481, doi:10.2166/wst.2008.238

AGT (Australian Groundwater Technologies), CSIRO Land and Water, Sinclair Knight Merz (SKM) and United water international (2002): Assessment of the potential of reclaimed water ASR in metropolitan Melbourne, 2002/12, Australian Groundwater Technologies, Adelaide

Almulla A, Hamad A, Gadalla M (2005) Aquifer storage and recovery (ASR): a strategic cost-effective facility to balance water production and demand for Sharjah. Desalination 174(2):193–204

ANZECC and ARMCANZ (2000) Australian and New Zealand guidelines for fresh and marine water quality—Volume 1: The guidelines (Chapters 1–7). 4, Australian and New Zealand environment and conservation council (ANZECC); Agriculture and resource management council of Australia and New Zealand (ARMCANZ), Canberra

Appelo CAJ, Postma D (1999) Geochemistry, groundwater and pollution. Balkema, Rotterdam

ASCE (2001) Standard guidelines for artificial recharge EWRI/ASCE 34-01. American society of civil engineers (ASCE), New York

Bahr J, Grundl T, Harrington G, Krohelski J, Werner M (2002) A review of aquifer storage recovery techniques, Wisconsin Department of Natural Resources

Baker MA, Valett HM, Dahm CN (2000) Organic carbon supply and metabolism in a shallow groundwater ecosystem. Ecology 81(11):3133–3148

Barbosa AE, Hvitved-Jacobsen T (1999) Highway runoff and potential for removal of heavy metals in an infiltration pond in Portugal. Sci Total Environ 235(1–3):151–159

Barclay A (2010) A sunburnt grain. Rice Today 9:12–17
BGR (2008) Groundwater and climate change: challenges and possibilities, Federal Institute for Geosciences and Natural Resources (BGR), http://www.bgr.bund.de
Bouwer H (2002) Artificial recharge of groundwater: hydrogeology and engineering. Hydrogeol J 10(1):121–142
Braskerud BC, Lundekvam H, Krogstad T (2000) The impact of hydraulic load and aggregation on sedimentation of soil particles in small constructed wetlands. J Environ Qual 29(6):2013–2020
Brown CJ, Hatfield K, Newman M (2006) Lessons learned from a review of 50 ASR projects from the United States, England, Australia, India, and Africa, increasing freshwater supplies. In: Proceedings of the 2006 UCOWR/NIWR Annual Conference, Santa Fe, New Mexico, USA, 18–20 July 2006
Bunn RA, Magelky RD, Ryan JN, Elimelech M (2002) Mobilization of natural colloids from an iron oxide-coated sand aquifer: effect of pH and Ionic Strength. Environ Sci Technol 36(3):314–322
Bürgisser CS, Cerník M, Borkovec M, Sticher H (1993) Determination of nonlinear adsorption isotherms from column experiments: an alternative to batch studies. Environ Sci Technol 27(5):943–948
Characklis GW, Dilts MJ, Simmons OD, Likirdopulos CA, Krometis L-AH, Sobsey MD (2005) Microbial partitioning to settleable particles in stormwater. Water Res 39(9):1773–1782
Cheremisinoff NP (2002) Handbook of water and wastewater treatment technologies. Butterworth-Heinemann, Boston
Datry T, Malard F, Gibert J (2004) Dynamics of solutes and dissolved oxygen in shallow urban groundwater below a stormwater infiltration Basin. Sci Total Environ 329(1–3):215–229
Day DM, Hart BT, McKelvie ID, Beckett R (1994) Adsorption of natural organic matter onto goethite. Colloids Surf A 89:1–13
Deletic A, Orr D (2005) Pollution buildup on road surfaces. J Environ Eng 131(1):49–59
Dempsey BA, Tai YL, Harrison SG (1993) Mobilization and removal of contaminants associated with urban dust and dirt. Water Sci Technol 28(3–5):225–230
Dillon P, Molloy R (2006) Technical guidance for ASR CSIRO land and water client report, http://www.smartwater.com.au/downloaddocs/Technical_Guidelines_for_ASR.pdf
Dillon P, Toze S, Pavelic P, Skjemstad J, Davis G, Miller R, Correll R, Kookana R, Ying G-G, Filderbrandt S, Banning N, Gordon C, Wall K, Nicholson B, Vanderzalm J, Le Gal La Salle C, Giber M, Ingrand V, Guinamant J-L, Stuyfzand P, Prommer H, Greskowiak J, Swift R, Hayes M, OcHara G, Mee B, Johnson I (2005) Water quality improvements during aquifer storage and recovery, Vol 1, Subsurface processes for water quality improvement, AWWARF
Dillon P, Pavelic P, Toze S, Rinck-Pfeiffer S, Martin R, Knapton A, Pidsley D (2006) Role of aquifer storage in water reuse. Desalination 188(1–3):123–134
Dillon P, Pavelic P, Page D, Beringen H, Ward J (2009) Managed aquifer recharge: an introduction, Waterlines report series No. 13, Australian government national water commission, http://www.nwc.gov.au
Dudding M, Evans R, Dillon P, Molloy R (2006) Report on broad scale map of ASR potential for Melbourne. SKM and CSIRO Land and Water, Melbourne
Ellis JB (2000) Infiltration systems: a sustainable source-control option for urban stormwater quality management? J Chart Inst Water Environ Manag 14(1):27–34
EPA Victoria (2003) State environment protection policy (Waters of Victoria). Publication 905. Victorian Environmental Protection Agency, Melbourne
Eriksson E, Baun A, Scholes L, Ledin A, Ahlman S, Revitt M, Noutsopoulos C, Mikkelsen PS (2007) Selected stormwater priority pollutants—a European perspective. Sci Total Environ 383:41–51
Foppen JWA, Schijven JF (2006) Evaluation of data from the literature on the transport and survival of *Escherichia coli* and thermotolerant coliforms in aquifers under saturated conditions. Water Res 40(3):401–426

Genç-Fuhrman H, Mikkelsen PS, Ledin A (2007) Simultaneous removal of As, Cd, Cr, Cu, Ni and Zn from stormwater: experimental comparison of 11 different sorbents. Water Res 41(3):591–602

Gerges NZ, Dillon PJ, Sibenaler XP, Martin RR, Pavelic P, Howles SR, Dennis K (2002) South Australian experience in aquifer storage and recovery. In: Dillon P (ed) Management of aquifer recharge for sustainability, Proceedings of the international symposium on artificial recharge of groundwater, 4th, Adelaide, Australia, 22–26 September 2002, 453–458

Ghosh SK, Schnitzer M (1980) Macromolecular structures of humic substances. Soil Sci 129(5):266–276

Greskowiak J, Prommer H, Vanderzalm J, Pavelic P, Dillon P (2005) Modeling of carbon cycling and biogeochemical changes during injection and recovery of reclaimed water at Bolivar, South Australia, in Water Resour Res, Vol 41, W10418, doi:10.1029/2005WR004095

Gu B, Mehlhorn TL, Liang L, McCarthy JF (1996a) Competitive adsorption, displacement, and transport of organic matter on iron oxide: I. Competitive adsorption. Geochimica Cosmochimica Acta 60(11):1943–1950

Gu B, Mehlhorn TL, Liang L, McCarthy JF (1996b) Competitive adsorption, displacement, and transport of organic matter on iron oxide: II. Displacement and transport. Geochimica and Cosmochimica Acta 60(16):2977–2992

Hatt BE, Deletic A, Fletcher TD (2006) Integrated treatment and recycling of stormwater: a review of Australian practice. J Environ Manag 79(1):102–113

Hatt BE, Fletcher TD, Deletic A (2007) Treatment performance of gravel filter media: Implications for design and application of stormwater infiltration systems. Water Res 41(12):2513–2524

Herczeg AL, Rattray KJ, Dillon PJ, Pavelic P, Barry KE (2004) Geochemical processes during five years of aquifer storage recovery. Gr Water 42(3):438–445

Ho Y-S (2003) Removal of copper ions from aqueous solution by tree fern. Water Res 37(10):2323–2330

Howard PH (2000) Chapter 12 Biodegradation. In: Boethling RS, Mackay D (eds) Handbook of property estimation methods for environmental chemicals environmental and health sciences. Lewis Publishers, Boca Raton

Howe C, Jones RN, Maheepala S, Rhodes B (2005) Melbourne water climate change study: implications of potential climate change for Melbourne's water resources. CMIT-2005-106, Melbourne Water, CSIRO Urban Water, CSIRO Atmospheric Research

Jardine PM, Weber NL, McCarthy JF (1989) Mechanisms of dissolved organic carbon adsorption on soil. Soil Sci So Am Jl 53(5):1378–1385

John DE, Rose JB (2005) Review of factors affecting microbial survival in groundwater. Environ Sci Technol 39(19):7345–7356

Johnson JS, Baker LA, Fox P (1999) Geochemical transformations during artificial groundwater recharge: soil–water interactions of inorganic constituents. Water Res 33(1):196–206

Jones GW, Pichler T (2007) Relationship between pyrite stability and arsenic mobility during aquifer storage and recovery in southwest central Florida. Environ Sci Technol 41(3):723–730

Kalbitz K, Wennrich R (1998) Mobilization of heavy metals and arsenic in polluted wetland soils and its dependence on dissolved organic matter. Sci Total Environ 209(1):27–39

Konikow LF, August LL, Voss CI (2001) Effects of clay dispersion on aquifer storage and recovery in coastal aquifers. Transp Porous Med 43(1):45–64

Kretzschmar R, Sticher H (1997) Transport of humic-coated iron oxide colloids in a sandy aquifer: influence of Ca2 + and trace metals. Environ Sci Technol 31:3497–3504

Kurniawan TA, Chan GYS, Lo W-H, Babel S (2006) Physico-chemical treatment techniques for wastewater laden with heavy metals. Chemi Eng J 118(1–2):83–98

Lattemann S, Höpner T (2008) Environmental impact and impact assessment of seawater desalination. Desalination 220(1–3):1–15

Lee H, Lau S-L, Kayhanian M, Stenstrom MK (2004) Seasonal first flush phenomenon of urban stormwater discharges. Water Res 38(19):4153–4163

Legret M, Pagotto C (1999) Evaluation of pollutant loadings in the runoff waters from a major rural highway. Sci Total Environ 235(1–3):143–150

Liebens J (2001) Heavy metal contamination of sediments in stormwater management systems: the effect of land use, particle size, and age. Env Geol 41(3–4):341–351

Makepeace DK, Smith DW, Stanley SJ (1995) Urban stormwater quality: summary of contaminant data. Crit Rev Environ Sci Technol 25(2):93–139

Maliva RG, Guo W, Missimer TM (2006) Aquifer storage and recovery: recent hydrogeological advances and system performance. Water Environ Res 78(13):2428–2435

Marks R, Chapman F, Lane S, Purdie M (2005) Parafield urban stormwater harvesting facility. Water 32(5):42–45

Marshall WE, Champagne ET (1995) Agricultural byproducts as adsorbents for metal ions in laboratory prepared solutions and in manufacturing wastewater. J Environ Sci Health, Part A 30(2):241–261

McCarthy JF, Williams TM, Liang LY, Jardine PM, Jolley LW, Taylor DL, Palumbo AV, Cooper LW (1993) Mobility of natural organic matter in a sandy aquifer. Environ Sci Technol 27(4):667–676

Mikkelsen PS, Weyer G, Berry C, Walden Y, Colandini V, Poulsen S, Grotehusmann D, Rohlfing R (1994) Pollution from urban stormwater infiltration. Water Sci Technol 29(1–2):293–302

Mitchell G (2004) Integrated urban water management: a review of current Australian practice. CSIRO, Melbourne

Mitchell GV, Mein RG, McMahon T (2002) Utilising stormwater and wastewater resources in urban areas. Aust J Water Resour 6(1):31–43

NHM RC (National Health, Medical Research Council) (2004) Australian drinking water guidelines. National Water Quality Management Strategy, Commonwealth of Australia

NWQMS (2007) Australian guidelines for water recycling: managing health and environmental risks (Phase 2): augmentation of drinking water supplies In: NWQMS (ed): Natural resource management ministerial council, environment protection and heritage council, Australian health ministers conference

OECD/OCDE (2000) OECD guideline for the testing of chemicals: adsorption—desorption using a batch equilibrium method, 106, OECD

Page D, Wakelin S, van Leeuwen J, Dillon P (2006) Review of biofiltration processes relevant to water reclamation via aquifers. CSIRO land and water science, Report 47/06

Pavelic P, Dillon P, Barry K, Gerges N (2006) Hydraulic evaluation of aquifer storage and recovery (ASR) with urban stormwater in a brackish limestone aquifer. Hydrogeol J 14(8):1544–1555

Pitt R, Clark S, Field R (1999) Groundwater contamination potential from stormwater infiltration practices. Urban Water 1(3):217–236

Po M, Kaercher JD, Nacncarrow BE (2003) Literature review of factors influencing public perceptions of water reuse. Technical Report 54/03, CSIRO Land and Water

Prommer H, Barry DA, Zheng C (2003) MODFLOW/MT3DMS-based reactive multicomponent transport modelling. Gr Water 41(2):247–257

Pyne RDG (1995) Groundwater recharge through wells: a guide to aquifer storage recovery. Lewis Publishers, Boca Raton

Read J, Wevill T, Fletcher T, Deletic A (2008) Variation among plant species in pollutant removal from stormwater in biofiltration systems. Water Res 42(4–5):893–902

Ryan JN, Gschwend PM (1994) Effects of ionic strength and flow rate on colloid release: relating kinetics to intersurface potential energy. J Colloid Interf Sci 164(1):21–34

Sansalone JJ (1999) Adsorptive infiltration of metals in urban drainage—media characteristics. Sci Total Environ 235(1–3):179–188

Sansalone JJ, Buchberger SG (1997) Partitioning and first flush of metals in urban roadway storm water. J Environ Eng 123(2):134–143

Sansalone JJ, Buchberger SG, Koechling MT (1995) Correlations between heavy metals and suspended solids in highway runoff: implications for control strategies. Transp Res Rec 1483:112–119

Schweich D, Sardin M (1981) Adsorption partition, ion exchange and chemical reaction in batch reactors or in columns—A review. J Hydrol 50:1–33

Stevenson FJ (1982) Humus chemistry: genesis, composition, reactions. Wiley, New York

Stumm W, Morgan JJ (1996) Aquatic chemistry: chemical equilibria and rates in natural waters, environmental science and technolology. Wiley, New York

Toze S (2006) Water reuse and health risks—real vs. perceived. Desalination 187(1–3):41–51

Vanderzalm JL, Le Gal La Salle C, Dillon PJ (2006) Fate of organic matter during aquifer storage and recovery (ASR) of reclaimed water in a carbonate aquifer. Appl Geochem 21(7):1204–1215

Victoria Stormwater Committee (1999) Urban stormwater: best practice environmental management guidelines

Walker DJ (2001) Modelling sedimentation processes in a constructed stormwater Wetland. Sci Total Environ 266(1–3):61–68

Walker DJ, Hurl S (2002) The reduction of heavy metals in a stormwater Wetland. Ecol Eng 18(4):407–414

Walsh CJ, Roy AH, Feminella JW, Cottingham PD, Groffman PM, Morgan IRP (2005) The urban stream syndrome: current knowledge and the search for a cure. J North Am Benthol Soc 24(3):706–723

Wendelborn A (2008) Zinc and copper behaviour during stormwater aquifer storage and recovery in sandy aquifers. PhD thesis, Monash University, Australia, http://arrow.monash.edu.au/hdl/1959.1/68715

Wong T (ed) (2006) Australian runoff quality—a guide to water sensitive urban design. Institution of Engineers, Australia

Zhao LYL, Schulin R, Nowack R (2007) The effects of plants on the mobilization of Cu and Zn in soil columns. Environ Sci Technol 41(8):2770–2775

Melbourne Water (2010) Weekly dam storage update, Accessed May 2010, http://www.melbournewater.com.au/content/water_storages/water_report/zoom_graph.asp

CSIRO Land and Water graphics (2010) Accessed May 2010, http://www.smartwater.com.au/projects/round2/asr/Pages/asr.aspx

Chapter 3
Hydrologic Balance of Citarum Watershed under Current and Future Climate

Rizaldi Boer, Bambang Dwi Dasanto, Perdinan and Delon Marthinus

Abstract As the biggest watershed in West Java, Citarum plays an important role in supplying water for many districts in the province. The Citarum watershed supplies approximately 7,650 million cubic metres of water per year ($m^3 \ a^{-1}$). Currently, approximately 78% of the extracted water is used for irrigation, 14% for industrial activities and electricity generation, and 8% for domestic consumption. Analysis of this watershed found it to be very vulnerable to climate change. It was found that all of the sub-districts already experience water deficit problems (i.e. not enough supply to meet their demands), particularly in the lower areas such as Kerawang, Bekasi and Purwakarta, even without a changing climate and if the level of water extraction from the streamflow was limited to 10% of the mean annual flow. In 2080, the water deficit for most of the sub-districts in this lower area would be even more severe. Increasing water extraction to 20% of the mean annual flow would not change the water status of these sub-districts. Consequently, conflicts among water users may be a serious problem for these regions in the future.

Keywords Climate change · Hydrology balance · Citarum watershed

R. Boer (✉) · B. D. Dasanto · Perdinan · D. Marthinus
Laboratory of Climatology, Department of Geophysics and Meteorology
Faculty of Mathematics and Natural Sciences, Bogor Agricultural University,
Gedung FMIPA Level 4 Wing 19 Kampus IPB Darmaga, Bogor, 16680, Indonesia
e-mail: rizaldiboer@gmail.com

Introduction

Citarum watershed is an important watershed in West Java, Indonesia. Water supply from this watershed is used for many purposes, including electricity generation, domestic consumption, irrigation water for agricultural uses and flushing Jakarta canals. The watershed supplies approximately 7,650 million cubic metres of water per year ($m^3\ a^{-1}$); approximately 5,750 million $m^3\ a^{-1}$ (75%) comes from Citarum watershed dams (Saguling, Cirata and Jatiluhur) and 1,950 million $m^3\ a^{-1}$ (25%) comes from other rivers (Perum Jasa Tirta II 2003). Currently, approximately 78% of the water is used for irrigation, 14% for industrial activities and electricity generation and 8% for domestic consumption. Water is not only supplied for irrigating districts within the watershed area, but also for districts outside the watershed. For example, Jakarta, the capital of Indonesia, receives clean water from Jatiluhur dam. The dam also supplies water for other districts outside the Citarum watershed such as Indramayu and Sukamandi (the main rice production centres of West Java).

Extreme climate events, climate change, and land use and land use cover (LULUC) change such as forest degradation, are expected to significantly impact the Citarum watershed by decreasing the water supply. During drought years associated with the El Niño phase of the El Niño Southern Oscillation Index (ENSO), irrigation supply from a great portion of the watershed decreased significantly causing widespread drought in irrigated rice paddies, particularly in the tail-end of the irrigation area. In West Java, the average area (over the period 1989–2004) suffering from drought during normal years was approximately 7,800 ha, while during ENSO years, this area increased sharply to 290,000 ha (Directorate of Plant Protection 2000).

It is very likely that due to the increasing demand for water supply, conflicts between water users may be a serious problem in the future. It is predicted that after 2010, the Citarum watershed might not be able to meet the water demand of its users (Hernowo 2001). Observations using historical data from 1896 to 1991 have shown that annual rainfall over the Citarum catchments and the corresponding water discharge have decreased by about 10 and 3.1 mm annually. The output of the ECHAM model suggests that rainfall in this region would continue to decrease in the future.

This study aims to assess the vulnerability of the watershed to current climate variability and to evaluate the status of the water balance of the watershed under the current and future climate using a number of climate change scenarios.

Fig. 1 Geographic characteristics of Citarum watershed

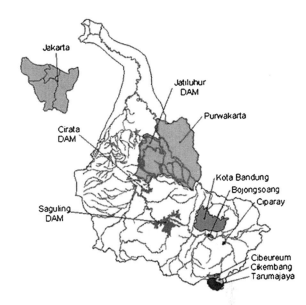

Methodology

Study Area

Citarum watershed is situated in West Java province, Indonesia. The catchment area is approximately 6,867 km² with a length of about 269 km. The highest geographic peak is approximately 1,700 m above sea level (ASL) and the lowest point is 0 m ASL. There are three dams in the watershed: Saguling in the upper part, Cirata in the middle, and Jatiluhur or Juanda in the lower part, as well as nine hydrological observation stations (Fig. 1). The main city is Bandung, which has a population of approximately 2.5 million people.

The annual rainfall is about 2,580 mm divided over two seasons–a rainy season (from November to April) that receives approximately 1,840 mm and a dry season (from May to October) that receives approximately 740 mm. The rainy season peaks around January. The average streamflow of Citarum river is 173 m³ s⁻¹. The Saguling and Cirata Dams are mainly used for electricity generation while the Jatiluhur Dam is mainly used for agriculture. Perum Jasa Tirta II (2003) predicted that the water demand by allocation in 2005 would be approximately 5,519 million m³ a⁻¹ (75%) for irrigation, 672 million m³ a⁻¹ (9%) for drinking water, 473 million m³ a⁻¹ (6%) for industry, 315 million m³ a⁻¹ (4%) for fisheries, 315 million m³ a⁻¹ (4%) for flushing Jakarta canals, and 100 million m³ a⁻¹ (1%) for electricity generation.

The three dams (Saguling, Cirata, and Jatiluhur) are interconnected. Outflow from Saguling Dam goes to Cirata Dam, outflow from Cirata Dam goes to Jatiluhur Dam, and outflow from Jatiluhur is used for the many purposes as described above and then finally flows out to the Java sea (Fig. 1). Inflow from other local rivers also supplies the three dams.

Streamflow Analysis

ENSO events are one of dominant factors that cause extreme climate conditions in Indonesia such as drought (ADPC 2000; Yoshino et al. 2000; Kirono and Partridge 2002; D'Arrigo and Wilson 2008). Water inflow into the reservoirs decreased significantly as a result of decreased rainfall during ENSO-related drought events. The behaviour of streamflow during these events was assessed using a simple correlation analysis between the Southern Oscillation Index (SOI) and streamflow data from one hydrology station (Cigulung-Maribaya station) with a long historical record (1953–2002). The five SOI phases from Stone et al. (1996) were used in the analysis. These phases are consistently negative (Phase 1), consistently positive (Phase 2), rapidly falling (Phase 3), rapidly rising (Phase 4), and neutral (Phase 5). The SOI values can be downloaded from http://www.dpi.qld.gov.au. When SOI falls rapidly and then remains negative, it indicates that an El Niño event is likely, which normally reduces rainfall in Indonesia, particularly in parts of South Sumatra, Java, and the eastern part of Indonesia. Conversely, when SOI increases rapidly and then remains positive, it indicates that La Niña is likely, which normally increases rainfall.

A graphical analysis showing the change in streamflow mean between normal, El Niño and La Niña years was also conducted. Total inflows from local rivers to each dam from 1986 to 2002 were obtained from Perum Jasa Tirta II were used in the analysis. This record covered five El Niño years (1987, 1991, 1994, 1997 and 2002), two La Niña years (1989 and 1998), and ten normal years.

Hydrology Balance Analysis

The purpose of the hydrology balance analysis is to assess the status of hydrology water balance at Citarum watershed under current and future climate. The analysis was conducted down to the sub-district level.

Hydrology balance is expressed in the following equation:

$$\text{Supply} = \text{Demand} + \text{Surplus}$$

If the demand is higher than supply, the surplus becomes negative and vice versa (if supply is higher than demand, the surplus becomes positive).

Water Supply

In the water supply analysis, the annual water supply (i.e. surface flow or discharge) of the watersheds was calculated using a simple linear regression between annual surface flow as dependent variable and annual precipitation as independent variable. This approach has been applied by previous work that assessed hydrologic balance in Indonesia (Pawitan 1996). The equation used for this study is expressed below,

$$V_Q = aP - b$$

where V_Q represents annual surface flow or streamflow (in mm), **P** represents annual precipitation (in mm), and *a* and *b* are constants. In this analysis, the Citarum watershed is divided into three regions–upper, middle and lower–with areas of 1,874 km^2, 2,477 km^2, and 2,517 km^2, respectively. The relationship between annual streamflow of local rivers and annual rainfall for each region was developed using the above regression equation. Because Cirata Dam receives water outflow from Saguling Dam, and Jatiluhur Dam receives water outflow from Cirata Dam (refer back to Fig. 1), the water supply equation for these two dams is:

$$V_Q = aP + b - I$$

where **I** represents the water outflow from the respective dams. As the streamflow data is in m^3 s^{-1}, these units were converted into annual streamflow measured as depth over the catchment (in mm). The calculation was conducted as follows:

$$V_Q = [I * (365 * 24 * 60 * 60)/A] * 1000$$

where **I** represents the average of annual inflow or outflow (m^3 s^{-1}) and **A** represents the area of the corresponding sub-watersheds (m^2). The rainfall data used in the analysis is taken from 26 stations (8 stations in the upper region, 7 stations in the middle region, and 11 stations in the lower region).

This analysis assumed that the maximum annual streamflow (VQ) that can be used as water supply should not be more than the minimum inflow. Based on 15-year data, it was found that the minimum inflows to the Saguling, Cirata, and Jatiluhur dams were approximately 21, 23 and 22%, respectively. As such, this study used two water supply scenarios equal to 10 and 20% of the annual streamflow (discharge).

Water Demand

Demand for water comes from three sectors: domestic use (urban and rural), industry, and agriculture. Water demand for domestic use was estimated by multiplying population size by water consumption per capita. Bina Program Cipta

Table 1 Freshwater use per capita in Indonesia

Population size according to city category	Water demand (l/cap/day)			Loss	Total
	Household use	Drinking water	Non-household		
>1,000,000	190	30	60	75	280
500,000–1,000,000	170	30	40	55	230
100,000–500,000	150	30	30	50	200
20,000–100,000	130	30	20	40	165
<20,000	100	30	10	30	125

Source: Bina Program Cipta Karya 1991

Karya (1991) stated that the level of water consumption could be categorized based on the population size of the city. The higher the population is, the higher the demand per capita (Table 1). Thus, water demand projections will follow population projections, assuming no changes in the patterns or management of demand.

For the industrial sector, water demand is estimated based on the size of each industrial area. The two water demand categories used by Bappenas (1991) were $0.55 \text{ l s}^{-1} \text{ ha}^{-1}$ (minimum) and $0.75 \text{ l s}^{-1} \text{ ha}^{-1}$ (maximum). The analysis in this paper used a water demand value of $0.65 \text{ l s}^{-1} \text{ ha}^{-1}$. Because precise data on the size of industrial areas was unavailable, this analysis assumed that the industrial area of each sub-district follows the proportion of the sub-district relative to the total area of the watershed. Using this approach, the industrial area of the Citarum watershed was estimated to be approximately 49,615 ha (0.3% of the total watershed).

For the agriculture sector, the dominant water use is for irrigation (for rice cultivation, specifically). The length of the irrigation season for rice in the two watersheds is between 90 and 150 days. The amount of water required is between 140 and 150 mm per month, equivalent to between 4,500 and 7,000 $m^3 \text{ ha}^{-1}$ per season. Thus, the total annual water demand for irrigation was estimated by multiplying the annual planting area by the demand. The annual planting data of irrigated rice was obtained from the Dinas Pertanian Propinsi Jawa Barat (Department of Agriculture for West Java Province) website (http://www.diperta-jabar.go.id).

Water Supply Scenarios

As the annual water supply is predicted using annual rainfall data, the scenario for water supply will follow rainfall scenarios. The rainfall scenarios were developed based on General Circulation Model (GCM) outputs under two emission scenarios, SRESA2 and SRESB2. Changes in rainfall under global warming varied considerably between GCMs. Two GCM models, CCSR and CSIRO, suggested that the seasonal rainfall would increase consistently over the period from 2020 to 2080

3 Hydrologic Balance of Citarum Watershed

Table 2 Percent change of seasonal precipitation of wet season (DJF) and dry season (JJA) from current rainfall in 2080 under the SRESA2 and SRESB2 scenarios

Scenarios	Months	CGCM2	CSIRO-MK2	CSM-1.3	ECHAM4	GFDL-R15b	MRI2	CCSR/NIES2	DOE-PCM	HadCM3
A2	DJF	-9.16	5.09	-0.23	-7.3	52.21	10.69	10.62	2.18	3.65
A2	JJA	-8.28	12.54	-12.63	-44.92	-23.83	5.61	-5.08	10.04	-27.38
B2	DJF	-6.52	-2.82	-7.56	4.02	-2.05	7.44	-0.25	-2.8	10.72
B2	JJA	-15.72	12.68	-10.01	-13.82	-41.68	7.85	-15.74	19.05	-14.51

Source: Unpublished data (Xianfu 2002)

under both scenarios, except for September–October–November (SON) rainfall. Analysis from the ECHAM4 and CGCM1 GCMs projected that the rainfall would decrease consistently. The HadCM3 GCM did not produce a consistent impact. HadCM3 suggested that December–January–February (DJF) rainfall might not change up to 2020, but would increase up to 2.5% from the baseline in 2050 and then decrease by 2% from the baseline in 2080. The two most interesting findings are (i) the SON rainfall might not change more than 5% from the baseline under the two emission scenarios, and (ii) rainfall during other seasons would increase or decrease up to 15% from the baseline in 2080. Analysis prepared by Xian Fu (2002) also found similar features (Table 2). Because the impact of global warming on Indonesian rainfall is not consistent among GCMs, hypothetical climate scenarios were used. The five rainfall scenarios adopted by this study were -20, -10, 0, +10 and +20% change from the mean rainfall value.

Water Demand Scenarios

Three scenarios were used for water demand. The first scenario is called the baseline scenario (a scenario developed based on data of historical trend and taking into consideration long-term government plan for 2025) and the other two scenarios were developed based on assumptions used in SRESA2 and SRESB2. Therefore, the rate of population growth at each sub-district for the latter two scenarios followed from the population growth rates used in the SRESA2 and SRESB2. Similarly, the development of industrial areas under the other two scenarios was assumed to follow the pattern of GDP growth rate under SRESA2 and SRESB2. Meanwhile, the development of agriculture area was assumed to be the same as that under the baseline scenario. This assumption was used because of the limited land available for the development of agriculture area. The historical data suggested that irrigated paddy area has decreased at a rate of approximately 0.5% per year. Hereafter, the other two scenarios used are referred to as SRESA2 and SRESB2. The result of projections for population growth rate, rice planting area and industry area are presented in Tables 3, 4 and 5, respectively.

Table 3 Projected population growth rate within Citarum watershed under baseline, SRESA2, and SRESB2 scenarios from 2005 to 2080

Scenario	Population growth rate (% per year)				
	2000–2005	2006–2010	2011–2020	2021–2050	2051–2080
Baseline[1]	1.67	1.52	1.40	1.21	0.88
SRESA2[2]	1.40	1.48	0.82	0.52	0.45
SRESB2[2]	1.30	1.38	0.76	0.45	0.34

(1) Average values over a number of districts in Citarum watershed
(2) Growth rates under these scenarios were developed based on growth rates used in IPCC scenarios (IPCC 2000). It was assumed that the maximum population density is 20,000 people per km^2 in a city (urban area), and 5,000 people per km^2 in rural areas

Table 4 Projected rice planting area within Citarum watershed from 2000 to 2080

District name	Rice planting area (ha/year)					
	2000	2005	2010	2020	2050	2080
Bandung City	4,465	3,422	1,992	500	500	500
Bandung	105,524	108,243	109,348	112,000	120,000	120,000
Bogor	88,185	92,010	95,280	102,000	120,000	120,000
Cianjur	114,415	99,687	88,707	80,000	60,000	60,000
Sukabumi	124,545	107,111	101,171	90,000	70,000	70,000
Subang	167,059	159,166	147,241	125,000	125,000	125,000
Sumedang	69,168	72,625	73,114	75,000	77,000	77,000
Garut	110,746	119,941	128,066	150,000	190,000	190,000
Purwakarta	28,886	33,565	36,820	45,000	63,000	63,000
Karawang	185,147	199,085	208,730	228,000	228,000	228,000
Bekasi	101,964	115,243	124,023	125,000	125,000	125,000

Source: Dinas Pertanian Jawa Barat 2002

Table 5 Projected industrial area within Citarum watershed under baseline, SRESA2 and SRESB2 scenarios from 2000 to 2080

Scenario	Industrial area (ha)					
	2000	2005	2010	2020	2050	2080
Baseline[1]	19,440	23,000	27,000	35,000	50,000	60,000
SRESA2[2]	19,440	23,000	28,000	38,000	52,000	65,000
SRESB2[2]	19,440	23,000	30,000	40,000	55,000	70,000

(1) Perum Jasa Tirta II 2003
(2) Growth rates under these scenarios were developed based on growth rates from IPCC scenarios (IPCC 2000)

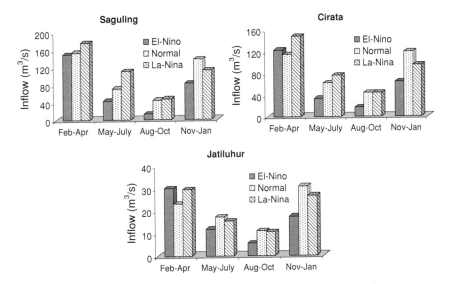

Fig. 2 Mean inflow to the three dams during El Niño, normal and La Niña conditions

Results of Analysis

Impact of ENSO on Streamflow

The impact of ENSO on inflows to the three dams was found to be significant in particular seasons. The impact on streamflow from February to April was statistically insignificant and the impacts of El Niño and La Niña were inconsistent. The impact of El Niño was clear for May–July, August–October and November–January streamflows, while the impact of La Niña was only clear for May–July inflow. The reduction of inflow to the three dams during El Niño years could be as much as 60% of normal inflow (Fig. 2).

Further analysis of the long-term historical streamflow data of Citarum watershed at Nanjung station also showed similar results. The impact of El Niño was significant only during May–July and August–September streamflows. The results of the regression analysis between the seasonal rainfall and SOI showed that May–July streamflow increased by 0.37 m^3 s^{-1} for every 10 unit increase in SOI, while August–October streamflow increased by 0.24 m^3 s^{-1} for every 10 unit increase in SOI (Fig. 3). The November–January and February–April streamflows were not significantly correlated with SOI.

The seasonal streamflow distribution developed from SOI phases one month before the season starts is shown in Fig. 4. The distribution suggests that the August-October streamflow distribution changes when the July SOI falls rapidly or is consistently negative, or when the July SOI increases rapidly or is consistently positive. The same relationship applies to November–January streamflow.

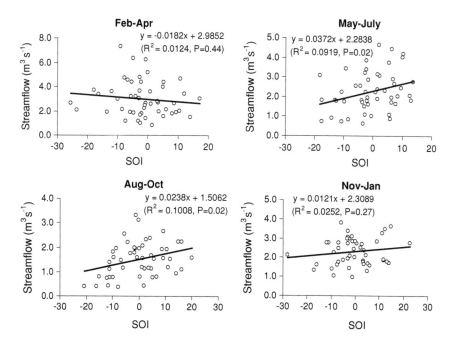

Fig. 3 Relationship between SOI and streamflow of Citarum at Nanjung Station

This analysis suggests that when the SOI phase in July falls rapidly or is consistently negative, the chance of high streamflow from August to October is low. On the other hand, if the SOI phase in July increases rapidly or is consistently positive, the chance of having high streamflow in August–October will increase. Similarly, the chance of having high streamflow from November to January will be low if the October SOI falls rapidly or is consistently negative. For example, the probability of having a November–January streamflow of least 2 m^3 s^{-1} when the October SOI falls rapidly or is consistently negative (Phase 1 + 3) is only 0.35, but when the October SOI increases rapidly or is consistently positive (Phase 2 + 4), the probability will increase to more than 0.60 (Fig. 4). This finding is consistent with a study conducted by D'Arrigo et al. (2009). They found a significant correlation between Citarum streamflow and Southern Oscillation Index (SOI) for June–September.

Water Supply

In the Citarum watershed, the relationship between annual rainfall and total annual streamflow is presented as simple linear regression equations (Fig. 5). Every 1,000 mm of rainfall yields 547 mm of streamflow for the upstream area, 736 mm for the middle area, and only 92 mm for the lower area.

3 Hydrologic Balance of Citarum Watershed

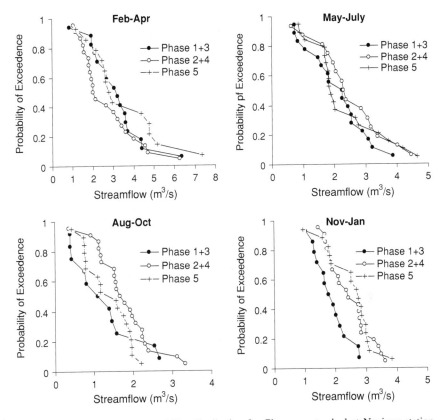

Fig. 4 Seasonal streamflow probability distribution for Citarum watershed at Nanjung station, associated with SOI phase of the previous month

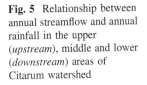

Fig. 5 Relationship between annual streamflow and annual rainfall in the upper (*upstream*), middle and lower (*downstream*) areas of Citarum watershed

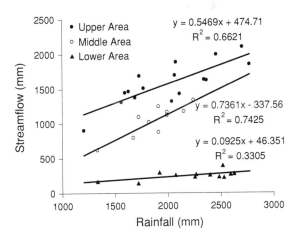

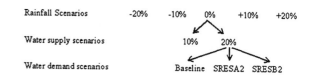

Fig. 6 Water balance scenarios

The middle and lower areas receive water primarily from the local rivers, as well as dams (see Fig. 1). The middle area receives outflow from Saguling Dam (Inflow Saguling-Is), and the lower area from Cirata Dam (Inflow Cirata-Ic). The outflow from Saguling Dam is approximately 710 mm per year (equivalent to 56 m³ s⁻¹) during normal years and 434 mm (34 m³ s⁻¹) during dry years. The outflow from Cirata Dam is approximately 1,532 mm (122 m³ s⁻¹) during normal years and 888 mm (71 m³ s⁻¹) during dry years; outflow from Jatiluhur Dam is about 2,505 mm (200 m³ s⁻¹) during normal years and 1447 mm (116 m³ s⁻¹) during dry years.

Water Balance

The water balance analysis consists of 30 scenarios: five rainfall scenarios, two water supply scenarios and three water demand scenarios (5*2*3 = 30). A diagram tree of the scenarios is shown in Fig. 6.

No Change in Rainfall Scenario

Under present climate (no change in rainfall) scenario, if increases in the volume of water extracted from the streamflow were limited to 10%, all sub-districts in the region would continue to have a water deficit problem, particularly in the lower areas of sub-districts in Kerawang, Bekasi and Purwakarta (Fig. 7). The water deficit in these sub-districts would total more than 60 m³ per year.

Under projections of demand changes, more areas in the sub-districts of Kerawang, Bekasi and Purwakarta would experience severe water deficits by 2020. In 2080, the water supply for most of the sub-districts in this lower area would be insufficient. Increasing the volume of water extraction by 20% would not change the water status of these sub-districts (Fig. 8). Therefore, these sub-districts are considered vulnerable areas. However, Fig. 8 also shows that by increasing the level of water extraction from 10 to 20%, the status of water balance in a number of sub-districts in Sukabumi and Purwakarta the middle areas would be in surplus most years except 2080. By 2080, all districts in the region would face serious problems with water scarcity.

3 Hydrologic Balance of Citarum Watershed

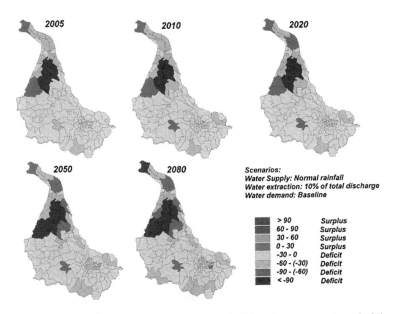

Fig. 7 Projection of water status with no change in rainfall and water extraction of 10% using baseline demand scenario by sub-district in the Citarum watershed

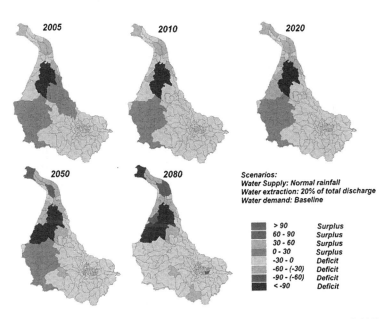

Fig. 8 Projected water status with no change in rainfall and water extraction of 20% using baseline demand scenario by sub-district in the Citarum watershed

Table 6 Percent change in rainfall under SRESA2 and SRESB2 using ECHAM model in West Java, Indonesia

Scenarios	Year	DJF	JJA
SRESA2	2020	0 to 5	−10 to −20
	2050	−5 to −10	−25 to −35
	2080	−5 to −20	−40 to −50
SRESB2	2020	+5 to −5	−10 to −20
	2050	0 to −5	−10 to −15
	2080	−5 to −10	−20 to −30

Change in Rainfall Scenario

As indicated by some of the GCM models, rainfall in West Java may change in magnitude by +5 to -50%. The ECHAM model projects mainly decreases in regional rainfall. The magnitude of the decrease will increase over time from 2020 to 2080 due to global warming (Table 6). Furthermore, rainfall decreases during the dry season are projected to be more pronounced than those during the rainy season. This projection is consistent with historical trends (Pawitan 2002).

By decreasing rainfall by 10 or 20%, and increasing the level of water extraction by 20%, the sub-district in Sukabumi would remain in surplus irrespective of water demand scenarios until the year 2010. If the increase in water extraction is minimized to 10%, all sub-districts would experience deficits similar to those shown in Fig. 7. However, if water demand scenarios followed SRESA2 and SRESB2, the number of sub-districts with a deficit of more than 60 million cubic metres (MCM) would be less. Further analysis showed that if water extraction increased by 10%, an increase in rainfall by 10 or 20% would not change the status of water deficits in the Citarum watershed significantly, irrespective of water demand scenarios. The condition would be similar to those with no change in rainfall (Fig. 7). However, if the level of water extraction were increased to 20%, the water status of most of the sub-districts within Citarum watershed would improve to a surplus (Fig. 9). Sub-districts in Sukabumi might experience a surplus until the year 2080. The CSIRO model suggests that rainfall in West Java will potentially increase up to 20% of the current rainfall under global warming.

Discussion

The results of the analysis suggest that the Citarum watershed is extremely vulnerable to current climatic conditions and, moreover, to future climate change. This finding is consistent with previous analysis conducted by The Directorate of Plant Protection (2000). The aforementioned study reported that West Java is the most vulnerable province to drought and flood events. During El Niño years,

3 Hydrologic Balance of Citarum Watershed

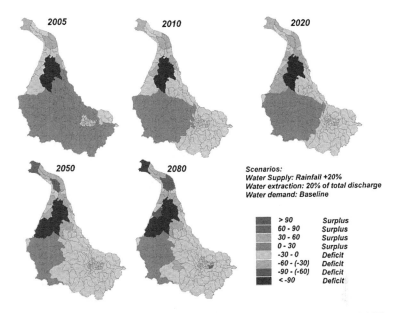

Fig. 9 Projected water status with 20% rainfall increase and water extraction of 20% using baseline demand scenario by sub-district in the Citarum watershed. Note: Water status for SRESA2 and SRESB2 was the same as the baseline

the area suffering from drought increased dramatically, while during La Niña years, areas susceptible to flood also increased significantly. D'Arrigo et al. (2009) found that there was a tendency for greater drought in Java during ENSO warm events (negative SOI).

Our study found that the occurrence of water deficit in the region under current climatic conditions ranges from one to three times over a 10-year period. The frequency of occurrence will potentially increase in the future, as suggested by a number of GCM models such as ECHAM and CGCM (CRU 1999) and based on historical trend data (Pawitan 2002; Kaimuddin 2000). Analysis of annual rainfall data for Citarum watershed over the period 1896–1994 indicated that the annual rainfall in this watershed has decreased at a rate of 10 mm/year. The mean annual rainfall was approximately 2,800 mm per year in the early 1900s, and decreased to about 2,350 mm by the 1990s (Pawitan 2002).

Furthermore, analysis of the long-term land use strategy of Bandung district indicated that the total area of rice paddy field is planned to increase from about 40,000 ha to 100,000 ha (Bapeda, 2002). When this plan is implemented, agricultural demand for water would increase significantly, while the available water supply from Citarum would not change. Under this condition, conflicts between water users might increase. In addition, if programmes for reforesting critical land are not achieved as planned, the area under forest cover might decrease. Consequently, the flood and drought risk in the districts within the Citarum watershed are likely to increase in the future.

Conclusion

Citarum watershed is an important watershed in West Java, Indonesia. Water supply from this watershed is used for many purposes, including electricity generation, domestic consumption, irrigation water and flushing the canal.

The results of the analysis suggest that the Citarum watershed is very vulnerable to current climatic conditions and, moreover, to future climate change. Under the current climate, the chances of experiencing a severe deficit problem are between one and three times over a 10-year period. This frequency is expected to increase in the future.

Under the scenario of no changing climate and if the level of water extraction from the stream flow were limited to 10% of the mean annual flow, it was found that all sub-districts within the Citarum watershed region, particularly in the lower areas in a number of sub-districts in Kerawang, Bekasi and Purwakarta, already have water deficit problems and do not meet their demands. By 2080, the water deficit problem would become even more serious. Increasing water extraction to 20% of the mean annual streamflow would not change the water status of these sub-districts.

Potential water deficit problems may trigger conflicts among water users in these regions in the future. Therefore, the long-term land use strategy for the Citarum watershed and its catchment areas, such as reforesting critical land, should be devised carefully in order to overcome or at least minimize the potential problems of water deficit in the region.

Acknowledgement The authors wish to acknowledge START/GEF/UNEP, who provides financial support to the study. The authors also acknowledge Eko and Mamnun for their assistant in preparing maps and data processing during the study and Jessica for editing the manuscript.

References

ADPC (2000) ENSO impact and potential forecast applications. in Indonesia. Extreme Climate Events Program, Asian Disaster Preparedness Center, Bangkok, Thailand, 90
Bapenas (1991) Perkiraan kebutuhan air industri sampai tahun 2020. In: Pengkajian Kebijaksanaan Strategi Pengembangan Sumberdaya Air Jangka Panjang di Indonesia. Badan Perencanaan Nasional, Jakarta
Bina Program Cipta Karya (1991) Perkiraan perkembangan penggunaan air di Indonesia menjelang tahun 2020. In: Pengkajian Kebijaksanaan Strategi Pengembangan Sumberdaya Air Jangka Panjang di Indonesia. Badan Perencanaan Nasional, Jakarta
CRU (1999) Climate change scenarios for Indonesia. Climatic Centre Research Unit, UEA, Norwich, UK. Available at www.cru.uea.ac.uk
D'Arrigo R, Wilson R (2008) El Nino and Indian Ocean influences on Indonesian drought: implications for forecasting rainfall and crop productivity. Int J Clim 28:611–616
D'Arrigo R, Abram N, Ummenhofer C, Palmer J, Mudelsee M (2009) Reconstructed streamflow for Citarum River, Java, Indonesia: linkages to tropical climate dynamics. Clim Dyn. doi:10.1007/s00382-009-0717-2

Dinas Pertanian Jawa Barat (2002) Jawa Barat dalam Angka. Biro Pusat Statistik Propinsi Jawa Barat, Bandung

Directorate of Plant Protection (2000) Impact of extreme climate events on crop management in Indonesia, Directorate General of Food Crops and Horticulture, Ministry of Agriculture Indonesia in collaboration with NOAA/OFDA. USA and Asian Disaster Preparedness Center, Bangkok

Hernowo (2001) Pengelolaan waduk Ir. H. Juanda di Jatiluhur. In: Prosiding Lokakarya Selamatkan Air Citarum 75:80

IPCC (2000) Emission scenarios, A Special Rerport of Working Group III of the IPCC, Intergovernmental Panel on Climate Change, Cambridge University Press

Kaimuddin (2000) Dampak perubahan iklim dan tataguna lahan terhadap keseimbangan air wilayah Sulawesi Selatan: Studi kasus DAS Walanae Hulu dan DAS Saddang, Disertasi Program Pasca Sarjana, IPB, Bogor

Kirono D, Partridge IJ (2002) The climate and the SOI. In: Partridge IJ, Ma'shum M (eds) Will It Rain?: The effect of the Southern Oscillation and El Niño in Indonesia. Queensland Government, Department of Primary Industry, Australia, pp 17–24

Pawitan H (1996) Keseimbangan Hidrology menurut Kabupaten di Indonesia. Laporan Penelitian ARMP, Bogor (unpublished report)

Pawitan H (2002) Long-term hydrological balance of Citarum Basin and Java island. Indones J Agric Meteorol 16:49–60

Perum Jasa Tirta II (2003) Data utama bendungan Jatiluhur, Perusahaan Umum Jasa Tirta II, Jatiluhur

Stone R, Hammer GL, Marcussen T (1996) Prediction of global rainfall probabilities using phases of the Southern Oscillation Index. Nature 384:252–255

Yoshino M, Urushibara-Yoshino K, Suratman W (2000) Agriculture production and climate change in Indonesia. Global Environ Res 3:187–197

Chapter 4
Towards a Database for an Information Management System on Climate Change: An Online Resource

V. Ramani Bai, S. Mohan and Reza Kabiri

Abstract The aim of the research study is to develop a resourceful database information and management system (DIMS) for climate change and coastal degradation. An important element of the DIMS is to have a coastal resource. This usually requires a great deal of data, and an information system can provide tremendous assistance in organizing, managing, understanding and reporting this information. Together, relational databases and geographical information systems (GIS) provide powerful tools for organizing and analysing environmental data. The climate change coastal resource information system has been designed to be simple, yet flexible. The database structure allows for variation in the level of detail provided for each variable and country. In addition to the ability to view, query and report monitoring data, the DMIS also allows users to display the data spatially using a GIS.

Keywords Database · Climate change · Coastal climate · Geographical information system · Metadata · Geodata · Meteorology

Introduction

Climate change data encompasses a very large array of data objects, with many available in a variety of data formats. It is important to make a database management system (DBMS) such that one can project and model the climate

V. Ramani Bai (✉) · S. Mohan · R. Kabiri
Department of Civil Engineering, Faculty of Engineering,
University of Nottingham (M) Campus, Jalan Broga,
43500 Semenyih, Selangor, Malaysia
e-mail: Ramani-bai.v@nottingham.edu.my

change parameters (Liu et al. 2009). This paper emphasizes the essential need for a total approach to climate change data management, particularly through the development and application of scenarios at the scale of regions to assess the response of natural (land and water) systems at the local level. The climate change data management initiative is the result of a link between ESRI (2002) and the climate change sciences community regarding climate change data representation and analysis challenges. Collaboration towards a development, using the common data model, helps to address the needs of the climate change community, and provides direction for ESRI software and tool development (Barnolas and Llasat 2007).

The ultimate goal of an ArcGIS climate change data management and model is to represent each of these data objects in a uniform manner, allowing their integrated analysis in the ArcGIS desktop environment. Some of the data considered in the climate change study includes:

1. Map backgrounds, geopolitical boundaries
2. Climatology and meteorology (rainfall, temperature, humidity, solar radiation, wind speed, wind direction, etc.)
3. Oceanographic observations and products (surface and deep)
4. Geological observations and products
5. Sociological (land use, population, etc.)
6. Surface properties (soils, soil moisture, etc.)
7. Biological observations and products (vegetation cover, agricultural production, etc.)
8. Space environment (STP, solar activity, magnetic fields, etc.)
9. Hydrology
10. Sub-surface water
    - Emission inventories (EPA, etc.)
11. Emergency management criteria, constraints

Problem Definition

This paper emphasizes the essential need for a total approach to climate change data management, particularly through the development and application of scenarios at the scale of regions to assess the response of land and water systems at the local level and their policy issues. Physical changes are implicated in the functioning of our regional climate. In addition, the biophysical and ecological functions of our environments also experience the impacts of climate change. Scientific evidence is increasingly indicating that environmental changes are occurring at all scales, as a result of climate change and climate variability. The research focuses on database management systems such as geographic distribution, the impact of climate change, etc.

Goals and Objectives

The concept is aimed for use by environmental sectors and for reporting to environmental authorities. The focus is on environmental data relevant to the state of climate change in the world. The objectives of this paper include:

a. Development of an integrated database management,
b. Share determination of the different ecosystems in environmental categories (global warming, Ozone depletion, etc.) in climate change study,
c. Environmental promotion of sensitive ecosystem and sustainable development, human health and improvement by database management,
d. Improvement of inventory and climate model.

Materials and Methods

This trend has been further developed with the progressive evolution and widespread use of the Internet and World Wide Web (WWW): with the expansion of Geographical Information Systems (GIS) onto the Internet, access to geographic data, and the manipulation of it, will become even easier for the environmental manager who may not necessarily be a computer applications specialist or scientist. The importance of being able to share scientific data has received increasing attention in the last few years from many researchers. Şahin and Kurum (2002) analysed by GIS the areas with higher landscape value in the impact assessment of dam constructions in the Seyhan-Köprû Hydroelectric Dam project proposal. The studies assessed the GIS tools and found them to be crucial for impact assessment and predicted that usage will dramatically increase in the near future.

A considerable amount of time and effort has already been directed towards the development of data models, data formats and spatial data infrastructures (SDI) to help overcome the problems associated with data sharing. Efficient implementation and monitoring of environmental measures related to climate change requires interoperable spatial information across national borders.

To effectively implement a GIS using the geodatabase, a solid database design and development must be put in place. In the GIS, the geodatabase provides a framework for geographic information and supports topologically integrated feature classes (Barnolas and Llasat 2007). Figures 1 and 2 refer to the methodology of development of the database system through this research.

The conceptual design illustrates the database organization and structure in tables and Unified Modelling Language (UML) diagrams that define the features and geographic representation of the required ArcInfo datasets and relationships. It also shows the user's view of the data within a database environment. This step basically involves the actual creation of the geodatabase tables from the abstract features defined in the study that is a skeleton of the final geodatabase. NIWA (2009) web information is also captured for database related to lakes, aquaculture and many other recent information on environment.

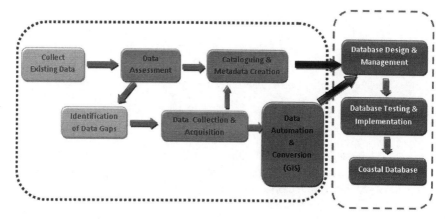

Fig. 1 Methodology of new database information management system

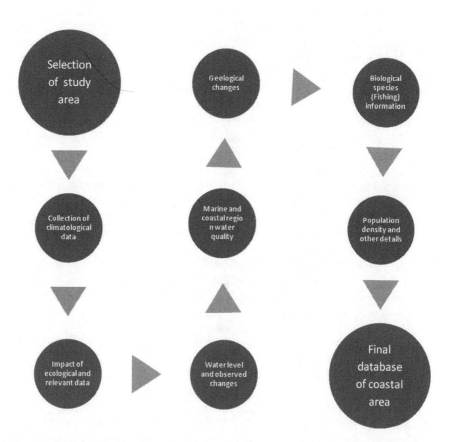

Fig. 2 Task flow in database system

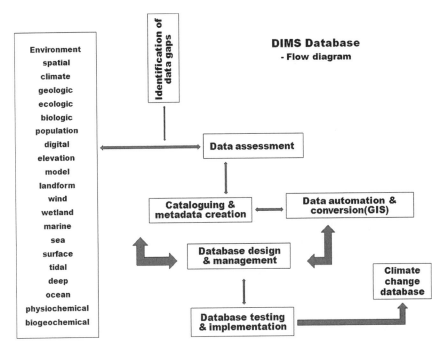

Fig. 3 GIS workflow diagram in DIMS

Metadata is a standard part of any GIS and is not an extension of functionality. The tools for creating and managing metadata are key technologies for sharing geographic information and building communities. This tool is in the core of ArcGIS software. Metadata makes spatial information more useful to all types of users by making it easier to document and locate data sets (Wilhelmi et al. 2005). It is a summary document providing content, quality, type, creation, and spatial information about a data set. It can be stored in any format such as a text file, Extensible Markup Language (XML), or database record. Because of its small size compared to the data it describes, metadata is more easily shareable. By creating metadata and sharing it with others, information about existing data becomes readily available to anyone seeking it. Metadata makes data discovery easier and reduces data duplication.

Next, design and implementation of a data-driven website and relational database consisting of climatic, geologic, ecologic, biologic (both quality and quantity) information, and spatial data will be done with GIS tools, as well as greenhouse gas (GHG) emissions and global warming to predictions of local impacts and system feedbacks (Liu 2009; Baker 2009). For proper development and use of the geodatabase, an accompanying data management plan will be developed. It will review important issues the environment must address such as the server configuration, user access, security, workflow and data location. The flow of work in creation of database information and management system (DIMS) is shown in Fig. 3.

Once the conceptual and logical geodatabase are developed, any of three methods below can be used to create the physical geodatabase:

- Migrating existing coverage/shapefile data into the geodatabase
- Creating a new geodatabase from scratch using ArcCatalog
- UML and computer-aided software

ArcGIS software uses the XML standard for metadata processing. Every feature dataset, feature class and attributes will be defined and created in this template, so the actual data layers can be placed into the geodatabase. In the GIS, the geodatabase provides a framework for geographic information and supports topologically integrated feature classes.

These data sets are stored, analysed and queried as layers similar to the coverage and shapefile models (Zeiler 1999). The system is based on web-based client–server architecture. The content management system will be handled by Joomla and application customization will be carried out using PHP to implement searchable interfaces. The web mapping server handles the linkage between spatial objects and non-spatial attribute data stored in a relational database. The web-based system will allow the users to interactively query, visualize data and analyse spatially through decision support tools.

Conclusion

From the layers used in climate change study, we derive a more comprehensive list of elements that are used to help analysis. It is here that we add attributes to features. This helps to determine which objects will be represented as points, lines and polygons, and also in the form of raster and tables. The approach provides a constant flow for the exchange of ideas, data and methods that works towards an end result that works in a GIS. In recent years, we have acquired more sources of information, including newspapers, showing some ordinary climate change that in the past could have gone unnoticed. This geodatabase will be created. Besides its wide geographic coverage, the climate change geodatabase also offers the advantage of containing updated information up to years included in scenarios. It is very important to have the information updated to the model.

In order to store, manage and analyse all the information available, a relational structure should be chosen. This geodatabase will be implemented on a GIS. It is a more efficient way to store information and to analyse it. In this way, it can be a helpful tool aimed at improving climate change assessment in addition to vulnerability information and others.

Supporting Information Available

Detailed set of database, products, events, key partners and other useful information are available on the Internet for public use. The materials are currently available on request at www.globalclimate-engine.org and will be provided free of charge once proper legal and policy matters are settled by end of the project period, which is June 2011.

Acknowledgments The authors acknowledge the research grant provided by the Asia Pacific Network, Japan, for their support and publication of the paper that has resulted in this article. The authors also thank the collaborating organizations of the research work for their fine co-ordination between the researchers.

References

Baker M (2009) Updated GIS database design: geodatabase model. Virginia, 30 pp
Barnolas B, Llasat MC (2007) A flood geodatabase and its climatological applications: the case of Catalonia for the last century. Nat Hazards Earth Syst Sci 7:271–281
ESRI (2002) Metadata and GIS. An ESRI White Paper. ESRI, pp. 1–13, www.esri.com
Liu J (2009) A GIS-based tool for modeling large-scale crop-water relations. Environ Model Softw 24:411–422
Liu DL, Mob J, Fairweather H, Timbald B (2009) A GIS tool to evaluate climate change impacts. In: Anderssen RS, Braddock RD, Newham LTH (eds) 18th World IMACS congress and MODSIM09 international congress on modelling and simulation. Modelling and simulation society of Australia and New Zealand and International association for mathematics and computers in simulation, pp 1936–1942
NIWA (2009) Fresh water biodata information system (FBIS). NIWA, New Zealand, available at https://secure.niwa.co.nz/fbis/index.do
Şahin Ş, Kurum E (2002) Erosion risk analysis by GIS in environmental impact assessments: a case study Seyhan Köprü dam construction. J Environ Manag 66(3):239–247
Wilhelmi O, Betancourt T, Boehnert J, Shipley S, Breman J (2005) Arcgis atmospheric data model. ArcGIS Data, 35 pp, available at http://www.gis.ucar.edu/sig/data_model/documents/ADM_CFD4.pdf
Zeiler M (1999) Modeling our world. ESRI Press, Redlands

Chapter 5
Assessing the Impact of Climate Change on Evapotranspiration and Soil Salinization

Mohammad Hassan Rahimian and Samaneh Poormohammadi

Abstract In this study a hypothesis was tested and evaluated: evapotranspiration of vegetation cover in dry saline waterlogged areas has a negative impact on salinization of soil resources due to pumping of saline underground water to the surface. For this purpose, reference, potential and actual evapotranspiration (ET) were simulated over the 2010–2039 time period via integration of Surface Energy Balance Algorithm for Land (SEBAL), MODIS remotely sensed satellite images and General Circulation Model (GCM)-derived data. Results of the study disproved the mentioned hypothesis; ET comparisons between green vegetated surfaces and bare abandoned lands represented the positive role of surface vegetation in alleviation of soil salinity in the studied conditions. This means that acceleration of the salinization rate is probable in a lack of vegetation cover condition, especially when the groundwater is saline-shallow.

Keywords Climate change · Evapotranspiration · SEBAL · HadCM3 · Salinization · Iran

Introduction

Iran, with about 25.5 Mha of slightly to moderately salt-affected soils and with 8.5 Mha of severe saline soils, is a country that is now faced with salinity problems in more than 20% of its area (Moameni et al. 1999). Due to different geologic,

M. H. Rahimian (✉) · S. Poormohammadi
RS & GIS Department, National Salinity Research Center (NSRC),
89195-315 Yazd, Iran
e-mail: mhrahimian@gmail.com; rahimian@insrc.org

climatic and hydrologic conditions in the country, different factors take part in soil salinization. These can be classified in two groups of natural and anthropogenic factors (Siadat et al. 1997). Environmental factors play the most important role on rate and spatial distribution of natural soil salinization. This type of salinization has affected the main portion of soil and water resources in the country and also agricultural yields, especially in arid, semi-arid and shallow saline groundwater environments. In such conditions, evapotranspiration (ET) is known as a main accelerative factor for salinization of soil resources.

This phenomenon (ET) is a combined process of transpiration from surfaces of plant leaves and evaporation from the soil in which the plants are growing. It counts the total volume of water lost through transpiration and evaporation processes and is expressed in cubic metres, or mm in a specified time period (Allen et al. 1998). Terminologically, three main types of ET can be defined: actual, potential and reference ET. Actual evapotranspiration (ET_a) is a process in which crop demand for water mainly controls, through crop leaves and stomata status, climatic conditions, environmental stresses, soil features and management practices. In contrast, potential ET (ET_p) is an indicator of crop-required water for healthy growth and productivity in non-stressed, disease-free and water-abundant conditions. Reference ET (ET_o) also demonstrates evaporative demands of the atmosphere and fully depends on climatic conditions. In fact, ET_o is known as a crop-independent term of ET. Among these different ET terminologies, ET_a seems to be paramount in illustration of wasted water through plant evapotranspiration, especially when the area is being under stress. It is also a good indicator for the illustration of CO_2 impact on stomatal conductance and photosynthesis and, therefore, crop productivity and water use efficiency (Le Houerou 1996). In areas with a shallow saline water table, actual ET, caused primarily by deep-rooted phreatophytes, is a significant factor for capillary rise and consequently salinity build-up hazard in the region. In contrast, when the soil is bare and also availability of water is limited to shallow groundwater, reference ET can demonstrate the power of ET for acceleration of capillary rise and movement of salts to the surface. Therefore, to investigate the effect of ET on salinization hazard in areas with a shallow saline water table, two conditions of bare soil and vegetated surfaces should be considered together with the help of ET_o and ET_a, respectively. Since ET is controlled by different climatic factors (e.g. temperature, sunshine, wind and humidity) it would be possible to simply monitor the impact of climate change on this phenomenon as well as the impact on soil salinization in areas with shallow saline groundwater.

At present, there are different approaches for determining actual ET. Examples of techniques used to estimate ET include water balance, lysimeters, micro-meteorological techniques such as Bowen Ratio and Eddy Covariance, and water table fluctuations (Scanlon et al. 2005). Information on ET in Iran is generally limited to meteorological station data used to estimate potential ET in different parts of the country. But none of these measurements provide estimates of actual ET. New procedures such as remotely sensed (RS) energy balance algorithms are other beneficial techniques for estimation of ET. The advantage of these techniques is that they provide a good illustration of actual ET in the region both spatially and temporally (Allen et al. 2003). These techniques are now available in Iran in addition to the availability of high, moderate or low resolution satellite images (Poormohammadi et al. 2010).

In this study, Surface Energy Balance Algorithm for Land (SEBAL) was used to determine actual ET in the satellite overpass times. Since data derived from the General Circulation Model (GCM) is also available, a link between RS techniques and GCM data would help towards a proper assessment of climate change impacts on actual ET in future decades. The main objective of this study is to assess the impact of climate change on potential and actual evapotranspiration of bare lands and vegetated surfaces and, consequently, their impacts on changes of soil salinity status in Azadegan plain, west of Iran, a region with severe salinity and waterlogging problems, which are the main hazards to its resources.

Materials and Methods

Study Area

The study area is Azadegan plain or Dasht-e-Azadegan (DA), one of the plains located in the lower part of the Karkheh river basin (KRB) in Khuzestan province, west of Iran. It lies between latitude 31°05' to 31°51' N and longitude 47°43' to 48°39' E (Fig. 1). The total area of DA is 334,000 ha, of which about 250,000 ha have been cultivated and used for agricultural purposes. Based on the statistics, the annual long-term precipitation in DA varies between 150 and 250 mm in different parts of the region. Meanwhile, annual evaporation is between 3,500 and 3,700 mm.

Methodology

Figure 2 shows the methodology used to estimate actual, potential and reference ET in the future period of time and, consequently, the impacts on soil salinization in two soil surface conditions of bare and green vegetated surfaces. As shown, in this study two main sources of data were used: historical and simulated data. Historical data includes T_{min}, T_{max}, sunshine, humidity and wind speed of the Bostan meteorological station from 1960 to 1989. In addition, a time series of satellite images was acquired and used as historical RS data. GCM data includes monthly data of the HadCM3 model for the projected period of 2010–2039 that resulted from GCM-runs for the Third Assessment Report (TAR) based on the IPCC[1]-SRES scenario of A2 and B2, respectively. HadCM3 is a well-known GCM

---

[1] Intergovernmental Panel on Climate Change.

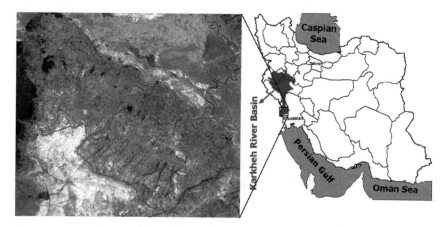

Fig. 1 Satellite image of Azadegan plain in the Karkheh river basin, Iran, and its geographical setting on the country map

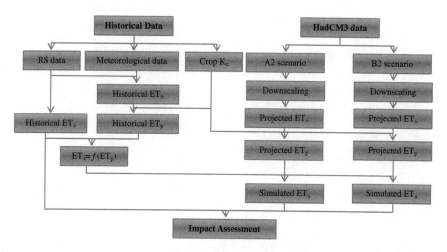

Fig. 2 Flow chart of the methodology used for impact assessment of climate change on evapotranspiration and soil salinization in Azadegan plain, west of Iran

model that is used worldwide (Bates et al. 2008). Key assumptions in A2 scenario are a very heterogeneous world, strengthening regional cultural identities, with an emphasis on family values and local traditions, high population growth, and less concern for rapid economic development. B2 emission scenario represents a heterogeneous world with less rapid and more diverse technological change but a strong emphasis on community initiative and social innovation to find local, rather than global solutions (IPCC 2010). In fact A2 and B2 are the most frequently used emission scenarios in the world.

To determine actual ET, a time series of remotely sensed images and also ancillary climatic data was prepared. RS data included 18 cloud-free images that were acquired from the Terra (MODIS) satellite during the wheat cropping season. Surface Energy Balance Algorithm for Land (SEBAL) was used to determine ET_a. This algorithm is a relatively new procedure for estimating actual ET at different scales, e.g. farm, catchments and basin levels, and is now used widely around the world and also in Iran for computing crop water requirements under standard and non-standard conditions. Generally, SEBAL uses digital images acquired by satellites that record thermal infrared radiation in addition to visible and near-infrared radiation and computes ET on a pixel-by-pixel basis for the instantaneous time of the satellite overpass. This algorithm predicts ET from the residual amount of energy in the form of latent heat flux (LE) and is equal to Rn-G-H, of which Rn is above-ground surface net radiation, H is sensible heat flux and G is the ground heat flux. Details about the SEBAL procedure and steps for running the algorithm are presented in Bastiaanssen et al. (1998a, b), Tasumi et al. (2000). In addition, reference ET was calculated through the Hargreaves–Samani procedure (Alizadeh 2004). Thereafter, potential ET of winter wheat, as dominant vegetation cover, was calculated by considering its coefficient (K_c), which was varied during the growth season (0.3–1.15). Afterwards, relationships between ET_p and ET_a were investigated to develop the best-fit model for determining actual ET via potential ET. This model was then used to project actual wheat ET in the future period of time. For this reason, General Circulation Model (GCM)-derived data was linked to the Hargreaves equation, first to project reference and potential ET and then to insert into the developed model for simulation of actual wheat ET over the next three decades. This procedure was accomplished for the 2010–2039 projection period using HadCM3-A2 and HadCM3-B2 data, respectively. Since HadCM3 data was available in a low spatial resolution of $2.5° \times 3.75°$ (latitude by longitude), a stochastic approach was used for downscaling HadCM3 data (both A2 and B2) in the projected period of 2010–2039 (Massah Bavani 2006).

Results and Discussion

Figure 3 shows the best-fit quadratic model between historical data of ET_a versus ET_p. As was mentioned before, potential ET has been estimated through the Hargreaves equation by considering the winter wheat coefficient (K_c). Fair correlations could be found in the model ($R^2 = 0.72$). This model was then employed to simulate wheat actual ET during the 2010–2039 time period. As shown in the Figure, for ET_p of less than 6 mm/day, a higher value for ET_a would result in higher ET_ps through this model. After this point, the ET_a values would have a descending trend. It is noticeable that ET_p of more than 6 mm/day explains the shifting climate to some drier conditions, lack of water for evapotranspiration and consequently decrease of ET_a, as shown in the model outputs. Therefore, it is anticipated that the lower ET_a would be projected in

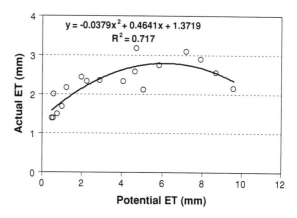

Fig. 3 Best-fit quadratic model for simulation of actual ET using potential ET in Azadegan plain, west of Iran

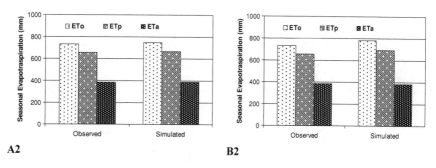

Fig. 4 Observed and simulated seasonal evapotranspiration (ET_o, ET_p, ET_a) for 2010–2039 using different scenario data (A2 and B2)

the future compared with the historical period of time, when the climate shifts to higher ET_p values.

Figure 4 shows averages of observed and simulated reference ET (ET_o), potential ET (ET_p) and actual ET (ET_a) in historical and projected periods of time, based on A2 and B2 scenario data, respectively. As was mentioned before, ET_a terminologically belongs to green vegetated surfaces and reflects the crop response to water availability, while ET_o illustrates the evaporative demands of the atmosphere and depends fully on climatic conditions of the region. ET_o can therefore represent the evaporation rate in bare soil and consequently explains the movement of salts to the surface.

As shown in this figure, both simulated ET_o and ET_p tend to higher values compared with observed values. In contrast, differences between observed and simulated values of ET_a for 2010–2039 seem to be negligible, both for A2 and B2 scenarios. This means that evaporative demand of the atmosphere would increase in the upcoming three decades, but green vegetated surfaces would prevent the wasting of water to the atmosphere. On the other hand, salinity of soil in vegetated surfaces tends to be constant, but salinity build-up hazard is expected for bare surfaces. In fact, the role of vegetation cover seems to be significant in preventing the salinization hazard in the upcoming three decades in the region under investigation.

Conclusion

Key deterministic factors in salinization of soil resources in the studied condition included climate, vegetation cover, salt stores, and geological and topographical features of the region. In addition, the roles of these factors in contributing to salinization were interrelated. Among these factors, fluctuations of climate and surface vegetation cover were found more probable as results of climate and land use changes, respectively. Among these, the role of vegetative surface seems to be paramount in alleviating salinity in a long-term period and in a global changing climate. This means that acceleration of the salinization rate is probable in a lack of vegetation cover condition. In this way, tactics and strategies to mitigate climate change impacts via a proper land use and sustainable environment programme must be implemented and supported by decision-makers and social, political and other related sectors.

References

Alizadeh A (2004) Soil, water, plant relationships, 4th edn. Ferdowsi University of Mashhad press, Iran, p 470
Allen RB, Pereira LS, Raes D, Smith MS (1998) Crop evapotranspiration (guidelines for computing crop water requirements). FAO irrigation and drainage paper, vol 56, p 300
Allen RG, Morse A, Tasumi M (2003) Application of SEBAL for western US water rights regulation and planning. ICID workshop on remote sensing of ET for large regions
Bastiaanssen WGM, Menenti M, Feddes RA, Holtslag AAM (1998a) A remote sensing surface energy balance algorithm for land (SEBAL). Part 1: formulation. J Hydrol 212–213:198–212
Bastiaanssen WGM, Pelgrum H, Wang J, Ma Y, Moreno J, Roerink GJ, van der Wal T (1998b) The surface energy balance algorithm for land (SEBAL). Part 2: validation. J Hydrol 212–213:213–229
Bates BC, Kundzewicz ZW, Wu S, Palutikof JP (eds) (2008) Climate change and water. Technical paper of the intergovernmental panel on climate change, IPCC Secretariat, Geneva, p 210
IPCC (2010) IPCC-DDC GCM data archive. http://www.mad.zmaw.de/IPCC_DDC/html/ddc_gcmdata.html. Accessed September 2010
Le Houerou HN (1996) Climate change, drought and desertification. J Arid Environ 34:133–185
Massah Bavani AR (2006) Risk assessment of climate change and its impacts on water resources, case study: Zayandeh Rud basin, PhD thesis, Tarbiat Modarres University, Iran, pp 189
Moameni A, Siadat H, Malakouti MJ (1999) The extent distribution and management of salt affected soils of Iran. FAO global network on integrated soil management for sustainable use of salt affected soils. Izmir, Turkey

Poormohammadi S, Rahimian MH, Taghvaeian S (2010) Applying remotely sensed energy balance methods in Iran, potentials and limitations. Remote Sensing and Hydrology Symposium, Jackson Hole, Wyoming, USA

Scanlon B, Keese K, Bonal N, Deeds N, Kelley V, Litvak M (2005) Evapotranspiration estimates with emphasis on groundwater evapotranspiration in Texas. Texas Water Development Board, Texas

Siadat H, Bybordi M, Malakouti MJ (1997) Salt-affected soils of Iran: a country report. International symposium on sustainable management of salt affected soils in the arid ecosystem, Cairo

Tasumi M, Bastiannssen WGM, Allen RG (2000) Application of the SEBAL methodology for estimating consumptive use of water and streamflow depletion in the Bear River basin of Idaho through remote sensing, appendix C: a step-by-step guide to running SEBAL, Final report, The Raytheon Systems Company, EOSDIS Project

Part II

Chapter 6
The South African Water Sector: On its Way Towards Adaptive Water Governance?

Elke Herrfahrdt-Pähle

Abstract The African continent is likely to be highly affected by the consequences of climate change. The ensuing projected changes in the ecological system such as decreasing water availability and higher amplitudes of droughts and floods require responses from the social system, i.e. adaptation measures. Even though climate modelling is still plagued by high uncertainty, models suggest a reduction of precipitation and runoff for South Africa. This poses a major threat for South Africa, which depends heavily on surface water and whose water resources are already under stress. Adaptive governance has been proposed as an approach that provides a way to cope with rising uncertainty and environmental change. Against the background of decreasing water availability and in addition to technical solutions, adaptive water governance structures can serve as an important element for increasing water use efficiency, adaptive capacity and resilience of the water sector. The first part of the paper takes a conceptual approach towards adaptive water governance. It suggests characterizing adaptive water governance by referring to and merging elements of good governance, water governance and adaptive governance. In the second part of the paper, some of the elements of adaptive water governance are examined with regard to their relevance for South Africa's water governance reforms. The paper concludes that, on the one hand, progress can be attested with regard to (1) a high level of flexibility built in the national water legislation, especially the National Water Act, (2) the provision of buffer capacities regarding institutions for solving water conflicts (redundancy), (3) the high level of participation of stakeholders at many levels and (4) the provision of equitable access to water. On the other hand, some (unintended) negative implications of these developments are documented. Among them are (1)

E. Herrfahrdt-Pähle (✉)
German Development Institute/Deutsches Institut für Entwicklungspolitik (DIE),
TULPENFELD 6, 53113 Bonn, Germany
e-mail: elke.herrfahrdt@die-gdi.de

the time-consuming processes of elaborating procedures and guidelines in the process of implementation, (2) the potentially negative effects of functional overlaps of CMA and regional offices of DWA, (3) effectively involving relevant stakeholders and (4) the lacking long-term sustainability of water infrastructure due to lack of capacity and capital.

Keywords Adaptation · Adaptive governance · Water governance · Good governance · Resilience · South Africa

Introduction

On the African continent, climate change impacts are expected to be especially severe. The Intergovernmental Panel on Climate Change (IPCC) states that "warming [in Africa] is very likely to be larger than the global, annual mean warming throughout the continent and in all seasons, with drier subtropical regions warming more than the moister tropics" (Christensen et al. 2007, p. 850). For South Africa, climate change models suggest that the eastern part of the country is likely to receive more rain, while the western parts become even drier (Lumsden et al. 2009). Due to its close interconnectedness with the climate system, the hydrological system is very sensitive to changes in the climate. A 20% decrease in precipitation might lead to a decrease of up to 70% of the drainage in some parts of the Orange-Senqu river basin, which serves as a major water source for irrigation agriculture (De Wit and Stankiewicz 2006). However, modelling the consequences of climate change is still afflicted with considerable uncertainties, especially concerning local projections of climate change impacts. Furthermore, uncertainties arise since past experience may often no longer provide reliable guidance for the future, and governance of natural resources is confronted by unprecedented situations (Pahl-Wostl et al. 2007).

These developments of the ecological system and increased uncertainties put the social system under pressure to change and adapt.[1] Climate change threatens recent achievements in development, thus underlining the urgency of action, i.e. adaptation, especially in developing countries. Economic and social development has so far assumed environmental stability and relied upon relatively stable environmental conditions. In the past, the focus of natural resource use has been on managing and governing for stability and efficiency. This produced good results in the short term while in the long term it led to resource degradation and decreasing ecological resilience (Olsson et al. 2006). Faced with uncertainty, the results

---

[1] A social system is defined as all man-made structures, relations and objects and encompassing social and economic aspects, while an ecological system is a system of interrelated and dependent organisms or biological units (Anderies et al. 2004). Interlinked and interdependent social and ecological systems constitute a social-ecological system (SES; Berkes et al. 2003).

of these approaches have been surprise and ecological and economic losses (Holling and Gunderson 2002).

At the same time, in the context of climate change, natural resource governance needs to reflect and be able to deal with uncertainties as well as gradual and abrupt change, i.e. it needs to become adaptive. It needs to be in accordance with the requirements of sustainable development, ensuring a balance of social, economic and environmental needs, not only today but also in the future. Sustainable use of natural resources in a changing environment is more likely if resilience and adaptive change become integral parts of natural resource and especially water governance. In this sense, building resilience can be termed a sustainable response to climate change (Tompkins and Adger 2003). With a view to natural resource governance and especially institutions, the problems of fit and interplay have been highlighted as important realms on the way towards sustainable resource governance regimes (Young 2002). The requirement of fit between an institutional setting and the natural resource it is addressing becomes even more important in a changing environment, highlighting the need to adapt. The problem of interplay, i.e. the interaction and coherence of the various institutions at different levels or sectors of natural resource governance regimes is also gaining importance, since climate change is a cross-cutting issue, which needs to be addressed coherently across sectors and administrative levels.

The need for adaptation is especially high in the water sector, which will be especially affected by climate change through changes in precipitation patterns and river runoff, increased evaporation from water bodies and evapotranspiration of plants. Technical measures for adaptation only constitute part of the solution. Firstly, climate change is likely to produce an impact on livelihoods high above any extent of technical progress. Secondly, already available technical solutions are unlikely to be affordable by the people hit hardest by climate change. The social system and especially the water sector need to develop strategies and structures enabling it to cope with the challenges of climate change, i.e. it needs to increase its adaptive capacity and resilience. With decreasing water availability and increasing demand, the need for effective governance, enabling conflict management and water allocation with the consent of the majority of the population, increases. Adaptive governance has been proposed as an approach that provides the means to cope with rising uncertainty and environmental change (Dietz et al. 2003; Folke et al. 2005).

This paper aims at developing a concept of adaptive water governance and applying it to the water sector. In the following section, such a concept is derived by combining the concepts of good governance, water governance and adaptive governance. The elements of adaptive water governance reflect the dimensions of sustainable development, as well as addressing the problems of fit and interplay. The third section provides an application of this concept to the South African context of water governance reform. Finally, conclusions are drawn with respect to the adaptability of the South African water governance regime.

The paper builds upon findings from research on the transition towards adaptive water governance regimes in the context of climate change in South Africa undertaken as part of the NeWater Project. It draws upon a literature review and several weeks of field research. The research was informed by a qualitative research paradigm. Empirical data on water governance reforms, governance structures and processes (especially regarding the elements of adaptive water governance described above) was mainly derived from 34 semi-structured interviews with policy-makers, water managers, stakeholders and scientists. Subsequently, data was updated and validated via contacts and discussions with colleagues from South Africa.

Towards a Concept for Adaptive Water Governance

A water governance regime that is able to cope with the challenges of climate change needs to, firstly, have a high level of resilience (Folke 2006). This includes flexibility to confront uncertainty and to react to surprise and changing climatic, hydrological and social conditions. It thus needs to be adaptive. Secondly, it should have the support and trust of the people and build upon a broad base of knowledge and experience. These issues are covered by the concept of good governance (Doornbos 2003). Finally, it should take the specific necessities of the water sector and the water cycle into account. This requirement is met by the concept of water governance (GWP 2003). A combination of adaptive governance, good governance and water governance is suggested as a way to capture the elements of adaptive water governance. In the following, the concepts are briefly outlined before they are merged to form a sustainable concept of adaptive water governance.

Adaptive Governance

"Adaptive governance is a process of creating adaptability and transformability in [social-ecological systems]" (Walker et al. 2004, p. 7).

Adaptive governance acknowledges that modern governance regimes increasingly face complexity, that governance resources change, and that surprise and new uncertainties are inevitable. However, the term remains ambiguous in the resilience literature (Jordan 2008). In this paper adaptive governance is defined as "the evolution of rules and norms that better promote the satisfaction of underlying human needs and preferences given changes in understanding, objectives, and the social, economic and environmental context" (Hatfield-Dodds et al. 2007, p. 4). Elements of adaptive governance most frequently mentioned in the literature are flexible and redundant institutions as well as multi-level, horizontal and polycentric governance structures (Folke et al. 2002; Folke et al. 2005; Olsson et al. 2006; Pahl-Wostl 2009).

At this point, the mutually reinforcing relationship between adaptive governance and adaptive management needs to be addressed. Adaptive management is an attempt to address uncertainties in everyday water resources management through the integration of feedback loops to accelerate learning, increased flexibility and perceiving policy as a set of controlled experiments (Mclain and Lee 1996; Folke et al. 2002). Adaptive management thus provides the basis for being able to take informed decisions on water allocation and distribution. Being a management tool for addressing uncertainties that remain in planning, it complements adaptive governance at the management level. Adaptive management relies on functioning institutions and institutional capacity and thus on adaptive governance (Medema and Jeffrey 2005). Water management takes place within and is influenced by the water governance system. It needs to be supplemented and facilitated by fitting (and possibly adaptive) governance structures that enable adaptive management of water resources.

Good Governance

"Good governance must underpin effective adaptation strategies" (AfDB et al. 2003, p. 24).

In the 1990s, international development agencies coined the term "good governance" as they found that shortcomings in development were linked to deficient governance structures and processes (Dolzer 2004). Good governance was identified as central to strong and equitable development and thus as a precondition for economic prosperity and political stability (World Bank 1992). Bad governance, on the contrary, often serves as a barrier to development, increases vulnerability and decreases the ability to react to change and adapt (GWP 2003). Good governance has a strong influence on social-ecological systems and their ability to react to disturbances (Walker et al. 2006, p. 7). In other words, good governance positively affects a system's ability to adapt to changing circumstances.

The aim of good governance is to improve the effectiveness and quality of governance processes, structures and outcomes, which is assumed to benefit the well-being of the citizens. "Governance can be seen as the exercise of economic, political and administrative authority to manage a country's affairs at all levels. It comprises the mechanisms, processes and institutions through which citizens and groups articulate their interests, exercise their legal rights, meet their obligations and mediate their differences" (UNDP 1997). Frequently mentioned elements of good governance include participation, transparency, accountability, effectiveness and equity. It promotes the rule of law and "ensures that political, social and economic priorities are based on broad consensus in society and that the voices of the poorest and the most vulnerable are heard in decision-making over the allocation of development resources" (UNDP 1997).

Sustainable development and building resilience are often connected to trade-offs, e.g. ecological resilience versus social resilience, economic versus ecological sustainability (Lebel et al. 2006). Each society needs to decide what kind of resilience it seeks to establish and towards which threats it wants to build resilience. Good governance facilitates these kinds of processes. A good governance approach ensures an equitable and broadly based discourse. It thus helps finding a broadly based consensus on how to deal with trade-offs, define priorities and how to position a society within areas of conflict.

The concept of good governance mainly focuses on providing political stability and economic prosperity, while not explicitly referring to the third dimension of sustainable development, namely ecological sustainability. Among other things, the ecological dimension is supplemented in the concept of water governance.

Water Governance

"Water governance refers to the range of political, social, economic and administrative systems that are in place to develop and manage water resources, and the delivery of water services, at different levels of society" (GWP 2003, p. 16).

Governance issues arise in almost all spheres of development and social life and thus also play an increasingly important role in formerly technically dominated spheres such as water management. In coherence with the principles of Integrated Water Resources Management (IWRM), water governance is defined as being (GWP 2003, pp. 27–29):

- *Open and transparent*: accessible and transparent policy formulation and implementation to increase confidence and trust in organizations and institutions of water allocation and distribution.
- *Inclusive and communicative*: enabling broad stakeholder participation to increase the quality and effectiveness of water policies, organizations and institutions.
- *Coherent and integrative*: inclusion of all users and uses and the coordination of related policies to adequately address problems of complex systems with a consistent, holistic and cross-cutting approach.
- *Equitable and ethical*: enforcement of justice, property rights and rule of law regarding access to and ownership of water to provide all citizens with equal opportunities to sustain their well-being.
- *Accountable*: ensure clear responsibilities and obligations as well as sanctions for the violation of rules.
- *Efficient*: apart from economic efficiency support social, political and environmental efficiency in water issues (e.g. minimize economic and environmental inefficiencies of water management such as over- or under-allocation or provide a socially accepted and affordable level of access to water resources and sanitation).

- *Responsive and sustainable*: orient policies and institutions on demand, the policies' impact and past experience. Implement these policies at the relevant level (principle of subsidiarity), keeping in mind present and future interests of water users.[2]

Water governance functions as a transmitter of impact from the social system to the ecological system and vice versa. Thus institutions as a basic component of water governance "link the society to nature, and have the potential to coordinate the human and natural systems in a complementary way for both ecological and human long-term objectives. They also have the potential to inhibit adaptive responses to ecosystem changes" (IHDP 1998). For example, water governance sets the incentives for the sustainable or unsustainable use of the water resources through the institutions of property rights. This function of water governance as interface implies that water governance and especially the institutional arrangement in the water sector is one of the spheres firstly coming under pressure of adaptation when changes of the ecological system occur. If water governance proves resistant to such adaptation pressure, the resulting (increased) mismatch between social and ecological systems is likely to exacerbate degradation and overuse of water resources.

Adaptive Water Governance

Based on the concepts of adaptive governance, good governance and water governance elements of adaptive water governance can be derived. These elements are grouped according to address the need for sustainable water governance and the problems of fit and interplay (cf. Fig. 1). The concept of adaptive water governance provides a point of reference for examining and assessing institutional change in water governance regimes.

The Problem of Interplay: Multi-Level and Horizontal Governance

The importance of vertical and horizontal interplay between institutions and organizations for sustainable and adaptive governance of natural resources has been frequently underlined (Young 2002; Folke 2006; Young 2006; Pahl-Wostl 2009). Horizontal interplay or governance refers to institutions that facilitate the cooperation and linkages (e.g. exchange of knowledge and information) among

---

[2] Water governance is, on the one hand, based on and significantly overlaps with good governance. On the other hand, it differs from good governance by not addressing issues of democratization, electoral systems and sovereignty (GWP 2003).

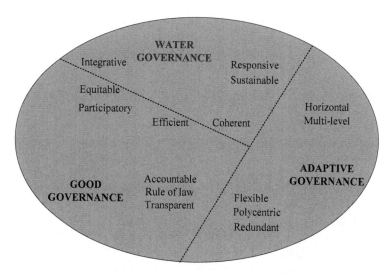

Fig. 1 Adaptive water governance. Source: own compilation

governance entities at the same level of administration (e.g. ministries, local water management organizations), as well as cross-sectoral and cross-disciplinary thinking (Hill 2005). It foresees that e.g. ministries concerned with water management tasks or local water management organizations, such as WUA, exchange information, thus allowing for new practices to spread quickly. This is increasingly important with regard to the transfer of responsibilities in decentralized governance systems. The newly empowered organizations need to cooperate to identify and share best practices.

The demand for *multi-level* or *vertical governance* structures within adaptive water governance (Folke et al. 2002; Pahl-Wostl 2009) originates from the observation that ecological systems rarely follow administrative boundaries and that ecosystems dispose of multiple levels. An effective resource management should thus mirror these multiple levels (Low et al. 2003) and possess governance mechanisms working across levels including state and non-state actors and integrating the local, provincial, national and international levels of governance and administration. Multi-level governance is based on the assumption that complex issues such as water governance can only be achieved through the integration, interaction and cooperation of the different levels of a governance system. However, multi-level governance does not necessarily imply the linear organization of administrative levels, which is more crisis-prone or susceptible to risk, since the failure of one element can break the system (Low et al. 2003).

The Problem of Fit: Polycentric, Flexible and Redundant Governance Structures

Adaptive water governance regimes should aim at improving the fit between the ecological and the social system (Young 2002). Evolving complex social-ecological systems (SES), which may be characterized by state changes and flips, require polycentric, flexible and redundant governance structures, management institutions and organizations in order to prevent mismatch and mismanagement.

Polycentric organizations and institutions "improve the fit between knowledge, action, and social-ecological contexts in ways that allow societies to respond more adaptively at appropriate levels" (Lebel et al. 2006, pp. 8–9). It is assumed that nonlinear polycentric governance structures are better geared to react to nonlinear ecological crisis if compared to linear governance structures (Low et al. 2003). For example, the impacts of climate change on water resources are likely to vary across a country, thus calling for individual adaptation measures in different regions. These are more likely to develop in the local context than within a hierarchical governance setting. Unlike strictly hierarchical systems polycentric governance structures dispose of multiple centres of power. "Polycentric governance systems can be characterized as complex, modular systems where differently sized governance units with different purpose, organization, spatial location interact to form together a largely self-organized governance regime. Polycentric governance systems are characterized by many degrees of freedom at different levels" (Pahl-Wostl 2009, p. 7). These degrees of freedom allow the development of locally appropriate institutions (Lebel et al. 2006).

Adaptive water governance requires *flexible institutions*, which offer mechanisms that provide for the adjustment of management procedures and management structures to new (environmental) conditions and new (scientific) knowledge (Folke et al. 2002). Flexible institutions permit and request feedback and monitor its implementation, i.e. they allow for learning from past experience, are able to quickly identify inefficient practices and rules and the need for changing them. However, too much flexibility may also turn out negatively. "Systems where change is not allowed will almost certainly generate surprise and crisis. Systems that allow too much change and novelty will suffer loss of memory" (Berkes et al. 2003, p. 376). On the one hand, the increasingly uncertain environment forces social systems and institutions to become more adaptive—that is, flexible and open to change. On the other hand, strong and reliable institutions are needed to establish and sustain a functioning water governance system. The task is therefore to find the appropriate balance between innovation and conservation.

The *redundancy* of institutions similarly requires finding such a balance. Redundancy influences system performance—under some conditions positively, under others negatively (Low et al. 2003). Contrary to mainstream economic thinking it is increasingly acknowledged that redundant structures do not necessarily lead to inefficiency but may even improve system performance—especially if system performance is not measured in terms of short-term output but rather in terms of long-term capacity to deal with risk and uncertainty and adapt to change.

In ecosystems, redundancy of species guarantees ecosystem resilience and stability and prevents ecosystem failure (Naeem 1998). Redundant species are not primarily needed to provide ecosystem functioning and services because they provide similar functions to the ecosystem as other species. But they are able to replace other species once these fail or become extinct. Thus, redundancy resembles one of the core principles of evolution and can potentially ensure the survival of the system in times of crisis.

Applied to social systems, the overlap and redundancy of institutions may increase the capacity of a system to diverse responses to a problem (Walker and Salt 2006). This implies that one institution can provide the same (or similar) function as a second one and replace it once the second one became ineffective through crisis and change. Redundancy may refer to the overlap of institutions or functions of organizations or the prevalence of similar subsystems (Holling et al. 2002; Low et al. 2003). The doubling of institutions and a modest overlap of functions support the spreading of risks and help to absorb disturbance (Folke et al. 2005). Moreover, governance structures that mirror and thus fit ecosystem structures are more likely to identify system failure and adequately respond to it (Low et al. 2003). Redundant functions and organizations may thus not contribute to system functioning under normal conditions but may provide relevant functions and information during unpredictable events, i.e. when the system reorganizes.

Redundant subsystems need to operate independently to have positive effects. The administrative unit compensating for another unit's failure should not be affected by that failure (Low et al. 2003). However, since the doubling of functions and subsystems is costly, there is a need to identify the optimal level of redundancy (Low et al. 2003). Since it is often difficult to identify the kind of redundancy in advance that will generate positive effects in a crisis, redundancy should be limited to those cases where it can be achieved with low costs.

Social Sustainability: Equity, Integration and Participation

Broad stakeholder *participation* on the basis of *equity* is an essential element of making water governance regimes adaptive (Pahl-Wostl et al. 2005; Biermann et al. 2009). This refers both to the involvement of relevant stakeholders in decision-making processes at different levels of administration (e.g. in the development of a water use strategy) and the equitable access to and use of water resources. The *integration* of all users ensures that different interests within society and interactions of a topic with other issues come to light and are discussed. Participation gives underrepresented groups the chance to raise their issues and claim their rights. Through this process, trust and a shared understanding can be built and social learning be fostered (Lebel et al. 2006). Through participation, different kinds of knowledge on ecosystem functioning and management practices (including local and traditional knowledge) can be integrated in management decisions.

Economic Sustainability: Efficiency and Coherence

The efficient use of natural, human and financial resources is a prerequisite of adaptive water resource governance, especially in the context of developing countries, which more often than not are not well endowed with any of them. One prerequisite for effective and *efficient* governance of resources is *coherent* decision-making (Sadoff and Muller 2009). This is especially important since water is a cross-cutting issue. Decisions at all levels and in many different spheres of governance affect the water resource. It is important, however, to keep in mind the inherent trade-offs between efficient governance, redundancy and an increased long-term vulnerability to uncertainty and surprise (Davidson-Hunt and Berkes 2003). This trade-off, which is related to what Walker has termed the difference between general and specific resilience (Walker et al. 2006), will be addressed in the section on synergies and trade-offs.

Political Sustainability: Rule of Law, Accountability and Transparency

Water governance organizations underlie the rules which have been formulated by legislature for the governance and management of water resources. Their actions should be authorized by law (García-Salmones 2009). Laws define their room of manoeuvre for managing water resources and the necessary cooperation with other state organs. The *rule of law* aims at preventing discretionary, ad hoc water resources governance. This is closely connected to upwards (towards superior bodies) and downwards (towards the public) *accountability*. Authorities that can be made responsible for their actions (and inaction) tend to pursue a more equitable distribution of benefits and thus enhance the adaptive capacity of a social system (Lebel et al. 2006).

Accountability is supported by *transparency*, i.e. the disclosure of organizational structures, water management procedures and strategies, as well as monitoring data (Young and Lipton 2006). Transparency, accountability and rule of law increase the predictability of system behaviour and create trust in the system and its functioning.

Ecological Sustainability and Responsiveness

Water governance should be organized with a view to the long-term *sustainability* of water use. This includes governing (surface water and groundwater) resources such that they are neither overused nor polluted and that enough water is provided to support ecosystems (so-called environmental flows). In a changing environment, *responsiveness*, i.e. the ability to recognize changes in the water system (e.g. decreasing water availability, overuse of groundwater resources or increasing water pollution), and taking timely and adequate measures to react to these changes gain importance. This includes measures of demand management rather than supply management.

Sustainable water governance should not only serve present water users but also take demands of future water users into account (GWP 2003). The emphasis on sustainability and responsiveness introduces the timescale to adaptive water governance. Past experience and current changes in the ecological system need to be monitored and taken into account. Long-term effects of present interventions need to be anticipated and (especially when this is not or not sufficiently possible) effects need to be monitored during implementation to be able to adjust measures, should they entail unintended negative effects.

Synergies and Trade-Offs Between the Elements of Adaptive Water Governance

Many of the elements of adaptive water governance described above are closely interlinked and difficult to examine in isolation. Some of the elements produce synergies, while trade-offs and tensions exist between others. For example polycentric governance may facilitate redundancy and experimenting because of a large number of independent units thus fostering resilience of the system (Warner et al. 2008; Huitema et al. 2009). Polycentric governance and redundancy in institutional set-up and structures, however, may come to the detriment of efficiency and coherence. Public participation on the one hand adds transparency for stakeholders, fosters social learning and improves water governance through knowledge exchange. On the other hand, it is often costly (in terms of human and financial resources), thus negatively affecting efficiency. Likewise, it might be difficult to arrange for participatory governance structures that allow for accountability (who is responsible for decisions taken with large involvement of stakeholders?) or to establish flexible institutions that remain sufficiently transparent.

From a short-term perspective, most of these tensions seem inevitable. With a longer-term perspective, which does not only take the functioning of the social but also of the ecological system into account, the elements of adaptive water governance are not necessarily mutually exclusive. The trade-off of participation and efficiency may at least partly dissolve when taking a longer-term perspective. Participation helps to produce resilient solutions that take local context and knowledge into account, are supported by stakeholders and tend to be easier to implement and sustain than decisions taken in a top–down manner. Likewise, redundant institutions or structures may prove useful in times of crisis. The aim of adaptive water governance should therefore be to find ways how to achieve one goal while not negatively affecting the others. In those cases where this cannot be achieved, good governance should provide for a societal negotiation process on finding an acceptable level of negative effects. An example is the trade-off between participation and accountability or the increasing costs of cooperation and coordination with increasingly redundant structures.

Adaptive water governance should not be understood as a "one-size-fits-all" approach, but rather as one leaving room for composing the elements and adjusting them to local needs. The optimal amount of one or the other element will differ in different cultural, social, political and economic contexts. The respective contexts also determine the relevance of a certain set of elements, which might not be the same in two countries. In addition, different stages of development may require the emphasis of different sets of elements, i.e. the composition and accentuation of elements in a particular case may change over time. For example the creation of trust through rule of law, transparency and accountability may be a precondition for meaningful participation. However, it is assumed that—regardless of the socioeconomic situation—a certain level of each of these elements is required for constituting an adaptive regime and the building of resilience. For example, the provision of a certain level of flexibility of institutions should be provided in order to assure the ability to react to disturbance. The "optimal" level of this flexibility can differ in different contexts though and should be determined through good governance.

Adaptive Water Governance Applied to South African Water Governance Reforms

South Africa is a country undergoing comprehensive transformation since the end of the apartheid regime in 1994. After decades of economic sanctions, domestic markets and foreign trade were liberalized and a parliament was democratically elected. In addition to this transformation, the country is prospected to be highly affected by environmental change. The South African water system in particular is highly vulnerable to climate change (De Wit and Stankiewicz 2006). The country is characterized by a relatively low mean annual precipitation [495 mm in 2007 (FAO 2009)] and a low conversion rate of rainfall to runoff while being highly dependent on surface water. As a consequence, water resources are highly stressed in some parts of the country. In the past, increasing demand was met through increasing the water storage capacity by building a highly sophisticated system of water transfer schemes and dams. Today, there are few adequate places left to build new dams (Muller 2002) and the variability of runoff is further increased through the impact of climate change, thus possibly rendering large, inflexible infrastructure less efficient. Apart from these economic limitations, several ecological objections exist against dams and water transfers. For example, the Greater Fish transfer scheme turned a previously seasonal river into a permanent one, providing excellent breeding conditions for the biting black fly. This insect turned from a seasonal occurrence into a pest, causing significant stock losses among cattle (Bohensky 2008). Thus, besides technical solutions, improvements of water governance regimes need to be explored. The prevailing structural engineering and supply management approach needs to be replaced by an approach which combines demand management and economic incentives for water saving (e.g. through pricing) with equitable access and ecological sustainability.

In the following, the elements of adaptive water governance are applied to the South African water sector. Even though adaptive water governance is no explicit goal of water governance reforms in South Africa, the current and future exposure of the country to climate change justifies the assessment of reforms using this concept. Examples are given that illustrate some elements of the concept without claiming to be exhaustive. The focus is on the problem of fit (especially flexible institutions and redundancy) and social sustainability (especially participation and equity), since most progress can be attested in these spheres.

Flexible Institutions: the National Water Act

Following the end of the apartheid regime, the South African social system has undergone paramount institutional change, beginning with a new constitution (Republic of South Africa 1996). In the water sector, the Water Services Act (Republic of South Africa 1997) and the National Water Act (Republic of South Africa 1998) that has been internationally praised for representing state-of-the-art water law (De Coning and Sherwill 2004) followed. The National Water Act foresees the transformation towards a holistic, decentralized and participatory approach to water management with the aim of increasing water use efficiency, securing equal access to water and sustainably using water resources.

Being conceptualized as a developing institution, the National Water Act provides several entry points for flexibility. Firstly, different parts of the legislation can be promulgated at different times. As a consequence, the National Water Act itself only provides a framework. Its concrete elaboration takes place during implementation and is authorized to the state administration in charge (in this case the Department of Water Affairs, DWA). Expressions such as "in a phased and progressive manner" allow for a phased implementation of the Act over time and space (Rowlston et al. 2000, p. 6). Thus, the largest part of the Act was promulgated in August 1998, but sensitive sections, such as the ones on licensing, the establishment of the Reserves and water use registration, were only promulgated in

October 1999 after they had been thoroughly designed, instruments had been developed and processes had been defined (De Lange 2004). With this approach it was acknowledged that the resources to implement comprehensive change in administrative and operational procedures were limited.

Secondly, several provisions of the Act include mechanisms allowing later adjustments. The Act itself and the National Water Resources Strategy are subject to reviews every five years. Likewise, water licences are reviewed every five years and may not be granted for a period longer than 40 years (Republic of South Africa 1998, Sect. 28). In the context of scarce water resources, the reallocation of water from relatively low productivity uses such as agriculture to uses with a higher productivity per unit of water such as industry are discussed in South Africa (Otieno and Ochieng 2004). Should there be political consensus on the issue, the flexibility of the water licensing procedure and the short term of water licences would allow for the implementation of such a change.

In the implementation process, however, the flexibility of the Act can also have negative repercussions. For example, the establishment of Catchment Management Agencies (Burt et al. 2008) was significantly delayed since the Act did not provide the necessary procedures, such as criteria for evaluating CMA proposals, which then had to be developed in the process. As a consequence, the assessment of the proposal for the Inkomati CMA was delayed for 18 months, thus alienating the involved stakeholders from the process (Brown and Woodhouse 2004; see also Herrfahrdt-Pähle 2010).

Redundancy: Functional Overlap of Institutions

In the South African context, institutions for solving water conflicts may serve as an example for potentially beneficial overlaps or redundancy. The National Water Act establishes a Water Tribunal as an "independent body with a mandate to hear and adjudicate appeals on a wide range of water-related issues, mainly against administrative decisions made by responsible authorities and water management institutions" (DWAF 2004b, p. 100). Appeals to the tribunal may relate to various aspects of water conflicts such as decisions on water allocation, licence applications, declarations of existing lawful water use or compensation for refused water rights.

Parallel to the Water Tribunal, another institution of conflict resolution has been acknowledged with the Traditional Courts Act (Republic of South Africa 2008). Traditional courts are foreseen to enhance "access to justice by providing a speedier, less formal and less expensive resolution of disputes" (Republic of South Africa 2008, p. 9). Thus traditional courts might serve as an alternative option to the Water Tribunal for solving water conflicts. In case of crisis, for example during a severe drought, traditional courts could serve as buffer capacity to unburden the Water Tribunal. It seems that the costs of the associated functional overlaps are relatively low since they are relatively inexpensive. The requirement of independence of the two institutions is also given since they operate at different levels and within different sets of norms and structures.

Another example of a functional overlap of institutions is linked to the transition from administrative to hydrological water management. Until recently, water was managed by province offices of the Department of Water Affairs (DWA). The National Water Act inter alia calls for the transition from a water management system based on riparian rights and administrative boundaries towards licensing of water use and the management along hydrological boundaries (Seetal 2005). This includes the restructuring of the water management bodies of the DWA and the introduction of Catchment Management Agencies. Even though they will not be completely replaced by CMA, the role of the province offices of DWA is not adequately referred to in most water governance documents. Often reference is made to the changing role of the DWA from a regulating and implementing agency towards a provider of policies and strategies (DWAF 2004b). It remains unclear what this means for the province offices, which were mainly involved in policy implementation and operation and maintenance in the past. The parallel existence of both CMA and province offices of DWA may easily result in functional overlaps. In this case, however, overlaps rather seem to be connected to a number of disadvantages, since both organizations act simultaneously, at the same level and in the same sphere of governance. While it is not clear so far how exactly responsibilities and tasks will be delineated between the two bodies, it seems likely that their coexistence will be associated with higher costs than benefits. In any case, this form of redundancy will require good coordination to avoid excessive inefficiency and ineffectiveness.

Participation of Stakeholders: Top–Down Versus Bottom–Up

Since the end of the apartheid era, participation and participatory democracy range high in all spheres of government. Accordingly, the National Water Act foresees mechanisms for participation at all levels. For example, the formulation of a new strategy for the water sector needs to be informed by a comprehensive participation process. Stakeholder workshops are conducted at province and municipality level. Due to lacking capacities and skills at all levels of water administration, these consultation processes often take several years.

However, it seems that participatory mechanisms have so far mostly been implemented in a top–down approach by government agencies and that the initiative and major input generally comes from the central government. For example, the implementation process of the CMA seems to be initiated and organized either by DWA's province offices or the DWA headquarters, thus leaving little room for stakeholder participation in the sense of impact on the design of these new organizations or their procedures. There seems to be a bias regarding participation. In strategically more important issues (such as the establishment of CMA, which touch upon the core of DWA competencies), participation seems to be less intensive than in others.

On the other hand, the dominant role of the government can be justified by the large imbalances that still exist between different stakeholder groups and which may cement inequality in water issues. Within Water User Associations (WUA), for example, the participation of previously disadvantaged persons is a topic of concern. They are often sidelined by white farmers and efforts need to be made to empower them to effectively take part in WUA decision-making (Karodia and Weston 2001). In this regard, it is the role of DWA to "level the playing field by stimulating information provision, social facilitation, dispute resolution, capacity building, etc". (Van Koppen et al. 2002, p. 15). The Minister has used her power several times and rejected proposals for CMA implementation processes because of the deficient representation of historically disadvantaged communities, smallholders and water users. "Only DWAF's [i.e. DWA's] strong, effective steering role in crafting CMAs that build upon local law can lead to inclusive, bottom–up river basin" (Van Koppen et al. 2002, p. 16). It will be difficult for DWA to determine the point in time when to withdraw from the process and to establish the necessary capacities with the underrepresented parts of the population. There is the risk that the domination of participatory processes through special interest groups (e.g. large-scale farmers) undermines the credibility of participatory mechanisms. De Villiers (2008) observes this trend and a resulting decline in participation with regard to ward committees.

Equity: Providing Equal Access to Water and Sanitation Services

Among the major concerns of the new political administration is the provision of the population with access to safe drinking water. Human and environmental water demand come first in the water law and there is a strong commitment to provide all citizens with access to drinking water under the slogan "some for all forever", which culminates in the constitutional right to water (Republic of South Africa 1996). To address the backlog in provision with access to water and sanitation from apartheid times, comprehensive programmes were put in place. Between 1994 and 2004, 13.4 million people received access to safe drinking water and 6.9 million people received access to basic sanitation (DWAF 2004a). Even though considerable progress was made by connecting large parts of the population to piped water in the 1990s, it has been stated that many of these infrastructure developments are inoperable today due to a lack of operational and maintenance capacity, inefficient cost recovery, the absence of institutional arrangements and vandalism (Mukheibir 2007).

The National Water Act foresees the implementation of the Reserve (consisting of a social and an ecological reserve) to prioritize human needs and environmental integrity of the system in relation to other water uses, e.g. by agriculture or industry. The "basic human needs reserve" guarantees a minimum of 25 l of water per person per day and the "ecological reserve" was established to assure sufficient provision of ecological flows (Hamann and O'Riordan 2000). Since 2001, the free basic water policy ensures that the basic human needs reserve is provided free of charge (to poor households). By 2007, more than 75% of the population received free basic water, which has contributed to achieving social equity (Muller 2008).

Conclusion

Governing water resources adaptively is becoming a crucial element of sustainable development because existing economic and social problems are increasingly exacerbated by the impact of climate change on water availability, especially in Africa. This paper sought to provide a way to characterize adaptive water governance in the context of sustainable development by referring to and merging elements of good governance, water governance and adaptive governance. Good governance is widely acknowledged as the basis for democratic and fair governance processes, both of which should also be guiding decision-making in the water sector. However, the concept of good governance appears not to be able to provide governance structures and processes with the resilience which is necessary regarding the challenges of a changing environment and especially climate change. Adaptive governance can complement good governance since it is geared towards increasing resilience, e.g. through flexible institutions. Both good governance and adaptive governance fall short of including the specific needs of the water sector. This element is provided by adding water governance, which among other things highlights the need for (ecological) sustainability in governing and managing water resources. Since the elements of adaptive water governance hold both synergies and trade-offs, the main challenge for future research will be finding ways of how to use synergies and minimize trade-offs.

In the second part of the paper, some elements of adaptive water governance were applied to the South African context of water governance reforms. The new water legislation and policy is largely in line with adaptive water governance. Elements such as ecological sustainability, institutional flexibility, participation and redundancy are addressed by the new water legislation through the provision for a social and ecological Reserve, CMA and the involvement of stakeholders. However, implementation of these ambitious legal frameworks is still weak. Little progress has been made with implementing ecological sustainability, e.g. by determining the demand for environmental flows. Similarly, the political will to curb water use and ensure sufficient environmental flows seems to be rather low. The examples on the flexibility of institutions, redundancy of institutional structures, participation of stakeholders and equitable access to water likewise provided a mixed picture concerning implementation. On the one hand, progress can be attested with regard to (1) a high level of flexibility built in the national water legislation, especially the National Water Act, (2) the provision of buffer capacities regarding institutions for solving water conflicts (redundancy), (3) the high level of participation of stakeholders at many levels and (4) the provision of equitable access to water. On the other hand, some (unintended) negative implications of these developments are documented. Among them are (1) the time-consuming processes of elaborating procedures and guidelines in the process of implementation, (2) the potentially negative effects of functional overlaps of CMA and regional offices of DWA, (3) effectively involving relevant stakeholders and (4) the lacking long-term sustainability of water infrastructure due to lack of capacity and capital. This shows the close integration of the elements of adaptive water governance and their trade-offs as well as the difficulty to find an appropriate level of each of these elements. Thus, depending on the context, more participation does not necessarily increase adaptiveness and resilience (e.g. if it takes too much time to generate meaningful results due to lack of capacities).

Despite these caveats, progress has been made towards a more adaptive way of governing water resources in South Africa. The National Water Act and its flexibility provide an important place of departure for adaptive water governance. However, despite this good starting point, most drawbacks have to be attested regarding implementation. This shows the long way from conventional water governance approaches towards more adaptive approaches. Current limitations such as lack of political will, capacities and skills have to be overcome to facilitate timely and comprehensive implementation and thus increase the resilience of the water governance regime. With current water governance reforms in South Africa, significant first steps have been made but many more have to follow to establish adaptive water governance in practice and find sustainable solutions for increasing water scarcity and uncertainty.

Acknowledgments The author is grateful to the numerous interview partners in South Africa for sharing their insights and time and to Steffen Bauer, Chinwe Ifejika-Speranza and Claudia Pahl-Wostl for constructive comments.

References

AFDB (African Development Bank), ADB (Asian Development Bank), DFID (Department for International Development), UK Directorate-General for Development, European Commission, BMZ (Federal Ministry for Economic Cooperation and Development, Germany), M.O.F.A. Development Cooperation, Netherlands, OECD (Organization for Economic Development), UNDP (United Nationals Development Programme), UNEP (United Nations Environment Programme), World Bank (2003) Poverty and climate change: reducing the vulnerability of the poor through adaptation, Berlin

Anderies JM, Janssen MA, Ostrom E (2004) A framework to analyze the robustness of social-ecological systems from an institutional perspective. Ecol Soc 9(1):18

Berkes F, Colding J, Folke C (eds) (2003) Navigating social-ecological systems. Cambridge University Press, Cambridge

Biermann F, Bestill MM, Gupta J, Kanie N, Lebel L, Liverman D, Schroeder H, Siebenhüner B (2009) Earth system governance: people, places and the planet, IHDP Report, International Human Dimensions Programme on Global Environmental Change

Bohensky EL (2008) Discovering resilient pathways for South African water management: two frameworks for a vision. Ecol Soc 13(1):19

Brown J, Woodhouse P (2004) Pioneering redistributive regulatory reform: a study of implementation of a catchment management agency for the Inkomati water management area, South Africa, Working paper series number 89. Centre on regulation and competition, Manchester, UK

Burt J, McMaster A, Rowntree K, Berold R (2008) Local institutions for water governance: a story of the development of a water user association and catchment forum in the Kat River Valley, Eastern Cape. Gezina, Water Research Commission

Christensen JH, Hewitson B, Busuioc A, Chen A, Gao X, Held I, Jones R, Kolli RK, Kwon WT, Laprise R, Magana Rueda V, Mearns L, Menéndez CG, Räisänen J, Rinke A, Sarr A, Whetton P (2007) Regional climate projections. In: Solomon S, Qin D, Manning M, Chen Z, Marquis M, Averyt KB, Tignor M, Miller HL (eds) Climate change 2007: the physical science basis, contribution of working group 1 to the fourth assessment report of the intergovernmental panel on climate change. Cambridge University Press, Cambridge

Davidson-Hunt IJ, Berkes F (2003) Nature and society through the lens of resilience: toward a human-in-ecosystem perspective. In: Berkes F, Colding J, Folke C (eds) Navigating social-ecological systems: building resilience for complexity and change. Cambridge University Press, Cambridge

De Coning C, Sherwill T (2004) An assessment of the water policy process in South Africa (1994–2003), Water Research Commission

De Lange M (2004) Water policy and law review process in South Africa with a focus on the agricultural sector. In: Mollinga PP, Bolding A (eds) The politics of irrigation reform: contested policy formulation and implementation in Asia, Africa and Latin America, Aldershot, UK, Ashgate

De Villiers B (2008) Summary of observations made during workshops on the review of provincial and local governments held under the auspices of KAS and DDP from 10–13 November 2008, Briefing paper. KAS (Konrad-Adenauer-Stiftung), Johannesburg

De Wit M, Stankiewicz J (2006) Changes in surface water supply across Africa with predicted climate change. Science 311:1917–1921

Dietz T, Ostrom E, Stern PC (2003) The struggle to govern the commons. Science 302:1907–1912

Dolzer R (2004) Good governance: Neues transnationales Leitbild der Staatlichkeit? Zeitschrift für ausländisches öffentliches Recht und Völkerrecht 64:535–546

Doornbos M (2003) Good governance. J Int Aff 57:3–17

DWAF (Department of Water Affairs and Forestry) (2004a) A history of the first decade of water services delivery in South Africa: 1994–2004, Department of water affairs and forestry, Republic of South Africa

DWAF (Department of Water Affairs and Forestry) (2004b) National water resource strategy: our blue print for survival, Pretoria

FAO (Food and Agriculture Organization of the United Nations) (2009) AQUASTAT: FAO's information system on water and agriculture

Folke C (2006) Resilience: the emergence of a perspective for social–ecological systems analyses. Glob Environ Chang 16:253–267

Folke C, Carpenter S, Elmqvist T, Gunderson LH, Holling CS, Walker B, Bengtsson J, Berkes F, Colding J, Danell K, Falkenmark M, Gordon L, Kasperson RE, Kautsky N, Kinzig A, Levin S, Mäler K-G, Moberg F, Ohlsson L, Ostrom E, Reid W, Rockström J, Savenije H, Svedin U (2002) Resilience and sustainable development: building adaptive capacity in a world of transformations, Swedish Environmental Advisory Council, Stockholm

Folke C, Hahn T, Olsson P, Norberg J (2005) Adaptive governance of social-ecological systems. Annu Rev Environ Resour 30:441–473

García-Salmones M (2009) Taking uncertainty seriously: adaptive governance and international trade: a reply to Rosie Cooney and Andrew Lang. Eur J Int Law 20:167–186

Hamann R, O'Riordan T (2000) South Africa's policy transition to sustainability: environmental and water law, Surrey, Water Policy International Ltd

Hatfield-Dodds S, Nelson R, Cook DC (2007) Adaptive governance: an introduction, and implications for public policy, ANZSEE conference. Noosa, Australia

Herrfahrdt-Pähle E (2010) South African water governance between administrative and hydrological boundaries. Clim dev 2:111–127

Hill H (2005) Good governance—Konzepte und Kontexte. In: Schuppert GF (ed) Governance Forschung: Vergewisserung über Stand und Entwicklungslinien. Baden–Baden, Nomos

Holling CS, Gunderson LH (2002) Resilience and adaptive cycles. In: Gunderson LH, Holling CS (eds) Panarchy: understanding transformations in human and natural systems. Island Press, Washington, DC

Holling CS, Carpenter SR, Brock WA, Gunderson LH (2002) Discoveries for sustainable futures. In: Gunderson LH, Holling CS (eds) Panarchy: understanding transformations in human and natural systems. Island Press, Washington, DC

Huitema D, Mostert E, Egas W, Moellenkamp S, Pahl-Wostl C, Yalcin R (2009) Adaptive water governance: assessing the institutional prescriptions of adaptive (co-)management from a governance perspective and defining a research agenda. In: Ecology and society 14:26

IHDP (International Human Dimensions Programme on Global Environmental Change) (1998) The problem of fit between ecosystems and institutions, International human dimensions programme on global environmental change, Bonn

Jordan A (2008) The governance of sustainable development: Taking stock and looking forwards. In: Environment and planning C: government and policy, vol 26, pp 17–33

Karodia H, Weston DR (2001) South Africa's new water policy and law. In: Abernethy CL (ed) Intersectoral management of River Basins: proceedings of an international workshop on "Integrated water management in water-stressed river basins in developing countries: strategies for poverty alleviation and agricultural growth," Loskop Dam, South Africa, International Water Management Institute, 16–21 October 2000, Colombo

Lebel L, Anderies J, Campbell B, Folke C, Hatfield-Dodds S, Hughes TP, Wilson J (2006) Governance and the capacity to manage resilience in regional social–ecological systems. Ecol soc 11:19

Low BS, Ostrom E, Simon C, Wilson J (2003) Redundancy and diversity: do they influence optimal management? In: Berkes F, Colding J, Folke C (eds) Navigating social-ecological systems: building resilience for complexity and change. Cambridge University Press, Cambridge

Lumsden TG, Schulze RE, Hewitson BC (2009) Evaluation of potential changes in hydrologically relevant statistics of rainfall in Southern Africa under conditions of climate change. Water SA 35:649–656

Mclain RJ, Lee RG (1996) Adaptive management: promises and pitfalls. Environ Manag 20:437–448

Medema W, Jeffrey P (2005) IWRM and adaptive management, NeWater report series, NeWater (New approaches to adaptive water management under uncertainty)

Mukheibir P (2007) Access to water—the impact of climate change in small municipalities, Energy Research Center. University of Cape Town, Cape Town

Muller M (2002) Inter-basin water sharing to achieve water security—a South African perspective, Pretoria, Department of water affairs and forestry, Republic of South Africa

Muller M (2008) Free basic water—a sustainable instrument for a sustainable future in South Africa. Environ Urbanization 20:67–87

Naeem S (1998) Species redundancy and ecosystem reliability. Conserv Biol 12:39–45

Olsson P, Gunderson LH, Carpenter SR, Ryan P, Lebel L, Folke C, Holling CS (2006) Shooting the rapids: navigating transitions to adaptive governance of social–ecological systems. Ecol Soc 11:18

Otieno FAO, Ochieng GMM (2004) Water management tools as a means of averting a possible water scarcity in South Africa by the year 2025. Water SA 30:120–124

Pahl-Wostl C (2009) A conceptual framework for analysing adaptive capacity and multi-level learning processes in resource governance regimes. Glob Environ Chang 19:354–364

Pahl-Wostl C, Downing T, Kabat P, Magnuszewski P, Meigh J, Schüter M, Sendzimir J, Werne S (2005) Transition to adaptive water management: the NeWater project, Institute of Environmental Systems Research, University of Osnabrück

Pahl-Wostl C, Jeffrey PJ, Brugnach M, Sendzimir J (2007) Adaptive water management: how to cope with uncertainty. Policy Brief No. 4, NeWater (New approaches to adaptive water management under uncertainty)

GWP (Global Water Partnership) (2003) Effective water governance, TEC background papers, Global water partnership, Stockholm

Republic of South Africa (1996) Constitution of the Republic of South Africa, Republic of South Africa

Republic of South Africa (1997) Water Services Act, 1997, Government Gazette, Cape Town, Republic of South Africa

Republic of South Africa (1998) National Water Act—Act No 36 of 1998, Pretoria, Republic of South Africa

Republic of South Africa (2008) Traditional Courts Act, Pretoria, Republic of South Africa

Rowlston B, Barta B, Mokonyane J (2000) Implementing new water law: a South African experience, XTH World Water Congress. Melbourne, Australia

Sadoff C, Muller M (2009) Water management, water security and climate change adaptation: early impacts and essential responses, TEC background papers number 14, Global water partnership, Stockholm

Seetal AR (2005) Progress with water allocation reform in South Africa, OECD Workshop on agriculture and water: sustainability, markets and policies, 14–18 November 2005

Tompkins EL, Adger NW (2003) Building resilience to climage change through adaptive management of natural resources, Tyndall Centre for climate change research

UNDP (United Nations Development Programme) (1997) Governance for sustainable human development, UNDP policy document

Van Koppen B, Jha N, Merrey DJ (2002) Redressing racial inequities through water law in South Africa: interaction and contest among legal frameworks, The commons in an Age of Globalisation. The ninth conference of the international association for the study of common property, Victoria Falls, Zimbabwe, International Water Management Institute Africa Regional Program

Walker B, Salt D (2006) Resilience thinking: Sustaining ecosystems and people in a changing world. Island Press, Washington, DC

Walker B, Holling CS, Carpenter SR, Kinzig A (2004) Resilience, adaptability and transformability in social–ecological systems. Ecol Soc 9:5

Walker B, Gunderson L, Kinzig A, Folke C, Carpenter S, Schultz L (2006) A handful of heuristics and some propositions for understanding resilience in social–ecological systems. Ecol Soc 11:13

Warner J, Wester P, Bolding A (2008) Going with the flow: river basins as the natural units for water management? Water Policy 10:121–138

World Bank (1992) Governance and development. World Bank, Washington, DC

Young O (2002) The institutional dimensions of environmental change. MIT Press, Cambridge

Young O (2006) Vertical interplay among scale-dependent resource regimes. Ecol Soc 11:27

Young KR, Lipton JK (2006) Adaptive governance and climate change in the tropical highlands of Western South America. Clim Chang 78:63–102

Chapter 7
Increase in Port Downtime and Damage in Vietnam Due To a Potential Increase in Tropical Cyclone Intensity

Miguel Esteban, Nguyen Danh Thao, Hiroshi Takagi and Tomoya Shibayama

Abstract It is currently feared that the increase in surface sea temperature resulting from increasing levels of greenhouse gases in the atmosphere could result in an increase in tropical cyclone intensity in the future. Although the economic consequences have been studied for a number of developed countries, very little work has been done on developing countries. The present paper attempts to indicate what are the likely economic effects of this, by using a Monte Carlo simulation that magnifies the intensity of historical tropical cyclones between the years 1978 and 2008. This tropical cyclone model is then coupled with a socio-economic model that attempts to provide a projection of the likely development course of the Vietnamese economy and society. The simulation shows how annual downtime from tropical cyclones could increase from 0.23 to 0.37% by 2085 which could cause the loss of between 0.015 and 0.035% of GDP growth per year (between 600 bn and 1,400 m USD after factoring in the likely growth in the Vietnamese economy by this time). The effect that this could have on port operations and a preliminary assessment on the potential for increases in direct damage due to high winds are also made, showing a typical 33 to 65% increase for the centre and north of the country.

Keywords Climate change · Typhoon · Tropical cyclone · Vietnam · Intensity increase · Productivity · Direct damage · Indirect damage · Port location

M. Esteban (✉) · N. D. Thao · H. Takagi · T. Shibayama
Department of Civil and Environmental Engineering,
Waseda University, 3-4-1 Ookubo, 51 Goukan,
4-kai, 7 shitsu, Shinjuku-ku, Tokyo, 169-8555, Japan
e-mail: esteban.fagan@aoni.waseda.jp

Introduction

The balance of scientific evidence now suggests that anthropogenic emissions of greenhouse gases are having an effect on climate, as highlighted in the Fourth Assessment Report of the Intergovernmental Panel on Climate Change, or 4th IPCC report (IPCC 2007). It is feared that as a consequence of global warming, the frequency and intensity of tropical cyclones will increase due to the warming of the sea, which is confirmed by an analysis of the last 30 years of satellite records of tropical cyclones (Webster et al. 2005). Elsner et al. (2008) also examined the trends in the upper quantiles of cyclone maximum wind speeds and found a significant upward trend for wind speed quantiles above the 70th percentile. However, the accuracy of satellite-based pattern recognition remains a matter of intense debate (Landsea et al. 2006). What is clear is that tropical cyclones can have devastating effects, especially in poor countries, such as the 1970 Bangladesh cyclone, where between 300,000 and 500,000 people perished (Landsea et al. 2006). In the US, the 2005 hurricane Katrina caused major damage and left more than 1,800 people dead, triggering a debate about whether such tragic events will occur more frequently in the future. More recently, Cyclone Sidr was one of the strongest cyclones ever recorded in the Bay of Bengal, making landfall in Bangladesh on 15 November 2007. Sidr slammed the highly vulnerable low-lying densely populated coastal areas of Bangladesh with heavy rain, winds of up to 215 km/h, and a significant storm surge (Shibayama et al. 2009).

However, the area most frequently affected by tropical cyclones is that of the western North Pacific Ocean, accounting for approximately one-third of these phenomena (Imamura and Van To 1997). In this area, even for developed countries like Japan, the economic damage due to these events is considerable and appears to be increasing with time. Although this could be due to an increase in people taking out insurance, between 1980 and 2008, eight of the ten costliest natural disasters in Asia were due to typhoons in Japan after 1998, according to data from the Munich Re website (2009). In the case of Vietnam, more than 70% of all natural disasters are due to typhoons, and yearly frequency and damage totals have been generally increasing (Imamura and Van To 1997).

There is a growing fear that the pace of increase in damage can reach a point where the costs of natural hazards may potentially outpace economic growth (Webersik 2010). However, attempting to analyse the economic damage due to tropical cyclones is not an easy task. The damage caused by these events can be divided into two components, the damage caused by the destruction of houses, infrastructure and other material objects (direct damage) and the loss in productivity due to temporary shutdown in production and because of prevention activities (indirect damage). Some research has been done on the possible effects of direct damage, such as that by Howard et al. (1972), Pielke (2007), Hallegatte (2007), all of whom agree that damage in the United States is exponentially related to the maximum storm wind speed. However, different authors propose different exponential powers.

Much of the existing research focuses on the physical damage of natural hazards without calculating the indirect effects of tropical cyclones. Hallegatte (2008) explains how the indirect costs include "business interruption in the event aftermath, production losses during the reconstruction period, and service losses in the housing sector". Other such losses could include increases in energy prices, loss of workers' income or increases in insurance premiums following the passage of major events, and Hallegatte (2008) explains how the total socioeconomic damage can be much larger than direct economic impacts. Very little research has so far gone into estimating the indirect economic damage of tropical cyclones and Hallegatte (2007) points out how additional research is necessary to understand these indirect impacts. This effect includes, for example, the loss in economic productivity due to downtime in the public transport system, important industries and commerce. Particularly, urban areas can be greatly affected by the passage of a tropical cyclone, with their inhabitants leaving work early and shuttering houses and shops before its arrival and taking shelter until it has passed. Hallegate (2008) found that the total losses due to a disaster affecting the area of Louisiana in the USA increase nonlinearly with respect to direct losses when the latter exceed $50 billion (for instance, when direct losses exceed $200bn, total losses are twice as large as direct losses). The model given by this author attempts to reproduce the disruption in production that takes place after the event, and is useful to model the effects of high intensity events. However, in many countries in the Asia-Pacific region, the majority of the tropical cyclone-related downtime is due to low-intensity but high-frequency events, where the downtime is directly related to the duration of the event. As the tropical cyclones grow larger, the number of hours that a given area of a country will be affected by them will increase in the future. Esteban et al. (2009a, b) thus analysed the effect of downtime on the economy due to an increase in the size of future tropical cyclones. They concluded that these events could deduct 0.7% of the annual GDP of the Taiwanese economy by 2085, and that additional investments would be required in Japanese Port infrastructure to remove potential bottlenecks in the export of goods.

To attempt to understand how tropical cyclones are likely to be affected by an increase in global temperatures, global climate models using powerful supercomputers have been carried out, as highlighted in the 4th IPCC report (IPCC 2007). This report highlights that although there is a general agreement that tropical cyclones are likely to increase in intensity, there is as yet no broad consensus on the future frequency of these events.

However, most of the studies of the economic effects of an as yet hypothetical increase in tropical cyclone intensity have focused on developed countries, ignoring the potentially far more devastating effects on developing countries. The present paper will try to apply some of the techniques used by previous authors to the more complex case of Vietnam. This country has suffered numerous tropical cyclones, which have caused considerable damage to the country due to flooding, storm surges and wind damage. Examinations of yearly frequency and damages caused by tropical cyclones indicate a relatively increase in the value of losses (Imamura and Van To 1997).

The model used is not without its problems, as will be shown in the course of the paper, but the results can highlight the tremendous challenge that Vietnam could be facing, and hence highlight the clear need for adaptation measures to be put in place. The model proposed is in fact composed of two parts, a tropical cyclone intensity model and a simple socioeconomic model. From the combination of these two models, an assessment can be made of the economic cost of an increase tropical cyclone related downtime to the productivity of urban areas and Vietnamese ports.

Methodology

The majority of comparative studies undertaken in the field of natural hazard studies use the country/year dyad as the unit of analysis. However, the study of human-environment systems should include geophysical variables, and hence the use of gridded sub-national-level data that can account for spatial variation is highly desirable. The economic impact of tropical cyclones in Vietnam depends on several factors such as the location of economic activity, number of storms, intensity of storms, and the topography of the affected region and other geographical attributes, such as land-use patterns. As all these factors vary geographically the authors propose a disaggregated computational approach to measure the potential economic damage caused by tropical cyclones under a climate change scenario for the year 2085. In Vietnam, as in most countries, the value of economic output varies geographically, and much of it is concentrated mostly around the city of Ho Chi Minh in the south and Ha Noi in the north. The present study expects that the current economic distribution in Vietnam will not change, and that future population growth will retain the current population distribution pattern. Such a simplistic study is perhaps less realistic than studies that account for socioeconomic future changes, but is easier to interpret.

The objective of this study is to attempt to quantify the indirect economic damage that an increase in tropical cyclone size will have in urban and port productivity in Vietnam. Also, a simple indicative assessment will be made on which parts of the country are likely to see increase in damage potential in the future. Although the indirect damage also consists of the effects after the passage of a tropical cyclone, these effects require a more significant study of each area concerned (see Hallegatte 2008). Although surveys are carried out following major typhoon events (see for example Takagi et al. 2009) statistics for direct economic damage for the country as a whole are quite difficult to obtain for the case of Vietnam, and hence it is difficult to place a direct monetary term on these. The methodology to calculate the geographical influence of the tropical cyclones is based on the work of Esteban et al. (2009a), and based on this methodology a tool to estimate the potential increase in damage will be developed. Though the method proposed in this paper is rather simple compared to other procedures [see (Hallegatte 2007), for example] it appears to be the only one available without generating large amounts of tropical cyclone tracks using a computer. Essentially procedures such as that of Hallegate (2007) calculate the damage based on the maximum central wind speed, which requires a much larger sample of tropical cyclone tracks to provide accurate results than the historical number of tropical cyclone tracks available. The simulation used in the current paper does not randomly generate the tropical cyclones; instead, they are each picked at random from a set of 831 historical tracks. The strength and size of each tropical cyclone are then randomly altered according to the expected future distribution of maximum wind speeds proposed by Knutson and Tuleya (2004). This allows the number of hours that an area can be expected to be affected by wind of certain strengths to be determined. The results calculated are for the year 2085, as the work of Knutson and Tuleya (2004) provides the expected distribution of maximum surface wind speeds for this year only. Furthermore, randomly generating tropical cyclones [(Hallegatte 2007), for example] and then applying a Monte Carlo simulation would result in a prohibitively long computational time. As a large number of historical tracks are available for the target region, and tropical cyclones generally follow the same general trajectories, by keeping the original historical tracks it is possible to obtain a solution relatively quickly.

The next sections will briefly explain some of the main aspects of the methodology used. First, many of the assumptions on which the model is based will be explained in detail. Then, the databases regarding tropical cyclones, GDP and demographics of Vietnam will be introduced, together with a projection of how these are expected to change in the future. After that will follow a discussion on the expected increase in tropical cyclone intensity by the year 2085. Finally, the procedure to compute the expected number of hours that a certain area is affected by winds of a certain strength will be presented based on the methodology of Esteban et al. (2009a).

Assumptions

The model of Esteban et al. (2009a) builds on a number of assumptions, which are crucial to understand its limitations:

1. Typhoon tracks, frequency and seasonal distribution of tropical cyclones will not change in the future. It is possible that future increases in sea temperature could make the tropical cyclone season longer and/or increase the frequencies of these events. A number of studies on tropical cyclone frequency in warmer climate have been made, but the results of these are contradictory, and are still regarded as inconclusive as indicated by the 4th IPCC report (IPCC 2007). A consensus statement from the 6th International Workshop on Tropical Cyclones of the World Meteorological Organization (2006) states how "although recent climate model simulations project a decrease or no change in global tropical cyclone numbers in a warmer climate there is low confidence in this projection". Thus, in

the absence of any clear guidance from other authors on this point, the assumption of keeping the routes and frequencies the same can be seen as the default starting point of any simulation to determine economic risks.
2. There is a general relationship between the maximum sustained wind speed and the size of the tropical cyclone. This point is not clearly established for the case of large typhoons (Knutson and Tuleya 2001), although Esteban et al. (2009a) provide some statistical analysis to back this assumption.
3. The topography and population distribution in Vietnam will not change dramatically in the future. The socioeconomic model assumes that there will be migration from the countryside to the cities, but the relative size of each city compared to each other will stay the same. It is assumed from the population projection data available for Vietnam that the population will remain stable between 2050 and 2085. For most developed countries, population has now been relatively stagnant for a fairly long time, and to assume that the same would happen for the case of Vietnam if it were to become a developed country is fair. No country has so far experimented significant population decrease due to population aging and falling fertility despite the relative stagnation in the population of most industrialized countries for a relative long time now. Although immigration has played a relatively large part in maintaining populations fairly constant, there is no reason to think that the population of poorer countries would migrate to Vietnam in the same way as has happened during the industrialization of other countries.
4. Any wind higher than 30 knots will result in the temporary shutdown of all human economic activities.
5. The likely onset of damage for tropical cyclones in Vietnam is a wind of 50 knots or greater. Below this value damage will not be significant.

By making these assumptions the model can be thought to be conservative, as it will provide a lower estimate of the possible consequences of climate change. If the tropical cyclone season were to become longer or the frequency of the storms increased, then this would exacerbate the results provided in the present paper.

Tropical Cyclone Data

The tropical cyclone data was obtained from the website of the Japan Meteorological Agency (2008), which provides best track data for tropical cyclones in the Western North Pacific and South China Sea between 1951 and 2008. Prior to 1977 there is no satellite data available and hence it was only possible to use the historical tracks after this date. Nevertheless, the 30 years of useful data still provides a total of 831 tracks of tropical cyclones, which cover the area well. The data shows, for each storm, snapshots at various intervals the storm geometry and wind speed. Figure 1 shows the best track paths of all typhoons in Vietnam between 1990 and 2007.

Fig. 1 Typhoon tracks in Vietnam (1990–2007)

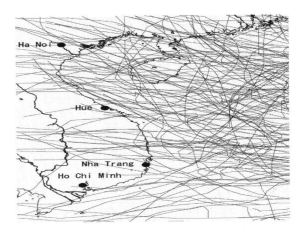

The influence of tropical cyclones, however, is not uniform throughout Vietnam. The country consists of three distinct topographical regions:

- The north region, characterized by mountainous or hilly terrain. Here the major rivers flow through the hills into flood plains and coastal plains;
- The central region, where the average distance from the coast to the mountain has a range of only 70 km;
- The southern region, comprising the Mekong river delta, characterized by a low-lying flat topography.

The principal spawning ground for the tropical cyclones that affect Vietnam is the western North Pacific Ocean, from where they usually travel though the Philippines to finally hit Vietnam or southern China. Thus, tropical cyclones are frequent in the northern and central regions but rarely affect the southern region (Imamura and Van To 1997). These authors state that the typhoon season in Vietnam starts in March and increases in intensity to reach a peak in October and then weakens progressively till the end of the year. The peak occurrences of typhoons typically shift from the north to the south as the season progresses, a pattern related to the seasonal change of water and air temperatures. Typically the country is affected by an average of 5.9 typhoons/year (Imamura and Van To 1997). The extensive coastline and low topography in the areas adjacent to the sea expose Vietnam to potentially devastating storm surges (recognized in the Vietnamese "National Strategy for Natural Disaster Prevention", 2007).

Population Projections

In order to assess the potential economic losses to each urban area, the population of each major city was estimated between 2009 and 2085. Most available future population projections for Vietnam involve a single national figure. However, since the purpose of this paper is to assess the potential effect of tropical cyclones

Fig. 2 Major urban population centres in Vietnam (2005)

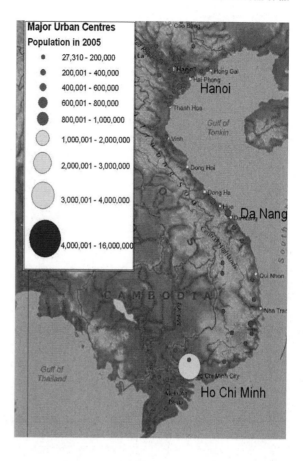

on each area of Vietnam, it was necessary to make an estimate for each city. Population estimates for present and future population (for the year 2050) were obtained from the medium variant for Vietnam from the United Nations Department of Economic and Social Affairs (2008), forecasting a population of 89 million for 2010 and 111 million by 2050. This data also estimates age distribution and fertility levels, and from it an assumption was made that the population would stabilize after 2050 and remain constant till at least 2085.

However, in a similar way to other countries in Asia, Vietnam is experiencing rapid urbanization. In 2004, the level of urbanization was 25.8%, but this is increasing at around 2% a year (To Lang 2005). The present simulation also included this increase in urbanization levels, although it is assumed to peak at around 80%, a similar level to what is seen in industrialized Asia nations such as Korea or Japan. This increase in urban population was redistributed throughout the cities keeping the proportion of population in each one similar to what it is today. The projections show how the population will stabilize by 2050, resulting in an increase of around 411% in the urban population. Figure 2 shows the current population of each major urban centre in Vietnam in the year 2005, Fig. 3 the

Fig. 3 Projection of population of major cities for the year 2085

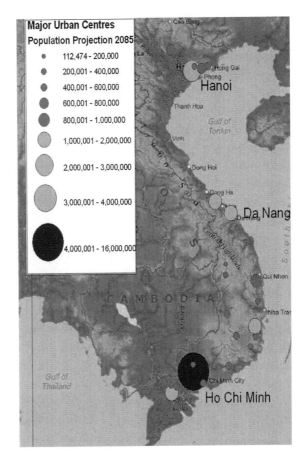

population projections for each city in the year 2085, with Fig. 4 showing the aggregated population projections for the whole country.

GDP Projections

To obtain an estimate for the GDP per capita of Vietnam, a study by Goldman Sachs (2007) was used as the starting point. This study gives an increase in GDP per capita from US$655 in 2006 to US$33,472 by 2050 using a gradually reducing rate of economic growth, (7.8% per year for up to 2015 and then reducing to 4% by 2050). This Goldman Sachs study (2007) only calculates population up to the year 2050, but as the climate change scenario model of Knutson and Tuleya is for certain CO_2 concentrations in the year 2085, it was necessary to extend this model to 2085. For the period 2050–2085, the authors thus choose a more conservative level of economic growth of 3% per annum, which would result in a GDP per capita of US$94,185 by 2085.

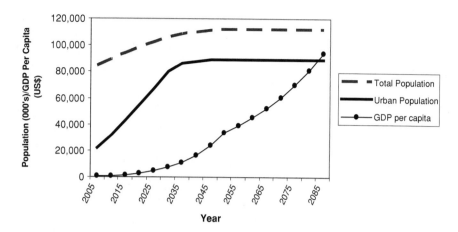

Fig. 4 Projected population growth and GDP per capita (US$) in Vietnam

Simulation Methodology

The Monte Carlo-based methodology described by Esteban et al. (2009a) was used as the basis to obtain the expected affected time $\hat{\vartheta}(c)$ that an area is under the influence of wind of a certain strength in one future year. In this method, the results are computed for two different conditions, one being the "present day" conditions (using the unaltered tropical cyclone records) and the other being the "climate change" conditions (where the tropical cyclone records are altered using the methodology described in the next sections). By doing so, it is possible to compare the future and present results and gain an insight into possible future economic consequences of an increase in tropical cyclone intensity.

The expected affected time can be defined as the sum of each of the values of time affected by storms for one year $\vartheta(c)$ for all the simulation runs divided by the number of simulated runs N, or,

$$\hat{\vartheta}(c) = \frac{\sum_{1}^{N} \vartheta(c)}{N} \qquad (1)$$

The reason to use a Monte Carlo simulation is that each simulation run produces completely different results and hence it is necessary to obtain an average of the expected result.

For each of the two conditions ("present day" or "climate change"), the simulation starts by generating a random number of tropical cyclones for each month of the year from the probability distribution parameters given in Table 1. These values were obtained by analysing the number of storms in the Western North Pacific and China Sea between 1971 and 2006, as published by the Japan Meteorological Agency. After the number of tropical cyclones in each month has

Table 1 Probability distribution functions of number of tropical cyclones per month

Month	Normal	Standard deviation
January	0.47	0.55
February	0.14	0.35
March	0.33	0.67
April	0.72	0.77
May	1.08	1.09
June	1.78	1.25
July	4.00	1.63
August	5.58	1.69
September	4.86	1.34
October	3.75	1.48
November	2.39	1.25
December	1.28	0.90

been generated, the simulation then selects for each cyclone in the month one random historical cyclone track from the record of all the tropical cyclones between 1978 and 2008. For the case of the "climate change" condition, the intensity of each tropical cyclone must be altered, as outlined in the following section.

Increase in Tropical Cyclone Intensity in the Year 2085

The assumptions regarding the increase in storm intensity in the year 2085 are derived from the work of Knutson and Tuleya (2004). These authors carried out 1,300 five-day idealized simulations using a high-resolution version of the Geophysical Fluid Dynamics Laboratory (GFDL) R30 hurricane prediction system. These simulations were carried out for a Surface Sea Temperature change of between +0.8 and +2.4 °C, which assume a linear +1% compounded yearly increase in CO_2 over a period of 80 years (up to the year 2085) in order to calculate the surface sea temperature. This +1% yearly increase means that CO_2 levels would reach 2.2 times the control value (that of 2004) by the year 2085. These authors acknowledge that other radiative forcing agents besides greenhouse gases may have important effects on the global climate, but quantification of their past and possible future forcing remains even more unclear than for greenhouse gases. Although these authors continuously refer to CO_2 and greenhouse gases, it appears that they only actually considered CO_2 emissions in their model.

These authors also note that surface sea temperature is not the only factor that affects the intensity of tropical cyclones. For example, other factors such as vertical wind shear can also play a crucial role, although how to correctly apply this is at present still under discussion. For this reason, Knutson and Tuleya (2004) choose not to include this effect, as they take the view from Emanuel (2000) that it is possible to obtain useful information on the relative distribution of intensities

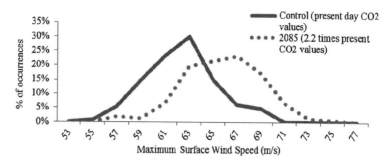

Fig. 5 Resolved inner-grid convection hurricane intensity simulation, after the work of Knutson and Tuleya (2004)

from knowledge of the potential intensity alone. Further discussion on this can be found in Chan (2006), Emanuel et al. (2008) or Vecchi et al. (2008).

Knutson and Tuleya (2004) computed histograms of the maximum surface wind speed for four different types of hurricane simulation, all of which result in an increase in both storm intensity and near-storm precipitation rates related to the increase in surface sea temperature. The method of Esteban et al. (2009a) simplifies the 2085 histogram into a probability distribution curve, and uses this to modify the intensity of historical storms as shown in Fig. 5.

For the case of the "climate change" condition, the computer simulation randomly generates an "intensity multiplier" from this probability distribution curve and multiplies it by the maximum wind speeds throughout the life of the historical storm. This intensity multiplier is normally greater than 1, resulting in a tropical cyclone of greater intensity than that of the historical record on which it is based, but it can also be less than one and result in a weaker storm. In this way, although the tracks of the tropical cyclones do not deviate from that of the historical norm, their intensities and shapes can be made to change slightly. The increased wind speed is then used to alter the radius of the tropical cyclone, as described in the next section. Note that for the case of the "present day" condition the computer uses a intensity multiplier of 1, meaning that the historical records of the tropical cyclone are unaltered.

Effect of Maximum Sustained Wind Speed on Radius of Tropical Cyclone

The data of the Japan Meteorological Agency provides radii for the sustained 30 and 50 knot winds at various time intervals. This data can be used to model the tropical cyclone as a circle representing the area which is affected by 30 or 50 knot winds or higher.

Esteban et al. (2009a) carried out an historical analysis of all tropical cyclone data between 1978 and 2008 to show how there is a positive relationship between the maximum wind speed and the radius of a storm of the shape:

$$R = b_0 + b_1 W_{max} + e \qquad (2)$$

where b_0 and b_1 are two parameters relating to the slope of the curve, R is the radius of the 30 or 50 knot winds, W_{max} is the maximum sustained wind speed and e is the error. For the case of 30 and 50 knot winds the following values for b_0 and b_1 were used:

$$R_{30} = 46.744 + 2.168 W_{max} + e \qquad (3)$$

$$R_{50} = -81.345 + 2.099 W_{max} + e \qquad (4)$$

Nevertheless, Esteban et al. (2009a) propose that investigate the sensitivity of a simulation of increased tropical cyclone size two different climate change scenarios should be investigated, a Scenario A with $b_1 = 1$ and a Scenario B with b_1 as shown on Eqs. 3 and 4.

Any scenarios with a b_1 higher that that shown in Eq. 3 and 4 would result in even greater time losses and thus goes against the conservative principle of the present paper.

For the data points where the maximum sustained wind speed is greater than 100 knots (1.78% of data), the relationship between maximum wind speed and size is not so clear. However, the effect of these points on the simulation is minor. For the case of downtime most of the time loss is not caused by the most intense tropical cyclones but by the more normal ones occurring with high frequency. However, for the damage computation this effect could be more important, though most of the damage events analysed in the present work were caused by the average strength tropical cyclones.

Both the 30 and 50 knot radius of wind speeds are significant to determine the downtime and damage that an area will suffer as a consequence of tropical cyclones, as will be shown in the following sections.

Computation of Wind Downtime and Minimum Damage Threshold

The assumption of the model used is that once wind speeds of more than 30 knots (55.56 km/h) are reached, it will disrupt many human activities. According to the Beaufort wind force scale, a moderate gale (over 27 knots) would make it difficult to walk against the wind. Anything over 34 knots ("fresh gale") would cause twigs to be broken and cars to veer on the road, and would result in a gale warning in places such as the UK and USA. From this point on, normal economic activity and transport are usually disrupted, and usually results in the shutdown of factories and commerce, and workers returning home early to take shelter. Winds of over 48 knots are classified as a "whole gale" or "storm", with trees broken off or uprooted, saplings bent and deformed and considerable tumbling of waves with heavy impact. At this point, it is clearly hazardous for any economic activity to take place and could be considered the point from which serious damage starts to take place. The present simulation hence assumes that any wind higher than 30 knots will result in the complete shutdown of the whole economy in the city. There are a number of problems with this assumption, of course, as human activities do not normally follow such pattern, and such a dramatic cut-off does not exist, though economic activity will clearly be disrupted starting from winds of 30 knots and leading to a complete shutdown much before 50 knots are reached. In a similar way, the threshold for events that start causing damage to infrastructure due to high winds will be placed at 50 knots, although damage is exponential in nature as highlighted by numerous authors (Howard et al. 1972; Pielke 2007; Hallegatte 2007, for example).

Estimation of Indirect Damage

The expected number of hours that each urban centre is affected by 30 knot winds was found using the above described model, as shown in Fig. 6. A comparison was then made between the two climate change scenarios and a control scenario where the strength of the tropical cyclones was kept identical to the historical average. This figure shows how the areas most affected by tropical cyclones are the centre areas around the city of Da Nang, with around 41 h lost each year in the control scenario, which would rise to 62 in Scenario B. For the case of Hanoi in the north, it would be 13 and 24 h respectively, while for the south of the country around Ho Chi Minh it would be around 4 and 7 h. Multiplying these by the population of each city Figs. 7, 8 and 9 are obtained. Here it can be seen how although Ho Chi Minh is not frequently affected, the large number of people living in it means that even infrequent disruption can lead to significant economic losses. The total extra cost to the economy due to an increase in tropical cyclone intensity could be between around US$900 m and US$1 400 m by 2085, although under the projections, Vietnam would be significantly richer by then. However, the economic damage caused by this effect could cut Vietnamese growth by between 0.015 and 0.035% of GDP by that time (for Scenarios A and B respectively)

Loss of Productivity of Ports

Ports are a crucial part of the infrastructure of heavily trade-oriented countries. Kawakami and Doi (2004) show how there is a direct correlation between the natural logarithm of the Real Port Capital Stock (RPCS) and the growth in Japanese GDP. The RPCS could be defined as the total value of all port

7 Increase in Port Downtime and Damage in Vietnam 115

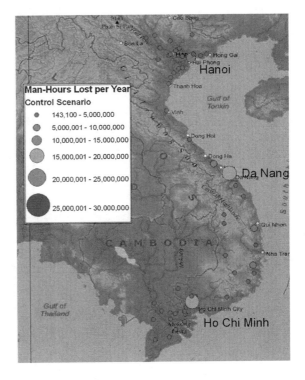

Fig. 6 Average number of hours lost in each urban centre

infrastructure and stock in Japan ports. These authors analysed the causal relationships between GDP, private capital, user transport cost and port capital and the magnitude of effects of port capital formation on private capital formation on GDP in a multivariate time-series framework. From their research it is clear that for the Japanese economy to grow there must be a continuous expansion of RPCS. Vietnam, although at a very early stage in its development, is currently attempting to build an export-oriented economy, and hence the development of port infrastructure is crucial (Thao 2009).

Effect of Climate Change on Port Operations

Thao et al. (2006) report that port operations in Vietnam are inefficient. There appears to be a number of reasons for this, such as shallow water depth and limited navigable time, poor navigation service, lack of well-maintained cargo-handling equipment, lack of trained port labour, inadequate supervision and management, and lack of incentives and unclear port charges. Thao et al. (2006) explain how for example in 1998 one state-owned operator reported it suffered a demurrage of 1,122 days from operating 21 general cargo vessels mainly because of waiting for high tides and poor cargo handling.

Fig. 7 Man hours lost in each major urban centre (control scenario)

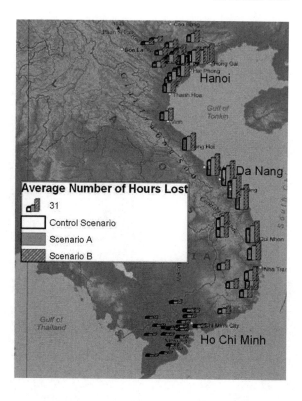

An increase in the number of hours when a port cannot operate could thus further exacerbate these problems by making port operators lose more frequently the precious windows of time when port operations are possible. Fortunately for Vietnam, most of the economic activity and major ports are located in the south, where high winds due to tropical cyclones are rarely an issue (Fig. 6). The ports in the centre of the country, around Da Nang, have relatively large capacities and deep berths, and are thus not as constrained, even if the amount of downtime could increase by around 50% by 2085. The north of the country, however, has a relatively large economy and currently few deep berths, and although potential increases in downtime are not as severe as in the centre of the country, they could create potential bottlenecks in the future. The present results thus highlight the need to improve the capacity of port infrastructure to take care of a potential increase in future downtime, and to improve the resilience of the port infrastructure and operations in general.

Estimation of Direct Damage

Analysing the direct losses due to typhoons is quite a complicated task as the impact of maximum wind speeds on damage is nonlinear, and physical damage

7 Increase in Port Downtime and Damage in Vietnam

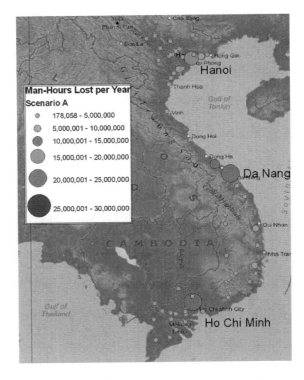

Fig. 8 Man hours lost in each major urban centre (Scenario A)

increases sharply with maximum winds. However, storms also vary in duration, and the period over which the wind is exerted also contributes to this failure. The analysis is further complicated by the fact that damage is likely to happen only after a certain threshold (nonlinear failure mode), which is reached by few storms. For the case of Vietnam, the amount of precipitation also plays a key part in the damage due to typhoons, as much of the damage is due to flooding (CCFCC 2009, as shown in Table 2). The statistics about the damage caused by natural disasters in Vietnam in the last ten years show that the number of people killed and the estimated economic losses were caused mostly by either independent flooding or those created by typhoons (CCFCC 2009).

Generally speaking, there is a strong relationship between the maximum wind speed and the damage that takes place (Howard et al. 1972; Pielke 2007; Hallegatte 2007). The model used, however, is not apt for estimating the damage in this way, as it relies on a limited number of historical typhoon tracks. As the maximum wind speed only occurs on a limited radius around the eye of the storm, to obtain a statistically significant result for the maximum wind speed, thousands of tracks would be needed. The only way to do this would be following a procedure similar to that shown by Hallegate (2007). However, there is a relationship between the maximum wind speeds and the size of the storm (Esteban et al. 2009a) and hence a storm of high wind speed will mean that a certain area would be affected by 50 knot winds for a longer period of time and hence this value can be

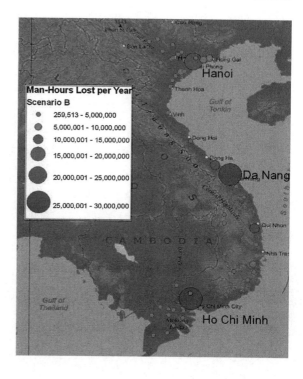

Fig. 9 Man hours lost in each major urban centre (Scenario B)

used as a proxy of the damage caused by a storm. The significance of 50 knot winds is that they can be described as the threshold of damage, and hence the calculation of the expected increase in these events can provide an indication of where damage is likely to increase in the future. Figure 10 shows the number of hours that each urban centre can be expected to be affected by winds of 50 knots for each scenario. According to the results and the likelihood of different natural hazards, several areas can be identified, although only the coastal areas are discussed below.

- The area around Ho Chi Minh City and the Mekong. This area at present is hardly ever affected by strong winds from typhoons, and the simulation shows this will not change in the future. This is not to say that damage will not necessarily increase, as the area is still vulnerable to storm surges, and there is the possibility that typhoon tracks might change in the future.
- East of southern Vietnam. This area is rather infrequently hit by strong typhoons, and wind speeds around this area are seldom a problem.
- Coastal areas of central Vietnam. The southern part of this area experiences around 0.3 h on average of 50 knot winds/year, and this is expected to increase by around 30–50% in the future (Fig. 10). The area around Da Nang is currently the hardest hit by tropical cyclones, which is currently affected on average 1.7 h/year by winds of 50 knots or greater, the highest rate in the whole country. This is expected to increase by between 24 and 47%, depending on the scenario.

7 Increase in Port Downtime and Damage in Vietnam

Table 2 Estimation of the loss caused by natural disasters from 1998 to 2008 (US$m)

Year	Typhoon	Flood	Whirlwind	Tropical low depression	Flash flood and landslide	Others
2008	126	661.212	11.697	0.485	5.818	
2007	347.564	344.381	4.098		1.769	
2006	1,105.29	7.05	22.682		25.332	
2005	270.527	31,613	4.549	23.144	21.566	
2003	3.551	50.757	1.394	42.644	1.013	
2002		833.114	2.701		22.956	
2001	51.312	160.947	1.764		4.5	6.248
2000	34.602	283.972	0.864	16.975	3.466	
1999	5.485	321.154	14.815		20.355	
1998	103.117	15.835	7.752		1.67	
Total	2,047.448	34,291.4	72.316	83.248	108.445	6.248

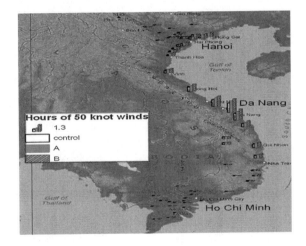

Fig. 10 Number of hours that urban centres will be affected by 50 knot winds per year

Hence, future damage potential is likely to increase in this area and it is crucial that adaptation measures are prioritized in this region. Moving north towards Dong Hoi, the rates of increase in 50 knot winds decrease slightly as compared to Da Nang.

- Coastal areas of Northern Vietnam. The north part of Vietnam is currently not as frequently affected by tropical cyclones, however some of the highest rates of growth can be seen in this region. In particular, around Thanh Hoa, just south of Hanoi, rates of increase of around 500% were computed (from 0.1 to 0.5 h for Scenario B).

The reason for this dramatic increase in the potential damage around Thanh Hoa probably lies in the fact that many typhoons turn north after exiting the Philippines and entering the China sea, eventually moving to the region around Hanoi and south China. However, if these typhoons become bigger and continue to follow the same path, although they might still not make landfall in Vietnam, they will start to have increasing effect in some of the Vietnamese regions due to their increased size.

Although the simulation gives an idea of where damage due to high winds is likely to increase, it does not really explain the economic effect on the population of the country. To understand this it is necessary to not only know the local geography and location of the population but also how developed and resilient each of these regions is against tropical cyclones. Hallegatte (2007) carried out a vulnerability assessment against hurricanes for the east coast of the United States, which supported the view of a wind-speed-to-loss relationship. However, this author does point out that large variability in recorded economic losses arises from hurricane characteristics other than wind speed (such as the amount of precipitation) and the precise landfall position and time. Also, he does mention how the vulnerability coefficients he proposed have the same problem as the normalized economic losses proposed by Pielke and Landsea (1998) and Pielke et al. (2006), namely that they do not take into account adaptation actions and hence probably overestimate current vulnerability. Hallegatte (2007) also tries to analyse the evolution of vulnerability, and states that "it seems that vulnerability evolves with time, probably toward a reduction of vulnerability in most but not all cases". Other authors, such as Yusuf and Herminia (2009) carried out a study of various countries in the Asia-Pacific and defined adaptive capacity (A_c) as the degree to which adjustments in practices, processes or structures can moderate or offset potential damage or take advantage of opportunities (from climate change). In equation form it can be written as

$$A_c = \text{f (socioeconomic factors, technology, infrastructure)} \quad (5)$$

Although this concept is very useful, for it to be used correctly it is necessary to correlate it to the economic damage after the passage of individual tropical cyclones. For the case of Vietnam this data is not easily accessed, and hence further analysis is not possible at this time. However, and despite the fact that damage increases exponentially with wind speed, what can be concluded at this point is that the damage potential is likely to increase *at least* as much as the increase in 50 knot winds, and hence for the hardest hit areas around Da Nang this damage potential could increase by 47% by 2085. In this case the damage potential could be seen as the damage level that would occur in a society where the $A_c = 0$. This clearly highlights the importance of adaptation measures to be put into place to reduce the levels of damage and increase the A_c of the country.

Implications and Discussion

There is now broad scientific evidence that climate change is a serious issue and that it could dramatically affect human economic systems. Its effects are likely to be felt more severely in developing than in developed countries. The Stern Report claims that "the overall costs and risks of climate change will be equivalent to losing at least 5% of global GDP each year, now and forever. If a wider range of risks and impacts is taken into account, the estimates of damage could rise to 20% of GDP or more" (Stern 2006). In the present simulation, the authors are essentially trying to move from the general approach used in the Stern Report into a more detailed assessment of the impacts for the case of Vietnam.

Climate models form an important tool to investigate the potential change in tropical cyclones. They contain hypotheses relating to how the climate system works, and yield fairly different results depending on these assumptions. The present model uses the results provided by Knutson and Tuleya (2004) to develop a methodology to evaluate the future economic consequences of an increase in tropical cyclone intensity. This method has a number of problems associated with it, such as not being able to take into account changes in typhoon paths. Although it could be possible to alter these routes using the work by other authors, much of this research is still inconclusive. Hence, the assumption of keeping the routes and frequencies the same can be seen as the default starting point of any simulation to determine the productivity and economic consequences of tropical cyclones.

The population and economic growth model should also be viewed with caution. Though it is based on reputable projections, attempting to predict the future development of a country is a complicated task. Also, the figures are only available for the entire country, and it is not clear that the assumption of keeping the distribution of population the same as the present one is valid, as countries rarely develop in such a simplistic manner. The projections assume that population and GDP growth with gradually ease in the future, which although typical in the historical development of other countries, should be nevertheless taken as an estimate.

The GDP per capita estimate for 2050 from Goldman Sachs (2007) is based upon current GDP per capita figures in US dollars, which are *not* adjusted for purchasing power parity (PPP). Thus, if the GDP estimates were adjusted for purchasing power parity, they would be much higher. Goldman Sachs (2007) predicts that Vietnam will be one of the "N-11" or Next Eleven nations to have a large economic and developmental boom—something which would likely alter the Purchasing Power of currency in the country.

The results overall suggest that damage will increase in the centre and north of the country and to a lesser extent in the south, and hence that further development in the south should be encouraged, as this area is not hit by tropical cyclones. However, this is clearly a simplistic prescription, as the present simulation does not take into account sea level rise or storm surges, which could leave large areas of the Mekong delta frequently inundated (Wassmann et al. 2004; MONRE 2009). The present work should thus be seen as a first step to quantify the likely changes in damage distribution in the country, and further future work should be done to expand the model by computing also the storm surge height.

Although the possibility of having stronger future tropical cyclones has been used by many to make the case for action on energy policies (climate change mitigation), Pielke (2007) argues that policy action should focus on reducing vulnerabilities, at least in the short term. In this respect, it is important that protection works such as those described by Imamura and Van To (1997) are continued to further improve the resilience of Vietnam against tropical cyclones. The present paper's approach to calculating direct damage is fairly simplistic due to the lack of accurate data. Nevertheless, and not withstanding that damage increases exponentially and not linearly with wind speed, it can provide a conservative estimation of the likely increase in the potential damage (assuming a society with no adaptation measures) that would occur if the damage progression was linear.

It can thus be seen that vulnerability is crucial to the direct damage that occurs in countries affected by tropical cyclones and the indirect damage that occurs after the tropical cyclone has passed [as shown by Hallegatte (2008) for the case of Katrina]. However, it is not a significant factor related to the indirect economic losses that occur during the passage of the tropical cyclone. If anything, an increase in adaptation capacity would increase this part of the indirect economic loss, as noted by Pielke (2007). Indeed, although wind speed is only one of the factors that influences damage and other indirect losses, it would be the major component in determining when people would choose to stop working and begin to take remedial action, as tropical cyclone alerts are based to a large extent on wind speeds. It should be noted thus that the current simulation still gives a conservative estimate of the effect of the tropical cyclones. Although it assumes that complete shutdown occurs when 30 knot winds occur, it does not take into account the "precautionary" cessation of activities that would occur before the storm arrives. Workers would leave work much earlier in order to return home to take shelter. If these losses were also to be included, the computed downtime could increase substantially. However, the estimation of this "precautionary" downtime is difficult to calculate and highly variable depending on the circumstances and personality of each individual. Thus, this effect is better ignored in the absence of better information and in order to keep the results of the paper conservative.

Conclusions

The objective of this research was to calculate the effect that a future potential increase in tropical cyclone intensity would have on economic damage to Vietnam. The paper uses as its basis the methodology of Esteban et al. (2009a), which in turn uses the results of Knutson and Tuleya (2004) for the estimation of the increase in tropical cyclone intensity in 2085. Essentially, the downtime and damage expected from tropical cyclones are computed for two different cases, namely the "present day" and "climate change" conditions.

The results show that economic downtime could increase from a national 0.23–0.37% by 2085, which could cause the loss of between 0.015 and 0.035% of GDP growth per year (between US$600 bn and 1,400 m considering the potential size of the economy in 2085). This downtime could also have a big effect on port operations, and the paper highlights how it is thus imperative to increase the capacity in ports (especially those situated in the north of the country, which have less deep berths) and the resilience in those in the centre (where damage is likely to increase) in order to prevent future bottlenecks. The downtime shown would probably not be significant for a developed country with enough capability to adapt to these effects, but could cause severe disruption in a developing country like Vietnam.

Regarding direct damage, an accurate economic figure could not be obtained due to the lack of accurate economic damage data and limitations in the model used. However, the authors could conclude that the potential increase in wind damage would increase throughout most of the country (with the exception of the south). However, the authors do highlight how this area faces increased flooding risks (MONRE 2009; Socialist Republic of Vietnam 2005), which, although not assessed in the current paper, are crucial to the development of the country. The potentially devastating effects that an increase in sea level and storm surge (due to tropical cyclones) could cause has been mentioned elsewhere in the literature, but not computed in detail. This effect should be further analysed in the future as this area produces a large part of the rice of Vietnam, and increasing flooding could ruin the lands and significantly reduce rice yields.

It is very difficult to estimate the direct damage to infrastructure and housing, though the paper showed how it is possible to compute a relative increase in potential direct damage. This level would vary through the country but was shown to be in the 33–65% range, although a dramatic 400% plus increase was computed for some coastal areas south of Hanoi. This highlights the important problems facing the country and the need to increase the resilience of communities in the future.

References

Chan JCL (2006) Changes in tropical cyclone number, duration, and intensity in a warming environment. Science 311:1713

Elsner JB, Kossin JP, Jagger TH (2008) The increasing intensity of the strongest tropical cyclones. Nature 455:92–94

Emanuel K (2000) A statistical analysis of tropical cyclone intensity. Mon Weather Rev 128:1139–1152

Emanuel K, Sundararajan R, Williams J (2008) Hurricanes and global warming: results from downscaling IPCC AR4 simulations. Bull Am Meteorol Soc 89(3):347–367. doi: 10.1175/BAMS-89-3-347

Esteban M, Webersik C, Shibayama T (2009a) Methodology for the estimation of the increase in time loss due to future increase in tropical cyclone intensity in Japan. J Clim Change, DOI 10.1007/s10584-009-9725-9. Available at http://www.springerlink.com/content/ u568p666t2h04075/

Esteban M, Webersik C, Shibayama T (2009b) Effect of a global warming induced increase in typhoon intensity in urban productivity in Taiwan. J Sustain Sci 4(2):151–163

Goldman Sachs (2007) Global economics paper no: 153, The N-11: more than an acronym, 28 March 2007

Hallegatte S (2007) The use of synthetic hurricane tracks in risk analysis and climate change damage assessment. J Appl Meteorol Climatol 46(11):1956–1966

Hallegatte S (2008) An adaptive regional input–output model and its application to the assessment of the economic cost of Katrina. Risk Anal 28(3):779–799

Howard RA, Matheson JE, North DW (1972) The decision to seed hurricanes. Science 176:1191–1202. doi:10.1126/science.176.4040.1191

Imamura F, Van To D (1997) Flood and typhoon disasters in Viet Nam in the half century since 1950. J Nat Hazards 15:71–78

IPCC (2007) Climate change 2007: the physical science basis contribution of working group i to the 4th assessment report of the intergovernmental panel on climate change. Cambridge University Press, Cambridge, UK and NY, p 996

Japan Meteorological Agency (2008) Available at www.jma.go.jp Accessed 1 June 2009

Kawakami T, Doi M (2004) Port capital formation and economic development in Japan: a vector autoregression approach. Reg Sci 83:723–732

Knutson TR, Tuleya RE (2004) Impact of CO_2-induced warming on simulated hurricane intensity and precipitation sensitivity to the choice of climate model and convective parameterization. J Clim 17(18):3477–3495

Knutson TR, Tuleya RE et al (2001) Impact of CO_2-induced warming on hurricane intensities as simulated in a hurricane model with ocean coupling. J Clim 14:2458–2468

Landsea CW, Harper BA, Hoaran K, Knaff JA (2006) Can we detect trends in extreme tropical cyclones? Science 313(5786):452–454

Ministry of Natural Resource and Environment (2009) Climate change, sea level rise scenarios for Vietnam, Report, Vietnam, p 34

Munich Re (2009) Natural disasters 1980–2008, 10 costliest typhoons ordered by insured losses. Available at http://www.munichre.com/. Accessed 28 April 2009

Pielke Jr RA (2007) Future economic damage from tropical cyclones: sensitivities to societal and climate changes. Philos Trans R Soc. doi: 10.1098/rsta.2007.2086

Pielke RA, Landsea CW (1998) Normalized hurricane damages in the United States: 1925–95. Weather Forecast 13:621–631

Pielke RA Jr, Gratz J, Collins D, Landsea CW, Saunders M (2006) Normalized hurricane losses in the United States: 1900–2005. Nat Hazards Rev 9(1):29–42

Shibayama T, Tajima Y, Kakinuma T, Nobuoka H, Yasuda T, Hsan RA, Rahman M, Islam MS (2009) Field survey of storm surge disaster due to cyclone Sidr in Bangladesh. In: Proceedings of coastal dynamics conference, Tokyo, 7–11 September 2009

Socialist Republic of Vietnam (2005) National report on disaster reduction in Vietnam, world conference on disaster reduction, Kobe, 18–22 January 2005

Stern N (2006) Stern review on the economics of climate change. HM Treasury, London

Takagi H, Thao ND, Cong LV (2009) Recent state of coastal areas in Vietnam. J Coast Zone Stud 21(4):113–119

Thao ND (2009) Sea-port system in the southern key area of Vietnam: present status and development strategies, 4th meeting of the Port-City University League, Yokohama, 21–22 May 2009

Thao ND, Trang DDT, Tuong TN (2006) The Study on Port Development in the Key Area of South Region of Vietnam. 3rd Vietnamese–Japanese students' scientific exchange meeting, VJSE 2006, November 2006, pp 153–156

The Central Committee for Flood and Storm Control (2009) Historical disaster database-summary of damages, Vietnam. Available at http://www.ccfsc.org.vn/ccfsc/. Accessed 1 June 2009

To Lang N (2005) Urbanization in Vietnam Features and Developing Trends in the Early 21st Century, Expert Group Meeting on Urban Sector Strategy Review, Managing Growth in Asia, pp 28–29 July 2005, ADB Headquarters

United Nations Department of Economic and Social Affairs, Population Division (2008) World population prospects: the 2008 revision population database. Available at http://esa.un.org/unpp/. Accessed 11 March 2009

Vecchi GA, Swanson KL, Soden BJ (2008) whither hurricane activity? Science 322, doi: 10.1126/science.1164396

Wassmann R, Nguyen XH, Hoanh CT, Tuong TP (2004) Sea level rise affecting the Vietnamese Mekong delta: water elevation in the flood seasson and implications for rice production. J Clim Ch 66:89–107

Webersik C (2010 forthcoming) Climate change and security: a gathering storm of global challenges, praeger security international series, security and the environment. In: Liotta PH (series ed), Praeger Publishers/PSI

Webster PJ, Holland GJ, Curry JA, Chang HR (2005) Changes in tropical cyclone number, duration, and intensity in a warming environment. Science 309(5742):1844–1846

World Meteorological Organisation (2006) 6th international workshop on tropical cyclones, San Jose, Costa Rica

Yusuf AA, Herminia F (2009) Climate change vulnerability mapping for Southeast Asia, EEPSEA special and technical paper, no. tp200901s1, January 2009

Chapter 8
Hydro-Economic Analysis for Water Resources Management in a Changing Climate

David Corderi Novoa

Abstract Water is potentially one of the most affected resources as climate changes. The impact of climate change in hydrologic regimes will affect water quantity, water quality and the occurrence of extreme events such as floods and droughts. Future water availability will also depend on the evolution of water use and water demands which are influenced by socioeconomic factors such as population growth, technological progress, infrastructure development, land use changes, and institutions. Climate change analysis for water resources management has traditionally focused on the physical impacts of different climate regimes on water availability with a limited representation of the socioeconomic dynamics within water systems. At the same time, economic analysis of climate change impacts on water resources has lacked a careful treatment of the physical aspects of water resource systems. Hydro-economic models have the potential to bridge the gap between these two types of analysis by integrating economic analysis into detailed water resource system analysis. A hydro-economic framework of analysis can be instrumental in studying how climate-related changes in water availability affect social welfare and how different policy interventions and investments can help reduce risks and take advantage of possible opportunities in the water sector resulting from climate change. This framework can be a useful tool for planners and decision-makers in the water management sector. Hydro-economic assessments integrate models describing the physical effects of climate change (e.g. hydrologic changes) with models describing engineering, economic and institutional responses at appropriate scales, providing a robust framework to study whether the broad range of water uses might remain compatible under future

---

D. C. Novoa (✉)
Agricultural and Resource Economics,
University of California, Davis One Shields Avenue,
Davis, CA 95616, USA
e-mail: dcorderi@primal.ucdavis.edu; david.corderi@aya.yale.edu

climate scenarios. We review case studies of different river basins around the world where applications of hydro-economic models were used to assess climate change impacts and adaptation strategies in water resources management. The cases under review address climate change adaptation strategies such as water allocation for agriculture and other uses in a river basin, water storage infrastructure expansion, flood control investments, and institutional barriers to adaptation.

Keywords Climate change · Adaptation · Water · Hydro-economic models · Integrated water resources management (IWRM)

Introduction

Water is potentially one of the most affected resources as climate changes. Knowledge and understanding continue to evolve about the nature, extent and distribution of the physical effects of possible climate change on water resources. According to the IPCC FAR (Bates et al. 2008) increases in global temperatures as well as changes in precipitation patterns (frequency and intensity) can be associated with changes in the hydrologic cycle and the subsequent impact on water availability from rivers, lakes and underground resources. The impact of global warming on water systems fed from snowmelt will also have an effect on water resources, since snow fields and glaciers act as natural reservoirs that balance water availability between winter and summer. Reduced glacier runoff could dramatically decrease river flows in the dry season, which will disrupt not only irrigated agriculture, but also human consumption and hydropower generation. Changes in the timing of runoff can also increase the risk of flooding in the wet season. In addition to the possibility of changes in water quantity and occurrence of extreme events, changing runoff patterns and temperatures may result in water quality effects such as increases in salinity or water pollution.

Future water availability will not only depend on the climate change effects on water supply but also on the evolution of water use and water demands which are influenced by socioeconomic factors such as population growth, technological progress, infrastructure development, land use changes, and institutions. For this reason, measures designed to ensure future water availability require integrated supply-side as well as demand-side strategies. Supply-side strategies generally involve infrastructure development such as additional storage capacity and improvements in conveyance infrastructure, or changes in the operations of a given hydraulic infrastructure system. Demand-side strategies focus on the distribution and efficiency of water uses. The development of water markets, the adoption of efficient irrigation technologies, and changes in land use are good examples of demand-side adaptation measures.

Climate change analysis for water resources management has traditionally focused on the physical impacts of different climate regimes on water availability (a supply-side approach) with a limited representation of the socioeconomic dynamics within water systems. At the same time, economic analysis of climate change impacts on water resources has lacked a careful treatment of the physical aspects of water resource systems. Hydro-economic models have the potential to bridge the gap between these two types of analysis by integrating economic analysis into detailed water resource system analysis. A hydro-economic framework of analysis can be instrumental in studying how climate-related changes in water availability affect social welfare and how different policy interventions and investments can help reduce risks and take advantage of possible opportunities in the water sector resulting from climate change.

Hydro-economic assessments of climate change integrate models describing the physical effects of climate change (e.g. hydrologic changes) with models describing engineering, economic and institutional responses at appropriate scales, providing a robust framework to study whether the broad range of water uses might remain compatible under future climate scenarios. This integrated hydrologic-economic-engineering approach allows for a comprehensive appraisal of adaptation strategies in a context of potential large-scale trade-offs between supply and demand across multiple water-dependent sectors with varied socioeconomic, environmental and institutional characteristics.

This paper is organized as follows: section one gives an overview of the underlying features of hydro-economic models; section two discusses examples of hydro-economic models that either incorporate climate change analysis explicitly or have the potential to be used for this purpose; section three concludes on the potential of using hydro-economic models to model adaptation to climate change in the water sector.

Hydro-Economic Models: Concept and Design

A Description of the Approach

Hydro-economic models represent regional scale hydrologic, engineering, environmental and economic aspects of water resources systems with a coherent framework. The main underlying philosophy of the models is that water systems perform economic functions, which needs to be accounted for when making decisions (Harou et al. 2009). These models have emerged as a useful tool for conducting economic analysis for integrated water resources management (IWRM). Hydro-economic models possess several distinguishing features that make them more adequate for climate change analysis than purely economic or hydrologic-engineering models.

Hydro-economic models can improve traditional hydrologic/engineering models by providing a better socioeconomic representation of the different water users in a water system. On one hand, this modelling framework introduces the economic concept of water demands from water users that respond to different water supply allocations as opposed to the traditional engineering approach of static water demands that can lead to the overestimation of climate change impacts. On another hand, these models allow for quantitative valuation of economic impacts of climate change for different water users. In other words, the models can evaluate the socioeconomic consequences of climate related changes in water supply using a bottom-up framework.

Traditional economic models for climate change analysis were also not useful for water sector planners due to the oversimplification of the water systems; these models provided top-down analysis that did not take into account the specificities of water resources management units. Hydro-economic models improve this traditional approach in economic analysis by considering hydrologic and engineering components such as water balance and relevant water supply infrastructure.

The typical structure of a hydro-economic model takes into account hydrologic and engineering features and represents them in a node-link mathematical network, where economic demands are represented as nodes and costs (or benefits) can be incurred on links. Schematic diagrams are a convenient way to illustrate the spatial features of a hydro-economic model and to indicate the relative locations of upstream and downstream water supplies and users. The hydro-engineer model structure depicts key physical characteristics of the natural and man-made water supply system, including tributaries, inflows, and return flows, diversion points, reservoirs, and basin imports and exports. The spatial structure characterizes where water enters the system, how it travels and is distributed, where it is used, and how it leaves the system.

Hydro-economic models link the investment decisions of water resource planning authorities, the water allocation decisions of water managers, and the water consumption decisions of water users together in a spatially and temporally differentiated framework that is consistent with the geophysical features of a water resources management unit. These models constitute a useful tool to inform decision-makers with respect to the economic aspects of climate change in the water sector in an integrated manner.

Examples of Model Design

There are many choices for designing hydro-economic models to study the effects of and responses to climate change in the water sector. Choosing the appropriate scale for modelling is instrumental to the analysis of climate change and depends on the water management question of concern. Most hydro-economic models use monthly or yearly time steps. Seasonality is normally captured in models that include agriculture, for example in the hydro-economic model of Howitt et al. (2006) or in Williams and Ray (1999). When questions such as long-term infrastructure investments need to be addressed, the temporal scale is divided in year time steps for a sufficiently long time horizon; see for example Callaway et al. (2007). The representation of space can vary considerably; most models reviewed in this paper are semi-distributed with different levels of spatial disaggregation. Hurd et al. (2004) study water management at the basin level taking into account different irrigation districts and urban areas located around the basin. Williams and Ray (1999) examine an irrigation canal where the locations of up to forty farms are considered.

Another important choice for hydro-economic model design is the type of model solver framework. Hydro-economic simulation models are used to examine and evaluate specific "what if" scenarios consisting of particular management decisions under a particular climate change scenario. Several models of this type are reviewed in this paper, for example Callaway et al. (2007) use a hydro-economic model to simulate the welfare effects of expanding the existing hydraulic infrastructure.

While simulation models can estimate the effects of specific adaptation options, optimization models can identify the most appropriate adaptation option. Optimization models typically employ simpler formulation of the system to identify "what is best". Lund et al. (2006) use the CALVIN model to find the optimal mix of adaptation options for California's water system. Both simulation and optimization models are complimentary to each other, since simulation models can test and refine optimization model results.

Analysing Adaptation in Water Resources Management with a Hydro-Economic Approach

Water and Agriculture

Agriculture accounts for 70% of freshwater withdrawals from rivers, lakes and aquifers—up to more than 90% in some developing countries. While rain-fed agriculture covers 80% of the world's cultivated land, and is responsible for about 60% of crop production, irrigated agriculture is responsible for 20% of crop production. Climate change impacts in the water cycle will have an important effect on the agriculture sector through higher temperatures and reduced precipitation, ultimately affecting water availability and crop yields.

Hydrologic-engineering analysis of climate change impacts and adaptation for agriculture water use, see for example Rosenzweig et al. (2004), combines a hydrologic model with a water engineering model and a crop model. This approach does not take into account that agriculture water demands may shift due to the economic response of farmers to altered water availability and yield changes on crops.

Economic models to study agriculture, water and climate change such as the Ricardian model (Mendelsohn et al. 1994) uses a cross-sectional econometric model to explore how farm values and net revenues vary across climatic zones and different conditions of water availability. These studies estimate response functions that are assumed to incorporate optimal adaptation. However, the studies were criticized for failing to properly take into account the effects of irrigation and other water supplies when analysing farm revenue impacts of climate change (Cline 1996; Darwin 1999). Kurukulasuriya and Mendelsohn (2007) and Mendelsohn and Dinar (2003), explored the importance of water availability within the Ricardian model by estimating the role of irrigation as an adaptation measure to both drier and hotter climate conditions in Africa. Structural Ricardian models have been used to estimate other demand side adaptation options using household data in developing countries; Mendelsohn and Seo (2007) analysed crop choice and livestock choice as adaptation options in agriculture in Latin America and Africa. These studies, however, do not approach climate change impacts and adaptation responses with a coherent representation of the water systems; cross-country data is used without controlling for hydrologic units such as a river basin.

Hydro-economic models such as the one by Howitt and Pienaar (2006) overcome the limitations of previous economic and hydrologic studies mentioned before. A Statewide Water and Agricultural Production (SWAP) model is constructed to study the economic aspects of adaptation to climate change in California agriculture taking into account water resource impacts. The SWAP model is integrated with a hydrologic and hydraulic infrastructure model (CALVIN) for California's water system to incorporate physical constraints on water availability in the most realistic manner. An agronomic component is further incorporated into the analysis; changes in crop yields and crop water demands are estimated and included as inputs to the agriculture sector model for a more accurate analysis.

The SWAP model is an economic optimization model that maximizes net returns to agricultural production for 21 regions in California subject to specific resource and production constraints. The production possibilities of a region are modelled through multi-input production functions which relate crop specific production with aggregate inputs such as land, water and capital. Estimated regional crop yield factors are used to scale the production functions and reflect agronomic potential given climate change. The model also allows state crop prices to be endogenously determined by introducing a demand function for each of the crops produced in California.

The SWAP model uses the region-specific production opportunities to derive an implicit water demand per region and combines it with each region's water allocation determined by a water distribution model (CALVIN), which would be the equivalent of the water supply per region. By using this supply–demand approach, it is possible to estimate the implicit value per unit of water taking into account the temporal and spatial dimensions of water scarcity. The shadow value of water represents the willingness to pay for an additional unit of water and depends on the type of crop grown and its region-specific price. This approach is useful not only for optimizing inter-regional water allocations in the agriculture sector but also for optimizing inter-sector water allocations in a water system with multiple users.

This hydro-economic model incorporates implicit agriculture water demands that respond to changes in crop yields and water allocations. This is used to study different adaptation possibilities for agricultural water demands. The economic implications of different water demand adaptation strategies at the farm level for two climate regimes are studied using this modelling framework. Three types of adaptation responses are estimated endogenously in the model. Farmers can adjust the total area irrigated, the combination of crops to be grown and the input mix in response to the supply of water. For example, farmers may choose to plant crops that have a higher return per unit of water, or substitute capital for water by investing in more efficient field-level water delivery systems, which reduces the amount of applied water without reducing crop yields.

The model is used to assess adaptation for two climate change scenarios that capture possible hydrologic extremes. Under the dry scenario, regions shift from water-intensive crops that have low economic value (pasture and rice) to high-value crops (vegetables) in order to maintain income. The demand for water for each crop increases to compensate for the higher temperature raising the agriculture sector water demand. Furthermore, the high-value crops have the largest increase in irrigation water since they constitute the highest value per unit of water. Under the wet scenario, the increase in water leads to an increase in low-value agriculture.

This hydro-economic analysis is useful for planning purposes since it can shed light on the potential welfare implications of different future climate scenarios for water availability. One of the main conclusions from the analysis is that it is more efficient to allocate water to high value agriculture in situations of low water availability. The model can also evaluate the optimal response to agriculture water shortages at the farm level. The assumptions made in the construction of the model tend to overestimate the level of adaptation and hence the sizes of the residual impact of climate change. In particular, the assumption of no transaction costs or institutional barriers for reallocating water from low value to high value uses and the assumption of perfect foresight with respect to future water conditions can significantly overestimate the mitigation of negative climate impacts.

Water and its Multiple Users

Besides being a critical input in food production activities, water is a vital resource for domestic uses, power generation, transportation, manufacturing processes, and recreation among others. This multisectoral nature of water resources development has also been analysed using a hydro-economic framework. Hydro-economic models have integrated and studied competing uses of water in a consistent framework, examining the economic welfare implications of alternative climate scenarios. Hurd et al. (1999, 2004) use a hydro-economic model to study the climate change impacts and adaptation for several water use sectors in four large river basins in the US. The model incorporates three types of consumptive uses—agriculture, municipal and industrial, and thermoelectric power—and five non-consumptive uses—hydropower, navigation, flood control, thermal waste heating, and water quality (divided into secondary and advanced wastewater treatment).

The study considers water reallocations among the water users in the basin as the main adaptation strategy. The focus is on surface water supplies and uses which are linked to climate variables through rainfall runoff models. Water allocation options for adaptation follow an economic optimization model which is subject to constraints that define the physical characteristics of the natural and man-made water supply system. This partial equilibrium framework optimizes water allocation to maximize the value of economic uses and services of available water supply in the river basin over both space and time.

Benefit functions for the different users are developed using valuation functions based on the concept of economic surplus (producer and consumer surplus), an indicator of willingness to pay for a given quantity of water. The authors construct linear economic demand curves for water consumers and use available data and methods to build non-consumptive demand functions. Net economic returns per use is calculated as total economic surplus from each activity that uses water as an input minus the costs of providing water, flood damages and diminished water quality. Using a river basin schematic the authors are able to identify water uses, net economic returns, water use levels and instream flows at each node of the watershed so that a spatially explicit optimization can be carried out. The model also studies the dynamic nature of water allocation by capturing the intertemporal trade-offs between the value of current and future water use as reflected in the reservoir storage decisions.

Using the above-mentioned components, the model estimates the marginal value or implicit price of water for every period and location based on the valuation functions for each benefit and cost in the system. The optimization framework then allocates water under the different climate change scenarios so that the marginal values of water are equalized. Projected seasonal runoff enters into each basin, and the model solves simultaneously for water allocations and implicit water prices for both consumptive and non-consumptive uses, reservoir storage and releases, and instream flows over a multiseason (three annual seasons), multiyear planning period.

Given the model set-up, adaptation to climate change is identified through a supply and demand framework as if there was a competitive market for water with no transaction costs and perfect information. Water relocations follow from the signals of changes in the marginal value of water which takes into account changes in water supply costs and in economic surplus for different water delivery quantities. Demand elasticities and marginal supply costs will determine water redistributions.

This hydro-economic model allows for an economic interpretation of the climate change simulation results in these four US river basins. Water allocations in the agriculture sector are significantly affected, which follows from the low marginal value of agriculture, its relatively low marginal supply costs and relatively high price elasticity of demand. Allocations to the municipal and industrial sector do not change much because of the sector's relatively inelastic demand schedule and the high marginal cost of supplying water to the sector. Furthermore, welfare impacts by sector and basin are calculated using this framework.

When it comes to non-consumptive uses, their dependence on streamflow makes them also particularly vulnerable to climate change; losses in these sectors cannot be allocated to those with the lowest marginal benefits in the same manner as agriculture bears the burden in the consumptive sector. Therefore, the magnitude of the welfare losses to non-consumptive sector tends to be greater than consumptive uses.

In addition to this, results suggest that there are significant differences in adaptation across river basins for a given climate scenario. This fact highlights the importance of the spatial scale, and the possibility of neglecting these differences if the model had a greater spatial resolution.

One of the disadvantages of modelling adaptation with such a wide spectrum of water use sectors is that their characterization tends to be more simplistic. For example, the agriculture sector demand included in Hurd et al. (1999) is not modelled with as much detail as in the SWAP model (Howit et al. 2006), where the derived demand for water is obtained taking explicitly into account endogenous changes in crop prices, changes in irrigation efficiency and shifts in cropping patterns.

Lund et al. (2006) use a hydro-economic model for evaluating climate change impacts and adaptation in California's inter-tied water system; this economic-engineering optimization model is also known as the CALVIN model. Compared to Hurd et al. (1999), CALVIN has fewer water use sectors (agriculture and water) but includes a more detailed representation of the agriculture sector as well as the water system; for example, groundwater is included in the model in addition to surface water. The model optimizes both water allocations and the operation of water facilities to study different adaptation options given a set of infrastructure and physical constraints. CALVIN finds the best water operations and allocations for maximizing regional statewide agricultural and urban economic value from water use and minimizing operating costs. The model considers broad mix of adaptation options such as system re-operation, conjunctive use, water reuse and desalination, water markets and water conservation.

Marginal values of water drive the allocations in the study of water allocations in the assessment of California's water system adaptation. The definition of marginal water value is reformulated into a scarcity cost indicator which measures users' willingness to pay for additional water when users' target demand is not fulfilled. By using this definition, the model's optimization target is to minimize the scarcity costs and operating costs of the water system. Operating costs include pumping, treatment, urban water quality, recharge, reuse, desalination and other variable operating costs for the system. Scarcity costs represent how much users would be willing to pay for desired levels of water delivery.

Agriculture water demands and marginal values are estimated using the SWAP model (Howitt et al. 2001); urban water demands are estimated using empirical household economic water demand curves adjusted for a number of factors. The model also takes into account environmental flows which are included as minimum instream flows by area. The CALVIN model calculates also the implicit costs of changing environmental flow requirements using the economic values of the system (urban, agricultural, hydropower and operations).

Using CALVIN, Lund et al. (2006) study the implications of a wet and a dry climate change scenario for California's water system. Results from the modelling exercise shed light on the economic desirability of different adaptation options as well as the magnitude and distribution of the climate change impacts and residual damages. A warmer and dry climate increases scarcity costs for water users and operating costs of infrastructure significantly (Medellin 2008). Water scarcity is shared across the different users, agricultural regions see larger increases in scarcity costs and water is transferred to urban regions due to the price-inelastic nature of their demand. Besides increased water market transfers, increased conjunctive use of groundwater and surface water, increased urban water conservation, and use of additional water reuse treatment arise as economically viable adaptation options for California's water system. In general, study results suggest that there will be a need for better water management given the increasing scarcity of water and infrastructure capacity.

Additional Applications of Hydro-Economic Models

New Infrastructure and Water Markets

Climate change adaptation in the water sector may require building additional infrastructure to store more water in cases where the seasonal distribution of runoff is altered significantly, or natural storage of water, such as glaciers, disappears due to temperature increases. Callaway et al. (2007) use a hydro-economic model to examine the economic desirability of the expansion of the water infrastructure system by assessing the economic benefits and costs of building a dam in the water system in the Berg river system in South Africa.

The analysis uses a simplified version of the Berg river basin and its water uses. A hydrologic model is coupled with an economic optimization model that considers urban and agricultural water uses in a spatially and dynamically explicit manner. The objective function includes the benefits such as the willingness to pay of the urban sector and the farm income of the agriculture sector as well as costs such as the short-run water delivery and operating costs, different farm costs, and the capital costs of building a dam.

The adaptation strategies considered in the study are structural measures such as building a dam (which is considered a long-run adaptation measure) and non-structural measures such as changes in operations and introducing efficient water market allocations (a short-run adaptation measure). The model is expanded to be able to calculate endogenously the optimal capacity of a reservoir and other structural works (water pumps). In other words, the model calculates the optimal size of the projected dam, which in turn determines the capital cost of the project to be included in the overall optimization framework. This model is a good example of how hydro-economic models can inform decisions between trade-offs such as long-term supply-side infrastructure investments and new institutional approaches such as introducing water markets.

The Berg river basin study develops a framework for economic analysis that is able to disentangle the effects from climate change such as decreased runoff from those of development such as the effect of population growth on urban water demand. By building alternative climate change and development scenarios, an exhaustive welfare analysis of adaptation strategies is carried out. This approach captures the uncertainty aspects of policy planning for climate change in a stylized manner by considering ex ante decisions and ex post outcomes. It also allows the authors to perform an economic analysis of no regret measures by comparing structural and non-structural adaptation measures. The concept of no regret is particularly important in the context of long-run investments that are irreversible or costly to change, or whose performance is climate sensitive such as a reservoir.

Results from the assessment suggest that both implementing efficient water markets and constructing a dam can mitigate some of the negative effects that climate change and population growth will have in the river basin's welfare. The implementation of efficient water markets has positive effects over the entire range of scenarios. When compared to other adjustment strategies, it is less risky and constitutes a no regret measure. Building a reservoir is relatively less effective than the water market adjustment, especially in very dry scenarios. The size of the reservoir (storage capacity) varies with the magnitude and the source of water scarcity (population growth or climate change). Higher urban demand growth makes markets more desirable than additional storage capacity because reducing consumption through high prices has a lower marginal cost than adding storage capacity for high levels of storage. Efficient water markets can also make a higher storage capacity more desirable by increasing the marginal present value of storage in some situations. Finally, the optimal size of the reservoir tends to follow a nonlinear pattern with respect to the severity of climate change.

Flood Infrastructure Investments

Another important problem arising from climate change effects in water resources is the intensification of extreme events such as floods. Zhu et al. (2007) study flood protection and adaptation of floodplains to climate change and urbanization in a basin in California. The hydro-economic approach tries to capture the intertemporal, spatial and uncertainty aspects of investments in flood protection. The focus is on levee design in the Sacramento river basin as a long term adaptation option. This study goes beyond standard climate change analysis of floods where changes in the frequency and extent of floods are analysed without an explicit treatment of the economic decisions regarding flood protection investments.

The authors study the economic implications of different levee setbacks and levee heights using an economic-engineering approach. The model integrates a probabilistic assessment and a hydraulic model of the river to analyse changes in flood frequency due to climate change and subsequent changes in the probability of levee failure. By studying levee failure and the possibility of overtopping flows, the hydraulic model is able to construct stage damage curves that allow for a spatially explicit analysis of floods. Cost and benefit functions were constructed based on available data such as levee construction costs and the economic value of land use in the floodplain. The authors study the economic trade-off of increasing levee setback for additional protection and decreasing the annual land value benefits due to land reclamation.

The dynamic programming model maximizes the value of the floodplain by choosing levee setback and height, taking into account the costs and the benefits associated with those decisions as well as the probability of flooding. The analysis uses a base case scenario with stationary climate and then constructs climate change and urbanization scenarios. Results suggest that there might be economic value to expand levee heights and setbacks over long periods of time. Sensitivity analysis is also performed and shows that greater damageable property and urban values encourages additional investments in levees. In addition to this, lower discount rates and high growth rates in levee construction costs tend to accelerate investments in levees towards the beginning of the planning period.

Institutional Aspects: Barriers to Climate Resilient Water Management

Hydro-economic models can also be used to study different institutional arrangements in the water sector, such as imperfect markets or the existence of transaction costs and their implications for water resource development strategies in a changing climate. Markets and institutions that are less than competitive tend to restrict and retard adjustments to water scarcity changes and hence lower the capability of adjusting to changes in water supply by precluding flexible responses to offset increased competition over fixed water supplies. Ray and Williams (1999) use a hydro-economic model to study the welfare implications of agriculture pricing policy and the efficiency and equity considerations arising from water theft along a watercourse used for irrigated agriculture in a drought-prone area in India.

The study area is known to have price supports for crops that are water-intensive and have high-value (sugarcane) in addition to heavily subsidized canal water for irrigation. Besides this, institutional arrangements for water allocations from the irrigation authorities make it possible for farmers to steal water from the canal. Given the hydro-geologic structure of the canal, farmer's location asymmetry can constitute an advantage or disadvantage that is exacerbated by institutionalized water theft and the pricing policies. Farmers located upstream make downstream farmers worse off due to water theft, which is further induced by the price incentives. This hydro-economic model improves upon standard economic analysis by taking into consideration the hydrologic structure of the canal and its implications for the economic welfare of the different water users located in different places around the canal.

The authors use a hydro-economic model to study the welfare implications of different pricing policies and different water allocation regimes for the households along the water course. The model of the watercourse has representative farms that only differ in their location along the irrigation canal. A relatively detailed representation of the hydrologic and water operating features of the canal is also embedded in the model constraints. The authors then establish three water allocation regimes and study their implications using economic welfare analysis.

Results suggest that water theft produces higher social costs compared to the other regimes. However, there is no clear advantage between efficient and equitable water allocations; the welfare outcome depends heavily on the pricing system in place. Furthermore, changes in agriculture pricing policies can improve situations of socially inefficient allocations of water.

Several lessons from this study can be applied to the analysis of climate change adaptation in water resources. Firstly, the pricing structure in the different sectors of the economy will have an important influence in future adaptation pathways. Changes in the pricing policy can be a mechanism to encourage adaptation to climate change. Secondly, equity considerations should be included in the study of adaptation to climate change; in this study, downstream farmers are significantly worse off due to their inability to access water resources. Finally, adaptation to climate change has important aspects of cooperation at the user level; hydro-economic models have the potential to study this aspect.

Conclusion

The analysis of water resources management under future climate change conditions has traditionally focused on the physical impacts of different climate regimes on water availability (a supply-side approach) with a limited representation of the socioeconomic dynamics within water systems. On the other hand, economic analysis of climate change impacts on water resources has lacked a careful treatment of the physical aspects of water resource systems. Hydro-economic models have the potential to bridge the gap between these two types of analysis by integrating economic analysis into detailed water resource system analysis.

Hydro-economic models link the investment decisions of water resource planning authorities, the water allocation decisions of water managers, and the water consumption decisions of water users together in a spatially and temporally differentiated framework that is consistent with the geophysical features of a water resources management unit. These models constitute a useful tool to inform decision-makers with respect to the economic aspects of climate change in the water sector in an integrated manner.

This paper has provided an overview of the hydro-economic model design and its applications to study climate change adaptation in the water sector. The studies reviewed touch upon some of the most important issues in water resources management and climate change such as water and agriculture, the competing demands for water in a river basin context, evaluating long-term water infrastructure investments, market and institutional issues in water allocation, and infrastructure investments to protect from changes in extreme hydrologic events such as floods. The examples reviewed suggest that hydro-economic models are a promising approach to water sector planning and decision-making in the context of climate change, improving upon purely hydrologic-engineering or economic models alone.

References

Bates BC, Kundzewicz ZW, Wu S, Palutikof JP (2008) Climate change and water. Technical paper of the intergovernmental panel on climate change, IPCC Secretariat, Geneva, 210 pp

Callaway JM, Louw DB, Nkomo JC, Hellmuth, ME, Sparks DA (2007) The Berg River dynamic spatial equilibrium model: a new tool for assessing the benefits and costs of alternatives for coping with water demand growth, climate variability, and climate change in the Western Cape. AIACC Working Paper No. 31

Cline WR (1996) The impact of global warming on agriculture: comment. Am Econ Rev 86:1309–1312

Darwin R (1999) The impacts of global warming on agriculture: a Ricardian analysis: comment. Am Econ Rev 89:1049–1052

Harou JJ, Pulido-Velazquez M, Rosenberg DE, Medellin Azuara J, Lund JR, Howitt RE (2009) Hydro-economic models: concepts, design, applications, and future prospects. J Hydrol 375(3–4):627–643

Howitt RE, Pienaar E (2006) Agricultural impacts. In: Smith J, Mendelsohn R (eds) The impact of climate change on regional systems: a comprehensive analysis of california. Edward Elgar, Northampton, pp 188–206

Hurd BH, Callaway JM, Smith JB, Kirshen P (1999) Economic Effects of Climate Change on U.S. Water Resources. In: Mendelsohn R, Neumann J (eds) The impact of climate change on the United States economy. Cambridge University Press, Cambridge, pp 133–177

Hurd BH, Callaway M, Smith J, Kirshen P (2004) Climatic change and U.S. water resources: from modeled watershed impacts to national estimates. J Am Water Resour Assoc 40(1):129–148

Kurukulasuriya P, Mendelsohn R (2007) Modeling endogenous irrigation: the impact of climate change on farmers in Africa. World Bank Policy Research Working Paper 4278

Lund JR, Zhu T, Tanaka SK, Jenkins MW (2006) Water resources impacts. In: Smith J, Mendelsohn R (eds) The impact of climate change on regional systems: a comprehensive analysis of California. Edward Elgar, Northampton, pp 165–188

Medellín-Azuara J, Harou JJ, Olivares MA, Madani K, Lund JR, Howitt RE, Tanaka SK, Jenkins MW, Zhu T (2008) Adaptability and adaptations of California's water supply system to dry climate warming. Clim Chang 87(Supplement 1):75–90

Mendelsohn R, Dinar A (2003) Climate, water, and agriculture. Land Econ 79(3):328–341

Mendelsohn R, Seo SN (2007) An integrated farm model of crops and livestock: modeling latin American agricultural impacts and adaptations to climate change. World Bank Policy Research Series Working Paper 4161, Washington, D.C., USA

Mendelsohn R, Nordhaus WD, Shaw D (1994) The impact of global warming on agriculture: a ricardian analysis. Am Econ Rev 84(4):753–771

Rosenzweig C, Strzepek K, Major D, Iglesias A, Yates D, McCluskey A, Hillel D (2004) Water resources for agriculture in a changing climate: international case studies. Glob Environ Chang 14:345–360

Williams J, Ray I (1999) Evaluation of price policy in the presence of water theft. Am J Agric Econ 81:928–941

Zhu T, Lund JR, Jenkins MW, Marques GF, Ritzema RS (2007) Climate change, urbanization, and optimal long-term floodplain protection. Water Resour Res 43:11

Chapter 9
Climate Change and Water Resource Availability: What to Do?

Eugenia Aloj, Mariagrazia De Castro, Michela Totàro and Anna Zollo

Abstract There is a widespread recognition that the world is facing a growing water crisis affecting the well-being of the poorest people. Rapidly growing populations, urbanization, agricultural intensification and climate change (such as global warming) all contribute to greater competition and scarcity of water resources. Despite massively increased provision of water facilities over the past few decades and the development of low-cost sustainable technical solutions to many aspects of water provision, millions still suffer from water-related diseases and the physical, social and economic burdens associated with scarcity. One of the keystones is represented by saving water that is being pursued through an ethic of consumption, as a challenge for environmental education of young generations. Teaching the younger generation water-saving means to embody the principles of sustainable development. In this paper, we also show the results of a general survey on the schools, carried out through monitoring and surveys that demonstrate how educating young people to a conscious use of water may represent a good model of sustainable development to face the global water crisis.

Keywords Climate change · Water resource · Water lack · Ethics of consumption · Environmental education · Education to consumption

Water Resource and Climate Change: A Background

There is a widespread recognition that the world is facing a growing water crisis affecting the well-being of the poorest people (European Environment Agency 2003; Kenney 2005). Rapidly growing populations, urbanization, agricultural

E. Aloj (✉) · M. D. Castro · M. Totàro · A. Zollo
Faculty of Economy, Sannio University, Via Calandra, 82100 Benevento, Italy
e-mail: galoj@email.it

intensification and climate change all contribute to greater competition and scarcity of water resources (IPCC 2007).

Water is change: drought, floods, hurricanes, increase of sea level, breaking of banks. The alteration of precipitations causes floods in some regions and drought in others: the effects are evident on the coasts, devastated by storms that are provoked by air superheating with a very high intensity. In dry areas, the same situation increases evaporation and drought. Climate change can affect the quantitative and qualitative status of water resources by altering hydrological cycles and systems (IPCC 2007). Changes in these variables lead to impacts on all the socioeconomic and environmental goods and services that depend on these variables directly or indirectly. Water is the main public gift. Once upon a time we thought of a water course as unlimited and the idea to protect water seemed a strange idea.

The importance of water to our life support system is widely recognized, as can be seen clearly in the international context (e.g. Agenda 21, World Water Fora, the Millennium Ecosystem Assessment and the World Water Development Report). Freshwater is indispensable for all forms of life and is needed, in large quantities, in almost all human activities.

Climate, freshwater, biophysical and socioeconomic systems are interconnected in complex ways, so a change in any one of these induces a change in another. Anthropogenic climate change adds a major pressure to nations that are already confronting the issue of sustainable water use. The challenges related to freshwater are: having too much water, having too little water, and having too much pollution. Each of these problems may be exacerbated by climate change. Water-related issues play a pivotal role among the key regional and local vulnerabilities. Therefore, the relationship between climate change and water resources is of primary concern and interest. In addition, the Intergovernmental Panel on Climate Change (IPCC) in the Contribution of Working Group II to the Fourth Assessment Report (2007) recognized that there is a strong connection between hydrologic cycle, climate change and use of water (IPCC 2007).

Another Global Problem Connected with Water Resource: The Price of Water

Another global problem with ethical and sustainability implications is the problem of water price, even though it is recognized as a public gift (OECD 1998).

"If you live in a slum in Manila, you pay more for your water than people living in London." That is the conclusion of a report from the United Nations Human Development Programme (Fig. 1) (UNHDR 2006).

The World Bank argues that the problem is not privatization itself, but that privatization is not being practised properly. Yet the market-based paradigm for such a vital resource has come under question. The aforementioned World Development Movement report noted that the goals of a responsible government (universal access), and the goals of a private company (profit, typically by

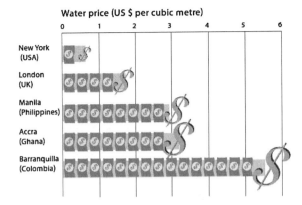

Fig. 1 The price of water (Source: Water Rights and Wrongs, UNHDR Youth Booklet, November 2006, p. 16)

providing access to those who can pay) imply that private sector efficiency for profit may not mean that the same efficiency will lead to universal access.

Certainly, there are cases where markets have provided innovative ideas and efficiency in management. This typically requires a market where people that can pay for the service. For *universal* access, however (which includes people who may not be able to pay, for a variety of reasons, and may require subsidies or assistance), a solely market-based privatization may be inappropriate.

Adaptation to Climate Change as Sustainable Water Management

Water managers have typically assumed that natural resource base is reasonably constant over the medium term and, therefore, that past hydrological experience provides a good guide to future conditions. Climate change challenges these conventional assumptions and may alter the reliability of water management systems. Management responses to climate change include the development of new approaches to system assessment and design, and non-structural methods through such mechanisms as the European Union Water Framework Directive.

In Table 1 we describe some adaptation options for water supply and demand.

Adaptation to climate change in terms of sustainable use of water requires an equilibrium between demand and availability. It means that demand could be reduced from suppliers using some measures such as the measure of consumption and the education and sensitization of users in terms of conservation of water.

Water availability could be increased by developing hydro-basins and transferring water from areas in which it is abundant to areas in which there is scarcity. Other measures to increase availability include the utilization of wastewater (such as depurated slurries to use when these is no need for high quality, e.g. watering golf courses) or the desalination of seawater. Finally, the reduction of losses in the distribution systems could increase availability.

Table 1 Some adaptation options (WGII)

Supply-side	Demand-side
Prospecting and extraction of groundwater	Improvement of water use efficiency by recycling water
Increasing storage capacity by building reservoirs and dams	Reduction in water demand for irrigation by changing the cropping calendar, crop mix, irrigation method and area planted
Desalination of seawater	Reduction in water demand for irrigation by importing agricultural products, i.e. virtual water
Expansion of rainwater storage	Promotion of indigenous practices for sustainable water use
Removal of invasive non-native vegetation from riparian areas	Expanded use of water markets to reallocate water to highly valued uses
Water transfer	Expanded use of economic incentives including metering and pricing to encourage water conservation

In consideration of the implications with climate change and with price, we understand that it is a problem of global management. Furthermore, Indian scientist and activist Vandana Shiva noted in a documentary that the water crisis is a human-created crisis only in the last two or so decades. In other words, it is not so much a water *shortage* crisis as a water *management* crisis.

So if we should respond to the question: "What does climate change have to do with wasting water?", we can say that wasting water could aggravate the availability of freshwater that is already a significant effect of climate change.

A Way for Sustainable Water Management: Environmental Education as Ethics of Consumption for Young Generations

One of the keystones is represented by saving water that is being pursued through an ethic of consumption, as a challenge for environmental education of young generations (Aloj et al. 2007a). Teaching the younger generation water-saving means to embody the principles of sustainable development. For this reason, in this paper we also show the results of a general survey on schools, carried out through monitoring and surveys that demonstrate how education of young people may represent a good model of sustainable development to face the global water crisis.

A need for a survey was born out of this consideration: in Italy, we have a lack of international, interdisciplinary water education programmes for formal and informal educators (Aloj et al. 2006, 2007b). For some years now, they have been assisted by the elaboration of different communication plans by public administration (such as the Department of the Environment) or public water services, but these plans are not integrated and hence isolated. Sometimes they are not for specific targets (such as citizenship or children, for example) but aimed at generic goals. But we know that environmental education holds the priority position in a formal education programme.

Fig. 2 Age of respondents

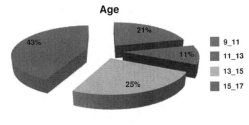

Fig. 3 Gender of respondents

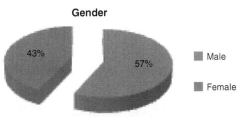

The relationship between water and climate change is not yet considered as a core area of environmental education, even if this relationship must be developed in the conception of environmental complexity because it establishes a link between the natural and the socioeconomic world (Palmer and Neal 1994). Environmental education should consider and emphasize the social implications of the water crisis. The effects of climate change will impact poorer parts of the globe disproportionately. In this case, environmental perception could support environmental education: environmental perception and then a set of perceptual processes, cognitive and affective, through which individuals acquire knowledge of the socio-physical environment and the information necessary for the development of cognitive schemes, whose maps are a particular mode (Bell et al. 2001; Mira Garcia and Real Deas Eulogio 2005).

This is the platform of our survey: monitoring the environmental perception of young generations, which is important to have a general outline supporting the initiation of an environmental education process and to pursue an increasing understanding of environmental conditions.

In March and April 2010, we carried out a survey in the schools of Benevento, Italy, interviewing 280 pupils between 9 and 17 years of age, asking pupils to express their opinions on the environmental perception of the problem of water connected with climate change, even connected with their own local community, as a main indicator of local sustainability.

Discussion of Results

We interviewed 43% of pupils aged between 15 and 17 years, 25% between 13 and 15 years, 21% between 9 and 11 years and 11% between 11 and 13 years (Fig. 2). In terms of gender, 57% were male, and 43% female (Fig. 3).

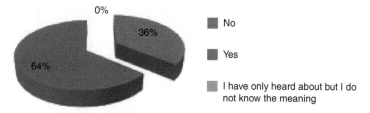

Fig. 4 Meaning of climate change

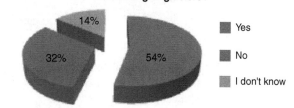

Fig. 5 Rise in temperature and melting of glaciers

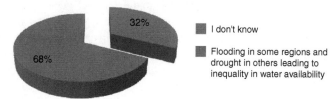

Fig. 6 Melting of glaciers

Referring to "climate change", 64% know the meaning of the term (Fig. 4) and 54% know that climate change is the reason for the melting of glaciers (Fig. 5) and consequently the change in equal distribution of water.

In order to deepen the line of questioning, we asked whether they knew the effect of the melting of glaciers and 68% know that it causes flooding in some regions and drought in others, leading to inequality in water availability (Fig. 6). Starting from this awareness we asked if they knew that, nowadays, although the global problem of water availability is known, we waste this precious resource and 50% of those interviewed are aware of this, even if 29% that think water is an unlimited resource for us and for future generations (Fig. 7).

Referring to the waste of water, 61% think that the problem is the lack of controls on industries and only 25% are aware that our own consumption is very wasteful (Fig. 8), and 61% are aware that we must do something to protect the

9 Climate Change and Water Resource Availability

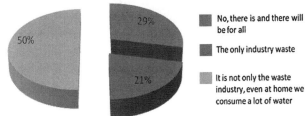

Fig. 7 Waste of water

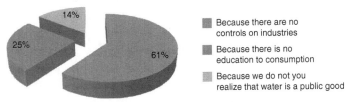

Fig. 8 Reasons of waste of water supply

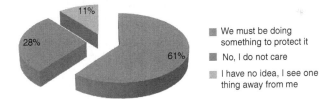

Fig. 9 Need for water for future

water supply (Fig. 9) as a right for future generations but, as shown in Fig. 10, 53% are not able to indicate how to do this.

In discussion of the overall results, we can say that there is awareness of climate change and of water availability as not interconnected global problems (Fig. 11). There is no awareness that climate change will impact on water availability (changes in temperature, precipitation patterns and snowmelt) and on water quality (changes in the timing, intensity and duration of precipitation). This is the confirmation of our starting point: environmental education needs environmental perception.

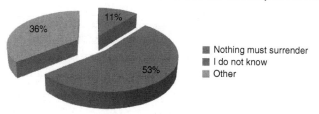

Fig. 10 Practical ideas for reduction of water consumption

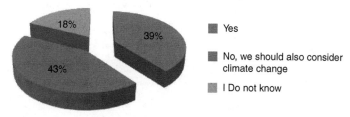

Fig. 11 Actions between reduction of water and climate change

Conclusions

In order to have the clearest picture of the situation deriving from our survey, we will briefly repeat the findings of the survey, concluding what can be interpreted from them.

The people interviewed are aware of the potential damage that climate change can do and are also aware of wasting the precious resource of water, but the two problems are not seen to be interconnected. We can conclude this because when we asked: "What can be done in practice?" they did not have a concrete solution.

There is no awareness that climate change will impact on water availability (changes in temperature, precipitation patterns and snowmelt) and in water quality (changes in the timing, intensity and duration of precipitation). We deduct from this that there is a need for environmental education to bring about environmental perception.

Environmental perception plays a strategic role within the framework of environmental education. Research and theories in environmental perception have primarily focused on the development and processing of mental representations by subjects to better interpret and understand their surroundings. In this specific topic, environmental perception should investigate the mental representations of the link between climate change and water as a cognitive map.

On the other hand, perception of environmental factors of local community should also guide the strategic water management behaviour in organizations. Management perception of the importance and impact of environmental factors should remove the gap in terms of environmental education and should focus not only on the physical characteristics of the environment but also on the capabilities of the subjects to orientate in it.

In this way, we can find rational solutions to cohabit with ecologic limits and we have to be prudent in setting new goals to bring attention to the reality, to the science and to the sensibility to protect common goods such as water.

References

Aloj E, De Castro M, Zollo A (2006) The role of scientific community for the actuation of the sustainable development strategies in the connection among the different stakeholders. In: Leal Filho W (ed) Innovation, education and communication for sustainable development. Peter Lang, pp 21–39

Aloj E, De Castro M, Zollo A (2007a) Educational actions in Italy to face the problem of climate change. In: Leal Filho W, Mannke F, Schimdt-Thomè P (eds) Information, communication and education on climate change–european perspectives. Peter Lang, pp 101–116

Aloj E, De Castro M, Zollo A (2007b) Higher education in environmental and ethical–social communication for companies: new challenges and new perspectives for the scientific community. In: Leal Filho W, Manolas IE, Sotirakou MN, Boutakis GA (eds) Higher education and the challenge of sustainability: problems. promises and good practice. Evrographics, pp 1–6

Bell PA, Greene TC, Fisher JD, Baum A (2001) Environmental psychology. Earl McPeck, USA

European Environment Agency (2003) Europe's water: an indicator-based assessment. Topic Report No. 1

Kenney DS (2005) In search of sustainable water management. Edward Elgar Publishing, UK

Kingsolver A (2010) L'acqua è vita. Natl Geogr 25(4):2–20

Mira Garcia R, Real Deas Eulogio J (2005) Environmental perception and cognitive maps. Routledge, London

Organization of Economic Cooperation and Development (1998) Household water pricing in OECD countries. OECD, Paris

Palmer J, Neal P (1994) The handbook of environmental education. Routledge, New York

Parry M, Canziani OF, Palutikof JP, van der Linden PJ, Hanson CE (eds) IPCC (2007) Climate change 2007: impacts, adaptation and vulnerability. Cambridge University Press, Cambridge

Water Rights and Wrongs (2006) UNHDR Youth Booklet Nov, p 16

Chapter 10
Adaptations to Climate Change and Variability Among Smallholder Farmers in Tanzania

Stephen Justice Nindi and David Gongwe Mhando

Abstract Tanzania is not an exception to the situation where climate change and variability (CC&V) are emerging threats to development, with adverse impacts on environment, human health, natural resources and physical infrastructure. CC&V has been one of the major limiting factors in agriculture production, resulting in food insecurity and low household incomes possibly due to over-dependence on rain-fed agriculture by the majority of people living in rural areas of Tanzania. Thus, understanding adaptation to CC&V impacts among rural communities is of paramount importance for effective interventions on adaptation strategies. Checklist questions were used to collect information on various adaptation strategies to CC&V impacts carried out by farmer groups in two villages of the Matengo. Spectacular adaptation strategies found in both villages were working in farmer groups for activities such as agroforestry, dairy cattle keeping, running of hydro-mills, beekeeping, and fish farming. Most of these activities ensured both livelihood diversification and landscape conservation under a changing environment. Unique to these farmer groups were their spontaneous expansions, largely owing to farmers' own inventiveness. This study, however, was only preliminary, hence calls for in-depth and longitudinal studies that could allow panel data collection and enable strategic intervention to the problems of CC&V.

Keywords Sub-Saharan Africa · Matengo highlands · Rural communities · Climate change and variability · Adaptation · Livelihood diversification · Landscape conservation · Smallholder farmers

S. J. Nindi (✉) · D. G. Mhando
Centre for Sustainable Rural Development,
Sokoine University of Agriculture (SCSRD),
P. O. Box 3035 Chuo Kikuu, Morogoro, Tanzania
e-mail: nindistephen@yahoo.com; snindi@suanet.ac.tz

Introduction

The impacts of climate change and variability (CC&V) have been the highest priority on the agenda in modern times. The impact of climate change especially to millions of poor households mostly residing in Africa is horrifying. To be precise, Africa, especially south of the Sahara, is one of the most vulnerable regions in the world to the impacts of climate change. Previous assessments (IPCC 2007) concluded that the region is particularly vulnerable to the impacts of climate change because of factors such as widespread poverty, recurrent droughts, inequitable land distribution and over-dependence on rain-fed agriculture. Nindi and Itani (2008) assert that the problems of CC&V are expected to escalate among many subsistence farmers in rural Africa where information flow systems are pathetic and technological changes are slow.

According to IFPRI (2009), agriculture and climate change are inextricably linked. Agriculture is part of the climate change problem, contributing about 13.5% of annual greenhouse gas (GHG) emissions (with forestry contributing an additional 19%), compared to 13.1% from transportation. Agriculture is, however, also part of the solution, offering promising opportunities for mitigating GHG emissions through carbon sequestration, soil and land use management, and biomass production. Climate change threatens agricultural production through higher and more variable temperatures, changes in precipitation patterns, and increased occurrences of extreme events such as droughts and floods. Thus, if agriculture is not included, or not well included, in the international climate change negotiations, resulting climate change policies could threaten poor farming communities and smallholders in many developing countries. The policies could also impede the ability of smallholders to partake in new economic opportunities that might arise from the negotiations.

Devereux and Maxwell (2001) reported that countries in East Africa, for instance, are already among the most food insecure in the world and CC&V will aggravate failing harvests. Having realized the gravity of CC&V, various conventions, protocols, policies, acts, frameworks, guidelines and programmes have been established and ratified and numerous studies conducted from global to local levels trying to understand the devastating impacts of climate change and take on board specific and transboundary initiatives to adapt and mitigate these impacts (EPMS and CEEST 2006).

Tanzania is not exceptional to this situation where CC&V is rapidly emerging as one of the most serious problems affecting many sectors and is considered to be one of the leading threats to sustainable development with adverse impacts on environment, human health, food security, economic activities, natural resources and physical infrastructure (IPCC 2007). Indeed, given the over-dependence on rain-fed agriculture by the majority of people living in rural areas of Tanzania, CC&V has been one of the major limiting factors in agriculture production thus resulting in natural resource vulnerability, food insecurity and low household incomes. The impacts of CC&V are manifested by floods, droughts, erratic rains and natural events. United Republic of Tanzania (URT 2005) revealed that famine resulting from either floods or drought has become increasingly common since the mid-1990s and is undermining food security. CC&V is likely to intensify drought and increase the potential vulnerability of the communities to future climate change (Hillel and Rosenzweg 1989), where crop production and livestock keeping are critically important to food security and rural livelihoods.

Certainly, the impacts of and adaptations to CC&V vary with geographical zones and households due to the varying distribution of opportunities in Tanzania, and households throughout the country are engaged in different and often multiple adaptation strategies (Liwenga 2003). Thus, understanding adaptations to CC&V impacts is of paramount importance so as to devise effective interventions that ensure both livelihood improvements and landscape conservation, which is also important to carbon sequestration under climate change. Thus, this study assessed climate change adaptations in two villages that have distinct climatic characteristics of Tanzania based on farmer groups' activities.

Materials and Methods

The study was conducted in the Matengo highlands, located in Mbinga district, Ruvuma region which is located in the southwestern part of the URT. The district lies between longitudes $34° 24'E$ and $350° 28'E$ and latitudes $100° 15'S$ and $110° 34'S$. Two villages, namely Kindimba and Kitanda, located 22 km apart from each other were purposely selected for this study. Kindimba village was selected because it is one of the oldest villages in the Matengo highlands and is located in the Mountain Area sub-ecological zone (about 2,000 m above sea level (masl)), while Kitanda was relatively a new village in the Rolling Hills North sub-ecological zone of the Matengo highlands (located between 900 and 1,500 masl) and inhabited by immigrants mainly from Mountain Area. Both villages have been severely affected by land degradation partly associated with CC&V and have interacted with other development partners in conducting activities that aim at landscape conservation and livelihood transformation.

Subsistence rain-fed agriculture is the mainstay of the Matengo farmers' economy in both villages. Farmers in the study villages largely cultivate coffee, tobacco and wheat as cash crops, whereas maize, beans and recently cassava serve as food-cum-cash crops. Except for coffee, which is planted on flat or bench terraces, other crops are planted on Matengo pits (ngolo). They keep a limited number of livestock, mainly cattle, goat, pigs and chicken. Their main system of livestock-keeping is either through tethering or free range and seldom have they conducted zero or indoor grazing.

This study used a cross-sectional design in which two sample villages were studied. Combinations of methods were used in order to elicit relevant information. Since adaptations at village level in these two villages largely followed activities by farmer groups, five farmer groups from each village were selected for interviews using simple random sampling approach. First, all farmer groups available in the village were listed (20 and 12 farmer groups in Kitanda and Kindimba, respectively), and then a simple random sampling was carried out to select five groups from each village. Two interviews, one in each village, were held with farmer group representatives. Each farmer group was represented by five members: the Chairperson, Secretary, Treasurer and two ordinary members making a total of 25 members per each interview in each village. Besides this, two focused group discussions (FGDs), one in each village, were conducted with the secretariat of farmer groups' umbrellas. Ten members of farmer groups' umbrellas from each village were involved in FGDs. Two sets of checklist questions were respectively used to collect information from farmer group representatives as well as for FGDs. Satellite image analysis for the periods of 1984, 1989, 1991, 1994 and 2000 were analysed for vegetation cover change detection for both villages and their results presented in tables to see if all adaptations are in line with sustainable resource management. Field observation and informal discussions were also part of the research process at all times. Information gathered from farmer groups' interviews, FGD, observations and informal interviews were subjected to content analysis.

Results and Discussions

Five farmer groups from each village (Kindimba and Kitanda) were involved in this study, as indicated in Table 1. During discussions with members of farmer groups, they noted various indicators of CC&V in their respective localities. These include severe drought of the early 1970s, recurrent droughts, El Niño rains of 1997–1998, change of rainfall season (onset and offset, durations, increasing dry spells), crop failures, increasing temperatures, outbreak of unusual pests and diseases such as maize leaf rust in the 1990s, increasing incidences of malaria, and declining soil fertility that forced replacement of maize with cassava. However, they mentioned failure of crops as another indicator of CC&V but they also linked it to other changing socioeconomic factors such as globalization of the market systems, which has taken place since the mid-1980s.

Their adaptation strategies to these CC&V include working in farmer groups, rural–rural migration and adoption of improved cooking stoves, intensive valley bottom cultivation and changes in eating habits. Other adaptations included intensifying agroforestry activities, beekeeping, fish farming, and adoption of new coffee clones, cultivating drought resistant crops such as cassava and engagement in non-farm activities. Most of these activities ensured both livelihood diversification and landscape conservation, which is also important in carbon sequestration. Indeed, farmers in these villages do adapt to climate change at household level but the most striking feature was their organization and working in farmer groups in the wake of adapting to CC&V. Unique to these two villages was that both farmer groups and farmer activities have registered spontaneous vertical and

Table 1 Farmer groups involved in interviews during survey

Villages	Farmer groups	Membership			Year established	Main activities
		M	F	Total		
Kitanda	Ujamaa	8	7	15	2002	1, 2, 3, 4, 5
	Familia	8	7	15	2003	1, 2, 3, 4
	Vumilia	7	5	12	2003	1, 2, 3, 4
	Ushindi	0	20	20	2004	5[a]
	Mundeki	6	2	8	2004	1, 2, 3, 5
Kindimba	Jiokoe	6	9	15	2003	1, 2, 3, 4
	Ondoa Umaskini	15	4	19	2003	1, 2, 3
	Women SACCOS	0	5	5	2005	5[a]
	Amani	9	12	21	2004	1, 2, 3, 4, 5
	Jikwamue	11	13	24	2004	3,4,5
	Total	70 (45.5)	84 (54.5)	154 (100)		

Numbers in parentheses are percentages
[a] Women Farmer Group whose activity was diary cattle keeping only
1 Fish farming, *2* tree planting, *3* beekeeping, *4* crop farming, *5* others (Chicken and pig husbandry, pasture production, manure and local brewing)

horizontal expansions, largely owing to farmers' own inventiveness. In subsequent sections, we will discuss the evolution of farmer groups and their activities aiming at adapting to CC&V.

Evolution and Dynamics of Farmer Groups in the Study Areas

With respect to interviews with farmer groups' representatives and FGD with the Group Union, effective farmer group formation in the study villages can be traced back to 2002 when one farmer group (Ujamaa) was formed in Kitanda village with the founding objectives of landscape conservation and livelihood diversification through fish farming, after observing a devastating decline in rural productivity due to changing environment. Development of farmer groups in the study villages received an impetus from the presence of the SCSRD Project (Centre for Sustainable Rural Development of Sokoine University of Agriculture), whose activities involved chanting out with farmers on possible methodology that can enable realization of sustainable rural development. The Project was implemented from 1999 to 2004 under collaboration of Sokoine University of Agriculture (SUA), Japan Cooperation Agency (JICA), Mbinga District Council and farmers in the Kindimba and Kitanda villages. Farmer groups involved in this study, their memberships, year of establishment and their subsequent activities are presented in Table 1.

Table 2 Number of farmer groups in Kindimba and Kitanda villages during and after SCSRD Project period

Villages	2002–2004 During Project period	2004–2005 Additional farmer groups after the Project period	Total number of groups	Main activities
Kitanda	12	8	20	Fish farming[a], beekeeping, tree planting, pasture, crop cultivation, poultry keeping, pig and dairy cattle keeping, hydro-mill machine, vegetables farming
Kindimba	8	4	12	Fish farming[a], beekeeping, tree planting, hydro-mill machine, local brewing

[a] Main founding activity

The subsequent parts present how farmers from these two villages continuously implemented activities spontaneously by themselves even after the end of the SCSRD Project. Inside these two villages, we can observe not only quantitative progress (increasing of the number of farmer groups), but also qualitative progress (diversifying and diffusing their activities within their communities and outside their villages) that is in line with adaptations to the changing environment.

Adapting to changing climate, farmer groups have enacted activities that have shown both qualitative and quantitative progress. Table 2 shows that in terms of qualitative and quantitative progress, farmer groups had increased in number and diversified their activities. For instance, farmer groups in Kitanda village have increased from one group in 2002 to 12 groups by SCSRD project termination in 2004 and to 20 groups in 2005 one year after the SCSRD project ended. A similar trend was also observed in Kindimba village. Their activities have diversified from original fish farming to pig farming, dairy cattle and beekeeping, tree planting and local brewing. While implementing these activities, farmer groups have always been involved in tree and fodder planting and refraining from setting bush fires and, on other occasions, villagers were restricted to cultivating near river courses so as to ensure natural vegetation restoration and hence landscape conservation, which is also important in adapting to changing environment.

It was further learnt during interviews that farmer group activities in the study area seem to have been diffusing throughout and even outside their communities. Group activities have been permeating from a few leading villagers, e.g. "relatively educated and/or the governing class/status", to "ordinary villagers". Certainly, this progress of spontaneous diversifying and diffusing of activities are essential for sustainable resource management because these activities of diversification and diffusion reflect villagers' ingenuity. Tamura (2006) described such quantitative and qualitative processes as "vertical expansion", implying that the expansion takes a hierarchical family or clan model wherein the founding group is viewed as a fore parent and subsequent groups are viewed as its descendants. However, on the basis of out-scaling of the group formation that went along with the expansion of farmer group activities in a ripple-like pattern to both the neighbouring and distant villages, the process is also worthy of the phrase "horizontal expansion". Some group leaders from Kindimba and Kitanda have been advising and assisting new groups from other villages at their own initiatives. The vertical and horizontal expansion processes are currently working closely together in the Matengo highlands, owing to originality from Kindimba and Kitanda villages. It is through these processes that farmers from distant villages are working together to improve their natural environment and livelihoods in the effort to adapt to climate change and variability.

Vertical Progress of Farmer Groups

Various vertical progresses were registered by farmer groups in the study villages. It was observed during the discussions with farmer groups' representatives and FGDs that robust vertical expansion of farmer groups occurred after a three-day farmers' exchange visit that took place in January 2004. About 150 farmers from both Kindimba and Kitanda villages were involved in this exchange visit (SCSRD 2004). On the first day of the exchange visit, Kindimba farmers visited Kitanda village, while on the second day, Kitanda farmers visited Kindimba village, and the third day involved discussions by participants from both villages. Group formation, group activities and management were the central matters under discussion. Demonstrations of farmer group activities depicting relationships between fish farming, tree planting and beekeeping as an integrated practical model for landscape conservation and livelihood improvement by already established groups attracted many participants.

Interviews with farmer group representatives revealed that after the exchange visits in 2004, many villagers requested the senior group leaders to include them in their groups. However, the group leaders declined the request because they were afraid the groups would become too big to manage. Instead, the consulted group leaders advised these villagers to form their own groups and promised to offer them managerial assistance and fish fingerlings for free. Such "free" assistance is not uncommon among the Matengo who have been practising reciprocal labour, "ngokela", for many years. There were some cases where some members from the old groups shifted to the new groups. Ten and seven farmer groups with diversified activities were respectively formed in Kitanda and Kindimba villages after the exchange visits of 2004, which is typical of what Tamura (2006) classified as vertical progress.

FGDs and interviews with farmer group leaders and representatives revealed another vertical development, in that all farmer groups in each village have established a closer connection to each other and formed a farmer groups' umbrella, which in Kitanda village is referred to as Groups Union and in Kindimba is known as Sengu Committee. Sengu is a traditional place where men used to assemble in the evening and discussed various issues to their lives including marriage and work plans for the purpose of improving lives in the family. The farmer groups' umbrella conducts monthly meetings, which involve the participation of all farmer group leaders, village government authorities, the agricultural extension officer and religious leaders, forming the so-called tripartite relationship (local people, religious leaders and government agents) in local resource management. They exchange their experiences and discuss challenges and ways forward. When necessary, the leaders of the umbrella visit a group in need of attention to discuss and advise on their challenges. The umbrella also managed the farmers' exchange visits inside the village. In Kindimba village, the umbrella (Sengu Committee) also manages the day-to-day running of the village hydro-mill project, including bookkeeping and accounts of the hydro-mill project.

A characteristic feature for all farmer groups both in Kindimba and Kitanda villages was the fish farming. Thus, fish farming has become a symbol of group activities for nearly all groups. Certainly, the impact of fish farming as a catalyst of formation of farmer groups and mushrooming of group activities came as a serendipitous event rather than a planned one. Some reasons can be attributed to this phenomenon. The growth of the fish is very rapid and thus the group output is seen within a short time. The making of fishponds also goes hand in hand with tree planting or keeping the micro-catchment vegetation intact to enable sustainable water flow to the ponds. The conserved micro-catchment can be used for other human-environmentally friendly purposes such as beekeeping and sustainable harvesting of mushrooms, medicinal plants and construction materials. Additionally, digging of fish ponds is perceived as not being so difficult by the Matengo people, who are used to labour-intensive traditional cultivation methods known as ngolo (a kind of ploughing to make four-tiered ridge farms by digging steep slopes). Matengo people have been using this type of cultivation method for more than 100 years (MWARP 1998) and they are applying similar hard-working spirit and energy to digging fishponds. The hard-working virtue of the Matengo is referred to by SCSRD (2004) as the "potential of indigenousness" to sustainable resource management.

Fingerling distribution to new groups also played a significant role in farmer groups' vertical expansion. It must be noted that the first fingerling installation in Kitanda village was done in November 2002 by one farmer group (Ujamaa) in one fishpond. Ujamaa group distributed 360 fingerlings to its seven group members by 2005. After two years, six members of this group managed to distribute about 4,300 fingerlings to emerging groups and individuals within and outside the village in Kindimba, Lipumba, Mtama and Kikolo villages. It was encouraging noting during discussion that one member of Ujamaa group, Mr Kevin Ndimbo, for instance, who initially received only 19 fingerlings, managed to distribute about 1,500 to other groups and individuals. As a result of active fingerling distribution, there were 210 ponds in Kitanda village and 75 fishponds in Kindimba village by the end of 2005. However, in Kindimba village, the first fingerling installation took place in July 2003.

Next to fish farming, social forestry activities were another fundamental activity that overlapped in many farmer groups from older to new groups. While several

Table 3 Some of the indigenous tree species planted in Kindimba and Kitanda villages by 2005

Local names	Botanical names	Major uses
Mdhambarau pori	*Syzygium guineense*	Conservation of water sources
Muwawa	*Khaya anthotheca*	Conservation of water sources
Msindano	*Pinus patula*	Timber, landscape conservation
Mwanga	*Pericopsis angolensis*	Timber, fuelwood supply, agroforestry
Mtela wa mbamba	*Rauvolfia caffra*	Medicinal, timber
Mzombo	*Brachystegia spiciformis*	Fuelwood, building poles
Mtumbati	*Pterocarpus angolensis*	Medicinal, agroforestry, timber
Mdonga	*Strichnos cocculoides*	Fruits, fuelwood, building poles
Mnyenda	*Bridellia micrantha*	Conservation of water sources
Mkaranga mti	*Macadamia spp*	Alternative cash crop to coffee
Mtanga	*Albizia schimperiana*	Soil fertility conservation and shade in coffee plantation

farmer groups received tree seeds and seedlings from the District Office and SCSRD, some groups collected indigenous tree seeds and/or seedlings from Miombo woodland forests and sowed or transplanted them to rehabilitate the degraded landscape. The total number of trees planted by farmer groups in Kindimba and Kitanda villages by 2005 was 10,500 and 3,500, respectively. Farmers regard the increase or diversification of farmer groups and activities (vertical and horizontal expansion) as a strategy of strengthening the vigour of their groups. The tendency of farmer group activities to directly integrate tree planting activities and micro-catchment vegetation preservation is vital not only for landscape conservation but also adaptive mechanisms to changing environment. Table 3 shows some of the tree species planted in Kindimba and Kitanda villages by farmer groups.

Indigenous knowledge is an important tool employed in the process of restoring the degraded landscape cover in the study area. Various micro-climates have different kinds of appropriate vegetations that revitalize their ecological potentials. Since local people have existed for a long time with the natural environment around them, they possess an indigenous knowledge resource on trees, which used to grow around catchments before land cover was degraded. This knowledge was very useful during the identification of the right tree species for conservation of water sources and degraded landscapes. Tree seedlings were distributed not only for planting within village water sources, mountain tops and open spaces, but also within and around river sources.

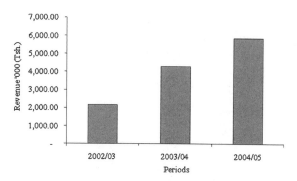

Fig. 1 Revenues (Tanzanian shillings) accrued from hydro-mill project at Kindimba village

Step Towards an Economic Stage

Farmer activities not only realized landscape conservation but they were also involved in income generation activities and accessed financial institutions, which is referred to as an economic stage to improve livelihoods. In Kindimba village, the farmer group umbrella (Sengu Committee) pointed out the success achieved by the village hydro-mill project, which is run collaboratively by all groups under the leadership of the umbrella group. Figure 1 shows revenues accrued from hydro-mill project in respective yeas and Fig. 2 presents the number of hydro-mill users, including those from neighbouring villages. It was learnt during informal discussion with district leaders that the Kindimba village hydro-mill project was the biggest sole project (in terms of number of users and revenues) manned by villagers themselves in the district.

The Sengu Committee elaborated an efficient system of record keeping of hydro-mill revenues against customers' attendance registry. The Committee has also opened a bank account for the project as an important step towards an economic stage. Revenues from this project were being used to articulate other development activities in the village such as providing financial support to the ongoing electrification project in the village and purchase of tree seeds and/or seedlings for landscape conservation. Expenses for the servicing of the hydro-mill machine are also borne by part of the revenues from the same machine. The milling charges are far cheaper compared to a diesel-propelled milling machine and the served income is used by households to cover other domestic needs, as reported by SCSRD (2004) and Tamura (2006), which is important to livelihood improvement.

Besides collaborative (whole village) economic transformation, some farmer groups such as Ushindi and Ujamaa in Kitanda village had also opened bank accounts, with the aim of acquiring bank loans for both the group and individual members as an important step towards an economic stage. Normally, after opening the group's bank account, the group members make weekly deposits which later can act as collateral. When each group member's deposits reach 5,000 Tanzanian shillings (Tsh.) (~US$4), each member can get a maximum loan of Tsh. 50,000 (~US$40) from the bank. This is a new type of the rural micro-finance

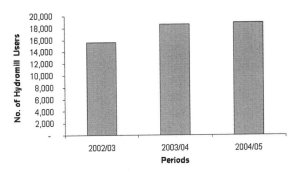

Fig. 2 Number of hydro-mill users at Kindimba village

arrangement and the villagers can access loans without any other collateral because their deposits in the group account guarantee them. For many years, financial institutions have been reluctant to give loans to poor people especially when collateral is missing (Barger 1989; Albee 1994). However, this new invention can permit many poor rural dwellers access to financial services and hence improve their livelihoods. For instance, Ushindi group in Kitanda has capitalized on the available loan system at the Savings and Credit Cooperative Society (SACCOs) and have used the loan to acquire a dairy cow. Farmer group savings from the selling of fingerlings, fish, vegetables and honey were other examples of economic diversification managed by farmer groups.

Horizontal Expansion of Farmer Group Activities

Discussion with farmer groups and FGDs noted that vertical expansion of farmer groups (in number and activities) in Kindimba and Kitanda villages also influenced the spontaneous formation of farmer groups even outside these two villages. By the same token, tree seeds and/or seedlings and fish fingerling distribution were extended up to the neighbouring and distant villages. Some of the outside villages were rather further than "neighbouring". For example, it took about a 7 h walk for Lipumba villagers to bring fingerlings from Kitanda to their village. Through such a process, horizontal expansion of farmer groups was realized, as discussed in subsequent sections.

Notably, villagers have been implementing these expansions mainly spontaneously. Neighbouring villagers and groups who wanted to get fingerlings, tree seedlings and advice received support from older groups mainly for free. On some isolated occasions, new groups had to buy fingerlings and seedlings from established groups. In October 2004, Vumilia group from neighbouring Lipumba village purchased 200 fingerlings from Ujamaa group. Vumilia group had other activities such as farming, beekeeping, tree planting and pig farming, but they wanted to diversify to fish farming as well. In this way, Ujamaa group increased group income and expanded fish farming activities to others. This was one of the ways to realize horizontal expansion of farmer groups' activities, which was a way of adapting to the changing environment.

The second type of horizontal expansion was observed through expansion of new groups and their activities to both neighbouring and distant villages under the assistance of older groups in Kindimba and Kitanda villages. For instance, Furaha group in the neighbouring Masimeli village (20–25 km away from Kitanda) was established in April 2005 with assistance from Ujamaa group of Kitanda village. It was in 2003 when one villager from Masimeli village (the present group leader), heard about fish farming in Kitanda. He visited the Ujamaa group leader in March 2005 and received some advice on how to form and manage a farmer group and how to start activities including fish farming. On returning to his village, he convinced his relatives, friends and neighbours to form a group, which is now known as Furaha. Their first activity was to establish a fishpond and purchase 250 fingerlings from *Ujamaa* group. Since Furaha group had no skills to start their activities they received "brotherly" advice from Ujamaa and other groups from Kitanda village. By the time of this survey, Furaha group was also conducting other activities such as crop, poultry and pig farming, and beekeeping and had established a tree nursery. All these activities were supported by the advice from Ujamaa group in terms of both skills and philosophy. The intention of Ujamaa group to help Furaha group was that the latter would be a keystone in Masimeli village and extend its activities within and outside the villages as well, hence ensuring livelihood improvement and conservation of degraded landscape due to changing climate.

The third horizontal expansion was to individuals who wanted to start fish farming or to add new or good fish pedigree to their existing ponds. Commonly, in Kindimba and Kitanda villages, individual farmers and households were also involved in making their own fishponds and some travelled long distances (40–50 km) near Lake Nyasa (or Malawi) to purchase "good" fish breeds or pedigree. Ujamaa group and other old groups always extended their advice and support to such individual household activities. Focus group and key informants' discussions contended that fish farming ensured food security, income to the households and conservation of village landscape.

So far, various farmer group adaptive activities to CC&V have been discussed. It is very important for farmers to initiate their own activities to adapt to a new environment and for outsiders to respect farmers' ways of doing things. Outsiders can act as "messengers" of farmers, carrying their opinion to upper radars where farmers cannot easily access. Most of these activities reflect the great potential to forest catchment conservation and livelihood transformation. Conservation of catchment environment serves to reduce the fluctuation in water levels and prevent soil erosion, and provides daily livelihood materials such as timber and medicine, and food materials such as honey, fruits and mushrooms (Nsenga et al. 2004). In addition, the hydro-mill machine project, fish cultivation, planting of useful tree species and promoting beekeeping are intended to serve as management approaches to these catchment forests. Intriguing to note is that these adaptive activities, even the advanced ones such as the hydro-mill project, have been initiated and

Table 4 Land cover changes (%) in Kitanda village extracted from satellite image analysis

Periods	Tree cover	Burnt area	Grass cover
1984	≥80	≤5	≤15
1989	≤70	≥5	≥25
1991	≤50	–	≥55
1994	≤40	≤5	≥55
2000	≤20	≥10	≥70

Table 5 Land cover changes (%) in Kindimba village extracted from satellite image analysis

Periods	Tree cover	Burnt area	Grass cover
1984	≤30	–	≥75
1989	≤40	–	≥60
1991	≤40	–	≥60
1994	≤45	≤5	≥55
2000	≤35	–	≥60

managed by the villagers themselves, although support from other development partners cannot be ruled out. In addition, the expansion processes have been supervised by the villagers themselves. This is a very important aspect in sustainable resource management and rural development in general, especially under the changing environment influenced by CC&V.

Impacts of Climate Change Adaptations on Land Cover Degradation

Sometimes adaptations or coping strategies can lead to severe repercussions on the landscape. The extent and trends of vegetation cover degradation were extracted from satellite imagery from the 1984, 1989, 1991, 1994 and 2000 periods and are presented in Tables 4 and 5, and Figs. 3 and 4, respectively. Values from Table 4 extracted from Fig. 1 show that land area covered by trees dropped from over 80% in 1984 to less than 20% in 2000 in Kitanda village. In contrast, in the same village, the grass-covered areas increased from less than 12% in 1984 to more than 70% in 2000, indicating unprecedented convention of tree cover to grassland, largely due to expansion of cultivation activities in these Rolling Hills due to decline in soil fertility in the Mountain Area.

Analysis of satellite images of Kindimba village (Mountain Area) presented in Table 5 show that tree cover seems to have stabilized for a long time, though at low level. Equilibrium had been reached through long-time habitation largely due to agroforestry practice in coffee farms and the establishment of eucalyptus woodland. Table 5 and Fig. 4 indicate that from 1994 tree cover showed slight downscale changes and the decline was associated to conversion of a few remaining fallows and woodland to crop fields and valley bottom cultivation after

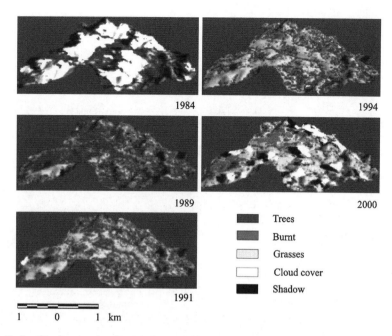

Fig. 3 Satellite images showing trends of land cover change in Kitanda village

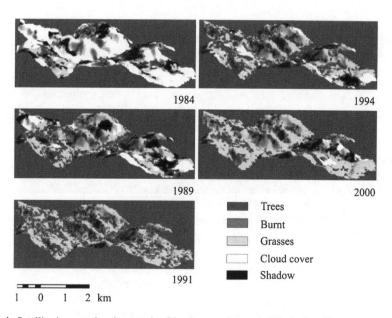

Fig. 4 Satellite images showing trends of land cover change in Kindimba village

Table 6 Percentage factors affecting tree cover in Kitanda village

Factors	Percentage
Opening of new farms	58.0
Fuelwood demand	21.0
Lumbering	12.5
Others	8.5
Total	100

changing climate and economic crisis in Matengo highlands since the 1990s. Lack of ample agricultural land for expansion within the Mountain Area could have contributed to low vegetation change within the sub-agroecological zone.

Reading from Table 4 and Fig. 3, one can conclude that land use in the Rolling Hills of the Matengo highlands has changed significantly, especially since the 1980s. Apart from changing climate, Nindi (2004, 2007) and Mhando (2005) also relates this situation to the economic destabilization faced by the Matengo farmers since the mid-1980s, especially after the introduction of trade liberalization of the coffee market, which led to the collapse of the Mbinga Cooperative Union (MBICU). They further reported that the collapse of the MBICU negatively impacted the ability of the Matengo farmers to access chemical fertilizers, which supported crop production in their discrete mountain farm plots. Interviews with farmer group representatives, FGDs and informal interviews confirmed that coffee economy (the major cash economy) has supported the Matengo and their agricultural systems for many years. At most of the times, MBICU ensured them with markets and a reliable input supply chain to farm level, hence stable farm production not only for coffee but also other food crops. Inputs bought from coffee money at several times supported production of food crops in their small highlands farm plots. With the biting of CC&V coupled with economic destabilization, farmers from Mountain Area had to invade adjacent frontiers in the Rolling Hills, and Kitanda village was not spared this invasion. Mhando (2005) contends that tumbling of the coffee production and its marketing systems from the 1980s led to the vast and abrupt creation of new farms on virgin land in adjacent frontiers in the Rolling Hills, where the use of agro-chemicals was not necessary.

FGD and informal discussions during the survey also confirmed that recent unprecedented deforestation of the Rolling Hills was largely due to the influx of Matengo immigrants from the Mountain Area after the collapse of the coffee economy, declined land productivity and invasion of new maize leaf rust disease in 1990s. Nindi (2004) revealed that in the new land, farmers from Mountain Area largely conduct extensive agricultural practices of slash-and-burn. He further pointed out that the practice of a slash-and-burn cultivation system on such steep and rugged slopes coupled with intense and erratic rains certainly provoked not only vegetation cleansing but also intensive soil erosion and sediment runoff to river valleys. Revelations from Kjekshus (1977), Pratt and Gwynne (1978), Pomeroy and Service (1986), and Itani (1998) show that soil erosion from uncoordinated agricultural practices is a serious problem in mountainous areas that experience frequent heavy rains. In such environments, the removal of vegetation can easily erode surface soil if the land is cultivated, erosion becomes more serious and the land may become barren within several years.

Indeed, prior to the mid-1980s, the Matengo had adaptive mechanisms to the changing environment of migration into frontiers, especially when the population was against the carrying capacity in the Mountain Area. Normally, after 3–4 generations, a Matengo extended family would allow some of the family members to migrate to "new land" and open up farms, and live their permanently, initially practising slash-and-burn agriculture and later resorting to their permanent intensive indigenous ngolo cultivation system (MWARP 1998; Nindi 2004). In the new land, they also followed their traditional land holding practice known as "ntambo" [a mountain ridge sandwiched between two river valleys owned and utilized in sustainable manner by extended family (Itani 1998)]. When the coffee economy was good, ntambo was normally bought from money saved from cash sales (MWARP 1998; Nindi 2004, 2007). Thus, the Matengo managed their natural landscape largely through the cash saved from coffee sales. The recent abrupt migration of the Matengo farmers from the Mountain Area to the frontiers did not follow this normal ntambo land holding system. The land tenure shifted from permanent ntambo land holding to the hiring of farm plots to be utilized for a few seasons only, and hence lacked the indigenous sense of resource ownership. Thus, both the abrupt invasion of Rolling Hills' frontiers and the changing land-holding system posed severe repercussions to vegetation cover and the livelihood of the people in the Matengo highlands.

It was learnt during FGDs that cutting down trees to open up new farms in the Rolling Hills contributed to forest degradation by 58%, as shown in Table 6. It was further noted that new farms in the Rolling Hills were opened up not only by the inhabitants of the Rolling Hills and immigrants from Mountain Area alone but also the urban-based farmer inhabitants from Mbinga Township. District officials lamented that fast population growth at Mbinga Township coupled with low food supply chain from adjacent rural areas also contributed to debilitating vegetation cover in the Rolling Hills. For instance, by 1978, the population of Mbinga Township was only 7,308 but reached 25,416 in 2002 (Mbinga District Council 1997; National Bureau of Statistics, 2003) and most of them (86%) were immigrants (Mbinga District Council 1997). Unfortunately, most of these immigrants and original inhabitants demanded land for cultivation as well as forest goods and services mostly from the adjacent Rolling Hills. This abrupt population increase in Mbinga town was partly due to an increased number of petty traders and artisan-mining activities that had concentrated in the district since the late 1980s. Economic liberalization also allowed free trading, a circumstance that brought a number of petty traders of second-hand clothing ("mitumba"), shops, kiosk, and restaurants in Mbinga Township, all demanding goods and services from the adjacent rural landscape.

According to Mbinga District Council (1997), about 64% of Mbinga town dwellers were farmer-based inhabitants and utilized farms and obtained their fuel wood and timber demand from the same Rolling Hills, consequently affecting tree cover in the area. Cool climate and lack of power supply complicated the situation further, as more fuel wood was needed for heating as well. Such an abrupt influx of immigrants in the non-electrified town of Mbinga possibly influenced deforestation in the woodland further. It could therefore be surmised from above that environmental degradation in the Matengo highlands is also a result of interplay of urban–rural relationships rather than an exclusive rural–rural based problem.

It was also revealed during FDs that most tree species ruined for lumbering were Muwawa (*Khaya anthotheca*), Mchai (*Harungana madagascariensis*), Mpuga (*Erythrophleum africanum*) Mgwina, (*Breanardia africana*) and Muwanga (*Pericopsis angolensis*). Tree species most affected by fuelwood demand include Mbuni (*Parinari curatellifolia*), Mnyonyo (*Syzygium cordatum*), Muwanga (*Pericopsis angolensis*), Mzombo (*Brachystegia spiciformis*) and Mteteleka (*Faurea speciosa*). One elder also commented that due to intense deforestation in the Rolling Hills, people now even make charcoal out of mango trees (*Mangifera indica*), previously not meant for that purpose.

It can be summarized that tree cover degradation under the influence of CC&V and trade liberalization has recently been very severe in the Rolling Hills area of Kitanda village than in the mountain area village Kindimba. The most influential factor in the fast deterioration of tree cover in the Rolling Hills could therefore be agreed as the tendency of opening up new food crop farms especially by mountain immigrants as well as Mbinga town dwellers who attend such farms on a temporal basis. This could be ascertained by the increased percentage of grass cover, which also represents food crop farms. However, the tendency of agroforestry cover in coffee farms would in future conserve the environment in the Matengo highlands, as was seen in the case of Kindimba village.

Conclusions

Undeniably, climate change and its impacts are very obvious and very serious, especially among poor farm families mostly residing in rural areas of Tanzania and the vulnerability to a rural production system cannot be tied to a single causality, but rather an interplay of factors and sometimes CC&V has only intensified the vulnerability of the changing rural landscape. In response to changing climate, the villagers have been developing and deepening their adaptation activities related to landscape conservation and livelihood diversification, both quantitatively and qualitatively, as a means of adapting to their changing environment. And in most cases, they have been doing these activities by themselves. Certainly, all areas and communities have their own peculiarities and identities, with sustainable adaptation strategies depending on their sense of value and behaviour or culture. Thus, we must find the possibility for sustainable adaptations within their sense of value/ behaviour or culture, which reflect their potential of indigenousness.

The impact of integrated farmer group activities is an eye-opener to new adaptation to landscape degradation influenced by CC&V and other socioeconomic factors. Thus, the effect of farmer group activities to adapt to changing environmental issues has revealed the new perspective of integrated resource management in the area. Through vertical and horizontal progress, many people and villages will be integrated into the natural resource management, which is the basis of their adaptation to changing climate. With such micro-catchment conservation strategies, larger areas can be conserved through integrating more farmer groups in a wider area. Integrated resource management also provides added value to the resource, hence ensuring its sustainability; e.g. tree planting and beekeeping can lead to careful management of tree and forest resources. With such committed spirits, time and financial resources, the Matengo highlands will without doubt be rehabilitated, improving livelihoods under changing climate conditions and other socioeconomic factors.

References

Albee A (1994) Support to women productive and income-generating activities, evaluation and research. Working paper series, number 1, UNICEF, Rome, 13 pp

Berger M (1989) Giving women credit: the strengths and limitations of credit as a tool for alleviating poverty. World Development 17(7):1017–1032

Devereux S, Edward S (eds) (2001) Food security in sub-Saharan Africa. Intermediate Technology Development Group (ITDG), London

EPMS, CEEST (Environmental Protection and Management Services and the Centre for Energy, Environment, Science and Technology) (2006) Poverty and climate change: a south north collaboration: summary of Tanzania vulnerability and adaptation to climate change, variability and extreme events, SSN Tanzania adaptation team

Hillel D, Rosenzweig C (1989) The greenhouse effect and its implications regarding global agriculture. In: Research bulletin number 724, Massachusetts University, prepared for the climate change team/ENV, The World Bank, Washington, DC

IFPRI (the International Food Policy Research Institute) (2009) agriculture and climate change: an agenda for negotiation in Copenhagen. In: Agriculture science and technology needs for climate change adaptation and mitigation, focus 16, brief 2, May

IPCC (Inter-governmental Panel on Climate Change) (2007) Climate change impacts, adaptation and vulnerability. In: Climate change working group II assessment report, accessed July 2009. Available at http://www.ipcc.ch/SPM6avr07.pdf/

Itani J (1998) Evaluation of indigenous farming system in the Matengo highlands, Tanzania, and its sustainability. Afric Stud Monogr 19(2):55–68

Kjekshus H (1977) Ecology control and economic development in east African history. Villiers Publications, London

Liwenga ET (2003) Food insecurity and coping strategies in semiarid areas: the case of Mvumi in central Tanzania. PhD thesis, Department of Human Geography, Stockholm, Stockholm Studies in Human Geography: ISBN9122020098 9789122020097

Mbinga District Council (1997) Mbinga town master plan, University College of Lands and Architectural Studies, University of Dar es Salaam, 81 pp

Mhando DG (2005) Farmers' coping strategies with the changes of coffee marketing system after economic liberalisation: the case of Mbinga district, Tanzania. Unpublished PhD thesis, Graduate School of Asian and African Area Studies, Kyoto University, Kyoto, 178 pp

MWARP (Integrated Agro-Ecological Research of the Miombo Woodlands in Tanzania) (1998) Sokoine University of Agriculture and centre for African area studies, Kyoto University, 413 pp

National Bureau of Statistics Tanzania (2003) population and housing census 2002. United Republic of Tanzania, NBS, Dar es Salaam

Nindi SJ (2004) Dynamics of land use systems and environmental management in the Matengo Highlands, Tanzania. Unpublished PhD thesis, Graduate School of Asian and African Area Studies, Kyoto University, Kyoto, 165 pp

Nindi SJ (2007) Changing livelihoods and the environment along lake Nyasa, Tanzania. Afric stud monogr Suppl 36:pp 71–93 (March)

Nindi SJ, Itani J (2008) Land use and environmental dynamics in the rural settings: local peoples adaptations to climate change. Paper accepted for publication during 3rd SADC/EU scientific conference at Lusaka Zambia (27–30 May, Cresta Golfview Hotel)

Nsenga JV, Mahonge CPI, Mtengeti EJ, Rutatora DF, Tamura K, Itani J, Araki M (2004) The SUA method: the case study from Mbinga. In: Rutatora DF, Tarimo AJP, Mattee AZ, Mtengeti EJ (eds) Perspectives and approaches for sustainable rural development in Africa. SCSRD, Morogoro, pp 333–357

Pomeroy D, Service MW (1986) Tropical ecology. Longman, Hong Kong

Pratt DJ, Gwynne MD (1978) Rangeland management and ecology in east Africa. Hodder and Stoughton, London

SCSRD (Sokoine University of Agriculture Centre for Sustainable Rural Development), JICA (Japan International Cooperation Agency) (2004) SUA method: concept and case studies, SCSRD and JICA, Dar es Salaam, 181 pp

Tamura K (2006) Monitoring and evaluation report on SCSRD/JICA activities, SCSRD, Morogoro, 49 pp

URT (United Republic of Tanzania) (2005) *Ruvuma region socio-economic profile*, 2nd edn, National Bureau of Statistics and Ruvuma Regional Commissioner's Office, Tanzania

Chapter 11
"Greening" Integrated Water Resources Management Policies for Tackling Climate Change Impacts: A Call for Sustainable Development

Nazmul Huq and Jean Hugé

Abstract This paper is a conceptual analysis of how "green" Integrated Water Resources Management (IWRM) can be the key to sustainable social and economic development under the regime of climate change. The prevailing IWRM principles developed by the Dublin and Rio statements (1992), the Millennium Assembly (2000) and the WSSD (2002) attach a narrow, technical and single purposive orientation to IWRM, which are inadequate in fighting against the impact of climate change. Standing on the brink of irreversible impacts of climate change, it is now most urgent to secure the water sector from the devastating impacts of climate change by appropriate responses through policies and acting accordingly. Integrating a climate-proofing approach to IWRM policies is termed as "green" IWRM. Analysis emphasizes that only "green" IWRM can promote water security under the changing climate. The paper outlines the strategies that should be taken to make the IWRM "green", which will further ensure sustainable use of freshwater, participation, gender balancing and equitable and efficient management fostering sustainable development.

Keywords Adaptation · Climate change · Greening · IWRM · Policies

N. Huq (✉) · J. Hugé
Bangladesh Centre for Advanced Studies, House 10,
Road 16A, Gulshan-1, Dhaka 1212, Bangladesh
e-mail: nazmulhuqrussell@gmail.com

Introduction

The word "crisis" is sometimes overused in development. But when it comes to water, there is a growing recognition that the world faces a crisis that, left unchecked, will derail progress towards the Millennium Development Goals (MDGs) and hold back human development (UN-DESA 2008). Throughout human history, progress has depended on access to clean water and on the ability for societies to harness the potential of water as a productive resource. Water for life in the household and water for livelihoods through production are two of the foundations for human development identified by the United Nations Development Programme Human Development Report of 2006 (UNDP 2006).

Climate change is arguably the most severe long-term threat to development facing this and future generations (Adger et al. 2003; Klein et al. 2008). By altering the hydrological cycle, climate change will exacerbate the water management problems that countries already face. Climate change will have significant, often dramatic, consequences: higher sea levels, more variable rainfall, more frequent and intense floods and droughts, and rapid desertification (GWP 2009). The consequences of water sector vulnerabilities for climate change are posing fundamental challenges to achieving the development aspirations. Water is already a scarce resource for reasons not directly associated to climate change: burgeoning population, excessive groundwater extraction, and industrial pollution, among others. Projections of the impact of climate change suggest that it would further exacerbate the water stress felt in many places around the world. As the causes of water quality and quantity deterioration become increasingly climate-challenged, it will be a daunting challenge for policy-makers to attribute responsibility to specific stakeholders for taking corrective action (David and Pandya 2009).

Better water management is thus essential for us to adapt to climate-induced changes in water resources. The future resilience (or vulnerability) of human communities to climate change-related impacts will depend on their success (Slootweg 2009; Bates et al. 2008). Since the early 1990s, the Integrated Water Resources Management (IWRM) concept has been emerging in the sustainable development and water resources management context. IWRM is perceived as the best approach to manage water resources in an efficient, equitable and sustainable manner.

Climate change and its devastating impacts are approaching at such a rapid pace that policy reconfiguration is urgently needed. The existing policies, for example the National Development Plan (NDP), land use, water use, transportation, forestry, agriculture, biodiversity and demography, etc., are all being readjusted to fight against climate change. Those readjustments are the necessary responses against the impacts of climate change (Eriksen et al. 2007; Klein 2008; Parry 2009). IWRM is regarded as the most effective approach for managing valuable water resources, but it also needs to be reorganized and readjusted to develop an equal interface against impending climate change impacts. This approach is identified as "greening" IWRM. The paper argues that the existing IWRM principle should be modified according to the response against climate change impacts in such a way that it would be responsive and provide an effective solution to global warming impacts.

Concept of IWRM

IWRM is an approach as well as a philosophy to support decision-making for managing complexity in the water sector. Water is the most scarce and valuable resource in the world; it is predicted that by 2025 around 3 billion people will be living under a water-stressed situation (GWP 2009). Global environmental changes add an additional burden on the water sectors. Since early 1990, environmental conventions adopted an integrated approach to manage scarce water resources in a coordinated manner for better use, which includes the allocation among competing human activities (Slootweg 2009). The Rio Earth Summit in 1992 is the major blow to push the IWRM concept embedded into policy and practice. In the twentyfirst century, IWRM has been widely accepted for the water management regime (Jonker 2007). The approach has been defined and conceptualized from academic, research and field experiences. The true meaning of IWRM is very much oriented to sustainable livelihood promotion through efficient uses of water resources.

Jonker (2007) and Merrey et al. (2005) describe IWRM from a livelihood point of view. They view the whole approach as a paradigm shift of traditional sectoral water resources management to an integrated manner where efficient management, allocation of water resources, sustainability of the water resources and support to human activities on the basis of equity are key. IWRM is a framework within which to manage people's activities in such a manner that it improves their livelihoods without disrupting the water cycle (Merrey et al. 2005; Jonker 2007).

The Global Water Partnership (GWP) defines the IWRM concept from the same point of view with some additional complements and components to make it functional into working for the practitioners (GWP 2009; Slootweg 2009). They take the essence of defining its guiding principles from the Dublin and Rio statements (1992), from the Millennium Assembly (2000), which gave rise to the Millennium Development Goals, and from the World Summit on Sustainable Development (2002) Plan of Action, which set a target for the preparation of IWRM and Water Efficiency plans. The guiding principles are:

- Freshwater is a finite and vulnerable resource, essential to sustain life, development and the environment.
- Water development and management should be based on a participatory approach involving users, planners and policy-makers at all levels.
- Women play a central part in the provision, management and safeguarding of water.
- Water is a public good and has a social and economic value in all its competing uses.

It also describes IWRM as: "Integrated water resources management is based on the equitable and efficient management and sustainable use of water and recognises that water is an integral part of the ecosystem, a natural resource, and a social and economic good, whose quantity and quality determine the nature of its utilisation" (GWP 2009).

Different views allow us to understand what an IWRM approach can offer for sustainable and efficient use of scarce water resources and for the betterment of humans and the planet. It can be summarized that IWRM is not just about managing physical resources; it is also about reforming human systems to enable all people to benefit from those resources.

IWRM and Climate Change

The impacts of climate change on the water sector are not linear; rather they are most complicated with multilevel interconnected impacts on society, community and ecosystems. Sea level rising, floods, lowering groundwater table, shortage of irrigation water, etc. will invoke settlement loss, crop damage, diseases, food insecurity, malnutrition, income loss and more poverty. The most threatening aspect is that the whole development track will lag behind the target. According to the recent MDG progress report, development targets cannot meet the expected targets, especially for low income countries (Eriksen et al. 2007; Klein et al. 2008; Parry 2009). Numerous reports, studies and researches establish the fact that impacts of climate change are playing the major role for this development apartheid. MDG target No. 3 of goal No. 7 clearly states that "Halve, by 2015, the proportion of the population without sustainable access to safe drinking water and basic sanitation". The crude reality of MDG goal No. 7 was shown to be that in 2009 already about 1.6 billion of the world's poor people were denied access to water and, by 2025, more than 2 billion will be added to this number (UN 2008). The majority of the water-stressed people will be the silent victims of climate change.

In this context, the challenges of water management will become increasingly important because there is general agreement that the supply of and demand for water resources will be substantially affected by climate change. The first key message is that, if the global energy habits are the focus for mitigation, the way of using and managing the water must become the focus for adaptation. One reason for this is that it is widely predicted that relatively small temperature changes of a few degrees will see average river flows and water availability increase by 10–40% in some regions while, in others, they will decrease by 10–30% (GWP 2005; Slootweg 2009). A further message is thus that changes in climate will be amplified in the water environment. The best approach to manage the impact of climate change on water is that guided by the philosophy and methodology of IWRM. The principles proposed by the IWRM methodology are based on equity-based efficient and effective management of the water resources and those principles are very much synergetic with the very core notion of sustainable development.

If the challenges of climate change for the world's water are not understood and addressed, the risk exists that the water supplies provided to the communities of a growing, urbanizing world, the infrastructures built to serve them and the industries and agriculture that supply and feed them will prove to be unsustainable (IFPRI 2009; Pahl-Wostl 2007). There are also broader dangers. If we fail to understand the interaction between climate change and water, other climate change strategies may actually aggravate the problems and increase the vulnerability of communities to both natural and man-made calamities (GWP 2005).

IWRM can be instrumental for both mitigation and adaptation strategies to climate change impacts if properly materialized. IWRM can play a vital role in mitigating the water-stressed climatic hazards, e.g. floods, lowering groundwater table, irrigation crisis and water-based livelihood vulnerabilities such as fishing. Effective planning to use IWRM for mitigating water-stressed hazards can eventually contribute to the holistic adaptation process and environmental sustainability that will boost the overall development trajectory especially to meet the MDG targets.

Greening IWRM: A Demand of the Time

Why "Greening IWRM"?

Policy greening is now one of the most oft-repeated pieces of jargon in policy domain. The unprecedented rate of climate change puts a strong impetus on greening the policies for all sectors. "Greening" refers to cross-cutting integration of principles in policies, plans and programmes. By nature, "greening" itself suggests a classical ecological focus and environmental sustainability (Hugé and Hens 2009). Encouraged by the essence of environmental sustainability, the paradigm has been shifted from "environmental policy greening" to "climate policy greening". As long as climate change impacts are concerned, the term "greening" also bears the synonymous explanation of the concepts "climate integration", "climate mainstreaming" and "climate proofing" (Klein 2008; Ahmad 2009). For climate change, it refers to the integration of climate-related vulnerabilities and mitigation and adaptation techniques into policies so that policy itself can be climate-resilient and when policies are put into action, those actions will also be climate-proof. It entails the whole spectrum of the penetration of climate resilient policies.

What is Green IWRM?

As a subset of climate policy, water policies also deserve to be "green" for better climate resilience. IWRM is an already established idea, concept and philosophy for managing water sustainably with special attention to environmental sustainability. However, to tackle the adverse impacts on the water sector and related cross-cutting sectors, it is time make another paradigm shift with "green IWRM". "Green IWRM" refers to the concept of cross-cutting integration of climate policies into major policies, programmes and actions related to IWRM. It also refers to the fact that the fundamental principles of IWRM would be embedded with policies and actions related to the fight against climate change.

The Global Water Partnership (2009) acknowledges that the prevailing mindset on water by promoting and applying the Integrated Water Resources Management (IWRM) approach is not enough for adaptation and mitigation for climate action. So far, IWRM is characterized by some overwhelming features such as narrow scope of implementation, more technical orientation, lack of capabilities of dealing with versatile issues, and being purpose-specific. So far, 68% of developed countries and 38% of developing countries have implemented IWRM but none of them use "outside the box" thinking to deal with the multi-dimensional complexities that are imposed by climate change (Jonker 2007; Mazvimavi et al. 2008; GWP 2009), although, in principle it is supposed to be efficient, effective, sustainable and equitable in nature. This fundamental lacking of existing IWRM raised the need to embrace IWRM with the principle of "green" that can effectively work for the sustainability of the water sector under a changing climate, keeping the human development as the most instrumental guiding principle (Merrey et al. 2005).

"Green IWRM" promotes a holistic approach to water management and recognizes that there are multiple pathways to building resilience against climate change. The methodology seeks to identify, and then to achieve trade-offs between, different water management objectives, including environmental sustainability, economic efficiency and social equity. It encourages the structured engagement of communities and sectors impacted upon by water into its management both to seek and promote "win–win" solutions but also to ensure that a better understanding of water constraints and challenges is developed and diffused into the society (GWP 2005).

Greening IWRM: A Process Approach

The strength of IWRM lies in its strong roots in the water sector and its subsequent extensive theoretical and practical knowledge of water-related issues. At the same time, this sectoral basis can be a point of weakness when issues beyond the sectoral boundaries have to be addressed. This has resulted in a call to think "outside the box", to see beyond what is common practice within the sector. In everyday practice, however, IWRM is not always effective in addressing the climate change adaptation challenge. In order to develop IWRM more effectively and responsively, various sources have suggested the linking of IWRM with Strategic Environmental Assessment (SEA) (Slootweg 2009).

The literatures suggest that in making IWRM climate-responsive, different steps can be followed. One of the processes is taking such an approach that will make convergence of different far-ended approaches into a single policy. The ingredients of this approach are as follows (UNESCO 2009; GWP 2009; Parry 2009; Slootweg 2009):

- A systems approach: assesses the linkages between, for example, humans and nature, water and the land, and the local and national perspective.
- An integrated approach: provides for a more coordinated and managed approach looking at the catchment and coastal level and surface and groundwater.
- A managed approach: strives to maximize water resources, minimize negative impacts and balance supply and demand.
- The stakeholder approach: examines the need for participatory decision-making at all levels, e.g. from government to individual. This will lead to a partnership approach and a sharing of common objectives.

Finally, the sustainable approach will focus on the necessity for equitable access to water resources. There are often compromises to be made between protection and use. This last thought appears to be a key one. To ensure water sector sustainability, there must be a balance between protection and rational use of the water. Different tools can be applied to bring the balance, such as developing policies and strict enforcement which, for instance, different countries have developed without significant tangible improvement (World Bank 2009a). However, practitioners suggest that improved basin management, community dialogue, trans-boundary cooperation, user sensitivity and attitudinal changes towards the precious water resource can bring prosperous development for bringing balance in protection and use. Moreover, more government attention, research and consensus among the stakeholders are deemed necessary (Cap-Net 2005; UNDP 2006; Cap-Net 2009).

Cap-Net (2005) also suggests a different approach for preparing IWRM which follows a sequential and cyclical approach. In this approach, there is plenty of space available to insert climate-proofing policies to make the IWRM green. The process approach suggested by the Cap-Net are as follows, showing the possible spaces to incorporate climate resilience policies.

Figure 1 shows a simple diagram of IWRM planning cycle. This is a very orthodox approach used for developing IWRM planning. To make it climate-proof and "green", some components can be added in each section of this cycle (Cap-Net 2005; Cap-Net 2009).

In the "Initiation" step, climate change impacts need to be integrated in the planning process. In advocacy towards policy-makers, the argument can be brought up that this will be instrumental for decision-makers to advance demand management strategies, which otherwise might be politically difficult to implement.

During the "Vision/Policy" phase, climate change adaptation is an additional element, not a replacement of IWRM goals. The overall aims of IWRM will remain the same.

In the "Situation Analysis" step, the use of climate information and impact analysis needs to be incorporated. Further, the adaptation/mitigation theme can be

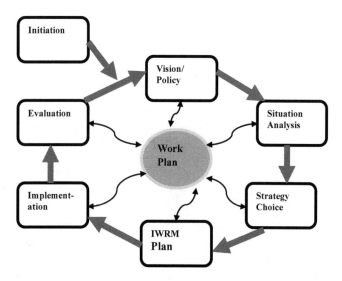

Fig. 1 The IWRM planning cycle. Source: Adapted from Cap-Net (2005)

brought out to suggest that the IWRM process should reduce the risk of adaptation options negatively impacting on the mitigation targets, and vice versa.

In the "Strategy Choice" phase, the anticipatory or "precautionary" approach can be introduced as the basis for strategies for IWRM.

Consider the roles of local authorities in adaptation strategies when drafting an IWRM plan. Legal frameworks, economics and health, and other variable conditional elements that have been analysed from the cornerstone for implementation of IWRM are decisive in how it contributes to climate change adaptation.

During evaluation, results must be measured against indicators, taking into consideration the adaptation measures proposed in the plan.

Throughout the process, stakeholder involvement is essential so that the results of the impacts assessment and strategic choice are owned by the implementing agencies.

The "greening" process of IWRM is not an easy one to implement. As stated earlier, so far only 68% of developed countries and 38% of developing countries have their IWRM plan. Studies conducted by Kramer (2007), Hugé and Hens (2007) on PRSP show that national development plans of developing countries are not green in principle to tackle environmental disasters, let alone IWRM. Klein (2008) conducted another portfolio study of ODA-funded projects and finds most donor-funded projects are not climate-proofed. In these circumstances, it is not easy to incorporate the greening policies into IWRM. Moreover, the developing countries who will be the first victims of climate change have a very low level of awareness in their policy level. The institutional strength is not enough to develop such realistic policies that will comply with the needs to tackle the climate stresses.

Greening IWRM: The Case of Bangladesh

Bangladesh, considered one of the most vulnerable countries to climate change impacts, is suffering tremendous water-related problems initiated by climate change (Ahmed 2006). Bangladesh recently prepared three documents related to water use, the National Water Policy 1999, National Water Management Plan 2005 and Poverty Reduction Strategy Paper 2008. All three documents, especially the first two, exclusively deal with the rational use of water resources and are based on the idea of traditional IWRM. It is a matter of great concern that despite inclusion of all of the aspirations of IWRM, those policies cannot make a proper response to the climate change problem. The detailed analysis reveals the truth that even though IWRM philosophy was the cornerstone of these policies, they are not enough to tackle the adverse impacts of climate change (GED 2008). As a result, countries' aspiration towards sustainable development and the positive achievements towards development are seriously at stake. Agricultural production is going down, livelihoods are becoming vulnerable and acute shortage of drinking water has gradually surfaced (MoEF 2008). This evidence suggests that IWRM policies need to be shifted towards a green policy, where climate-proofing development would be ensured. The ministry should take the lead to save the countries' most vulnerable resources and coordinate the efforts that are being taken to reduce the vulnerabilities from climate change impacts. For better response to the impacts of climate change there is no alternative than to go for "climate-proofing" development, which can be started from preparing climate-proofing action plans and policies and acting accordingly. It is thus necessary to incorporate the basic principles of developing climate-proof policies into action. If not, all the development efforts might go in vain. From this aspiration, it is time, albeit rather late, to start making the development and management of the water sector "climate proof", developing "green" IWRM and inserting the policy as a cross-cutting issue in all development intervention.

Conclusion

The whole discussion is aimed at developing the consensus that the present common interpretation of IWRM is not sufficiently compatible to fight against climate catastrophe. Much more comprehensive policies have to be developed which will be green in nature and sufficiently capable to face climate change effectively. At the same time, it is equally necessary to recognize that to make this paradigm shift, the core values of IWRM cannot be erased. IWRM is the best possible solution that works for sustainable use of scarce vital water resources. Thus, it is an urgent call to make "green" IWRM policy that will hold its core components at its centre and save the water resources for the people and ecosystem of the planet. However, most importantly political reluctance, lack of awareness and economic hindrance are working against stimuli for a green IWRM. As long as sustainable development is the dream, making efficient and equitable use of water based on the philosophy is of the utmost importance. IWRM policies thus should be directed towards accommodation of green policies so that climate-proofing notions can be ensured and water efficiency can be reached.

The paper attempted to provide a conceptual approach of how green IWRM can be possible, acknowledging the different shortcomings that exist. Besides this, scholarly literature on this issue is not yet developed, although this issue deserves much more attention than it has at present. However, it is always better to act fast, especially standing on the brink of climate catastrophe, otherwise the whole notion of sustainable development will never be achieved.

References

Adger W, Huq S, Brown K, Conway D, Hulme M (2003) Adaptation to climate change in the developing world. Prog Dev Stud 3:179–195
Ahmed A (2006) Climate change in Bangladesh—a synthesis, climate change cell, department of environment, Dhaka
Ahmad IH (2009) Climate policy integration: towards operationalization. United Nations Department of Economic and Social Affairs, New York
Bates B, Kundzewicz Z, Wu S, Palutikof J (eds) (2008) Climate change and water, IPCC technical paper IV, Intergovernmental Panel on Climate Change, Geneva
Cap-Net (2005) Integrated water resources management plans—training manual and operational guide, International Network for Capacity Building in Integrated Water Resources Management, Available at http://www.cap-net.org/TMUploadedFiles/FileFor67/IWRM_Plan.doc
Cap-Net (2009) IWRM as a tool for adaptation to climate change. International Network for Capacity Building in Integrated Water Resources Management, Available at www.cap-net.org
David M, Pandya A (2009) Troubled waters—climate change, hydropolitics, and transboundary resources. Henry L Stimson Center, Washington DC
Eriksen SE, Klein RJ, Ulsrud K, Næss LO, O'Brien K (2007) Climate change adaptation and poverty reduction: key interactions and critical measures. Norwegian Agency for Development Cooperation, Oslo
GED (2008) Moving ahead—national strategy for accelerated poverty reduction II (FY 2009–11), General Economic Division, Planning Commission, Government of the People's Republic of Bangladesh
GWP (2005) Climate change adaptation and integrated water resource management—an initial overview, Stockholm
GWP (2009) Global Water Partnership Strategy, Stockholm, 2009–2013
Hugé J, Hens L (2007) Sustainability assessment of poverty reduction strategy papers. Impact Assess Proj Apprais 25(4):247–258
Hugé J, Hens L (2009) The greening of poverty reduction strategy papers: a process approach to sustainability assessment. Impact Assess Proj Apprais 27(1):7–18
IFPRI (2009) Climate change—impacts on agriculture and costs of adaptation. International Food Policy Research Institute, Washington DC
Jonker L (2007) Integrated water resources management: the theory—praxis—nexus, a south african perspective. Phys Chem Earth 32(15–18):1257–1263
Klein R (2008) Mainstreaming climate adaptation into development—briefing note for the European parliament temporary committee on climate change. Stockholm Environment Institute, Stockholm

Klein RJ, Kartha S, Persson Å, Watkiss P, Ackerman F, Downing TE et al (2008) Adaptation: needs, financing and institutions. Stockholm Environment Institute, Stockholm

Kramer AM (2007) Adaptation to climate change in poverty reduction strategies. United Nations Development Program, New York

Mazvimavi D, Hoko Z, Jonker L, Nhapi I, Senzanje A (2008) Integrated Water Resources Management (IWRM)—from concept to practice. Phys Chem Earth 33(8–13):609–613

Merrey D, Drechsel P, Penning de Vries F, Sally H (2005) Integrating 'livelihoods' into integrated water resources management: taking the integration paradigm to its logical next step for developing. Reg Environ Change 5:197–204

MoEF (2008) Bangladesh climate change strategy and action plan. Ministry of Environment and Forest, Dhaka

Pahl-Wostl C (2007) Transitions towards adaptive management of water facing climate and global change. Water Resour Manage 21:49–62

Parry M (2009) Climate change is a development issue, and only sustainable development can confront the challenge. Clim Dev 1:5–9

Slootweg R (2009) Integrated water resources management and strategic environmental assessment—joining forces for climate proofing. Netherlands Commission for Environmental Assessment, Istanbul

UN (April 2008) United Nations Department of Public Information. Accessed 31 May 2010, available at http://www.un.org/millenniumgoals/2008highlevel/pdf/newsroom/Goal%207%20FINAL.pdf

UN-DESA (2008) The millennium development goals report, New York

UNDP (2006) Beyond scarcity: power, poverty and the global water crisis. Human development report, United Nations Development Programme, New York

UNESCO (2009) Climate change and water—an overview from the world water development report 3: water in a changing world, United Nations World Water Assessment Programme, Perugia

World Bank (2009) Convenient solutions to an inconvenient truth: ecosystem-based approaches to climate change. The World Bank, Washington

Chapter 12
Urbanization and Flood Vulnerability in a Peri-Urban Neighbourhood of Dakar, Senegal: How can Participatory GIS Contribute to Flood Management?

Audrey Maheu

Abstract Land and water are intricately connected and the inability to manage one component will necessarily impact the other. As urban sprawl grows out of control of authorities in developing countries, poor populations are forced to settle in hazard zones. Thus, comprehending the urbanization process is key in understanding flood vulnerability. This research uses the urbanization process as a starting point to understand the social and environmental dynamics behind floods in a peri-urban neighbourhood of Dakar, Senegal. Participatory GIS (PGIS), land use change analysis, interviews, transect walks and ethnographic observations were used to understand settlement in flood-prone areas. Limited access to risk knowledge was identified as a contributor to flood vulnerability. We argue that although risk awareness existed among peri-urban communities, an uneven access to knowledge led to settlement in risk-prone lowlands. We demonstrate risk awareness among the community by looking at the influence of three variables on the urbanization process: landforms, land availability and land price. This paper concludes with a critical analysis of PGIS regarding analytical and empowering outcomes of this methodological approach.

Keywords Participatory GIS · Participation · Flood · Vulnerability · Peri-urban · Urbanization

We argue that although risk awareness existed among peri-urban communities, an uneven access to knowledge led to settlement in risk-prone lowlands. We demonstrate risk awareness among the community by looking at the influence of three variables on the urbanization process: landforms, land availability and land price.

A. Maheu (✉)
International Development Research Centre (IDRC),
7 Des Frênes, Vaudreuil-sur-le-lac, QC J7V 8P3, Canada
e-mail: amaheu@idrc.ca

More than half of the world's population is currently living in urban areas, and by 2050, the world urban population will have almost doubled that from now, reaching 6.4 billion (United Nations 2007). Urban areas in Africa and Asia are expected to absorb most of the anticipated population growth, with "full urbanization" already completed in Europe, North America, and most of Latin America. This widely recognized and expected increase in urban population hides another reality: the growth and transformation of peri-urban areas. Studies in Latin America and Asia show that peri-urban areas absorb nearly all "urban" population growth, a trend likely to continue in Africa (da Gama Torres 2008). This poses a particular challenge as peri-urban environments "suffer some of the worst consequences of urban growth, including pollution, rapid social change, poverty, land use changes and degradation of natural resources" (United Nations 2007).

More than a de-densifying periphery around the city, the peri-urban interface is characterized by a heterogeneous mosaic of natural, agricultural and urban ecosystems affected by the flows (material, energy, people) between the urban and rural systems (Allen 2003). Institutional fragmentation, rooted in the urban–rural dichotomy between space and inner-city, as well as rural migration movements, often lead to the uncontrolled sprawl of peri-urban areas. Peri-urban areas are typically characterized by the existence of informal economies, often based on illegal activities such as clandestine slaughterhouses, or the intensive use of agrochemicals in agriculture and mining (Allen 2003). In such environments, urbanization is devoted from conscious planning and settlement precedes the extension of basic services implying large public health concerns. In fact, the problems of peri-urban areas are similar to those of low-income urban settlements: limited access to water and sanitation, waste accumulation, concentration of poverty, etc. Land speculation and the private appropriation of land in peri-urban areas further exacerbate the marginalization of the poor within the city's suburbs.

Slums often form at cities' edges, where they have been shown to increase the risk of natural disasters for the entire city (ISDR 2009). More often than not, these informal settlements develop in hazardous areas, where land is inexpensive, leaving the slum inhabitants particularly vulnerable to natural disasters. Illustrating this are the chronic problems of fires, landslides and floods in informal settlements at the edge of cities (Aragón-Durand 2007; Murray 2009; O'Hare 2005).

The problem of Floods at the Peri-Urban Interface

Hydro-meteorological hazards pose the biggest challenge for both developed and developing countries in the face of climate change (Thomalla 2006). In developing countries, flooding has become a recurrent problem in numerous peri-urban areas as populations settle in flood-prone areas. Already in the 1970s, flooding was a persistent problem in squatter settlements in the suburbs of Kuala Lumpur (Aiken 1975). The Chalco Valley at the outskirts of Mexico City has been suffering from "chronic flooding" since the late 1980 s (Aragón-Durand 2007). In Bamenda,

Cameroon, 20% of the population in informal settlements live in flood plains and urban expansion on hill slopes has triggered devastating flash floods (Acho-Chi 1998).

Flooding is not only the result of the occupation of hazard zones, it is also amplified by concurrent environmental degradation in peri-urban areas. For example, the loss of vegetation due to an increase in impervious surfaces was identified as an aggravating factor in the flood pattern in Dhaka, Bangladesh (Dewan 2008). Floods in Mumbai in 2005 led to a ban on plastic bags with garbage-clogged drains blamed for the magnitude of the disaster (Chatterjee 2005). Current flooding problems are likely to be exacerbated by both climate change and the booming urban growth in developing countries, making flooding an urgent issue to address in peri-urban areas, especially given the acute vulnerability of the poor populations concentrated in these areas.

GIS as a Diagnostic Tool

In the wake of the multiplication of flood events, urban planners and geographers have tapped the potential of Geographical Information Systems (GIS) to study flood events and to design solutions. GIS is thus a widely used diagnostic tool for understanding the causes behind the observed rise in flood events. A large number of studies have been produced on the dynamics of land use change and their relationships to floods. (Hara 2005) showed that land use changes following rural to urban transitions have reduced the volume of floodwater retention in peri-urban Bangkok, leaving both formal and informal settlements more vulnerable to flooding in this new landscape. The increase in impervious areas in suburban areas has been put forward as an explanation for flash floods (Romero 2004; Zhang 2008). In the Pearl river delta, China, for example, the filling of water bodies and the clearing of vegetation associated with urban sprawl impaired the buffering capacity of the ecosystem, thus triggering floods (Zhang 2008). These various GIS studies typically conclude that uncontrolled urban growth is the root cause behind the detrimental land use changes. Many have warned against the destruction of crucial ecological services when urban growth is not carefully planned (Romero 2004; Yu 2009).

While GIS has provided valuable insight in understanding the physical causes of flood events, it cannot *explain* the anthropogenic factors involved (Pelling 1999). Here, social sciences and political ecology can complement GIS analyses by investigating the social changes that triggered a modification of the physical landscape. Aragón-Durand (2007), for example, researched the socio-historical context behind flood risk generation at the outskirts of Mexico City. Texier (2008) showed that the marginalization of the poor by the municipal government was at the core of the 2007 flood disaster in Jakarta. These studies demonstrate that an understanding of the social construct of flood hazards is necessary in order to have a complete picture of the dynamics behind flood events. Unfortunately, GIS practitioners typically focus solely on the physical phenomena behind flooding, thereby providing only an incomplete picture of the problem.

GIS as a Prescriptive Tool

GIS in flood management has also been used to provide a "scientific input to planning" (Webster 1994). As such, hydrodynamic models have been integrated in GIS to map the behaviour of floods (Pradhan 2009; Vojinovic 2009). Along the same lines, multicriteria analyses based on physical attributes of the landscape (elevation, slope, land cover, distance from channel, depth to groundwater table, etc.) were designed to determine hazard zones (Fernandez 2010; Yalcin 2004) These different applications do not focus explicitly on peri-urban areas and usually look at the city as a whole, although more comprehensive approaches coupling flood hazard maps and urban growth models are being developed (Correia 1999). GIS is also put to use to map social vulnerability by investigating the spatial distribution of socioeconomic conditions of populations (Morrow 1999). Of course, these social analyses are often combined with a physical vulnerability assessment in order to obtain a complete picture of the spatial distribution of vulnerability (Bizimana 2010; Chakraborty 2005). In a positivist approach, GIS is seen as a prescriptivist and prioritization tool, the main purpose of which is to answer the "where" question, e.g. "Where are floods likely to occur?" and "Where are flood prevention measures required?" This approach is fundamentally different from the first use of GIS in flood management described above, which applies GIS to understanding the root causes leading to flood events. This difference in GIS usage is not peculiar to flood management and the tendency to prioritize isolated problem-solving measures as opposed to investigating the root of a problem has been a cause of tension in GIS research for some time (Ramsey 2009).

The main problem behind the prescriptive approach lies possibly in the prescriptions themselves, which are often difficult to put in place logistically and which meet strong public opposition. The positivist approach to land use planning has in fact been criticized in the planning community (Hodge 1991). Cities in developed countries are learning the hard way that expert-led land use planning has its limits. GIS-based studies looking at new landfills' location are perhaps the best example: although a location appears optimal according to a set of physical and social criteria, the failure of the siting process due to public opposition has become ubiquitous in North America (Lober 1995). GIS is a powerful tool for understanding the territory: however, it provides little help in the design of solutions and, most importantly, their acceptance by stakeholders.

Alternative GISs: PPGIS, PGIS, and CiGIS

Some may argue that GIS as a tool for flood management is inherently confined to the above-mentioned limits. However, the development of new fields in GIScience may be exactly what the doctor ordered. As GIS technologies rapidly spread to a variety of fields in the 1990s, a debate arose over the theoretical and societal aspects of GIS. In that context, GIS in relation to society became a field of research of its own—commonly referred to as the "GIS and Society" movement.

Of particular interest here is the criticism of GIS as an elitist system that reinforces top-down management. These critics condemned the technocratic avenue taken by GIS and argued for the recognition of spatially differentiated knowledge and the re-engagement of GIS with communities. As a result, new fields in GIScience emerged: Public Participation GIS (PPGIS), PGIS, and Community-integrated GIS (CiGIS).

Whereas the GIS and Society movement is generally concerned with theoretical and epistemological aspects of GIS (the whether and why questions), the just mentioned new fields are concerned with the "how" questions, e.g. "How can we put these new theories into practice?" (Sieber 2006). Thus, while alternative GISs all embrace the GIS and Society discourse, each has taken a slightly different avenue. Public participation GIS (PPGIS) aims at making GIS "available and accessible to all those with a stake in official decisions" (Schroeder 1996) and focuses on access to GIS technologies by grassroots organizations. PGIS has typically been critical of the overrepresentation of the privileged sectors of society and reaches out to marginalized populations. PGIS is strongly connected to the participatory movement in development, and the majority of applications labelled as PGIS take place in developing countries. Finally, in reaction to PPGIS, the goal of which is to develop applications controlled by the communities themselves, Community-integrated GIS (CiGIS) takes an alternative path: while fostering community participation in the GIS process, it also encourages expert-driven implementation in reaction to fundamental barriers in local capacity (Weiner 1999). Although these alternative GISs hold different values, they all share the common goal of working towards broader participation in GIS applications. To facilitate communication, this paper will adopt PGIS as an umbrella term for the different ways of implementing an interface between community and GIS.

PGIS: a Definition

The use of the PGIS label has been broad and inconsistent across applications (Schlossberg 2005), making it hard to produce a formal definition of this new field. By and large, PGIS has been understood as differing from traditional GIS in two aspects: (1) the type of information fed into the system and (2) the source of that information (Dunn 2007). With regard to the first aspect, the scope of the information feeding analysis in PGIS is much broader including narratives, fuzzy information, experiential knowledge and socially-differentiated information. The integration of qualitative information within a GIS framework has brought questions on two levels: GIS *with* qualitative methods and GIS *as* qualitative method. Discussions on GIS *with* qualitative methods look into the technical challenge of interpreting and representing different forms of spatial knowledge while research on PGIS *as* a qualitative method investigates the epistemologies of PGIS as a knowledge production strategy (Elwood 2006). As for the second aspect defining

PGIS, contrary to traditional GIS, PGIS necessarily involves citizen participation, with a special consideration of marginalized communities.

Participation in PGIS

A definition of PGIS cannot dodge the question of participation which has been at the core of discussions around PGIS. The notion of participation has been debated within the development community for a number of years and the requirements entailed by the participatory label have grown significantly over the years. In reaction to the utter failure of top-down projects, the mainstreaming of participatory approaches in the 1980s left a bitter taste for many in the development community. In particular, the wave of participatory rural appraisals (PRA) transformed participation into a technical fix to complex development problems (Hickey 2004). A radical change ensued that promoted commitment to the output of participation rather than to innovative ways of implementing it. Lengthy debates over the past decades have left many proposals on the table regarding the definition of participation. The definition gaining the most currency sees participation as a transformative process that strives for the achievement of full citizenship by disempowered populations (Hickey 2004). According to this vision, participation needs to be embedded in the broader political process rather than focusing on local initiatives (Williams 2004). However, setting such high standards for participation from the outset leaves little recognition of participatory approaches on the lower end of the spectrum and their potential as stepping stones towards a full transformation. The debate around participation remains wide open and although participation has been regarded as an imperative in development, "the 'trendiest' activity (at least among academics) is [still] the withering critique of participation" (Kesby 2007).

It is argued here that while mere tokenism is not an acceptable path for PGIS, the contribution of different levels of participation should be appreciated. Community-based GISs emerged in recognition of the limits of "overparticipation" (Weiner 1999). Kyem's (2001) work in Ghana showed that granting overwhelming responsibilities to communities who are not ready for them can actually have a negative impact on social capacity. In fact, climbing on top of the participation ladder should not be the objective of PGIS. Instead, it should be recognized that some forms of participation are more suitable for certain environments than others. Hence, the definition of participation adopted here is not based on any "level" of participation but is understood as a process inclusive of all stakeholders and in which the stakeholders represent all individuals "who are affected by, bring knowledge or information to, and possess the power to influence a decision or program" (Schlossberg 2005). While inclusion of the public at large is not considered here, the potential for doing so should not be dismissed (see Schlossberg 2005 for a discussion on the topic).

The Benefits of Participation

Access to Differentiated Information

The benefits of participation in PGIS can be understood by looking at the two different (and some would argue conflicting) understandings of participation. On the one hand, participation has been understood as a means of obtaining differentiated information. With this approach, participation has been used for ground-truthing or as a source of new information (Koti 2006; Harris 2002; Sliuzas 2003). The core objective of these projects is the production of knowledge and the literature abounds with cases of PGIS applied to produce risk and vulnerability assessments (Kienberger 2007; Nethengwe 2007). One of the benefits of participation is thus a better understanding of a given issue by capturing new sources (e.g. experiential knowledge) and forms (e.g. qualitative) of information.

The Potential for Empowerment

On the other hand, participation can be viewed as an end in itself in PGIS. While some projects aim at increasing citizen involvement (Al-Kodmany 2000), education, and communication (Mills 2008), others implement PGIS in order to promote empowerment (Bauer 2009; Stonich 2002). The common denominator here is that the GIS output is seen merely as an accessory to social change at large. In disaster management, PGIS has been used to impart a better perception and management of risk (Barclay 2008). Instead of using a top-down education approach where stakeholders are "taught" the elements of risks within a community, PGIS allows stakeholders to structure their relationship to the territory and gain a better understanding of the level of risk they are subject to. This process has empowering qualities as stakeholders reflect on their current standing and on how they can improve their conditions (D'Aquino 2004). PGIS can also help communities consolidate their message to be heard by authorities, building on the power of GIS as a communication tool. A number of cases exist in which PGIS has been used as an advocacy tool (Maantay 2002; Stonich 2002).

The PGIS process of producing spatial information has been acknowledged as empowering itself (D'Aquino 2004; Jordan 2002; Dunn 2007). Borrowing from Freire's concept of conscientization (Freire 1970), PGIS encourages social and political action on a collective level by inducing reflection on living conditions and the development of a critical worldview. However, others have argued that "participation in the creation of knowledge does not necessarily give power to those involved in, and affected by decision-making" (Aitken 1995). This argument was made in reaction to sweeping claims on the potential of achieving empowerment through PGIS, claims which lacked empirical support (Corbett 2005).

Objectives

The above discussion on participation and PGIS provides an overview of this emerging field in addition to showing the potential of PGIS to complement GIS with regard to flood management. This research project investigated flood dynamics in Dakar's suburbs using a PGIS approach. The aim of this paper is to (1) discuss social and environmental dynamics behind floods in Dakar's suburbs, with a focus on understanding the settlement process in flood-prone areas and (2) explore new ways of doing GIS and critically analyse the contribution of participation to research and to participants.

Study Area

The suburbs of Dakar have suffered severe flooding over the last ten years, with the department of Pikine being the most severely hit. While the floods seemed to be exceptional events at first, they have since established themselves as a recurrent phenomenon with worsening consequences every year. The flood problem in Dakar's suburbs has its origins in the exponential population growth since the 1970s, when an enduring drought throughout the country triggered massive rural migration towards urban centres (Goldsmith 2004). This large influx of people resulted in an erratic occupation of land, especially in the suburbs of Dakar.

The geomorphology of Dakar's suburbs features highlands, composed of coastal dunes and Ogolian dunes, a set of inland dunes running parallel along the southwest–northeast axis and lowlands composed of interdunal depressions in between Ogolian dunes known as the Niayes region. The lowlands are humid environments where the water table comes to the surface. These areas were traditionally used for agriculture given their easy access to water. During the ongoing dry years, the lowlands dried up, thus freeing some urgently needed space that then was quickly occupied. Although these lowlands had dried out, the water table remained near the surface.

Recently, the Sahel region has seen a rise in precipitation (Ali 2009). Although portrayed as exceptional, this recent increase in rainfall is most likely a return to normal conditions after 40 years of drought (Sene 2002). Unlike many flood events across the world that are triggered by catastrophic amounts of water unleashed onto a city (through rain or storm surges), the flood problem in Dakar is the aftermath of a prolonged period of climate perturbation. Recent wetter years raised the water table closer to previous levels (<2 m below surface) and now little precipitation is required for lowlands to become saturated, leaving its occupants flooded (Mbow 2008).

Our research on these flood events was conducted in Diamalaye, an informal neighbourhood in the districts of Malika and Keur Massar located at the edge of the department of Pikine (Fig. 1). While districts are the smallest government units, neighbourhoods and their neighbourhood representatives are credible local authorities and, in fact serve as an interface between formal authorities and the population (Tall 1998). Thus, the neighbourhood provides an appropriate scale to

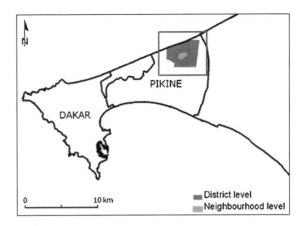

Fig. 1 Location of the study area at district and neighbourhood levels. The district level study area overlaps the districts of Malika and Keur Massar. The neighbourhood level study area corresponds to the neighbourhood of Diamalaye

understand social dynamics. A larger area was also considered to frame the flood problem within a wider context. Spatial analyses were conducted at both neighbourhood and district levels (Fig. 1).

This study area was selected for its landscape and for its peri-urban nature and the intense urbanization process currently underway there. Diamalaye displays the various physical features that are typical of Dakar's suburbs. Approximately half of the neighbourhood spreads across highlands and the other half across lowlands, providing subsets representative of the suburbs. Moreover, floods were late to come to the neighbourhood of Diamalaye. While the rest of the district began experiencing floods in 2005, floods only came to the neighbourhood of Diamalaye in 2009. Still, as the lowlands were being flooded in the surrounding neighbourhoods, people continued to settle in the lowlands in Diamalaye. Overall, the neighbourhood of Diamalaye provides the perfect setting for understanding the settlement process in lowlands (flood-prone areas) versus highlands.

The neighbourhood of Diamalaye is undergoing intense urbanization. Diamalaye is a young neighbourhood that emerged in the late 1970s and that truly exploded in the late 1990s. A 2007 survey shows that only 27% of the 2,000 inhabitants are native to the neighbourhood (Gaye 2008). The anticipated shutting down of the M'beubeuss landfill, which borders Diamalaye, also contributes to land speculation and intensifies urbanization pressure. On that account, Diamalaye presents the perfect conditions for studying how environmental and social dynamics of flood events unfold in the urbanization process. It will also offer useful information for managing urbanization in the context of city perimeter that is prone to flood problems.

Methodology

Similar to CiGIS studies (Koti 2006, 2010; Weiner 1999), this project used a variety of analytical methods to study the social and environmental dynamics underlying flood events in Diamalaye. The methodology comprised expert and participatory analyses that together formed the PGIS.

The participatory analysis took various forms. Four mapping workshops with an average of five participants were held in the summer of 2010. A local facilitator helped organize and conduct the workshops and served as an interpreter (French–Wolof). Each workshop targeted specific groups: young men, young women, farmers and neighbourhood representatives. Young people were targeted as they demonstrate an ease of communication with the researcher, which facilitated insight into social dynamics. The group of farmers was chosen due to the peculiar impact of floods on their livelihoods. During the three workshops with the above-mentioned groups, participants were asked to map flooded areas and flood risk within their neighbourhood and to describe flood impacts on their lives and adaptation strategies. Workshops were open-ended and included discussions on various aspects of neighbourhood life. The fourth workshop was held with neighbourhood representatives, including the chief neighbourhood representative and elders. During this workshop, the urbanization process was discussed along with the land acquisition process and land tenure issues. Interviews ($n = 6$) with key informants and local authorities were also conducted along key themes identified during the mapping workshops. Information gathered through workshops and interviews was then fed into the PGIS.

This research is also based on an ethnographic approach. For a period of four months, the researcher lived in an adjacent neighbourhood, undertook numerous walks across the neighbourhood and made observations. This data was also entered in the PGIS. While not participatory per se, this was part of a strategy to access differentiated information in keeping with basic PGIS principles.

The research project also included expert analyses going along traditional GIS methods. Land use changes were mapped at both neighbourhood and district levels through the manual classification and comparison of aerial photographs from 1954 (1:50,000), 1979 (1:60,000), 1997 (1:20,000), 2003 (1:10,000) as well as a 2009 SPOT image. Using this time series, conditions previous to (1954), during (1979), and after (2003 and 2009) the drought could be compared and analysed. A DEM of the region was also produced for the region using both surveying data from an ongoing project of Université Laval as well as contour maps (1:10,000) from the *Direction des travaux géographiques et cartographiques du Sénégal*.

Using the 1954 aerial photograph, a map of geomorphological features was also produced to delineate lowlands and highlands. The DEM was used to corroborate this map. Through manual classification, four features were mapped: coastal dunes, Ogolian dunes, the Niayes areas and water/lakebed. This latter category represents waterbodies from 1954 but given that water levels fluctuated during the drought, the category was named "water/lakebed". Spatial analysis was conducted to compare urban development against geomorphology features (Fig. 2). This was a step towards understanding the settlement process in lowlands.

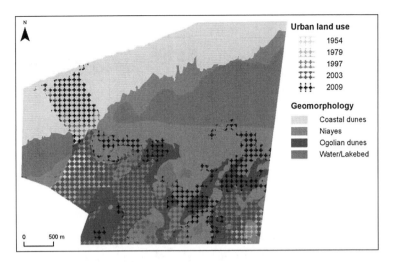

Fig. 2 Intersection between geomorphological features and urban development at the district level. Ogolian dunes correspond to highlands and Niayes correspond to lowlands

Results

The Urbanization Process

The flood hazard in Dakar's suburbs has its roots in the urbanization of lowlands. Land use changes between 1979 and 2003 show the steady invasion of lowlands during drought years (Fig. 3). During the same period, water levels and humid vegetation in lowlands receded, freeing up space for occupation. This goes along research by (Mbow 2008) on the adjacent district of Yeumbeul, where the land cover change induced by drought brought about urban development in lowlands.

Land cover change is only part of the reason behind occupation of lowlands. While physical processes allowed for the occupation of lowlands, land use and land cover changes are essentially driven by people's responses to opportunities (Lambin 2001). We argue that although risk awareness existed among peri-urban communities, an uneven access to knowledge led to settlement in risk-prone lowlands. We demonstrate risk awareness among the community by looking at the influence of three variables on the urbanization process: landforms, land availability and land price.

The Influence of Landforms

It is generally assumed that the drought modified the landscape significantly enough to blur the flood risk associated with lowlands, thus prompting the occupation of lowlands (Lo 2000; Mbow 2008). We hypothesized that if this

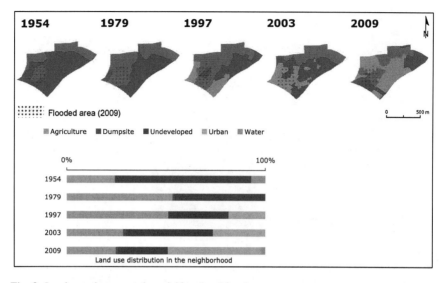

Fig. 3 Land use changes at the neighbourhood level

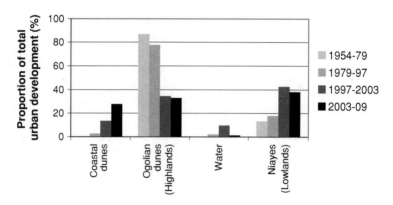

Fig. 4 Where urban development occurs: the geomorphological distribution of urban development for a given time period. Analysis at the district level

assumption holds true, urban development will then occur regardless of landforms during periods of drought, as water levels and vegetation then receded from lowlands, leaving little apparent differences between lowlands and highlands.

To verify this hypothesis, we identified which landforms were the sites of new urban development (Fig. 4). The drought period was set between 1960 and 2000, mainly for lack of more precise parameters. Most climatologists consider the Sahel drought as beginning between 1960 and 1970 and while the drought is still not over, climatic research has observed an increase in precipitation beginning in the early 2000 s which points toward the end of the drought (Aguiar 2009; L'Hôte 2002; Sene 2002).

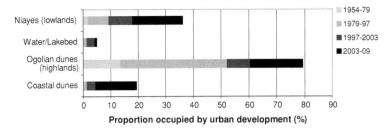

Fig. 5 Urban development as a proportion of total land available for a given geomorphology category. Analysis at the district level. Urban development corresponds to *new* urban land use that originated during the time period indicated. Ogolian dunes correspond to highlands and Niayes correspond to lowlands

Figure 4 shows that between 1954 and 1997, urban areas preferentially developed on highlands at the district level. Even during the drought, highlands were favoured over lowlands for urban development. The same trend is observed at the neighbourhood level. The neighbourhood of Diamalaye first developed on top of an Ogolian sand dune and only later did settlement spread over to lowlands. The preference of highlands for settlements, a phenomenon occurring worldwide, can be explained by various factors such as climate, diseases, defence mechanisms and agricultural diversity. While the preference for highlands is not necessarily a demonstration of more flood awareness per se, it does reflect how this landform is considered more suitable for settlement than lowlands. Interestingly, between 1997 and 2009, as precipitations slowly returned, highlands and lowlands hardly differed with regard to urban development. Then, as urban pressure built up during this period, land scarcity could have played a role in the gradual urbanization of lowlands. However, as it will be discussed below, land availability was not a strong determinant of the urbanization of lowlands.

The Influence of Land Availability

Dakar's flood problem is usually framed within the rapid population growth of the city and it was suggested that the occupation of lowlands is the result of land scarcity (Mbow 2008). We tested these hypothesis by distinguishing the urban development per total land available for a given landform (Fig. 5). The hypothesis is that urban development in lowlands did not begin until highlands available for development became scarce. An important assumption here is that unoccupied land is available for acquisition, which usually holds true as landowners tend to validate ownership by laying down at least crude foundations. Figure 5 shows that the occupation of lowlands started even as only 53% of highlands (Ogolian dunes) were occupied. Spatial evidence shows that the occupation of lowlands is not merely the result of land scarcity as the peri-urban district develops.

The Influence of Land Price

As land availability did not show to be a strong determinant of the occupation of lowlands, the influence of a third variable, land price, was investigated. Through interviews and workshops, cheaper land price was identified as the main motivation for settlement in the lowlands of Diamalaye. Similarly, participants in mapping workshops repeatedly associated lowlands with lower land value throughout the district. Land price impacts the urbanization process at two levels: it acts as a migration attractor and as a situational determinant.

The cheaper cost of living in the suburbs through subsidized services, an attractive fiscal system, and a lack of speculation bringing down land and property value have been identified as important drivers of urban sprawl in developed countries (Ewing 2008). Urban sprawl remains poorly researched in developing countries, even though urbanization has become a prominent force in the reconfiguration and economic development of these regions. While the literature on West African urbanization has focused mainly on rural-to-urban migration, the contribution of migrants as a driver of urban expansion is being revisited (Beauchemin 2004). In Ouagadougou, peri-urban neighbourhoods are teeming with young inner-city migrants who are attracted to these districts by government employment plans aiming to relieve the congested downtown area (Boyer 2010). Inner-city migration plays an important role in the demographics of the urban expansion of Dakar, with second or third generation rural migrants attracted towards the periphery in order to gain access to home ownership (Vernières 1973; Legros 2003). As land prices go up in the city and in the proximate suburbs, peri-urban locations are the only remaining option, as emphasized in workshops and interviews. Cheaper land prices drive urban sprawl to the periphery by attracting new migrants.

Cheaper land prices attract new migrants towards the periphery and this same driver influences the choice of location for a parcel of land within the peri-urban environment. At both neighbourhood and district levels, lowlands were associated with land prices that are lower, often less than half the price of parcels located on highlands. Land value reflects a large set of environmental variables, road development, proximity to amenities, provision of services, human as well as environmental risks. Even in informal land markets, prices are not set arbitrarily and typically follow economic principles (Adams 2004). The price difference not only explains the specific motivation for settlement in lowlands during drought years, but also shows that landowners recognized differences in environmental conditions significant enough to impact land value. Still, in the midst of the drought period, lowlands and highlands showed little distinct physical differences (water and vegetation had receded from lowlands) and both areas reached very similar development levels (e.g. proximity and availability of services). Just as urban development was privileged on highland landforms, the differential price treatment between lowlands and highlands suggests risk awareness among the community.

Risk Awareness and Marginalization

Spatial patterns of urban development point toward risk awareness among the local population. Interviews with key informants and local authorities corroborate this finding by acknowledging that most landowners knew about the land's unsuitability for residential development. However, this finding goes against repeated claims of flood victims stating that they were completely unaware of flood possibilities. In the neighbourhood of Diamalaye, many pointed to the pond that had turned into a soccer field during dry years as evidence that the flood risk was impossible to recognize. How can such different perspectives on flood risk arise? We argue that the problem did not lie within the presence of risk knowledge as demonstrated above but with the unequal access to that knowledge among the population.

Risk mapping exercises with the population showed that now that the hydro-meteorological conditions are turning back to normal, the relationship between flood risk and lowlands is well understood. Thus, at present, settlement in lowlands is not the result of a misconception of risk. The uneven distribution of knowledge is part and parcel of other disparities occurring in the Diamalaye. A distinct gap persists between the lowland and highland portions of the neighbourhood regarding access to services and political representation. Figure 6 shows the uneven electricity delivery within the neighbourhood leaving lowlands' dwellers to deploy cable extensions (up to 500 m) to hook up to power sources outside of the neighbourhood. Along the same trend, Fig. 6 also shows the uneven spatial distribution of neighbourhood representatives. Despite their informal status, the neighbourhood chief and his representatives play an important role in the daily activities of the neighbourhood. The current composition of local representatives meant little attention was given to issues arising in the lowlands' portion of the neighbourhood, including the issue of floods.

These disparities demonstrate the differential treatment of lowlands and highlands within a same neighbourhood, corroborating (Agrawal 1999) conceptualization of a community as a non-homogeneous entity. Limited access to knowledge is a key contributor to vulnerability among poor urban dwellers regarding natural hazards (Blaikie 1994). Environmental knowledge is thus a form of power, while conversely, the lack of access to knowledge exacerbates marginalization. Similarly, the uneven distribution of services and political power are also indicative of exclusionary practices and can be considered spatial manifestations of marginalization. Lowlands dwellers were typically found to lack relational qualities prior to their arrival in the neighbourhood. Most of them were introduced to the neighbourhood through outsiders of the community: friends from a neighbourhood adjacent to Diamalaye or word of mouth about available land. More importantly, lowlands dwellers lacked any kinship relationship with the main original landholding families. Nevertheless, although chaotic in appearance, the urbanization process in the informal neighbourhood of Diamalaye is controlled by social relationships and informal institutions thus transposing marginalization and exclusion throughout the spatial development of the neighbourhood.

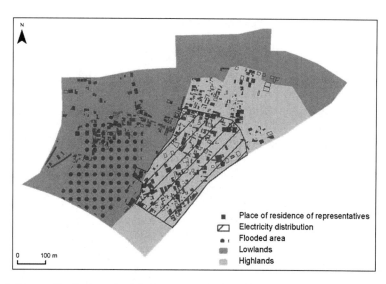

Fig. 6 Uneven distribution of political representation and service distribution at the neighbourhood level

The disconnection between a heterogeneous community and its homogeneous leaders often results in the exclusion of the most vulnerable regarding development goals of a community. For example, flood mitigation or adaptation was not identified as a priority by the population in a participatory development project currently underway in the neighbourhood. As Pelling (1998) observed in Guyana, vulnerability differentials and the perceived non-inclusiveness of the political space "encourage individuals to withdraw from the participatory system and concentrate on the lower efficiency but greater control to be found in investing in household or family-based coping mechanisms." When investigating adaptation strategies, individual structural measures such as sand filling in houses, localized road raising and informal diversion works were the most common coping mechanisms, along with family relocation during flood periods. The perceived non-inclusiveness of political space has downstream repercussions on the neighbourhood's development and is translated into neighbourhood-dwellers investing in individual strategies rather than relying on community-based action. This then prevents the building of social capital between lowlands and highlands groups, a shortcoming that is at the heart of flood vulnerability as it impedes the flow of information on risk.

PGIS: The Intersection between Intervention and Research

The use of participation within this research required the negotiation of participation as an intervention (linked to empowerment) and as an analytical method (as a source of information in a perspective of fundamental research).

The following section discusses how PGIS shapes and contributes to research on flood management as these two levels within the urbanization context.

Participation and Empowerment

The perceived lack of openness of the political space in the neighbourhood prevents the building of social capital between the two social groups in the lowlands and highlands. Building social capital would facilitate the flow of risk knowledge, which would promote the empowerment of marginalized groups at the same time. Social capital has been identified as a key element of adaptive capacity in the face of natural hazards and climate change (Adger 2003). While its importance has been recognized in the field of natural resources management, developing pathways to supplement social capital has proven difficult (Pelling 2008). As discussed above, PGIS can help develop collective ownership of an environmental problem, thus suggesting its potential for social capital building.

The success in building social capital varies depending on the way in which a PGIS is implemented. In this project, we set up the mapping workshops according to already established association regarding age, gender and occupation. We also sought to facilitate communication by creating a comfortable setting conducive for discussion. Furthermore, the workshops mainly targeted flood victims given the topic of this research. Overall, we concluded that the implementation of the PGIS reinforced social bonds within existing groups, as participants shared common experiences and a common fate. However, results demonstrated that the lack of social bridges between the lowlands and highlands impeded the flow of risk knowledge. We suspect that the PGIS could have contributed more to empowerment had it been implemented to develop social bridges between the lowlands and highlands, as opposed to only reinforcing social bonds between flood victims, as was the case in this project. Therefore, experience from this research shows that the design of PGIS implementation can have a significant influence on the building of social capital, and by extension on empowerment outcomes.

Participation and an Improved Access to Knowledge

One of the alleged benefits of a PGIS is to allow access to local knowledge, which is difficult to achieve through a conventional GIS approach given the source and form of local knowledge. This claim is based on numerous assumptions.

First, local knowledge in PGIS is often framed as a collective resource and assumed to be evenly distributed within the community. Such an assumption is the remnant of the conceptualization of the community as a homogeneous unit. Research by Agrawal and Gibson (1999) in natural resources management criticized the concept of community as "small, integrated groups using locally evolved norms to manage resources sustainably and equitably." Instead, they proposed a new framework where the community is understood through local politics and institutions. Their research triggered a series of articles framing communities within a political space, mainly focused on the distribution of power (Agarwal 2001; Klooster 2000; Nygren 2000). At the same time, concepts of local knowledge and indigenous technical knowledge (ITK) were being implemented in natural resources management (Agrawal 1995; Gadgil 1993). Despite the enthusiasm for these new concepts and the political focus following research by Agrawal and Gibson, certain aspects, such as the question of knowledge and its distribution remained unscrutinized. In this research, we showed that knowledge deserves to be meticulously investigated when trying to understand hazard vulnerability. Access to knowledge, in this case risk knowledge, is at the roots of flood vulnerability in the neighbourhood of Diamalaye. A PGIS can be an efficient tool to access local knowledge; however, the distribution and modalities of access to this knowledge must be carefully considered within the PGIS framework. As this research demonstrates, knowledge is not always evenly distributed and while a PGIS enables the mapping of knowledge distribution, it still requires a conscientious and targeted approach in order to obtain a comprehensive picture.

Second, many people assume that relevant authorities or institutions will make use of local knowledge, given the beneficial aspects. The knowledge flow (or lack thereof) is considered to be at the heart of the problem. In this sense, a PGIS is often interpreted as a way to instil or restore communication between the local population and authorities. However, in a context of development where weak institutions often prevail, no assumptions can be made regarding the outcomes this information will actually bring. In this project, the production of spatial information was beneficial as it helped bring attention to the flood problem and power asymmetries in this specific neighbourhood. Therefore, the translation of this knowledge into action by local authorities remains uncertain as a PGIS is subject to the same struggle of bringing research to policy-making.

Third, the central assumption of a PGIS in accessing local knowledge is that sustainability had been impeded by a lack of information. This project shows that it is not the lack of information that prevented sustainable urban planning but rather its uneven distribution among the population coupled with inefficient planning authorities. Access and distribution of information played a central role in creating the flood hazard in the study area. A PGIS can be used to target this specific problem and be used as a means to distribute this information by establishing channels to share geographical and environmental knowledge. However, more information is not a panacea for reaching sustainability; it only informs policy-makers during the actual policy-making process. Instead of concentrating on establishing vertical communication (between local population and authorities), a PGIS could be used to stimulate horizontal communication (between community members). In the context of weak institutions, such an investment in the knowledge flow within civil society is of utmost importance.

Conclusion

Spatial patterns unveiled by traditional GIS were used as an entry point to understand urban expansion. A PGIS implemented through mapping workshops as well as interviews and ethnographic observations revealed a geography of exclusion that exacerbates flood vulnerability. Therefore, any initiative tackling the flood hazard will need to acknowledge the heterogeneity of the community and exclusionary processes at play. Assistance and mitigation plans will need to expand actions beyond current institutional structures, even traditional informal ones, to successfully engage with marginalized and vulnerable flood victims. For example, surveys of flood victims and compensation distribution carried out by the state should try establishing direct communication channels with flood victims rather than going through indirect channels (e.g. informal institutions) as is currently the case.

At the methodological level, a PGIS constitutes an excellent tool for gaining access to local knowledge. However, assumptions regarding knowledge distribution and the practical benefits this additional information brings should be carefully considered. A PGIS helped to unveil the geography of exclusion within the neighbourhood of Diamalaye. (Schmitt 1996) already raised the question of "systems for prospective exploration" rather than "systems for control" for GISs. In that spirit, participation contributes an added value by allowing a broader framing of land and water management problems and by supporting a freer investigation of spatial and non-spatial processes.

The use of a PGIS showed that the tool could contribute to empowerment through building social capital. However, the way in which a PGIS is implemented plays a significant role in its efficacy. PGISs should target specific social bonds and/or bridges to ensure empowerment outcomes rather than building social capital in general.

References

Acho-Chi (1998) Human interference and environmental instability: addressing the environmental consequences of rapid urban growth in Bamenda Cameroon. Environ Urban 10(2):161–174

Adams J, Antwi A (2004) Urban myth vs. economic explanation: An empirical analysis of land purchasing decisions in Accra. World Rev Sci Technol Sustain Dev 1(2):192–208

Adger WN (2003) Social capital, collective action, and adaptation to climate change. Econ Geogr 79(4):387–404

Agarwal B (2001) Participatory exclusions, community forestry, and gender: An analysis for South Asia and a conceptual framework. World Dev 29(10):1623–1648

Agrawal A (1995) Dismantling the divide between indigenous and scientific knowledge. Dev Change 26(3):314–439

Agrawal A, Gibson C (1999) Enchantment and Disenchantment: The Role of Community in Natural Resource Conservation. World Dev 27(4):629–649

Aguiar LA (2009) Impact de la variabilité climatique rcente sur les ecosystems des niayes du Sénégal entre 1950 et 2004, unpublished PhD thesis, Université du Québec à Montréal

Aiken SR, Leigh CH (1975) Malaysia's emerging conurbation. Ann Assoc Am Geogr 65(4):546–563
Aitken SC, Michel SM (1995) Who contrives the 'real' in GIS? Geographic information, planning and critical theory. Cartogr Geogr Inf Sci 22(1):17–29
Ali A, Lebel T (2009) The Sahelian standardized rainfall index revisited. Int J Climatol 29(12):1705–1714
Al-Kodmany K (2000) GIS in the Urban Landscape: reconfiguring neighborhood planning and design processes. Landscape Res 1:5–28
Allen A (2003) Environmental planning and management of the peri-urban interface: Perspectives on an emerging field. Environ Urban 15(1):135–148
Aragón-Durand F (2007) Urbanisation and flood vulnerability in the peri-urban interface of Mexico City. Disasters 31(4):477–494
Barclay J, Haynes K, Mitchell T, Solnana C, Teeuw R, Darnell A, Crosweller HS, Cole P, Pyle D, Lowe C, Fearnley C, Kelman I (2008) Framing volcanic risk communication within disaster risk reduction: finding ways for the social and physical sciences to work together. Geol Soc Lond Special Publ 305:163–177
Bauer K (2009) On the politics and the possibilities of participatory mapping and GIS: using spatial technologies to study common property and land use change aong patoralists in Central Tibet. C Georaphies 16:229–252
Beauchemin C, Bocquier P (2004) Migration and urbanization in francophone West Africa: an overview of the recent empirical evidence. Urban Stud 41(11):2245–2272
Bizimana JP, Schilling M (2010) Geo-information technology for infrastructural flood risk analysis in unplanned settlements: A case study of informal settlement flood risk in the Nyabugogo flood plain, Kigai City, Rwanda In: Showalter PS, Lu Y (eds) Geospatial techniques in urban hazard and disaster analysis, Springer Netherlands, London, pp 99–124
Blaikie P, Cannon T, Davis I, Wisner B (1994) At risk: natural hazards, people's vulnerability, and disasters. Routledge, London
Boyer F (2010) Croissance urbaine, staut migratoire et choix résidentiels des ouagalais. Vers un insertion urbaine ségrégée. Revue Tiers Monde 1(201):47–64
Chakraborty J, Tobin GA, Montz BE (2005) Population evacuation: Assessing spatial variability in geophysical risk and social vulnerability to natural hazards. Nat Hazards Rev 6(1):23–33
Chatterjee P (2005) Mopping up in Mumbai. Lancet 366(9488):795
Corbett JM, Keller CP (2005) An analytical framework to examine empowerment associated with participatory geographic information systems. Cartographica 40(4):91–102
Correia FN, Saraiva MDG, Da Silva FN, Ramos I (1999) Floodplain management in urban developing areas, Part II. GIS-based flood analysis and urban growth modelling. Water Resour Manag 13:23–37
da Gama Torres H (2008) Social and environmental aspects of peri-urban growth in latin American Megacities United Nations Expert Group Meeting on Population Distribution, Urbanization, Internal Migration and Development
D'Aquino P (2004) Pour une expertise participative qui accompagne l'émergence de territoire citoyens du local vers le global. CIRAD, available at http://hal.archives-ouvertes.fr/hal-00157746_v1/
Dewan AM, Yamaguchi Y (2008) Effect of Land Cover Changes on Flooding: Example from Greater Dhaka of Bangladesh. Int J Geoinformatics 4(1):11–20
Dunn CE (2007) Participatory GIS: a people's GIS? Prog Hum Geogr 31(5):616–637
Elwood S (2006) Critical issues in participatory GIS: deconstructions, reconstructions, and new research directions. Trans GIS 10(5):693–708
Ewing RH (2008) Characteristics, causes, and effects of sprawl: a literature review. Urban Ecol 5:519–535
Fernandez DS, Lutz MA (2010) Urban flood hazard zoning in tucuman province, argentina, using GIS and multicriteria decision analysis. Eng Geol 111:90–98
Freire P (1970) Pedagogy of the oppressed. Seabury, New York
Gadgil M, Berkes F, Folke C (1993) Indigenous Knowl Biodiversity Conserv 22(2–3):151–156

Gaye A (2008) Enquête socio-économique de Diamalaye de *2007:* rapport d'analyse, Institut Africain de Gestion Urbaine

Goldsmith PD, Gunjal K, Ndarishilanye B (2004) Rural-urban migration and agricultural productivity: the case of senegal. Agric Econ 31:33–45

Hara Y, Takeuchi K, Okubo S (2005) Urbanization linked with past agricultural landuse patterns in the urban fringe of a deltaic Asian mega-city: A case study in Bangkok. Landscape Urban Plan 73(1):16–28

Harris T, Weiner D (2002) Implementing a community-integrated GIS: perspectives from south african fieldwork. In: Craig W, Harris T, Weiner D (eds) Community participation and geographic information systems. Taylor & Francis, London, pp 246–258

Hickey S, Mohan G (2004) Toward participation as transformation: critical themes and challenges. In: Hickey S, Mohan G (eds) Participation: From tyranny to transformation?. Zed Books, London, pp 3–24

Hodge G (1991) Planning Canadian Communities. Nelson Canada, Scarbrough

ISDR (2009) Global Assessment Report on Disaster Risk Reduction. United Nations, Geneva

Jordan G (2002) GIS for community forestry user groups in Nepal: Putting people before technology. In: Craig W, Harris T, Weiner D (eds) Community participation and geographic information systems. Taylor & Francis, London, pp 232–245

Kesby M (2007) Spatialiasing participatory approaches: the contribution of geography to a mature debate. Environ Plan A 39(12):2813–2831

Kienberger S (2007) Assessing the vulnerability to natural hazards on the provincial/community level in Mozambique: the contribution of GIScience and remote sensing In: The 3rd international symposium on geo-information for disaster management, Toronto: Canada

Klooster D (2000) Institutional choices, community and struggle: A case study of forest co-management in Mexico. World Dev 28(1):1–20

Koti F, Weiner D (2006) (Re)Defining Peri-urban Residential Space Using Participatory GIS in Kenya. Electron J Inf Sys Dev Ctries 25(8):1–12

Koti FT (2010) Confronting sociospatial exclusion on the fringe of Africa's cities using participatory GIS: Lessons from Athi River town, Kenya. Afr Today 56(3):63–82

Kyem PAK (2001) Power, participation, and inflexible institutions: an examination of the challenges to community empowerment in participatory GIS applications. Cartographica 38(3):5–17

Lambin EF, Turner BL, Geist HJ, Agbola SB, Angelsen A, Bruce JW et al (2001) The causes of land-use and land-cover change: moving beyond the myths. Glob Environ Change 11:261–269

L'Hôte Y, Mahé G, Somé B, Triboulet JP (2002) Analysis of a Sahelian annual rainfall index from 1896 to 2000; the drought continues. J Hydrol Sci 47(4):563–572

Legros O (2003) Le gouvernement des quartiers populaires: production de l'espace et regulation politique dans les quartiers non réglementaires de Dakar (Sénégal) et de Tunis (Tunisie), unpublished PhD thesis, Université François-Rabelais de Tours

Lo PG, Diop MB (2000) Problems associated with flooding in Dakar, western senegal: influence of geological setting and town management. Bulletin Eng Geol Environ 58:145–149

Lober DJ (1995) Resolving the siting impasse: modeling social and environmental location criteria with a geographic information system. APA J 61(4):482–495

Maantay J (2002) Mapping environmental injustices: pitfalls and potential of geographic information systems in assessing environmental health and equity. Environ Health Perspect 110(2):161–171

Mbow C, Diop A, Diaw AT, Niang CI (2008) Urban sprawl development and flooding at yeumbeul suburb (Dakar-Senegal). Afr J Environ Sci Technol 2(4):75–88

Mills JW, Curtis A (2008) Geospatial approaches for disease risk communication in marginalized communities. Prog Community Health Partnersh Res Educ Act 2(1):61–72

Morrow BH (1999) Identifying and mapping community vulnerability. Disasters 23(1):1–18

Murray MJ (2009) Fire and ice: unnatural disasters and the disposable urban poor in post-apartheid johannesburg. Int J Urban Reg Res 33(1):165–192

Nethengwe NS (2007) Integrating participatory GIS and political ecology to study flood vulnerability in the limpopo province of south africa, PhD edn, West Virginia University, Morgantown

Nygren A (2000) Development discourses and peasant-forest relations: natural resource utilization as a social process. Dev Change 31(1):11–34

O'Hare G, Rivas S (2005) The landslide hazard and human vulnerability in La Paz City Bolivia. Geogr J 171(3):239–258

Pelling M (1998) Participation, social capital and vulnerability to urban flooding. J Int Dev 10(4):469–486

Pelling M (1999) The political ecology of flood hazard in urban guyana. Geoforum 30:249–261

Pelling M (2008) Shadow spaces for social learning: a relational understanding of adaptive capacity to climate change within organisations. Environ Plan A 40(4):867–884

Pradhan B, Shafiee M, Pirasteh S (2009) Maximum flood prone area mapping using RADARSAT images and GIS: kelantan river basin. Int J Geoinformatics 5(2):11–23

Ramsey K (2009) GIS, modeling, and politics: on the tensions of collaborative decision support. J Environ Manag 90:1972–1980

Romero H, Ordenes F (2004) Emerging urbanization in the southern andes: environmental impacts of urban sprawl in santiago de chile on the andean piedmont. Mt Res Dev 24(3):197–201

Schlossberg M, Shuford E (2005) Delineating 'Public' and 'Participation' in PPGIS. J Urban Reg Inf Sys Assoc 16(15):26–28

Schmitt E, Brassel K (1996) From GIS for control to GIS for creative exploration GIS and Society workshop, specialist meeting report, South Haven

Schroeder P (1996) Criteria for the design of a GIS/2 specialists' meeting for NCGIA initiative 19: GIS and society

Sene S, Ozer P (2002) Évolution pluviométrique et relation inondations-événements pluvieux au sénégal. Bulletin de la Société géographique de Liège 42:27–33

Sieber R (2006) Public participation geographic information systems: a literature review and framework. Ann Assoc Am Geogr 96(3):491–507

Sliuzas R (2003) Opportunities for enhancing communication in settlement upgrading with geographic information technology-based support tools. Habitat Int 27:613–628

Stonich SC (2002) Information technologies, PPGIS, and advocacy: globalization of resistance to industrial shrimp farming. In: Craig W, Harris T, Weiner D (eds) Community participation and geographic information systems. Taylor & Francis, London, pp 259–269

Tall SM (1998) La decentralization et le destin des délégués de quartier à Dakar (Sénégal). In: Bulletin de l'Association Euro-Africaine pour l'Anthropologie du Changement Social et du Développement, 15, http://apad.revues.org/567

Texier P (2008) Floods in Jakarta: when the extreme reveals daily structural constraints and mismanagement. Disaster Prev Manag 17(3):358–372

Thomalla F, Downing T, Spanger-Siegfried E, Han G, Rockstrom J (2006) Reducing hazard vulnerability: toward a common approach between disaster risk reduction and climate adaptation. Disasters 30(1):39–48

United Nations Population Fund (2007) State of the world population 2007: unleashing the potential of urban growth

Vernières M (1973) Campagne, ville, bidonville, banlieue: migrations intra-urbaines vers dagoudane pikine, ville nouvelle de dakar (Sénégal). Cahiers ORSTROM, série Sciences Humaines 10(2/3):217–243

Vojinovic Z (2009) Supporting flood disaster management with numerical modelling and spatial mapping tools. Int J Geoinformatics 5(4):33–40

Webster CJ (1994) GIS and the scientific inputs to planning. part 2: prediction and prescription. Environ Plan B: Plan Des 21(2):145–157

Weiner D, Harris T (1999) Community-integrated GIS for land reform in south africa

Williams G (2004) Toward a repoliticization of participatory development: political capabilities and spaces of empowerment. In: Hickey S, Mohan G (eds) Participation: from tyranny to transformation?. Zed Books, London, pp 92–109

Yalcin G, Akyurek Z (2004) Analysing flood vulnerable areas with multicriteria evaluation. XXth ISPRS Congress Istanbul, Turkey

Yu K, Wang S, Li D, Li C (2009) The function of ecological security patterns as an urban growth framework in beijing. Shengtai Xuebao/Acta Ecologica Sinica 29(3):1189–1204

Zhang H, Ma W, Wang X (2008) Rapid urbanization and implications for flood risk management in hinterland of the pearl river delta, China: the foshan study. Sensors 8:2223–2239

Chapter 13
Vulnerability of Andean Communities to Climate Variability and Climate Change

Harry Diaz, Rosa Garay-Fluhmann, Julia McDowell, Elma Montaña, Bernardo Reyes and Sonia Salas

Abstract There is mounting evidence that climate change will increasingly impact large areas of Latin America, affecting people's livelihoods and important natural resources such as water. These impacts will make rural people disproportionately more vulnerable, given their dependency on natural resources and their exposure to other stressors, such as globalization and restricted fiscal policies. Climate change, however, could also bring new opportunities, such as the expansion of cultivated areas. The paper presents the results of a 3-year study focused on rural vulnerabilities to climate variability in three Andean watershed basins: the Mendoza Basin in Argentina, the Choquecota Basin in Bolivia and the Elqui Basin in Chile. Following the vulnerability approach, the presentation discusses and compares (a) the present exposures, sensitivities and adaptive capacities of different rural producers in the three basins to present and past climate variations and their impacts on local water resources; and (b) the instituional challenges faced by these rural producers in the context of the present climate variability and expected future changes in climate conditions.

Keywords Vulnerability · Andean communities · Exposure · Adaptive capacity · Sensitivity · Institutional framework · Governance

H. Diaz (✉) · R. Garay-Fluhmann · J. McDowell · E. Montaña · B. Reyes · S. Salas
Canadian Plains Research Center, University of Regina,
Regina, SK S4T 6N8, Canada
e-mail: harry.diaz@uregina.ca

Introduction

There is mounting evidence that the impacts of global warming will increase, producing variations in local weather patterns and water supplies, disturbing ecosystems and soil landscapes and impacting on economic production and social conditions. The last report of the Intergovernmental Panel on Climate Change (IPCC 2007) indicates that climate change will impact large areas of Latin America. Important changes in rainfall patterns, and increases in temperatures and the intensity and severity of extreme climate events have already been observed, with negative impacts for people's livelihoods (WGCCD 2006).

Based on various climate scenarios, the IPCC estimates that by the year 2,100, mean regional temperatures will rise between 2 and 6°C. Projected impacts of these increases in temperature involve the displacement of forests, reduction of the extent and volume of glaciers, loss of agricultural soil, biological imbalances and increasing pest intensity, sea-level rise, further changes in precipitation, salinization and desertification, and water scarcities. The IPCC expects that by 2020, several million people in Latin America will experience water stress due to climate change, facing critical problems with drinking water supply and sanitation (Magrin et al. 2007; see also, WGCCD 2006). Thus, anticipated climate changes will seriously impact regional development, affecting everyone. However, it will disproportionately affect the livelihoods of rural people, given their dependency on natural resources and the extent to which they are already exposed to other stressors, such as globalization and restricted fiscal policies.

In this context there is an increasing need for an appropriate understanding of existing climate vulnerabilities and adaptive capacities of local rural populations. Expanding our knowledge about present and past climate impacts and vulnerabilities is essential not only for an effective management of present risks but also for the development of adaptive capacities able to deal with the future challenges of climate change in the region. The development of an adaptive capacity—the capacity of governments and civil society to combine strengths and resources to manage risk—increases the ability of rural regions and communities to reduce the present and future adverse impacts of climate change and maximize its benefits.

This paper examines the vulnerabilities of rural actors in three Latin American watersheds: Mendoza in Argentina, Choquecota in Bolivia, and Elqui in Chile. The paper is based on a 2-year interdisciplinary project, "Coming Down the Mountain: Understanding the Vulnerability of Andean Communities to Hydroclimatologic Variability and Global Environmental Change", which was carried out with the financial support of the Inter-American Institute for Global Change Research. The project, completed in December 2009, sought to identify and characterize the vulnerabilities of rural actors to climate variability and climate-induced water problems in each one of the basins and to evaluate the capacities of water governance institutions to reduce these actors' vulnerabilities. In addition to the collection and analysis of secondary data, the study involved approximately 260 in-depth interviews oriented to gain insights from agricultural producers and representatives of water governance organizations in the three countries (further information about the project can be found on the website www.climaycomunidades.org).

Based on the project results, this paper focuses on the capacities of governance institutions to reduce the vulnerability of rural producers in the three basins. It begins with a brief discussion of the concepts of vulnerability and institution, which framed the methodological and theoretical approach of the project. A brief description of the three basins follows, with special consideration given to the climate and productive conditions that characterize each one of the watersheds. The next section focuses on a comparison of the exposures and adaptive capacities of rural actors in the three basins and the role played by water governance institutions in the development of these capacities. Finally, the paper offers some insights about the role of institutional conditions in determining the adaptive capacities of local people.

Water Governance, Vulnerability and Adaptation to Climate Change

The project adopted a vulnerability model (see Fig. 1), which guided the research approach and analysis in the three basins, facilitating the process of comparison. Following the IPCC, vulnerability was defined as the degree to which a system, such as a rural community or a farm, "is susceptible to, or unable to cope with, adverse effects of climate change, including climate variability and extremes", IPCC 2001, p. 995).

Vulnerability, in this definition, is a function of the exposure/sensitivity and the capacity of a community to adapt to climate stress (Liverman 1994; Wisner et al. 2005; Handmer et al. 1999). *Exposure-sensitivities* refer to the interaction of both the characteristics of the system and a stimulus. They reflect the manner in which a system experiences conditions to which it is sensitive (Smit and Wandel 2006). The actions taken to ameliorate risks and capitalize on opportunities are considered *adaptive strategies*. The system's ability to employ adaptive strategies reflects its *adaptive capacity*. In this context, the most vulnerable systems are those most likely to be exposed to climate change impacts and are more sensitive to perturbation, with a limited capacity for adaptation (Adger and Kelly 1999).

The vulnerability of a system, such as a rural community, is not a function of climate alone, but rather, it is the result of multiple social conditions including environmental, social, economic and political factors. These multiple factors are interlaced in the dynamics of community vulnerability and the consequent development of adaptive strategies to reduce vulnerability (Smit and Wandel 2006). It is the combination of biophysical and social conditions, usually structural conditions (such as access to resources or the existence of social capital), which makes people more or less exposed to climatic variability and events.

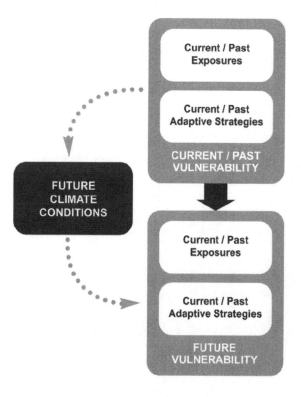

Fig. 1 The vulnerability approach model

What is the role of institutions in the context of vulnerability? The Third Report of the IPCC has identified a set of determinants of adaptive capacity that "influence the occurrence and nature of adaptation and thereby circumscribe the vulnerability of systems and their residual impacts" (IPCC 2001, p. 893). These determinants include the existence of economic resources, technology, information and skills (including human capital), infrastructure, equitable social relations, and well-developed institutions such as government bodies. The IPCC considers that established institutional conditions, such as the existence and availability of insurance mechanisms or water conservation programmes, facilitate the management of climate-related risks reinforcing the adaptive capacity of the population (IPCC 2001, pp. 896–897). Accordingly, countries with well-developed institutions have a greater adaptive capacity than countries with less effective institutional arrangements.

Institution is defined as a persistent, reasonably predictable arrangement, process, custom or organization structuring aspects of the political, social, cultural or economic transactions and relationships in a society (Henningham 1995) and it could range from highly formalized settings to informal arrangements. Rural settings are places where formal and informal institutions coexist and interact. Regional and local governments, church organizations, and forms of social capital such as networks for mutual support, become structured into complex sets of relationships that contribute or disrupt the stability, viability, and in our case, the adaptive capacity of rural people (Adger 2003; Halpern 2005; Dale and Onyx 2005). In addition, rural settings, like any other human setting, function within larger institutional political systems that link them with the larger society. These political systems, which are central to adaptive capacity given their purposeful mandate, permanency, social acceptance and legal basis, pervade the lives of the community members by imposing a body of regulations, rules, processes and resources on communities.

An important political institutional cluster is governance, which refers to "the patterns by which public power is exercised in a given context" (Jenkins 2002, p. 485). More specific to this paper, water governance is an organizational form of public power that defines the organization and management of the interrelationships between society and water resources, determines how much water may be used, by whom, and under which conditions, and assesses the impacts of society upon water resources in the natural environment. As expected, water governance assumes a diversity of forms, ranging from a highly centralized system to one that encompasses a diversity of public and non-public organizations. The dominant tendency, however, has been to define governance as a more inclusive system than solely government, defining it as form of political power that encompasses laws, regulations, public organizations, and includes those sectors of civil society that participate, interact with, or influence the management of water resources (Hurlbert et al. 2009a, b). This understanding of governance is the one that informs the paper's discussion of the role that formal institutions play in influencing vulnerability to climate and climate-related water stress of the rural population.

The Study Areas

The three basins selected for the study are mainly agricultural regions, although the Mendoza and the Elqui Basins have a more complex economic base where mining, industry and services complement a very dynamic agricultural industry.

The Bolivian watershed selected for the study is the Choquecota river basin (all the information provided in the paper is found in McDowell and Zeballos 2008). This basin covers about 100 km^2. The Choquecota river originates in the glaciers of Mururata mountain, descending quickly through the valley until it finally reaches a low altitude of 3,200 m. Within this basin, researchers selected the Municipality of Palca, which is located approximately 20 km southeast of the city of La Paz, as the study area. The municipality sits in a valley at the western base of the Illimani and Mururata mountains in the Andes Mountain Range. It has a total area of 743 km^2, of which 53.4% are highlands (above 4,000 m), wetlands, grasslands and rocky hills that are unsuitable for agriculture. Only 27% of the territory is appropriate for agriculture, especially in the lower parts of the valley.

The local climate is characterized by a humid and a dry season each year. The sun radiation can be very intense, especially on the northeast-facing slopes, which can generate strong winds during the day. Local temperature varies greatly depending upon altitude. The average annual temperature in the municipality is 12.2°C, with an average highest temperature of 22.4°C and a lowest average of 5°C. The average annual rainfall for the municipality is 557 mm, much of which falls shortly and intensely from November to March (410.9 mm). The past 20 years indicate fairly stable patterns in precipitation levels. The driest years—1982, 1983, and 2005—were also the hottest. On the other hand, June and July are the driest and coolest months.

The population of the municipality of Palca was approximately 14,000 people in 2001. They live in either of the two small rural towns of the area, Palca and Cohoni, or in over 60 smaller agricultural communities. Most of the active population, close to 70%, is occupied in agricultural activities. The rest are either part of a small cottage industry confined to the town of Palca or seasonally employed in the cities of La Paz or El Alto. A large segment of the local population, approximately 78%, lives in extreme poverty with limited access to water and sanitation services, low levels of education, and with a severe deficit in access to basic healthcare. An important characteristic of the local population is the strong indigenous identity of its inhabitants. Nine out of ten people identified themselves as Aymara, and 85% of the population indicated that Aymara was their first language, while only 14.75% indicated the same of Spanish. Furthermore, over one-third of the local population only speaks Aymara.

Local agriculture could be defined as traditional peasant agriculture. Families are the central unit of social and political life and organized in community settlements, each with an organizational structure and a system of governance framed in the indigenous Aymara culture that has predominated in the region for centuries. A large amount of local production is for self-subsistence and the rest is traded mostly in local markets. Most of the units of production are *minifundios*—small parcels that are own by a family. These small landholdings—ranging from 100 m^2 to half a hectare—are characterized by an intensive use of family labour and a limited use of modern technology. They generate very low family incomes, which forces family members to seek alternative sources of income by seasonally migrating to the large cities or mines. Each family has several of these parcels, distributed at different altitudes or *pisos ecologicos*, which allows for a diversified production of crops—such as potatoes, beans, oats and vegetables—using the ecological advantages of different altitudes. Communities have established a complex system of *trueque*, the bartering of goods, where products produced at different altitudes are traded among families from the different communities, facilitating their nutritional diversification.

The Argentinean research site is the Mendoza river basin, in the province of Mendoza. The river originates in the Andes of central Argentina, where it collects the snowmelt at an altitude of 6,000 m above sea level, and rapidly descends into the plains of the western *pampas*, at 600 m above sea level, where its joins the San Juan river. It is mostly an agricultural region, but in the western part of the basin, close to the Andes's foothills, is the city of Mendoza with a population of approximately one million, the main administrative and service centre of the watershed.

The basin covers approximately 19,000 km² and its climate is characterized by the altitude, its large distance from the Atlantic Ocean, and the presence of the Andes mountains that act as a barrier to the humid winds of the Pacific. In particular, altitude is the main factor in determining climate conditions. The climate is mainly arid in the lower areas but becomes increasingly humid in the high altitude of the Andes, where the Mendoza river is born. As expected, the average temperature changes with altitude, increasing from the west to the east. The annual average temperature in the city of Mendoza is 17.1°C but in the month of January the average reaches 32.2°C in the same area. Precipitation also varies with the altitude, from an annual average of 272 mm in the highlands to 90 mm in the plains, but high levels of evapotranspiration create the semi-arid and arid conditions that predominate in the watershed.

The regional economy is predominantly agricultural. Given the regional climatic conditions, irrigation is essential for the economy. There is a well-established irrigation infrastructure in the basin that uses mostly river water, with a large number of storage dams able to guarantee the irrigation needs. Underground water is used mostly to complement the use of surface water, especially during drought periods. Most of the population of the region—approximately 800,000—lives either in the city of Mendoza or in a group of oases that are the result of irrigation. These oases occupy no more than 3% of the area of the province of Mendoza, but over 90% of the economic activities of the area are concentrated in them. The main crops grown are grapes (almost 50% of the cultivated area), vegetables, olive and fruit trees, and grass. A more marginal agriculture is found in the drylands, *el desierto*, those areas with no access to irrigation and, accordingly, highly dependent on precipitation.

The Argentinian basin is very different to its Bolivian counterpart. It is characterized by a very dynamic, modern economy, where the agricultural sector plays a significant role. The region has a strong wine industry—a very dynamic sector that orients its production towards national and international markets—as well as a strong horticultural sector. It is, however, a region characterized by contradictions, tensions and conflicts that emerge around the use of water resources. Water scarcity is a characteristic of the region, mostly due to an increasing demand for the resource and a lack of efficiency in its use. There are significant tensions between urban and rural uses, about allocations among different economic sectors, between upstream and downstream users, and among agricultural producers. This last tension assumes a spatial contradiction between the oases, where a modern agriculture predominates, and the drylands, where subsistence agriculture predominates (Montaña et al. 2005). In other words, the Mendoza river basin is a dry watershed with an unequal distribution of power upon the scarce water resources.

The Chilean watershed, the Elqui river basin (ERB), shows similar characteristics to the Mendoza Basin. The Elqui Basin is the northernmost basin of the Coquimbo region in northern Chile. It spans 9,675 km², from the Andes Mountains to the Pacific coast. As in the case of the Mendoza river, the Elqui river originates in the high Andean mountains, where it feeds from the high snow accumulation at an altitude that ranges from 3,000 to 6,000 m, and descends through a narrow valley to the gentler slopes of the coastal plateaus.

The climate of the ERB is strongly influenced by the presence of the Pacific Ocean, especially by the high pressure of its anti-cyclonic system that moves north during the winter, bringing precipitation one or two months of the year to the valley. Average precipitation is 100 mm per year, although it may double or triple during the wet El Niño phase of the El Niño Southern Oscillation (ENSO), and be half the average or less during the dry La Niña phase of ENSO. Eight to ten months without precipitation is a common feature of the valley, but at higher altitudes the basin experiences a much higher average precipitation than in the valley (Fiebig-Wittmaack et al. 2008). There seems to be, however, a decrease in precipitation over the past century from between 150 and 180 mm per annum in the early 1900s to the currently experienced 100 mm. In the last 25 years, the decreasing precipitation trend has levelled off, and even shows a slight recovery, but it is still among the most pronounced decreases in all of Chile, and it represents a risk that affects both human and natural systems (Cepeda et al. 2009). In these terms, the region, as it is the case of the Mendoza Basin, has a dry climate adjacent to a major mountain system and landscapes at risk of desertification and it is highly dependent on runoff from the Andes—a primary source of water for irrigation (Cepeda et al. 2004). The Chilean Basin has a well-developed irrigation system that provides water to an area of approximately 20,000 ha.

The region is heavily engaged in agriculture and mining, two activities that impose increasing pressure upon scarce water resources. As in the case of the Mendoza Basin, a process of agricultural differentiation has characterized the development of the Elqui Basin since the late 1970s. Agro-industrial companies and modern landowners have transformed the valley: cereals, pasture land and orchard production gave way to vineyards, citrus and avocado plantations that have higher input costs associated with them and a need for sophisticated irrigation systems. Traditional farmers have either been marginalized into pockets of small agriculture that contain most of the poor rural households in the area or they have become seasonal labourers for large agricultural operations (Bodini and Araya 1999). In this context, access to water and improved technologies is unevenly distributed. Most agro-industrial operations have the financial resources and know-how to secure the latest technologies to reduce the risks associated with water stress. Small producers and traditional farmers, on the other hand, have little access to water resources and tend to mainly use traditional irrigation systems that are not as efficient, such as flood irrigation (Diaz et al. 2009).

This brief description of the three basins shows a significant difference among the basins related to the process of differentiation among agricultural producers. The Bolivian basin is characterized by a high degree of homogeneity among producers, which seems to be the result of the predominance of an ethnic culture where community—as an institutional form—is central to the lives of the producers. The cases of Argentina and Chile, in opposition, represent a process of modernization characterized by an increasing bipolar process of differentiation of the agricultural units: on the one hand, modern, large units that orient their production to highly profitable crops and, on the other hand, a precarious, marginal agriculture characterized by small units of production.

There are also significant differences among the institutional systems that exist in the three countries and their presence in the watersheds. In comparison with the other two countries, Bolivia is perhaps the weakest in terms of the presence of its public institutional system in the studied basin. Few central government institutions are active in the region and the municipal government lacks the technical and institutional capacity to provide substantive assistance to agricultural producers in the region. There are also some differences between Chile and Argentina. While Argentina is a federal country where provincial public organizations have some degree of autonomy from the central government, Chile is characterized by a centralized government and by regional institutions dependent on a central decision-making government. Despite this difference, both countries have more developed and established institutional systems compared to Bolivia, with a strong presence at the level of the basin. In both cases, water governance institutions have played a strong role in shaping the existing adaptive capacity in the basins, although with some limitations.

The Vulnerability of the Basins

As expected, agricultural producers and rural households are exposed and sensitive to a variety of climate variability-related events. A variety of climate events, such as hail, frost, changes to the precipitation and temperature patterns, and droughts, are the most relevant exposures identified by the respondents in the three basins. Different producers have different sensitivities to each of these exposures, either because of the nature of their productive systems (sensitivity of crops to high temperatures) and/or limited access to resources that could contribute to reduce their sensitivities (lack of access to irrigation), or location within the basin (upstream/downstream).

Droughts are hydroclimatic hazards that affect the three basins. Droughts are less relevant on the Choquecota Basin, but many local producers noted in the interviews an increase in abnormal precipitation rainfalls. According to them, the rain season has shortened during the last 20 years. In previous years, precipitation extended from September to April but in recent years, rains do not start until the last days of October and end by March. No less concerning for the Bolivian producers is the lack of accumulation of snow in the mountains, which provides the necessary water for irrigation during the dry season (normally from May to August). Less water in the dry season creates the potential for conflict among the producers, especially among the downstream communities. In the case of Argentina and Chile, droughts are a phenomenon that is a natural and recurrent characteristic of the regional climate. As an example, the Elqui Basin had, between 1915 and 2003, 11 years of extreme drought (annual precipitation of less than 30 mm) and 16 years of moderate drought (annual precipitation of between 30 and

60 mm) (Diaz et al. 2009, p. 33). In both basins, droughts affect not only the supply of water to the local population and agricultural producers—with devastating effects on their production—but they also amplify the negative impacts of aridity on soil erosion and ecosystems. In addition, historical and political conditions limit the access to water resources of many agricultural producers, creating socially related water scarcities.

An excess of rain is also problematic. In the case of Argentina, an abnormally wet season creates problems for the fruit, especially grapes. Rain could affect the development of vines and grapes or bring plagues that affect both the quantity and the quality of the harvest. In the case of Chile and Bolivia, an excess of rain, particularly in ENSO years, increases the risk of mudslides and floods due to topographic conditions characterized by steep valley slopes. In the two basin communities, crops and irrigation channels located in the foothills are particularly exposed to mudslides.

Frost and hail are also exposures that affect agricultural producers in the three basins. Argentinean producers noted that the phenomenon of hail has increased in recent years, while producers in Bolivia state that hail and frost are less predictable now. Extreme cold spells and hot days are also defined as an exposure, mostly because they impact on the yield and quality of a variety of crops, reducing the profitability of the farm.

As expected, sensitivity to all these exposures is unevenly distributed. Sensitivities to climate are uneven in terms of geographical location, such as the case of downstream Argentina producers who are more vulnerable to water scarcities as a result of a lack of control of the upstream use of the resource, or by virtue of differential access to natural, social and economic resources that could reduce exposure to climate events, as is the case of Bolivia and Chile where the precarious location of rural housing or agricultural parcels in risk-prone areas highly increases rural people's exposure to mudslides and floods. Thus, vulnerabilities to climate are not isolated from other conditions that inform the livelihoods of agricultural producers. Moreover, climate variability is not always the determinant factor that places agricultural producers in an extremely vulnerable situation but rather it is the interaction of the climate disturbance with the inherent social conditions of the producers that defines the degree of their vulnerability. Accordingly, a variety of conditions such as distribution of wealth, social marginalization, differential access to resources and access to healthcare play a role in affecting social vulnerability (Adger and Kelly 1999). In the case of Bolivia, a central determinant to the producers' sensitivities to climate is the predominance of the *minifundio* and its associated conditions, such as limited income and resources. In Chile and Argentina, where regional agriculture is well integrated into national and international markets, the main determinant of the sensitivities is related to market conditions, increasing concentration of regional economic resources and limited income alternatives.

Vulnerabilities to climate are defined not only by exposure and sensitivity to climate events, but also by the adaptive capacity of producers. The research findings show that an adaptive capacity already exists in the three basins, a capacity that has been the product of a historical process of adjusting to climate variability. Its effectiveness—as is the case of sensitivity to climate—is obviously related to producers' differential access to social, economic, natural and institutional resources.

In the case of Bolivia, most forms of adaptation have emerged as a product of the direct local experience of climate and in direct relation to coping mechanisms to other forms of stress. Some of these adaptive strategies are based on old customs that have strong cultural roots, such as the use of *pisos* (farming terraces), access to local community-based social capital networks, or forms of collective works based on reciprocity. Others are more religious in nature, such as the ceremonies offered to Aymara deities. Other strategies are more recent in time and are a response to the challenges imposed by the limited opportunities offered by the *minifundio*. Given the limited opportunities that small land parcels offer to increasing and/or diversifying production to secure family consumption, the generation of alternative sources of income, such as obtaining city jobs or seasonal migration to the mines, has become a new normal.

What is relevant in the Bolivian case is that the existing adaptive capacity is based on local resources and capacities and more or less uniformly distributed within the local community. As indicated before, the presence of the public institutional system in the Choquecota Basin is relatively weak. Few government institutions are active in the region. The most important entities are non-state institutions, such as the local agrarian union, which concerns itself not only with agricultural issues, but also with the general oversight of communal well-being. Local state institutions are weak, regional institutions have little direct influence in the region, and relevant national institutions are only recently gaining the capacity to support communities. The municipal government suffers from accusations of chronic corruption and it has limited technical and institutional capacities to provide substantive assistance to agricultural producers in the region. Its support is sporadic and only takes place in situations of emergency (disaster relief). Important institutional actors in the area of climate change research and programme implementation in Bolivia are the foreign aid agencies. In addition to implementing their own development programmes linked with climate change adaptation and mitigation, foreign aid funds several government programmes and occupies many arenas where local governments should have jurisdiction. As both donors and implementers, foreign aid organizations have a great deal of power in terms of setting the agenda for climate research and response in Bolivia but with a limited presence at the regional level.

In the cases of Chile and Argentina, there is also an historical process that has informed the development of a regional adaptive capacity to climate and other stressors. The fundamental difference with the Bolivian situation is that Chile and Argentina have more developed and established institutional systems, with a strong presence at the level of the basin. Thus, the existing adaptive capacity in these two countries has been considerably shaped by a variety of government programmes and political decisions oriented to improve the conditions of productive systems.

The construction of irrigation infrastructure in the Mendoza and Elqui watersheds has been a central element of government policies aimed to regulate river flows and water storage. Adaptations to secure reliable water resources during periods of scarcity have been primarily achieved through the construction of dams and irrigation systems and, more recently, by introducing more efficient forms of irrigation, such as drip irrigation. This has been accompanied by administrative decisions regarding the distribution of water resources and the supply of drinking water; other measures have contributed to the development of an adaptive capacity that reduces a variety of water scarcities and other climate or climate-related risks. In addition, in recent years a large number of policies and programmes oriented to support the production of agricultural products for exportation and access to new market opportunities associated with free trade have been developed, as in the case of fresh fruit and avocados in the Elqui valley and wine in the Mendoza watersheds.

This adaptive capacity, however, is unequally distributed. Argentinian and Chilean respondents recognized the existence of a variety of supporting government programmes (credits, technological dissemination, and others) but many of them considered those programmes to be biased to the most productive and organized economic sectors (large producers). Small producers, especially dryland farmers and ranchers, have limited access to many of the irrigation and productive programmes, which forces them to adopt other strategies to cope with problems, such as limiting the amount of cultivated land, reducing the number of animals, income diversification, seasonal migration and the use of underground water. These strategies, however, do not always allow for an expansion of the adaptive capacity to face more drastic crises. Larger agricultural producers, on the other hand, have access to larger resources that facilitate the implementation of more reliable forms of adaptation, such as the use of more efficient irrigation techniques, adoption of varieties more resistant to extreme weather conditions, effective management of water resources, access to market information, and others. In these terms adaptive capacity follows power distribution patterns.

The capacity of governance to sustain and increase the adaptive capacity of agricultural producers is limited not only by its bias in support of specific producers but also by the weakness of state policies, decision-making processes and program management in relevant areas. As an example, climate governance is relatively weak in all three countries. They all have agencies focused on climate change, but with limited resources and no long-term perspective. Bolivia is perhaps the weakest country in terms of institutional knowledge and capacities, with limited climate and water data at the national and regional level. Argentina has a more established institutional network in the area of climate change, but is still weak at the level of the province. Chile is, perhaps, the most advanced country in terms of an institutional response. It has established a National Plan of Adaptation and Mitigation and some sector-based plans, but its coordination and effectiveness are still limited. Although adaptation has been defined as a component of the Chilean policy approach to climate change, most of the efforts in recent years have been focused on understanding the potential impacts of climate change.

In those terms, in the three countries climate, policy integration, i.e. the mainstreaming of climate into a variety of areas, is non-existent and the formulation of programmes oriented to support the most vulnerable local agricultural producers is still a challenge. Significant limitations also exist in terms of the coordination of the agenda and activities of government agencies, data collection and dissemination, regional resources, and policy flexibility to adopt a reliable and planned support for local adaptive capacities.

The existing adaptive capacity in the three countries has been developed in the context of the present and past range of climate variability. Climate change, however, brings a new, expanded range of climate variability that the existing adaptive capacity cannot face.

We do not have a clear sense of what the future climate change scenarios for the Choquecota Basin could be, other than the very general information available in the chapter on Latin America in the 2007 IPCC Report (Magrin et al. 2007). Water availability, as a reduction of the glaciers, seems to be an issue that could potentially affect agricultural producers. Forecasts for the other two basins are already available and they provide similar trends: rise in temperature, rising of the zero-degree isotherm with the consequent reduction of snow accumulation, reduction of precipitation (although we are less certain of this for the Chilean case), increasing evapotranspiration, changes in river flows, and increasing frequency of extreme climate events (Boninsegna and Villalba 2007 as cited in Montaña 2008, pp. 36–37; Fiebig-Wittmaack et al. 2008; Diaz et al. 2009). These changes create significant risks for the dominant agricultural activities in the basins, such as increasing water scarcities and loss in productivity and reduction in the yield and quality of grapes and fruit. Given the unequal distribution of adaptive capacity, some households and producers will be affected more severely by future climate change than others. On the other hand, these scenarios could also create opportunities, such as new crops and the possibility of having earlier harvests of vegetable crops and fruit. Maximizing opportunities and reducing risks, however, will require enhanced adaptation in terms of the awareness of various agro-climatic trends, for example, and of their implications for current and future possible agricultural products. This, of course, needs to be complemented with an increase in the adaptive capacity of governance institutions in the basins in order to reduce the vulnerability of local producers. This involves not only an effort to develop a proper climate integration approach to the development and implementation of policies, but also to develop measures oriented to reducing the unequal present conditions in terms of the distribution of adaptive capacity.

Conclusion

The case studies discussed above provide an interesting opportunity to discuss the institutional conditions that influence adaptive capacities to present and future climate. The distinctions between Bolivia and the other two countries show the relevance of having a well-developed formal institutional system able to reach all corners of the country. There is no doubt that rural people in Choquecota have learned how to deal with a multitude of stressors that impact their lives. Social capital, cultural-based networks, use of terraces, different and combined use of family labour, and other mechanisms are expressions of an existing coping capacity. These current strategies are, however, largely reactive and they only lead to a partial alleviation of the communities' problems. Local adaptive strategies without the support of a larger institutional system have limitations, especially in terms of developing a robust adaptive capacity able to deal with the new range of climate variability presented by future climate change scenarios.

Argentina and Chile have, on the other hand, well-developed, stable institutional systems with the capacity to provide adequate support to agricultural producers. The existence of a robust irrigation infrastructure in both basins is a good example of the ability of the formal institutional system to reduce the risks of water scarcities. However, the existence of a strong institutional system is no guarantee of an adaptive institutional capacity. The cases of Argentina and Chile show institutional deficiencies in managerial abilities and a structural weakness to provide resources in a fair and equitable manner to all producers, making some of them more vulnerable than others to climate and non-climate stressors.

There is clearly the need to develop a better understanding of what institutional efficiency is in relation to climate change. Given the multidimensional nature of rural vulnerability, where climate interacts with many other stressors, an inclusive institutional approach that emphasizes the links between adaptation and development priorities is needed. The interrelations between climatic, economic, and social vulnerabilities require an approach able to strengthen the general sustainability of the rural population. It is impossible to foster agricultural resiliency with ad hoc approaches; rather, we need to plan and act across traditional sectors, issues and political boundaries, and integrate ecosystem-management, disaster reduction, and social and economic development measures.

References

Adger N (2003) Social aspects of adaptive capacity. In: Smith J, Klein R, Huq S (eds) Climate change, adaptive capacity and development. Imperial College Press, London

Adger N, Kelly PM (1999) Social vulnerability to climate change and the architecture of entitlements. Mitigation and adaptation strategies for global climate change 4:253–266

Bodini H, Araya F (1999) Visión geográfica global. In: La Región de Coquimbo. Espacios y recursos para un desarrollo sustentable, Seminarios-Talleres, Agosto, 1998. Centro de Estudios Regionales, Universidad de La Serena, La Serena

Boninsegna J, Villalba R (2007) La oferta hídrica en los oasis de Mendoza y San Juan. Los escenarios de cambio climático y el impacto en los caudales. Jornadas de Investigación en Recursos Hídricos, CELA-ICA. Mendoza, 27 de septiembre de 2007

Cepeda J, Fiebig M, Morales H, Salas S (2004) Description of the Elqui River Basin, IACC project working paper, 2. www.parc.ca/mcri/pdfs/papers/iacc002.pdf

Cepeda J, Zuleta C, López-Cortes F (2009) Síntesis Los Sistemas Naturales de la Cuenca del Río Elqui en el Contexto del Cambio Climático. In: Cepeda J (ed) Los sistemas naturales de la cuenca del Río Elqui (Región de Coquimbo, Chile): Vulnerabilidad y cambio del clima. Ediciones Universidad de La Serena, La Serena, pp 327–369

Dale A, Onyx J (eds) (2005) Social capital and sustainable community development: a dynamic balance. UBC Press, Vancouver

Diaz H, Hadarits M, Barrett-Deibert P (eds) (2009) IACC final report December 2009. Comparative study of Dryland River Basins in Canada and Chile, Regina, CPRC Press. www.parc.ca/mcri/pdfs/papers/IACC_Final_Report_e.pdf

Fiebig-Wittmaack M, Perez-Valdivia C, Lazo E (2008) Aspectos Climáticos del Valle del Elqui. In: Cepeda-Pizarro J (ed) Los sistemas naturales de la cuenca del Río Elqui (Región de Coquimbo, Chile): Vulnerabilidad y cambio del clima. Ediciones Universidad de La Serena, La Serena, pp 33–55

Halpern D (2005) Social capital. Polity Press, Cambridge

Handmer J, Dovers S, Downing T (1999) Societal vulnerability to climate change and variability. Mitigation and adaptation strategies for global change 4:267–281

Henningham J (ed) (1995) Institutions in modern societies. Oxford University Press, Melbourne

Hurlbert M, Corkal D, Diaz H (2009a) Government and civil society: adaptive water management in the South Saskacthewan River Basin. In: Marchildon G (ed) A dry Oasis. Institutional adaptation to climate on the Canadian plains. CPRC Press, Regina

Hurlbert M, Diaz H, Corkal D, Warren J (2009b) Climate change and water governance in Saskatchewan, Canada. International Journal of Climate Change Strategies and Management 1(2):118–132

IPCC (Intergovernmental Panel on Climate Change) (2001) Climate change 2001: impacts, adaptation, and vulnerability technical summary. A report of working group II of the intergovernmental panel on climate change, Cambridge University Press, Cambridge

IPCC (2007) Climate change 2007: impacts, adaptation, and vulnerability. Contribution of working group II to the fourth assessment report of the intergovernmental panel on climate change. Cambridge University Press, Cambridge

Jenkins R (2002) The emergence of the governance agenda: sovereignty, Neo-liberal Bias, and the politics of international development. In: Desai V, Potter R (eds) The companion to development studies. Arnold, London

Liverman D (1994) Vulnerability to global environmental change. In: Cutter S (ed) Environmental risks and hazards. Prentice Hall, Englewood Cliffs

Magrin G et al (2007) Latin America. In: Climate change 2007: impacts, adaptation, and vulnerability. Contribution of working group II to the fourth assessment report of the intergovernmental panel on climate change. Cambridge University Press, Cambridge

McDowell J, Zeballos G (2008) "Caso Bolivia: Cuenca de Palca" project document. www.climaycomunidades.org

Montaña E (2008) Central andean foothill farmers facing global environmental change. IHDP Update 2:36–40

Montaña E et al (2005) Los Espacios Invisibles Subordinación, Marginalidad y Exclusión de los Territorios no irrigados en las Tierras Secas de Mendoza, Argentina. Región y Sociedad 32:3–32

Smit B, Wandel J (2006) Adaptation adaptive capacity and vulnerability. Global environmental change 16:282–292

WGCCD (Working Group on Climate Change and Development) (2006) Up in smoke? Latin America and the Caribbean. New Economic Foundation, London. www.neweconomics.org/publications/smoke-latin-america-and-caribbean

Wisner B et al (2005) At risk. Natural hazards, people's vulnerability and disasters. Routledge, New York

Chapter 14
Water Management Issues in Southern Tunisia Under a Climate Change Context

Nizar Omrani and Dieter Burger

Abstract In the southern part of Tunisia, oases sustain an extreme arid climate. Their tributary to the underground water resources definitely weakens their viability. The continuous decrease in the aquifer level, coupled with the salt contamination risks coming from drainage water, make water management more complex in this particular water shortage context. The recurrent creation of newer private parcels has required more water allocation and the irrigation system capacity became unable to satisfy the water demand. This situation is exacerbated by the climate change impacts that are becoming increasingly evident. These regions experience more consecutive droughts while they have less access to reliable water resources. In order to face these challenges, water resources management should be improved; more efficient water management is required. Moreover, the farmers' current practices have to include more water saving to meet the full capacity of the irrigation systems. This paper focuses on the preparedness of the irrigated sector within southern Tunisia to cope with the impacts of climate change. Our approach emphasizes the technical constraints to the feasibility of such new practices, as well as the socio-economic conditions prevailing within the southern Tunisia oases.

Keywords Tunisia · Climate change · Water · Management · Efficiency · Oases · Farmers' practices

N. Omrani (✉) · D. Burger
Institute of Arid Regions (IRA), Route El Jorf,
Km 22,5, 4119 Medenine, Tunisia
e-mail: Nizar.Omrani@ira.rnrt.tn

Introduction

Located at the southern border of the Mediterranean Sea, Tunisia is typified by a sweet-tempered climate; the mean rainfall is estimated to be 207 mm/year, with the minimum equalling a third of this value and the maximum trebling it (Fig. 1).

Beyond the variability in time, important differences occur in the spatial distribution. The mean annual rainfall varies from 594 mm in the north, 296 mm in the centre, to 156 mm in the south and less than 100 mm in the extreme south. Nearly 80% of the precipitation is concentrated between October and March. The annual evaporation ratio varies from 1.200 mm in the north to 1,800 mm in the south (Aquastat 2005).

In southern Tunisia, the oases cover an area of 23,000 ha between the provinces of Gabes, Gafsa, Kebili and Tozeur. These perimeters are intensively cultivated and particularly known for their important biodiversity. There are three cultivations layers that characterize agriculture at the oases from the ordinary irrigated perimeters. De Haas (2002) defined the oasis as "agricultural sites in arid environment where agriculture is normally impossible without irrigation".

Under an arid climate (less than 150 mm/year), the southern Tunisia oases still represent the main pillar of economic development in this region. The main agricultural production is date palm, which has a high added value and provides the country with a promising exportation sector.

Nevertheless, the main water resources available are underground and are hardly renewable water. National forecasts are expecting increasing pressure on these water resources. As a concrete consequence of climate change impacts, the drought periods are expected to be longer and prolonged over one or more years.

Cross-border Water Management Issues

In the context of water shortage, expected to be more acute with the impact of climate change, the main water resources exploited in the southern country remain the deep aquifers.

The only potential water resources available are those provided by the SASS (North-Western Sahara Aquifer System). It deals with a sedimentary basin holding huge underground water volumes and extended over an area of 1 million km^2 across three countries: Algeria (700,000 km^2), Libya (250,000 km^2) and Tunisia (80,000 km^2). During previous decades, water mobilization from this basin has considerably enhanced from 0.6 in 1970 to the current level of 2.5 million km^3.

Nevertheless, regarding climate aridity, the recharge of the aquifers in this basin remains low, less than 1 million km^3 across the entire basin area. These resources are currently used from about 8,800 drillings. They are distributed respectively as 6,500 in Algeria, 1,200 in Libya and 1,100 in southern Tunisia.

Fig. 1 Annual rainfall in Tunisia (adapted from Ben Mechlia 2006)

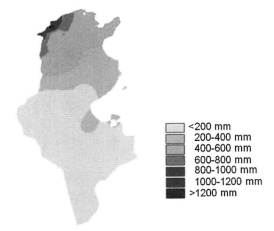

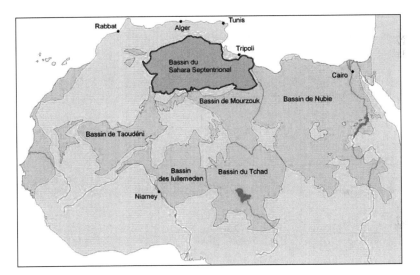

Fig. 2 Localization of the North-Western Sahara Aquifer System in northern Africa (adapted from Schmidt et al. 2006)

There are mainly two deep aquifers that are built by the SASS (Fig. 2): the Complex Terminal (CT) aquifer, with a depth ranging between 100 and 500 m, and the Continental Intercalary (CI), with a depth that could reach 2,800 m. The CI aquifer supplies nearly 80% of the irrigation and drinking water for these regions (Kamel et al. 2006); the mean salinity between the two aquifers varies between 2.5 and 5 g/l (Prinz and Loeper 2008).

Table 1 Water amount value targets of southern Tunisia (adapted from MAREH/EAU XXI, MAREH 1998; SAPI 2005)

Year	1996	2010	2020	2030
Irrigated surface (ha)	46,000	49,000	50,490	52,035
Irrigation water consumption (m^3/ha/year)	11,000	9,500	8,167	7,022
Irrigation water demand (million m^3/year)	506	466	412	365

Southern Tunisia Water Policy Milestones

At the national scale, it seemed evident that national policy should focus more on water demand management as well as efficiency enhancements for both the drinking water sector and the irrigation sector (improving distribution efficiency in the irrigation network from 66% to 80%). In the absence of such tendencies, the objective of 400,000 ha of total irrigated surface could not be met by 2010.

In the case of the southern Tunisia oases, the irrigated area is expected to increase from 46,000 ha in 1994 to 52,035 ha after 2030. For this purpose, the main orientation of the long-term national water strategy (EAU XXI) is the decrease in the irrigation water demand. Indeed, this approach focuses on efficiency improvement and targets to reach the average of 365 million m^3 in 2030, having been more than 506 million m^3 in 1996 (SAPI 2005), (Table 1).

The main challenges to meet for irrigated agriculture in general and particularly the oases ecosystems remain climate change. The impact of such a phenomenon is expected to be more severe on the water resources.

Indeed, national prospective studies attribute a decrease of nearly 28% in underground resources until 2030 (Fig. 3). Production in drought periods will observe a decrease of 50%, which equals 800,000 ha for rain-fed agriculture. These impacts will be seen in livestock, which will decrease by 80%, in the centre and southern parts of country (OSS 2009).

Facing these risks, Tunisia has elaborated a national adaptation strategy to decrease these impacts; the evolution of the mean climate indicators will be taken into account for the future natural resources management plan. The alert systems for both floods and drought events are already established, with a network of climate and hydrological stations across the country.

Water resources protection remains the focal point of such a strategy; the enhancement of efficiency was supported by the institution by several measures. In terms of drinking water, the hydraulic network was subject to an integral assessment that emphasized the commitment of the population to water saving practices. In the irrigated sector, the improvement of intensification within farmers' parcels is still to be improved in order to save the water resources. Cultivation of high-added-value crops should be intensified. Moreover, the contribution of the irrigated sector to total agricultural production needs to be enhance and attempt nearly 50% in the long term.

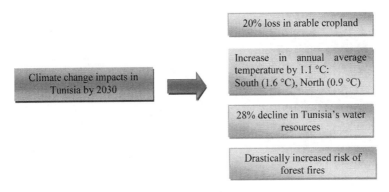

Fig. 3 Main expected climate change impacts on agriculture by 2030 (GTZ 2007)

For further economical efficiency, the government undertook a progressive disengagement in the management of water resources and required more involvement from the farmers' groups to protect the hydraulic infrastructures as public facilities. The growing participation of private investors, e.g. in the geothermic sector, helps to promote water value and minimizes the risks of water wastage.

The national programme of water saving was reinforced by the allocation of financial grants to introduce water-saving equipment within farmer parcels. About 40, 50 and 60% have been attributed to important, middle and small exploitations respectively (Hamdane 2004).

It is evident that all these major efforts passed by the government should be sooner relieved by the farmers' commitment. The current behaviour still shows major water consumption inside parcels, with the applied water amount being more than the effective crop requirement. Traditional irrigation methods are still widely used within farmers parcels. The absence of any field levelling and the over-application of water during irrigation, up to treble the amount of the real crop requirements, causes water losses (Mechergui and Van Vuren 1998).

The first prospective studies conducted by OSS experts expect a considerable decrease in artesianism in the extreme south of the country (Mamou 2009). In the Nefzaoua region oases, Zammouri et al. (2007) simulated three scenarios of pumping strategies from the CT aquifers. The main results emphasized a common impact being water quality deterioration across the whole Nefzaoua region.

Technical Constraints

There are several constraints prevailing within the oasis systems. There are relevant water losses that still occur along the distribution system, starting from the pumping station until the parcel entrance. These losses especially affect connections between *surface canal* transect connections which don't present good

Fig. 4 Conversion of irrigation techniques into more efficient systems

resistance to high temperature and often break out, inducing huge water losses. Due to such dysfunctions, the water flow becomes lower and contributes to higher water temperatures throughout the distribution network. Moreover, having such water losses through the water distribution causes a relevant and recurrent delay in the irrigation duration and could create a context of tension between the farmers' groups that share the resource.

Furthermore, the main water distribution to the parcels remains the surface canal, the conversion into water-saving techniques such as sprinklers or drip irrigation require the adaptation of such techniques in terms of water discharge, and the suitable water flow that could be delivered through such techniques (Fig. 4).

Regarding the irrigation water quality, which varies between 2 and 5 g/l, the use of drip irrigation is permanently exposed to warping risks and needs to be closely monitored before being widely implemented within parcels.

Irrigation with such water quality is still highly risky as soil salinization became evident (Fig. 5). On the other hand, the sandy soil with low retention capacity that characterizes these perimeters causes relevant water infiltration through the drainage network.

The problem is more acute in the oases supplied by artesian wells on the CI. In the absence of suitable proportions, the water mixture is between CT and CI water. Indeed, when the geothermic water coolers facilities, called "air cross systems", stop during electricity cuts, only high-temperature (60–70°C) artesian CI water runs within the network, which obliges the farmers to let them directly run off to drainage canals to avoid the crops burning.

In addition to the irrigation feasibility constraints, water management within the southern Tunisia still has a real lack of cultivation system water requirements data. The amount of water currently applied does not take into account the accurate water needs; water wastage is high, so the three cultivations layers that typify the oases remain absent. Such behaviour doesn't militate water resource valorization; the effective cost of the water mobilization is also partially recovered.

Fig. 5 Soil salinization occurs in absence of adequate farming practices

As crucial consequences of the depletion of water resources, the quality deterioration threatens the availability of drinking water.

The implementation of the desalination stations in the main regions (Gabes, Djerba, Zarzis) is expected to be generalized throughout the whole of southern Tunisia. With urbanization and the expansion of the tourism and industrial sector, more water conflict situations will then be addressed (Kassah 2009).

It seems evident that such a context aggravates more frequent and longer drought periods, as well as a rise in the mean temperature, threatening the survival of these ecosystems. The implementation of viable options for adaptation to an uncertain future is crucial.

The Illegal Oasis Extension

The oasis extension is one of the most important current challenges to face within the coming decades. The multiplication of private parcels always occurs at oasis peripheries. Their water consumption is very high, nearly double that of the public irrigated area. This induces important water wastage and significant drainage water amounts (Fig. 6). The water surplus provides the shallow water table that rises up and leads to the phenomenon of waterlogging (Prinz et al. 2005).

As concrete impacts of these extensions, the first prospective studies led by OSS experts also expect a considerable decrease in the water table charge in the extreme south of the country (Mamou 2009).

To limit the multiplication of private parcels, the development sector is called to restore attractive water delivery conditions (in terms of water flow and available

Fig. 6 Drainage water collected downstream in the southern Tunisia oasis

Fig. 7 Illegal well in private parcel at downstream oasis in southern Tunisia

discharge). The several subsidies that the development authorities allow for water-saving equipment within farmers' parcels seems to be a conclusive approach.

In order to tackle this illegal extension, radical changes in the main stakeholders' behaviour dealing with water management in this part of the country should be reconsidered. This context addresses the need for strengthening the role of education, knowledge and capacity development in the sustainable water management of water resources. The importance of awareness raising for all the farmers, as well the southern country population is crucial.

The farmers' groups are exhorted to halt the implementation of private wells (Fig. 7) and their role as a water resource oasis guarantor is expected to be greater.

Furthermore, the legislative regulation should be stricter in order to eradicate the illegal oasis extension. There is an obvious need for redefinition of the institutional framework for the creation of penal and financial penalties as a concrete dissuasion tool.

Conclusion

With regard to its limited water resources (quota of 480 m^3/capita/year), Tunisia has begun to implement a national water resources mobilization strategy. The prospective plan focused on water saving by essentially managing the demand for the main sectors and especially agriculture, which consumes more than 83% of the total water resources. Such a policy aims to gain control of more than 95% of the available water resources by 2016.

Nevertheless, similarly to the Maghreb countries region, Tunisia is also strongly affected by the impacts of climate change. Since 2005, within the framework of the Tunisian–German cooperation, an assessment analysis managed by the GTZ aimed to identify the impacts of climate change in Tunisia with a focus on agriculture and water resources development vulnerability.

This two-year process provided the country with comprehensive references and national forecasts that have led to the development of the national adaptation strategy to climate change.

Particularly in the arid southern Tunisia region, where agriculture still plays a key role in development, the main incomes are substantially provided by oasis production.

These ecosystems had also been identified as strongly vulnerable to the future impacts of climate change. Their tributary to the underground water resources, which are sustaining a chronic depletion, is compromising their sustainability for the coming decades.

Under an acute water shortage context, these perimeters are expected to experience more difficulties in managing less reliable underground water resources. The main issue that determines the survival of irrigated agriculture in such conditions will definitely be the promotion of optimal water management solutions. There is a crucial need for such measures to be implemented to meet the coming challenges of climate change impacts on irrigation water reliability.

In those regions of southern Tunisia where irrigated areas are sustaining permanent desertification risks, a deeper understanding and a better assessment of the available water resources management is required. The collection of accurate data regarding this aspect will provide the water policy with future options and facilitate effective decision-making in order to meet various societal needs and overcome the risks of water resources degradation.

As the pressure on water demands is expected to become more severe with climate change, the uncertainty in food security, as well as the water resources availability, has become the key issue for the national water policy.

The research is already engaged in the study of irrigation sector efficiency enhancement. The study case of the adaptation of drip irrigation and the sprinkler method to palm irrigation is being carried out in arid regions. The institution of reliable references on water consumption is also being undertaken.

Recourse to non-conventional water resources should be strengthened. Irrigation with drainage water of the fodder crops as well as the forests trees which has been implemented around the oases allowed the creation of a green cordon around them. Such a technique still assumes a significant role in the protection of the irrigated lands from permanent desertification risks. It also contributes to decreasing the pressure on water resources for irrigation and drinking purposes.

It would also be an innovative option to valorize the brackish water for fish farming production, which could be attractive for investors, as the drainage water is being made available in significant amounts and free of charge.

In the rapidly changing and ever more complex context of water resources management, these numerous good practices will need to be achieved. Indeed, resistance to climate change impacts will require pressing attention and more efficient irrigation practices, the associative mode of the water management and the introduction of higher-added-value cultivation in the irrigated sector.

Such an approach aims to provide the water policy with adequate options and facilitate effective decision-making to enhance local capacity and preparedness to cope with climate change issues.

References

Aquastat (2005) http://www.fao.org/nr/water/aquastat/countries/tunisia/indexfra.stm

Ben Mechlia N (2006) Water conservation measures in Tunisia, WASAMED project meeting, Malta, 3–6 May 2006

De Haas H (2002) Agricultural transformations in the Maghreb and the role of multidisciplinary research, In: Kroll A (ed) Water, land, agriculture and policies in the mediterranean, notes, workshop. Brussels, institute for prospective technological studies, JRC and European Commission, 19–20 April 2001

GTZ (2007) Factsheet on the development of a strategy for adaptation to climate change in the Tunisian agricultural sector

Hamdane A (2004) La modernisation des systèmes irrigués en Tunisie. Note technique du ministère de l'agriculture et des ressources hydrauliques, p 18

Kamel S, Dassi L, Zouari K (2006) Hydrogeological and hydrochemical approach of hydrodynamic exchanges between deep and shallow aquifers in the Djerid basin (Tunisia). Hydrol Sci J 51:713–730

Kassah A (2009) Oasis et aménagement en zones arides: Enjeux, défis et stratégies. Actes de L'atelier SIRMA « Gestion des ressources naturelles et développement durable des systèmes oasiens du Nefzaoua », 25–27 February 2009, Douz, Tunisia, CIRAD, Montpellier, France, p 6

Mamou H (2009) Ressources en eau et développement agricole dans le sud tunisien Workshop projet SIRMA Gestion des ressources naturelles et développement durable des systèmes oasiens de Nefzaoua. Douz, Tunisia 25–27 February

MAREH (1998) EAU XXI. Stratégie du secteur de l'eau en Tunisie à long terme 2030, p 97

Mechergui M, van Vuren G (1998) Improved irrigation efficiencies in Tunisian oases. ILEIA, Newsletter July

OSS (2009) The system aquifer of septentrional Sahara, the commun management of a transborder basin, Drafts serie 1:56

Prinz D, Loeper T (2008) Nutzung fossilen Grundwassers in der tunesischen Oasenwirtschaft. Effizienz–Notwendigkeit–Sinnhaftigkeit. BGR, Hannover 21 February

Prinz D, Chaban B, Kastl A (2005) New approaches in Oasis water management—Experiences from North Africa. Proceedings, XII world water congress of IWRA—Water for sustainable development—Towards innovative solutions, New Delhi, 22–25 November 2005

SAPI (2005) Assistance spéciale pour l'exécution du projet, Projet d'amélioration des périmètres irrigués des oasis du Sud de la Tunisie (TS-P10), Final report

Schmidt G, Manfred H, Soefner B (2006) Investigations on regional groundwater systems in North-East Africa and West-Asia

Zammouri M, Siegfried T, El Fahem T, Kriaa S, Kinzelbach W (2007) Salinization of groundwater in the Nefzaoua Region Tunisia: results of a regional-scale hydrogeologic approach. Hydrogeol J 15:1357–1375

Chapter 15
Climate Change and its Impacts on the Livelihoods of the Vulnerable People in the Southwestern Coastal Zone in Bangladesh

Md. Afjal Hossain, Md. Imran Reza, Sania Rahman and Imrul Kayes

Abstract Bangladesh is globally considered one of the most vulnerable and exposed countries to climate change (Climate change and Bangladesh Department of Environment, Government of People' Republic of Bangladesh. Climate Change Cell, Dhaka, 2007). There is evidence of prominent increases in the intensity or frequency of many extreme events such as flood, land erosion, heat waves, tropical cyclones, intense rainfall, tornadoes, drought, storm surges, salinity intrusion, etc. which cause loss of livestock, damage to pasturelands, increase fodder scarcity, destroyed shelters, decreased production, increased management costs to incidence of diseases, etc. in Bangladesh. This paper therefore intends to do three things: (1) it shall identify the extreme climatic hazards, vulnerabilities and risks; (2) it shall find out the impacts of climatic hazards on the livelihood of the vulnerable people; and (3) it shall propose some possible strategies for reducing the vulnerability to the climatic hazards. The present paper is intended as a concept paper to deal with the impact level assessment on livelihoods due to climate change. The method has followed both qualitative and quantitative approaches in the southwestern coastal zone in Bangladesh and used secondary data and information. The livelihood and income of a large population depends on the natural resource base and most of the poor people often live in marginalized lands and areas more prone to natural disasters. Climate change means that many natural disaster-prone areas will become more prone due to increased frequency and intensity of disasters. Drought-prone areas will become hotter and drier, with less predictable rainfall; flood frequency and intensity along onset and recession will be changed in future; the nature of cyclone and storm surges will be different from the historical trend. All of these together will change crop yields and affect many poor people's livelihoods. Agriculture yields have been decreased and cropping pattern has been changed in recent years. Adverse

Md. Afjal Hossain (✉) · Md. Imran Reza · S. Rahman · I. Kayes
382/1 North Kazipara, Mirpur, Kafrul, Dhaka, 1216, Bangladesh
e-mail: robinpln@gmail.com

impacts of climate change are likely to reduce availability and deteriorate quality of water for domestic use. Moreover, climate change is likely to increase the prevalence and infection of vector- and water-borne diseases such as malaria and dengue fever, cholera and dysentery, etc. Degradation of biodiversity will reduce the availability of many traditional medicines which may affect poor and rural people who depend more on natural resources for medicine as well as income and food. Sea level rise (SLR) will drastically affect the poor people who are in coastal area and flood plain zone in Bangladesh. However, many actions undertaken to address the baseline or contextual risks in Bangladesh are also synergistic with the so-called adaptations that might be required as climate change impacts manifest themselves.

Keywords Climate change · Livelihood · Vulnerability

Introduction

Bangladesh is globally considered one of the most vulnerable and exposed countries to climate change (Climate Change Cell 2007). Bangladesh is geographically exposed to a multitude of adverse impacts of climate change, because of its location in the tropics, in the delta of three of the world's biggest rivers and its flat low-lying deltaic topography. The country has low adaptive capacity due to its extreme poverty (World Bank 2000). The Stern Review and IPCC 4th Assessment Report both state that climate change will have adverse impacts on people's health, safety and livelihoods, with the "poorest people in the poorest countries expected to suffer first and foremost". Predicted climate change will create barriers to future poverty reduction and reverse many of the important socioeconomic gains made by developing countries.

Sea level rise will affect the vast coastal area and low-lying river estuary zones of Bangladesh. The livelihoods of coastal communities and the natural environment of the coastal zones will be submerged by the anticipated sea level rise. The Mangrove forest locally called Sundarbans is the most important ecosystem of the country on which 10 million people's subsistence depends. People will be forced to move out from their homes, will have to take refuge on dry lands and change their way of earning (Sarwar 2008); this will result in a total change of their way of life.

There is evidence of prominent increases in the intensity or frequency of many extreme events such as flood, land erosion, heat waves, tropical cyclones, intense rainfall, tornadoes, drought, storm surges, salinity intrusion, etc. in Bangladesh. Such impacts pose additional risks for already vulnerable communities striving to combat poverty and achieve sustainable development. Therefore, climate change poses a serious and additional threat to the region's poor farmers and rural communities who live in remote, marginal areas. Impacts of such disasters range from hunger and susceptibility to disease, to loss of income and human livelihoods. Climate change is in fact emerging as the pre-eminent development issue in Bangladesh.

The present paper is intended as a concept paper to deal with the impact level assessment on livelihoods due to climate change, with both qualitative and quantitative approaches in the southwestern coastal zone in Bangladesh and the use of secondary data and information. This paper therefore shall do three things: (1) it shall identify the extreme climatic hazards, vulnerabilities and risks; (2) it shall find out the impacts of climatic hazards on the livelihood of the vulnerable people; and (3) it shall propose some possible strategies for reducing vulnerability to the climatic hazards.

Key Hazards, Vulnerabilities and Risks Due to Climate Change

Bangladesh is one of the poorest countries in the world and is a member of the least developed countries group. It has a population of 139.2 million (UNDP 2006), making it one of the most densely populated countries, and the present population growth rate is 1.7% in 2004. Though growth rate has decreased significantly, the high base population makes the need for development efforts that much more challenging. Climate change comes as an additional and huge burden on an already stressed economy and ecosystem. Bangladesh's geological and spatial location makes its highly populous and extended coastal lands and islands extremely flat, dynamic and vulnerable.

Current Climate

Bangladesh has a humid, warm, tropical climate. Its climate is influenced primarily by monsoon and partly by pre-monsoon and post-monsoon circulations. The southwest monsoon originates over the Indian Ocean and carries warm, moist and unstable air. Besides monsoon, the easterly trade winds are also active, providing warm and relatively drier circulation.

In Bangladesh there are four prominent seasons, namely, winter (December to February), Pre-monsoon (March to May), Monsoon (June to early October), Post-monsoon (late October to November). The general characteristics of the seasons are as follows:

- Winter is relatively cooler and drier, with the average temperature ranging from a minimum of 7.2–12.8°C to a maximum of 23.9–31.1°C. The minimum occasionally falls below 50°C in the north, although frost is extremely rare.
- Pre-monsoon is hot with an average maximum of 36.7°C, predominantly in the west for up to 10 days, very high rate of evaporation, and erratic but occasional heavy rainfall from March to June. In some places the temperature occasionally rises up to 40.6°C or more.
- Monsoon is both hot and humid, brings heavy torrential rainfall throughout the season. About four-fifths of the mean annual rainfall occurs during monsoon.

The mean monsoon temperatures are higher in the western districts compared to those for the eastern districts.
- Post-monsoon is a short-living season characterized by withdrawal of rainfall and gradual lowering of night-time minimum temperature.

Sea Level Rise

The SAARC Meteorological Research Council (SMRC) carried out a study on the recent relative sea level rise on the Bangladesh coast. The study used 22 years of historical tidal data from the three coastal stations. It is predicted that around 10% of the Bangladesh area will be inundated by 45 cm and over 21% of the country areas confined in the coast Bangladesh will be inundated by 1 m sea level rise (IPCC 2001). This will reduce agricultural productivity, alter livelihoods and risk food security and climate-induced marine migration. Put simply, 30 years of development investment will be wiped out by a 30 cm sea level rise.

Saltwater Intrusion

Changes in surface water salinity due to a 30 cm and 1 m rise in sea level were investigated by considering the topography and contours of the area and present streamflow patterns. Changes in the surface water salinity pattern due to a 30 cm rise in sea level revealed that the present dry season saline front (2 dS/m) is expected to move 30 km to 50 km north, affecting most of Khulna, Jessore, Barisal, Patuakhali and Noakhali (greater) districts and parts of Faridpur and Comilla districts. With a 1 m rise in sea level, the saline water front will move far north on the northeastern side of Bangladesh. Most of Jessore, Faridpur, Comilla and part of Dhaka (greater) districts will be affected by saline surface water intrusion. Most of Barisal, Patuakhali, Sundarbans, Bhola, Hatia and Sandwip will be directly inundated by saline/brackish water or will have a serious saline waterlogging problem. The drinking water supply of major cities such as Dhaka, Chittagong and Khulna will be affected by salinity. On the western side of Bangladesh, the saline water front will move close to Kushtia and Pabna districts. With a 1 m rise in sea level, it may be expected that cyclonic surges will penetrate further north into the country.

Increased Intensity of Extreme Events

The geographical setting of Bangladesh makes the country vulnerable to natural disasters. Every year, one or more natural calamities upset people's lives in some parts of the country. The major natural hazards include flood, cyclone and storm

Table 1 Historical record of cyclones formed in the Bay of Bengal

Year	Maximum wind (km/h)	Surge height (m)	Casualties	Year	Maximum wind (km/h)	Surge height (m)	Casualties
Nov 1904	62–88			Nov 1970	224	4.55	500,000
Oct 1905	62–88			Dec 1973	111	5.1	1,000
Sep 1919	120		3,500	Dec 1981	167	4.55	72
May 1923	89–117		6	Nov 1983	135	1.5	300
May 1926	89–117		2,700	May 1985	154	4.55	10,000
May 1941	89–117	3.03–3.64	5,000	Nov 1986	110	0.61	60
Oct 1947	89–117		500	Nov 1988	161	4.4	5,683
Oct 1960	129	6	9,450	Apr 1991	225	7.6	138,882
Oct 1960	193	6.6	5,149	Jun 1991	110	2.5	300
May 1961	161	5	11,468	Nov 1992	50		
May 1961	161	6.5		Apr 1994	210	4.85	184
Oct 1962	93			Nov 1995	210		650
May 1963	193	6	22,000	May 1997	230	4.55	155
Oct 1963	81			Sep 1997	150	3.05	67
May 1965	161	3.7	19,279	May 1998	150	2.44	
Dec 1965	184	3.6	3,000	Nov 1998	90	2.44	
Oct 1966	139	6.67	850	Nov 2007	220	4.5	3,000

Source: SMRC 2003 and Wikipedia

surge, flash flood, drought, tornado, earthquake, riverbank erosion and landslide. Floods and cyclonic storm surges are major killers, as well as being a cause of most direct and indirect impediment to economic development. Agriculture probably suffered more than non-agricultural sectors.

Increasing Cyclones and Storm Surges

One of the predictions of climate change is the intensification of extreme weather events such as cyclones and associated storm surges. The Bangladesh coast is vulnerable to recurrent cyclones. The increase in the intensity of wind velocity is expected to incur greater losses to vulnerable communities and ecosystems. The Bay of Bengal is an ideal breeding ground for tropical cyclones and depressions. The funnel-shaped configuration of the coastline of Bangladesh produces the catastrophic ravages of cyclones and storm surges. During pre-monsoon and post-monsoon periods, disastrous tropical cyclones form in the Bay of Bengal. In Bangladesh alone, about 40% of the total number of global storm surges are recorded. Over the last 30 years, different scales of cyclones have been affecting the country with loss of valuable lives and property. Table 1 shows that cyclones usually struck this country in the months of April, May and October, November and December (Fig. 1).

In the 1991 "super cyclone" (which was a storm of exceptional intensity with wind velocities up to 225 km/h), a large number of deaths occurred mainly in three

Fig. 1 Cyclone pathways in Bangladesh

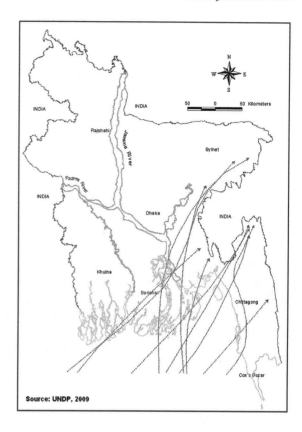

coastal districts in Bangladesh: Chittagong (79,697 dead and 2,600 injured) Cox's Bazar (51,147 dead and 133,000 injured) and Noakhali (8,878 dead and 995 injured). The economic losses alone from the cyclone were estimated at US$2.4 billion. The loss of life was, however, substantially less than the 1.2 million people killed by the 1970 cyclone. The loss of human life further declined due to large investment and awareness-raising activities after the 1991 cyclone.

Climate Change Impacts on Coastal Flooding

The low-lying costal zone in Bangladesh is located between the extensive drainage network of the Ganges–Brahmaputra–Meghna river system on one side, and tidal and cyclonic activity from the Bay of Bengal on the other. The coastal embankments paradoxically also tend to block efficient drainage of freshwater on the other (land) side at times of excess rainfall and riparian flooding.

The increased snowmelt from the Himalayan permafrost, due to increases in temperature, will force more water to flow through the Ganges, Meghna, Brahmaputra river systems and their river networks. This will create additional

Fig. 2 Flood prone area in Bangladesh

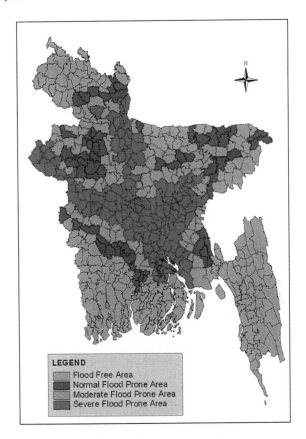

flooding extending over the central floodplain of Bangladesh. Furthermore, the additional flow will bring with it sediments which will make the shallow riverbed even shallower. This will result in a lower capacity of the riverbed to flow out water rapidly, thus increasing the probability of enhanced flooding and erosion of riverbanks (Fig. 2).

Flood is a regular natural disaster occurring in Bangladesh, entailing huge damage to the economy. Four main types of natural floods occur in Bangladesh: (Table 2)

The devastating floods of 1987, 1988 and 1998 inundated more than 60% of the country. The 1998 flood alone caused 1,100 deaths, inundated nearly 100,000 km^2, rendered 30 million people homeless, damaged 500,000 homes and caused heavy losses to infrastructure. In 2004, floods inundated 38% of the country (MoEF (Ministry of Environment, Forest) (2005)).

But over the last ten years, the country has been experiencing early, late or prolonged floods. NAPA-Bangladesh has also identified the erratic behaviour of major disasters, including flood. About 1.32 m ha of cropland is highly flood-prone. Besides this, crops, perennial trees and livestock are damaged by flood every year. In two severe flood years of 1974 and 1987, the shortfall in production

Table 2 Different types of flood occurring in Bangladesh

Type of flood	Causes of occurrence	Time/duration	Tentative affected area
Flash flood	Runoff during exceptionally heavy rainfall occurring in neighbouring upland areas	Pre-monsoon months of April and May	The foot of the northern and eastern hills of Bangladesh
Rainwater flood/ Monsoon flood	Heavy rainfall occurring over floodplain and terrace areas within Bangladesh	April–May, June–August	In the southwestern part of the country
River flood	Snowmelt in high Himalayans, heavy monsoon rainfalls over the Himalayans, the Asam Hills, Ganges floodplains	April–May and June–September	Catchments areas of three major rivers
Coastal flood	In case of major cyclones the entire coastal belt is flooded. Coastal areas are also subjected to flooding	Tidal flood occurs from June to September	Southwestern coastal areas

Source: Ahmed 2006

Table 3 Broad adverse impacts of major floods during the last 50 years

Year	Impact
1954 floods	Affected 55% of country
1974 flood	Moderately severe, over 2,000 deaths, affected 58% of country, followed by famine with over 30,000 deaths
1984 flood	Inundated 52,520 km^2, cost estimated at US$378 million
1987 floods	Inundated over 50,000 km^2, estimated damage US$1.0 billion, 2,055 deaths
1988 floods	Inundated 61% of country, estimated damage US$1.2 billion, more than 45 million homeless, between 2,000 and 6,500 deaths
1998 floods	1,100 deaths, inundated nearly 100,000 km^2, rendered 30 million people homeless, damaged 500,000 homes, heavy loss to infrastructure, estimated damage US$2.8 billion
2004 floods	Inundation 38%, damage US$6.6 billion, deaths 700, affected people nearly 3.8 million

Source: MoEF (Ministry of Environment, Forest) (2005)

from the trend was about 0.8 and 1.0 million Mt of rice, respectively. During 1984, flood affected both Aus and Aman rice crops and the shortfall was about 0.4 million Mt. The following table shows major floods and its impacts: (Table 3)

Increased Drought

Drought is primarily an agricultural phenomenon that refers to conditions where plants are responsive to a certain level of moisture stress that affects both the vegetative growth and yield of crops. The farmers' concern on drought is when

Fig. 3 Kharif drought
(T. Aman) in Bangladesh

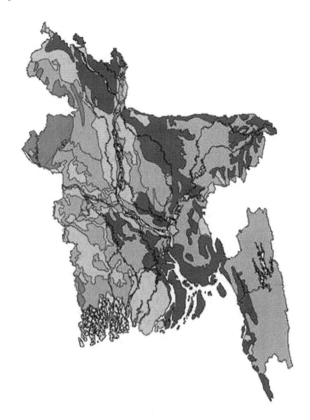

crop production gets hampered due to lack of rain and soil moisture. According to the local people of the drought-affected areas, the intensity and longevity of drought are increasing over time. The western part of the country is receiving less rainfall, averaging some 1400 mm as against the national average of about 2150 mm. As a consequence, susceptibility to and severity of drought in the western districts are much higher than elsewhere. Drought of different intensities in Kharif, Rabi and pre-Kharif seasons cause damage to 2.32 million ha of T. Aman and 1.20 million ha of Rabi crops annually. Yield reductions due to drought vary from 45 to 60% in T. Aman and 50 to 70% in Rabi crops in very severe drought situations. (Fig. 3)

Key Impacts and Vulnerabilities

The geographical location and socio-demographic features of Bangladesh make it one of the most vulnerable countries to climate change, variability and extreme events. Its long coastline, vast low-lying landmass, high population density and nature-dependant traditional agricultural practices would be impacted upon due to climate change. It is likely that vulnerability of the disadvantaged and poor community would be worse than the non-poor and better-off strata of the society. The issue of climate change in relation to vulnerability and adaptation for Bangladesh has been assessed through several studies.

Crop Agriculture and Food Security

Various studies indicate that a temperature rise of 1 to 20°C in combination with lower solar radiation causes sterility in rice spike lets. High temperature was found to reduce yields of HYVs of Aus, Aman and Boro rice in all study locations and in all seasons. The effect was particularly evident at a rise of temperature by 40°C. Climate changes, especially in temperature, humidity and radiation, have great effects on the incidence of insect pests, diseases and microorganisms. A change of 10°C changes the virulence of some races of rust infecting wheat (DAE 2007).

The production of crops in Bangladesh is constrained by too much water during the wet season and too little during the dry season. Presently, the total irrigated area is 4.4 million ha which is more than 50% of the potentially irrigable area of 7.12 million ha cultivated area (DAE 2007). This area is being irrigated through surface and groundwater resources. As a result, the groundwater table in Bangladesh is declining at a rapid rate, causing STWs to become non-operational in many parts of the country during the dry period.

It was noticed that a temperature increase of 40°C would have a severe impact on food-grain production, especially for wheat production. On the other hand, carbon dioxide fertilization would facilitate food-grain production. A rise in temperature would cause a significant decrease in production of 28 and 68% for rice and wheat respectively (DAE 2007). Moreover, doubling of atmospheric concentration of CO_2 in combination with a similar rise in temperature would result in an overall 20% rise in rice production and 31% decline in wheat production (Karim et al. 1999).

The apparent increase in yield of Boro (dry season rice crop generally grown under irrigated conditions and including high-yielding varieties) and other crops might be constrained by moisture stress. It is feared that moisture stress would be more intense during the dry season, which might force the Bangladeshi farmers to reduce the area for Boro cultivation. A shortfall in food-grain production would severely threaten food security of the poverty-ridden country (Karim et al. 1999).

Under a severe (40°C temperature rise) climate change scenario, the potential shortfall in rice production could exceed 30% from the trend, while that for wheat and potato could be as high as 50 and 70% respectively (Karim 1996). Under a moderate climate change scenario, the crop loss due to salinity intrusion could be about 0.2 Mt (Habibullah et al. 1998). The loss of production due to such effects may be relatively higher compared to that under floods. The effect of low-flow on agricultural vulnerability is considered to be much less intense compared to other effects. The ultimate impacts of loss of food-grain production would increase import of food which would require spending hard currency.

Livestock

In addition to affecting human beings, natural disasters cause tremendous sufferings for the livestock population of Bangladesh. Livestock suffer large-scale death in cyclonic storm surges (Haider et al. 1991). Prolonged flood can also cause death of livestock through a number of direct and indirect mechanisms (Ahmad et al. 2000). During droughts, livestock in Bangladesh do not suffer death, but lack of water increases their vulnerability to diseases. Since climate change would increase susceptibility to natural disasters, as mentioned in earlier sections, the anticipated toll on the livestock sector would be quite high (Ahmed 2005; GoB 2005).

The sufferings of livestock in the coastal zone are much higher than in other parts of the country. Field observations clearly suggest that livestock density is relatively low in the coastal areas, particularly in the southwestern parts of the country. The local elderly informed that gradual increase in salinity also increased competition for freshwater resources, and the livestock suffered the brunt of such a calamity (RVCC 2003). Animals used to have the least access to freshwater sources during the dry season. Due to drinking of poor-quality water, these animals fall victim to diseases, which reduce their economic efficiency (draught power, milk production, etc.). Lack of grazing land and the proliferation of shrimp areas are also identified as potential reasons for decreasing population of livestock in coastal areas. It may therefore be concluded that the livestock sector would also be vulnerable to the adverse impacts of climate change.

Coastal Zone

Several studies indicate that coastal zone vulnerability would be acute, due to the combined effects of climate change, sea level rise, subsidence, changes of upstream river discharge, cyclone and coastal embankments (World Bank 2000). Four key types of primary physical effects, i.e. saline water intrusion, drainage congestion, extreme events and changes in coastal morphology, have been identified as key vulnerabilities in the coastal area of Bangladesh (World Bank 2000).

The effect of saline water intrusion in the estuaries and into the groundwater would be enhanced by low river flow, sea level rise and subsidence. Pressure of the growing population and rising demand due to economic development will further reduce the relative availability of freshwater supply in future. The adverse effects of saline water intrusion will be significant on coastal agriculture and the availability of freshwater for public and industrial sectors will fall.

The combined effect of higher seawater levels, subsidence, siltation of estuary branches, higher riverbed levels and reduced sedimentation in flood-protected areas will impede drainage and gradually increase waterlogging problems. This effect will be particularly strong in the coastal zone. The problem will be aggravated by the continuous development of infrastructure (e.g. roads) further reducing the limited natural drainage capacity in the delta. Increased periods of inundation may hamper agricultural productivity, and will also threaten human health by increasing the potential for waterborne disease.

Impact on Mangrove Ecosystem

Sundarbans in coastal southern Bangladesh will be exposed to several of the above risks, particularly sea level rise, saline intrusion and intensive extreme weather events. Given enough time, the mangrove under threat and rapid change is expected to readjust and recolonize if space and time permit. But the demographic pressures in areas north of the Sundarbans would not permit the requisite space. The rate at which climate change-related sea level rise and saline intrusion is likely to take place is going to be much faster than the rate at which the mangrove ecosystem will be able to readjust. This will result in the reduction of species and biodiversity, as well as a decrease in the areas of the mangrove forest with all the concomitant consequences.

Settlements and Infrastructure

About half the population of Bangladesh falls below poverty line. Under the prevailing socioeconomic circumstances, it is easily understandable that the poor do not have good-quality houses. Moreover, natural disasters often take a huge toll on poorly built houses and sanitation infrastructure (Ahmad et al. 2000). Human settlements are, therefore, highly vulnerable to climate change-induced floods and cyclonic storm surges. Floods, especially the high-intensity floods, often devastate physical infrastructure such as road networks, educational centres, market places, administrative buildings, etc. (Nizamuddin et al. 2001; Siddiqi 1997; Siddique and Chowdhury 2000). The telecommunications network was torn off during the cyclone of 1991 and the entire coastal belt was disconnected for weeks.

Climate change-induced high-intensity events pose huge threats to existing physical infrastructure. Damage to national motorways due to flood alone is estimated at 1,011 and 3,315 km by 2030 and 2050, respectively. The corresponding damage to embankments is estimated at 4,271 and 13,996 km by 2030 and 2050, respectively. The aggregated damage figures for health centres and hospitals due to floods, cyclones, sea level rise and salinity intrusion is estimated at 1,682 and 5,212 km, respectively, for the above two time horizons (BRTC-BUET 2005).

Table 4 Relation between disaster and livelihood vulnerability

Type of disaster	Livelihood vulnerability
Flood	Reduce livelihood options due to loss of agriculture, illness, restriction of movement
Drought	Lack of job opportunity due to lack of agricultural activities, illness due to extreme heat, etc.
Cyclone and storm surges	Limited fishing time in the Bay, loss of life, and illness

Source: UNDP Human Development Report, 2007

Impact on Livelihoods

Whatever happens to the climate, and subsequently to various other sectors, is important, for the main reason that it affects the lives and livelihoods of the people. Climate change is expected to have major physical impacts on agriculture, industry, infrastructure, disaster, health and energy and consequently on people's livelihoods in terms of employment, income and consumption (including food security). Various groups in society will experience the impacts in various degrees depending upon their initial economic conditions (poor or non-poor), location (coastal or non-coastal, rural or urban) and gender.

The impacts on livelihood due to climate change depend on the nature and severity of the physical impacts relating to agriculture, water availability and quality, disaster-proneness, hospitability of the physical environment due to rising temperature and changing water regimes to pathogenic activity and coastal inundation. Climate change impacts on livelihood thus become a challenge of development under the most adversarial changes in dynamics of nature.

Part of the vulnerability will be due to water shortages for agriculture. But there are other areas where water-related vulnerability may increase. Some of this would be related to health and disaster. On the other hand, extensive waterlogging that is being experienced now may exacerbate, creating major problems for the livelihood of a poor person, all of whose land may be submerged permanently.

Both flooding and drought may increase in frequency. Particularly floods may be more devastating, creating major problems of livelihood and macroeconomic dislocations, slowing growth and pushing people down to the poverty line. Also, if cyclones and storm surges increase in frequency and intensity, the potential losses to life and livelihood would be most severe. The following table illustrates the relationships between different types of disaster and livelihoods (Table 4).

The health problems due to climatic factors, such as temperature rise and degrading water quality and shortage, will increase the likelihood of cholera, diarrhoea, dysentery, malaria and typhoid, and also involuntary foetus abortion, in the coastal areas due to rising salinity leading to hypertension. Increased food insecurity will exacerbate the problems further by causing more widespread malnutrition. Unfortunately, these are little calibrated or not enough to be superimposed on to socioeconomic trends to refine the livelihood impacts.

Taking a livelihood analysis approach to potential impacts of climate change, it is clear that the most vulnerable groups within each community are the poorest among them and even within the poor groups the most vulnerable are the women, children, elderly and the sick. It is therefore quite likely that the adverse impacts from climate change will fall disproportionately on these most vulnerable groups within the country as a whole, as well as within each vulnerable region of the country.

Vulnerability of Small Farmers in the Coastal Zone

A recent study on coastal area vulnerability and livelihood relationship revealed that physical or natural vulnerabilities are strong in regard to small farmer's livelihoods all over the coastal zone. Particularly cyclone/tidal bore, waterlogging/ drainage congestion, various types of flooding, sand deposition and soil salinity became the major physical hindrances to farming and production. These are also found as the major causes of sudden crop damage to small farmers. Seasonal attack by rats and insects was another important vulnerability which led to massive crop damage in many districts.

Physical vulnerability varies across the coastal region. For example, in Khulna region, waterlogging, soil salinity, lack of cultivable land are significant; in Barisal, region tidal flooding, cyclone/tidal bore, salinity, siltation, rats/insects are found to be the most adverse physical or natural vulnerabilities; and in Chittagong region, cyclone/tidal bore, different types of flood, deterioration of soil fertility due to salinity are found to be the major physical vulnerabilities.

Wage Labour in the Coastal Zone

Rural wage labourers are more vulnerable to economic and social factors. Lack of employment opportunities and low wage rate seem to be the major vulnerability factors among the rural wage labourers. Physical vulnerabilities such as lack of cultivable land, climate change stress, etc. that has an effect on agriculture ultimately converts into lack of employment opportunities for the wage labourers (Fig. 4).

Changing Lives

Changes have already been visible in the fact that saltwater from the Bay of Bengal has penetrated inland along tributary channels. The impacts of these observed changes have been significant on agriculture, particularly on food production and security. Climate change here is a day-to-day reality, as it would entail an additional 17 million Bangladeshis being displaced from their homes and farms

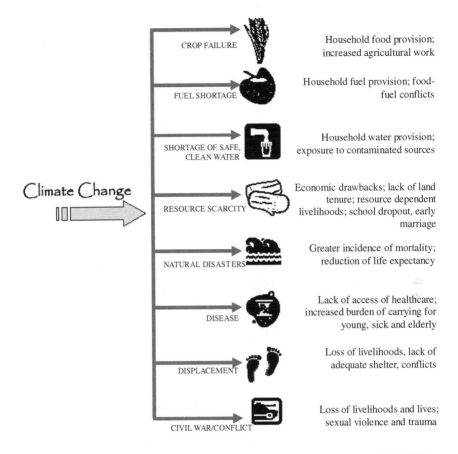

Fig. 4 Impacts of climate change on human security

by 2030 (UNEP 1989 in Sarwar 2008). Their lives would be plunged into darkness. These "climate refugees" would have to leave the region to find work in the cities in Bangladesh or neighbouring India.

Losing Livelihood: Gradually

The problem has been compounded because the flow of the rivers during the dry season has fallen, offering less resistance to the increasing power of the sea. With saltwater penetrating further inland and deeper into the groundwater, climate change is the root cause of the problems for gradually destroying from traditional cultivation and all available livelihoods. The mangrove forest is a breeding ground for fish, shrimp and crab, thus providing livelihoods for hundreds of thousands of people living in the area. Among the other economic activities, collecting seed shrimps or fry shrimps is the main livelihood of many landless men and women. Collection of seed shrimps (*Panaeus monodon*) is one of the main sources of earning for the small and landless fishermen and women of this area. These shrimp larva are highly sensitive to the pH condition of water, which mainly depends on the salinity of water and temperature. Shrimps are destroyed by invaded saline water and also because they are losing their natural habitats as mangrove forest is being destroyed. Due to the reduced abundance of fry over the past several years and a shortening of the fry season, many fry collectors are moving towards alternative employment. This is the common story of thousands, as when one man found himself unable to feed his two children, he moved to Dhaka, the capital of Bangladesh, where he pedals a rickshaw tricycle through the heavy traffic and takes in a swelling slum.

Lost Livelihood: Sudden

A category four storm SIDR hit the Sunderbans and southwestern coast of the country on 15 November 2007. According to the Disaster Management Centre in Bangladesh, the cyclone resulted in severe damage and 3,363 deaths, 8.9 million affected people, 2,472,944 acres of crop damage and an estimated US$3.1 billion economical loss (GoB 2008). Spiral winds of 220 km/h with 5-m water surges devastated much of the country's coastal belt, leaving almost 1.5 million homes damaged or destroyed. The total value of damage to housing is BDT 57.9 billion (US$839 million), representing more than half of the total damage and losses of all sectors (GoB 2008: 17). Dwelling houses blew away, trees were uprooted, standing crops demolished, roads and livestock washed away. Almost one-third of the Sunderbans has been totally demolished. A large number of people who earn their livelihood by engaging in fishing or fish-related business have literally been made paupers. The landless men and women and fishermen have lost their boats and nets, the shrimp cultivators their stock of shrimp, and the fish traders their business. As many of the households had complex livelihoods before the cyclone, combining several overlapping activities, some were simultaneously affected by losses in crops, livestock, fisheries, commerce, industry or wage employment. The most vulnerable groups were landless labourers, and marginal farmers with very little land and no other sustainable source of income (GoB 2008).

The concomitant losses to livelihood, trade and industry in these costal communities have not yet recovered, as the BBC documented 100 days after the cyclone hit. The BBC interviewed one Union chairman (Union is the lowest administration in Bangladesh) in the southern community of the Sundarbans, who described their main problem is the homeless, as they lost homes and were forced to live in the open. Although the international community promised to build their houses, by that time they had not yet done it. There were no shops to buy foods, all having been destroyed. They had no food. They were forced to live on handouts. They had to wait until the next crop harvesting sessions.

Table 5 Causes of impacts, vulnerable areas and impacted sectors

Climate and related elements	Critical vulnerable areas	Most impacted sectors
Temperature rise and drought	Northwest	Agriculture (crop, livestock, fisheries)
		Water
		Energy
		Health
Sea level rise and salinity intrusion	Coastal area Island	Agriculture (crop, livestock, fisheries)
		Water (waterlogging, drinking water)
		Human settlement
		Energy
		Health
Floods	Central region Northeast region Char land	Agriculture (crop, livestock, fisheries)
		Water (urban, industry)
		Infrastructure
		Human settlement
		Health
		Energy
Cyclone and storm surge	Coastal and Marine zone	Marine fishing
		Infrastructure
		Human settlement
		Life and property
Drainage congestion	Coastal area Urban Southwest	Water (navigation)
		Agriculture (crop)

Source: UNDP Human Development Report, 2007

After one year following the cyclone, the BBC revisited the same area and were told by development workers in the area that around 80% of people were still living in tents or temporary dwellings. As the government decreed a ban on entering the forest, hoping for natural regeneration of the forest, people have not found their livelihoods. As fishermen lost their boat and nets, people were forced to change their occupation to one in which they had no competence, resulting in less income than before. Many have left for nearby cities and the capital to find work.

The point is the Sundarbans' great value was evident once again as it protected local communities from huge damage. In this case, the importance of the Sundarbans as a natural barrier to reduce wind velocities (protecting embankments and settlements) was evidenced (GoB 2008). The Sundarbans is Bangladesh's guardian angel as far as storms and tidal waves are concerned. If there was no Sundarbans, the cyclone would have caused more havoc in the southern districts.

Table 6 Intensity of impacts on different sectors due to climate change

Sectoral vulnerability context	Physical vulnerability context							
	Extreme temperature	Sea level rise		Drought	Flood		Cyclone and storm surges	Erosion and accretion
		Coastal inundation	Salinity intrusion		River flood	Flash flood		
Crop agriculture	+++	++	+++	+++	+	++	+++	−
Fisheries	++	++	+	++	++	+	+	−
Livestock	++	++	+++	−	−	+	+++	−
Infrastructure	+	++	−	−	++	+	+	+++
Industries	++	+++	++	−	++	+	+	−
Biodiversity	++	+++	+++	−	++	−	+	−
Health	+++	+	+++	−	++	−	++	−
Human settlement	−	−	−	−	−	−	+++	+++
Energy	++	+	−	−	+	−	+	−

Source: MoEF (Ministry of Environment, Forest) (2005)

Summary of the Key Impacts

From the above findings, the impacts of climate change and variability on biophysical systems and consequences are evident. It is also found that the coastal zone, northwestern zone, central region and piedmont plain are most susceptible to existing climate variability and anticipated future climate change. A summary of causes of impacts, vulnerable areas and impacted sectors are given (Table 5). Intensity of impacts on different sectors due to changes of vulnerability context is given in Table 6.

Adaptation Strategies for Reducing Vulnerabilities Due to Climate Change

A careful analysis of the above definitions suggests that all these refer to adjustments in a system in response to climatic perturbation, but they also indicate differences in scope, application and interpretation of the term adaptation. Adaptation can refer to climate change, to change and variability, or just to climate. Adaptation can be in response to adverse effects of climate variability and change, which refer to past, actual or anticipated conditions, changes or opportunities.

The implications of high-intensity floods cannot be overemphasized in Bangladesh. Management of floods in future will remain a major challenge, especially in view of further densification in increasingly flood-vulnerable lands (Ahmed et al. 1998). Community efforts to cope with floods can benefit tremendously from the issuance of early warnings. Improvement of the current flood warning system and dissemination in a people-friendly manner are thought to be highly potential adaptation options for the future (Ahmed 2005).

Removal of impediments of drainage (dredging/re-excavation of choked rivers/khals, drainage canals), construction of drainage structures (culverts, bridges and regulators), rehabilitation of structures such as roads, embankments, etc. should be considered as adaptation measures towards facilitating drainage and reduce flood-related vulnerability (Ahmed et al. 1998). Pumping out water to remove water-logging, especially in polder areas, has already been practised, and is likely to be considered an adaptation option for the future. Multi-purpose cyclone shelters and flood shelters should be built in FVAs (Choudhury et al. 2003; GoB 2005). In recent years, community-based flood management practices had shown high potential, which could also be considered as an important modality to adapt to climate change-induced floods (Ahmed 2004).

For drought management, making water available to offset moisture deficit appears to be the major adaptation modality (Karim 1996). However, the creation and recreation of water storage systems (ponds, khals, reservoirs, etc.)—operated and maintained by vulnerable communities—needs to be given due emphasis (World Bank 2000). Capacity building for advanced irrigation techniques could also be considered as an important adaptation option in order to conserve available water resources. Resuscitation of surface water bodies including silted-up rivers and rivulets should be given due priority in order to maintain water bodies even during the dry season for irrigation purposes (Ahmed et al. 1998).

Deaths arising from cyclones and associated tidal bores (both human and livestock) could be minimized by maintaining the Cyclone Preparedness Programme, and further strengthening the programme by means of building new MCSs, killas and other facilities along the coastal zone (Mahtab 1989; Ali 1999). The polders which might be at risk of inundation due to rising sea levels and/or by invigorated tidal waves should be identified and rationalized, in order to enhance their efficiency towards safeguarding lives, crops and properties (Ahmed 2005). NAPA for Bangladesh proposed community-focused coastal afforestation as a priority adaptation measure to reduce climate hazards (GoB 2005).

In addition to adaptation in the water resources sector, one must consider adaptation in the agricultural sector. The gravity of the issue and its importance on people's livelihoods deserve special treatment, which is why the potential adaptation options in agriculture are discussed separately in the following section.

According to WB, the risk associated to human health in tropical developing countries is one of the salient risks of climate change (World Bank 2000). Bangladesh's current vulnerability to outbreaks of cholera and other waterborne and diarrhoea diseases such as dengue or dysentery needs to be given due importance in view of increasing risk potentials caused by climate change-induced drainage congestion and standing water. Treating pathogen-laden water with a mixture of lime, bleaching powder and alum, as provided in Ahmad et al. (2004), should be given due importance to avoid a large-scale outbreak of waterborne diseases. Inadequate provisions for drinking water in saline-affected regions add to people's vulnerability, which needs to be given high priority towards designing a national adaptation programme (Ahmed 2005). Providing saline-free drinking water should be considered as an immediate adaptation in view of current as well as future health risks (Ahmed 2004). The pressure on the availability and access to safe water, in particular during the dry period, and the increasing reliance on groundwater are an additional threat.

Awareness needs to be increased among illiterate and poor people, especially along the drier western parts of the country, to combat heat stress-related health disorders. Improved cyclone and flood shelters, with increasing capacity and coverage, are likely to reduce overall death tolls in the case of climate change-induced high-intensity disastrous events. Similarly, building relatively stronger houses by low-cost retrofitting along the cyclone-affected coastal regions could save lives as well as assets (RVCC 2003). Safe use of carbolic acid would reduce susceptibility to snake bites in flooded regions. Use of oral rehydration saline for treating diarrhoea patients will continue to save lives. Another major adaptation proposed for human health involves improving the healthcare system, which is needed anyway to address the current human health situation. These improvements could significantly reduce the risks to human health from climate change (World Bank 2000). Thus, the benefits of improving healthcare are likely to be even greater when avoided health impacts of climate change are accounted for.

Conclusion

The livelihood and income of a large population depends on the natural resource base and most of the poor people often live in marginalized lands and areas more prone to natural disasters. Climate change means that many natural disaster-prone areas will become more prone due to increased frequency and intensity of disasters. Drought-prone areas will become hotter and drier, with less predictable rainfall; flood frequency and intensity along onset and recession will be changed in future; the nature of cyclone and storm surges will be different from the historical trend. All of these together will change crop yields and affect many poor people's livelihoods. For example, the 2007 floods inundated 32,000 km^2 in area, destroying over 85,000 houses and destroying or partially damaging approximately 1.2 million acres of crops. Total estimated loss was over US$1 billion.

It is likely that natural disasters will damage more houses and will cause temporary migration. It may also require children to help more with household tasks, leaving less time for schooling. Malnourishment and diseases also impair learning. Extreme climate change-related disasters threaten school buildings and educational materials. For example, cyclone Sidr caused huge damage to school buildings and wiped out teaching materials.

Adverse impacts of climate change are likely to reduce availability and deteriorate quality of water for domestic use. Direct climate change effects include increases in mortality and illness associated with heat waves, particularly among the elderly and the urban poor. Women and children are particularly vulnerable to extreme weather events. For example, when the 1991 cyclone hit Bangladesh, 90% of victims were women and children.

Climate change is likely to increase the prevalence and infection of vector- and waterborne diseases such as malaria and dengue fever, cholera and dysentery, etc. Children and pregnant women are particularly susceptible to such diseases. Climate change will probably cause a decline in the quantity and quality of drinking water, which is a prerequisite for good health. Malnutrition, the main cause of ill health among children, could also be exacerbated due to declining natural resource productivity and inadequate supply of food.

Changes in temperature and rainfall distribution, and sea level rise and salinity intrusion are likely to change ecosystem characteristics and shift ecosystem boundaries. Climate change also poses a greater survival threat than the destruction of many natural habitats including coral reefs. Degradation of biodiversity will reduce the availability of many traditional medicines which may affect poor and rural people who depend more on natural resources for medicine as well as income and food.

Sea level rise (SLR) will drastically affect the poor people who are in coastal areas and floodplain zones in Bangladesh. The disasters of climate change are somewhat obvious. It is already visible in the costal area in this delta. Climate change is not some "distant fury"; its widespread consequences are already being felt in costal areas especially among local communities around the Sundarbans with livelihood losses caused by gradual and sudden climatic changes. It is not a matter of the distant future. Climate change has effects on living conditions in ways that violate human rights. The lives of millions have changed due to global warming. Especially in these local communities, climate change has had multiple consequences on the mangrove forest. The mangrove forest in the Sundarbans is linked with millions of people's livelihoods. Their livelihoods have been destroyed by climate change-induced natural hazards. Climate change is not only destroying the mangrove forest, it affects the Sundarbans "ecological services".

Many projected climate change impacts, including sea level rise, higher temperatures and evapotranspiration losses, enhanced monsoon precipitation and runoff, potentially reduced dry season precipitation, and an increase in cyclone intensity, would in fact reinforce many of these baseline stresses that already pose a serious impediment to the economic development of Bangladesh. By the same token, many actions undertaken to address the baseline or contextual risks in Bangladesh are also synergistic with the so-called adaptations that might be required as climate change impacts manifest themselves. With regard to structural adaptations, such as coastal embankments and salinity reduction, even though it is true that many of these measures have already been integrated in development projects and policies in Bangladesh, there remains an ongoing challenge with regard to their durability and sustainability. Structural adaptations therefore need to be matched by efforts to facilitate financial and institutional adaptation—sustained interest on the part of the government and donors, and the participation of local populations to help monitor and maintain infrastructural projects.

References

Ahmad QK, Chowdhury AKA, Imam SH, Sarker M (eds) (2000) Perspectives on flood 1998. The University Press, Dhaka

Ahmed AU (2004) A review of the current policy regime in Bangladesh in relation to climate change adaptation, CARE-Bangladesh, under reducing vulnerability to climate change (RVCC) project, Khulna

Ahmed AU (2005) Adaptation options for managing water related extreme events under climate change regime: Bangladesh persectives. Balkema Press, London

Ahmed AU, Alam M, Rahman AA (1998) Adaptation to climate change in Bangladesh: future outlook in vulnerability and adaptation to climate change for Bangladesh. Kluwer, Dordrecht

Ali A (1999) Vulnerability of Bangladesh coastal region to climate change with adaptation option. Bangladesh Space Research and Remote Sensing Organization (SPARRSO), Dhaka

BRTC-BUET (2005) Final report of the sectoral working group on industry and infrastructure, Bureau of research, testing and consultation (BRTC), Bangladesh University of Engineering and Technology (BUET), Dhaka

Choudhury AM, Quadir DA, Neelormi S, Ahmed AU (2003) Climate change and its impacts on water resources of Bangladesh. In: Muhammed A (ed) Climate change and water resources in South Asia. Asianics International, Islamabad

Climate Change Cell (2007) Climate change and Bangladesh Department of Environment, Government of People' Republic of Bangladesh. Climate Change Cell, Dhaka

Department of Agricultural Extension (DAE) (2007) Disaster risk management in agricultural project, Supported by comprehensive disaster management program (CDMP), UNDP

GoB (2005) National adaptation programme of action (NAPA), Final report: November 2005, Ministry of Environment and Forest, Government of the People's Republic of Bangladesh (GOB), Dhaka

GoB (2008) Cyclone Sidr in Bangladesh damage, loss and needs assessment for disaster recovery and reconstruction Government of Bangladesh, Ministry of Environment and Forest, Government of the People's Republic of Bangladesh (GOB), Dhaka

Habibullah M, Ahmed AU, Karim Z (1998) Assessment of foodgrain production loss due to climate induced enhanced soil salinity. Kluwer, Dordrecht

Haider R, Rahman AA, Huq S (eds) (1991) Cyclone'91: an environmental and perceptional study. Bangladesh Centre for Advanced Studies, Dhaka

IPCC (2001) Climate change 2001 impacts adaptation and vulnerability. In: McCarthy JJ, Canziani OF, Leary NA, Dokken DJ, White KS (eds) Contribution of working group II to the third assessment report of the intergovernmental panel on climate change. Cambridge University Press, Cambridge, p 1032

Karim Z (1996) Agricultural vulnerability and poverty alleviation in Bangladesh. In: Downing TE (ed) Climate change and world food security, NATO ASI series, 137. Springer, Berlin

Karim et al (1999) Climate change vulnerability of crop agriculture: a case study in vulnerability and adaptation to climate change for Bangladesh. Kluwer, Dordrecht

Mahtab F (1989) Effect of climate change and sea level rise on Bangladesh. Commonwealth Secretariat, UK

MoEF (Ministry of Environment, Forest) (2005) Bangladesh national adaptation programme of action (NAPA), Dhaka

Nizamuddin K, Khuda ZRMM, Ahmad N (2001) Women-headed households displaced by river bank erosion: problems and strategies of survival. Disaster Research Training and Management Centre, Dhaka

RVCC (2003) Report of a community level vulnerability assessment conducted in Southwest Bangladesh, a report prepared by the reducing vulnerability to climate change (RVCC) Project, CARE Bangladesh, Dhaka

Sarwar G (2008) Impacts of sea level rise on the coastal zone of Bangladesh. Lund University, Sweden

Siddiqi MH (1997) Floods: problems and prospects. Paper presented at the workshop on observance of the international decade for natural disaster reduction 1997, organized by the disaster management bureau, Ministry of Disaster Management and Relief, the Government of the People's Republic of Bangladesh (GoB), Dhaka

Siddique QI, Chowdhury MMH (2000) Flood'98: losses and damages. In: Ahmad QK et al (eds) Perspectives on flood 1998. The University Press, Dhaka

SMRC (2003) The vulnerability assessment of the SAARC coastal region due to sea level rise: Bangladesh case. SMRC Publication, Dhaka

UNEP (1989) Impacts of sea level rise on the coastal zone of Bangladesh, Master thesis, Lund University, Sweden (in Sarwar G (2008))

UNDP (2006) Human Development Report 2006, Dhaka

World Bank (2000) Bangladesh: climate change and sustainable development, Report No. 21104-BD, Rural Development Unit, South Asia Region, The World Bank (WB), Dhaka

Chapter 16
Conservation and Management of Water Resources for Livelihood Sustainability in Mountainous Regions of Uttarakhand, India

Anil Kumar

Abstract The inhabitants of the middle and upper regions of the Himalayan regions of Uttarakhand State in India struggle for their survival, food security and sustainable livelihoods due to various topographical and socioeconomic constraints. Agriculture, which is largely rain-fed (90%), is the mainstay for their livelihoods, but various climatic, geographical and socioeconomic constraints have led to a dismally low agricultural productivity in the region. For generations, the inhabitants have been depending on the natural water springs and streams for meeting their day-to-day water needs for drinking and domestic uses, irrigation, animal consumption, etc. Most of the perennial springs and streams have become seasonal or have dried up for want of recharge due to climatic variability and other natural and man-made hazards. Efforts are made to develop, conserve and manage the available water resources, namely overland flow, runoff through rooftops of houses, and natural water springs by making suitable low-cost storage structures. The field studies have revealed that construction of a brick-cemented tank to store spring water and filtered roof water for drinking and household needs, in combination with a dug-out farm pond lined with 0.25 mm thick low-density polyethylene sheet to collect the overflow from the cemented tank and the overland flow, is a technically feasible and economically viable option.

Keywords Water resources · Livelihood · Sustainability · Mountainous region · Water harvesting · Spring recharge

---

A. Kumar (✉)
Department of Soil and Water Conservation Engineering,
G. B. Pant University of Agriculture and Technology, Pantnagar,
263 145 Uttarakhand, India
e-mail: anilkumar_swce61@yahoo.co.in

Introduction

Water is undoubtedly the single most important and integral part of all the facets of the mountain environment. Mountains are the major source of all the natural resources including forest, land, water, animal, minerals, etc., and they are called life-givers to the biotic means, not only to the inhabitants residing in this region but also to the inhabitants downstream. The Himalayas have given birth to many perennial rivers and streams for the survival of living beings in the downstream regions of most of the northern states of India, but the inhabitants of the middle and upper reaches in the Himalayan region struggle for their own survival for want of adequate water resources at their disposal, food security and sustainable livelihoods due to various topographical and socioeconomic constraints. For generations, the inhabitants of the Himalayan region of Uttarakhand State in India have been depending on the natural water springs and streams for meeting their day-to-day water needs for drinking and domestic uses, irrigation, animal consumption, etc. The topographical and human settlement pattern in this region is such that a large section of the population lives on upper reaches of hill slopes, where the only available sources of water are rainfall and existing natural water springs. Today, most of the perennial springs and streams have become seasonal or have dried up for want of recharge due to various environmental hazards. In the remote areas, the womenfolk walk several kilometres to fetch a head-load of water for drinking and domestic uses. Water-conserving trees such as oak (*Quercus leucotrichophora*), Utis (*Alnus nepalanses*), etc. have been reducing in number, which has subsequently reduced outflow from the natural springs. Valdiya and Bartrya (1991) investigated that in the past five decades the discharge from natural springs has reduced from about 25 to 75% and subsequently, the flow in rivers has also reduced drastically. The potential of the existing natural water springs in the region has been completely overlooked in the so-called modern era.

At national level, there is wide scope to beneficially utilize the available rainwater in the zone of 1,000–2,500 mm annual average rainfall. The middle and high hills of Uttarakhand, in general, fall under this category, where about 90% of the population in the hilly areas earn their livelihood from agriculture and animal husbandry, and are still in the subsistence class characterized by extremely limited capital resources and consistent use of traditional means of crop production. In the hilly areas, drinking water has the highest priority, followed by cattle feeding and irrigation. Therefore, it is essential not only to ensure an adequate supply of water, but also to maintain its quality for the different types of demand of the people for their survival. In small cities and towns, government agencies have provided the facility of drinking water at the cost of millions of rupees by pumping water from valley areas, but villages in remote hilly areas are still out of their reach. The demand for water has been persistently increasing with the growth of population, agriculture and industry; meanwhile, the per-capita availability of water has been decreasing every year.

Although most hilly areas receive good annual rainfall, its intensity and distribution is quite erratic and causes severe drought spells which hamper the growth of timely sown crops due to lack of soil moisture at critical times. This situation forces the farmers to risk their crops at the germination and ripening stages of growth. Frequent and long dry spells retard the growth, size and yield of important fruit crops such as apple, plum, peach, apricot, etc. The monsoon runoff, if allowed to flow freely, causes soil erosion from arable and non-arable lands, but if used wisely and judiciously can be a boon for improving crop production in the rain-fed areas. In order to minimize the adverse effects of water stress, particularly at the productive stages of crop growth, the conservation of rainfall in soil profiles and providing irrigation through runoff/spring flow harvesting in ponds or tanks at suitable locations are the only ways out to solve drinking-water problems as well as to enhance the productivity of rain-fed agriculture on high and medium hills. Several authors have emphasized runoff harvesting to eliminate the ill effects of droughts and low productivity in the arid and semi-arid foothill areas in the country (Chitranjan and Rao 1986; Grewal et al. 1989; Oswal 1994). The studies conducted by Kumar (1992) suggested the feasibility of cost-effective low-density polyethylene (LDPE) lined dug-out small ponds for irrigation purposes in the mid-Himalayan region. If proper irrigation facilities are assured, vegetable crop production has great potential to raise the economic standard of hill farmers. Off-seasonal vegetables (pea, potato, cauliflower, cabbage, etc.) can be produced on a large scale and sold at high prices in plain areas. Assured irrigation can also promise cultivation of pea and potato crops twice a year.

The need for harnessing the vast potential of Himalayan water resources has been felt to mitigate the effect of water scarcity not only in mountains but also in the plain areas of the country. A study was conducted to understand the hydrologic behaviour of existing perennial natural water springs and compute the water yield and storage potential of these springs along with harvesting of huge surface runoff, which otherwise causes severe soil erosion in the middle and high hills, to fulfil the needs of the people for drinking water, domestic uses, animal consumption and irrigation, using cost-effective materials and techniques. Realizing the fact that global warming is a reality and water resources are limited, the obvious solution to meet the water demands of the present and future lies in harnessing the available resources in the most efficient and effective way to optimize their use.

Study Area

This study was conducted at the Hill Campus, an off-campus of G. B. Pant University of Agriculture and Technology, located at Ranichauri (78° 2′ E longitude and 30° 15′ N latitude), at an altitude of about 2,000 m above mean sea level, in Garhwal region of Uttarakhand State in India. Two perennial natural water springs were selected, namely Hill Campus (HC) spring located at an altitude of about 2,000 m, and Fakua spring located at about 1,800 m above mean sea level and at a distance of about 500 m from Hill Campus spring. The weekly measurements of the discharge of these springs have been taken regularly since

January 1995 by a volumetric method using a container of 5-l capacity and a stopwatch. During the rainy season, the measurements were taken twice a week to monitor the effect of frequent rains. Each measurement was taken three times and a mean discharge rate was calculated to avoid any observational error. The rainfall data was collected from the meteorological observatory located at Hill Campus, Ranichauri. The mean annual rainfall is about 1,176 mm, ranging between the minimum of 4 mm in November and the maximum of 246.8 mm in August. The soil properties indicate a little more fractions of silt and clay in Hill Campus soil as compared to Fakua soil.

Hydrologic Analysis of Water Resources

In the mountainous areas, water is available in three forms, namely surface runoff, runoff through rooftops of houses and the discharge from natural water springs. Since conventional irrigation of all rain-fed lands is not feasible in undulating hilly terrains, there is tremendous scope for improving the productivity of these areas through rainwater harvesting. Since water harvesting practices are location specific, there are different water-harvesting practices that have evolved locally, broadly classified as:

i. in situ retention of rainfall on the land itself by biologic or agronomic and/or engineering or mechanical measures;
ii. surface runoff harvesting and its storage;
iii. storage and management of discharge from natural water springs;
iv. diversion of surface/sub-surface water into the storage structure; and
v. rooftop water harvesting.

The surface runoff, which can be estimated using various methods (e.g. curve number method) on the basis of past rainfall data and land use, is suitable for irrigation only. Runoff through rooftops can be estimated using a reasonable value of runoff coefficient for different types of roof, and may be utilized for human consumption and domestic uses after proper filtration. The flow from natural water springs, which is well suited for drinking purposes, can also be estimated using past records of discharge. The spring water can be stored in closed cemented tanks for the purpose of drinking, sanitation and other domestic uses. The optimum size of a storage structure depends on the amount of runoff expected, water demand for various purposes and the benefit–cost ratio for the proposed harvesting system. In this study, the spring flow and surface runoff data are analysed separately for the purpose of water harvesting and storage. The flow duration curve, which is a plot between the rate of discharge and the per cent of time the flow was equalled or exceeded, was useful in water resource planning for evaluating the dependable flow from a spring. By arranging the mean monthly discharge data in descending order of magnitude, the plotting position or the probability of exceedance corresponding to a discharge rate is computed by Weibull's formula (Subramanya 1995).

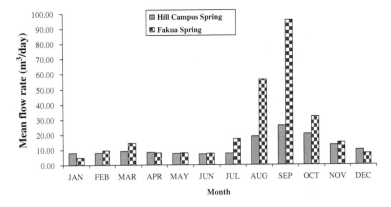

Fig. 1 Mean monthly discharge from the natural water springs

Computation of Minimum Required Storage for Spring Flow

The discharge pattern for the Hill Campus spring and the Fakua spring, shown in Fig. 1, indicates that the mean monthly rate of spring flow varied from 7.1 m^3/day (June) to 25.76 m^3/day (September) with an average of 11.96 m^3/day for the Hill Campus spring; and 4.83 m^3/day (January) to 95.45 m^3/day (September) with an average of 22.96 m^3/day for the Fakua spring. The variability in monthly average flow volumes of Hill Campus and Fakua springs was very high, indicating more variation in the flow of the Fakua spring as compared to that of the Hill Campus spring. Therefore, these springs were categorized as variable-discharge springs, having more than 100% variability in their mean monthly discharges. The flow pattern revealed the fact that both the springs produced the maximum discharge rate in the month of September (last monsoon month in India), while the minimum was in June (just before the onset of monsoon). This pattern indicates continuous recharging of both the springs during monsoon months. The flow duration curves for these springs revealed that at 50 and 80% probability levels, the Hill Campus and Fakua springs had 8.5 and 5 m^3/day, and 10.80 and 2 m^3/day mean flow rates, respectively, which indicates the availability of 80% dependable flows from the Hill Campus spring and Fakua spring at the rate of 5 and 2 m^3/day, respectively, which is insufficient to fulfil the drinking water needs of even a few families depending solely on these springs without proper storage arrangement for continuous water supply.

The computation of minimum storage required to fulfil the variable demands of the users can be done by using the flow-mass curve or by arithmetical computations without using the curve. Assuming that the storage reservoir is full at the beginning of the dry period (when inflow rate is less than the demand rate), the maximum amount of water drawn from storage is the cumulative difference between the supply and demand volumes (column 7) from the beginning of the dry season. The arithmetical computations are given in Tables 1 and 2 for the Hill Campus spring and Fakua spring, respectively.

Table 1 Computation of minimum required storage corresponding to a variable daily demand rate for Hill Campus spring

Month	Mean flow rate (m³/day)	Flow volume (m³)	Demand rate (m³/day)	Demand volume (m³)	Difference between flow and demand volumes (m³)	Cumulative volume	
						Excess demand (m³)	Excess flow (m³)
Jan	8.08	250.48	7	217	33.48		1769.79
Feb	7.78	217.84	9	252	−34.16	−34.16	
Mar	9.23	286.13	10	310	−23.87	−58.03	
Apr	8.36	250.8	10	300	−49.2	−107.23	
May	7.39	229.09	10	310	−80.91	−188.14	
Jun	7.1	213	10	300	−87	−275.14	
Jul	7.69	238.39	8	248	−9.61	−284.75[a]	
Aug	18.6	576.6	5	155	421.6		421.6
Sep	25.76	772.8	5	150	622.8		1044.4
Oct	20.32	629.92	7	217	412.92		1457.32
Nov	13.21	396.3	7	210	186.3		1643.62
Dec	9.99	309.69	7	217	92.69		1736.31

[a] Minimum required storage = 284.75 m³

Table 2 Computation of minimum required storage corresponding to a variable daily demand rate for Fakua spring

Month	Mean flow rate (m³/day)	Flow volume (m³)	Demand rate (m³/day)	Demand volume (m³)	Difference between flow and demand volumes (m³)	Cumulative volume of	
						Excess demand (m³)	Excess flow (m³)
Jan	4.83	149.73	7	217	−67.27	−67.27	
Feb	9.47	265.16	9	252	13.16		13.16
Mar	14.53	450.43	10	310	140.43		153.59
Apr	7.93	237.9	10	300	−62.1	−62.1	
May	8.08	250.48	10	310	−59.52	−121.62	
Jun	7.64	229.2	10	300	−70.8	−192.42[a]	
Jul	17.11	530.41	8	248	282.41		282.41
Aug	56.13	1740.03	5	155	1585.03		1867.44
Sep	95.45	2863.5	5	150	2713.5		4580.94
Oct	32.03	992.93	7	217	775.93		5356.87
Nov	14.69	440.7	7	210	230.7		5587.57
Dec	7.57	234.67	7	217	17.67		5605.24

[a] Minimum required storage = 192.42 m³

It is evident from the tables that no storage is required for Hill Campus spring, if the demand rate is 7.1 m³/day or less (minimum flow rate in column 2) of Table 1; and correspondingly 4.83 m³/day for Fakua spring. The variable monthly demand rates were arbitrarily assumed for different months based on the daily

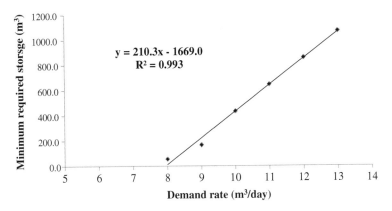

Fig. 2 Minimum required storage versus daily demand rate for Hill Campus spring

domestic water needs of the inhabitants. The storage required (column 7) of the Tables shows the cumulative of the consecutive values with the "−" sign (i.e. for excess demand) in column 6. The maximum of the cumulative value for excess demand (column 7) indicates the volume of excess demand which is to be met through storage only. Column 8 indicates the excess flow from the storage. Because of wide variations in mean monthly discharge, the Fakua spring will be able to fulfil more water demands of the people with the same storage than the HC spring. The importance of the stored water was realized during the lean spring flows in summer months when the tank was continuously filling during nights and people utilized the water during day time. The required volume of clean drinking water from the springs can be stored in a single or cluster of cement-concrete tanks for proper distribution throughout the year. The excess flow from springs during monsoon months and overflow from the main tank along with the surface flow can be collected in the nearby dug-out farm ponds.

A regression analysis was done to find the relationship between uniform daily demand and minimum storage required to ensure uninterrupted water supply throughout the year (Figs. 2 and 3).

A few arbitrarily chosen values of demand rates were considered (e.g. 8–13 m³/day for Hill Campus spring, and 5–13 m³/day for Fakua spring). The relationship between the required volumes of minimum storage and demand rate for Hill Campus and Fakua springs were developed as the best-fit linear relationships as given below:

Hill Campus spring:

$$S_h = 210.3\, D_h - 1669.0 \quad (R^2 = 0.993)$$

Fakua spring:

$$S_f = 55.3\, D_f - 318.5 \quad (R^2 = 0.915)$$

where, S_h and D_h are the minimum required storage (m³) and uniform demand rate (m³/day), respectively for Hill Campus spring; and S_f and D_f are the required

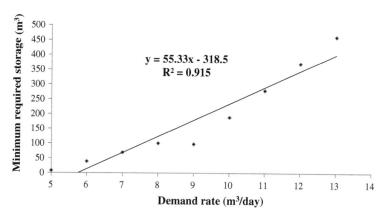

Fig. 3 Minimum required storage versus daily demand rate for Fakua spring

storage (m³) and uniform demand rate (m³/day), respectively, for Fakua spring. Based on these equations, the minimum required volume of storage can be estimated to fulfil a uniform daily demand of water for these springs.

Design of Pond for Runoff Harvesting

The size of water storage structures is decided according to water availability and topography of the location, where structure is to be constructed. In order to be cost-effective, the site should be located where maximum storage volume is obtained through minimum earthwork. The capacity of the storage depends mainly on the availability of relatively flatter land on which these structures could be made, and the runoff/spring flow passing through that point. The small and scattered land holdings on different terrains can only permit the construction of small water storage tanks upstream from a cluster of fields to facilitate irrigation through gravity flow. Based on the experiments conducted at the research station and nearby areas to evaluate the feasibility and economic viability of lining materials, it was found that a low-density polyethylene (LDPE) sheet of 0.25 mm thickness is sufficiently strong and economical for runoff harvesting for irrigation purposes.

The construction of dug-out pond includes digging of a truncated reverse-pyramid-shaped pit with 1:1 side slopes (Fig. 4). The depth of the pond is restricted to 1–1.5 m. The inner surfaces of the pond are plastered with 5 cm thick mud plaster so that the sheet is properly stuck to the surfaces. A single piece of LDPE sheeting of 0.25 mm thickness at the required size is placed with its corners properly folded and ends buried on all sides at the top of the pond. Another 10–15 cm layer of mud mixture of excavated soil and wheat straw or chopped dry pine needles in a 4:1 ratio is plastered on the sides and bottom. While harvesting only the surface runoff, a small silt retention trench of 1 × 0.5 × 0.5 m size is dug at the entry point to the main pond so that debris and suspended particles along

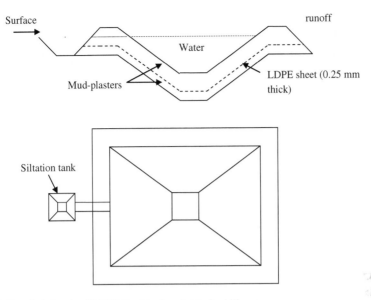

Fig. 4 Detailed sketch of LDPE-lined tank suitable for hilly areas

with overland runoff are settled down and relatively cleaner runoff water may enter the main pond. After plastering, the pond is left to dry for 1–2 days. The cracks developed during drying, if any, are filled with mud mortar. The runoff is allowed to enter the pond with low velocity so that the mud-plaster remains intact. Evaporation losses from the pond can be minimized by spreading on the water surface a small quantity of burnt engine oil. The construction cost of the LDPE-lined tank comes out to be 150 rupees (US$3)/$m^3$ of water stored, which is much less than the cemented tanks of the same capacity costing more than 1,000 rupees (US$20)/$m^3$ of water stored. Another advantage of the LDPE-lined pond is that this system can be constructed, repaired and maintained by the farmers themselves at a reasonably low cost, as the only material to be purchased from the market is the LDPE sheet. As a precaution, the LDPE sheet should not be exposed to sunlight for long durations as the sun's ultraviolet rays can damage the sheet. The useful life of such ponds is normally 20–25 years, which can be further extended if special care and maintenance is ensured. Water from these tanks is taken by siphoning through rubber pipes to irrigate the crops at lower elevations through gravity flow. As an integrated approach in hilly areas, all the available water resources can be combined in such a way that a cemented tank is used to store spring water and runoff from rooftops through a filter arrangement for drinking and domestic uses, while the overflow from this tank and overland surface runoff may be stored in the cost-effective and easily maintainable LDPE-lined dug-out ponds at lower elevations (Fig. 5). In this way, the water resources are utilized to the maximum extent and all the needs of the farming communities are also met simultaneously.

Fig. 5 Appropriate water harvesting model for hilly areas

Rejuvenation of Natural Water Springs

The survival and livelihoods of the people living in the middle and high hills is directly connected with the existence and development of natural water springs in mountainous regions. Studies have revealed that the spring discharge is mainly controlled by geological structures, land use pattern, and the nature and extent of vegetal cover in the recharge zone of the springs. Infiltration of rainwater can be enhanced by using biological and engineering measures and subsequently the spring discharge will be increased. A study was undertaken to rejuvenate and enhance the discharge of the existing natural water springs with the specific objective to increase the recharge potential of natural water springs using vegetative and engineering measures in the recharge zone of the springs. The vegetative measures included planting of water-conserving forest species such as oak (*Quercus leucotrichophora*), utis (Nepalese alder, *Alnus nepalensis*), willow

(*Salix alba* and *Salix babylonica*), etc. as per the recommended practices, along with the existing vegetation. Concurrently, the engineering measures in the form of staggered contour trenches of $1.0 \times 0.5 \times 0.5$ m (depth) size with a side slope of 1:4 at a spacing of 1 m along the approximate contours, and 12.5 m horizontal interval/cm of expected runoff were adopted in the recharge zone of the springs. The results revealed that the discharge from the springs increased significantly after adopting the biological and engineering measures. The contour trenches showed an immediate effect on spring discharge due to enhanced recharging of the spring catchments, whereas the trees had a delayed but long-term effect on spring recharge. Using these measures, the disappearing and dead springs were rejuvenated.

Conclusions

- In order to develop and conserve water resources in the Himalayan region on a sustainable basis, appropriate water-harvesting tanks and ponds must be constructed at suitable sites to meet the crucial needs of drinking, domestic uses, animal consumption and irrigation. This will ensure an adequate and better water supply for the villagers and reduce drudgery on farm women.
- An integrated approach must be adopted in hilly areas to capture all the available water resources such as spring water, runoff from rooftops for drinking and overland runoff for various uses for the household and irrigation.
- This study clearly indicated that even the small flows of perennial water springs can generate a huge volume of storage for beneficial uses in the middle and high hilly areas. Development of such facilities at suitable locations will save a huge amount of money being spent on water supply to hilltops from deep valleys using multi-stage water lifting devices.
- Large-scale adoption of biological and engineering measures for rejuvenating the existing natural water springs will greatly enhance their discharge and help towards the emergence of new springs at suitable locations with active participation and cooperation from governmental and non-governmental organizations.

References

Chittaranjan S, Rama Mohan Rao MS (1986) Runoff harvesting and recycling on vertisols for increasing crop production. In: Gupta RK, Khybri ML (eds) Soil Conservation in India. Jugal Kishore and Company, Dehradun, pp 188–191

Grewal SS, Mittal SP, Agnihotri Y, Dubey LN (1989) Rainwater harvesting for management of agricultural droughts in the foothills of northern India. Agric Water Manag 16(4):309–322

Kumar A (1992) Development and conservation of water resources in Garhwal Himalaya. J Soil Water Conserv 47(2):249–250

Oswal MD (1994) Water conservation and dryland crop production in arid and semi-arid regions. Ann Arid Zone 33(2):95–104

Subramanya K (1995) Engineering hydrology. Tata McGraw-Hill, New Delhi

Valdiya KS, Bartrya SK (1991) Hydrological study of springs in the catchment of Gaula river, Kumaon lesser Himalaya. Mt Res Dev 11(3):239–258

Chapter 17
Ports and Climate Change: Building Skills in Climate Change Adaptation, Australia

Melissa Nursey-Bray and Tony Miller

Abstract The debate on climate change in relation to the ports and shipping sector has largely focused on their impacts rather than the question of adaptation and vulnerability. Limits to current levels of adaptation to existing climatic variation are demonstrated by the impacts and costs to society associated with extreme events such as floods, ice storms, droughts and hurricanes. Assessing vulnerability, and therefore the resilience of social-ecological systems, such as ports, needs to take account of three dimensions: (i) real or potential impacts on the system, (ii) the systems' ability to cope and adapt to these impacts and (iii) the extent to which coping capacity may be constrained by environmental or societal conditions. In this context, this paper reflects on the results of a training needs analysis conducted within the ports and shipping industry in Australia, in order to explore what might be most relevant issues for the sector and its implications for future adaptation strategies. The paper reflects on the issues raised and argues that the key to building adaptive capacity is implementation of training packages focused on vulnerability assessments.

Keywords Ports · Shipping · Climate change · Vulnerability · Needs analysis · Training · Management

---

M. Nursey-Bray (✉) · T. Miller
University of Adelaide, North Terrace Campus,
North Terrace, Adelaide, SA 5005, Australia
e-mail: Melissa.Nursey-Bray@adelaide.edu.au

Introduction

Climate change, its impacts and solutions are at the forefront of a collective global consciousness. Instituting adaptation frameworks that can assist professionals to respond to climate change, yet be tailored to specific sectoral needs, is crucial. Professionals working at the "coalface" of the climate change challenge need to build new skills and create innovative solutions in social and political contexts. However, there are a number of challenges to creating management frameworks that address climate change while meeting the needs of different sectors.

Using the ports and shipping sector as a case study, this paper reports on the experience of implementing a training needs analysis of the ports industry in Australia in relation to climate change.

This needs analysis was part of a project designed to develop climate change adaptation curricula for ports professionals, and part of the process we adopted was to try and determine what core training needs were. However, this process also revealed interesting trends in relation to the adaptive capacity and perception of the industry, which form the basis of this paper. The project is funded by the Department of Climate Change and Energy, Australia, and was conducted while the authors were working at the National Centre for Marine Conservation and Resource Sustainability, Australian Maritime College, UTAS, Tasmania, Australia. The results are now incorporated within a draft online certificate in climate change adaptation training.

While obvious to some, it is nonetheless worth noting that the ports and shipping sector is very diverse. While our summary of the industry is done by using the term "ports and shipping", we acknowledge that this is a fairly rudimentary term that does not capture the nuances of the industry. We do, however, use it for convenience, especially as we ensured that we undertook to gain an understanding within our review of the experiences and needs of the sector across the board.

Methodology

The training needs analysis was undertaken to: (i) develop an understanding of what the industry training needs were and (ii) what content and style of delivery to adopt for the ports sector. Table 1 shows the methodology we adopted to undertake this process. We conducted a desktop survey of ports across the world, but in particular focused on the ports of Tasmania, Queensland and Victoria, Australia as our case studies.

Why Ports?

Current climate change science has arrived at four important conclusions: (i) that warming of the climate system is "unequivocal", (ii) the study of palaeo-climatic data supports the view that this warming is unusual and comparatively accelerated

Table 1 Methodological framework for conducting a needs analysis

Typology of needs	Definition	Methods used
Expressed needs	Those stated by industry representatives themselves	Desktop survey Workshops
Indicative needs	Those indicated by community/sector characteristics	Data sources such as the ABS
Normative needs	Those drawn from applying benchmarks	Performance indicators for industry
Comparative needs	Those suggested by comparison to others such as other sectors, i.e. local government, fisheries	Expert workshops Individual interviews Focus groups

and shows that where warming did occur, so did sea level rise, (iii) that the increase in temperatures that have been documented since the mid-20th century are "very likely" to have been human induced, and (iv) that continued emissions outputs will cause further warming (Steffen 2009; IPCC 2007).

Climate change is anticipated to have a major impact on ports. An international study by the OECD of 136 cities with over 1 million people (in 2005) into the vulnerability of ports across the world highlights the importance of this issue (Nicholls et al. 2007). Key findings from this report indicate that already large populations are exposed to coastal flooding in port cities, and that across all cities approximately 40 million people (0.6% of the global population, or roughly 1 in 10 of the total port city population in the cities (within the project study) are exposed to a 1 in 100-year coastal flood event. The study also found that the top ten cities (in 2005) that have the most exposed populations include: Mumbai, Guangzhou, Shanghai, Miami, Ho Chi Minh City, Kolkata, Greater New York, Osaka-Kobe, Alexandria and New Orleans (Nicholls et al. 2007). The sheer number of ports, combined with Australia's geographical size, means that any climate change impact on ports in Australia will have flow-on effects for maritime transport, food security and environmental protection.

Extreme weather events will include chaotic, heavy precipitation, high wind loads, increased wave action and storm surges. These events will lead to a variety of impacts for ports including: increased coastal runoff and siltation (requiring ports to generate increased GHG emissions from more frequent dredging); increasing bioavailability of entrained heavy metals and other pollutants; increased high wind and temperature stoppages under OH&S laws, with consequential delays to berthing and cargo-handling operations; and coastal flooding requiring increased coastal hardening. Extreme wind speeds are likely to require engineering upgrades to piers, berths, wharf moorings, container gantries and other cargo-handling equipment.

Most ports are already operating at their limits as far as channel utilization capacity goes, and margins for error have been reduced to a bare minimum. Adverse weather conditions will require shipping to go outside the boundaries of Permanent International Association of Navigation Congresses (PIANC) guidelines for channels. Present resources may not be adequate to facilitate safe movements, resulting in down time. Pilots and tug operators will probably call for the safety margin in channels and swinging basins to be significantly improved by (i) reducing length and beam restrictions or widening channels; (ii) restricting transits of vessels with high wind loads.

Extreme weather and changing climatic conditions may also force the creation of new ports and marinas as commercial fish stocks migrate further southward and trading patterns change in keeping with greater focus on greenhouse gas emissions. Development of new ports are likely to occur in response to proximity of bulk export resources as well as the conflicts with urban development restricting existing port expansion requirements.

Rising ocean and air temperatures coupled with increasing ocean acidity are likely to increase corrosion, biodeterioration and biofouling and also create new opportunities for biological invasions associated with shipping. New environmental conditions coupled with ships spending increased time in port waters due to more frequent storms will result in greater potential for the establishment of species in ports.

It is highly likely that these weather changes will drive changes to the regulatory framework, including: tightened planning regulations for coastal developments and spoil dumping, and changes to the weather components of OH&S legislation. Other changes will probably include upgraded requirements for harbour tug capacity, changes to draft and air-draft restrictions, and possible moves towards cold-ironing. Regulatory restrictions on port emissions will probably lead to significantly slower transit speeds in port waters and the need for greater fuel efficiency in cargo-handling systems and operations. The use of alternative fuels, such as LNG, in ships to reduce greenhouse gas emissions will, in turn, require new infrastructure for bunkering operations.

Moreover, the ports industry will not only be subject to climate impacts but is also seen as an important point source for ongoing emissions. CO_2 is the major greenhouse gas emitted by ships. SO_x emissions have a potential cooling effect and/or create local climate disturbance. The sulphur content of marine fuels will decrease due to measures being adopted by IMO. NO_x, HC, CO, PM, CH_4, N_2O, HFCs, fugitive VOCs, POPs and others have a minor role in global warming and will need to be accounted for. SO_x, NO_x, HC, CO, PM, fugitive VOCs, POPs and others have significant local air quality impacts. Although most air quality emissions only play a minor role in global warming, the increased community awareness of emissions due to the prominence of the global warming issue, and the potential for climate change to affect the dispersion of air quality emissions, will emphasize the need to accurately quantify air quality emissions. There is also the potential for greenhouse gas mitigation measures to affect levels of air quality emissions. Emissions sources at ports include ocean going vessels, harbour vessels, cargo handling equipment, locomotives and vehicles (EPA 2009).

Climate change also introduces another level of complexity to the operating environment in which port authorities and government plan Australia's port infrastructure capacity. With approximately 30,000 commercial vessel calls a year (Ports

Australia 2009), ports are a critical element of Australia's trade infrastructure. Although there is great diversity in the delivery of current capacity in our ports, ranging from small, regional ports such as Flinders Island to major ports such as Melbourne and Sydney, all contribute to Australia's position as a global trader, with nearly 800 million mass tonnes of throughput in 2007–08 alone (Ports Australia 2009). Over 99% of Australia's exports and imports, by mass, are carried by sea, and expectations are that Australia's trade will continue to grow: international container trade alone is predicted to nearly treble by 2020 to almost 12 million TEU. It is therefore critical that Australian ports remain effective for the industries they serve.

Results of the Training Needs Analysis: Ports and Shipping

How does this review of the issues for ports compare with industry understanding and perception of what the issues are? The following section sums up the range and breadth of these issues (Table 2).

This summary shows that in relation to the physical impacts of climate change in ports there is a diversity of opinion, both within the industry and the literature about it, about what the needs are in responding to the problem. To clearly identify appropriate adaptation and mitigation strategies, this diversity needs to be incorporated. For example, different components of the industry will have different needs. A climate change response strategy within a single port, may in fact, then, be a composite of many different types of adaptation and mitigation mechanisms, appropriate to the different needs of the sub-components of the whole. Within Australia, different ports are in fact looking at different strategies. The Port of Melbourne, for example, is developing a climate change policy, and the Port of Sydney is developing a climate change risk assessment and has instituted the practice of considering climate change mitigation factors in development applications.

This summary also shows that there is a diversity of issues confronting professionals working in the ports area. Interestingly, while there was variability in emphasis about these issues across different ports, (i.e. for one port, infrastructure might be of greater importance than navigation), there was little conflict within the sector about the issues per se. While our review did not cover the relationship between port personnel and other stakeholders, future research might reflect on whether the resolution of some of these issues and needs may not only affect others outside the industry but provide an opportunity to engage with others in a profitable manner, hence avoiding potential conflict. It would also be worth reflecting on the question of at what level training or skills development might occur. It is important to distinguish between different levels of interest in training, and who might undertake it.

Reflection on the results of stakeholder workshops from a related project examining vulnerability and ports is, however, instructive (Nursey-Bray et al. in review). In this case, potential for conflict between ports in Australia, originating from the differing perceptions of port personnel over the nature of the climate change threat,

Table 2 Summary of issues raised in needs analysis of ports and shipping in relation to climate change

Impact	Need
Sea level rise	• Conduct specific risk/hazard and vulnerability assessments to determine how these climate change hazards will affect the shipping industry
	• Durability of port coastal infrastructure needs examination, monitoring to determine its resilience to sea level rise and associated impacts
	• Development of the appropriate adaptation strategies (retreat, protection) to reduce infrastructure impacts
Managing extremes	• Risk assessment/hazard training needed
	• Raising structures and services above expected inundation levels
	• Building a dyke or levee to keep the rising sea back
Warmer seas	• Increase quarantine measures against the risk of introducing invasive marine species
	• Develop and adapt strategies for the monitoring and management of both ballast water and anti-fouling/hull fouling systems
Change in wind conditions	• Potentially increase birthing time at docks causing economic impacts for shipping companies and time delays for ports
	• Invest in boat designs that handle increases in wave height as result of warming
	• New locations for terminals
	• New emergency responses to enable stability when departing and entering marinas and terminals
Changes in sea chemistry	• Monitoring strategies for increased salinity levels
	• Adapt coastal structures that may be affected by corrosion or degradation
Navigation	• Need to understand the varied affects and impacts climate change will have on the sectors related to navigation
	• Need to consult the stakeholder groups (port, coastal, offshore and vessels) that operate within the industry
Infrastructure	• Increased understanding of methods for reducing impacts to existing and future marine and coastal infrastructure
	• Risk assessment
	• Ongoing research capability
	• Retrofitting
	• Need new and efficient climate-sensitive building designs and methods for old and new buildings
	• Relocation of marinas and wharfs
Vessel impacts	• Alternate routes for entering ports and adjusting docking procedures
	• Research to develop methods for reducing turbulent seas reducing port closure allowing vessels to dock or depart safely

(continued)

Table 2 (continued)

Impact	Need
Port closures	• Examine gaps in knowledge and improve methods and designs for the implementation of adaptation measures to reduce these disruptive impacts
Reduction in emissions (from vessels, handling of bulk products, and transport operations undertaken by port customers)	• Fuel efficiency measures or alternate fuel and energy source • Ship/port operational changes for improved efficiency and fuel savings • Market-based instruments to encourage behavioural change
Stakeholder consultation	• Training professionals in how to consult with community, stakeholders and partners

Industry recommendations based on needs analysis
- Appropriate adaptation training needs to be provided to maritime professionals so they can develop clear objectives, strategies and actions
- Train professionals in adaptation responses, policy requirements and implementing, risk assessment and response training and managing and conserving coastal regions and their infrastructure
- Management plans regarding infrastructure and operations should be adapted to allow for protection at all levels from climate impacts
- The use of adaptation tools such as Geographical Information Systems (GIS) to monitor any changes in sea level rise or topography of coastal areas in order to monitor vulnerable areas and adapt any plans accordingly
- Buildings that are in danger of being impacted by climatic events should be retrofitted to the highest standards
- Establish communication channels between scientists, stakeholders and maritime professionals. This will include creating partnerships with key research agencies, including CSIRO
- Coastal development needs much stronger regulation and monitoring
- Identify priority areas for action on adaptation and mitigation. Development of a climate change strategy to assist ports in developing actions, goals, strategies and targets for mitigating against and adapting to the impacts of climate change
- Identify any future information requirements and fund research studies to reduce any gaps in knowledge

and therefore need to respond to it. It was clear in this instance that some port managers evinced a higher order of scepticism than others. Moreover, those who were concerned about the impacts of climate change expressed a high conviction in the ability of ports to respond to change because they "always had". Thus, for them, the need for training, delivered by external parties, was not necessarily prioritized.

Taking the Next Step: Managing for Change...

This training needs analysis has implications for port personnel, not least in establishing what level of adaptation and mitigation needs to be implemented at what level. For example, an appropriate training programme for a port worker will not be the same as that for a port manager. There are clear lines of management, delegation and responsibility that need working out, as well as the funding investment required to build adaptation and mitigation options at all levels. Port authorities and the heads of shipping industries often prepare management plans for specific areas such as infrastructure, vessels and port operations. These plans are intended to guide decisions about how these areas should be managed in a balanced way and may include detailed recommendations for setbacks for development, as well as designating specific areas for particular uses. In the case of climate change, the analysis highlighted that management options will now need to incorporate adaptation and mitigation strategies to reduce the impacts of climate change to the region.

While our review highlights a number of issues that will affect the industry and merit inclusion into a climate change adaptation training course, we argue that there are two areas where priority attention is vital. The first is the need to invest in research, development and implementation of adaptation strategies for the following:

1. the vulnerability of ports to extreme weather events;
2. increased monitoring of coastal zone and infrastructure;
3. implications of coastal erosion, sea level rise and increased storminess;
4. looking at potential dredging requirements; increasing the height of dykes;
5. relocation of ports and industries;
6. examining impacts of increased storminess on vulnerable shipping routes;
7. adapting existing building codes to ensure that long-term infrastructure will be buffered against future climate risks;
8. updating of disaster management strategies;
9. implications of coastal erosion, sea level rise and increased storminess.

The second need is to train ports and shipping professionals in risk and vulnerability assessment techniques. It is important that maritime industry planners and professionals are able to rate the probability of climate change impacts in order to assess the social, ecological, political, economic and legal consequences of that particular impact. The use of risk and vulnerability assessments allows professionals to identify areas in their industry most susceptible to climate impacts in order to establish priorities and incorporate these into the strategic planning process (Nursey-Bray and Ferrier 2009). The ability of planners, managers and professionals to be competent in the use of information collected during a vulnerability/risk assessment to estimate the consequence, probability and resulting risk of specific climate change impacts to systems in a planning area will become vital in coming years.

There is much experience in this area to draw upon, as the term vulnerability has multiple definitions. Adger (2006), and Cutter et al. (2003) provide good overviews of its history and intellectual precedents. Füssel (2007), for example, outlines a suite of approaches within vulnerability research including (i) classical approaches, (ii) risk-hazard approaches (Downing and Patwardhan 2004), (iii) political economy approaches (Adger and Kelly 1999), (iv) pressure and release model which defines risk as a product of hazard and vulnerability (Blaikie et al. 1994), (v) integrated approaches which combine the above such as

Cutter's model, which combines exposure to hazards with the interaction with the social profile of communities (Cutter et al. 2003), and (vi) the resilience approach, based on the idea of exploring vulnerability in the context of social-ecological resilience (Folke 2006).

Füssel (2007) notes that the United Nations outlines four groups of vulnerability factors that are relevant in the context of disaster reduction (and of relevance to the ports sector): physical factors (i.e. external stressors to which people/property are exposed), which describe the exposure of vulnerable elements within a region; economic factors, which describe the economic resources of individuals, populations groups and communities; social factors, which describe non-economic factors that determine the well-being of individuals, population groups and communities, such as the level of education, security, access to basic human rights and good governance; and environmental factors, which describe the state of the environment within a region.

All of these factors describe properties of the vulnerable system or community rather than of the external stressors. However, they cannot be simply adopted; professionals need additional training and expertise to undertake such assessments. This is best summed up by a comment from the submission presented by the Port of Enfield in Adelaide, South Australia, to a Federal government inquiry on climate change and coastal communities arguing that "regional coastal vulnerability assessments (incorporating economic, social, and environmental impact assessment) be undertaken in accordance with a nationally consistent framework, with the flexibility to include particular issues of local interest or concern" (Port of Enfield 2008, p. 3).

Summary: Responding to Change

This paper reported on a training needs analysis conducted on the ports industry in Australia. The analysis highlighted that (i) the industry needs to make some key investments in research so as to ensure the knowledge about how to address climate change issues are addressed, and (b) that the industry would benefit from training in vulnerability assessments so as to identify key issues, and build ongoing adaptive capacity. In this way ongoing resilience will buffer the ports and shipping sector against the potentially sudden perturbations of climate, and the ongoing stresses caused by climate change.

The ports and shipping industry is not a stranger to change, nor responding to it over time. However, there is less time now to adapt, and a greater imperative to build resources from within to cope with external stressors. Both mitigation and adaptation strategies are required in order to effectively respond to the idea of climate change, and investments in the research and training needed to build understanding of vulnerability over time. Then the industry will build the skills essential to ensuring it has ongoing capacity to address climate-induced risks in the future.

References

Adger WN (2006) Vulnerability. Glob Environ Chang 16:268–281
Adger WN, Kelly M (1999) Social vulnerability to climate change and the architecture of entitlements. Mitig Adapt Strateg Glob Chang 4:253–266
Blaikie P, Cannon T, Davis I, Wisner B (1994) At risk: natural hazards, people's vulnerability, and disasters. Routledge, London
Cutter SL, Boruff BJ, Shirley WL (2003) Social vulnerability to environmental hazards. Soc Sci Q 84:242–261
Downing TE, Patwardhan A (2004) Assessing vulnerability for climate adaptation. In: Lim B, Spanger-Siegfried E (eds) Adaptation policy frameworks for climate change: developing strategies, policies, and measures, chapter 3. Cambridge University Press, Cambridge
Folke C (2006) Resilience: the emergence of a perspective for social-ecological systems analyses. Glob Environ Chang 16:253–267
Füssel H-M (2007) Vulnerability: a generally applicable conceptual framework for climate change research. Glob Environ Chang 17:155–167
IPCC (2007) Climate change 2007: the physical science basis. contribution of working group 1 to the fourth assessment report of the intergovernmental panel on climate change. In: Solomon S, Qin D, Manning M, Chen Z, Marquis M, Averyt KB, Tignor M, Miller HL (eds). Cambridge University Press, Cambridge
Nicholls RJ, Hanson S, Herweijer C, Patmore N, Hallegatte S, Corfee-Morlot J, Château J, Muir-Wood R (2007) Ranking port cities with high exposure and vulnerability to climate extremes: exposure estimates, In: OECD environment working papers, Number 1, OECD Publishing
Nursey-Bray M, Ferrier T (2009) Risk assessment and local government, Tasmania: applying an inter-disciplinary approach to climate change adaptation. In: Filho W, Mannke F (eds) Interdisciplinary aspects of climate change. Peter Lang, Frankfurt, pp 245–265
Nursey-Bray M, Hewitt C, Francis J, Brooks B, Wright J, Lawrence N, Haugstetter H, Blackwell B, Campbell M Ports, climate change and risk: adaptation as an evolving risk management strategy (in review)
Port of enfield (2008) Submission to the commonwealth house of representatives standing committee on climate change, water, environment and the arts, Inquiry into climate change and environmental impacts on coastal communities
Ports Australia (2009) Australia's Port Industry. http://www.portsaustralia.com.au/port_industry/. Accessed 4 September
Steffen W (2009) Climate change 2009: faster change and more serious risks, department of climate change, Australia. http://www.anu.edu.au/climatechange/wp-content/uploads/2009/07/climate-change-faster-change-and-more-serious-risks-final.pdf

Chapter 18
Preliminary Report on the Socioecology of Rural Groundwater Markets in the Gingee Watershed

Susan Varughese and K. V. Devi Prasad

Abstract According to the latest World Bank report, India is the largest groundwater user in the world and accounts for approximately 60% of the irrigated agriculture in the country. In the hard rock and semi-arid areas of peninsular India, uncontrolled overexploitation of groundwater is leading to a continuous decline in the water table, driving farmers to participate in and increase their dependence on informal markets for irrigation water. This paper examines the existence, structure and role of groundwater markets in the Gingee river watershed in the Tamil Nadu and Puducherry region of India. The state government's policy of providing free electricity, subsidized bore installation schemes and promoting water-intensive crops has led to a drop in the water level in successive years. The large landholders have the resources to dig bores at greater depths; small landholding farmers depend greatly on surface water and rainfall, leading to exacerbation of inequalities. With changes in climate and rainfall variability during the last few years, and also the ineffectual surface water irrigation system, the whole agriculture structure has received a setback.

Keywords Groundwater · Water markets · Agriculture · Canal · WEMs · Tamil Nadu

Introduction

Today, groundwater accounts for more than half of the total irrigation in India. The agriculture sector is the largest consumer of water, with 90% of the total available water being used for irrigation. Agriculture in India accounts for as much as 85%

S. Varughese (✉) · K. V. Devi Prasad
Department of Ecology and Environmental Sciences, Pondicherry University,
Kalapet, Puducherry, India
e-mail: susan.ecology@gmail.com

W. Leal Filho (ed.), *Climate Change and the Sustainable Use of Water Resources*,
Climate Change Management, DOI: 10.1007/978-3-642-22266-5_18,
© Springer-Verlag Berlin Heidelberg 2012

of the total annual withdrawals. The scarcity of water resources along with the rapid increase in the demand for water has resulted in overexploitation of aquifers. It is often stated that the availability of cheap water to the agricultural sector has tended to encourage its pre-emption for a low-value, high-volume use and has encouraged its waste and profligate consumption (Narain 1997).

Indian villages on the whole have traditionally relied on local rainfall for meeting their water requirements. When crop production was practised mostly under rain-fed conditions, the development of irrigation was mainly through exploitation of surface water and irrigation and was primarily protective and extensive. The increasing use of well and bore well irrigation has resulted in a rise in net irrigated area, but at the same time has also led to overexploitation of groundwater (Janakarajan 1993).

Water is one of the main inputs in modern agriculture and determines, to a large extent, the level of agricultural production and income. Use of groundwater in irrigation provides a higher degree of control to the farmer than canal irrigation does, in most areas. The practice of selling groundwater appears to have been prevalent in many parts of the Indian subcontinent even under traditional water extraction technology. The earliest formal reference to water-selling can be traced to the mid-1960s (Patel and Patel 1969). In many other parts of the Indian subcontinent, especially Pakistan, the Indian states of Punjab, Haryana, western and central Uttar Pradesh, West Bengal, Assam and in Bangladesh, and in some southern pockets, water transactions in some form or other have existed since the 1940s.

A water market in the Indian context is an informal arrangement in which individuals having access to water extraction mechanisms trade their water rights with each other or to outside parties. There is "renting" of water between neighbouring farmers and the seller sells some portion of his water, for a brief period of time, without any legal conditions. Although the volume of water sold is not metered, the buyer and seller have good information on the amount exchanged (Dinar et al. 1997). The key feature that distinguishes water markets from other forms of irrigation institutions such as a commonly owned tank is that water pumped is assumed to be the property of the pumper (Shah 1993). Informal water markets emerged in order to bridge the gap between the demand and supply of water, and these markets continue to exist in many parts of the country (Venkatachalam 2008).

Over the years, groundwater markets have come to play a significant role in India's groundwater economy (Bhatia et al. 1995). The rising share of groundwater in irrigation is attributed largely to its higher productivity compared with canal or tank irrigation (Shah 1993). In the hard-rock water-scarce areas of Tamil Nadu, the evidence in relation to the development of water markets is mixed. Guhan and Mencher (1983) found active and developed water markets operating in Iruvelpattu village in Ramanathapuram district of Tamil Nadu (Shah 1993). Swaminathan and Kandaswamy (1989) in their study in the Avinashi and Perianaickenpalyam taluk of Coimbatore district found the absence of water-selling; while Copestake (1986) found active water markets in the Shivaganga block near Madurai. He found evidence of water sales from wells as well as tanks.

Researchers and academics are divided on the equity and efficiency impacts of groundwater markets. In economically efficient resource allocation, the marginal benefit from the use of the resource should be equal across sectors. Equity of water allocation is particularly concerned with the "fairness" of distribution across economically disparate groups in a society or across time and may not be compatible with efficiency objectives (Dinar et al. 1997). Some favour water markets on the grounds of equity in resource distribution, in terms of access to water to farmers irrespective of their social and economic status (Rosegrant and Binswanger 1994; Meinzen-Dick and Sullins 1994). Groundwater markets make the resource accessible to those who cannot afford it. Water markets increase the use of installed pumping capacity, thereby improving the economic efficiency of private bore irrigation (Meinzen-Dick and Sullins 1994). Those who oppose water markets argue that it amounts to favouring the rich over the poor by monopoly rents, leading to worsening of income distribution (Janakarajan 1993; Singh 2002). Narain (1997) noted that in the absence of well-defined rights and an institutional structure in water, the development of groundwater markets could stimulate aquifer depletion while at the same time creating a powerful monopoly of water lords. Both arguments hold good under different situations. The former may be true in water-abundant regions and the latter in water-scarce regions.

Dwindling water tables often catalyse the process of "competitive deepening" of modern water extraction mechanisms (WEMs) as each user chases a declining water table. Over the years, the area irrigated by tanks in Tamil Nadu is decreasing while the area irrigated by wells is increasing (Palanisami and Meinzen-Dick 2001). Both underpricing of water and lack of cost recovery mechanisms in government-managed irrigation systems have resulted in poor operations and maintenance (Bandara 2005). Economic imperatives force farmers to choose "cash" crops that are generally water intensive over "food" crops which are less water-demanding, as the former provide greater economic return per unit of land. Tail-end farmers do not get canal supplies and groundwater is not recharged, which leads them to spend additional money on pumping groundwater from lower depths (Shah 2004).

Kajisa and Sakurai (2005) studied the nature of groundwater markets by examining the impacts of bargaining relationships and of output-sharing contracts on efficiency and equity, using household data from Madhya Pradesh, India. In a study conducted by Dubash (2000), in two villages in Gujarat, he concluded "that the markets are 'informal' should not be equated with an assumption that they are unregulated". He also reasoned that while spatial characteristics, land patterns and water depth exert a powerful shaping influence on groundwater markets, the actual form of outcome depends on socioeconomic factors such as the distribution of land ownership, access to credit and caste. He pointed out the fact that analysis of specific village characteristics is required to understand these path-dependent outcomes.

Agriculture in Tamil Nadu and Puducherry largely depends on southwest and northeast monsoons. Taking advantage of the southwest monsoon, large-scale coverage of oilseeds, pulses and cotton crops are taken up under rain-fed cultivation. The northeast monsoon rains are also very crucial for recharge of sub-soil water. In a climate change scenario, the decreasing groundwater level and the erratic rainfall pattern will present new challenges and uncertainties to the whole irrigation system.

The purpose of this paper is to examine the presence of groundwater markets in the Gingee watershed. The paper deals with the preliminary analysis of the extent of water market development, the nature of the transactions and who participates in such transactions, with special reference to the Veedur command area in the Villupuram district of Tamil Nadu and Mannadipet commune of Puducherry. The paper attempts to document the dependence of farmers on informal water transactions for irrigation in the canal command areas.

Study Area

The area under study forms a part of the Gingee watershed. This watershed comprises of parts of Tamil Nadu and Puducherry,[1] extending from north latitude 11° 55′ 00″ to 12° 33′ 00″ and east longitude 79° 18′ 00″ to 79° 52′ 00″. The total geographical area of the basin is 2250 km^2. The watershed lies on the drainage basin of the Gingee river, also known as "Varahanadi", which crosses Puducherry diagonally from northwest to southeast. The river is basically ephemeral, flowing only during rains and floods. Even during rains, due to interdiction of the flow for irrigating the upper reaches, the river flows only intermittently.

The Veedur Main Canal takes off from the saddle dam beyond the left flank of Veedur through two sluices. The channel feeds a total irrigated area of 1,280 ha, of which 400 ha are in Puducherry region. Under the Veedur dam command area, sixteen villages are officially the beneficiaries of the dam. Of these sixteen, eleven fall under Tamil Nadu and five under Puducherry.

Methodology

Six of the sixteen villages were chosen for the study, two from each head, middle and tail end, located in the Villupuram district of Tamil Nadu state and Mannadipet commune of Puducherry Union Territory, respectively. The grounds for choosing these villages were that all of them are beneficiaries of the Veedur dam canal command area, and would have been expected to use the canal system as the main source of irrigation. The existence and extent of water transaction under such a scenario would be an indication of the dependence and importance of groundwater in irrigation. It also provides an insight on how these transactions differ along the head, middle and tail reaches within the command area.

---

[1] Formerly known as Pondicherry, the name of Union territory was changed to Puducherry by an Act of Parliament in 2006.

A distinctive feature of many of the canal irrigation systems of India is that the farmers in command areas of the canal depend on pumping groundwater for irrigation (Shah 2009). Choudhury (2007) studied eight irrigation systems and found that tail-end farmers usually receive much less water than they need for growing crops. He found that even in the canal command area, less than 58% of farmers rely on canal water for irrigation.

The present study was conducted at two levels. A focus group discussion was first conducted in all the villages to get a general idea about the villages, their main source of irrigation, changes in the cropping pattern, nature of water transactions, sources of irrigation and the causes of the changes in the cropping patterns and irrigation sources. A detailed individual questionnaire survey was later conducted at households chosen randomly in all the villages. A total of 138 interviews were carried out from the six villages, taking 10% of the population subject to a maximum of 50. The survey was conducted from August 2009 till January 2010.

Structure and Role of Groundwater Markets

Selling and buying of groundwater is a common practice in most of the villages, though the manner in which water markets function varies across different regions. A study conducted in four villages of the Banaskantha district of Gujarat concluded that groundwater irrigation is a major source of livelihood not only for well owners, but also for water buyers and farm labourers (Kumar et al. 2004).

In the present study, the existence of groundwater markets was found in all the villages except one. In all cases where water transactions occurred, it was found that the selling and buying of water has been a common practice for generations and is based on the ownership of wells. The villages surveyed under the present study had similar climates but different geologies. The main monsoon season and the major rainfall in this part of the country occur during the northeast monsoon. The geomorphology of the Puducherry region comprises mainly alluvial and flood plains, whereas the Tamil Nadu part consists of hard bedrock in the middle reaches and clay and sandstones in other parts. The farmers in the area practise irrigation, either through canal, tank or personal water extraction mechanisms (WEMs), depending on the availability of canal/tank water. Farmers without any WEMs either depend on rainfall or buy water from other farmers at an hourly basis or on an output-sharing contract basis.

Table 1 gives a comprehensive outline of the market structure of the six villages. The head and middle reaches are in the Tamil Nadu state, whereas the tail end comes under Puducherry. As can be seen from the table, the water transactions perform an important role in the irrigation of crops at the tail end and at the head end. There are few water transactions in the mid-reaches of the canal, mainly due to the geology of the area. These middle-reach villages have hard bedrock at a depth of around 40–60 m. There are no bores and existing wells do not go beyond a depth of 30–60 m in the majority of cases. Most of the existing wells do not have

Table 1 Market structure in the study area

		Number of agricultural households	Number of well owners (non-sellers)	Number of well owners (water sellers)	Number of water buyers	Number of well owners who are also water buyers	% of respondents involved in water transactions
Head	Veedur	33	14	3	0	2	6.06
	Pombur	30	13	5	7	1	26.67
Mid	Nemili	20	7	2	0	0	0
	Thollamur	20	5	0	0	0	0
Tail	Lingareddy-palayam	20	7	5	7	0	36.84
	Suthukeny	15	5	1	9	0	60

water during the dry season and those with water in their wells have just enough to irrigate their own fields. There is no surplus left for any water transactions to occur.

Among the six villages, Thollamur did not have any instances of a respondent selling or buying water over many years, which was also confirmed during focus group discussions. In Thollamur, the villagers depend mainly on canal and rain water and raise only one crop a year. The situation worsens during years when rains fail or rainfall is below normal. During normal monsoons the water from the canal and tank is enough for three months. For the rest of the year, villagers work as labourers in the granite mining industry in and around the area or they go to neighbouring villages or states as agricultural labourers. It was also observed that most of the respondents sold some part of their land to the granite industry, as it was more profitable than cultivating it.

The number of well owners in Veedur was higher compared to other villages. Being the first village from the dam, it also enjoys assured water availability for longer periods. The availability of free electricity to all farmers in Tamil Nadu could be one of the reasons for the proliferation of bore wells. During the survey, only five farmers were found to be using diesel pumps. Also, Table 1 shows that among the two head-end villages, there were more water transactions in Pombur than in Veedur. This is because Pombur is the first beneficiary village in line, and canal water is available for five to six months.

However, Pombur has to wait for the water to come through the canal, which passes two villages, and then fill up the Pombur tank from where it is diverted to different fields. Therefore, to reduce their dependence on canal water, people engage in informal water transactions. Also, as Varahanadi has irregular water flows, water is not available at all times. This, together with the unpredictable rainfall patterns, has generated water trading.

The tail-end villages were more active in water selling and buying, as indicated by the number of water sellers and buyers among the respondents. The lack of canal water over many years and the subsidies provided by the government to install new bore wells have played a significant role in the existing state of affairs in these villages.

Table 2 Well ownership by land ownership

	Land owned(acres)	0 to ≥ 2.5	< 2.5 to ≥ 5	< 5 to ≥ 10	> 10
Veedur	well owner	4	4	4	7
	no well	9	5	0	0
Pombur	well owner	2	7	6	4
	no well	8	3	0	0
Nemili	well owner	1	5	3	1
	no well	4	4	2	0
Thollamur	well owner	0	2	3	0
	no well	13	2	0	0
Lingareddypalayam	well owner	4	2	3	3
	no well	5	2	0	0
Suthukeny	well owner	0	1	1	4
	no well	9	0	0	0
	TOTAL	59	37	22	19

There have been cases where a water seller also became a water buyer for some parcels of land. Only three such cases were observed in the area where the WEM owner was also a buyer of groundwater from other sellers. It was observed in all these three cases that the owner's land was away from the command area of their own wells.

The ownership of water is closely associated with land ownership in India. The chances of being a well owner increases with an increase in the land owned. Table 2 provides data on well ownership by land ownership. The respondents in the area were classified under four different categories of land ownership, namely marginal (less than 2.5 acres), small (between 2.5 and 5 acres), medium (between 5 and 10 acres) and large (more than 10 acres). The majority of respondents were marginal landowners and owned the lowest number of wells. Most marginal farmers were totally dependent on the canal system and rainfall for cultivation and were also partly dependent on large well owners to provide water when there was no water from the canal. Nearly all the medium and large farmers have their own wells.

In terms of cropping patterns, although paddy still continues to be the major crop in the area, casuarina has become an important competitor in terms of acreage. As a result of the problems associated with labour availability, difficulties in getting access to water at the right time, erratic rainfall and low profit, most respondents shifted to casuarina as it required low amounts of water and labour and yielded higher income per acre. It also gives farmers a chance to earn money from other sources.

Table 3 gives a picture of the major crops in the study area and the acreage under each crop in all the six villages. Most of the farmers grow paddy just after the northeast monsoon (December to March). Where water is available through canals or private wells, paddy is grown two to three times a year. The land under paddy cultivation is higher in Veedur and Pombur compared to the other villages. According to the respondents, the main reason for this difference is the availability

Table 3 Village-wise acreage under major crops

Village name	Paddy	Sugarcane	Casuarina	Black gram	Cotton	Groundnut	Other crops
Veedur	73.9	33	75	18	10	36	11.1
Pombur	67.1	65.3	22	15	4	14	16
Nemili	22.5		37	17		8	8
Thollamur	39		22.5			2	2
Lingareddypalayam	17.9	42.5	72.5			3	
Suthukeny	30.9	46.3	25.7				

of water for a longer period in the head end and the production of paddy per acre is higher from dam water than groundwater. Respondents also stated that paddy under canal irrigation produces around 25–30 quintals per acre as compared to 18–25 quintals per acre under groundwater irrigation. Though the production per acre was not as much as in the head end, it was observed that paddy was cultivated two to three times in Puducherry villages, mainly due to good-quality soil and assured water supply from bores.

Sugarcane was the third most important crop after paddy and casuarina. Sugarcane is a highly water-intensive crop and, as Table 3 shows, none of the respondents from the hard bedrock middle-reaches grew sugarcane. Focus group discussion revealed that Pombur especially leads in terms of acreage of land under sugarcane as a result of state government subsidies and encouragement of drip irrigation scheme in this village. It is generally believed that the production of sugarcane is higher per acre under drip irrigation as compared to normal irrigation. In Puducherry, the presence of a sugar mill in Lingareddypalayam since the 1980s has gradually led to a shift from paddy to sugarcane cultivation. The government schemes favouring sugarcane cropping and subsidized bore installation schemes have encouraged sugarcane production and today most farmers prefer growing sugarcane as per-acre return is higher. Both sugarcane and casuarina were found to be more profitable and less labour intensive than paddy.

Among the legumes and pulses, groundnut accounted for around 7% of the total cultivated area in the study villages, followed by black gram, sometimes used as an intercrop with casuarina or groundnuts. Many dryland crops, such as Ragi and Bajra, grown previously, were either not grown any longer or confined to a few farmers with small acreages. Cotton was grown mostly as a rain-fed crop in the head-end villages. Other crops grown included the eucalyptus tree, bananas, fodder grass and vegetables such as lady's finger, brinjal and chillies.

Wells are the principal source of irrigation in Tamil Nadu and during 2007–2008, the net area irrigated by open wells and tube wells/bore wells accounted for about 55.6% of the total net area irrigated. The net area irrigated by wells during 2007–2008 was highest in Villupuram district, which accounted for 10.8% of the total net area irrigated by wells in the state (Season and Crop Report 2007–2008, Department of Economics and Statistics, Tamil Nadu). In Puducherry, during 2008–2009, the net area irrigated by tube wells was 64%, with canals accounting for only about 35% of the total net irrigated area (Season and Crop

Table 4 Comparative village dependence on surface and groundwater (number of farmers)

Village name	Surface water	GW + surface water	GW wells	Rented GW only[a]
Veedur	13	12	7	2
Pombur	7	4	15	7
Nemili	10	6	4	0
Thollamur	15	1	4	0
Lingareddypalayam	0	0	12	7
Suthukeny	0	0	6	7

[a] These farmers depend only on water bought from well owners for irrigation. They are also included in total groundwater users

Report 2008–2009, Government of Puducherry). Table 4 gives an overview of the dependence of each study village on surface and groundwater.

The main irrigation sources in the study area were private bores and wells. The farmers in the head and middle reach of the canal irrigated with dam and tank water for three to four months after the northeast monsoons. Table 4 provides a general overview of the dependence of respondents on surface and groundwater. It shows that groundwater played a major role as a source of irrigation with surface water irrigation absent in Lingareddypalayam and Suthukeny as a result of their location at the tail end of the main canal.

The socioeconomic conditions of these villages played a major role in their well ownership and dependence on various irrigation sources. Approximately 32% of the total respondents depended entirely on surface water for irrigation and 16% used both surface and groundwater, with the major users located at the head end of the canal. Illegal withdrawal of water through pipes attached to engines in main canals and smaller channels was also observed during the field visit to the head-end villages. Such withdrawals further reduced the chances of water reaching the tail end of the canal system. Groundwater accounted for around 35% of the total water used, with 16% of the respondents depending only on rented water from well owners. Since rented water is also groundwater, approximately 50% of respondents depended on groundwater sources for their irrigation needs. Groundwater is not used if canal water is available and if rainfall is normal. However, as this was not the case most of the time, pressure on the aquifers has increased rapidly, resulting in a lowering of water tables.

In 2005, an initiative was taken up by the Tamil Nadu government to develop the basin command area and increase the efficiency of irrigation systems with the help of the World Bank. The main objective of this venture, known as the "Irrigated Agricultural Modernisation and Water Resources Management" (IAMWARM) Project, was rehabilitation and modernization of anicuts,[2] flood banks, supply channels and tanks in the Varahanadi sub-basin. Under this project, the existing irrigation facilities such as the system tanks, non-system tanks and dams of the basin are being modernized to harness the full benefit of the available water

---

[2] Anicut is a Tamil word meaning "dam".

Table 5a Details of water buyers in the study site

	Veedur	Pombur	Nemili	Thollamur	Lingareddypalayam	Suthukeny
Max. no. of sellers/buyers	1	3	–	–	2	1
Output share[a]	(1/3)	(1/3)	–	–	(1/3,1/4)	(1/3,1/4)
Hourly rate[b]	–	(20–50)	–	–	–	–
Seasonal contract[c]	–	(1,000–7,000)	–	–	–	–

[a] The value in brackets denotes the fraction of share of the total produce given to sellers, either in cash or crops

[b] The number in brackets denotes the hourly rate of buying water in rupees. The rate varies from 20–50 rupees

[c] The number in brackets denotes the amount in rupees paid for the whole season for a crop. The amount varies according to the type of crop from a minimum of 1,000 rupees to a maximum of 7,000 rupees

potential. This project is expected to increase agricultural productivity and provide benefits to the farmers. Work is still taking place to line the main canal with cement and stones for efficient transfer of water and reduction of seepage losses. Once the whole project is completed, we will know if there is any improvement in the efficiency of the canal system and if the benefit is passed to the tail-enders.

Water Transactions

Three types of market contracts for water transactions were observed in the study area, namely, fixed charge per season, flat charge per application of water, and output sharing contracts. Among these, output sharing contracts were the most common. This contract type is mainly used for paddy where the water seller takes a third of the crop produce. In the case of sugarcane and casuarina, the water seller receives a quarter of the cash value of total production.

Flat charge contracts were found mainly in the hard rock areas in the middle reaches of the canal, where farmers cultivated once yearly after the north-east monsoons. Water was bought only for the last phase of the crop season and for a short period. Tables 5a and b provide data on water buyers and sellers in the six villages. In Table 5a, the first row gives the average number of sellers per buyer, which is around two. Buyers of water generally buy from the same sellers.

In the case of the tank irrigation systems of South India, the water market works in the later part of the crop season. The water is supplied on an hourly basis with rates varying from 20 to 50 rupees an hour depending on the crop period and the demand. Normally, about two to three buyers are covered under single well-owners (Palanisami and Suresh Kumar 2004). Evidence from the state of Bihar and the Ganga–Brahmaputra basin in India shows that cash transactions (Mukherji 2004; Shah 1991; Shah and Ballabh 1997; Fujita and Hossain 1995) are very common in water markets. Palanisami (2009), found the practice of non-cash

Table 5b Details of water sellers in the study site

	Veedur	Pombur	Nemili	Thollamur	Lingareddy-palayam	Suthukeny
No. of sellers	3	5	2	–	5	1
Max. no. of buyers/sellers	4	6	5	–	10	5
Output share[a]	(1/3)	(1/4)	–	–	(1/3 and 1/4)	(1/3 and 1/4)
Hourly rate[b]	(50–60)	(30–50)	(25–30)	–	(50)	–
Seasonal contract[c]	–	(2,000–10,000)	–	–	–	–

[a] The value in the bracket denotes the output share of the total produce with the seller
[b] The numbers in the bracket denote the minimum and maximum amount in rupees paid to the seller under the hourly rate system
[c] The numbers in the bracket denote the minimum and maximum amount paid in cash for a season. It varies according to different crops

contracts in the form of share-cropping in Tamil Nadu, particularly in tank and canal command areas. Janakarajan (1993), in his study of water markets in Tamil Nadu, found cases where water buyers offered labour services such as operating pumps and irrigating well owners' fields.

Seasonal contracts were found only in Pombur, where a fixed proportion of crops was agreed upon for the whole season, the proportion varying by crop, the highest being sugarcane and the lowest pulses and legumes. Two to three cases were observed where water sellers sold water "free of cost" but with the exception that the buyer would offer in return some other kind of service in the future. The hourly rate for water buying was used mostly for groundnuts, black gram and other legumes, as water was needed only for a short time during the last phase of cultivation. As Puducherry villages cultivated only paddy, sugarcane and casuarina, water transactions were only in the form of output-sharing contracts.

In Table 5b, water transactions are viewed from the water seller's standpoint in the study villages. The maximum number of water buyers per seller was higher than water sellers per buyer. This indirectly points to the dependence of small farmers on water transactions for irrigation purposes. Many water sellers were found to transact under all the three types of contract with their different water buyers.

The water sellers interviewed in Pombur and some sellers in Lingareddypalayam took a quarter of the total produce in terms of crops, including paddy. However, this was not common for all sellers. During personal interviews, it was found that the shares in output were increased because of competition with nearby sellers. The sellers and buyers of water had a maximum distance of 500 to 800 m between their fields. Most of the time, water was transported through PVC pipes and sometimes through small channels.

The prevalent social and ecological conditions of groundwater markets points to the fact that even though the groundwater market is not a formal institution, it plays a major role in providing access to water to marginal and small farmers, who cannot afford to have their own wells. It also points to the need to accept the existence and role of such an informal institution and to make relevant policies for both surface and groundwater use in these areas. Climate change will affect the water availability of the country. Aquifers would respond more slowly to climate change as compared to surface water systems. With a change in the climate, groundwater will become a crucial and threatened natural resource.

Conclusions

Water scarcity is becoming one of the major areas of concern in India. With increasing demand for water, its availability is decreasing. The highly unpredictable rainfall, added to increased population expansion, has led to an increase in competition between different uses and users for the limited supplies of water in many areas. Many farmers are moving out of agriculture to find more rewarding and secure jobs in nearby towns and cities while others work as agricultural labourers in nearby villages.

Groundwater irrigation is central to India's small farmers and rural poor. In the present study, groundwater ranks as the most important source of irrigation, with both head- and tail-end farmers depending on it. Around 50% of farmers depend on groundwater for irrigation. Water-intensive crops such as paddy and sugarcane take the lead in terms of acreage. With labour shortages and decreasing water availability, there is a gradual shift towards casuarina. Many are moving out of agriculture altogether.

The majority of farmers in the study area fall into the marginal category with meagre funds to invest in their own water extraction measures. They depend on groundwater for irrigation which they rent from large farmers. Even those in the canal command area depend upon groundwater as their main source of irrigation.

The increasing scarcity and competition for water across different sectors call for a more efficient, equitable and sustainable water allocation policy. The presence of informal institutions such as groundwater markets should be officially recognized and regulated in such a way as to check the overexploitation of aquifers and to provide opportunities for small and marginal farmers to access to water. Government policies which promote water-intensive crops and provide subsidies for installing bore wells need to be changed and will require strong political will to achieve this change.

Re-formalizing India's water strategy and a change in the administrative approach towards water management issues are needed to meet the challenge of climatic change. A more holistic and long-term approach is needed to protect the region's aquifers from irreversible damage.

Acknowledgments Susan Varughese acknowledges the guidance and help received from Mr Ravi Bhalla, Mr Dhandapani and Mr Alexander for providing logistical support for the fieldwork and Pondicherry University for research support. The authors thank the anonymous referees for their very useful suggestions.

References

Bandara KRN (2005) A national water policy for Sri Lanka, Concept paper, Interim national water resources authority. Water Resources Research

Bhatia R, Cestti R, Winpenny T C (1995) Water conservation and reallocation: best practiced cases in improving economic efficiency and environmental quality. World bank paper, UNDP/World bank water and sanitation program

Choudhury N (2007) Irrigation service delivery in canal systems: a study of eight canal systems in India. Int J Rural Manag 3:127–148

Copestake JG (1986) Finance for wells in a hardrock area of southern Tamil Nadu. ODA/NBARD research report no. 11

Dinar A, Rosegrant MW, Meinzen-Dick R (1997) Water allocation mechanisms: principles and examples. World bank policy research working paper 1779

Dubash NK (2000) Ecologically and socially embedded exchange: 'gujarat model' of water markets. Econ Political Wkly 35(16):1376–1385. Available at http://www.jstor.org/stable/4409175

Fujita K, Hossain F (1995) Role of the groundwater market in agricultural development and income distribution: a case study in a north-west bangladesh village. The Dev Econ 33(4):442–463

Guhan S, Mencher P J (1983) Iruvelpattu revisited: II. Econ Political Wkly 18(24):1063–1074. Available at http://www.jstor.org/stable/4372209

Janakarajan S (1993) Economic and social implication of groundwater irrigation: some evidence from South India. Indian J Agric Econs 48(1):65–75

Kajisa K, Sakurai T (2005) Efficiency and equity in groundwater markets: the case of Madhya Pradesh India. Environ Develop Econ 10:801–819

Kumar MD, Singhal L, Rath P (2004) Value of groundwater: case studies in Banaskantha. Econ Political Wkly 39(31):3498–3503. Available at http://www.jstor.org/stable/4415346

Meinzen-Dick R, Sullins M (1994) Water markets in Pakistan: participation and productivity. EPTD discussion papers 4. International food policy research institute (IFPRI), pp 1–69

Mukherji A (2004) Groundwater markets in Ganga-Meghna-Brahmaputra basin: theory and evidence Econ Political Wkly 39(31):3514–3520. Available at http://www.jstor.org/stable/4415349

Narain V (1997) India's water crisis: avenues for policy and institutional reform. TERI Inform Monitor Environ Sci 2(1):1–6

Palanisami K (2009) Water markets as a demand management option: potentials, problems and prospects. In: Saleth RM (ed) Strategic analyses of the national river linking project (NRLP) of India. Promoting irrigation demand management in India: potentials problems and prospects, vol 3. International Water Management Institute, Colombo, Sri Lanka, pp 47–70

Palanisami K, Meinzen-Dick R (2001) Tank performance and multiple uses in Tamil Nadu, South India. Irrigation and Drainage Syst 15:173–195

Palanisami K, Suresh Kumar D (2004) Study about suggestions for tank water sharing in kappiyampuliyur, vakkur tanks in Villupuram district and chengam tank in Thiruvannamalai district. Report submitted to Institute for Water Studies, Chennai, India

Patel SM, Patel KV (1969) Economics of Tubewell Irrigation. CMA Report, Indian Institute of Management, Ahmedabad

Rosegrant MW, Binswanger HP (1994) Markets in tradable rights: potential for efficiency gains in developing country water resources allocations. World Develop 22(11):1613–1625

Government of Tamil Nadu, Public Works Department, WRO (2007) Rehabilitation and modernisation of anicuts, Flood banks, Supply channels and all tanks coverd by Varahanadhi sub basin Under irrigated agricultural modernisation and water resources management (IAMWARM) project, Government of Tamil Nadu, Available at http://iamwarm.gov.in/dpr-pdf/Varahanadhi.pdf

Season and Crop Report, Tamil Nadu, 2007–2008. Department of Economics and Statistics, Chennai. http://www.tnstat.gov.in/seasonandcropreport2007-08.pdf

Shah T (1991) Water markets and irrigation development in India. Indian J Agric Econ 46(3):335–348

Shah T (1993) Groundwater markets and irrigation development—Political economy and practical policy. Oxford University Press, Bombay

Shah T (2004) Water against poverty: livelihood oriented water resource management. In: Mollinga P (ed.) Water for food and rural development: approaches and initiatives in South Asia. Sage Publications India Pvt. Ltd., New Delhi, pp. 38–68

Shah T (2009) Climate change and groundwater: India's opportunities for mitigation and adaptation. Environ Res Lett 4:1–13

Shah T, Ballabh V (1997) Water markets in North Bihar: six village studies in Muzaffarpur District. Econ Political Wkly 32(52):A183–A190. Available at http://www.jstor.org/stable/4406234

Singh D (2002) Groundwater markets in fragile environments. Indian J Agric Econ 57(2):180–196

Swaminathan LP, Kandaswamy P (1989) Groundwater development and its consequences in coimbatore district—Tamil Nadu. Workshop on efficiency and equity in groundwater use and management. IRMA, Gujarat

Venkatachalam L (2008) Market based instruments for water allocation in India: Issues and the way forward, Managing water in the face of growing scarcity, Inequity and declining returns: Exploring fresh approaches. Paper presented in the IWMI-Tata water policy program, pp 498–512

Chapter 19
Impact of Climate Change on Water Supply and Food Security in Seraro District, Southern Ethiopia

Fikadu Reta Alemayehu

Abstract While the issue of climate change and its potentially devastating impact is generally agreed among researchers and policy-makers worldwide, the level of exposure to the problem varies regionally. Sub-Saharan Africa is considered one region likely to be most affected by the impacts of climate change. However, there is limited evidence indicating the specific magnitude of the impact within and between countries in the region. This study used a quantitative and qualitative research design to collect data on agricultural production, food security, child nutrition and water supply in Seraro district of Oromia region, southern Ethiopia. A total of 418 households participated in the study and detailed discussions were done with the farmers. The result of the study showed that nutritional status of the children is far from the WHO standard and that shortage of drinking water both for household and livestock is one of the most important constraints in the study area.

Keywords Climate change · Water supply · Food security · Child nutrition · Southern Ethiopia

Introduction

While the issue of climate change and its devastating impacts is generally agreed among researchers and policy-makers worldwide, the level of exposure of different regions of the world varies (Chhibber and Laajaj 2008). At the same time, the available evidence about the degree of impact and necessary adaptation measures

F. R. Alemayehu (✉)
Faculty of Life Sciences, University of Copenhagen,
Frederikssundsvej 305, 1tv., 2700 Brønshøj, Copenhagen, Denmark
e-mail: fikadureta@gmail.com

is limited and inconclusive. For example, sub-Saharan Africa is considered to be a region most affected by the impacts of climate change. However, there is limited evidence indicating the specific magnitude of the impact within and between countries as well as the necessary actions to be taken to deal with the impact. Kung'u (2007) state that world hunger and poverty are among the leading challenges of the twenty first century. In recent years, climate change and rising food prices have added another level of complexity to existing problems of hunger and malnutrition in developing countries (Cline 2007; Burroughs 2007). On the other hand, in most developing countries, economic progress has been uneven and associated with limited improvement in the well-being of mothers and children in rural areas. For example, in India, which is one of the world's fastest-growing economies, 47% of children are malnourished, with child malnourishment in the poorest states increasing from 54% in 1991 to 60% in 2001 (IIPS 2007). Despite increased economic growth in the past few years, Ethiopia is one country experiencing periodic problems of drought, food shortage and child malnutrition (IMF 2008; FAO 2009). Globally, population growth, a growing economy, together with the problem of climate change, are putting increasing pressures on water availability and affecting the lives of the 900 million people without access to adequate water. Of these, 340 million are found in sub-Saharan Africa (Baron 2009). Agriculture is the backbone of people's livelihoods in sub-Saharan Africa, including Ethiopia, and issues related to the availability of water, food security and child nutrition are crucially important in planning adaptation measures (FAO 2009). Thus, this study uses a quantitative and qualitative research design to examine agricultural production, food security, child nutrition and water supply in Seraro district of Oromia region, Southern Ethiopia.

Method

The study area was one of Southern Ethiopia's remote districts, where the problems of crop failure and child malnutrition are widespread. From 15 December 2009 to the end of January 2010, data was collected from 418 farming households, which included detailed discussions with farmers. A stratified cluster sampling technique was used to select sample respondents for the study. Data was collected using a household survey questionnaire, focus group discussion and interviews, anthropometric measurements of mothers and children less than 5 years old and secondary data sources. The quantitative data was analysed using STATA 10 software and WHO Anthro software was used to analyse children's anthropometric data. Nutritional status of children was presented in Z scores (number of standard deviations that an observed anthropometric value of a child is below or above the population mean of the WHO reference population).

Results

Socioeconomic Characteristics of the Respondents

All respondents in the household survey were mothers aged from 17 to 45 years with a mean age of 29 years. Almost all (95.7%) were married while 3.1% were widowed. Family size ranged from 3 to 21 members with a median of six. A significant number (25.8%) of the respondents' husbands had more than one wife. About 82% of mothers and 72% of husbands were illiterate. Most respondents (82.5%) came from the Oromo ethnic group and 97% were Muslim. Agriculture was the major economic activity for the majority of respondents (88%) with 12% of respondents engaged in other economic activities such as labouring, selling firewood and small-scale trading. Only 21% of respondents owned a radio while 71% of households had one or more kinds of livestock, of which cattle and chicken are the most common. Almost all (98.6%) mothers did not use any family-planning methods.

Agricultural Production and Food Security

94.5% of respondents had some agricultural land, of whom 85% owned their own land. The reported mean size of agricultural land was ½ hectare with a minimum size of ¼ hectare and a maximum of six hectares. Maize, teff, sorghum and potato were the most common crops grown in the area. The mean amount of maize produce, which is the major cereal in the area, during the harvest season at the time of data collection was 480 kilograms per household with production ranging from 100 to 5,000 kilograms. Concerning sufficiency of produced food, 93% of respondents reported that annual food production was insufficient to meet their needs. As a result, 91% reported they had to buy food either by selling domestic animals or by looking for other income-generating opportunities. About 86% reported they depended on purchased food for an average of 6 months per year. Due to the crop failure problems since the 2007–2008 harvest season, 76% of respondents reported they had received periodic food aid for the last two years for about seven successive distributions during the most stressful months (June, July and August).

Water Availability Drought

All respondents reported they had faced shortages of water/rainfall during the previous harvest season. When asked the reasons for the drought, 89% said they did not know while only 7% reported it might be due to high population and land degradation. 90% of respondents considered prayer to be the main solution to drought, with 4% stating they did not know what the solution was and only 5% mentioning planting trees as a solution. 22% reported flood problems during the previous harvest season, while 97% reported that water availability had decreased during the last 10 years. 87% considered drought and shortage of rainfall as the causes of reduced water availability and 92% said they did not know of any options to increase water supply other than prayer since rainfall was provided by the supernatural. 96.7% of respondents reported that malaria had decreased over the past 10 years as a result of the drying of ponds and decrease in waterlogging.

Nutritional Status and Health of Children

Anthropometric measurement of children from two to five years of age in the study area indicates that about 70% were stunted (they were short compared to their ages) with a height-for-age Z-score measure less than -2, and about 8% were wasted (having lower weights compared to their heights) with weight-for-height Z-score less than -2. In addition, 21% of the children had a mid-upper arm circumference (MUAC) for age Z-score of less than -2. (Figs. 1, 2)

In a multivariate regression analysis, MUAC-for-age Z-score of children from fathers having an additional wife was independently lower by 0.22 ($p < 0.05$) than those children whose fathers did not have an additional wife. In addition, children from relatively food-insufficient households had lower MUAC-for-age values by 0.05 ($p < 0.05$).

About 34% of children had signs of sickness at the time of data collection. About 43% of mothers had experienced deaths of one to five children with an average of 1.75 child deaths per mother. Among those who experienced child deaths, more than half had death of one child and 30% of them experienced deaths of two children. Diarrhoea is the major reported reason for child deaths. Increase in the number of child births by one is independently associated with increase in the number of child deaths by 0.34 ($p < 0.0001$). In addition, increase in the maternal level of education by one year and increase in the size of agricultural land by ¼ hectares is independently associated with a decrease in number of child deaths by 0.5 ($p < 0.005$ and $p < 0.0001$), respectively.

Focus Group Discussion

The focus group discussion was conducted in three groups: a group of men, a group of women and a group with both sexes. After a detailed discussion about the most important development needs of the villages, the women agreed that drinking water was the most important, since each day they had to travel on foot at least five kilometres to get water from the Bilate river. The men gave priority to food aid

Fig. 1 Weight-for-age Z-score of the children compared to WHO standard

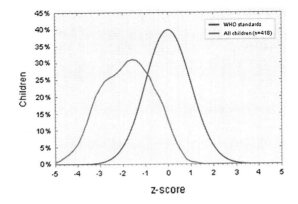

Fig. 2 MUAC-for-age Z-score of the children compared to WHO standard

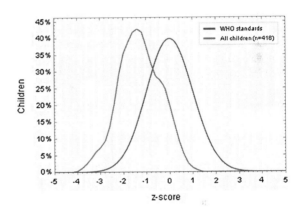

over that of drinking water. They said that since last season's crop production was poor, they were afraid that they would experience hunger again. They argued that if they received food aid, children and mothers would be able to bring water as usual. Both groups agreed that children in general were not in a good health condition. The villagers also agreed that it was the Government's responsibility to provide solutions, including providing food aid.

Discussion

The study shows that there is high problem of food insufficiency in the study area and almost all respondents had faced crop failure during the previous harvest season. Some of the demographic characteristics of the villagers such as number of child births and additional wife contributed to household food insufficiency and child mortality. The number of months of food purchase from market following exhaustion of food store from own produce was negatively associated with mid-upper arm circumference measure of the children. As the number of purchases of food from market increased by one month, children's MUAC-for-age Z-score measure decreased by 0.12 ($p < 0.01$). This indicates the importance of food prices and local agricultural production in improving food insecurity and child mortality. The villagers explained that food shortage forces them to buy food from the local market by selling their livestock and even renting their agricultural lands which in turn leaves their family in a more difficult situation. They also stated that some of the young generation migrate to urban areas to look for jobs because the agricultural production faces shortage of rain and also there is limited agricultural land. The other biggest challenge faced by the villagers was that they were using river water for drinking. The impact of shortage of safe drinking water can be seen indirectly from the high rate of child malnutrition, child deaths and the fact that diarrhoea was the major reason for child deaths. The association between unsafe drinking water and child mortality and morbidity has been widely documented in the literature.

The findings of this study give support to the United Nations development programme (UNDP 2008) report concerning fighting climate change and to a study by Ludi (2009) on climate change, water and food security. Both studies indicate that the 2008 food prices crisis resulted in increased numbers of malnourished children and left the poor in a state of high food deficit. The UNDP report warned that the achievements in fighting poverty and malnutrition can slow down or even be reversed due to climate change impacts on food production, availability of water, nutrition and community health. It is projected that by 2080, the impact of climate change could result in an additional 600 million people being exposed to malnutrition and 1.8 billion people facing shortages of water (UNDP 2008).

This finding can be a good example of the complex interaction between problems in local-level perception of climatic issues such as causes of drought, demographic problems such as large family size and population pressure, together with poor agricultural production and productivity, food prices, poor environmental health and sanitation, and drought. Thus, strategies aimed at tackling the impact of climate change and improving rural livelihoods in countries like Ethiopia need to pay attention to the above-mentioned points. In addition, local-level beliefs and the role of religion should be considered, since the farmers mentioned supernatural power as the major solution to their problems.

Conclusion

The southern Ethiopian study population is facing severe problems of food insecurity and shortage of water as a result of crop failure due to repeated drought. In addition, demographic factors also contributed to food insecurity and a high prevalence of child malnutrition. Safe drinking water is also a priority to help the community and save the lives of children suffering from malnutrition and death.

References

Baron C (2009) Water in Africa: availability and access. Civiliz Gov Water in West Africa 359:33–56
Burroughs WJ (2007) Climate change: a multidisciplinary approach, 2nd edn. Cambridge University Press, Cambridge, UK
Chhibber A, Laajaj R (2008) Disasters, Climate change and economic development in sub-saharan africa: lessons and directions. J African Econ 17(2):ii7–ii49
Cline RW (2007) Global warming and agriculture: impact estimates by country. Centre for Global Development and Petersen Institute for International Economics, Washington, DC
FAO (2009) The state of food and agriculture. Food and agriculture organization of the United Nations. Rome, Italy
IMF (International Monetary Fund) Report (2008) Accessed 28 October 2009, available at http://www.allafrica.com/stories/200903060610.html
International Institute for Population Sciences (IIPS) and Macro International (2007) National Family Health Survey (NFHS-3), 2005–06, Key Findings, IIPS, Mumbai, India
Kung'u JB (2007) Food security in africa: the challenges of researchers in the 21st century. In: Bationo A et al (eds) Advances in integrated soil fertility management in sub-saharan africa: challenges and opportunities. Springer, Nairobi, pp 105–113
Ludi E (2009) Climate change, water and food security. In: ODI Background Notes, Overseas Development Institute (Odi), London, pp 1–8
UNDP (2008) Fighting climate change—Human solidarity in a divided world. UNDP, New York

Chapter 20
Climate Change and Rural Water Supply Planning in Nigeria

Salisu Lawal Halliru and Da'u Abba Umar

Abstract This paper is aimed at providtsaveing a synthesis of what we currently know about climate change, especially its potential impacts on rural water availability in Nigeria. This is because climate change is known to have impacts on the quantity of water available as well as the geographical and seasonal distribution of this water through its significant impact on precipitation amount, its distribution and duration, as well as the rates of water loss by evaporation. This paper also joins others before it to call for more research on climate change and water resource availability for the sake of our rural population, who are the most vulnerable to climate change phenomena when it comes to water shortages. Finally, this paper gives recommendations on how this trend should be averted locally through educating the rural people that they are partly the construct of their misfortunes, mainly through bush burning and deforestation.

Keywords Climate change · Rural · Water supply · Planning

Introduction

Throughout much of the developing world (including Nigeria), freshwater usually comes in the form of seasonal rains. Although there is variability and unreliability in its occurrence with time (temporal) and with space (spatial), it is still the only source of annual recharge for both surface and sub-surface water storages.

S. L. Halliru (✉) · D. A. Umar
Department of Geography, Staff Secondary School Federal College of Education Kano, Kano state, P.M.B. 3045 Nigeria
e-mail: lhsalisu09@gmail.com

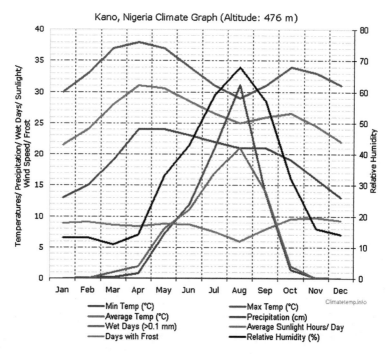

Fig. 1 Climate conditions in Kano, Nigeria (2009, NIMET)

This happens through the processes of runoff, streamflow and infiltration, since the falling precipitation ultimately sinks into the ground (as soil moisture and groundwater recharge) or runs off (as surface sub-surface flow) into the oceans, seas or some inland lakes, or is intercepted and temporarily stored (as interception and surface storage) before being eventually evaporated and transpired (as evapotranspiration), as is clearly demonstrated in the process of the water cycle (Ayoade 1988).

Because of this increased variability in precipitation, resulting in floods in humid areas, and decrease in precipitation resulting in drought in the savannah and semi-arid areas, the characteristics of the component of the hydroclimatological systems in the different ecological zones is altered, with their consequences on the availability of water resources. In its recent annual rainfall prediction, the Nigerian Meteorological Agency (NIMET) said that rainfall is expected to fall slightly below its normal level over a large portion of the country in 2009. This is likely to create water shortage in lakes, dams and rivers. The annual rainfall is expected to vary from 400 to 1200 mm in the northern half of the country to an increase from 1200 to 2800 mm in the southern half, implying high surface runoff (NIMET 2009) (Fig. 1). These conditions would eventually lead to changes in management strategies in order to balance water supply and demand through conservation efforts (Briscoe et al. 1990). This statement is especially true among the rural Nigerian population who were living under the tyranny of water resource scarcity.

Table 1 Population distribution type

Population distribution type	Community size	Population (million)	% of total
Urban	> 20,000	45.0	38
Small towns	5,000 to 20,000	40.0	33
Rural	< 5,000	35.0	29

Source. Field study, 2010

Population

Nigeria is the most populous country in Africa with an estimated population of 120 million and average density of about 130 persons per km^2. The population has been growing at an estimated average of 2.9% per annum (Table 1). The best estimate of the population's distribution is given in Table 1.

Climate Change

It is important to point out that the global climate, or climate of any part of the Earth for that matter, has never been static. Variability is an inherent attribute of climate. What is crucial is the degree of variability that climate is subjected to, as well as the duration of such variability. Minor fluctuations or variations constitute not more than a noise in climatic series and humans can easily adapt to such minor variations. However, when fluctuations in climate constitute significant departures from the normal climate or become prolonged to constitute a new climate state, then there are problems of adjustment and the environment, and humans and their socioeconomic activities become very vulnerable.

Climatic variations occur on various temporal scales, varying from a few decades to millions of years. Different nomenclatures are used to describe climatic variations depending on the timescales within which such variations occur. Thus, we have such terms as climatic trends, climatic cycles and climatic change. There are also general terms such as climatic fluctuations and climatic variability, which emphasizes the inherently dynamic nature of climate.

Climate change is therefore a situation when climatic variations or fluctuation over a long period of time occur to produce a shift or a change in the type of climate prevailing over an area. In other words, climate change represents a significant difference between two mean climatic states or climatic normals with a significant impact on the ecosystem (Ayoade 2003; Smith 1996).

Changes in weather and climate have been known to profoundly affect water resources, and thus increase human vulnerability to infection. Climate change will bring overall less rainfall and also heavier individual rainfall events in a single day. We depend on water for drinking, washing and cleaning, for agricultural needs (such as irrigation), for industrial needs (such as food processing), and for transportation, fishing and recreation. Pre-existing problems will only get worse with climate change.

Fig. 2 Point of rural water sources

Nigeria's low-lying coastline makes the country prone to sea-level water intrusion into coastal freshwater resources as climate change brings with it a rise in sea level that will seriously affect our coastline. Coastal erosion and flooding is not uncommon, and contributes to the "polluting" of freshwater systems. Nigerians do not enjoy an adequate water supply. This problem is more prominent and devastating in the northern areas of the country with their limited sources of water and harsh weather conditions (Fig. 2).

Water Availability and Sources

About 70% of the Earth's surface is water, but most of that is ocean. By volume, only 3% of all water on Earth is freshwater, and most of this is largely unavailable because about three-quarters of all freshwater is locked away in the form of icecaps and glaciers located in polar regions far removed from human habitation. The water found in lakes, rivers and those at shallow underground level are the easily accessible freshwater and are only 1% of the total water on Earth. Only this amount is regularly renewed by precipitation and its sustainability is threatened by the alarming climate change phenomena. Therefore, in all, only one-hundredth of one per cent of the world's total water supply is considered accessible for human use (Barabas 1986).

Globally, between 12.5 and 14 billion m^3 of water are deemed available for human use on an annual basis. This amounts to about 9000 m^3 per person per year, as estimated in 1989 (Alaba 2001).

By the year 2025, global per-capita availability of freshwater is projected to drop to 5100 m^3 per person, as another 2 billion people join the world's population. Even then, this amount would be enough to meet human needs if it were to be distributed equally among the world population. But the availability of freshwater supply is not distributed evenly around the globe throughout the season or from year to year, usually because of the inherent variability of climate coupled with the topical climate change issue that aggravates these disparities. In some cases, water is not where we want it, nor in sufficient quantities. In other cases, we have too much water, in the wrong place, and at the wrong time (Malin 1989, 1991).

For instance, about 20% of the global average rainfall each year is accounted in the Amazon basin (a vast region with fewer than 10 million people: a tiny fraction of the world's population). Similarly, the Congo river and its tributaries account for about 30% of the entire African continent's annual runoff, but the watershed contains only 10% of Africa's population. However, the disparities are starker when it comes to access to water (Postel 1997).

Freshwater Resources and Their Management

By the middle of the century, annual average river runoff and water availability are projected to increase by 10–40% at high latitudes and in some wet tropical areas, and decrease by 10–30% over some dry regions at mid-latitudes in the dry tropics, some of which are presently water-stressed areas. By 2020, around 75 million people are projected to be exposed to increased water stress due to climate change. If coupled with increased demand, this will adversely affect livelihoods and exacerbate water-related problems (Working Group 11 Fourth Assessment Report, IPCC 2007).

Rural Water Supply Planning

Water supply is the provision of water for drinking, domestic uses and irrigation; its availability is controlled by global water distribution (Oteze and Foyose 1998).

Water supply planning, therefore, is the collection, analysis and interpretation of hydrologic and geohydrologic data, followed by field reconnaissance surveys, geophysical investigation supported by test drilling and physical examination of cores samples (Oteze and Foyose 1998). These were done to isolate the water-bearing horizons (or aquifers), estimate their water resources potentials and establish the quality of the groundwater: water for rural people in the developing world should have easily applicable, minimum possible treatment. The types of treatment for water purification depend on the physical, biological and chemical characteristics of the water to be used. In rural areas where a population of 5,000 or more is spread over relatively short distances, and in villages with over 500 households, a small-scale safe water supply should be provided. To ensure convergence between water scarcity, demand availability and supply of water to the rural population, drinking water directly from its source should be treated before supplying and or distributing it to various locations according to population demand criteria, since untreated water may contain many bacteria, causing diseases such as typhoid fever, dysentery, gastroenteritis, hepatitis and amoebic dysentery.

The whole idea of water supply planning depends on the availability or amount and sources of water, which is directly controlled by hydrological cycle—by extension the climatic system and processes that dictate where and when rain will fall. Thus, water resource planning is aimed at achieving an orderly development of water resources to meet present and future demands.

Rural Water Supply and the MDGs

There are still at least 1.1 billion people across the world who do not have access to safe drinking water. Many of these people live in rural areas and are among the poorest and most vulnerable to be found anywhere in the world. In sub-Saharan Africa, 300 million people have no access to safe water supplies—approximately 80% live in rural areas (Fig. 2). Therefore, significantly increasing the coverage of rural water supply in Africa is fundamental to achieving many of the internationally agreed Millennium Development Goals (MDGs). Without safe water near to dwellings, the health and livelihoods of families can be severely affected; children's education suffers as the daily tasks of survival take precedence over all other concerns (MacDonald et al. 2010).

Facing Up to the Water Crisis

Water is the defining link between climate and agriculture. But even without climate change, we are in serious trouble. Competing demands combined with mismanagement of this critical resource means that water availability has become an urgent issue facing rural people (and other users) the world over. And typically, the most extreme shortages are experienced by those least able to cope with them—the most impoverished inhabitants of developing countries. Climate change will exacerbate an already critical situation. Water is already a critical limitation for many poor people (Moorhead 2009).

Impacts of Climate Change on Rural Water Supply

The hydrological cycle, driven by solar energy, involves water changing in form and moving from one storage to another. Consequently, climate change will lead to the intensification of the global hydrological cycle with consequences for the major world's water storages. The changes in the volume (magnitude), timing and distribution of water resources would necessitate changes in management strategies in order to balance water supplies and demand, since the hydrological cycle comes with no guarantees for humankind (Postel 1998).

Various methods have been employed to assess the impact of climate change on hydrology and water resources at various geographical scales varying from the globe to the river catchments. They include the following, among others:

- Statistical analysis of long-term concurrent variation in runoff and meteorological elements, especially air temperature and precipitation over past periods.
- Studies of the hydrological consequences of past periods of very warm or cold, wet or dry conditions.
- The use of methods of water balance over a long period of time to assess the impact of increased temperature and decreased precipitation on run off and soil moisture storage.
- The use of General Circulation Models (GCMs) of the atmosphere to obtain changes in the climatic and hydrological characteristics of a large area using different greenhouse gas emission scenarios.
- The use of deterministic hydrological conditions using input of climatic data including outputs from GCMs.

The methods above have their strengths and weaknesses when used to assess the impact of climate change on hydrology and water resources (see IPCC 1990b).

From the result of several studies conducted using the above approaches, here are the extracts that correspond to Nigeria's situation:

River catchments will be very sensitive to even small changes in climatic conditions, especially those in arid and semi-arid regions where the annual runoff is highly variable, as is the case in our river catchments in the Sudano-Sahelian region of Nigeria, which, according to Olofin (1987) stretches from 12 N to the international boundary of the Niger Republic (Fig. 3a, b).

Global warming is likely to lead to changes in runoff extremes both high and low. In other words, there are likely to be very high flows and very low flows within the year (Fig. 4).

There will be increases in the rates of evaporation/evapotranspiration, which will lead to reduced soil moisture storage and reduction in the total annual volume of runoff, particularly in areas where precipitation remains unchanged or decreases in amount or fails to increase in amount large enough to offset the increases in the rates of evaporation (IPCC 1990a).

There may be a decrease in infiltration rates where raindrop size increases or rainfall intensities increase, with consequent adverse effects on soil moisture, groundwater recharge and groundwater levels.

There will be changes in demand for water resources both domestic and agricultural; water consumption will increase with increasing dryness and/or heat, all things being equal. The radiational index of dryness RW/LP (where RW is the net radiation of a wet surface, P is precipitation and L is the latent heat of

Fig. 3 a Political map of Nigeria b Map of Nigeria showing the drainage pattern

condensation) provides an indirect measure of water consumption. The greater the value of the index the higher the water consumption rate (IPCC 1990b).

Regions that depend on unregulated river systems (as the case with most of rural Nigeria) will be more vulnerable to hydrological changes induced by climate change (IPCC 1990c; Umolu 1995).

With these impacts of climate change on water availability and distribution, the need for water supply planning especially in rural areas has arisen if the water supply crisis is to be averted. The current population growth has also exacerbated the problem of water scarcity. Over the years, access to clean water has been a key indicator of measuring the quality of life in nations across the world. Access to clean water has been proven to have a strong relationship with longevity, since clean water does not only reduce incidence of waterborne diseases, but is also a vital requirement in attaining high standards of sanitation.

Fig. 4 Impact of climate change on rural rivers

Conclusion

Water resource planning essentially resolves into three issues: the extent of available water resource, the future requirements of water for various purposes, and how these can be met. The most fundamental thing is water's availability: how can the diminishing resources be conserved or improved? This calls for action on what causes climatic change, especially the contribution wade by rural dwellers, whether deliberate or accidental. One such activity is deforestation, which reduces the most essential tropospheric sinks of carbon (the vegetation), which is the highest contributor of greenhouse gas concentration (50–60%). Today, carbon concentration in the troposphere is approaching 400 ppm and it is predicted to reach 450 ppm by the year 2050 if we do not cut down emissions at local, national, regional and international levels (Umar 2000).

At the rural (village) level, what is significant to note is that most rural dwellers were unaware of the danger of their actions and inactions. For instance, locally, the rural farmers were of the habit of setting fire to the bush thereby contributing to atmospheric carbon concentration and at the same time aggravating species attrition.

Besides bush fire, the activities of the rural dwellers to cut down trees recklessly (deforestation) may be poverty driven, but even then they should have been replaced or reforested, for the fact that vegetation ameliorates the microclimate, stabilizes soils, plays a role in interception and, above all, absorbs excess carbon that contributes immensely to global warming. This warming intensifies heat and encourages evaporation and at the same time contradicts rainfall reliability as a result of this change in climate. The Intergovernmental Panel on climate change indicates that in the 1980s, deforestation alone accounted for about 1.6 billion tonnes of carbon emissions (Umar 2000). It has been stated elsewhere that "man is the architect of his environmental misfortune", as this climate change is partly the result of human activities at various geographic locations.

Recommendations

Nigeria in general and rural areas in particular need a "blue revolution" in the first instance to conserve and manage the little available water resource in the face of growing demand. In another outlook, factors affecting climate, especially negatively, should henceforth be stopped (e.g. bush fire and deforestation as commonly practised by the rural populace). Besides this broad recommendation, the local, state and federal authorities should:

- Educate rural households to use water wisely and about the dangers involved during water shortages.
- Provide a legal framework to guide the activities of industries sited in rural areas so as to ensure no toxic waste contaminates water bodies.
- Punish non-law abiders through taxes.
- Involve qualified companies while awarding contracts on rural water supply, i.e. those capable of assessing, harnessing and distributing the available water accordingly.
- Educate the rural people about the significance of climate change on water resource availability, temporally and spatially.
- Improve upon the available water supply system, such as dams, reservoirs, inter-basin transfer for both single and dual purposes.
- Educate the rural people about the danger of their local actions such as bush burning and deforestation, which contribute immensely to changes in climate as we know it.
- Involve the supply of water through desalinization of ocean water and saltwater for public consumption, especially in the rural coastal population.
- Improve water re-use systems, especially where industries are located in rural areas. This involves sequential water use (e.g. household → industrial → agriculture).

Acknowledgments We are very grateful to Dr (Mrs) R. J. Muhammad for her support and encouragements at all times, directly or indirectly, Muhammed A. Muhammed, Musa Sule and the entire Management of Federal College of Education Kano, who have in one way or other contributed to the success we have achieved nationally and internationally. We are greatly indebted to the Education Trust Fund (ETF) for funding to attend an International Conference in Cairo and F. C. E. Kano in Nairobi to mention but a few. Thanks also to Associate Professor A. I. Tanko, Department of Geography, Bayero University Kano, Nigeria, Professor S. U. Abdullahi (former VC A. B. U. Zaria) Ibrahim Ahmad Gundutse Jamilu Gambo, Murtala Muhammed Adogi, Dr M. L. Mayanchi. All errors and omissions, and all views expressed, remain solely our responsibility.

References

Alaba AO (2001) Economics of water health and household, labour market participation, a final report. African Economic Research Consortium, Kenya

Ayoade JO (1988) Tropical hydrology and water resources. Macmillan, London

Ayoade JO (2003) Climate change: a synopsis of its nature, causes, effects and management. Vantage, Ibadan

Barabas S (1986) Monitoring natural water for drinking water quality. World Health State Q 39(1):32–45

Briscoe J et al (1990) Towards equitatable and sustainable rural water supplies; a contingent study in Brazil. The World Bank Rev 4(2):15–34

IPCC (1990a) Climate change: the IPCC scientific assessment. Cambridge University Press, Cambridge

IPCC (1990b) Climate change; the IPCC impacts assessment. Australian Government Publishing service, Canberra

IPCC (1990c) Climate change: the IPCC response strategies. WMO/UNEP, Geneva

IPCC (2007) Working Group 2 contribution to the fourth assessment report to the intergovernmental panel on climate change summary for policy makers and technical summary (IPCC 2007), pp 11–13

MacDonald A et al (2005) Groundwater and rural water supply in Africa Brief note IAH Burdon. Ground water network a simplified hydrogeological map of Sub-Saharan Africa http://www.iah.org/downloads/occpub/IAH_rural water

Malin F (1989) The massive water scarcity now threatening Africa, why isn't it being addressed? Ambio 18(2):112–118

Malin F (1991) Rapid population growth and water scarcity: the predicament of tomorrow's Africa. Macmillan Press, Ltd, London

Moorhead A (2009) Climate, agriculture and food security: a strategy for change. The Consultative Group on International Agricultural Research (CGIAR), Washington, pp 22–23

NIMET (2009): Annual rainfall prediction for the year 2009. Daily Trust, 20 February 2009

Olofin EA (1987) An aspect of physical geography of Kano state and related human aspect. University press, Kano

Oteze GE, Foyose EA (1998) Regional development in hydrology of the Chad Basin water resources. Water Resources Journal of the Nigerian Association of Hydrologist 1(1):9–29

Postel S (1997) Facing water scarcity. Norton, New York, pp 17–191

Postel SL (1998) Water for food production: will there be enough in 2015? Bioscience 48:629–637

Smith K (1996) Environmental hazard: assessing the risk and reducing disasters. Routledge, London

Umar DA (2000) Activities of Kano state afforestation programme (KNAP) in Bichi Local Government (BSc thesis) Geography Department, Bayero University Kano-Nigeria

Umolu JC (1995) Global climate change: impact on energy

Chapter 21
Impact of Climate Change in Bangladesh: Water Logging at South-West Coast

Md. Moniruzzaman

Abstract Bangladesh is a densely populated, agriculture-based country and is recognized as one of the areas most vulnerable to the impacts of global warming and climate change. This is due to its unique geographic location, dominance of floodplains, low elevation, high population density, high levels of poverty, and overwhelming dependence on nature for its resources and services. The country experiences severe flood and cyclone events and, in recent years, water logging has become a catastrophic problem along the coast. These coastal areas play important economic and environmental roles in the country. The present paper attempts to show the extent of water logged areas, caused by sea level rise and the sectoral impacts of settlement, agriculture, health and education in the south-western coastal areas of Bangladesh. A multi-temporal analysis method has been used with remote sensing (LandSat 1975 and LandSat 2010) data. SRTM data has been used to visualize the water logged areas attractively. To the Government of Bangladesh, NGOs, policymakers, planners and other interested parties it is important to measure and monitor present issues and to predict the future impacts of climate change. This will help to facilitate effective management and is particularly important where a large number of people are threatened. In this regard the present study is expected to be useful and will have strong implications in coastal planning and other climate change adaptation measures.

Keywords Global warming · Sea level rise · Floodplain · Remote sensing

---

Md. Moniruzzaman (✉)
Department of Geography and Environmental Studies,
Rajshahi University, Rajshahi, Bangladesh
e-mail: moniralvi@gmail.com

Md. Moniruzzaman
GI Technology, Muenster University, Muenster, Germany

Introduction

Global climate is a continuous changing phenomenon. The earth has long been subjected to sequential glacials, interglacials and warm periods, and some parts of it have been warmer and cooler at various times in the past (Bhatti et al. 2005). In glacial periods global temperature was 4–5°C less and interglacial periods 1–2°C more than the temperature we are facing in the present time (Harris 2004). The average temperature near the earth surface was increased by about 0.74°C over the past century and eleven of the last twelve years rank among twelfth hottest years on record since 1850. And the second half of the twentieth century was the warmest fifty abbreviate in the last 1,300 years (IPCC 2007). Like temperature, precipitation, the other important factor of climate, has changed significantly its trend to upward with most of the increase occurring in the first half of twentieth century. In high latitudes of both hemispheres the trend of precipitation has increased and across the tropical regions the trend has decreased (National Climatic Data Center 2008). The past feature was totally natural in origin when population size on the surface of the planet, earth, was minimal, but now humans are also responsible significantly to make variation in global climate along with the nature (Harris 2004).

Most of the observed increasing globally average temperatures are very likely due to the observed increasing human produced greenhouse gas concentration. The climate model project summarized by Intergovernmental Panel on Climate Change (IPCC 2007) indicates that the average global surface temperature will likely rise from 1.1 to 6.4°C during the twenty-first century.

Scientists now believe that the green house effect has been intensified by extra greenhouse gases that are released by humans. Human activities since the industrial revolution have increased the concentration of various greenhouse gases like carbon dioxide (CO_2), methane, ozone, chlorofluorocarbon (CFC) and nitrous oxide (Mastrandrea and Schneider 2005). The biggest factor of present concern is the increase in the CO_2 levels (Fig. 1). The 2005 level was highest over the last 650,000 years and the same rise in the CO_2 concentration at the end of ice ages took over 5,000 years while even a bigger rise has happened only at industrial age. Fossil fuel burning has produced approximately three-quarters of the increase of CO_2 from human activities over the past 20 years. Land use changes such as deforestation in the tropics, paddy rice farming, wetland changes and land fill emission among other factors are leading to the higher methane atmospheric concentration. Agricultural activities including the use of fertilizers lead to higher nitrous oxide concentration (Southwest Climate Change Network 2009).

Impact of Climate Change In Bangladesh

Ganges delta, the largest river delta of the world, was formed by the confluence of the major three rivers of Bangladesh, *Ganges, Jamuna* and *Meghna* and hundreds of their tributaries (Rashid 1991). The country's population density is the highest

Fig. 1 CO_2 and global temperature. (Source: zFacts.com (2007))

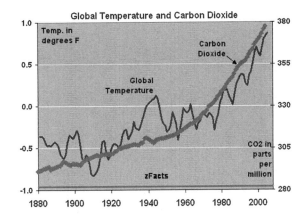

Table 1 Major cyclones affecting Bangladesh since 1960

Year	Death	Year	Death
1960	3,000	1970	300,000
1960	5,149	1985	11,069
1961	11,466	1988	2,000
1963	11,520	1981	138,000
1965	19,279	2001	3,064

Source: European Community (1998)

in the world and almost half of the total people live below the poverty line. Bangladesh can be the example of several kinds of natural calamities such as flood, cyclone, draught, earthquake, landslides and water logging in the present time (Climate Change Cell 2007). Every year hundreds of villages and many cities are inundated by floods damaging agricultural fields. Many people, in some years even millions, die along the coastal areas due to devastating cyclones (Table 1) which are considered as negative impact of global climatic change (Banglapedia 2006). Space Research and Remote Sensing Organization (SPARSO) of Bangladesh produced a map showing the cyclone affected areas of Bangladesh (Fig. 2).

Almost every five years Bangladesh is affected by droughts which shake crop life cycles and north western regions are particularly vulnerable to this affect. Although in monsoon period a substantial amount of rain occurs but some years in some parts the amount is not sufficient for agriculture. Moreover, land degradation, livestock population, employment and health sectors are also affected considerably by this hazard. During the last 50 years, droughts happened in Bangladesh 19 times. Among all, droughts of 1951, 1961, 1975, 1979, 1981, 1982, 1984, 1989, 1994, 1995 and 2000 year were severe (FAO 2007).

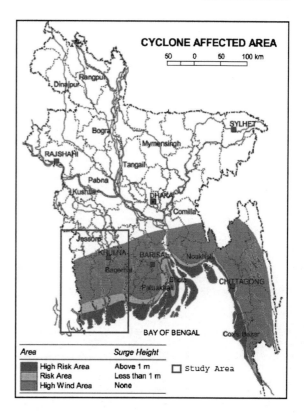

Fig. 2 Cyclone affected area of Bangladesh. (Source: Dhakadailyphoto 2007, adopted from SPARSO, Bangladesh)

The Coastal Zone of Bangladesh

Bangladesh is basically a riverine country and a few of the world's major rivers passing through the country helped to form as a deltaic land (IUCN 1994). The destination of all rivers is to meet to the Bay of Bengal which locates to the south of Bangladesh. The country's big coastal area is 710 km long, which covers 19 districts out of 64 and 147 *thanas* [subdistrict which is also known as upazila (Fig. 3, Islam 2006)]. This coastal zone reserves 32% of the total area and 28% of total population of Bangladesh and covers an area from the shore of 37 to 195 km (Islam 2006).

According to the closer to the sea the zone is divided into exposed and interior coast. The *thanas* which are facing the southern sea coast or river estuaries are known as exposed coastal zone and the total number of such exposed zones is 48, whereas interior coast which are located behind the exposed coast do not face sea coast or riverine estuaries and they are 99 in numbers (Fig. 3). The exposed coast zones are more vulnerable to climate change such as cyclone, sea level rise, tsunamis and water logging.

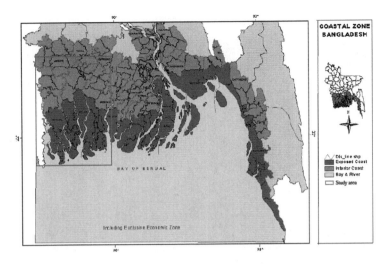

Fig. 3 Coastal zones of Bangladesh (Source: Islam 2006)

Remote Sensing and Water Logging

Remote Sensing is an extremely useful tool to detect natural or human induced disasters among others such as floods, cyclone, water logging. This tool has been widely using for mapping and monitoring waterlog both in urban areas and coastal zones. Different images of satellites are different and complement each other with respect to spatial, temporal and spectral resolution (Lelkes et al. 2002). Choubey (1997) used IRS-IA LISS data of 1989 to delineate water logged areas in the Tawa command areas of India during pre and post-monsoon period and he obtained both water logging areas and areas are sensitive to water logging. This study also suggest that periodic assessment of water logging using remotely sensed data should be carried out at regular intervals. Landsat Multi-spectral Scanner data of different years have been used to detect water logging and salinization areas of India's arable land (Dwivedi 2006). The study shows a significant decrease of soil fertility due to salinity over 15 years.

Objectives

Keeping the aforementioned problems in mind the research has accomplished with a view to find out the impact of climate changes. The main objectives are as follows:

(a) to determine the actual extent of water logging at south-west coastal areas of Bangladesh using landsat images of 1975 and 2010;
(b) to delineate different sectoral impacts of water logging such as agriculture, health, settlement and education.

Fig. 4 Study area

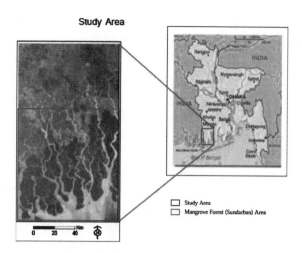

Study Area

The present study was conducted in the south-west coastal area of Bangladesh, Satkhira, Khulna, Bagerhat and Jessore Districts (Fig. 4), where a large number people are living. Total number of population is 2,520,961; 1,950,557; 2,586,771 and 1,620,216 in Khulna, Satkhira, Jessore and Bagerhat districts respectively (BBS 2007). Annual average temperature is maximum 35.5°C, minimum 12.5°C; annual rainfall 1,710 mm. The soil of this region is mainly alluvial floodplain. Main rivers are *Kobadak, Sonai, Kholpatua, Morischap, Raimangal, Hariabhanga, Ichamati, Betrabati* and *Kalindi-Jamuna* (Rashid 1991). The largest mangrove forest "*Sundarbans*" is also in this area. Sundori, Goran, Geoa etc. are the main trees of this forest.

Methodology

Data Collection and Software

In this study data were collected from different secondary sources. For identifying water logging areas two satellite images were used. One of them was Multispectral Scanner (MSS) in the year of 1975 and the other one was Thematic Mapper (TM) of 2010 which were taken from the USGS Global Visualization Viewer (GloVis). Band 1 and band 2 are visible bands and bands 3 and 4 are near infra-red (IR) of MSS scanner. For TM visible range is from band 1 to band 3, band 4 is near IR and band 5–7 are short wave IR (inforterra 2010). All information is summarized in Table 2.

Table 2 Image information

Sensor	Path/Row	Acquisition date	Spectral range (μm)	Bands	Pixel resolution (m)
MSS	148/44, 148/45	15 Dec., 1975	0.5–1.1	1, 2, 3, 4	60
TM	138/44, 138/45	10 Jan., 2010	0.450–2.35	1, 2, 3, 4, 5, 6,7	30

The acquisition month generally has to be taken into account when comparing multi-temporal data, especially in areas of strong seasonal change of water bodies. In this study Landsat imagery of dry season in Bangladesh (from November to March) were considered to identify the actual water logged areas as additional water from rain can be found in the rest months of the year to ensure that those water logged areas were not flood. Moreover, seasonal effects on vegetation are not so strong in the study area which can be mostly neglected.

SRTM data of 90 meter resolution were downloaded from the website of http://srtm.csi.cgiar.org/ and used for accurate visualization of water logging areas and to make a map of vulnerable places of the study area. Sectoral effect data have been used which were collected from government organizations, NGOs and different offices. ERDAS Imagine, ArcGIS and Global Mapper software were used to analysis and find out the results.

Methods of Water Logging Analysis

To find out water logged areas in the south-west coast of Bangladesh both images were subset first according to the study area where band 1, band 2, band 3 and band 4 were considered. Band 5 and band 7 of TM were not considered for spectral comparability of bands 1–4 of MSS and TM. And then carried out a supervised classification using parallelepiped non-parametric rule (Schanze et al. 2004). The image classification is a process of categorizing all pixels or regions of the image into classes intended to represent different physical objects or types. The output of such process is considered as a thematic map rather than image (Richards and Jia 2006). Unsupervised and supervised are the main two methods of image classification. In the unsupervised classification method training samples or user's previous knowledge on land cover of the study area is not necessary (CCRS 2005), whereas supervised classification method refers users training classes and a priori knowledge of the area. It rests upon using suitable algorithms to label the pixels in an image as representing particular ground cover types, or classes (Richards and Jia 2006).

The parallelepiped classifier is known as box decision rules. This classification is based on the ranges of values within the training data to define the regions within a multidimensional data space (Campbell 2002). Principally this classifier uses the class limits and stored in each class signature to determine if the class will receive the given pixel or not. The limits of class indicate the dimensions (in standard deviation units) of each part of a parallelepiped surrounding the mean of the class in feature space. The given pixel may fall or may not fall to the specified class and even some pixels are owned by more than one class (Richards and Jia 2006). The main advantage of the parallelepiped classification is that it is easy to leave areas unclassified. It might also be possible to set certain limits for classifying using nearest neighbour or maximum likelihood, but in the parallelepiped classification a box is defined based on statistical parameters, and only pixels within the box are assigned to the desired class.

However, the classification was accomplished to separate water bodies from the rest land cover areas, such as agricultural fields, forest and vegetation, roads and bare soils, urban areas, etc. Signatures, main hub of supervised classification, were collected from feature spectral images. Feature space layers help to explore spectral properties of image. ERDAS Imagine software allows user to create a scatterplot where the data file values of one band have been plotted against the data file values of another band. This can only be done in a two dimensional histogram, but theoretically data has multi-dimensions of spectral space (Spatial Analyst Laboratory 2009).

Three samples of 3 × 3 pixel size in both images were defined representing water bodies (Appendix, Tables A.1 and A.2) and then scattergrams were checked to examine if these three samples were located in spectral space. A polygon was defined in the scattergram that include all the three samples. This polygon therefore defines the spectral range of desired class "water bodies". The selected polygons were checked in the images with image alarm. Image alarm shows the pre-classification of the image data. Signature file with the defined colour can be checked on the image to be classified with this operation. This process can be done repeatedly to get a better classification result (LLU—ESSE21 2010).

These two classified images are then subtracted, classified image of 1975 were subtracted with union operation from classified image of 2010 with an aim to find out the changes in water logging situation in the study area. Image Subtraction is the extraction of information of differences between two images of a particular location.

This difference can be increased or decreased in the average intensity level (CCRS 2005). The resultant image was then rescaled where water bodies were layered as 1 and rest is 0. Rescaling of image converts one data type to another such as from 16 bit to 8 bit or so on. It decreases the file size changing the image contrast (Campbell 2002). This rescaling, however, is very much useful for detecting particular object and to overlay on other phenomena. To create the final water logging map opacity were used for non water bodies to make transparent so that water bodies can be overlaid. However, it was then overlaid on SRTM image which was made as $45°$ solar azimuth for proper visualization. ERDAS Imagine 9.1 was used for the above all operations. But for taking layout or final map ArcGIS 9.3 was used. The following flow chart (Fig. 5) shows the overall method mentioned above.

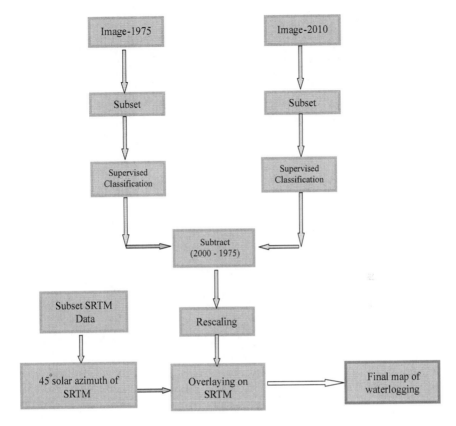

Fig. 5 Flow chart of the methodology

Results and Discussion

Water Logging Areas

Considering the methods mentioned above, the two images were classified with particular view to see the actual changes in water bodies of the coastal area. However, after supervised classification of MSS and ETM+ images it can be seen that in 1975 total area of water bodies was about 310,341 ha including rivers, lakes and parts of Bay of Bengal. This had increased to 492,759 ha in 2010. Here additional 182,418 ha area of water bodies existed in 2010 (Fig. 6). This can be identified as water logged areas influenced by sea level rise. In 1975, there was no impact of sea level rise and no stagnant water was found in the study area. Only some regular water bodies such as rivers, canals, ponds, lakes and wetlands existed.

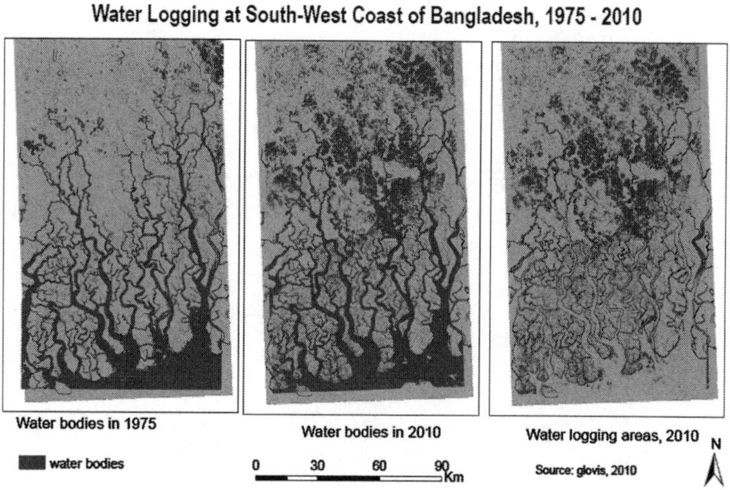

Water Logging at South-West Coast of Bangladesh, 1975 - 2010

Water bodies in 1975 | Water bodies in 2010 | Water logging areas, 2010

Fig. 6 Water logging in 2010

Bangladesh is characterized by a tropical monsoon climate with wide seasonal variations in rainfall. In the monsoon period, when every year lot of rain occurs, water levels become higher and rivers play a very important role to flow this rain water out to sea. Over the last 40 years normal flow of rivers in Bangladesh, especially in the south-west area, has become hindered by the high water level in the lower reaches caused by sea level rise. Water now stagnates in the area of 35 km north of the Bay of Bengal. As sea level is continuously rising, sea water is also adding to the rain water and the logging process is becoming faster and wider spread.

Sectoral Impact

Salinity Intrusion

The south-west coastal area, formed with fertile land by deltaic actions, is an important part of the country where a large number of people are living. Due to sea level rises the rate of saline water intrusion into this area has increased. Saline line has entered up to 20 to 35 km into the mainland since 1967 (Fig. 7). It is estimated that every year, on average, about 1 km area is coved with saline water (SRDI 1998a, b, c). *Monirampur, Keshabpur, Jhikorgacha* of *Jessore* District and *Tala, Kalaroa* and *Ashashuna thanas* of *Satkhira* districts are mostly affected by this sea salt water.

With the moving of salinity line northward soil in that area is becoming salty. From the map (Fig. 8) we can see that the area covered by salty soil in 1997 is

Fig. 7 Salinity line at south-west coast

considerably bigger than that in 1967 in the south-west part of Bangladesh. Both dimension and intensity have increased substantially (SRDI 1998a, b, c).

Impact on Agriculture

Agriculture is the main hub of Bangladesh economy where more than 60% of the total population depends on agriculture and the south-west agricultural zones are about one-sixth of the total area of the country. The total agricultural production in the study area is decreasing due to either water logging or salinity intrusion. In *Keshabpur* thana of *Jessore* district, the area most affected by water logging, agricultural lands decreased by 30% in the 20 years until 2008, in *Monirampur* thana the rate of decrease is 16%, *Tala* and *Kalaroa* *thanas* are comparatively less affected but currently, the degree of water logging is increasing with the potential to cause serious agricultural damage (Table 3, Fig. 9).

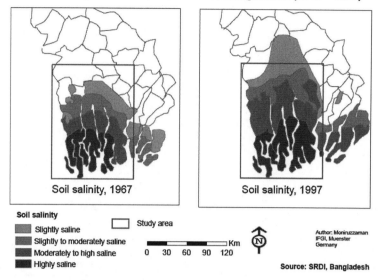

Fig. 8 Soil salinity at south-west coast

Table 3 Agricultural impact of different years

	1988 Total arable land (ha)	1998 Total arable land (ha)	2008 Total arable land (ha)
Jhikorgacha	28,230	26,909	23,370
Monirampur	41,343	38,465	34,478
Keshabpur	29,509	26,512	20,543
Tala	25,807	25,021	24,608
Kalaroa	21,258	20,387	17,710

Source: Agricultural Extension Office (2009a, b)

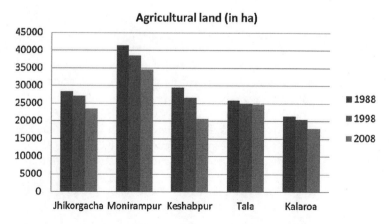

Fig. 9 Agricultural impact of different years

Impact on Population and Settlement

Bangladesh is one of the most densely populated countries in the world. Inhabitants are distributed throughout the country's rural and urban areas. There are no big cities in the south-west coastal areas of the country except Khulna Metropolitan Area. People generally live in the villages beside their agricultural fields. During the last four decades some settlement areas have disappeared due to severe water logging. From the following two tables (Table 4) we can see that 12,855 families were affected in 2008. Some of them have totally lost their houses while some are only partially affected. People are forced to live inhumanly with their remaining domestic animals, such as cows and goats, on the streets, high lands, in schools or shelters (Figs. 10, 11).

Education and Health

The education system in this area has totally collapsed due to factors related directly to water logging. First, as schools are being used as a shelter, teachers cannot take classes for teaching. Although it has been seen that some immediate make-shift schools are made on the street formed by polythene and paper which are vulnerable to the medium to strong wind. Secondly, transportation systems have broken down. Children and teachers from outer parts can no longer go to their schools or colleges. In some cases students use local small boats, but those are few in number and they have to spend much time and energy to arrive there. Thirdly, some children are busy collecting diets for themselves and even for their family members. Parents like to send their children to receive breads, rice, biscuits, and water from different shelters or aid centers. In 2006, total 121 primary and secondary schools were fully inundated and 80 schools were partially inundated (Fig. 12, Table 5).

Along with education, overall health condition is also concerned. People are living in unhygienic environments where there is no drinking water or sanitary latrine. In some areas all tube wells, the only source of drinking water, have gone under water, some partially inundated tube wells are even supplying salty water (Table 5). Some ponds where people used to collect their water for drinking purpose and other domestic uses have also become salty. People from those areas are suffering from water borne diseases like diarrhoea and cholera. Some NGOs opened mother and child care centers in few places, but most of them are either being used as shelter or are inundated wholly by the water logging. At the moment one severe problem has arisen, salty stagnant waters are getting more and more dirty day by day and people are suffering several skin diseases (Figs. 13, 14).

Table 4 Affected families in Jessore and Satkhira District

Jessore and Satkhira District

Upazila	Affected families	No. of shelter	No. of families took shelter
Jhikorghacha	988	5	261
Monirampur	1,428	6	323
Keshabpur	3,176	9	571
Tala	3,842	15	511
Kalaroa	3,421	7	742

Source: Disaster Forum (2008)

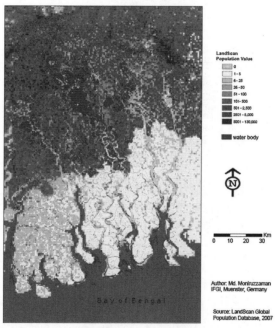

Fig. 10 Population and water logging

Fig. 11 Families affected by water logging in Satkhira District. Source: ClimateWire (2009)

Fig. 12 An affected primary school in *Keshabpur Upazilla*. Source: Khan (2008)

Table 5 Affected schools, tube wells and health care centers

	Fully inundated	Partially inundated	At the risk of inundation
Primary schools	81	55	28
Secondary schools	33	25	14
Tube wells and other sources of drinking water	13,838	9,443	4,665
Mother and child health care facilities	13	11	–

Source: LCG Bangladesh (2006)

Fig. 13 An affected secondary school in *Keshabpur Upazilla*. Source: (Khan 2008)

Fig. 14 Women are busy for collecting drinking water in *Kalaroa upazila*. Source: The Daily Star (2008)

Conclusions

The current study has shown water logging areas due to climate changes with remote sensing analysis which could be useful for the proper management in the affected areas. Land use pattern and land use modeling of the study area could be shown by remote sensing analysis (Skidmore 2002) but further research is necessary in this regard. Although salinity impact on soil and water what is considered a big sectoral impact for both environment and economy, has shown here using GIS application rather than Remote Sensing monitoring.

Mitigation and adaptation are two major steps which are necessary to face this problem in Bangladesh. Since, the country is so poor, it does not have enough money, technology and resources to take actions for mitigation. A big barrage across the coastal belt can be effective to protect excessive water coming from the sea level rise and salinity water intrusion as well into the land but since the deltaic land is connected to the southern sea by a number of rivers, it is totally impossible to manage all the big rivers mouths. Moreover, it is also necessary to drainage out huge rainwater occurred in the monsoon period into the sea. So, coastal barrage can make another prolonged problem.

People are thinking to leave their ancestors' home and migrate into another places, very few of them have already found their new destinations. But the whole Bangladesh is full of people; it is difficult to find out places where they can be accommodated. International migration policy, particularly to the developed countries, is also a big hindrance for them unless a new international agreement for climate refugee is going to make. Economic powerful countries with huge land resources like USA, Germany, Australia, and Canada can consider taking refugee people made by climate change for rehabilitation.

Climate change is not only environmental problem but also social and economical crisis. The question is why the people who have very less contribution to create such a big problem but they supper the most. It is well established that the industrialized countries of the world are emerging huge amount of CO_2 into the atmosphere and making the globe warmer. So, to resolve this worldwide problem they should cut down their fuel consumption, decrease the uses of natural resources and other high ambitious life style.

Last and but not least way for the people who are affected to the stagnant water is adaptation. They need a large amount of money to cope with the climate change impact. A poor country like Bangladesh has a few capacities to build fund for this hazardous people. International community should enlarge their helping hand with big budget and also technology and resources. In 2009, the climate conference held in Copenhagen, Denmark, Bangladesh claimed as the most vulnerable countries of the world due to climate change and demanded 15% of the total fund built by the developed nations as 15% of the total affected population of the world live in the country.

Appendix

See Tables A.1 and A.2

Table A.1 Digital number of water bodies in TM image

Sample- 1									
Blue green	21	21	21	20	21	21	20	20	20
Green	24	23	20	21	20	20	22	21	21
Red	18	14	14	15	13	12	15	13	14
Near-IR	4	4	4	4	4	3	3	3	4
Sample- 2									
Blue green	13	13	14	12	12	13	13	13	12
Green	14	14	15	14	13	15	13	12	14
Red	8	9	10	9	10	10	8	8	9
Near-IR	1	1	2	1	1	1	1	1	1
Sample- 3									
Blue green	12	12	11	13	13	12	12	12	13
Green	14	13	12	13	14	12	12	14	11
Red	10	10	7	9	9	9	10	10	11
Near-IR	2	3	4	3	3	3	2	2	5

Table A.2 Digital number of water bodies in ETM+ Image

Sample- 1									
Blue green	75	77	79	76	76	77	76	78	77
Green	62	61	60	59	62	61	61	61	62
Red	50	50	51	51	48	50	51	50	50
Near-IR	26	25	26	27	26	26	27	27	27
Sample- 2									
Blue green	74	76	75	76	74	76	74	74	74
Green	59	61	62	61	61	62	58	61	61
Red	51	51	51	53	52	50	53	50	53
Near-IR	28	29	27	28	27	28	27	30	27
Sample- 3									
Blue green	87	85	83	87	84	85	88	87	84
Green	70	69	73	69	72	71	70	72	72
Red	69	70	71	70	73	70	68	68	70
Near-IR	31	31	31	30	30	31	30	30	30

References

Agricultural Extension Office (2009a) Jessore, Bangladesh
Agricultural Extension Office (2009b) Satkhira, Bangladesh
BBS (Bangladesh Bureau of Statistics) (2007) District population of Bangladesh, Bangladesh
Bhatti J, Lal R, Apps MJ, Price MA (2005) Climate change and managed ecosystems. CRC press, Boca Raton
Campbell JB (2002) Introduction to remote sensing. The Guilford Press, New York
Choubey VK (1997) Detection and delineation of waterlogging by remote sensing techniques. J Indian Soc Remote Sens 25(2):123–139
Climate Change Cell (2007) Climate change and Bangladesh, Department of Environment, Government of the People's Republic of Bangladesh
Disaster Forum (2008) New flood in Kobadak basin villages, Mohammadpur, Dhaka
Dwivedi RS (2006) Study of salinity and waterlogging in Uttar Pradesh (India) using remote sensing data, National Remote Sensing Agency, Hyderabad
EU (European Community) (1998) Cyclone shelter preparatory study (CPSP) stage I: feasibility study, Draft Final Report; Supp. Vol. 3, Community Development Studies—Part A, supp vol 8; environmental analysis, field visits and other notes. EC/Sener Ingenieria Sistemas SA., 1996
Harris F (2004) Global environmental issues. Wiley, Chichester
IPCC (Intergovernmental Panel on Climate Change) (2007) Climate change 2007—impacts, adaptation and vulnerability: contribution of working group II to the fourth assessment report of the IPCC. Cambridge University Press, Cambridge
Islam MR (2006) ICZM initiatives and practices in Bangladesh. In: Krishnamurthy R et al (eds) Integrated coastal zone management (ICZM)—the global challenge, New Delhi
IUCN (1994) Mangroves of the Sundarbans: Bangladesh, vol 2. The IUCN Wetlands Programme, IUCN, Gland, Switzerland
Khan SA (2008) Water logging and primary and secondary education at Keshabpur Upazilla. An unpublished dissertation for the degree of master in disaster management, BRAC University, Bangladesh

Lelkes M, Csornai G, Wirnhardt C (2002) Natural disaster monitoring by remote sensing in Hungary: water logging and floods in the 1998–2001 period. In: Observing our environment from space: new solutions for a new millennium. Swets and Zeitlinger, The Netherlands

Rashid H (1991) Geography of Bangladesh. University Press, Dhaka

Richards JA, Jia X (2006) Remote sensing digital image analysis. Springer, Berlin

Schanze J, Zeman E, Marsalek J (2004) Flood risk management: hazards, vulnerability and mitigation measures. Springer, The Netherlands

Skidmore A (2002) Environmental modelling with GIS and remote sensing. Taylor and Francis, New York

SRDI (1998a) Coastal area and water salinity map of Bangladesh (1967 and 1997). Soil Resources Development Institute (SRDI), Dhaka

SRDI (1998b) Soil salinity map of Bangladesh (1973), Soil Resources Development Institute (SRDI), Dhaka

SRDI (1998c) Soil salinity map of Bangladesh (1997), Soil Resources Development Institute (SRDI), Dhaka

Internet Resources

Banglapedia (2006) Global warming. Available at http://www.banglapedia.org/httpdocs/HT/G_0137.HTM. Accessed on 12 Nov 2009

CCRS (Canada Centre for Remote Sensing) (2005) Glossary of remote sensing terms. Available at http://www.ccrs.nrcan.gc.ca/glossary/index_e.php?id=650. Accessed on 12 Jan 2010

CAST (Center for Advanced Spatial Technologies) (2010) Landcover classification and mapping. Available at http://www.cast.uark.edu/gap/chap2.htm. Accessed on 13 Jan 2010

ClimateWire (2009) Facing the specter of the globe's biggest and harshest mass journeys. Available at http://www.eenews.net/special_reports/bangladesh/. Accessed on 12 Oct 2009

Dhakadailyphoto (2007) Cyclone affected area. Available at http://2.bp.blogspot.com/_ZPf1HUBNpK4/R9KAM08GLkI/AAAAAAAAHHI/8EEnE_hS90g/s1600-h/Cyclone.jpg. Accessed on 2 Dec 2009

Environmental Protection Agency (2009) Coastal zones and sea level rise. Available at http://www.epa.gov/climatechange/effects/coastal/index.html. Accessed on 25 Nov 2009

FAO (Food and Agriculture Organization) (2007) Climate variability and change: adaptation to drought in Bangladesh. Available at http://www.fao.org/docrep/010/a1247e/a1247e00.htm. Accessed on 29 Nov 2009

Global Warming Art (2009) Global sea level rise risks. Available at http://www.global-warmingart.com/wiki/File:Global_Sea_Level_Rise_Risks_png. Accessed on 2 Dec 2009

inforterra (2010) Satellite band designations and principal applications. Available at http://www.infoterra.co.uk/data_sat_banddesc.php. Accessed on 11 Feb 2010

Jones P, Salmon M (2008) Climate research unit: data, temperature. Available at http://www.cru.uea.ac.uk/cru/data/temperature/. Accessed on 3 Dec 2009

LandScan (2008) LandScan population data 2007. Available at www.ornl.gov/landscan/ registration for population data collection in November 2008

LCG Bangladesh (2006) Children vulnerability in waterlogging 2006. Available at www.lcgbangladesh.org/.../2006%20Report_SCF%20UK_Children%20Vulnerability%20in%20Waterlogg%202006.pdf. Accessed on Oct 2009

Liu H (2006) Image band transformations. Available at http://geog.tamu.edu/~liu/courses/g661/spectral_trans.pdf. Accessed on 3 Feb 2010

LLU—ESSE21 (2010) Land use/land cover classification module. Available at http://resweb.llu.edu/rford/courses/ESSC5xx/docs/ESSE21_clssfn.pdf. Accessed on 12 Feb 2010

Mastrandrea MD, Schneider SH (2005) Global warming. In: World book online reference center. World Book, USA. Available at http://www.nasa.gov/worldbook/global_warming_worldbook.html. Accessed on 20 Nov 2009

National Climatic Data Center (2008) Global warming. Available at http://www.ncdc.noaa.gov/oa/climate/globalwarming.html. Accessed on 21 Nov 2009

Southwest Climate Change Network (2009) Science daily. Available at http://www.southwestclimatechange.org/news/feed/sciencedaily. Accessed on 20 Nov 2009

Spatial Analyst Laboratory (2009) Software. Available at http://sal.ocean.washington.edu/tutorials/erdas/qtips.html. Accessed on 13 Jan 2010

The Daily Star (2008) Food, water crisis in 85 Kobadak basin villages. Available at http://www.thedailystar.net/story.php?nid=57700. Last accessed 17 Nov 2009

US Geological Survey (2003) Using the landsat 7 enhanced thematic mapper tasseled cap transformation to extract shoreline. Available at http://egsc.usgs.gov/isb/pubs/ofrs/03-272/OFR03-272.pdf. Accessed on 12 Feb 2010

United Nation Environment Programme (2009) Climate change science compendium 2009. Available at http://www.unep.org/compendium2009/. Accessed on 15 Nov 2009

Wikipedia (2009) Sundarbans. Available at http://en.wikipedia.org/wiki/Sundarbans. Accessed on 10 Nov 2009

zFacts.com (2007) Evidence that CO_2 is cause. Available at http://zfacts.com/p/226.html. Accessed on 02 Dec 2009

Part III

Chapter 22
Characterizing Adaptive Capacity in Water Governance Arrangements in the Context of Extreme Events

Margot Hill

Abstract Climate change impacts on precipitation patterns, glacial retreat and associated changes in runoff regimes are observed from the Alps to the Andes. In order to manage future uncertainty as climate impacts on water resources further unfold, it is vital to better understand adaptive capacity and how it may be developed. Governance is an issue at the heart of the water discourse, with effective water governance seen as essential to building adaptive capacity in communities to manage future climatic uncertainty and stress. Governance and institutional components are more generally seen as key determinants of adaptive capacity, yet there has been relatively little empirical verification of indicators at the local and regional levels, as well as in the water sector. This study aims to contribute to the literature on adaptive capacity in the water sector, through the empirical and analytical development of more robust indicators of adaptive capacity relating to governance and institutions. The paper discusses how extremes can be an effective illustration of one type of climate uncertainty, in which to explore and assess the plasticity and adaptive capacity of the water governance system. Research is based on literature review, stakeholder interviews and statistical analysis of climatic extremes.

Keywords Adaptive capacity · Climate change · Water governance · Extreme events · Switzerland

---

M. Hill (✉)
Research Group on Climate Change and Climate Impacts, Site de Battelle/D 7, Chemin de Drize, 1227 Carouge GE, Switzerland
e-mail: margot.hill@unige.ch

Introduction

For most people around the world, climate change will be experienced through changes in local hydrological patterns (Parry et al. 2007). Moreover, mountainous areas, commonly considered "water towers" of the world, are at the forefront of warming patterns. Climate impacts on glacier retreat, precipitation patterns (seasonality and snow line) and associated changes in runoff regimes are already observed from the Alps to the Andes, and model projections suggest a continuation if not heightening of current trends (IPCC 2007b). Climate change is projected to affect the function and operation of existing water infrastructure and institutions, but it is recognized that current frameworks may not be robust enough to cope (Bates 2008). Effective adaptation and building adaptive capacity, therefore, is seen as crucial to managing water resources under the future uncertainty of changing climatic conditions.

Water, Climate Change and Adaptation

In his seminal book *On the Origin of Species*, Darwin famously noted: "It is not the strongest of the species that survives, nor the most intelligent that survives. It is the one that is the most adaptable to change". This observation perfectly elucidates how humans have always had to adapt to change, including climatic and meteorological variation. So what is different now? Why do we worry so much about society's ability to adapt to future variation in the 21st century? The answer to this can be found by looking at the speed of current climatic change, and the complex geopolitical-environmental context within which it is and will take place. As the different planetary systems move towards threshold or boundary conditions, which lie outside society's range of experience, resilience and the ability to adapt decline (Rockström et al. 2009). Current rates and type of change have meant that a more concerted effort must be placed on creating an enabling environment for adaptive capacity to accelerating rates of change in today's more complex and interconnected world.

The retreat of mountain glaciers is one of the clear indications that certain sub-systems of the earth are moving out of their relatively stable holocene state, and into an age becoming known as the anthropocene (Rockström et al. 2009). The movement towards such boundary or threshold situations have led some within the water sector to suggest that the concept of "stationarity" is dead, and that the lessons of the past can no longer guide how we manage our future (Milly et al. 2008). This implies that a shift in how we plan and manage water resources is needed, which respects non-stationary conditions and increased levels of uncertainty. To meet these challenges, the water resources and research community have in recent years focused more heavily on better understanding adaptive processes.

Some of the key concepts in this growing body of literature are vulnerability, adaptive capacity and resilience. *Vulnerability* is the degree to which a system is susceptible to, and unable to cope with, adverse effects of climate change, including climate variability and extremes (Parry et al. 2007). Vulnerability is comprised of a number of components including exposure to impacts, sensitivity and the capacity to adapt (Adger and Vincent 2005). *Adaptive capacity* is the capacity of actors, individuals and groups to respond to, create and shape variability and change in the state of the system (Chapan et al. 2009). It can be characterized by preconditions necessary to enable adaptation, including social and physical elements, and the ability to mobilize these elements (Nelson et al. 2007). For the purpose of this article, *resilience* is the ability of a socioecological system to absorb disturbances while retaining the same fundamental structure, function and identity, including the capacity to adapt to stress and change, through either recovery or reorganization in a new context (Parry et al. 2007; Chapan et al. 2009).

There is also wide recognition of the importance of institutions and governance mechanisms for building adaptive capacity and resilience (Nelson et al. 2007; Brooks et al. 2005; UNECE 2009), the role that public policy plays in fostering adaptive capacity at local and regional levels (IISD 2006), as well as the need for flexible approaches (UNECE 2009). In the preceding decade, adaptive capacity has become a more mainstream concept. As the IPCC (2007a) stated, climate change implies a speed and magnitude of change, which poses risks that are beyond the human experience and potentially the boundaries of coping ranges. Yohe and Tol (2002) suggest that determinants of adaptive capacity have a key role in defining the potential the boundaries of coping ranges. Determinants of adaptive capacity have been defined as including a variety of system-, sector-, and location-specific characteristics (IPCC 2001):

- the range of available technological options for adaptation,
- the availability of resources and their distribution across the population,
- the structure of critical institutions, the derivative allocation of decision-making authority, and the decision criteria that would be employed,
- the stock of human capital including education and personal security,
- the stock of social capital including the definition of property rights,
- the system's access to risk spreading processes,
- the ability of decision-makers to manage information, the processes by which these decision-makers determine which information is credible, and the credibility of the decision-makers themselves, and
- the public's perceived attribution of the source of stress and the significance of exposure to its local manifestations.

Adaptive Capacity

Building adaptive capacity, by cultivating or contributing to the presence of the above factors in communities, improves the ability of systems to be become resilient to surprises and longer-term changes (IISD 2006). The determinants of adaptive capacity lay the foundations for a number of different features and principles, which are seen as useful indications of a system's adaptive capacity, which will be discussed later in this paper. A number of theoretical discourses have developed, such as adaptive management, adaptive co-management and adaptive governance, in the quest for resilience in the face of uncertainty and climate change. Many of these theories focus heavily on the idea of "learning by judicious doing" (Holling 1978), which represents a departure from the more traditional approach of rigid and irreversible planning and anticipatory management to a concept of policy experimentation.

Adaptive governance is seen to meet the call for dealing with increased uncertainty and change, arising from the "growing number of failures among current approaches and increasing vulnerability of social-ecological systems" (Olssen et al. 2006, p. 1). Within the specific context of river basins, Pahl-Wostl et al. (2007) noted that more attention needs to be devoted to understanding and managing the transition from current management regimes to more adaptive regimes that "take into account environmental, technological, economic, institutional and cultural characteristics of the basin" (p. 49). Similarly to the adaptive management field, the concepts of learning by doing, social learning and scenario planning have become popular as a means of operationalizing the need for flexibility and better integration of social and ecological factors. These approaches are seen as a response to the challenge of "creating governance structures that are flexible and robust in the face of uncertainties and inevitable surprises" (TwinGo 2010, p. 3).

This paper presents the theoretical and empirical development of research presented at Climate 2009 on vulnerabilities in the water governance arrangement of the Rhone Basin in the canton of Valais, Switzerland. It builds on this previous research, discussing the challenges in developing indicators to better characterize and assess adaptive capacity. It presents a synthesis of the current state of indicators on adaptive capacity, and discusses the challenges arising from the mostly normative characterization of adaptive capacity. It uses results from earlier research combined with literature review to identify and theoretically develop characteristics of adaptive capacity. It then sets out a means to further test and develop these using the empirical setting of past extreme events and presents some initial results from preliminary stakeholder interviews. It suggests that the use of both qualitative and quantitative research techniques could lead to a more integrated understanding of physical and institutional attributes of the system.

Assessing Adaptive Capacity

The assessment of adaptive capacity is inextricably linked with that of adaptation. While the assessment of adaptation actions tend to be addressed within a framework of whether the outcome of such actions are equitable, effective and legitimate, there are also significant questions about not *how we adapt*, but rather *whether we can adapt*. The concept of adaptive capacity is used as a point of departure to determine measurable indicators that "could sustain comparable analyses of the relative vulnerabilities of different systems located across the globe and subject to a diverse set of stresses that lie beyond their control" (Yohe and Tol 2002, p. 25).

Such indices can be either qualitatively or quantitatively based, generated through formulaic or discursive data, but are critical for the management of risk in relation to climate change impacts. Engle and Lemos (2010, p. 3) note that "decision makers are interested in identifying and nurturing specific system characteristics that will increase adaptive capacity and resilience". The identification of determinants and indicators of adaptive capacity provide a broad suite of characteristics, among which governance and institutional processes are deemed particularly important for the development of adaptive capacity, reduction of vulnerability and prevention of overt and lasting damage from climate change (Nelson et al. 2007; Brooks et al. 2005).

A number of studies have highlighted that there are significant challenges in assessing adaptive capacity (Engle and Lemos 2010; Smit et al. 2000) since "adaptive capacity is latent in nature (…) it can only be actually measured after it has been realized or mobilized" (Engle and Lemos 2010, p. 5). However, the study of adaptation to climatic events can to a certain degree overcome the pre-impact intangibility of adaptive capacity. Smit et al. (2000) show that by studying system responses to past climate variability (which tends to be experienced through the nature and frequencies of extremes), it is possible to identify attributes of the system which were key to either successful or failed responses. Such studies of adaptation to extreme events have highlighted the importance of institutions and governance mechanisms for the capacity or inability to deal with change (Brooks et al. 2005; Engle and Lemos 2010).

Background

Case Areas

The two case studies chosen from the ACQWA case regions are the Rhone Basin in the canton of Valais, Switzerland and the Aconcagua Basin in Valparaiso, Chile. This paper will focus on the Swiss case area, in which the water governance framework is characterized by the "principle of subsidiarity", which defines the

Fig. 1 View of Valais, depicting the high peaks and low, highly cultivated valley floor. Available at http://valais.ialpes.com/gastronomie/gastronomie-valais.htm

decentralized implementation of public policy and law at the cantonal and municipal levels. The upper Rhone flows through the valley floor of the Valais. The area is characterized by an inner alpine arid climate, with large discrepancies in precipitation between higher and lower elevations. While water conflicts have been rare, periods of increased competition have been experienced more recently. The region has traditionally needed to cope with a range of events from flooding to low precipitation. Climate change impacts in the Swiss alpine areas will lead to enhanced glacial retreat, melting of permafrost, as well as the changes in vegetation and precipitation (Figs. 1, 2).

Higher winter temperatures and a more marked increase in summer temperatures will be noted. Precipitation will also be higher and more intense in winter, but much reduced in summer months (Häberli and Beniston 1998; Fig. 1). The largest source of vulnerability from climate change is likely to come from changes in the intensity or frequency of extreme events, such as heat waves (winter and summer), heavy precipitation events and drought (Beniston et al. 2007). Increased glacial melt also is leading to an increase in flood risks and other natural hazard events (OcCC 2008) (Fig. 3).

Increased flooding and extreme precipitation events are compounded by an increase in risk exposure due to infrastructure/housing development in vulnerable areas which are currently seen as "safe" due to technical interventions. Temperature increases at alpine elevations raise demand for water uses such as artificial snow making and summer cooling/drinking water leading to complex management shifts, compounded by changes in seasonality. There have already been examples where a lack of planning for drinking-water supply has led to sectoral issues between hydropower use, tourism use and drinking-water supply (Reynard 2000), as well as tensions between the hydropower and agricultural sector which arose in the 2003 summer heat wave.

The impacts of climate change may also compound the reduction in ecological status of many surface waters in Switzerland. Of 65,300 km of surface waters in Switzerland, 10,600 km have been considerably altered through technical projects, thereby impairing their ecological functions (FOEN 2009). Hydropeaking

Fig. 2 Districts of Canton Valais. Artist: Tschubby, 16 August 2004

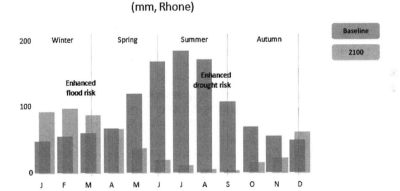

Fig. 3 Adapted from Beniston et al. (2007, p. 228). It represents the difference in seasonal distribution of runoff for the river Rhone between baseline values (1961–1990) and projected values for A2 SRES scenario (800 ppm) by 2,100. The bars aligned the *left* represent baseline runoff, showing typical seasonal flows (high runoff in summer, low in winter) for an alpine regime. The bars aligned to the *right* show the 2,100 projection, with increased runoff during early spring (increasing flood risk) and decreased runoff in mid- to late summer (increasing drought risk)

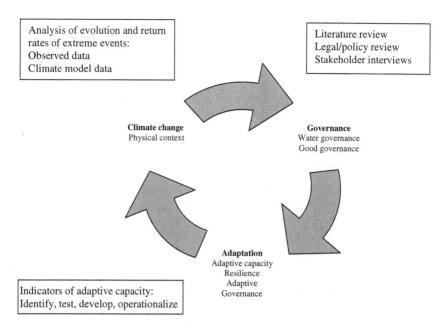

Fig. 4 Conceptual framework with associated methodological steps

(artificial high and low flow phases) also impacts rivers, in that they regularly dry up from over extraction of water, with damaging impacts on aquatic ecosystems.

Methods

An initial governance assessment, based on the Striver/Brahmatwin methodology (Allan 2008) was undertaken in order to understand the current vulnerabilities in the water governance system (Hill 2010). The Striver assessment drew on a number of good governance criteria, including transparency, accountability and participation, combining it with a set of IWRM-related sections. However, given the future uncertainty due to global change, deeper analysis of the system's adaptive capacity sought to contribute to the development of the governance assessment. To this end, extensive literature analysis identified key components of adaptive capacity to develop a further set of indicator questions for measuring adaptive capacity. Stakeholder interviews at the federal level and cantonal level in the Valais were used to critically appraise some of the normative assumptions within the governance and institutional determinants of adaptive capacity (Fig. 4).

Additionally, in order to ground the development of these indicators in empirical evidence, past examples of extreme climatic events will be used to explore particular problems experienced under climate extremes. The case studies of extreme events allow for deeper investigation into the potential performance of the governance system under climate change impacts. These case events in each case area will serve as reference points of climate variability and as useful indications for the impact of extremes in a future climate. Return period analysis was used to understand the current development of such events (trend analysis) and to project the likelihood of such events happening under climate change scenarios (return period analysis). Further stakeholder interviews in each case area, at the local and cantonal level, are then used to develop and refine the indicators, as well as provide important insights into the adaptive capacity of the water governance systems.

Review of Indicators of Adaptive Capacity

Within the water sector, there is a general call for all new water management measures to be climate-resilient. Adaptation strategies therefore aim at reducing vulnerability, including the possibility of increasing adaptive capacity (UNECE 2009). Water managers and scientists are progressively looking to integrated water resources management (IWRM) and adaptive management to help mitigate not only governance failures of the past, but also uncertainty in the future (Ingram 2011). IWRM places more emphasis on collaborative governance and the recognition of the multiple values of water, and is seen as one means to increasing capacity of water management in the face of climate change. Institutional capacity is also seen as a critical requirement in effective adaptation, particularly in the clarity of roles and responsibility of individual authorities, especially in extreme event situations (UNECE 2009).

In the literature on good governance, and therefore in the governance assessment itself, adaptive capacity to climate change tends to be assumed if indicators of good governance are adequately met. Results from the governance assessment in the Swiss case area showed that despite the water governance system performing well under the three initial indicators (accountability, transparency, participation), serious concerns were raised about its ability to cope and adjust to a changing climate and rising competition on water use, mainly through the lack of integration across different geographies and users (Hill 2010). Issues also arose in interviews which suggested that a correlation between "participation and decentralization" and greater adaptive capacity should not be taken as a normative assumption. The following section discusses why these presumptions should not be assumed and suggests how other indicators could be further developed and tested to determine adaptive capacity.

Tools and concepts used to measure the validity of outcomes of adaptive actions can also be employed to assess underlying states beneficial to the development of adaptive capacity. A number of determinants of adaptive capacity have been identified within the climate change impacts, adaptation and vulnerability literature (CCIAV). To recap, common factors considered determinants can be categorized into the following groups: economic resources, technology, information and skills, infrastructure, institutions, equity, social capital, and collective action (Engle and Lemos 2010; Eakin and Lemos 2006; Yohe and Tol 2002;

IPCC 2001). However, empirical verification of the merit of these norms for building adaptive capacity is sparse, particularly within the water sector (Engle and Lemos 2010; Wilbanks and Kates 1999).

Accountability, Participation, Transparency

The four indicators of the STRIVER assessment—accountability, participation, transparency and IWRM—were not specifically designed to measure adaptive capacity, but were rather shaped in the context of good governance for IWRM. However, these indicators also play different roles in other adaptive capacity assessments (Hurlbert 2008; Engle and Lemos 2010; Iza and Stein 2009). Accountability, participation and transparency are often considered key principles in adaptive capacity. A recent IUCN report (Iza and Stein 2009) refers to different process principles in the discussion on reforming water governance, which are requisite to provide an enabling environment, including transparency, accountability and participation. Their definition of participation broadens out from more than just consultation in decision-making to involvement in multi-stakeholder platforms and decision-making at the lowest appropriate level. It is considered these elements of participation could effectively raise levels of awareness, co-management and citizen initiatives, all components deemed necessary for fostering effective water governance capacity as well as sources of resilience in social-ecological systems.

IWRM and Integration

IWRM is currently held up as the ideal framework for managing water in an integrated and sustainable way that would enhance the system's resilience to cope with the impacts of climate change on water resources. However, despite the concept's use in addressing the need for water governance processes to effectively and equitably manage the fair distribution and protection of the resource, it has weaknesses in terms of complexity, uncertainty and adaptive capacity (Timmerman et al. 2008). Timmerman et al. suggest that in addition to recognizing multiple uses of water, multiple sources of knowledge and information should also be integrated into management systems.

Olsson et al. (2006) explore the different features that contribute to the resilience of social-ecological systems in the face of change (in the context of adaptive co-management). Their criteria do not follow the neat normative categories of many of the other studies into adaptive capacity, but provide some useful insights into governance-related criteria which can provide an enabling environment for enhanced resilience to environmental shocks and stresses. They suggest that an "enabling legislation that creates social space for ecosystem management" is requisite for the building of resilience. As vague as this may be, it deems that in order for resilience to be fostered, the institution of law should ensure that ecosystems and the environment are factored in as a relevant stakeholder. Not only should sectoral actors be integrated into legislation relating to resources (water in this case) but institutions also need to take account of ecosystem needs. This concept finds resonance with the element of integration and recognition for the non-economic uses of water within an IWRM context.

Leadership, Trust, Commitment

Olsson et al. (2006) use the criteria of "vision, leadership, and trust", which share some normative properties with accountability, in terms that an unaccountable system will not generate trust amongst its citizens. However, there is no reason to equate vision or leadership with the same norm, but both could be seen as requirements for the necessary political will requisite to foster proactive responses to climate change and develop relations across different networks and levels of decision making. Folke et al. (2005) also suggest that vision, trust and innovative leadership can provide key functions for adaptive governance, e.g., "building trust, making sense, managing conflict, linking actors, initiating partnerships, compiling and generating knowledge, mobilizing broad support for change".

The importance of these elements of leadership in building collaboration and resolving conflicts is underlined by its role as a key component in bridging interests and stakeholders and to a certain extent driving realization of other principles of adaptive governance. Leadership can be seen as an abstract concept, which can be highly subjective to personal opinion. Additionally, strong leadership may not always have a positive correlation with principles of adaptive governance, but it may be inferred that meeting the other principles of adaptive governance may not be as possible without the presence of leadership. Linkages may also exist with accountability, resources, networks, transparency and participation. Engle and Lemos (2010) also discuss the indicator "commitment", which refers to the belief held by the different stakeholders that the institutional and governance structures in place are adequate for managing the resource as effectively and efficiently as possible.

Experience

Engle and Lemos (2010) note that more experience would correlate with a greater ability to deal with everyday events, as well as extremes, in an effective and efficient way. While experience can broadly be deemed as relevant, just as with the concept of leadership, precise measurement of this principle is very abstract. However, although an actor may have many years of experience, preconditioned ideals or values may subject his/her decisions to preconceived notions, which may or may not still be relevant for changing conditions. UNECE (2009) highlight the importance not just of career experience, but also fostering experience through training and simulation exercises on a regular basis.

Resources

Olsson et al. (2006) propose "funds for responding to environmental change and for remedial action; capacity for monitoring and responding to environmental feedback" as indicators which both relate to the importance of human and financial resources for ensuring effective capacity for monitoring systems, enforcing laws and responding to extremes or feedbacks. The importance of information- and knowledge-sharing, not just in itself, but across different levels of stakeholders and decision-makers is touched upon through criteria 5 and 6 ("information flow through social networks; combination of various sources of information and knowledge"). These criteria are also relevant for the creation of the appropriate level of public perception (Yohe and Tol 2002) for adaptation through sense-making and collaborative learning (Olsson et al. 2006). Engle and Lemos (2010) also comment that levels of financial and human capital are critical for overall success of an organization or governance structure. Yet, while more resources (financial and human) may increase the capacity of the system, it is how these resources are applied and organized that may be more important. Less could mean more. There may also be linkages with experience, networks, accountability, transparency and decentralization.

Networks and Connectivity

Folke et al. (2005) explore the social elements of adaptive governance, which can enable adaptive ecosystem based management in the context of abrupt change. "Connectivity across Networks" refers to connectivity across individuals, organizations, agencies and institutions through bridging organizations. Networks capture the various institutional levels and relationships involved with river basin management. Folke et al. (2005) also suggest that adaptive co-management requires more flexible social networks, which may be more innovative and responsive than bureaucracies in times of rapid change. It is assumed that the greater the networking and connectivity between groups and stakeholders involved in the management processes, the greater the adaptive capacity (Engle and Lemos 2010). Just as in the critique of participation, connectivity alone may not imply a willingness to cooperate, which is requisite for systems to be adaptive (UNECE 2009).

Predictability—Flexibility

Flexibility is to be taken as the antithesis of irreversibility. This indicator is repeated across a number of the studies on adaptive capacity. The UNECE comments that "the capacity to adapt requires flexibility. As a result, measures that are highly inflexible or where reversibility is difficult should be avoided" (UNECE 2009, p. 78). In institutional terms, it refers to an ability to bend, but not break, and to learn iteratively, incorporating lessons learnt through experience efficiently and effectively (Engle and Lemos 2010). This concept of *iterative adaptive governance/learning by doing* is a key element of adaptive management and governance (Olsson et al. 2004; Pahl-Wostl et al. 2007). The assumption is that the greater the flexibility of rules (legislation, institutions), the greater the adaptive capacity (Engle and Lemos 2010). However, there is a struggle here between flexibility for adaptive management, and the need for certainty (Iza and Stein 2009; Tarlock 2009) or predictability (Hurlbert 2008) within the law. Predictability suggests that all laws and regulations should be applied fairly and consistently. The assumption is that consistency in application of the law will enhance adaptive capacity. The IUCN (Iza and Stein 2009) use a similar concept in the process principle of "certainty", which rests upon the rule of law in terms of both predictability and enforceability. This would of course be dependent upon laws also reflecting principles of ecological integrity, equitable access for all and linkages between land and water resources. Otherwise, rigidity in the application of "bad" laws and policies would diminish adaptive capacity.

Knowledge and Information

The UNECE (2009) cite the importance of supporting training and response systems with climate and hydrological information systems, which are "capable of delivering early warnings in a timely and efficient manner" (UNECE 2009, p. 42). Folke et al. (2005) relate the idea of knowledge with the creation of an iterative learning environment. There are therefore important links with *flexibility* through the process of learning by doing. The goal here relates to an improved understanding of the dynamics of the whole system so that an understanding is established for how to manage periods of rapid change. The interpretation of knowledge is also highly linked with how to effectively deploy scientific information across different networks or levels of decision-making for the management of resource issues in the context of change. Engle and Lemos (2010) also refer to the linkage of using scientific knowledge and information with the building of adaptive capacity, but add to the concept the importance of equality of decision-making and knowledge use (in terms of power distribution among stakeholders and access to technical knowledge). Nelson et al. (2007)

also suggests that the ability to maintain a response capacity is predicated in part on the capacity for learning.

Decentralisation

Decentralization and subsidiarity (Hurlbert 2008) refers to the delegation of responsibility and authority of water management to the lowest feasible level. Devolved decision-making means that a system would be "presumably, better able to recognize and respond to unforeseen circumstances" (IISD 2006, p. 119). There a theoretical link here to the IWRM component "Basin/Watershed Approach", as well as to Olssen et al. (2004) concept of enabling legislation that creates social space for ecosystem management. Yet, while a system may be highly devolved, this does not imply that there are ecological based units of decision-making.

Discussion

This list of indictors captures the development in the analytical field of adaptation and vulnerability in the preceding decade. However, it is equally recognized that there has been fairly minimal empirical verification of the correlation between different principles and indicators of adaptive capacity, particularly at local and regional scales, and more so within the water sector (Engle and Lemos 2010; Wilbanks and Kates 1999). There are a number of analytical challenges relating to the different principles and indicators of adaptive capacity listed above. The next section discusses these and proposes how further theoretical and empirical research could contribute to improving our understanding of which elements in which circumstances may create an enabling environment for adaptive capacity.

Normative Principles Versus Open Indicators

Much of the discussion around governance issues in adaptation and adaptive capacity has a strong normative edge. Normative principles such as accountability and participation tend to denote a stronger bias towards the researcher's analytical framework. More open indicators such as knowledge and levels of decision-making are less prescriptive and therefore predisposed to be more iteratively developed through the research process, both the theoretical and empirical exploration. While this distinction should be recognized, and normative bias to the analytical framework should be avoided where possible, it should not be seen as a major impediment to the development of more robust indicators. However, for the sake of the iterative development of indicators within this research, more prescriptive and normative indicators employed within the STRIVER assessment have been replaced by more open indicators. Additionally, it is difficult to make a priori judgements on issues to do with levels of financial and human resources, leadership or even experience. Instead, these indicators should be explored in an a posteriori framework, to investigate correlations with adaptive capacity across different scales (sectors, geographies, political levels).

Process Versus Outcome

There is a difference between the process indicators as described in many of the studies, and the more outcome associated determinants in others. Requirements such as "enabling legislation that creates social space for ecosystem management" (Olsson et al. 2004) and "institutional capacity" (UNECE 2009) can be seen as requisite for an enabling environment for adaptive capacity, but also as an outcome of sufficient adaptive capacity. A key issue is therefore how questions relating to enabling legislation and institutional capacity could be integrated into more open indicators. Or, are such concepts in fact outcomes of indicators such as "levels of decision-making and networks", and therefore should not be separately tackled within the adaptive capacity assessment per se? More specifically regarding "institutional capacity", one could perhaps infer that if indicators such as transparency, knowledge, networks, resources, decentralization/subsidiarity and experience are met, then institutional capacity should be strengthened, therefore it could be taken as an output.

Similarly, the issue of "process versus outcome" is pertinent to IWRM. While IWRM is not considered an indicator, its component parts could be seen as useful determinants of adaptive capacity. An indicator for "integration" could encapsulate a key element of IWRM. The initial Swiss assessment showed that its prime weakness in the face of future change was the lack of integration across different sectors, levels and geographies, indicating that a lack of these components may decrease adaptive capacity of the system, thereby increasing its vulnerability to external shocks and environmental change. Normative prescriptions could be avoided by not suggesting that an ideal level or type of integration pre-exists, but that different levels and types may enable adaptive capacity in varying sectors or geographies. Additionally, considering that numerous studies (including the Swiss governance assessment) have shown that "a substantial gap exists between promise and practice" (Ingram 2011, p. 2) in IWRM, it would be make more sense to focus on how different types of integration rather than IWRM per se contribute adaptive capacity, than testing normative assumptions based on the criteria of IWRM.

Finally, the concept of environmental integrity or ecological system resilience (Nelson et al. 2007) appears regularly as a key determinant for adaptive capacity in the adaptive management discourse. Since the capacity of aquatic ecosystems to produce many of the goods and services on which societies depend is rapidly declining, the provision of water for nature/nature as a buffer can be seen as a key determinant of adaptive capacity in a system under stress. If the biological component of the system is already under stress, then adapting to more extreme conditions may be limited. Principles purported within the adaptive governance literature are linked with achieving these outputs, but again the question arises of how to define the relationship between ecological integrity and resilience with adaptive capacity.

Decentralization, Participation and Governance Modes

Preferences concerning the right mix of modes of governance (hierarchy/state, market/private and decentralization/civil society) are rife within the literature on adaptation and vulnerability, despite the recognition by many that what matters is that prescriptions fit contexts (Ingram 2011). The focus on full participation and decentralization in water management as desirable norms is reflected across a broad swath of the literature (Hurlbert 2008; UNECE 2009; World Bank 2002; UNDP 1997; Nelson et al. 2007). However, other studies note the fact that decentralization and participation per se are not a priori requirements for better management and enhanced resilience. Berkes in Nelson et al. (2007, p. 409) suggests that "the balance of evidence shows that neither purely local level management nor purely higher level management works well by itself" (p. 239), and Lemos and Agrawal (2006) highlight the development of emerging hybrid, multilevel and cross-sectoral forms of environmental governance. Hill (2010) suggests that there may be a limit to the level of devolvement, and that it can only be effective when combined with requisite levels of experience and resources as well as a propensity for stakeholders to work across the other levels of decision-making.

Ingram (2011, p. 8) adds that "participation is no panacea for water conflicts". Other studies such as Iza and Stein (2009, p. 8) elaborate that other factors such as coordination across levels, rather than pure participation and decentralization hold significant importance. Results from earlier work (Hill 2010) also support these findings, which suggests the need to look beyond prescriptive norms such as participation and decentralization and subsidiarity, to more exploratory indicators which allow examination of causal relationships between different indicators and adaptive capacity within different sectors as well as governance regimes.

Knowledge and Information

In a number of studies the indicator of transparency is pinpointed as fundamental to good governance and adaptive capacity. However, drawing on studies and publications in the resilience framework and the wider climate dialogue, it might be worth broadening out from the normative prescription of transparency to a more thorough exploration of the contribution that different forms of knowledge and/or information play in enhancing resilience. By looking at knowledge as well, we therefore refer not just to scientific information and data (hydrological models, climate models, economic statistics, etc.), but can also recognize the potential importance of local and indigenous knowledge. A recent report from Switzerland comments on the need to take account of and integrate traditional knowledge in climate data systems (Lugon 2010).

An awareness of the need for climate services also recently arose out of the 3rd World Climate Change Conference in Geneva (WCC-3 2009), which refers to the provision of climate information (both current climate variability and recent and future climate change) (Lugon 2010). It also calls for better management, communication and understanding of this information so that resource managers and the public alike can actually generate knowledge out of the wealth of data and information available. The HEID report comments that while today, people are likely to be inundated with information, often "the hurdles are not the hard science but the communication" (Lugon 2010, p. 64). It also notes that climate information per se is not enough; to be truly valuable it needs to be integrated with socio-economic and other environmental data. It is therefore important to investigate not just what kind of information decision-makers are getting, but also how they use it, with whom they share it and how relevant it is to the problem they need to resolve.

Empirical Application

Using Extreme Events to Explore Tentative Indicators

Following on from the discussion above, a more refined list of tentative indicators has been proposed, which will be empirically explored through stakeholder interviews in each case region. They are: knowledge; networks; levels of decision-making; integration; predictability/flexibility; experience; resources; leadership (Table 1). The indicators and sub-criteria draw on current understanding and the different indicators in the discipline of adaptive capacity, adaptive governance and adaptive management, as well as the discourse on integrated water resources management. They have been supplemented with understanding based on theoretical exploration and results from the initial governance assessment within the Valais case area.

Previous studies assessing the adaptive capacity of water law to climate change have utilized case studies of water stress events to provide insights into important modifications in the institution of water law which will increase adaptive capacity (Hurlbert 2009). The IPCC (2007a) also recognizes that "empirical knowledge from past experience in dealing with climate-related natural disasters such as droughts and floods (…) as well as longer term trends in mean conditions, can be

Table 1 Level 1 operationalization of tentative indicators to be further explored and developed in interview

Tentative indicators	Sub-criteria
Knowledge	Right to information; communication/public perception; spatial planning; access to scientific/environmental information; exchange of data and information; integration of scientific expertise; quality of scientific information; use of traditional and local knowledge
Networks	Access to participation; selection of non-state actors; level of influence; type of participation; stage in the political process; social networks; professional networks; willingness to cooperate
Levels of decision-making	Ecological based units of decision-making; institutional arrangements
Integration	Geographical integration; sectoral/uses integration; political integration
Flexibility/predictability	Consistency in rule of the law; rigidity of legal provisions; iterative elements of law/institutions
Resources	Financial resources; quantity/quality of human resources; organization of resources; independence/impartiality of experts
Experience	Training and development; years of experience
Leadership	Political commitment; facilitating role; initiation of partnerships; support mobilization; linking of actors; trust amongst stakeholders

particularly helpful in understanding the coping strategies and adaptive capacity" (p. 138). This study therefore is utilizing case studies of past extreme events relevant to the two basins, which can serve as reference points of climate variability and as useful indications for the impact of extremes in a future, warmer climate.

In order to move beyond the characterization of adaptive capacity to be able to assess which specific approaches are associated with higher adaptive capacity, it is necessary to establish some qualification and quantification of the extreme event impacts. Therefore, to contribute to the characterization of adaptive capacity, the impacts and evolution of extremes were analysed at six climate stations in the Valais (Ulrichen, Sion, Visp, Zermatt, Gd. St. Bernhard and Fey). Summer heat waves were characterized by analysing temperature and precipitation data from June–July–August (JJA). Extreme precipitation events were characterized through August–September–October (ASO) precipitation records. Table 2 depicts a general overview of the types of case events used for the assessment of adaptive capacity, and the associated impacts for each event.

Climate model data from the ACQWA project to calculate return periods as per A1B and B2 emissions scenarios. Initial results from the stations at Visp and Zermatt are detailed below (Fig. 5).

Additionally, impacts data will be gathered at the regional and local level on alterations in water allocations as well as in water reserves. This will be done through both interview and physical data. Figure 3 represents projections of the development of enhanced drought and flooding risk in the Rhone basin. The graph suggests that summer months could experience enhanced drought situations through reduced glacial mass and precipitation, while in winter months increased

Table 2 Details of impacts on separate categories of water goods and services per each extreme event

Geomorphological impacts	Impacts on water goods and services Beniston (2005); Beniston et al. (2007)				
	Non-domestic consumption	Energy	Environmental flows	Transport and absorption of wastewaters	Domestic consumption
Summer heat wave Permafrost degradation and reduction in the cohesion of slope material, acceleration of glacier retreat	Heat stress, enhanced evaporation, droughts, soil-moisture depletion, acceleration of glacier retreat (which can also lead to increased river levels)	Changes in hydro-power supply because of seasonal shifts in the filling of dams	Damage to eco-systems through excessive heat and drought. Increased glacial retreat can leads to river level rising	Reduced flow increasing concentration of chemicals in water ways. Increased glacial retreat can leads to river level rising	Drinking water is always prioritized
Extreme precipitation Slope instability events in mountain regions, increased frequency and severity of floods	Sediments deposited in large quantities on agricultural lands, irrigation canals and streams—reductions in agricultural production	Erosion, discharge and sedimentation rates—damage to hydropower infrastructure		Increased flows causing overflow of wastewaters into water ways	

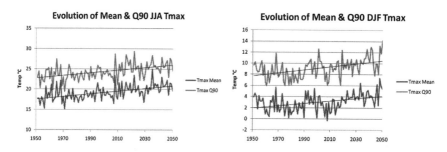

Fig. 5 Evolution of temperatures at Visp according to A1B scenario for summer temperatures (*left*) and winter temperatures (*right*)

intense precipitation periods could not only impact on flood risk, but a wide range of geomorphological processes such as landslides and rock falls.

Stakeholder interviews were conducted at the local and regional level in the Valais, in order to better characterize and assess the mobilization of adaptive capacity in response to the case events. The semi-structured interviews consist of questions relating to the different indicators, lasting 45–90 min with a range of individuals with expertise and experience in water and natural hazards policy and management at the local and cantonal level. Three sub-case areas were identified in the Valais, consisting of six communes that were representative of the different sectoral interests as well as the different microclimates in the Valais. The qualitative data from these interviews serves to provide greater insight into the indicators and will be used to operationalize the criteria of the indicators, as well as assess its mobilization in response to the events. A discussion of preliminary results from interviews in the Swiss case area is found in the following section.

Discussion: Key Issues for Adaptive Capacity

The predominant issue in the Valais canton is flooding. While there has been long-term experience of floods and related hazards, experts interviewed noted that the volume of water has significantly increased since the 1990 s, due to an increase in runoff from glacial melt. This has led to increased damage from floods and associated geological hazards. Interestingly, during the major heat wave in 2003, the Valais experienced increased runoff from glacial melt into the waterways. This served to buffer issues from the lack of precipitation during the event, and actually led to a situation where water levels in the Rhone, its lateral rivers, as well as connected groundwater sources reached high water marks. However, this trend is not projected to last (see Fig. 3). Some experts suggested that they are currently experiencing the peak of contribution from glacial melt, and in the near-term future, a decline will be observed.

Interviewees did point out that there have been increasing numbers of communes which have experience scarcity situations, predominantly in springtime or during winter months in the ski resorts. This tends to be dependent on a number of factors, such as amount of precipitation during winter, when the snowmelt begins, amount of precipitation falling in spring (e.g., April), which all impacts on how quickly the springs can recharge after winter.

With reference to the indicators, issues were raised with data *integration* across the different politico-administrative levels and disciplinary-sectoral foci. While there is little integration of climate change and ecosystem concerns (beyond a subconscious awareness) into the cantonal and communal offices so far spoken with, there are overriding legal and financial incentives for local actors to take these issues into better account (for more information on the Neuefinanzausgleich, see Hill 2010). Additionally, the Third Rhone Correction (Hill 2010) could be seen as one of the first attempts to better integrate different sectors and ecological and climatic concerns in water management in the Valais.

In some cases, such as Les Bagnes (the municipality which includes the tourist village Verbier), the municipality has had experience of over-demand in peak times (rather than scarcity) and issues of a lack of interconnection between neighbouring villages. Thirty to forty years ago, all the water resources of these villages were separate, so if they had issues of supply during a certain period, they had to deal with this alone. This has been managed by connecting the resources, so that where villages have only one source, they can now be integrated into the network, accessing resources higher up the mountain, or with other villages that have plentiful water. Creating a larger and more connected network of supply has allowed Les Bagnes (and in another case, Crans Montana) to better manage the peak demand during the winter tourism season, when flows are at their lowest.

With respect to *knowledge and information*, while there is a considerable amount of effort in improving monitoring and observation networks, access to information is seen as complex and uncoordinated. Actors at the cantonal level asserted that this was recognized and efforts were on going to generate central databases for hydrological information, but improving connectivity across different groups and administrative levels was a challenge.

The relevance of *networks* in the response to these events is highlighted by stakeholder comments that when extreme situations do arise, there is intensive contact with other stakeholders. This comes in different forms at the different levels. For example at the commune level, as part of their risk planning strategies, there is an "alarm group", which includes a number of different actors, some of whom are involved in water issues on a full-time basis, some of whom are volunteers. While a rich response network was discussed in relation to hazards (flooding, avalanches), relatively little exists across different stakeholder/sectoral groups for water use, outside of the different conventions between the municipalities and ski-lift owners (who manage the artificial snow production) or the hydropower companies.

The irrigation groups, known as *Geteilschaft/Gemeinschaft; Consortages*, which co-govern the water irrigation canals (*Suonen/Bisses*) (Reynard 2008), still exist in the Valais. There are currently about 706 km of canals in the Valais, with many being reinstated and repaired along with the accompanying organizations, not only for tourist value, but also as a means to foster solidarity in maintaining infrastructure and minimizing costs as the local level.

Collaboration exists across different stakeholders, notably in the domain of research and monitoring collaborations. For example, hydropower companies partner with the federal research institutes to develop projections of impacts on glacial melt from climate change. The cantonal administration relies on private consultancies to measure, collect and provide data to them on water quality and quantity. At the local level, communal authorities use private consultancies to outsource aspects of EIAs and other reports/studies they are required to submit to the canton. This aspect concerning networks is also tied into the resources indicator, since at the local level these deliverables are outsourced because resources are limited. Additionally, cantonal authorities noted that financial resources at the commune level tended not to be sufficient to cover damage costs from flooding events such as the 2000 event in Baltschieder, near to Visp. In these situations, the canton or federal government needs to step in and cover a certain amount of the costs.

The relationship between the different *levels of decision-making* in is highly interesting for a number of indicators. Cantonal authorities consistently clarify that they can only take a "hands off" approach, using incentives and guidance to suggest water management methods on the lateral rivers. The Rhone, on the other hand, remains the responsibility of the canton. Interestingly, while the federal government has passed a Federal Ordinance on Water Provision during times of emergency, it has not been brought into effect in the Canton Valais.

Additionally, it is clear that no one person is really responsible for "water management" as a whole. At the cantonal level, people have their individual responsibility, with intermittent collaboration across the different offices. At the local level, the communal authority is generally responsible, with different competencies for different individuals. Since the communes tend to be so small, people know who they need to go to.

With regards to *flexibility*, one key area is the length of hydropower concessions. For example, in the communes of Les Bagnes, these will not to be renegotiated until 2040. Currently in times of stress, the municipal industrial service must informally ask the hydropower companies if it would be possible to do an exchange of water, which is possible at their "goodwill". For an actual renegotiation of the terms of the concession, they will need to wait until 2040.

Discussions pertaining to the influence of *experience* of past events were particularly interesting and rich. The role of prior experience in extremes and impacts of natural hazards was considered highly important in determining the preparedness for extreme events. The importance of prior events, such as the 1993 and 2000 floods, in the collective consciousness should therefore be potentially considered as important as professional experience and training. More generally, the knowledge and experience of historical shifts in the climatic and socioeconomic situation, and the fact that the Valais is the driest canton in Switzerland, has meant that the irrigation infrastructure is well developed. This means that the Valais is perhaps better positioned perhaps than other areas in Switzerland for coping with drier, more stressed periods through the adaptations that stakeholders (mainly agricultural) have had to implement over the past decades (better irrigation technology and resurgence of the consortages) and centuries (development of the canals and consortages since the 13th century).

There is perhaps a need to focus on the broader implications of this in the planning for climate change impacts in the water sector, and in general. Since climate change impacts are seen to be pushing us beyond our range of experience, one question is how to build social resilience in the absence of prior experience of similar issues. This is particularly relevant for the tourist sector, since unlike the agricultural actors, the experience and knowledge of managing past change and uncertainty does not stretch back as far.

Leadership did not generate much discussion in interviews, except for the role of a certain individual in the cantonal administration for natural hazards, who is responsible mainly for overseeing avalanche preparedness. However, in the case of Les Bagnes, one can identify the important role the central organization, SIB, has played in creating a more integrated network for water provision. It may, therefore, be worth expanding on the understanding of *leadership,* so that it encompasses the leading role a centralizing local institution can play, not just individual leadership.

Issues of scale are relevant not only to *levels of decision-making,* but to other indicators as well, notably *networks* and *integration.* Politico-geographical scale is highly relevant to where water issues are experienced in the Valais and where they are not. In some villages, drinking water and irrigation water are taken from the same sources (mainly groundwater), which means that while one village may have no issues with conflicting uses (refer to the Le Châble example above), the next-door village will experience water scarcity during peak times of use, or low periods of precipitation. Scale plays a relevant role here on two counts. Firstly, the municipal level micromanagement means that neighbouring villages can experience highly contrasting management problems. Secondly, in this particular case, altitudinal and geographic scale directly affected the variety of sources Valaisanne communes can rely on.

The interviews showed that in practice there is a considerable amount of crossover between some of the indicators (i.e., *networks—knowledge—resources; resources—experience; integration—networks*), which will need to be addressed. It may also be required to rethink the underlying criteria behind an indicator such as *experience,* as to whether collective experience is a more driving factor than individual training, which could also be a feature of *resources.* Finally, two of the indicators, *leadership* and *resources* proved the most difficult in terms of eliciting rich responses as to their relevance in both managing the resource and the events.

Conclusion

This paper set out to discuss some of the issues in developing a deeper understanding of how to characterize and assess adaptive capacity to climate change in water governance arrangements. The discussion of the current state of indicators on adaptive capacity suggests that there are a number of challenges in present assumptions, yet identifying and refining characteristics of adaptive capacity through interviews also proves testing.

The initial results from interviews in the Swiss case area suggest that is necessary to rethink some of the characteristics of the indicators. While further work needs to be done to refine and operationalize all the indicators, two of them need particular attention and possibly reassessment. Partly as expected, *leadership* is perhaps conceptually too vague to have significant analytical relevance for assessing the governance system. It would perhaps be better placed in disciplines more concerned with anthropology or broader social sciences. With regards to the *resources* indicator, it is particularly difficult to be able to meaningfully define whether the level of financing or human resources, training or general education is sufficient. It is also unclear how this could be more universally operationalized to be comparatively meaningful across the different case areas and beyond. Other studies do not seem to provide many answers here, since a great deal of data and indicators on *resources* tends to lie at the national scale (Brooks et al. 2005; Engle and Lemos 2010).

Future work will apply the methodology in the Chilean case area (which was delayed due to the earthquake) in order to generate a comparative analysis into the relevance of these indicators in a different socioeconomic and climatic context. This research does not aim to provide a broad panacea of adaptive capacity insights and recommendations for building adaptive governance, but instead to use comparative local experiences to better operationalize adaptive capacity, from which we can build our understanding of how to reduce vulnerability to future uncertainty and climate change. It is vital to provide not only local guidance, but to use comparative cases to see where a more holistic understanding of adaptive capacity itself can be built, and how relevant universal indicators can be to local issues such as water governance.

Additionally, future work must focus on better understanding the outcome of the reaction to and management of these events. This could allow us to infer whether change (in this case extreme events) is being managed in a way that enhances adaptive capacity or leads to a more negative outcome (Chapin et al. 2009). Future work will also encompass deeper analysis of the qualitative data, with the tool Atlas-TI, as well as using richer climatological and hydrological model data from the ACQWA project to build a better picture of climate impacts on which to map the vulnerabilities and adaptive capacity of the governance system.

The issue of scale is still problematic in the adaptive capacity literature. Empirical application in the contrasting case regions will address not only issues of verification, but equally of scale. While there is a growing literature on adaptive governance and adaptive management with respect to water resources, there is still a tendency for the respective disciplines of social science and climatology to remain within their respective niches. This is to the detriment of a more holistic understanding of challenges from climate change and the exploration of potential responses to these threats or possibly opportunities. In this study, it has been attempted to explore ways of bridging this gap.

While this attempt may be a crude initial effort, it is hoped that it is a contribution to the understanding of how researchers need to start not only understanding the different languages of a multitude of different disciplines, but also should employ the tools associated with them within their own research to gain a more comprehensive understanding of the issues which are being researched. Finally, adding this extra layer of detail to the indicators will add clarity and a deeper understanding to their performance and the reporting of the results. While normative indicators can be useful, more analytical empirical studies need to assess how different determinants may contribute to building resilience across scales and sectors, in different forms of climatic extremes.

Acknowledgments The author would like to acknowledge all those who took the time to be interviewed for this research. This work was supported by the ACQWA project coordinated by the University of Geneva under EC contract 212250.

References

Adger WN, Vincent K (2005) Uncertainty in adaptive capacity in C. R. Geoscience 337:399–410
Allan A (2008) Governance assessment methodology. Centre for Water Law, Policy and Science, University of Dundee, Scotland
Bates BC, Kundzewicz ZW, Wu S, Palutikof JP (eds) (2008) Climate Change and Water. Technical paper of the intergovernmental panel on climate change, Switzerland IPCC Secretariat, Geneva
Beniston M (ed) (2005) The Risks Associated with Climate Change in Mountain Regions. Global change and mountain regions: an overview of current knowledge
Beniston M, Stephenson DB, Christensen O, Ferro CAT, Frei C, Goyette S, Halsnaes K, Holt T, Jylhä K, Koffi B, Palutikof J, Schöll R, Semmler T, Woth K (2007) Future extreme events in European climate: an exploration of regional climate model projections. Clim Chang 81:71–95
Brooks N, Adger WN, Kelly PM (2005) The determinants of vulnerability and adaptive capacity at the national level and the implications for adaptation. Glob Environ Chang 15:151–163
Chapin FS, Kofinas GP, Folke C (eds) (2009) Principles of ecosystem stewardship: resilience-based natural resource management in a changing world. Springer, London
Eakin H, Lemos MC (2006) Adaptation and the state: Latin America and the challenge of capacity-building under globalization. Glob Environ Chang 16:7–18
Engle N, Lemos MC (2010) Unpacking governance: building adaptive capacity to climate change of river basins in Brazil. Glob Environ Chang 20:4–13
FOEN (2009) Environment Switzerland 2009, Federal Office of the Environment (FOEN) Bern, Switzerland, accessed June 2009, available at http://www.bafu.admin.ch/publikationen/publikation/01039/index.html?lang=en&show_kat=/publikationen/00003
Folke C, Hahn H, Olsson P, Norberg J (2005) Adaptive governance of social-ecological systems. Annu Rev Environ Resour 30:441–473
Häberli W, Beniston M (1998) Climate change and its impacts on glaciers and permafrost in the alps. Ambio 27(4):258–265
Hill, M. (2010): "Converging threats: assessing socio-economic and climate impacts on water governance". Int J Clim Chang Manag Strateg, accepted

Holling CS (1978) Adaptive environmental assessment and management. Wiley, London

Hurlbert M (2008) An analysis of trends related to the adaptation of water law to the challenge of climate change: experiences from Canada, presented at Climate 2008, 3–7 November 2008, online

Hurlbert M (2009) The adaptation of water law to climate change. Int J Clim Chang Strateg Manag 1(3):230–240

IISD (2006) Designing policies in a world of uncertainty, change and surprise: adaptive policy-making for agriculture and water resources in the face of climate change, International Development Research Centre and The Energy and Resources Institute, Winnipeg, Manitoba and New Delhi, India

Ingram H (2011) Beyond universal remedies for good water governance: a political and contextual approach. In: Garrido A, Ingram H (eds) Water, Food and Sustainability, Routledge

IPCC (2001) Climate Change 2001: Synthesis report. A contribution of working groups I, II, and III to the Third assessment report of the intergovernmental panel on climate change. Watson RT, the Core Writing Team (eds) Cambridge University Press, Cambridge, New York, p 398

IPCC (2007) Climate change 2007: synthesis report. Contribution of Working Groups I, II and III to the Fourth assessment report of the intergovernmental panel on climate change, IPCC, Geneva, Switzerland

IPCC (2007) Summary for Policymakers. In: Parry ML, Canziani OF, Palutikof JP, van der Linden PJ, Hanson CE (eds) Climate change 2007: impacts, adaptation and vulnerability. Contribution of Working Group II to the fourth assessment report of the intergovernmental panel on climate change, Cambridge University Press, Cambridge, UK, pp 7–22

Iza A, Stein R (eds) (2009) "RULE–Reforming water governance", IUCN. Gland, Switzerland

Tarlock D (16 July 2009) Seminar, UNESCO Centre for Water Law, Policy and science, University of Dundee, Scotland

Lemos MC, Agrawal A (2006) Environmental governance. Annu Rev Environ Resour 31: 297–325

Lugon R (2010) Climate information for decision making: lessons learned from effective user-provider communication schemes. The Graduate Institute, Geneva, Switzerland

Milly PCD, Betancourt J, Falkenmark M, Hirsch RM, Kundzewicz ZW, Lettenmaier DP, Stouffer RJ (2008) "Stationarity is dead: whither water management?". Science 319:573–574 1 February 2008

Nelson DR, Adger NW, Brown K (2007) Adaptation to environmental change: contributions of a resilience framework. Annu Rev Environ Resour 32:395–419

OcCC (2008) Das Klima ändert—was nun? Der neue UN-Klimabericht (IPCC 2007) und die wichtigsten Ergebnisse aus Sicht der Schweiz, OcCC—Organe consultatif sur les changements climatiques, Bern

Olsson P, Folke C, Berkes F (2004) Adaptive comanagement for building resilience in social-ecological systems. Environ Manag 34(1):75–90

Olsson P, Gunderson LH, Carpenter SR, Ryan P, Lebel L, Folke C, Holling CS (2006) Shooting the rapids: navigating transitions to adaptive governance of social-ecological systems. Ecol Soc 11(No. 1 (18)): accessed January 2010, available at http://www.ecologyandsociety.org/vol11/iss1/art18/

Pahl-Wostl C, Gupta J, Petry D (2007) Transitions towards adaptive management of water facing climate and global change. Water Resour Manag 21:49–62

Parry ML, Canziani OF, Palutikof JP et al. (2007) Technical Summary. In: Parry ML, Canziani OF, Palutikof JP, van der Linden PJ, Hanson CE (eds) Climate change 2007: impacts, adaptation and vulnerability. Contribution of Working Group II to the fourth assessment report of the intergovernmental panel on climate change, Cambridge University Press, Cambridge, 23–78

Reynard E (2000) Cadre institutionnel et gestion des ressources en eau dans les Alpes: deux études de cas dans des stations touristiques valaisannes. Swiss Political Rev 6(1):53–85

Reynard E (2008) "Les Bisses du Valais. un example de gestion durable de l'eau?" in Lenaniques 69:1–6

Rockström J, Steffen W, Noone K, Persson A, Chapin S, Lambin EF, Lenton TM, Scheffer M, Folke C, Schellnhuber HJ, Nykvist B, de Wit CA, Hughes T, van der Leeuw S, Rodhe H, Sörlin S, Snyder PK, Costanza R, Svedin U, Falkenmark M, Karlberg L, Corell RW, Fabry VJ, Hansen J, Walker B, Liverman D, Richardson K, Crutzen P, Foley JA (2009) A safe operating space for humanity. Nature 461:472–475 24 September 2009

Smit B, Burton I, Klein R (2000) An anatomy of adaptation to climate change and variability. Clim Chang 45:223–251

Timmerman JG, Pahl-Wostl C, Möltgen J (eds) (2008) The adaptiveness of IWRM; analysing European IWRM research. IWA Publishing, London UK ISBN: 1843391724

TwinGo (2010) Synthesising research for adaptive water governance. Twin2Go newsletter 1:1–4 January 2010

UNDP (1997): *Governance for Sustainable Human Development*, accessed January 2009, available at http://mirror.undp.org/magnet/policy/

UNECE (2009) Guidance on water and adaptation to climate change, United Nations Economic Commission for Europe Geneva, Switzerland

WCC-3 (2009) Better climate information for a better future, Conference Statement, accessed May 2010, available at http://www.wmo.int/wcc3/documents/WCC-3_Statement_Ex_Summary_04-09-095PM.pdf

Wilbanks TJ, Kates RW (1999) Global change in local places: how scale matters. Clim Chang 43(3):601–628

World Bank (2002) Toward more operationally relevant indicators of governance (Prem Notes. No. 49, December 2002). Accessed February 2009, available at http://www1.worldbank.org/prem/PREMNotes/premnote49.pdf

Yohe G, Tol RSJ (2002) Indicators for social and economic coping capacity–moving toward a working definition of adaptive capacity. Glob Environ Chang 12:25–40

Chapter 23
The Contribution of Rural Development Programmes in Mitigating Greenhouse Gas Emissions in Italy

Rocío Dánica Cóndor, Marina Vitullo, Domenico Gaudioso and Marina Colaiezzi

Abstract The Health Check reform, reinforcing the Common Agricultural Policy, emphasizes the role of agriculture in climate change, facing issues related to climate change, renewable energies, protection of biodiversity, water management, innovation, and dairy production. The CAP Health Check targets were included in the Rural Development Programmes (RDPs); a preliminary qualitative assessment identified that 14 RDPs include climate change targets and eight RDPs include the six Health Check targets (MIPAAF (2009): Analisi dei PSR sulle nuove sfide del Health Check—17/09/2009. Rete Rurale Nazionale 2007–2013). This paper aims to make an initial quantitative assessment of the impact of agriculture greenhouse gas (GHG) emission reduction measures in Italy, taking into account the 21 Italian RDPs. The conceptual framework of this work was based on a multiple criteria approach, which considers the analysis of diverse GHG emission reduction measures for the agriculture sector, and multiple rural development *axis–measures–actions* from RPDs. Main activities were found under measure 121 (farm modernization) and 214 (agri-environment). A more rational use of nitrogen fertilizers can result in an effective mitigation action (reduction of N_2O emissions). Concerning emissions of CH_4 from manure management, the adoption of biogas recovery could be relevant; while other specific incentives for electric power production are also supporting methane emission reduction (<1 MW special fee and >1 MW green certificates).

R. D. Cóndor (✉) · M. Vitullo · D. Gaudioso · M. Colaiezzi
Dipartimento stato dell'ambiente e metrologia ambientale,
Istituto Superiore per la protezione e la ricerca ambientale (ISPRA),
Via Vitaliano Brancati 48, 00144, Rome, Italy
e-mail: rocio.condor@isprambiente.it

Keywords Climate change · Mitigation · Rural development programmes · Common agricultural policy

Introduction

In the last years, climate change issues have been addressed in different fields, such as the agriculture sector. In particular, the Common Agricultural Policy (CAP) and the Health Check reform has directly tackled specific issues related to climate change, renewable energies and water management, among others. In March 2010, the European Commissioner for the Environment called for a *Common Agricultural and Environmental Policy*, saying that the CAP should improve sustainability, soil quality, water quality and efficiency.

Right now, the CAP is based on two main pillars, the so-called Pillar 1 supporting farmers' income through market interventions and direct payments, and Pillar 2 supporting the development of rural areas, which take the form of Rural Development Programmes (RDPs). A first balance and analysis of the Italian situation related to rural development and the mechanism that transfers funds from direct payments to rural development measures is available for Italy (Sotte 2009). Currently, the 2007–2013 programming period contains a basket of policies with different objectives, management and intervention methods, beneficiaries and stakeholders. These policies range from structural, historical heritage, environmental conservation and biodiversity, to policies for training, information and assistance; also forest policies are included as well as those enhancing food quality; then there are policies for diversification to those for the quality of life in rural areas.

In Italy, the agriculture greenhouse gas (GHG) emission inventory (ISPRA 2010) and projections (Ministry for the Environment, Land and Sea 2009) has been improved thanks to research studies. Researches were concentrated on improving inventories from the methodological point of view, and efforts were oriented to develop country-specific emission factors, among other activities. Furthermore, there was a need to better understand the role of the CAP in the GHG emission trend and projections. For this purpose, ISPRA has followed a step-by-step research approach. Initially, the impact of Pillar 1 of the CAP was studied; secondly, the assessment of Pillar 2—Rural Development Programmes (RDPs)—was performed; and thirdly, we expect to follow up and update information from RDPs after the incorporation of the Health Check targets. Results of the last two activities will be described in this paper.

The objective of the article is to present the results obtained from the assessment of GHG mitigation measures described in RDPs from Italy. Two stages are evidenced: the first one assessed 21 RDPs based on information available by the end of 2008. RDPs in the period 2007–2013 were assessed only for climate change mitigation targets. The second stage considered RDPs recently approved under the

Health Check reform; thus, climate change and water management targets were considered. For this last phase selected RDPs were revised.

The outline of the article will provide information on recent GHG emissions from the agriculture sector in Italy; GHG policies and measures from this sector will be also presented. A brief overview of the state of the art of RDPs and climate change issues is provided, and the methodological approach is described. Results and discussions are provided, and finally main conclusions are presented.

Greenhouse Gas Emissions

The United Nations Framework Convention on Climate Change (UNFCCC) was ratified by Italy in 1994 through Law No. 65 of 15 January 1994. In the framework of the EU Burden Sharing Agreement and under the Kyoto Protocol, Italy has committed to reduce its GHG emissions by 6.5% below base-year levels (1990) over the first commitment period (2008–2012). On 1 June 2002, Italy ratified the Kyoto Protocol through Law No. 120 of 1 June 2002.

In order to comply with national and international commitments, the national GHG emission inventory is compiled and communicated annually by the ISPRA (ex-APAT) to the competent institutions, after endorsement by the Ministry for the Environment, Land and Sea. As a party to the UNFCCC and the Kyoto Protocol, Italy is committed to develop, publish and regularly update national GHG emission inventories as well as formulate and implement programmes to reduce these emissions. Therefore, Italy annually submits to the UNFCCC secretariat and the European Union's Greenhouse Gas Monitoring Mechanism the national GHG emission inventory through the compilation of the *Common Reporting Format* (CRF)[1] and the *National Inventory Report* (NIR) (ISPRA 2010). According to the guidelines provided by the UNFCCC, Parties should use for estimations the methodologies reported by the Intergovernmental Panel on Climate Change (IPCC).[2] The 2006 IPCC Guidelines for national greenhouse gas inventories has merged the agriculture and LULUCF (Land Use, Land Use Change and Forestry) sectors into the Agriculture, Forestry and Land Uses (AFOLU) sector. Up to now, Parties are not obliged to report according to the 2006 IPCC Guidelines since guidelines have not been formally approved by the UNFCCC. However, Italy has already included most methodological improvements from the 2006 IPCC Guidelines.

As required by IPCC guidelines, methane (CH_4) and nitrous oxide (N_2O) emissions are estimated for the agriculture sector. For Italy, the following emission

---

[1] Available at http://unfccc.int.

[2] IPCC Guidelines are: "Revised 1996 IPCC guidelines for national greenhouse gas inventories", "2000 Good Practice Guidance and Uncertainty Management in National Greenhouse Gas Inventories", "2003 Good practice guidance for land use, land use change and forestry for LULUCF" (GPG for LULUCF).

categories are included in the agriculture sector: enteric fermentation (4A), manure management (4B), rice cultivation (4C), agricultural soils (4D) and agricultural residues burning (4F). For the LULUCF sector, removals (CO_2) and emissions (CO_2, CH_4, N_2O) occur as a result of changes in land uses and forestry.

Total GHG emissions, in CO2 equivalent, excluding emissions and removals from LULUCF, have increased by 4.7% between 1990 and 2008, varying from 517 to 541 CO_2 equivalent million tons (Mt), whereas the national Kyoto target is a reduction of 6.5%, as compared to the base-year levels, within the period 2008–2012. The most important greenhouse gas, CO2, which accounts for 86.4% of total emissions in CO_2 equivalent, shows an increase of 7.4% between 1990 and 2008. In 2008, 6.6% of the Italian GHG emissions, excluding emissions and removals from LULUCF, originated from the agriculture sector, which is the second source of emissions after the energy sector, which accounts for 84%. For the agriculture sector, GHG trends from 1990 to 2008 show a decrease of 11.6% due to a reduction in the number of animals and cultivated surface. The LULUCF sector is responsible for 87.3 Mt of CO_2 removals from the atmosphere in 2008 (ISPRA 2010).

GHG Policies and Measures

The law that ratifies the Kyoto Protocol in Italy prescribed the preparation of a National Action Plan on GHG emissions. This action plan was adopted by the Interministerial Committee for Economic Planning (CIPE) on 19 December 2002 (No. 123).

GHG emission projections, which include mitigation policies and measures, are also part of international commitments and reporting. Information on how countries are implementing policies and measures is periodically presented through the National Communication reports to the UNFCCC. Italy recently presented the Fifth National Communication (5NC) to the UNFCCC (Ministry for the Environment, Land and Sea 2009). This report provides indications of future trends in GHG emissions and removals, given the current national circumstances.[3] In this context, energy and non-energy projections are prepared. Non-energy projections include the estimation of GHG trends for the agriculture, waste, industrial process and LULUCF sectors. Projections are provided for 2010, 2015 and 2020. In Italy, agricultural emission projections are estimated using the same methodologies utilized for the preparation of the GHG emission inventory, ensuring consistency between estimates. In Table 1, projections for the agriculture sector are shown.

---

[3] Parties shall report projections: "with measures" (currently implemented and adopted policies and measures); "without measures" (excludes all policies and measures implemented, adopted or planned after the year chosen as the starting point for the projection); and "with additional measures" (encompasses planned policies and measures) projections.

Table 1 Greenhouse gas emission projections from the agriculture sector (Mt CO_2 eq.)

GHG $MtCO_2$	1990	1995	2000	2005	2006	2007	2010	2015	2020
Enteric fermentation	12.18	12.27	12.17	10.84	10.63	11.03	10.75	10.70	10.52
Manure management	7.38	7.07	7.14	6.86	6.65	6.85	6.75	6.71	6.62
Rice cultivation	1.56	1.66	1.38	1.47	1.48	1.52	1.47	1.49	1.52
Agricultural soils	19.44	19.34	19.24	18.03	17.86	17.79	17.68	17.32	16.84
Field burning of agricultural residues	0.02	0.02	0.02	0.02	0.02	0.02	0.02	0.02	0.02
TOTAL	40.58	40.35	39.94	37.24	36.63	37.21	36.66	36.24	35.52

Source: Ministry for the Environment, Land and Sea 2009

With respect to 1990, the years 2010, 2015 and 2020 are expected to see a further reduction of 2, 3 and 5%, respectively.

In the previous Italian National Communication, policies and measures for agriculture related to the rationalization of the use of nitrogen fertilizers (reduction of N_2O emissions), and the recovery of biogas from animal waste (reduction of CH_4 emissions) were considered ("with additional measures"). However, for the 5NC, these measures were included in the trend scenario due to their adoption.

In general, most Parties cited other objectives in addition to climate change behind the implementation of agricultural measures and policies, and despite the lack of specific climate policies for the agricultural sector, most Parties reported decreases in emissions (UNFCCC 2003). In 2008, EU-15 emissions from the agriculture sector decreased by 12% with respect to 1990 (EEA 2010, p. 369). In the last four years, the Italian agriculture sector has verified an important annual GHG emission reduction with respect to 1990, from 8% (2005) to 12% (2008).

Mitigation measures and policies adopted in Italy for the agriculture sector, as described above, are in line with most Annex-I countries from the UNFCCC. The last available report from the UNFCCC describes that policies and measures directed at agriculture seek to: reduce N_2O emissions through manure management; reduce N_2O emissions from agricultural soils through optimized nitrogen fertilizer use; and reduce CH_4 emissions through changes in livestock management. This report also highlights as the most important policies, in terms of frequency of use and mitigation effects: fiscal incentives (either direct or within the context of agricultural market reform); and regulations (e.g. the EU Nitrates Directive) to a lesser extent (UNFCCC 2007, p. 37). In fact, a recent report from the European Environment Agency described that in the agriculture sector, very little emission reductions are projected from existing and additional measures for 2010 and 2020. The agriculture sector is also the sector where the least absolute and relative reductions are expected (EEA 2009, p. 13). Member States report almost no quantified GHG reductions expected from the CAP and other policies to reduce N_2O emissions from soils, in contrast with Commission estimates (EEA 2009, p. 52).

Climate Change and Rural Development Programmes

In the last years, the agriculture sector has contributed to the reduction of GHG emissions, but this sector is still called to intensify its efforts to reduce GHG emissions in the framework of global EU strategy related to climate change.

Agriculture in Europe is determined by the CAP. Therefore, drivers on the reduction of GHG emissions are linked to a wide range of policy objectives, such as market reforms (CAP Pillar 1); rural development (CAP Pillar 2); environmental pollution directives (e.g. the European Community Nitrates Directive and IPCC Directive); and other sustainable agronomic practices (organic farming, integrated agriculture, etc.) (UNFCCC 2003: 44). In Europe, Pillar 1 of the CAP had a strong impact on GHG emissions from agriculture. Specifically, the milk quota system leads to a strong reduction of animal numbers in the dairy sector (EEA 2010, p. 367). In Italy, between 1990 and 2008, the number of dairy cattle and non-dairy cattle decreased by 31 and 15%, respectively. For instance, this reduction led to a decrease in CH_4 cattle emissions (4A source) by 15%; cattle source represents 78% of 4A and 24% of total agricultural emissions. The animal number reduction (dairy cattle) has also led to an increase in productivity, reduction of CH_4 emissions per unit of milk produced, and reduction in cows' fertility (Coderoni and Cóndor 2010). A recent study analysed the impact of the CAP (from 1990 to 2007 reforms) for the Italian GHG agricultural emission trend by source category (4A, 4B, 4C, 4D and 4F). Results show that in Italy is in line with the rest of Europe; the CAP reform has driven GHG trends in the reduction in the number of animals, and for the increase in productivity (Coderoni and Cóndor 2010).

Rural development policy has adopted 4 axes and 41 measures (pillar 2 of the PAC). Axis 1 is about increasing competitiveness in the rural economy, especially in farming and forestry. Axis 2 focuses on conserving the landscape and environment, while axis 3 is related to the improvement of the quality of life in rural areas and helping the diversification of the rural economy. Axis 4 is aimed at helping local people to develop their own communities. RDPs need to express their choice for adopting each of the measures, how they will be financed, the way of access and management, and their priorities. In Italy, 21 RDPs are available (one for each region and autonomous province), representing 24% of the total RDPs in the European Union. This means that while the regionalization policy implements the principle of subsidiarity, on the other hand, it certainly implies a considerable additional administrative burden at regional, national and EU level, taking into account the articulated procedures for programming, approval, and monitoring and evaluation of the programmes (Camaioni and Sotte 2010: 45).

Regarding climate change challenges, in the European Union, in Pillar 2 of the PAC, the main GHG emission reduction activities are predominantly or exclusively supported by two rural development measures: farm modernization (measure 121) and agri-environment (measure 214). Some other activities support the modernization of farms through energy-efficient equipment and buildings, and

Table 2 Assessment of Health Check targets in Rural Development Programmes from Italy

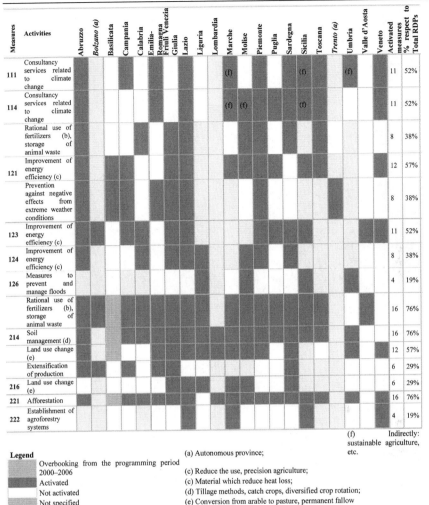

promoting biogas production (European Commission 2009: 20). In Italy, in a recent report from the Ministry of Agriculture—National Rural Network, the following climate change activities from the Health Check reform are given: reduction of the input of fertilizers, storage and use of animal waste for the production of biogas, biomass for the use of energy and all the interventions to increase energy efficiency. These are GHG emission reduction measures considered useful under the Kyoto Protocol (MIPAAF 2010). In Table 2, a qualitative assessment of climate change targets are presented. We have complete information for the 21 RDPs (information is based on the initial assessment presented by the Rural Network). Further analysis will be provided in the discussion section.

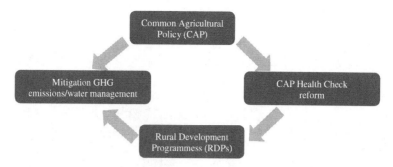

Fig. 1 Conceptual framework of this research

Methodology

The conceptual framework of this work was based on a multiple criteria approach, which considers the analysis of diverse GHG emission reduction measures for the agriculture sector, and multiple rural development *axis–measures–actions* from RDPs. The aim is to assess from a qualitative and quantitative (if possible) point of view the contribution of RDPs to climate change mitigation targets in Italy. Figure 1 gives a graphical representation of the conceptual framework. Two phases are evidenced: (i) *ex-ante assessment* performed at the end 2008, and (ii) *ex-post assessment*, which was executed in 2010 after RDPs included Health Check reforms.

A detailed analysis of RDPs was performed in order to get information related to climate change and water management activities. RDPs were downloaded from the Italian National Rural Network 2007–2013 site (www.reterurale.it).

For the ex-ante assessment, data and information contained in the 21 Italian RDPs 2007–2013 by the end of 2008 were considered. On the basis of the qualitative and quantitative figures reported for different measures (in the various pillars), an assessment of the impact of the scheduled policies and measures on climate change issues was carried out. Main findings were summarized in a matrix reporting the main challenges (i.e. reduction of CH_4 emissions from manure management, reduction of N_2O emissions from agricultural soils, increase of soil organic matter in agricultural soils, increase of carbon removals in forest lands, etc.) of the RDPs and an impact evaluation, for each activated measure.

The analysis of the effect on the atmosphere of the measures envisaged under the RDP was carried out through a detailed analysis of the relationships which may exist between the individual actions (different types of intervention) and the qualitative and quantitative aspects characterizing that environmental component.

The analysis refers to the specific objectives identified by the National Rural Development Plan for axis 2 measures relating to reducing emissions of GHG: (1) expansion of biomass and biofuels production; (2) carbon sequestration from forests and agricultural land; (3) reduction of GHG emissions; and (4) reduction of ammonia (NH_3) emissions. Under objective 3, two main types of actions, resulting in an emission reduction, are distinguished: the promotion of energy saving and renewable energy sources, and consequently a reduction in fossil fuel consumption and in carbon dioxide emissions, and the rationalization of agricultural and animal husbandry practices, leading to a reduction in emissions of CH_4 and/or N_2O.

Based on this approach, the analysis of the impact on the atmosphere has considered the following topics:

(a) climate change

    (a1) expansion of biomass production and biofuels
    (a2) carbon removal by forests and farmland
    (a3) promotion of energy saving and renewable energy sources
    (a4) rationalization of agricultural and animal husbandry practices

(b) air quality

    (b1) reduction of ammonia emissions,

without neglecting the overlaps between them and the frequent synergies with regard to the effects of measures.

In assessing the relationships between measures (analysed systematically and with specific reference to the individual actions included in each measure) and the atmosphere in its various aspects, the analysis evaluated:

- the relationship (positive or negative impact or neutrality) between actions under the plan and environmental issues according to the sustainability criteria set out above;
- the type of relationship (direct or indirect);
- the extent of the relationship (high, medium, low magnitude);
- the transience or permanence of the effects;
- the timescale (short, medium and long term);
- the scale of action (area of influence).

The relationship between specific measures and the environmental component also considered: likely development in the absence and presence of the measure, reference indicators, comparison with environmental protection objectives, mitigation or compensation measures, evaluation of alternative options, and the structure and adequacy of monitoring networks and actions to be taken to monitor the implementation of the measure.

In the case of significant effects (positive or negative), the possible synergy of various interventions has also been separately evaluated.

For the ex-post assessment, a text analysis revised RDPs, after the adoption of the Health Check and the European Recovery Package. The main objective was to identify new and possible quantitative information from selected RDPs. Six regions were chosen, considering their representativeness on animal and agricultural production and their contribution to national GHG emissions. Main contribution to GHG emissions for the agriculture sector is given by: Emilia Romagna, Piedmont, Veneto and Lombardy regions. However, Puglia and Campania were also considered in the detailed assessment. Climate change and water management targets information was collected from RDPs. In 1990, 54% of GHG agricultural emissions was represented by four regions in Italy: Lombardy (20.8%), Emilia Romagna (12.1%), Piedmont (11.0%) and Veneto (10.5%). In 2005, the regional contribution increased to 57%, as follows: Lombardy (23.4%), Emilia Romagna (11.5%), Veneto (11.2%) and Piedmont (10.7%) (Cóndor et al 2008).

Results and Discussions

Ex-ante Assessment

In this paper, only the results of two GHG mitigation activities are shown (CH_4 from manure management and N_2O from agricultural soils). The analysis was performed for the 41 measures; in this paper focus is given to axis 1 and axis 2. In Table 3, the ex-ante evaluation matrix for the 21 Italian RDPs is shown. The impact of activated measures on the reduction of CH_4 emissions from manure management are evidenced. Results show that most RDPs have activated measures mainly in axis 1. Major impact was given by the Campania, Calabria and Umbria through measure 131 (support farmers to face EU regulations). Emilia Romagna region addressed measure 215 for animal welfare.

In Table 4, the ex-ante evaluation matrix for the 21 Italian RDPs is shown, with a focus on the first two axes, concerning the impact of different activated measures on the reduction of N2O emissions from agricultural soils. A relevant impact is mainly addressed in axis 2 under measure 214 (agro-environmental payments). In many cases, RDPs disaggregate measure 214 in several actions which target GHG emission reduction. Further description of these actions is given in the ex-post section of this paper. In general, the most common target, highlighted in RDPs, is the reduction of the nitrogen surplus.

Ex-post Assessment

A first analysis is obtained from Table 2, where main climate change challenges were identified for selected rural development measures (13 RDPs were considered by the Rural Network). In order to have a complete panorama of the Health Check reforms regarding climate change challenges, we have incorporated information from the remaining RDPs (Liguria, Marche, Molise, Puglia, Sicilia, Umbria, Valle D'Aosta, Veneto). Two main results can be evidenced: 38 and 76% of RPDs include, under measure 121 and measure 214, actions such as the rational use of fertilizers and the storage of animal waste. However, other activities are also described such as the improvement of energy efficiency, measures to prevent and

Table 3 Ex-ante evaluation matrix on the reduction of CH$_4$ emissions from manure management

manage floods, afforestation, land use change, the establishment of agroforestry systems, and the consultancy services related to climate change.

The ex-post assessment has found that RDPs gave information on NH$_3$ and GHG emissions. Some of them identified the main source of GHG emissions, due to the use of fertilizers in agriculture and livestock production. The new revision of RDPs identified actions, and, in some cases, also specific indications of compliance under rural development measures. This is commonly found for the action on integrated production. The most relevant GHG mitigation activity for the agriculture sector defined in RPDs will be implemented through the reduction in the nitrogen surplus, which confirms the ex-ante assessment. In some RDPs, there is a clear reduction target on surplus reduction with respect to the current situation. Another important finding is the correlation evidenced by RDPs, between climate change and water management actions. Other mitigation targets are highlighted, such as the increase of carbon sequestration and the production of renewable

Table 4 Ex-ante evaluation matrix on reduction of N$_2$O emissions from agricultural soils

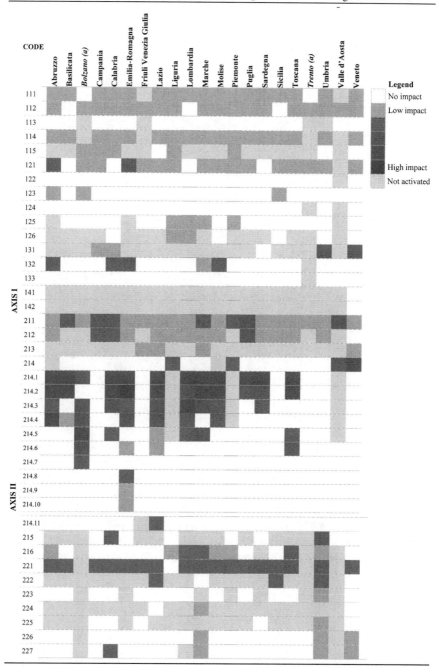

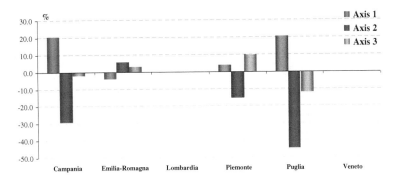

Fig. 2 Changes in financial balance between axes post Health Check

energy actions. Relevant information from selected RDPs is given in later sections of this paper. Attention is given to measures which target climate change and water management. The description of selected RDPs also includes a context presentation of each region in order to evidence the different territorial situations.

The financial balance between axes has been modified differently in the six regions under examination, as a consequence of Health Check reform. In Fig. 2, changes in regional balances between axes are reported; in Puglia and Campania reallocation of funds have resulted in an increase of the axis 1 share, at the expense of axis 2. Minor changes were performed by Piedmont or Emilia Romagna. Remaining regions have not changed their cost attributions to each rural development measure.

We performed estimations of GHG agricultural emissions considering findings from RDPs, based on the disaggregation, at regional level, of the national inventory for the last year available (2008, submission to UNFCCC, 2010). For the estimations, it was considered, on the basis of RDPs assessments, that Emilia Romagna, Piedmont and Puglia regions contribute an annual GHG emission reduction of 1% from 2009 to 2013. It was also assumed that Veneto and Lombardy regions have at least 0.5% reduction for the same period. Results show that national GHG emissions for the agriculture sector could be 35.54 Mt CO_2 eq. for 2010 and 34.83 Mt CO_2 eq. for 2015. Taking into account the above-mentioned findings and assessing the contribution to RDPs to the national GHG inventory, a further reduction of 3 and 5% with respect to the trend values from the 5NC (see Table 1) can be estimated.

Campania

Agriculture absorbs a large amount of underground and surface waters being used for irrigation purposes. The irrigation network covers 14.69% of the utilized agricultural area (UAA) and is currently serving 69% of agricultural land that can be irrigated. Overall the availability of water for irrigation is not a problem since the needs are satisfied by the availability of the resource. Rationalization in the use of water and technological improvements on water-saving methods and irrigation systems are foreseen. The approach in saving water is also crucial insofar as the food sector is concerned and investment for the use of wastewater for processing is considered a strategic need. Therefore, additional allocations will be provided to measures 121, 123 and 124 to meet this challenge. Regarding the climate change challenge, RDPs intend mainly to contribute to a reduction of CO_2 emissions by operations that will be financed under modernizations of the farms. It has been noted for measure 121 that a significant reduction of CO_2 emissions is expected (almost 25%) by the reduction of the fuel used in heating systems for greenhouses, dryers and stem production in on-farm milk processing plants.

It is stressed that the absorption of CO_2 provided by forests and the potential of carbon sinks in the soils can be exploited. In particular, the analysis underlines the contribution of carbon sequestration given by agricultural practices such as minimum work of land, precision agriculture, green cover, set-aside, organic farming, etc.; priority additional allocations will be given to measures 121 and 123.

Regional strategies are oriented towards the significant development of renewable energies from biomass of agroforestry origin in the light of sustainability. Therefore, the contribution that agriculture and forestry could give to the energy production from agroforestry biomass is important and the support planned in the approved RDP is deemed adequate to face these priorities. The emissions of greenhouse gases will be reduced by substituting fossils fuels with renewable sources of energy, which will improve the quality of air and contribute to combating climate changes, too.

Emilia Romagna

This region is based on intensive cultivation, which is translated into a high and increasing demand for water resources. In 2007, agricultural production showed a reduction in the production linked to yields. More demand on water resources for agricultural production led to competition for the use of water for agricultural and civil purposes. Farmers should pay more to get more irrigation, with the consequence of a higher water abstraction that leads to negative results. For this reason, measures oriented to save and use water resources efficiently are highlighted for this region (RER 2009a). Priority activities are focused on the reduction of the high pressure of agriculture, which leads to high NH_3 and GHG emissions. Thus, the reduction of loads of nitrogen fertilizers, and rationalization in the use of animal waste are prioritized. Importance is given to short rotation of poplars and biogas production from animal waste (RER 2009b).

For climate change challenges, three axes are involved. For axis 1, two main actions are encouraged, farm modernization in dairy production, and the protection of regional water resources due to climate change. Axis 1 includes measure 121 to incentive investment for the production of renewable energies, and measure 123 incentives on the recovery and waste disposal of sub-products from agro-industries also with energetic aims.

For axis 2, more financial resources are given to agro-environmental measures, biodiversity, and environmental quality in order to overcome climate change and for the protection of water resources (quantity and quality) (RER 2009b). Under measure 214, specific actions, related to climate change challenges, such as integrated production, organic farming and increase of organic carbon content are prescribed.

Integrated production action aims to reduce in average the use of macro-elements by 30–45% (N, P and K). Other advantages are obtained because of the method and the application period (time and split of fertilizer application), which determines a low relay to the groundwater by 40% of N and 60% of P (RER 2009c). Integrated production also contributes to water management through: (i) soil management practices (tillage method, catch crops, diversified crop rotation), and (ii) technologies for water saving (e.g. efficient irrigation systems). Specific indications for the use of fertilizers are given (integrated production): (i) define the quantitative maximum distribution of macro-nutrients by culture; (ii) define the timing and procedure for fertilizer distribution based on the climatic characteristics; and (iii) the rational use of liquid animal waste according to the best time for distribution for an efficient absorption of nutrients. For irrigation purposes (integrated production), indications encourage the adoption of a simplified water balance to define the time and the maximum amount of distribution (based on farm pedology and meteorological conditions) (RER 2009c, p. 319). Organic farming is an action aimed at the reduction of agricultural inputs, the conservation of soil fertility and the conservation of water resources. This action also contributes to water management through similar practices described for the integrated production (RER 2009c, p. 326). Organic soil content action (measure 214) considers the conservation of the soil resource through increasing the surface with organic fertilization. This action has a specific indication for the limitation in the use of mineral fertilizers corresponding to 40% of nitrogen needed for the culture (RER 2009c, p. 334).

Axis 3 considers the construction of plants for the production and distribution of bioenergy (measure 311). A specific action considers the production of renewable energies, use and sale of energy and heat (maximum 1 MW). Micro-plant for the production of biogas, which can obtain energy and electricity (includes cogeneration, but they only acceptable if it is demonstrated the CO_2 balance is positive). The result indicator shows that energy produced from this action could be equal to 5,955,524 KWh (RER 2009c, p. 399).

For the 2007–2013 period, the region estimates an annual reduction of GHG emissions of 48,000 t CO_2 eq. through the reduction in the use of nitrogen fertilizers, afforestation activities and the production of renewable energies. This is approximately 1% of total agricultural GHG emissions reduction per year (RER 2009b, p. 205).

Lombardy

The most important agricultural sector of the region is cow's milk, which accounts for nearly 33% of the total agricultural output. The composition of UAA is the following: 74.1% arable, 22.3% permanent pastures, 3.9% permanent crops. GHG emissions are concentrated in the lowlands, where intensive agriculture and livestock production is located.

Axis 1 considers the adjustment of irrigation infrastructures for water saving and rationalization is described. Measure 121 will support investments on irrigation equipment for water saving and for renewable energy production. Support will be given to investments for biogas production, using organic waste (maximum power = 1 MW, energy to be used on farm and energy plant dimension proportional to the farm energy consumption). For axis 2, activities will support "green territorial systems" for water purification, creation of ecological corridors, and countryside valorization, and promote low-impact agricultural and biomass production (measures 216, 221, 223). Specific actions with measure 214 will support integrated production and organic farming. Measure 221 will encourage the production of forest biomass to favour the conversion of agricultural production to sustainable woody production (GHG reduction). For axis 3, the promotion of renewable energy production (and linked services) will be promoted. Support will be received through measure 311 (diversification into non-agricultural activity), and measure 312 (support for business creation and development).

For climate change, this region specifically invests: in farms (i.e. renewable energy, manure storage), afforestation and actions to prevent forest fires, training and farm advisory services focusing on climate change issues. Some of the additional operation could have positive effects in terms of climate change mitigation (i.e. biogas in dairy farms). Currently, only 33% of farms in animal production have covered storage, as a consequence, there is the necessity to favour interventions to cover storage system for biogas production (Regione Lombardia 2009).

Piedmont

The forest sector is the most representative economic sector in this region. The prioritized agro-food sector is mainly composed of arable land and rice. Most of the Italian rice production comes from this region. Other relevant sectors are wine, fruit, milk and beef production. After approving the RPD with the Health Check reform, additional resources for climate change issues have been allocated to measures 111, 121, 123 and 124 for the improvement of energy efficiency and thus for reducing CO_2 emissions. Regarding the water management issue, the priority of the region is to safeguard water quality, and to improve water management by rationalizing livestock manure and wastewater treatment. Additional Health Check resources have been allocated to measures 111, 121 and 123, aiming to finance structural investments necessary for improving the efficiency of nitrogen fertilizer use, installations for wastewater treatments on farms, and for water-saving technologies.

Regarding climate change, for axis 1 the region focused on the improvement of efficiency in rural infrastructures. Interventions are expected to improve the multifunctionality and efficiency of the irrigation network, mainly through the reduction of losses in order to decrease agricultural water consumption (measure 125) (Regione Piemonte 2009, p. 206). Through measure 111, the region supports training courses in order to reduce GHGs and to adapt to climate change, measure 121 will encourage energy saving, and measure 123 and measure 124 the improvement of energy efficiency. Water management challenges are tackled under measure 111 through training courses on the improvement of water management; measure 121 for the improvement of the capacity to use water more efficiently and to improve the capacity to store water; and measure 123 for the improvement of the capacity to use water more efficiently.

The objective to reduce GHG emissions and acidificant pollutants is supported by the following measures in axis 2: agro-environmental payments (214), animal welfare payments (215), non-productive investments (216 and 227), afforestation of agricultural land (221), the first agroforestry system in agricultural land (222), afforestation in non-agricultural land (223), environmental forestry payments (225) and restoring forestry potential (226); and through a bioenergy programme. The most significant action to face climate change mitigation challenges is focused on the increase of carbon sequestration in biomass and soils and the production of energy from agricultural and forest biomass and other renewable energies. Activities are oriented to support agronomic practices, animal production, and animal waste management. These actions operate in synergy with those finalized to conserve water resources, in particular, through animal waste treatment (Regione Piemonte 2009, pp. 211–218).

For the water resources challenge, in axis 2, actions are related with improving the efficiency of irrigation resources, reducing the pollution of water because of agricultural chemical inputs, reducing the level of nutrient surplus, and reducing the negative effect of animal waste and wastewater.

This region has a specific target for GHG emissions that is estimated to be an annual reduction of 1% with respect to 1990 until 2020. Two impact indicators that are described by this region are indirectly tackling climate change: water quality improvement through the reduction by 13% of the baseline value of the gross nitrogen balance (71 kg N ha^{-1}), and the increase in the production of energy from renewable energies (+9.2 ktoe).

Puglia

Agricultural lands cover more than 73% of the territory. In the last years, the UAA has notably decreased, while the number of small farms (with UAA smaller than 5 ha) has increased. Almost 20% of the regional UAA is irrigated; nevertheless, water availability in rural areas of Puglia does not satisfy the needs of the farmers; in addition, the use of abusive wells is increasing. To address the new challenge of water management, additional funding will be used to reinforce measure 125, to support the improvement/modernization of the distribution network of treated wastewater for irrigation, in order to save water. The regional irrigated area will not be increased.

In the framework of diversification of farms and environment protection investing in the production of energy from renewable sources (mainly biomass) has been proposed; the increase in production is expect to be 20%.

The region is affected by processes of desertification, soil erosion and loss of organic matter. The priority of intervention will be addressed by the new measures 214/5, green cover of arboriculture land that will increase carbon sequestration and promote the sustainable use of soil, avoiding desertification and erosion, as well as measure 214/6, stubble leaving and/or ploughing, which will also increase the content of organic matter as a means of climate change mitigation (reducing desertification processes).

Puglia region assessed a reduction of emissions from the agricultural sector of about 7% for the 2007–2013 period, while an increase of 5% in removals is expected. Concerning ammonia emissions from the agricultural sector, a reduction of 15% is foreseen.

Veneto

Agriculture and forestry cover 75% of the territory. During the last few years, there has been a tendency towards reduction of the UAA and forest areas. Air quality is affected by ammonia emissions that cause acid rains with a negative effect on forests, biodiversity and soil quality. This problem is derived from livestock manure and agricultural activities (fertilizers).

For axis 1, it is described that agriculture and forestry could contribute with energy production from agroforestry biomass (reducing GHG emissions). Therefore, additional allocations are provided to improve the processing of forestry biomass (measure 122), and to transform forestry biomass into energy for the forestry sector's needs (measure 123).

Main climate change challenges are presented in axis 2. Under measure 214, specific actions are as follows: action b for the improvement of the quality of soil in order to improve the content of organic matter and nutrients in the soil, reduce nutrients such as nitrogen and phosphorus and decrease the usage of inorganic fertilizers; action c for supporting organic farming to reduce inputs and undertaking commitments foreseen in Reg. 2092/91 that would benefit biodiversity and water quality; and action g for water resources protection (ground and surface water) in the perspective of implementation of Directive 2000/60, in order to reduce GHG emissions due to conversion of arable land to grassland. For the forestation of agricultural land grant and premium scheme (measure 221), the aim is to increase area (except in mountains) under forestry and contribute to climate change mitigation; to provide a sustainable source of timber for energy, to improve water quality in NVZs, to increase biodiversity and create/improve the natural environment, and to protect soil against erosion. Measure 222 and measure 223 will operate for afforestation and the establishment of agroforestry systems, which have the potential for reducing N_2O emissions and increasing carbon sequestration. Measure 223 considers the conversion of agricultural land into forest/agroforestry system for the protection and improvement of water quality. Measure 227 addresses climate change challenges with interventions planned to improve the original conditions of natural forests.

This region describes that activities to tackle climate change challenges are focused on increasing carbon sinks. The analysis presents the impact on carbon sequestration also deriving from agricultural practices such as minimum work of land, precision agriculture, green cover, etc. With the Health Check reform, a new action that considers climate change and water management issue has been inserted (measure 214i: conservative agriculture) (Regione Veneto 2009, p. 628).

Conclusion

This paper summarizes an initial attempt to quantify the contribution of RDPs from Italy to the climate change mitigation challenge, and to highlight the relationship with the water management challenge. A quantification of the impact of 41 axis–measures–actions from the 21 RDPs was performed. The matrix analysis of regions, rural development measures and GHG mitigation targets shows a moderate impact of activated measures on reduction of CH_4 emissions from manure management under axis 1 (measure 121). Regarding axis 2, mainly under measure 214 (environmental payments), a high impact is expected for the reduction of N_2O emissions from actions oriented to reducing the nitrogen surplus.

For the ex-post assessment, carried out on six selected RDPs, a more specific analysis was performed for collecting information regarding climate change and water management targets, aiming to highlight the regional peculiarities adopted by regions in facing both targets. Regions tend to provide descriptive information, while other regions supply additional quantitative information. Each region has considered a different set of actions under each type of rural development measure.

Most actions oriented to climate change/water management challenges are addressed in axis 1 and axis 2 from the RDPs. A direct correlation between actions which can have an effect on both challenges is evidenced (e.g. Emilia Romagna region), finding positive synergies. In many cases, measures from axis 1 and measures from axis 2 converge and incentivize both challenges through diverse actions. For example, integrated agricultural production contributes with water management objectives through soil management and water-saving technologies.

Specific rural development measures were focused on: measure 121 commonly used to incentivise production of renewable energies (Emilia Romagna), energy-saving and efficient use of water (Piedmont, Campania) and biogas production (Lombardy, Campania), and measure 214 with integrated production and organic farming actions. Local peculiarity has used RDPs to prioritize actions, such as for the integrated production to rationalize the use of fertilizers (Emilia Romagna), increase organic soil content, organic farming and water resource protection (Veneto region), and avoid desertification and erosion processes (Puglia region). On the other hand, there has also been the case of actions targeting biodiversity conservation and conservation of high natural value (HNV), which are indirectly facing climate change challenges because they are encouraging organic production methods, the extensification of production and sustainable management of forests (Piedmont region).

Three regions out of the six selected RDPs have estimated a possible GHG reduction target achievable with the implementation of RDPs (Emilia Romagna, Piedmont and Puglia). Emilia Romagna and Piedmont regions assumed that the contribution of regional GHG mitigation is an annual reduction of 1% per year. The detailed analysis of RDPs allows us to estimate the contribution of the activated measures on GHG emission reduction for the agricultural trends. A further 3% (2010) and 4% (2015) of reduction could be achieved with the implementation of RDPs measures. In this context, the monitoring and evaluation of the different actions, which will be implemented under the RDPs, is fundamental to assess the impact of the measures for the mitigation of GHG emissions.

References

Camaioni B, Sotte F (2010) Un primo bilancio della politica di sviluppo rurale in Europa. Rivista Agriregionieuropa 6(20):41–45

Coderoni S, Cóndor RD (2010) Emissioni di gas-serra dall'agricoltura italiana e gli effetti della Politica Agricola Comune, submitted to QA-Rivista dell'Associazione Rossi-Doria

Cóndor RD, Di Cristofaro E, De Lauretis R (2008) Agricoltura: inventario nazionale delle emissioni e disaggregazione provinciale, ISPRA Rapporto tecnico 85/2008, Rome. http://www.apat.gov.it/site/it-IT/APAT/Pubblicazioni/Rapporti/Documento/rapporto_85_2008.html

European Commission (2009) Commission staff working document. The role of European agriculture in climate change mitigation, SEC(2009) 1093 final. http://ec.europa.eu/agriculture/climate_change/sec2009_1093_en.pdf

European Environment Agency—EEA (2009) Greenhouse gas emission trends and projections in Europe 2009. Tracking progress towards Kyoto targets, EEA Report N 9/2009

European Environment Agency—EEA (2010) Annual European community greenhouse gas inventory 1990–2008 and inventory report 2010. Submission to the UNFCCC Secretariat, EEA Report N 6/2010

ISPRA (2010) Italian greenhouse gas inventory 1990–2008. National inventory report 2010, ISPRA Rapporto tecnico 113/2010, Rome. http://www.apat.gov.it/site/it-IT/APAT/Pubblicazioni/Rapporti/Documento/rapporto_113_2010.html

Ministero delle Politiche Agricole, Alimentari e Forestali—MIPAAF (2010) Contributo tematico per l'aggiornamento del PIANO STRATEGICO NAZIONALE relativamente alla nuova sfida Cambiamenti Climatici Versione 1. Rete Rurale Nazionale 2007–2013, Italy

Ministry for the Environment, Land and Sea (2009) Fifth National Communication under the UN framework convention on climate change, Italy, Nov 2009. http://unfccc.int/resource/docs/natc/ita_nc5.pdf

Regione Emilia Romagna—RER (2009a) Programma di sviluppo rurale dell'Emilia Romagna 2007–2013 Analisi del contesto socio economico, dell'agricoltura e dell'ambiente. Integrazione novembre 2009. Direzione Generale Agricoltura, Regione Emilia Romagna, Italy

Regione Emilia Romagna—RER (2009b) Programma di sviluppo rurale dell'Emilia Romagna 2007–2013 I PARTE Sostenibilità ambientale, qualità sociale e distintività del territorio rurale dell'Emilia-Romagna Valori per tutta la società, fattori competitivi per un'agricoltura organizzata e innovativa, protagonista delle filiere agroalimentari e dei mercati globali'. Versione 4, Direzione Generale Agricoltura, Regione Emilia Romagna, Italy

Regione Emilia Romagna—RER (2009c) Programma di sviluppo rurale dell'Emilia Romagna 2007-2013. II PARTE Sostenibilità ambientale, qualità sociale e distintività del territorio rurale dell'Emilia-Romagna. Valori per tutta la società, fattori competitivi per un'agricoltura organizzata e innovativa, protagonista delle filiere agroalimentari e dei mercati globali. Versione 4. Direzione Generale Agricoltura, Regione Emilia Romagna, Italy

Regione Lombardia (2009) Programma di sviluppo rurale 2007–2013, Rev 3 del 13/11/09. Regione Lombardia Agricoltura, Lombardia, Italy

Regione Piemonte—RP (2009) Regione piemonte. Programma di sviluppo rurale PSR 2007–2013 ai sensi del regolamento (CE) n. 1698/2005. PARTE I Testo adottato con DGR n. 2-9977 del 5 novembre 2008 (modifiche anno 2008) e integrato con modifiche Health Check al 10 dicembre 2009. Piemonte, Italy

Regione Veneto (2009) Programma di Sviluppo Rurale per il Veneto 2007–2013, Regione del Veneto, Giunta Regionale. Veneto, Italy

Sotte F (2009) La politica di sviluppo rurale 2007–2013. Un primo bilancio per l'Italia, Gruppo 2013 Quaderni, pp 264. http://www.gruppo2013.it/working-paper/Documents/La%20politica%20di%20sviluppo%20rurale%202007-2013.pdf

United Nations Framework Convention on Climate Change—UNFCCC (2003) National Communications from parties included in Annex I to the Convention. Compilation and synthesis of third national communications. Addendum. Policies and measures (FCCC/SBI/2003/7/Add.2; 29 May 2003). http://unfccc.int/resource/docs/2003/sbi/07a02.pdf

United Nations Framework Convention on Climate Change—UNFCCC (2007) Compilation and synthesis of fourth national communications Addendum. Policies, measures, past and projected future greenhouse gas emission trends of Parties included in Annex I to the Convention (FCCC/SBI/2007/INF.6/Add.1; 27 November 2007). http://unfccc.int/resource/docs/2007/sbi/eng/inf06a01.pdf

Chapter 24
"Climate Proofing" Water Resources Development Policy: The Evidence from Bangladesh

Nazmul Huq, Zakir Hossain, Rashedul Hasan and Al Mamun Azad

Abstract The water sector of Bangladesh is instrumental in supporting the livelihoods of millions of people. Over the last few decades, water resources in Bangladesh have come under the severe threat of climate change impacts and the first effects of climate change are already being felt. The Intergovernmental Panel on Climate Change (IPCC) univocally confirms the future water sector vulnerabilities for Bangladesh in its Fourth Assessment Report. However, the water resource development policies of Bangladesh have insufficiently acknowledged the issue of climate change impacts on its economy and livelihood. The paper aims to examine the level of climate policy integration in newly developed water-related policies, i.e. National Water Policy (NWP), National Water Management Plan (NWMP), Poverty Reduction Strategy Paper (PRSP). The approach is conceptualized as "climate proofing" policy (integration of climate change as a cross-cutting issue). A descriptive analysis of "climate proofing" policy is presented based on a literature review. The paper finds that Bangladesh still cannot make its water policies "climate-proof" considering the high level of climate vulnerabilities. Furthermore, the lack of a wider contextual understanding of the issue has appeared as one of the major impediments. Subsequently, the paper outlines strategies for developing pro-adaptive, responsive and "climate-proof" water policies in order to sustain this valuable resource from climate change impacts. Improved coordination among the stakeholders and convergence of key policy documents has been deemed instrumental to make the policies "climate-proof" and sustainable. In addition, potential bottlenecks of developing such climate-responsive water policies occupy significant discussion in the paper.

N. Huq (✉) · Z. Hossain · R. Hasan · A. M. Azad
Bangladesh Centre for Advanced Studies, House 10,
Road 16A, Gulshan 1, 1212, Dhaka, Bangladesh
e-mail: nazmulhuqrussell@gmail.com

Keywords Bangladesh · Climate proofing · Climate change · Impacts · Policies · Vulnerabilities · Water resources

Introduction

Climate change is one of the most formidable long-term challenges faced by the global community. The word "crisis" is sometimes overused in the social dimension of development. But when it comes to water, there is a growing recognition that the world faces a crisis that, left unchecked, will derail progress towards the Millennium Development Goals (UN-DESA 2008) and hold back human development. Throughout history, human progress has depended on access to clean water and on the ability of societies to harness the potential of water as a productive resource. Water for life in the household and water for livelihoods through production are two of the foundations for human development (UNDP 2006). Water is probably the world's most intensely contested natural resource. Water is not only a necessity for life but also a critical input in nearly every type of economic activity, from farming to energy production (CCCD 2009). United Nations Human Development Report 2006 states that by 2025 up to 3 billion people will be living in a water-stress situation (UNDP 2006). Climate change, the gravest threat for sustainable development, often exacerbates the water problems with possible multiplier negative impacts on lives and livelihoods. Climate change is arguably the most severe long-term threat to development facing this and future generations. By altering the hydrological cycle, climate change will exacerbate the water management problems that countries already face (Bates et al. 2008). The paper argues that a "climate proofing" water policy is an urgent call for managing sustainably the scarce water resources under the threat of climate change. The issue of mainstreaming or adaptations of climate change policies are becoming more and more crucial for all development stakeholders for "climate-proof" development. That underpins the shift of assessment paradigm to "environmental sustainability" to "climate integration".

The paper is an assessment of the major policy documents of Bangladesh related to water management and development to examine the level of "climate proofing". Bangladesh is one the most vulnerable countries to climate change impacts and its water resource sector is considered one of the most affected sectors, which triggers other vulnerabilities. The paper uses three major policy documents for the assessment, i.e. National Water Policy (NWP), National Water Management Plan (NWMP) and Poverty Reduction Strategy Paper (PRSP), to examine the level of "climate proofing".

The aspiration of the paper is interdisciplinary and multidimensional, as both water-related vulnerabilities and policy documents are intertwined with an innumerable number of issues. On the one hand, climate change-related water vulnerabilities are considered to be the gravest threat not only to humankind but also to the existence of a living planet and, on the other hand, the documents are the guiding principles for nations on the road to sustainable development. As the apex guiding documents, those documents must have guiding indications as to how the water–related climate vulnerabilities can be tackled sustainably. Therefore, the paper promises to seek a level of interaction between PRSP and adaptation to climate change.

Climate Change Impacts on Water Resources and Impacts on Development in Bangladesh

As stated earlier, the water sector is one of the most vulnerable sectors to the impacts of climate change. An enormous number of assessments, reports, findings and scientific findings endorse the statement with a very high level of confidence. It is estimated that by 2050, if present trends continue, Bangladesh's annual incremental food production would be decrease significantly, in monetary terms the equivalent of 1% of total national GDP (MoEF 2008).

The IPCC forecasts that global warming will result in sea level rises of between 0.18 and 0.79 m, which could increase coastal flooding and saline intrusion into aquifers and rivers across a wide belt in the south of the country, although most of the area is protected by polders. Rainfall is predicted to become both higher and more erratic, and the frequency and intensity of droughts are likely to increase, especially in the drier northern and western parts of the country (Pender 2008; IPCC 2007b).

Considered one of the most vulnerable countries to climate change impacts, Bangladesh is facing tremendous negative impacts on its development aspirations. Floods, tropical cyclones, storm surges and droughts are likely to become more frequent and severe in the coming years. The water sector is projected as one of the vulnerable sectors, putting poor people's livelihoods at stake through the various erratic patterns of behaviour, described below:

- Increased temperatures in the water bodies of Bangladesh may also impact on fisheries.
- In terms of impact on livelihoods and the economy, the most affected by a rise in surface water temperatures would be Bangladesh's coastal shrimp-farming industry, for if the temperature goes above 32°C, the small shrimp fries would have high death rates (Pender 2008).
- Higher rainfall in the Ganges basin is projected (Stern 2006) in the monsoon period, which is likely to lead to more frequent and severe floods from swollen rivers, while less rain in the winter will mean less water in rivers in the dry season affecting river-fed irrigation, industry, fisheries, travel by launch/ferries and increasing salinity around the coast (Tanner et al. 2007).
- Agricultural production decrease significantly due to saline water, abrupt rainfall and unusual flooding and by 2050, food security will be a serious concern for the marginalized people (Ahmed 2006).

- Water sector impacts on livelihood become a challenge of development under the most adversarial changes in the dynamics of nature. Water-related impacts will threaten the significant achievements Bangladesh has made over the last 20 years in increasing incomes and reducing poverty, and will make it more difficult to achieve the MDGs (DoE 2007; MoEF 2008; GED 2008).
- Serious and recurring floods have taken place in 2002, 2003 and 2004. Cyclones originating in the Bay of Bengal have been noted to have decreased since 1970 but the intensity has increased (Mirza 2002).
- The frequency of monsoon depressions and cyclones formation in Bay of Bengal has increased (Pender 2008).
- Water shortages have been attributed to rapid urbanization and industrialization, population growth and inefficient water-use, which are aggravated by changing climate and its adverse impacts on demand, supply and water quality (Bates et al. 2008).
- Saltwater from the Bay of Bengal is reported to have penetrated 100 km or more inland along tributary channels during the dry season (Ahmed 2006).
- The precipitation decline and droughts have resulted in the drying up of wetlands and severe degradation of ecosystems (Pender 2008).
- The food security of the whole country would be severely at risk due to shortage of available water for irrigation, resulting in a decline in agricultural production (MoEF 2005).
- The very high of likelihood of spreading waterborne diseases that will increase the death toll if left unchecked (GED 2008).

"Climate Proofing" Policies

The term "climate proofing" is somewhat synonymous with the term "mainstreaming" of climate policies into development policies and action. Mainstreaming of climate policies into sectoral development plans helps to develop "climate proofing" policies and actions. Klein et al. (2007) state that "climate proofing" is the integration of policies and measures that address climate change into development planning and ongoing sectoral decision-making so as to ensure the long-term sustainability of investments as well as to reduce the sensitivity of development activities to both today's and tomorrow's climate. The term "climate proofing" also indicates that climate policies are integrated in the policy documents (Sperling 2003). As part of the whole development discourse, sometimes "climate proofing" development refers to "adaptation-friendly" development, especially for the countries which are suffering from adaptation deficit.

Climate change is real and its adverse impacts are already being felt. At the same time, the increasing impetuses on sustainable development push for making the policies and actions climate-proof so that they can reduce the vulnerabilities of climate change in different sectors and work towards the overall goal of sustainable development.

National Policies and Responses to Climate Change

Water is central to the way of life in Bangladesh and the single most important resource for the well-being of its people. It sustains an extremely fragile natural environment and provides a livelihood for millions of people (MoWR 1999). This vital resource sector has experienced poor management in the past and Bangladesh is experiencing the result of bad management in different forms including sedimentation, waterlogging, regular flooding, groundwater lowering and disruption of production, among others. The climate change phenomenon, especially seen in the rapid melting of the Himalayan glaciers (IPCC 2007a), puts additional severe negative impacts on the whole development aspirations of the country. Recently, the government of Bangladesh prepared a sector-based development approach. Among them, the NWP, 1999 and NWMP, 2001 were developed to regulate and manage the water resources efficiently and the 2nd PRSP, 2008 was the apex guideline for implementing the development policies over the next few years (2008–2011). This paper will carry out a thorough assessment of those documents from a "climate lens". It will address to what extent these documents are "climate proofing" and responsive to the climate change reality. In the age of rapid climate change and as a most vulnerable country, it might be expected that the climate change issue will be necessarily reflected by those documents for "climate proofing" development.

Methods

The assessment provides an overview of the integration of adaptation to climate change and climate risks management within a subjective qualitative framework. One of the limitations that the study finds is an inadequate amount of literature on Climate Policy Integration (CPI) (Ahmad 2009) and proxy indicators were used to evaluate the "climate proofing" performances of the three selected documents. In addition, through review of some pertinent studies done by Bojö et al. (2004); Huge and Hens (2009); Kramer (2007); Prowse et al. (2009); and Klein et al. (2007) suggested a subjective qualitative framework of analysis.

Coinciding in a subjective approach of assessment, this research has used eight indicators to examine the "climate-proof" approach of the documents. Those indicators are adapted from the researches mentioned above. The indicators are as follows (Table 1):

Identification of water-related challenges due to climate change. How the documents identify the upcoming water sector-related challenges posed by climate change.

Identification of water-related vulnerable sectors and communities and solutions. Issues related to a clear identification of sectors and communities due to climate change impacts. The vulnerabilities of the water sector and water sector-dependent communities are given preference.

Table 1 The performance of NWP in regard to the indicators set for justifying a "climate proofing" document

Name of the document	Indicators						
	Identification of water challenges due to climate change	Identification of sectors and vulnerable communities	Identification of solutions for water related vulnerable sectors and communities	Range of adaptation options considered	Enabling proactive strategies	Attention to process of mainstreaming adaptation	Total
NWP	0	1	1	1	0	0	3

Range of adaptation options considered. Identification of a set of adaptation projects, priorities and a coherent institutional capacity for the implementation phase.

Enabling proactive strategies. To what degree the documents are responsive to enable making proactive strategies and decisions as far as water sector vulnerabilities are concerned.

Attention to process of mainstreaming adaptation. This is a very important issue to consider when examining a "climate proofing" document. How much attention is paid to mainstreaming the adaptation process into development activities needs to be considered.

The analysis method is a combination of the works conducted by Bojö et al. (2004) and Kramer (2007). The assessment works are very much qualitative in nature and where those two are combative, they provide a basis of comparison among the case studies. The assessment of the documents across the selected indicators is based on a qualitative judgement. All variables received a score with respect to each document's merit, comprehensiveness, sensitivity to water sector vulnerabilities to climate change, time of preparation and necessity of inclusion:

0 = Not mentioned
1 = Identified but not elaborated
2 = Identification or elaboration of the concept

Though the assessment does not intend to be scientifically precise, it is a good indication of the level of developing "climate proofing" policies and actions. The interpretation of the scores is as follows:

0–4 = Little or no progress in the integration of developing "climate proofing" water policies for climate change and climate risk management
5–8 = The document has a growing level of awareness and understanding of the value and requirements of "climate proofing" water policies, mainstreaming, and recognizes the need for action. It may also have decided to take action
8–12 = It refers to an intermediate stage, where the document is developing plans and tools to address the requirements of "climate proofing" water action to climate change.

Results and Discussion

National Water Policy

NWP of Bangladesh was prepared in 1999 with the aim address the objectives of improved water resources management and protection of the environment. The policy envisaged that it would help to guide both public and private actions in the future to ensure optimal development and management of water, which benefits both individuals and the society at large (MoWR 1999). In its declaration, it remarked that the policies were designed to ensure continued progress towards fulfilling the national goals of economic development, poverty alleviation, food security, public health and safety, decent standard of living for the people and protection of the natural environment (ibid.).

To examine the document from a "climate proofing" perspective, it has undergone the indicators described in the methodology section. It is very surprising that the whole NWP does not even mention the term "climate change". The document identified six objectives for better water resources development and management and also acknowledges that the water resources of Bangladesh are facing severe environmental stress but climate change is not regarded as the threat to the water sector. The document also allocated a full section, namely "Water for the environment" and at the beginning of the report there is another section, "River basin management". It might be expected that the issue of climate change would attract considerable attention at least in those two special sections. Unfortunately neither of the sections has any clue of climate change vulnerabilities to the water sector.

As a result, the document does not meet any of the objectives by which this research paper identifies whether a document is "climate proofing". However, this document identifies some of the vulnerable communities and sectors which may suffer from water-related problems. As stated in objective two, "to ensure the availability of water to all elements of the society including the poor and the underprivileged and to take into account the particular needs of women and children" this document points out the poor, underprivileged, women and children as vulnerable groups and communities. Besides, in the sections "River basin management", "Water for agriculture" and "Water for the environment", it reports some potential natural vulnerabilities and disasters such as floods, lowering groundwater table, biodiversity loss, wetland destruction, shortage of irrigation water but the linkage of climate change with those identified disasters is not at all apparent. As potential remedial action, this document proposes some policies, but again the threat of climate change was not taken into consideration. It proposes to develop a NWMP for preparing detailed action against those vulnerabilities Table 1.

The score of the assessment is 3, which indicates that in terms of climate policy integration, this document is really very poor and cannot guide the water resource sector under a changing climate.

National Water Management Plan

Parallel to the NWP, Bangladesh embarked upon the preparation of the NWMP (2005). In the course of this, a development strategy for NWMP was adopted by Bangladesh in 2001. NWMP was prepared to provide guidelines for implementing water sectoral projects and actions at national and regional level in accordance with the changing global and national demands. This plan is divided into three phases; short phase (2001–2005), medium phase (2006–2011) and long phase (2011–2025) and will be revised after each five years (MoWR 2001).

The NWMP stands on eight building blocks for rational, optimal and sustainable use and development of water resources of Bangladesh. It is a comprehensive action plan consisting of detailed hydrological zone action plans, project portfolios, budget and future guidelines. The major building blocks for developing responsive efficient water management identified by NWMP are (MoWR 2001):

- Institutional development
- Enabling environment
- Main rivers
- Towns and rural areas
- Major cities
- Disaster management
- Agriculture and water management
- Environment and aquatic management

As mentioned earlier, this is a much more comprehensive action plan, which tries to give the necessary attention according to the geophysical and institutional needs of each individual hydrological zone. For the first time, this plan mentions the threat of climate change for water sector development. Ironically, this is the only additional development that this plan has made compared to the NWP, with regard to the climate change issue. Like the NWP, there is no extended discussion on climate change, its impacts on the water sector of Bangladesh and how it can be adapted to. The document provides sufficient directions for developing water resources through institutional developments, enabling environment, disaster management, agricultural management and environment management.

However, the necessary components that can make a "climate proofing" action plan are significantly missing. By analysing the investment portfolios, programme implementation phase and budgeting, it has been found that only under the block "disaster management" are there some projects which can meet the indicators of a "climate proofing" plan. But the total budget allocation for implementing the disaster-related water management projects is only 5% of the total investment cost.

Among other things, the NWMP puts special emphasis on institutional development and water sector development in urban and rural areas. Though there is no straight connection with the climate change reality, the last three building blocks make significant strides toward adapting too the changing environmental realm. That characteristic makes this plan somewhat "climate proofing", as some adaptation

Table 2 The performance of NWMP in regard to the indicators set for justifying a "climate proofing" document

Name of the document	Indicators						Total
	Identification of water challenges due to climate change	Identification of sectors and vulnerable communities	Identification of solutions for water related vulnerable sectors and communities	Range of adaptation options considered	Enabling proactive strategies	Attention to process of mainstreaming adaptation	
NWMP	1	1	1	1	1	1	6

measures are being considered for natural disasters, institutional strengthening and mainstreaming the water-sectoral problems as national development issues. But this can be considered a very minor level of mainstreaming considering the geo-spatial and economic vulnerabilities of climate change to Bangladesh Table 2.

The overall score 6 indicates that the document is considering that climate change can be a potential threat for development aspirations that this action plan wishes to implement. For this document, it is somewhat difficult to draw a straight-line conclusion as despite the lack of a direct link with climate change impacts, there are some projects, actions and plans which may be effective in the fight against climate change and help to adapt the water sector from further vulnerabilities.

Poverty Reduction Strategy Paper

Bangladesh is implementing its third PRSP, lasting 2008–2011, after finishing the first PRSP during the 2005–2008 period. Since 2003, Bangladesh has followed the PRSP approach of development replacing the previous national five-year development plan. The present PRSP provides a comprehensive overview and guidance for economic growth and development. In essence, PRSP (2008–2011) is the localized context of the MDGs; however, environmental concerns are heavily dominant throughout the document.

The document attaches special care to the water resources of Bangladesh. It dedicates a separate chapter to water resources management, considering water as the single most vital resource for Bangladesh. Moreover, water resources management comes across the five strategic blocks and five supporting blocks as a cross-cutting issue. The chapter related to water resources management can be summarized as follows:

> The vision for the water resources sector is acquisition of water resources, optimizing various uses, and managing water resources for poverty reduction and sustainable development for securing the lives and properties of the people, especially the poor from water-related disasters. Participatory planning and implementation of water sector programmes/projects will be a continuing process.

Table 3 The performance of PRSP in terms of its "climate proofing" commitment

Name of the document	Indicators						Total
	Identification of water challenges due to climate change	Identification of sectors and vulnerable communities	Identification of solutions for water related vulnerable sectors and communities	Range of adaptation options considered	Enabling proactive strategies	Attention to process of mainstreaming adaptation	
PRSP	2	2	1	2	1	2	10

The strategies of water resources development and management will be grouped under six major heads: (i) the main rivers of Bangladesh will be developed for the expansion of the multipurpose use of water resources, management for navigation, erosion control, and development of hydropower; (ii) flood protection and storm-water drainage measures will be undertaken with the rehabilitation and maintenance of existing FCD and FCD/I systems in a participatory manner and protection of rural and urban areas from floods; (iii) disaster management programmes will include provisions of cyclone protection, early warning and forecasting systems with adequate lead time, flood proofing shelters, control of riverbank erosion, drought management and rationalization of groundwater resources and climate change adaptation; (iv) adequate provisions will be made for water management for agriculture through public sector irrigation development and flood management and drainage; (v) protection of the natural environment and aquatic resources will be ensured with water pollution monitoring and control, water management for fisheries and ecologically sensitive areas and raising awareness of all stakeholders to support environmental measures; and (vi) development of institutions in the water sector will be ensured. (GED 2008).

The brief summary above gives a positive impression about the climate sensitivity of the PRSP document in regard to water resources management. The document provides a correlation between water resources and climate change. It shows how poverty can be more vulnerable due to mismanagement of valuable water resources. It also exhibits how impacts of climate change can worse the prevailing vulnerabilities of climate change and lead livelihoods to an endangered condition. In its environmental section, a range of adaptation options are considered for climate change and adapting water resources to changing climate is one of the pioneering adaptation sectors.

At the same time, the PRSP calls for the immediate implementation of the projects identified by NWMP. The policy matrix of the PRSP clearly defines the task of different authorities related to water resources management and climate change in a coordinated manner. It is important to note that both structural and non-structural policies for water sector management from the impacts of climate change are included.

The overall score that PRSP evaluation matrix achieves is 10 which means the PRSP documents can be regarded as a "climate proofing" document (Table 3). Still some lacking appears in terms of solution identification of vulnerabilities and enabling proactive adaptation strategies but as an apex guiding document for moving forward this is indeed a satisfactory performance.

Conclusion

The general essence that the above three documents reveal is not quite satisfactory. Bangladesh is a country which is considered one of the most vulnerable countries of the world to climate change impacts (Huq and Ayers 2007). The water sector of Bangladesh also holds huge potentiality to fight against climate change impacts. It can be a tool for community coping mechanisms which can further amplify the community-based adaptation system. From this aspiration, it is now the time—albeit late—to start action for making the development and management of the water sector "climate-proof". Considering the nature of the impacts of climate change on the water sector, developing a climate-sensitive development plan is now imperative. Without mainstreaming climate change into water development policy and planning, Bangladesh cannot achieve sustainable development. As the key referral strategies of national development as well as water development, PRSP, NWP, NWMP should be "green" in both planning and action. A mainstreamed climate change strategy should include measures that address the underlying factors of vulnerability to climate change, particularly greater understanding of the relationship between climate change and poverty. There are several entry points for the mainstreaming of climate change into the processes, e.g. higher level of coordination among the stakeholders, convergence and cohesion of development policies. A high degree of awareness on mainstreaming climate change can be a major starting point for integration of climate change policies into the policy guidance.

References

Ahmed A (2006) Climate change in Bangladesh—a synthesis. Climate Change Cell, Department of Environment, Dhaka

Ahmad IH (2009) Climate policy integration: towards operationalization. United Nations Department of Economic and Social Affairs, New York

Bates B, Kundzewicz Z, Wu S, Palutikof J (eds) (2008) Climate change and water: IPCC technical paper IV. Intergovernmental Panel on Climate Change, Geneva

Bojö J, Green K, Kishore S, Pilapitiya S, Reddy RC (2004) Environment in poverty reduction strategies and poverty reduction support credits. The World Bank, Washington

CCCD (2009) Closing the gaps: disaster risk reduction and adaptation to climate change in developing countries. Commission on climate change and development, Stockholm

DoE (2007) Climate change and Bangladesh. Climate Change Cell, Ministry of Environment and Forest, Dhaka

GED (2008) Moving ahead–national strategy for accelerated poverty reduction II (FY 2009–11). General economic division planning commission government of the people's republic of Bangladesh, Dhaka

Hugé J, Hens L (2009) The greening of poverty reduction strategy papers: a process approach to sustainability assessment. Impact Assess Proj Apprais 27(1):7–18

Huq S, Ayers J (2007) Critical list: the 100 nations most vulnerable to climate change–IIED opinion paper. The international institute for environment and development (IIED), London

IPCC (2007a) Summary for Policymakers, in Climate Change 2007: the physical science basis contribution of working group I to the fourth assessment report of the intergovernmental panel of climate change. Cambridge University Press, Cambridge

IPCC (2007b) Summary for Policymakers, in Climate Change 2007: impacts, adaptation and vulnerability. Contribution of working group II to the fourth assessment report of the intergovernmental panel on climate change. Cambridge University Press, Cambridge

Klein RJ, Eriksen SE, Næss LO, Hammill A, Tanner TM, Robledo C et al (2007) Portfolio screening to support the mainstreaming of adaptation to climate change into development assistance. Clim Chang 84(1):23–44

Kramer AM (2007) Adaptation to climate change in poverty reduction strategies. United Nations Development Program (UNDP), New York

Mirza MM (2007) Global warming and changes in the probability of occurrence of floods in Bangladesh and implications. Glob Environ Chang 12(2):127–138

MoEF (2005) National adaptation programme of action (NAPA). Ministry of Environment and Forest, Dhaka

MoEF (2008) Bangladesh climate change strategy and action plan (BCCSAP). Ministry of Environment and Forest, Dhaka

MoWR (1999) National Water Policy. Ministry of water resource, Government of Bangladesh, Dhaka

MoWR (2001) National water management plan. Ministry of water resources, Government of Bangladesh, Dhaka

Pender JS (2008) What Is climate change? And how it will effect Bangladesh?. Church of Bangladesh social development programme, Dhaka

Prowse M, Grist N, Sourang C (2009) Closing the gap between climate adaptation and poverty reduction frameworks. Overseas development institute (ODI), London

Sperling F (ed) (2003) Poverty and climate change–reducing the vulnerability of the poor through adaptation. AfDB, ADB, DFID, EC DG Development, BMZ, DGIS, OECD, UNDP, UNEP and the World Bank, Washington

Stern N (2006) Report of the stern review: the economics of climate change. HM treasury, London

Tanner T, Hassan A, Islam K, Conway D, Mechler R, Ahmed AU et al (2007) ORCHID: piloting climate risk screening in DFID Bangladesh–detailed research report. Institute of Development Studies, University of Sussex, Sussex

UN-DESA (2008) The millennium development goals report 2008. United Nations, New York

UNDP (2006) Beyond scarcity: power, poverty and the global water crisis. Human development report, United Nations Development Programme, New York

Chapter 25
Developing an Environmental Sustainability Toolkit to Integrate Climate Change Issues in Development Cooperation

Tom Waas and Jean Hugé

Abstract The KLIMOS research platform on climate change and development cooperation is an inter-disciplinary and inter-university research platform working on climate change adaptation and mitigation in development cooperation. Within KLIMOS, our team is working on the integration of environmental sustainability issues in development interventions in Belgium's southern partner countries, through the development of an environmental sustainability toolkit. The toolkit consists of a screening instrument and a database that need to be used together when preparing, monitoring and/or evaluating development interventions. This allows development practitioners in North and South to mainstream key environmental sustainability aspects such as climate change, but also biodiversity, desertification and deforestation, into the different aid modalities (projects, budget support, programmes, national strategies, etc.). A number of case studies will be conducted to test the toolkit in different contexts and to familiarize future users with the approach. In the context of this paper, we will describe the genesis of the toolkit, as well as the contents and the process of the proposed first version (toolkit 1.0). Finally, we will analyse the potential of the toolkit in the Sahel partner countries of the Belgian fund for food security through a brief case study in Benin. The paper concludes with some thoughts on how to fine-tune the toolkit methodology with the integrated water resources management (IWRM) approach, which is used successfully by non-governmental organizations in the region. The case study will allow us to bridge the gap between "abstract" climate change mainstreaming goals and the local realities in the South, by enriching the toolkit approach with IWRM insights.

T. Waas · J. Hugé (✉)
Biology Department (DBIO/APNA), KLIMOS Research Platform on Climate Change and Development Co-operation, Vrije Universiteit Brussel,
Pleinlaan 2, 1050 Brussels, Belgium
e-mail: Jean.Huge@vub.ac.be

Keywords Environmental sustainability toolkit · Integrated water resources management · Development co-operation · Benin · Belgium

Introduction

Humanity faces far-reaching anthropogenic environmental problems, putting increased pressure on socioeconomic and institutional systems. Climate change, biodiversity loss, deforestation, desertification, poverty, no or limited access to safe drinking water, lack of basic sanitation and healthcare, food insecurity are expected to become ever more prevalent, despite international commitments to tackle these problems (such as the millennium development goals). The dramatic state of the environment and its vast and threatening impact on human well-being is firmly stated by the United nations environment programme (UNEP 2007):

> Imagine a world in which environmental change threatens people's health, physical security, material needs and social cohesion (…) Some people experience extensive flooding, while others endure intense droughts. Species extinction occurs at rates never before witnessed. Safe water is increasingly limited, hindering economic activity. Land degradation endangers the lives of millions of people. This is the world today.

"Global change" thus forces us to think innovatively and to develop practical responses as soon as possible. Worldwide, "sustainable development" is considered to be the overarching concept that will provide the best way to address the complex interrelated environmental and human development challenges of the twenty first century. Even if sustainable development became widely used in international politics, the concept is often (mis)used, without real understanding.

Moreover, while many interpretations circulate, the concept has clear interpretational limits and one can distinguish between useful and trivial or meaningless interpretations (for example, unlimited and continued "sustainable" economic growth) (Lélé 1991). In order to clarify the meaning of the concept, we turn towards the first key documents defining the concept. Pursuant to the global political endorsement of sustainable development at the United nations conference on environment and development (UNCED), in 1992 in Rio de Janeiro, we adopt the report "our common future" (Brundtland report) (WCED 1987) as a reference. We distinguish between the report's sustainability "mission statement and its operational definition".

According to the former, sustainable development is "development that meets the needs of the present without compromising the ability of future generations to meet their own needs." It contains within it two key concepts:

- the concept of "needs", in particular the essential needs of the world's poor, to which overriding priority should be given; and
- the idea of limitations imposed by the state of technology and social organization on the environment's ability to meet present and future needs.

The latter defines sustainable development as follows: "In essence, sustainable development is a process of change in which the exploitation of resources, the direction of investments, the orientation of technological development, and institutional change are all in harmony and enhance both current and future potential to meet human needs and aspirations."

Similarly, the Intergovernmental panel on climate change (IPCC) fourth assessment report—climate change a major focus of greening tools—adopts the operational definition of "our common future" (Verbruggen 2007).

Within the context of this paper and this toolkit, we emphasize the most marginalized aspect of sustainable development in the cooperation sector, being the ecological/environmental pillar, hence the use of the term "environmental sustainability". Environmental sustainability represents only one essential part of sustainable development. Hugé and Hens (2007) define it as "the ability to maintain the qualities that are valued in the physical environment". These include the ecosystem services underpinning human life and the various linkages between these services and the socioeconomic bases of society. The ecological dimension is thus emphasized, even if firmly embedded in the broader definition of sustainability.

Development cooperation is an important pillar of international development politics to combat poverty in developing countries. While environmental protection and human development are inherently linked, the field of development cooperation only recently explicitly started to introduce sustainability considerations in its activities, as such contributing to the environmental sustainability goal (goal seven) of the millennium development goals (MDGs). The rise of climate change issues on the international political agenda (IPCC 2007), the visible and projected consequences and costs of climate change in the South (Stern 2006), as well as the acknowledgement of the interrelationships between food security, vulnerability, aid efficiency and poverty reduction have all contributed to create a sense of urgency for environmental integration.

This evolution triggered the interest of the development community to design and apply well-adapted environmental integration/sustainability tools. With regard to climate change, this led to the mushrooming of so-called "climate proofing tools", while different other types of impact assessment (environmental impact assessment, strategic environmental assessment and sustainability assessment) are being improved and applied in the field of development cooperation (OECD 2006). Countries such as Sweden and Germany and Belgium (this paper) are actively developing climate integration tools to improve the quality of their development interventions in the South.

This paper focuses on the development of policy-supporting tools, aimed at improving the way development interventions (policies, plans, programmes, projects...) are designed, implemented and evaluated.

While actions in favour of sustainability may range from shifting individual consumption patterns and implementing new business models to a re-organization of the economic system, innovative governance instruments for sustainable development are widely regarded as key factors of success in fostering sustainability. Governance, defined by Petschow (2005) as "the sum of many ways in which individuals and institutions manage their common affairs" can be interpreted in various ways, but it always involves a variable degree of cooperation between official government institutions and external partners such as the general public, businesses and/or non-governmental institutions. The term "sustainability assessment" can be seen as an overarching concept, and can be a powerful instrument fostering sustainable development. It is defined as "an ex ante evaluation of a proposal, aiming at the integration of sustainability principles at all levels of decision-making by adapting the initial proposal in order to make it coherent with aspirational sustainability principles agreed upon by a range of stakeholders". The theoretical advantages of sustainability assessment are manifold. Assessing the sustainability of a policy proposal before its implementation provides decision-makers with an overview of the impacts of that particular proposal on the economy, on society and on the environment. This kind of exercise highlights the strengths and weaknesses of a proposal by taking into account various spatial and temporal scales. Through a thorough multidisciplinary analysis of the proposal, potential risks and conflicts are detected in an early stage, allowing to manage and to adapt the proposal adequately. The involvement of stakeholders fosters consensus and facilitates the solution of potential conflicts. Ultimately, sustainability assessment allows decision-makers to make fully informed decisions, without jeopardizing the freedom of political decision-making (Gibson 2005).

Influenced by the international development discourse on climate change, by the scientific advances in sustainability science, and by national pressure groups, the main actors of the Belgian official development cooperation (DGDC and BTC) have repeatedly expressed their commitment to a better environmental sustainability mainstreaming in their interventions (among others, at the March 2007 International conference on climate change and development cooperation in Brussels and in the Minister of development cooperation's policy note). In order to turn this commitment into reality, the KLIMOS Research platform for climate change and development cooperation is developing an environmental sustainability toolkit (from here on, we will refer to it as "the toolkit"). This paper discusses the methodology followed to develop the toolkit, the toolkit's first version and a pilot application.

The Environmental Sustainability Toolkit: Methodology

The "Environmental sustainability toolkit or toolkit" aims to be a user-friendly and intervention-oriented tool that will be incrementally developed to support the integration of environmental sustainability considerations, including climate change (adaptation and mitigation), in the development intervention of Belgium and developing countries.

Therefore, an "action research" approach was adopted, continuously integrating implementation and evaluation in a participatory way, involving various stakeholders. This process will be backed by different capacity-building initiatives and knowledge sharing through active involvement in relevant (inter)national networks. Action research entails that a team of professional researchers works together with stakeholders to define problems to be examined, to co-generate relevant knowledge about these problems, to learn and execute social research techniques, to take actions and to interpret the results of actions based on what they have learned (Greenwood and Levin 1998).

When developing an action-oriented toolkit, one needs to concentrate on two key issues: the contents of the toolkit as such (e.g., which criteria will be used to assess the environmental sustainability of the planned interventions); and the process aspects of actually implementing the toolkit and embedding it in existing and/or future institutional arrangements and daily practice. Taking this into account, the following steps were followed to develop the toolkit:

1. overview of existing "greening" tools aimed at environmental and/or climate change integration in development interventions;
2. interactive meetings with future users to guarantee a needs-based approach;
3. development of a first version of the toolkit (Toolkit 1.0);
4. planning of a pilot application of the toolkit's first version on a well-defined development intervention focused on "integrated water resources management".

This paper will discuss the results of these three main project sections that have been realized so far (spring 2010).

Overview of Existing "Greening" Tools in Development Cooperation

In order to build on existing experiences, the KLIMOS research team first compiled an overview of existing tools for environmental integration in development cooperation. This overview was published as a KLIMOS working paper; we here provide a summary, without introducing every analysed tool, instead focusing on the categories of tools. While the developed tools have a different focus and target different groups, they all aim to "green" development. Greening refers to the cross-cutting integration of environmental sustainability principles in policies, plans, programmes and projects. Some tools focus on adaptation to climate change, others on biodiversity and yet others on environmental sustainability aspects as a whole.

The overview aims to introduce to the reader a range of tools with varying scope, origin, objective and methodology. This paper draws upon an analysis of publicly available documents, as well as upon scientific literature and NGO materials.

Instead of focusing on each separate tool, we propose a series of possible categorizations which will allow potential users to define what they expect from (a combination of) tools. Indeed, "greening" development aid will be facilitated by using (parts of) operational instruments developed by other agencies (cross-fertilization) and through collaboration with other agencies and institutions. Cross-fertilization and collaboration can considerably reduce time and efforts invested in the development of operational measures and contribute to greater donor harmonization, as is called for by the Paris declaration on aid effectiveness (Gigli and Agrawala 2007).

All of the following categorizations are merely indicative. The categorization of tools cannot be absolute. Although many of the presented tools have target audiences and stated objectives, their use and application is context-dependent. As most of the tools are openly accessed, they can all be used in a wide variety of ways.

Categorization According to the Tool's Objectives

We propose three categories:
1. Information tools

These tools, databases and platforms provide a wide variety of audiences with environmental, climate and/or vulnerability related information. Most of these tools are open source. They are not decision-making tools, but provide development practitioners with inputs that can be utilized for risk management and adaptation management processes, as well as in the design process of any development intervention. A number of international organizations provide country environmental profiles. These are reports that include the analysis of the country's environmental situation, current policies, institutional capacities and environmental cooperation experience with clear recommendations for the integration of the environment during preparations of country strategies.

2. Tools targeting intervention design

These so-called integration tools have been developed to screen projects, programmes and policies and to develop priorities. They are tailored toward specific decision-making processes and aim at integrating environmental sustainability issues throughout a stepwise process approach. These tools often entail the use of both quantitative models and stakeholder consultation. Most of these tools focus on a set of entry points (annual negotiation sessions, international conferences, participation in PRSP preparation, preparation of country strategies, sector planning reviews, etc.) in various aid modalities (country programmes, sector programmes, projects).

3. Frameworks and guidelines

These tools provide a general approach to environmental sustainability integration and are centred around principles and best practices. They are not specifically tailored to specific interventions such as the tools under (2).

Categorization According to the Decision-making Level and to the Aid Modalities

As the majority of the presented tools are to be used by development agencies' staff in a variety of situations, it is useful to categorize the tools according to their use in the existing aid modalities architecture. To that end, we follow the OECD's (2009) concept of entry points: stages in the intervention cycle where environmental sustainability issues can be integrated. This categorization is combined with a focus on the decision-making level: the national, the sectoral or the local level.

Categorization According to Main Target Audience

Various tools target different audiences, ranging from development agencies' staff in the North, field officers, non-governmental organizations, officials in the partner countries, grassroots communities, etc. Some tools are more general and hence do not have a specific target audience.

Conclusion on "Greening" Tools

This overview of existing greening tools in the field of development cooperation provides a source of inspiration for the development of an environmental sustainability toolkit targeting the actors of the Belgian development cooperation. It is clear that the development and the use of greening tools, and in particular climate-related tools, is a very dynamic field. Keeping the existing tools in mind will contribute to the development of a state-of-the art toolkit, tailored to the needs of its future users.

The method that was followed to design the environmental sustainability toolkit's first version was inspired by the combination of scientific expertise and participation (see next section). It is in line with the general "philosophy" underpinning the future application of the toolkit: involving the future toolkit users in its development is necessary as successful implementation of environmental sustainability measures will require the consent of all actors of decision-making in order to be successful. The method of development is rooted in the idea of the key importance of participation in sustainability thinking and acting. We refer to Kasemir et al. (2003) for a detailed description of the theory and practice of participation in sustainability science, as these scholars' research provided a source of inspiration for our research.

Interactive Meetings with Future Users

In order to develop a toolkit tailored to the needs of its future users, a series of interactive meetings was arranged. The future users of the toolkit are broadly composed of three categories: officials from the Belgian development cooperation

(Directorate-general for development cooperation and Belgian technical cooperation); non-governmental organizations (in North and South); and officials from the partner countries' ministries. In the project's first phase (2009–2010), meetings were set up with the users in Belgium:

- Meetings with the various sections of the DGDC (D0: Management; D1: Governmental programmes; D2: Humanitarian and food security programmes; D3: Non-governmental programmes; D4: Multilateral and European programmes; D5: Awareness-raising programmes).
- Meetings with the "attachés" (officers working in the Belgian partner countries in the south).
- Meeting with the Belgian technical cooperation (the Belgian development agency).
- Meeting with a number of Belgian NGOs at a seminar.

Although an exhaustive overview of the results of these meetings lies outside the scope of this paper, we hereby list some key questions raised:

- Will the toolkit be integrated within the existing project management cycle?
- Will the toolkit be user-friendly and not use too many indicators?
- How much time and effort will it cost to comply with the toolkit?
- Will the toolkit be user-friendly for non-environmental experts?
- Are there guarantees for harmonization with other donors?
- Will the toolkit slow down or even block decision-making?
- Will the toolkit be updated regularly?
- Will country—and region—specific information be available?
- Which dimensions of sustainable development will be included?
- ...

It is clear that the future users had many question on the concrete design and application of the toolkit. The meetings were always very lively and the attitude of all stakeholders was constructive, even though some healthy scepticism was regularly encountered. The KLIMOS research team tried to take all questions into account, and this resulted in the design of a first version of the toolkit, Toolkit 1.0 (see next section).

Toolkit 1.0

The toolkit's first version, "Toolkit 1.0" consists of two main elements, which should be used together. The user starts on a the homepage of the toolkit (access to the site will be made public in 2011), where, next to a brief introduction, two sections are visible:

- the environmental screening tool
- the database (under the heading "search for information")

The environmental screening tool provides a stepwise guide that should be used in all circumstances when applying the toolkit. This is shown in Box 1 below.

Box 1: Key steps of the environmental screening tool (part of the environmental sustainability toolkit)

Aim
The aim of the "Environmental screening tool" is to include and consider environmental considerations in development cooperation.
Such an analysis is indispensable because environmental sustainability and human development are interdependent. Environmental sustainability is an essential condition to meet the needs of present generations without compromising the ability of future generations to meet their own needs, in particular the needs of the world's poor. Meeting basic human needs and poverty alleviation on the other hand is an essential condition to realize environmental sustainability and to keep within the environmental limits of the planet. Therefore, during the different stages of the policy cycle environmental considerations should be addressed to reduce environmental risks and vulnerabilities of a project, and to counteract or minimize a projects impact on the environment.

Method
To accomplish an environmental analysis, an "Environmental screening tool" was developed. The tool is part of the "Environmental sustainability toolkit" (www.kuleuven.ac.be/klimos) which next to it, consists of a database with relevant and accessible environmental information for consultation to complete the "Environmental screening tool".
The "Environmental screening tool" consists of a set of *guiding questions* to assess a project's environmental risks and vulnerabilities, and impacts.
To use the tool the following steps should be followed:

Step 1: Complete the questionnaire
Step 2: Summarize the answers by listing the key issues for each completed question
Step 3: Draw one of the following three overall conclusions:

⇒ Accept

⇒ Minor changes are required

⇒ Major changes are required

Step 4: Paste the summary and overall conclusion to the evaluation dossier of the project and add the completed "Environmental screening tool" in appendix.

This list of steps is then followed by a questionnaire, which is different for each priority sector of the Belgian development cooperation. These sectors are:

- agriculture and food security;
- education;
- health;
- infrastructure.

For each sector, the questionnaire consists of eight sections. In Toolkit 1.0, only climate change is dealt with. In the future versions of the toolkit, biodiversity and land use will also be covered, in order to ensure a multidimensional interpretation of environmental sustainability. For now, however, the toolkit is limited to climate change issues. Sections 1 and 2 focus on the impact of climate change on the intervention and on the possible adaptation measures (which are listed). Sections 3 and 4 focus on the impact of the intervention on climate change, and on the possible mitigation measures (which are listed). Sections 5, 6, 7 and 8 focus on management and legislation:

Box 2: Key sections of the questionnaire of the environmental screening tool
1. Are there any climate change risks that might affect the intervention, such as …
2. If the intervention might be affected by climate change risks, are appropriate adaptation measures taken, such as …
3. Does the intervention have any expected impact on climate change, through …
4. If it is expected that the intervention will have an impact on climate change, are appropriate mitigation measures taken, such as …
5. Does the development partner has the capacity to deal with climate change (skills, financial resources …)?
6. Are climate change issues addressed in implementation, monitoring, follow-up and evaluation?
7. Does the intervention require any assessment according to the national legislation of the partner country, in which climate change issues could be addressed?
8. Is there any climate change legislation that might influence the intervention?

In order to correctly complete the environmental screening tool's questionnaire and in order to complete the stepwise approach in a satisfactory way, a vast amount of information is presented in the database. The information can be accessed through four entry points. These are:

- Country (each of the 18 Belgian partner countries' NAPA, PRSP, etc.—if available …).
- Sector (education, health, agriculture, …).
- Intervention (project, programme, budget support …).
- Theme (for now: only climate; in the future: also land use and desertification).

By following the stepwise approach of the environmental screening tool and thanks to the questionnaire, the users are guided towards a policy-supporting approach aimed at enhancing the quality of the design of their interventions. Any background information needed to apply the environmental screening tool can be accessed at any time through the regularly updated database.

Planned Pilot Application: Integrating IWRM in the Toolkit

The overview of existing tools and the interactive meetings with the future users resulted in a first version of the toolkit, presented above. In order to test if the developed tool is applicable on real-life development interventions, we applied it on a case study.

Officials from the Belgian development cooperation's fund for food security proposed to test the developed Toolkit 1.0 on a development project in Benin, called the *Projet d'Appui à la Croissance Economique Rurale* (PACER—project to support rural economic development). The project focuses on the development of the production chain of rice, tubers and grapefruit in various carefully selected regions of Benin. As the pilot application of the toolkit needs to be conducted in close collaboration with the officials involved in its design and implementation, the timing had to be adapted to the agendas of both researchers and officials. This means that until now (autumn 2010), the assessment has no field-testing part; this pilot assessment has been a desk study so far. For now, we will delve deeper into one of the components of the project, namely the focus on "integrated water resources management" (IWRM). Toolkit 1.0 currently focuses on climate change, without taking other dimensions of environmental sustainability into account. This pilot testing not only allows us to test how well the toolkit functions in a practical action-research context, but it also allows the KLIMOS research team to fine-tune the toolkit's contents by gaining insight on IWRM in a development cooperation context. IWRM fosters a sustainable and holistic perspective on water use and acknowledges the wide range of functions and services provided by water basins. In practice, this means that organizations fostering IWRM work together with local stakeholders to identify risks of degradation of the water resources (both in terms of quality and quantity) so as to implement protective measures such as protecting the drinking water sources, reforestation and protection of river banks, etc. This needs to be done while taking into account the different functions and services of the local water resources, ranging from consumption to cleaning, biodiversity protection, fishing, microclimatic regulation, etc.

Thus, IWRM is a key aspect of environmental sustainability, and should hence be an integral part of the toolkit. The linkages between IWRM and climate adaptation are evident, and just like climate change adaptation, water resources management has experienced an important paradigm shift. From an engineering approach based on controlling environmental problems with technical solutions, water management has shifted to a more holistic approach considering human attitudes and collective behaviours as an integral part of the management response (Pahl-Wostl et al. 2008). Many climate change adaptation responses proposed in Toolkit 1.0 are a combination of technical solutions (which are of course needed and particularly useful) and management/behavioural solutions, which emphasize social learning aspects. The design and application of the toolkit itself is also a learning process, and it might be very useful for the Belgian development actors to build on the IWRM experience of specialist actors such as the NGO Protos, to multiply the capacity-building and learning effects of early toolkit uses. Next to the paradigmatic and learning similarities between climate change adaptation and integrated water resources management, the geographic concentration of Belgium's partner countries in severely water-stressed countries in the Sahel belt further points to the need for a pooling of knowledge and experience.

Ultimately, although the toolkit is not meant to be the overarching response to any development problem—that would admittedly be overly ambitious and unnecessary—in the future it is to provide a thorough answer to any development practitioner aiming at improving the design of his interventions. It thus needs to consider as many aspects of environmental sustainability as possible. The fruitful experiences with IWRM can be very helpful to inform the further development of the toolkit. One could, for instance, think of a "good practice" section, as well as a stronger uptake of linkages between climate change issues and water issues in the Sahel region.

Conclusions

Climate change and other environmental problems are threatening the achievement of development commitments, and more importantly, are directly threatening vulnerable populations in the South. The development community has embraced the concept of sustainable development for some time now, but implementing sustainable development and realizing its objectives is still a challenge.

Developing integrated instruments to mainstream environmental sustainability concerns (and climate change issues in particular) in development cooperation interventions is now a priority for many development actors, as shown by the plethora of existing tools. Embarked on the process of developing an environmental sustainability toolkit for the actors of the Belgian development cooperation, the KLIMOS research team used an action research methodology to ensure adequate participation of all concerned stakeholders. The delivery of a first version of the environmental sustainability toolkit is the result of the first research phase. The toolkit is intended to be used in the different phases of the development intervention. We therefore follow the widely known "Plan–Do–Check–Act" approach to ensure that the toolkit method will be used iteratively and will gradually lead to an improvement of the quality of decision-making in the field of development cooperation.

In order to enhance the quality of the toolkit's contents and applications, a pilot-testing phase is planned. However, even before the testing phase, it is clear that successful development approaches can be an important source of inspiration to further develop the toolkit. Integrated water resources management has already been used for years, in particular by NGOs, and a lot can be learned from these experiences. The social learning fostered by IWRM, the multidimensional approach to management, the combination of scientific and participatory knowledge and the focus on adaptation to changing environments are important parallels with the climate change adaptation challenges faced in vulnerable regions such as the Sahel. The KLIMOS research team advocates the uptake of these experiences in the toolkit and will work towards a better integration of successful examples of more environmentally sustainable development cooperation in the next few years. This will hopefully lead to a more sustainable development in the future.

References

Gibson R (2005) Sustainability assessment—criteria and process. Earthscan
Gigli S, Agrawala S (2007) Stocktaking of progress on integrating adaptation to climate change into development co-operation activities, COM/ENV/EPOC/DCD/DAC(2007)1/FINAL, OECD, Paris
Greenwood DJ, Levin M (1998) Introduction to action research—social research for social change. Sage Publications, London
Hugé J, Hens L (2007) Sustainability assessment of poverty reduction strategy papers. Impact Assess Proj Apprais 25(4):247–258
IPCC (2007) Summary for policymakers. In: Climate Change 2007: impacts, adaptation and vulnerability. Contribution of Working Group II to the fourth assessment report of the Intergovernmental panel on climate change. Cambridge University Press, Cambridge
Kasemir B, Jäger J, Jaeger CC, Gardner MT (2003) Public participation in sustainability science—a handbook. Cambridge University Press, Cambridge
Lélé S (1991) Sustainable development: a critical review. World Dev 19(6):607–621
OECD (2006) Applying strategic environmental assessment. good practice guidance for development co-operation. OECD, Paris
OECD (2009) Integrating climate change adaptation into development co-operation. Policy Guidance, OECD Paris
Pahl-Wostl C, Tabara D, Bouwen R, Craps M, Dewulf A, Mostert E, Ridder D, Taillieu T (2008) The importance of social learning and culture for sustainable water management. Ecol Econ 64:484–495
Petschow A (2005) Environmental policy integration–greening sectoral policies in Europe. Earthscan, London
Stern N (2006) The stern review on the economics of climate change. Cambridge University Press, Cambridge
Verbruggen A (2007) Glossary. Climate change 2007: mitigation of climate change, IPPC. Cambridge University Press, Cambridge
WCED (world commission on environment, development) (1987) Our common future. Oxford University Press, Oxford
UNEP (2007) Global environmental outlook—environment for development. United Nations Environmental Programme. Progress Press, Valetta, Malta

Chapter 26
Spatio-Temporal Variation of Drought Severity in the Sudano-Sahelian Region of Nigeria: Implications for Policies on Water Management

Maruf Sanni, Theophilus Odeyemi Odekunle and Francis Adeyinka Adesina

Abstract The African Ministerial Council on Science and Technology agreed that drought is one of the major environmental problems that is afflicting the African continent, with its consequent negative impacts on the socioeconomic development of the affected areas. It is within this context that this study assessed the spatio-temporal variation of drought severity in the Sudano-Sahelian region (SSR) of Nigeria, which is highly vulnerable to drought. The study analysed spatio-temporal variation of drought severity in the region between 1961 and 2006 using the Standardized Precipitation Index (SPI) and Principal Component Analysis (PCA). It was established that most of the extreme droughts of the highest magnitude were recorded around the 1970s and 1990s and that there were indications that this would continue in the future. Results also identified sub-regions characterized by a different level of drought risks in the region. The study concluded by making appropriate policy recommendations for effective water management for the region.

Keywords Sudano-Sahelian region · Drought severity · Nigeria · Water · Policies · SPI · PCA

M. Sanni (✉)
National Centre for Technology Management, Obafemi Awolowo University,
Ile-Ife, PMB 012, Osun State, Nigeria
e-mail: marufsanni@yahoo.com

T. O. Odekunle · F. A. Adesina
Department of Geography, Obafemi Awolowo University, Ile-Ife, Osun State, Nigeria

Background

Over the years, people living in the arid and semi-arid lands have been in a continuous struggle with extreme events such as droughts and a wide range of other natural hazards. Many of these events could be of short duration, high intensity and with great impacts. It is against this backdrop that the African Ministerial Council on Science and Technology concluded that drought is one of the major environmental problems that is afflicting the African continent, with its consequent negative impacts on the socioeconomic development of the affected areas. Major scientific research activities have even linked global warming with extreme weather patterns such as droughts. It is usually difficult to define drought because the meaningful threshold of significant moisture deficiency is a function of the water use being impacted (National Academy of Sciences (NAP) 1986). As a matter of fact, there is no universal definition of drought. Wilhite and Glantz (1985) reviewed more than 150 definitions of drought. A critical assessment of these definitions showed that drought is unique to a region. For the purpose of this study, however, drought is considered as an interval of time, generally months or years in duration, during which the actual moisture supply at a given place is consistently less than the climatically expected or climatically appropriate moisture supply (Palmer 1965). Meanwhile, drought severity as it relates to public water systems is necessarily a function of human actions and/or inactions as well as the magnitude and duration of the individual hydrologic event (National Academy of Sciences (NAP) 1986). In view of this, it has been advocated that more effective research activities and abatement measures be directed toward the challenges of water management during drought events.

Droughts represent a problem of long-term common concerns in the Sudano-Sahelian region (SSR) of Nigeria. Drought severity during the 1960s and 1970s (popularly called Sahelian desiccation) has caused a wide range of impacts across the SSR of Nigeria (Nicholson et al. 2000). These impacts generated widespread scarcity of potable water as well as depleted supplies for irrigation, industry and domestic uses. One of the sectors of the economy that was severely affected was the agricultural sector, where there was significant reduction in livestock population. For instance, the drought impact during that period resulted in the loss of millions of metric tonnes of millet and sorghum worth millions of US dollars (Oladipo 1995). The aftermath of the effects of these droughts is also seen in biological productivity, as well as the regenerative capacity of the ecological systems, migration of active labour force and consequent reinforcement of rural poverty (Watts 1983; Anyadike 1987). In relation to rural poverty, the rural poor in the SSR, constituting the majority of the population, are the most vulnerable because they have few assets to call upon in order to cope with extreme events such as prolonged droughts (Thomas 2008). Meanwhile, vulnerability has been defined as the degree to which production and livelihood systems are susceptible to, or unable to cope with, adverse effects of climate change, including climate variability and extremes (IPCC 2001). To buttress the level of exposure of this region to the risk of extreme events, (UN-HABITAT 2006) has estimated that there are 199 million slum-dwellers in sub-Saharan Africa (SSR inclusive) with little or nothing to cope with the impacts of extreme events. In addition to the problem of poverty, the rural poor in this region also depend on the quality and functioning of the agro-ecosystems that they live in through the provision of goods and services such as food, water, fodder, fuelwood, and other materials. Therefore, any disruption in these ecosystems ultimately affects their survival. Another source of concern in this region is the high increase in population. Dynamic increase in population at the rate of 3.5% in sub-Saharan Africa, as compared to the global rate of 1.7%, has led to a lot of pressure on land and other natural resources (Naginder and Kundzewicz 1997). Consequently, it becomes imperative to provide for this ever increasing number of people. Meanwhile, this increase also poses danger to already over-stressed water infrastructures in the region. Moreover, the ability of governments and international relief agencies to deal with droughts in the region is also constrained by the absence of reliable data, information networks, professional and institutional capacities. In addition to this, (Naginder and Kundzewicz 1997) have reported that hydrological networks and observing stations in this region are in decline and the databases to assess drought and desertification risks and reduce their impacts in Africa are not adequate. Meanwhile, it is interesting to note that even though drought also hits the developed countries, it does not kill them. This has been made possible because of their ability to cope and implement efficient strategies such as sharp water-price rises, bans on watering lawns or car washes, etc. during drought episodes. Unfortunately, such strategies are difficult to implement in the SSR because of inadequate structures, institutions, legislation, markets, awareness and education (Naginder and Kundzewicz 1997). Glantz (1976) also corroborated this fact on the situation of Sahelian desiccation where he stated that even a six-month forecast of weather would not have been sufficient to avert the drought impacts in the region, due to the lack of institutional and human capabilities to manage the problems.

Today, the world's drylands are facing a greater water crisis, and droughts of lesser magnitude are now resulting in greater impacts. Even regions with normal precipitation are not left out of the challenges of water shortages, which are now widespread in both developing and developed nations (Wilhite 2005). Studies have shown that most of the strategies related to development (in most drylands), such as irrigation, improved land use, urban and rural water supply, range and livestock management, agricultural extension, energy (fuelwood), and community development, are inherently and intrinsically related to water (Naginder and Kundzewicz 1997). Drought management usually poses a serious challenge to water resources and usage. Although droughts predominate in arid and semi-arid regions of the world, their frequency and severity in the SSR regions have been on the increase. Scientists have shown that many of the world's arid and semi-arid lands would face increasing temperatures with climate change (Hulme 2001) and, more importantly, disruptions in their hydrological cycles, resulting in less and more erratic rainfall that could exacerbate the already critical state of water scarcity and conflicts over water allocation and usage (Thomas 2008). For instance, studies

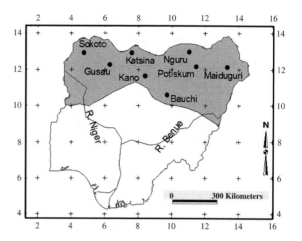

Fig. 1 Map of Nigeria, showing the Sudano-Sahelian ecological zone of Nigeria

have shown that an increase in temperature in this region could mean that recharge would be much reduced and that groundwater exploited would not be replaced at the same rate. Issues such as public health and economic development have brought improved water access to the global scene and attracted the attentions of policy-makers, water professionals, international bodies and non-governmental organizations. In view of these, the United Nation's Millennium Development Goals (MDGs) related to water and sanitation access are now on the agendas of many governments and NGOs (World Water Assessment Program (WWAP) 2006). It is apparent, therefore, that any strategy that resolves the water problem has the capability of solving many other challenges in the region and subsequently, action plans for other issues can easily be determined and effective policy successfully implemented. It is therefore important to assess spatio-temporal variation of droughts in the region in order to evaluate and anticipate drought risk and implement effective water management policies. The ability to determine the onset of droughts and map their severity will be a significant development in mitigating the negative effects of drought, combating desertification and ultimately promoting economic development in the advent of global warming. It is within this context that this study intends to assess the spatio-temporal variation in the SSR of Nigeria which is highly vulnerable to drought and climate change impacts, and suggests appropriate policies for effective and efficient water management strategies.

Study Area

Location and Size

The area chosen for this study is the SSR of Nigeria which lies roughly between Longitude 4° and 15° E and latitudes 11° and 13° N and comprises most of Borno,

Yobe, Kano, Zamfara, Katsina, Kaduna, Jigawa, Kebbi, and Sokoto states (Fig. 1). It occupies approximately 26,159 km².

Climate

In the SSR of Nigeria, rains fall mostly from June to September with the rest of the year virtually dry. The mean annual rainfall is between 1,016 mm in the wettest part (southernmost part of Sokoto state), less than 508 mm or less in the driest part (northern part of Borno state) (Ojanuga 1987) of the region with coefficients of variation ranging from 15 to 30% (Sivakumar 1989). Low annual rainfall is not necessarily as critical a constraint in crop production; rather it is the irregular occurrence of rainfall (Sivakumar and Wallace 1991). There is a 10% probability of receiving rainfall of 50 mm and 400 mm annually at the northern and southern border, respectively (Breman and De Wit 1983). The mean annual temperature is between 26°C and 28°C (Ojanuga 1987).

Variability in rainfall has been identified as a normal part of the climate of the SSR of Nigeria. An average departure of 20–40% from the annual mean rainfall has been estimated (Janowiak 1988). Mortimore (2000) identified the key source of this variation (and of drought) to be the seasonal rainfall. D'Amato and Lebel (1998) and Chappell and Agnew (2008) have recognized the importance of the number of storm events, rather than the amount of rain, to be more responsible for much of the rainfall variation in the region. This led to the conclusion that the migrations of the Intertropical Discontinuity (ITD)—the mechanism that brings rain to the SSR of Nigeria (Omotosho 1990)—are not the primary causes of the droughts. In other words, they were of the opinion that drought is the result of dynamic conditions in the weather system. Another study by Nicholson (1985) and Ba et al. (1995) also concluded that drought in the Sahel is apparently more clearly linked to reduced convective activity. The Nigerian climate is classified on a world scale as Af, Aw, and BS (Trewartha 1968). Af, Aw, and BS are Tropical wet, Tropical wet-and-dry, and Semi-arid climates, respectively (Trewartha 1968; Ojo 1987). The SSR could be classified under BS. The most obvious climatic feature of this category is that potential evaporation and transpiration exceed precipitation. The tropical desert climate is influenced by upper air stability and subsidence, which is the result of the presence of the subtropical high pressure zone. Relative humidity is normally low, averaging 10–30% in interior locations. Although the northward migration of the ITD and its associated winds appears to be the most popularly accepted of the factors known to promote rainfall in the SSR of Nigeria, precipitation is very low in quantity and very infrequent in distribution, both temporally and spatially (Adedokun 1978; Lamb 1982). However, there are opposing views that believe that rainfall deficit could be experienced even when there is sufficient incursion by the mT air mass. For instance, Nicholson and Chervin (1983) and Odekunle (2010) established that major rainfall fluctuations in the SSR appear to be primarily linked to factors modifying the intensity and frequency of disturbances. However, the other opinions (Winstanley 1973; Kraus 1977) ascertained that the inter-annual variability observed is significantly generated by the shift in the mean positions of the ITD. In the same light, Adedokun (1978) has explained the effect of strong cold-water upwelling in the Gulf of Guinea, which could bring about a strong walker circulation, which in turn strengthens the mT air mass, and push the ITD significantly further north of the equator, resulting in heavy precipitation in the SSR of Nigeria. The opposite trend has also been noted when there is weak cold-water upwelling, which brings about a weak Walker circulation. This phenomenon is usually associated with a weak mT air mass and limited ITD incursion into the SSR region, consequently bringing about limited precipitation in the SSR of Nigeria (Adedokun 1978). There are some opinions that SSR precipitation in general, and Nigeria in particular, is also modulated, to some extent, by the complex feedback mechanisms mainly due to deforestation/land-cover change (Brovkin 2002; Zeng 2003; Brooks 2004).

Methodology

Data Collection

The climate data required for this study includes monthly and annual rainfall totals. Data on these parameters was collected over a uniform period of time for all the rainfall stations in the SSR of Nigeria. These locations include Sokoto, Kano, Katsina, Nguru, Maiduguri, Bauchi, Gusau, and Potiskum (Fig. 1). The data was sourced from the archives of the Nigerian Meteorological Services, Oshodi, Lagos. The length of the data spanned a period of 45 years (1961–2006).

Data Analysis

A consistent framework for the description of drought is essential for any study of this magnitude. This study adopted the climatology definition of drought and used the six-month Standard Precipitation Index (SPI) to determine the drought severity index. A number of other different indices of drought have been developed to quantify drought, each with its own strengths and weaknesses. Some of them are Palmer Drought Severity Index (PDSI—Palmer 1965), Deciles (Gibbs and Maher 1967), Crop Moisture Index (CMI—Palmer 1968), Food and Agriculture Organization (FAO) Water Satisfaction Index (WSI—Frere and Popov 1979), Surface Water Supply Index (SWSI—Shafer and Dezman 1982), Agro-Hydro Potential (AHP—Petrasovits 1990), Vegetation Condition Index (Kogan 1990), Index of Moisture Adequacy (IMA—Sastri 1993), Vegetation Condition Index (Kogan 1990), and Soil Moisture Deficit Index (Narasimhan and Srinivasan 2005). Two of

Table 1 Classification of drought intensity using SPI

SPI Values	Drought intensity
2 and more	Extreme wet
1.5–1.99	Very wet
1–1.49	Moderately wet
−0.99–0.99	Near normal
−1−−1.49	Moderate drought
−1.5−−1.99	Severe drought
−2 and less	Extreme drought

Source McKee et al. (1993)

the most commonly used are PDSI and SPI. There are several other methods which can be used in drought analysis. SPI is a drought monitoring index developed by McKee et al. (1993) to quantify precipitation deficit for multiple timescales. The index uses long-term precipitation record for a desired period of time which will be fitted to a probability distribution. If a particular rainfall event gives a low probability on the cumulative probability function, then this is indicative of a likely drought event. The classification system shown below in Table 1 is used to define drought intensities resulting from SPI values.

Determination of Spatial Variation of Drought Severity

Another technique which can be used to identify spatially homogeneous drought regions is Principal Components Analysis (PCA) (Peter 1990). Studies such as those of Dyer (1975), Barring (1987) and Ehrendorfer (1987) have already established that PCA can be used to identify clusters of stations or spatial units that behave similarly with respect to a particular climatic variable. This study employed PCA to analyse spatial variation of drought severity in the region. The variable used is the drought severity magnitude in the period of 1961–2006. The values of drought magnitude are better than that of the raw SPI values because SPI values only give the probability of occurrence of a determined SPI value, which will be the same for the different areas represented by the PCA (Dracup et al. 1980). As a result of this, the index cannot be used for estimating the spatial differences in drought risk. However, the temporal patterns of drought can differ due to the temporal succession of SPI values. The loadings generated by the PCA can be mapped to show the spatial pattern of the characteristic under investigation (Peter 1990). The study used the rotated component matrix because the un-rotated one does not give physically interpretable patterns when dealing with meteorological or climatological data (Richman 1981, 1986). Thus, the principal components clusters of climatic divisions that behave similarly with respect to the various drought types or magnitude were identified and mapped (Peter 1990).

Discussion of Results

Assessment of Drought Severity in the SSR of Nigeria Between 1961 and 2006

The evaluation of drought severity in the SSR of Nigeria showed that most of the drought severity with the highest magnitude occurred between the 1970s and 1990s. It was also observed that the individual rainfall stations under examination experienced severe drought of highest magnitude at different periods during the years under study. For instance, Sokoto station experienced its highest magnitude of severe drought during 1987; severe drought of high magnitude was experienced in Katsina in 1996; that of Kano was in 1973; Nguru was in 1991; Maiduguri was in 1983; severe drought of high magnitude was experienced in Bauchi in 1985; Gusau was in 1984; while that of Potiskum was in 1982 (see Figs. 2, 3, 4, 5, 6, 7, 8, and 9). Observation of the duration of the drought severity showed that there is an increase in the occurrence of drought severity in 1973, the 1980s and 2000s. In Sokoto, 1987 recorded the longest duration of drought severity (Fig. 2a). Further insight into the analysis showed that there was a persistent drought severity for six months during this year. This is followed by 1973 and 1985, which experienced drought severity for a period of five months each, respectively. These periods coincided with the periods when West African Sahel experienced most of the extreme droughts. It was observed that this region experienced decline in rainfall of more than 20% (Hulme et al. 2001). These same periods also saw the death of hundreds of thousands of people and millions of animals (Glantz 1976, 1996). This particular change in climate in the region has since been referred to as Sahelian desiccation (Hulme 2001; Chappell and Agnew 2008). Some explanation could be adduced to explain this phenomenon. Recent studies have established that the Sahelian desiccation was a product of long-term climate variability driven by changes in patterns of global sea surface temperature (Giannini et al. 2003). Giannini et al. (2003), argued that dry conditions in the Sahel occur during periods in which the southern hemisphere oceans and northern Indian Ocean are warmer than the remaining northern hemisphere oceans, and that it is a shift to such a pattern of global temperature distributions that is responsible for the turn towards aridity in the Sahel from the late 1960s. Giannini et al. (2003) identify SSTs as the principal driver of Sahelian rainfall variability, which they model successfully for the period between 1930 and 2000 using only SST forcing. The model also represents land–atmosphere interaction via moisture feedbacks. They identify warming in the Indian Ocean as "the proximate cause for the negative rainfall trend observed in the Sahel from the late 1960s to the 1980s", which, they suggest, in combination with a more intermittently warmer than average eastern equatorial Atlantic, reduced the land–ocean temperature contrast that is crucial in monsoon dynamics, causing the deep convection associated with monsoon rainfall to migrate southwards (Brooks 2004).

26 Spatio-Temporal Variation of Drought Severity

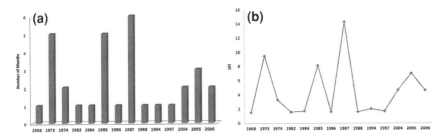

Fig. 2 a duration of drought severity in Sokoto between 1961 and 2006 b Magnitude of drought severity in Sokoto between 1961 and 2006

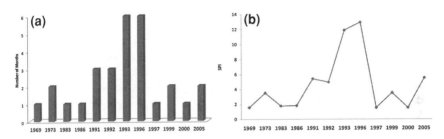

Fig. 3 a Duration of drought severity in Katsina between 1961 and 2006 b Magnitude of drought severity in Katsina between 1961 and 2006

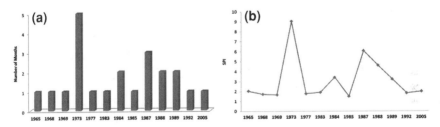

Fig. 4 a Duration of drought severity in Kano between 1961 and 2006 b Magnitude of drought severity in Kano between 1961 and 2006

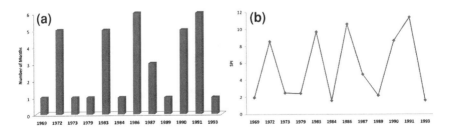

Fig. 5 a Duration of drought severity in Nguru between 1961 and 2006 b Magnitude of drought severity in Nguru between 1961 and 2006

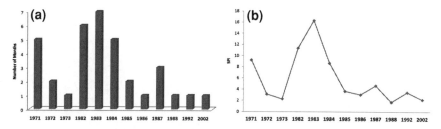

Fig. 6 **a** Duration of drought severity in Maiduguri between 1961 and 2006 **b** Magnitude of drought severity in Maiduguri between 1961 and 2006

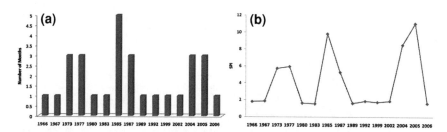

Fig. 7 **a** Duration of drought severity in Bauchi between 1961 and 2006 **b** Magnitude of drought severity in Bauchi between 1961 and 2006

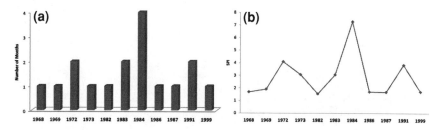

Fig. 8 **a** Duration of drought severity in Gusau between 1961 and 2006 **b** Magnitude of drought severity in Gusau between 1961 and 2006

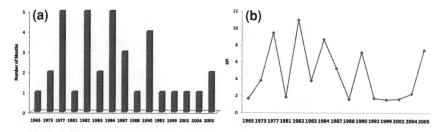

Fig. 9 **a** Duration of drought severity in Potiskum between 1961 and 2006 **b** Magnitude of drought severity in Potiskum between 1961 and 2006

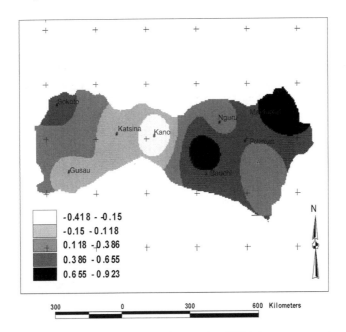

Fig. 10 Spatial distribution of rotated correlation matrix of PCA (component one)

Spatial Variation of Drought Severity in the Sudano-Sahelian Region of Nigeria

The analysis of PCA generated three main components. These components explained 73.63% of the total variance. The first component explained 34.43% of total variance, and represented mainly the drought evolution of the eastern region of the study area (Fig. 10). The second component (22.93% of total variance) mainly represented the western region of the study area (Fig. 11). The third component (16.27% of total variance) could be attributed to both the western and eastern parts of the study area (Fig. 12). The first two aforementioned patterns showed distinct spatial distribution, while the last one overlapped between the first two components. The implication of these results is that the areas with higher magnitude on the maps are those regions with higher drought risk.

The analysis of the spatial variation of drought severity indicated that rainfall-producing mechanisms in the different rainfall stations analysed could be as a result of local factors. For instance, analysis of the components one (which explained 34.43% of the variation in the drought magnitude) established that the area towards the eastern part of the SSR was homogeneous. This is because the map clearly showed that this section of SSR was highly coherent. Another important feature of this map was that this area was also susceptible to high drought risk. The same pattern could also be seen in the western part of the region. In all, the analysis of the spatial variation revealed that three zones could be

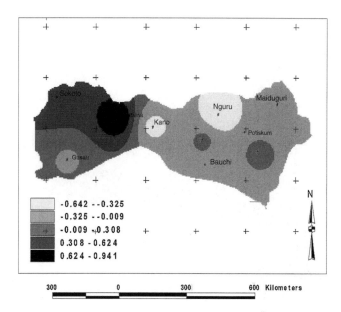

Fig. 11 Spatial distribution of rotated correlation matrix of PCA(component two)

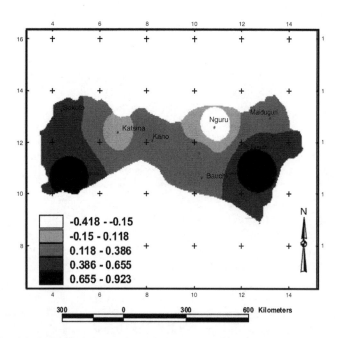

Fig. 12 Spatial distribution of rotated correlation matrix of PCA(component three)

delineated in the SSR in terms of drought risk and that areas to the extreme west and eastern section of the region were the areas which were most vulnerable. The factor which could explain the spatial variation of drought severity is the feedback mechanism associated with interactions between the atmosphere and the land surface in the mobilization and transport of dust (MacLeod 1976). The region under study is the largest source of atmospheric dust in the country. Atmospheric dust can affect the vertical temperature profile of the local atmosphere, affecting its stability and ability to support the convection necessary for rainfall generation. For instance, where dust is present as an elevated layer, cooling near the Earth's surface due to reduced incoming solar radiation can combine with heating in the dust layer due to absorption on outgoing longwave radiation, reducing the vertical temperature gradient, thus increasing atmospheric stability and inhibiting convection (Brooks 2004).

Conclusion and Policy Recommendations

Conclusion

This study examined drought severity in the SSR of Nigeria between 1961 and 2006. Drought severity was assessed in terms of temporal and spatial variations. It was observed that there is the tendency for drought severity for Katsina, Nguru, Bauchi, and Gusau to increase, while that of Sokoto, Kano, Maiduguri and Potiskum indicated a decrease in the nearest future. As a result of this, policy-makers both at the federal and state levels should be aware of this and put appropriate policies in place to cope with droughts in such areas. Within this context, people living around Katsina, Nguru, Bauchi, and Gusau are particularly vulnerable to the impact of drought, especially the local farmers. This is based on the fact that drought severity showed a tendency to increase in the future.

This study also established that even though the SSR is categorized as an ecological zone, the region is different in terms of exposure to drought risk. This is coming on the heel of findings from the spatial analysis of the drought magnitude. The spatial analysis revealed that the region could be divided into three sub-regions, where the extreme western and eastern sections of SSR have the highest level of exposure to drought risk. The implication of this revelation is that the policy-makers should be wary of sharing interventions equally to the region whenever there is drought occurrence. They should be aware that the region is not homogeneous in terms of drought risk.

Policy Recommendations

Drought severity analysis is an important study in semi-arid West Africa because of the risk of droughts and the constraint it imposes on the local populace. In a rain-fed agricultural region of SSR in Nigeria, an efficient water management system is very important. Looking at the fact that drought risks vary greatly in this region, it is imperative that research activities and policies are focused on building and strengthening capacity around local initiatives. These policies should be in line with those promulgated by the African Conference on Water Resources, as well as the results of the focused research activities at the local levels. Pertaining to water resources, communities should carry out water resources assessment both at the state and local levels for effective policy implementation. These assessments should be supported by both the international bodies and non-governmental associations. Any policy that would be implemented must involve all the stakeholders in the community: local and private sectors, and governments. More importantly, all legislations regarding water resources management should be enforced at all levels.

References

Adedokun JA (1978) West African precipitation and dominant atmospheric mechanisms. Archiv fur Meteorologie Geophysik und Broklimatologie Serie A 27:289–310
Anyadike RNC (1987) Measures for containing the effects of drought on the environment. In: Ofomata GEK, Ikpeze NI (eds) Austerity and the Nigerian society. UNN-Etukokwu Press, Onitsha, p 367
Ba MB, Frouin R, Nicholson SE (1995) Satellite-derived interannual variability of west African rainfall during 1983–88. J Appl Meteorol 34:411–431
Barring L (1987) Spatial patterns of daily rainfall in central Kenya application of principal component analysis common factor analysis and spatial correlation. J Climatol 7:267–289
Breman H, de Wit CT (1983) Rangeland productivity and exploitation in the sahel. Sci 221: 1341–1347
Brooks N (2004) Drought in the African sahel: long-term perspectives and future prospects. Tyndall Centre Working Paper no 61
Brovkin V (2002) Climate-vegetation interaction. J de Physique IV 12:57–72
Chappell A, Agnew CT (2008) How certain is desiccation in west African sahel rainfall (1930–1990). J Geophys Res 113, D07111, doi:10.1029/2007JD009233
D'Amato N, Lebel T (1998) On the characteristics of the rainfall events in the sahel with a view to the analysis of climatic variability. Int J Clim 18:955–974
Dracup JA, Lee KS, Paulson EG Jr (1980) On the definition of drought. Water Res 16:297–302
Dyer TGJ (1975) The assignment of rainfall stations into homogeneous groups: an application of 373 principal components analysis. Quart J Royal Meteorol Soc 101:1005–1013
Ehrendorfer M (1987) A regionalization of Austria's precipitation climate using principal components analysis. J Climatol 7:71–89
Frere M, Popov GF (1979) Agro meteorological crop monitoring and forecasting. Plant production and protection paper 17, FAO, Rome
Giannini A, Saravanan R, Chang P (2003) Oceanic forcing of sahel rainfall on interannual to inter-decadal time scales. Sci 302:1027–1030
Gibbs WJ, Maher JV (1967) Rainfall deciles as drought indicators. Bureau of meteorology bulletin no 48, Commonwealth of Australia, Melbourne
Glantz M (ed) (1976) The politics of natural disaster. Praeger Publishers, New York
Glantz M (1996) Drought follows the plough cultivating marginal areas. In: Ribot et al (eds) Climate variability climate change and social vulnerability in the semi-arid tropics. Cambridge University Press, Cambridge, pp 125–128

Hulme M (2001) Climatic perspectives on sahelian desiccation: 1973–1998. Glob Environ Chang 11:19–29

Hulme M, Doherty R, Ngara T, New M, Lister D (2001) African climate change 1900–2100. Clim Res 17:145–168

IPCC (2001) Third assessment report—climate change 2001. The third assessment report of the intergovernmental panel on climate change, IPCC/WMO/UNEP

Janowiak JE (1988) An investigation of interannual rainfall variability in Africa. J Clim 1:241–255

Kogan FN (1990) Remote sensing of weather impacts on vegetation in non-homogeneous areas. Int J Remote Sensing 11(8):1405–1419

Kraus EB (1977) The seasonal excursion of the Intertropical convergence zone. Mon Weather Rev 105:1052–1055

Lamb PJ (1982) Persistence of subsaharan drought. Nat 299:46–48

MacLeod NH (1976) Dust in the sahel cause of drought. In: Glantz M (ed) The politics of natural disaster: the case of the sahel drought. Praeger Publishers, New York, pp 214–231

McKee TB, Doesken NJ, Kleist J (1993) The relationship of drought frequency and duration to time scales. Preprints, 8th Conference on Applied Climatology, Anaheim, 17–22 January, pp 179–184

Mortimore M (2000) Profile of rainfall change and variability in the Kano-Maradi region, 1960–2000. Working paper 25 series drylands research Crewkerne, Somerset

Naginder SS, Kundzewicz ZW (1997) Water, drought and desertification in Africa. In: Proceedings of Rabat symposium titled sustainability of water resources under increasing uncertainty, IAHS Publ no 240, April 1997

Narasimhan B, Srinivasan R (2005) Development and evaluation of soil moisture deficitindex SMDI and evapotranspiration deficit index ETDI for agricultural drought monitoring. Agric For Meteorol 133:69–88

National Academy of Sciences (NAP) (1986) Drought management and its impact on public water systems: report on a colloquium sponsored by the water science and technology board, National Academies Press, Washington

Nicholson SE (1985) African rainfall fluctuation in 1850 to present spatial coherence, periodic behaviour and long-term trends. Third conference on climate variations and symposium on contemporary climate 1850–2100, American Meteorological Society, Los Angeles, January 1985

Nicholson SE, Chervin RW (1983) Recent rainfall fluctuations in Africa interhemispheric teleconnections. In: Street-Perrot A, Beran M, Ratcliffe R (eds) Variations in the global water budget. D. Reidel Publishing Company, Dordrecht, pp 221–238

Nicholson SE, Some B, Kane B (2000) An analysis of recent rainfall conditions in West Africa, including the rainy seasons of the 1997 El Niño and the 1998 La Niña years. J Clim 13:2628–2640

Odekunle TO (2010) An assessment of the influence of the inter-tropical discontinuity on inter-annual rainfall characteristics in Nigeria. Geographical research doi: 10.1111/j.1745-5871.2009.00635.x

Ojanuga AG (1987) Characteristics of soils of the semi-arid region of Nigeria. In: Ecological disasters in Nigeria: drought and desertification (Proc Nat Workshop, Kano, December 1985). Federal Ministry of Science and Technology, Lagos, Nigeria, pp 64–77

Ojo O (1987) Hydroclimatic consequences of climatic events in West Africa: the lessons of the Sahelian droughts. In: Rodda JC, Matalas NC (eds) Water for the future: hydrology in perspective, IAHS Publ no 164. IAHS Press, Wallingford, pp 229–238

Oladipo EO (1995) Some statistical characteristics of drought area variations in the savanna region of Nigeria. J Theor Appl Climatol 50(3–4):147–155

Omotosho JB (1990) Onset of thunderstorms and precipitation over northern Nigeria. Int J Climatol 10:849–860

Palmer WC (1965) Meteorological drought. Research paper no 45, US Department of Commerce Weather Bureau, Washington, pp 1–56

Palmer WC (1968) Keeping track of crop moisture conditions nationwide: the new crop moisture index. Weather 21:156–161

Peter TS (1990) Spatial patterns of multiple drought types in the contiguous United States: a seasonal comparison. Clim Res 1:13–21

Petrasovits I (1990) General review on drought strategies. Transactions of the 14th Congress on irrigation and drainage, Rio de Janeiro, International Commission on Irrigation and Drainage (ICID), vol 1-C, pp 1–12

Richman MB (1981) Obliquely rotated principal components: an improved meteorological map typing technique. J Appl Meteorol 20:1145–1159

Richman MB (1986) Rotation of principal components. J Climatol 6:293–335

Sastri ASRAS (1993) Agricultural drought management strategies to alleviate impacts: examples from the arid and sub humid regions of the Indian Subcontinent. In: Wilhite DA (ed) Drought assessment, management, and planning: theory and case studies. Kluwer Academic Publishers, Dordrecht, pp 65–86 Abstract-EconLit

Shafer BA, Dezman LE (1982) Development of a surface water supply index (SWSI) to assess the severity of drought conditions in snowpack runoff areas. In: Proceedings of the Western snow conference, Colorado State University, Fort Collins, pp 164–175

Sivakumar MVK (1989) Agroclimatic aspects of rainfed agriculture in the Sudano-Sahelian zone In: Soil crop and water management systems of rainfed agriculture in the sudano-sahelian zone (Proc Int workshop, January 1987), ICRISAT Sahelian Centre, Niamey, pp 17–38

Sivakumar MVK, Wallace JS (1991) Soil water balance in the Sudano-Sahelian zone: need, reliance and objectives of the workshop. In: Sivakumar MVK, Wallace JS, Renard C, Giroux C (eds) Soil Water Balance in the Sudano-Sahelian zone (Proc Niamey Workshop, February 1991). IAHS Publ no 199, IAHS Press, Wallingford, pp 3–10

Thomas RJ (2008) Opportunities to reduce the vulnerability of dryland farmers in Central and West Asia and North Africa to climate change. Agric Ecosyst Environ 126:36–45

Trewartha GT (1968) An introduction to climate, 4th edn. McGraw-Hill, New York, p 408

AT UN-HABIT (2006) The state of the world's cities report 2006/7. Earthscan, London

Watts M (1983) Silent violence: food, famine and peasantry in northern Nigeria. University of California Press, Berkeley, p 687

Wilhite DA (2005) Drought and water crises: science, technology, and management issues. Taylor and Francis Press, Boca Raton

Wilhite DA, Glantz MH (1985) Understanding the drought phenomenon: the role of definitions. Water Int 10:111–120

Winstanley D (1973) Rainfall patterns and general atmospheric circulation. Nat 245:190–194

World Water Assessment Program (WWAP) (2006) Water, a shared responsibility: the 2nd UN world water development report. UNESCO, Paris

Zeng N (2003) Drought in the sahel. Sci 302:999–1000

Chapter 27
Climate Change and Sustainable Management of Water Resources

Golda A. Edwin and G. Poyyamoli

Abstract In recent times, several studies have shown that climate change is likely to have a significant impact on the availability of freshwater resources. Freshwater-rich regions across the globe are projected to face water scarcity if current reserves are not managed effectively. Traditionally, the Puducherry region has been well endowed with large freshwater reserves, but demand for water has already increased manifold over the years due to an increase in agriculture, industrialization, urbanization, population and economic development. This has resulted in water scarcity and water quality problems in some regions. At present, the hydrological cycle is being modified at a rapid pace due to the overexploitation of water resources, changes in cropping pattern, land use, groundwater depletion, seawater intrusion, pollution and water pricing models. Apart from these, there have also been observed changes in the increase of average temperature, humidity and coastal erosion. It is recognized that sustainable water resources development and management is an important and urgent issue to be taken up seriously. Therefore, an objective assessment of the availability of water resources in the context of the future water requirements, impacts of climate change and its variability is very crucial for sustainable development. This paper examines in detail the potential for sustainable management of freshwater resources within the constraints of climate change.

Keywords Climate change · Adaptation · Mitigation · Policy-making · Water resources management · Water quality · Sustainable management

G. A. Edwin (✉) · G. Poyyamoli
Department of Ecology and Environmental science,
Pondicherry University, Puducherry 605014, India
e-mail: Golda_Edwin@yahoo.in

Introduction

Puducherry is a Union Territory in south India covering an area of 492 sq km^2 with 45 km of coastline stretching along the Bay of Bengal and to some extent along the Arabian Sea. The territory also has 675 km^2 of inshore waters. Figure 1 shows the four geographically unconnected regions of the Union Territory: Puducherry, Karaikal, and Yanam on the Bay of Bengal and Mahe on the Arabian Sea.

Puducherry, like many regions, is prone to natural hazards such as floods, cyclones and earthquakes. In the past, the northeast monsoon has wreaked havoc in many parts of the region by inundating a vast area of land along the coastal stretches. The heavy rains and cyclones due to the monsoon sometimes cause severe damage to life and property. Unchecked growth in the population during the past decade and a blooming economic activity have put severe stress on the available water resources.

With an estimated population of over 1 million, the region generates wastewater of about 50 million litres per day (MLD), with the entire amount discharged untreated into the sea through backwaters and creeks. Puducherry hosts a lot of water-intensive industries such as paper, alcoholic beverages, chemicals and pharmaceuticals. The total treated wastewater discharged from industries is over 7 MLD. Sewage wastewater contains mostly biochemical oxygen demand (BOD) compounds, nutrients and bacteria. Industrial wastewater mainly contains mostly suspended matter, BOD compounds and traces of oil and gas.

Water management is a critical issue here as the water-intensive industries and agriculture are mainly dependent on the groundwater for their needs. Because of this, the water tables have started falling throughout the region due to excessive exploitation of groundwater resources. Recent research has also revealed that the stress on the water table is resulting in the intrusion of seawater, especially in some coastal areas in the Puducherry region. Existing surface irrigation systems suffer from some inherent weaknesses that need rectification. Several of the tanks and ponds, formerly used for storing rain and surplus river water and irrigating a major proportion of the cultivated area, are already into a state of disrepair. The efforts to revive these systems will have to be speeded up considerably. Figure 2 shows the municipalities and communes of the Puducherry region.

In recent years, the government and the local NGOs have implemented various traditional, disconnected projects to promote sustainable development and improve the conditions of water resources. The predicted climate change over the coming years is likely to add greater stress to the already affected water resources in the region, including some areas that are currently well endowed. The stresses are likely to involve changes in the frequency of extreme events as well as gradual changes in mean annual net resources. Thus cooperation, proper planning and sustainable water resources management are required to address the impacts of climate change and meet the current and future demands in a sustainable way.

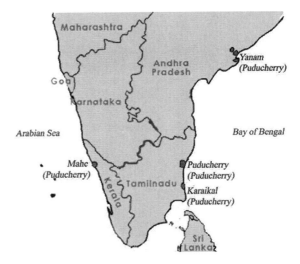

Fig. 1 Location map showing the four unconnected regions of Puducherry

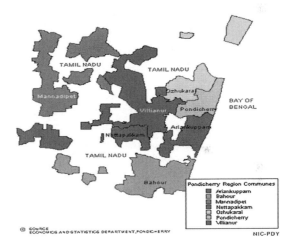

Fig. 2 A close-up view showing the major municipalities and communes in the Puducherry region

Challenges of Climate Change on Water Management

Water is the primary medium through which climate change will impact the environment, ecosystems, economies and people. It is widely accepted that integrated water resources management (IWRM) should therefore be an early focus for adaptation to climate change.

The response to the challenges posed by the new climate change conditions should be through adaptation and mitigation. Complexity, vulnerability and uncertainty are key issues to be addressed. It is in this setting that policy-makers and water managers need to tackle an increasingly scarce resource that varies greatly in space and time (Jones 1999).

Proper planning and management of water resources, comprehensive measures for the reduction of carbon and water footprint, prediction of unavoidable impacts of climate change, cooperation and coordination of actions should be among the priorities. The flexibility supported by the IWRM processes will therefore need to effectively respond to change and be capable of adapting to new ecological, economical, social and environmental conditions.

In Puducherry, the economic sectors that are mostly affected are agriculture, tourism, energy, fisheries and infrastructure, which depend heavily on climate, weather and environmental conditions. Successful adaptation to the impacts of climate change will depend not only on effective national action plans and regulations, but also on the ability of current water management strategies practised in the municipalities and communes to be integrated into affected sectoral policies such as agriculture, energy, tourism, fisheries, and infrastructure.

In the past century, in India, there has been an increase in mean annual temperature, increase in maximum temperature, increasingly trendless monsoon rainfall, decadal departures in summer monsoon rainfall are found above and below the long-time average alternatively for three consecutive decades. This has profoundly affected the environment, economy and society. Therefore, cooperation is required among the different stakeholders to concentrate on diverse measures to address the change.

On the one hand, infrastructural measures such as the construction of dams, levees, drainage canals, sewer networks, dykes, and desalination plants should be built considering the impacts of climate change. On the other hand, solutions such as crop choice (salt–and drought-resistant crops), crop efficiency, biofertilizers, decentralized demand management, water pricing, water markets, watershed management, artificial recharge, rainwater harvesting, flood proofing, and retention measures, knowledge management, decision support system and insurance should be worked upon to comprehensively address the challenges.

The challenges that face us can be better comprehended by analysing them under the areas of physical, social and economic dimensions.

Physical Dimension

Changing Rainfall Patterns

The average annual rainfall at Puducherry is 1254.4 mm. The region receives rainfall from both the northeast and southwest monsoons, with annual rainfall spread over a period of 8 months. The southwest monsoon brings 29% of the annual rainfall from June to September and the northeast Monsoon brings 63% of the annual rainfall during October to December (Puducherry City Development Plan 2007).

Table 1 Percentage deviation of the actual rainfall received from the normal rainfall for different seasons in 2009

Region	Winter period (January–February)			Hot weather period (March–May)			Southwest monsoon (June–September)			Northeast monsoon (October–December)		
	Actual	Normal	Deviation (%)	Actual	Normal	Deviation (%)	Actual	Normal	Deviation (%)	Actual	Normal	Deviation (%)
Puducherry	18	46	-60.9	224	64.3	248.4	140	378.5	-63	1238	849.4	45.7
Karaikal	44.5	66.5	-33.1	227.4	68.4	232.5	117.2	258.7	-54.7	1055.7	993.2	6.3
Mahe	0	10.4	0	272.8	279.7	-2.5	2036	2729.8	-25.4	485.6	367.2	32.2
Yanam	0	27.1	0	17.2	70.1	-75.5	688	691.3	-0.5	242.1	450.3	-46.2

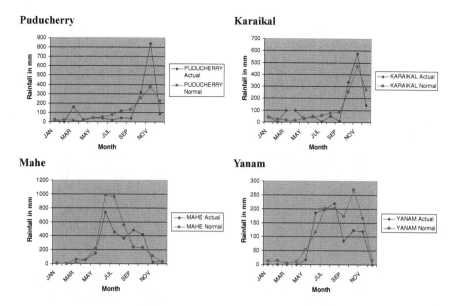

Fig. 3 Actual and normal rainfall month-wise in the four regions of the Union Territory for the year 2009

Table 1 shows the percentage deviation of the actual rainfall received from the normal rainfall for different seasons in the Union Territory for the year 2009 and Fig. 3 shows the month-wise actual and normal rainfall for the year 2009 in all the four districts of the Union Territory. The rainfall pattern is changing at random and deviation from the norm is too much and unpredictable. Although it is difficult to make good predictions about the rainfall and storms, what needs to be understood is that it is even more difficult to predict the impact of changing temperature and rainfall on water availability from rivers, aquifers and lakes.

As the majority of the water-users in this region are farmers, it is critically important to predict the amount and timing of rainfall. Excessive exploitation of groundwater in combination with climate change makes the situation difficult. In many places, groundwater wells are already contaminated, unprotected or close to becoming dysfunctional due to a lowering of the groundwater table and due to low and poor maintenance. Those wells, which serve as the basic water supply, will not be able to supply water in times of disasters and emergencies. They will either be contaminated or dried up.

The increased variability in rainfall results in change of pattern in recharging the aquifers which is the major water source for the economy. Since mostly it is a humid area, there will be a decrease in groundwater recharge because more frequent heavy rain will result in the infiltration capacity of the soil being exceeded, thereby increasing surface runoff. With such scenarios the relationship between the amount of rainfall received and the amount of water available in rivers, lakes and aquifers becomes a complex one (Stakhiv and Pietrowsky 2009).

Fig. 4 Humidity (%) in Puducherry from 1984 to 1996. Source: hydrology project

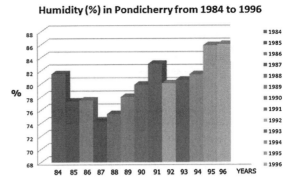

A change in rainfall pattern could lead to intense and short-lived precipitation in coastal urban watersheds, which would lead to calamitous flooding. When accompanied by sea level rise, which itself is a consequence of climate change, excessive flooding of urban settlements and industrial zones is a certainty, endangering the lives of the citizens and threatening infrastructure. In Puducherry, the development is not uniform across all municipalities and communes and some areas have poor drainage infrastructure, which is a major bottleneck in addressing the impacts.

Projected sea level rise and excessive groundwater extraction in areas close to the coast also increase the salinity problems in water supplies. The quantity and the quality of groundwater depends always on recharge conditions, which are controlled by factors such as annual precipitation, soil properties, land surface characteristics and vegetation cover.

Annual Runoff and Drainage System

As a result of drier ground and increased evaporation, water percolations into the aquifers or runoff to the rivers have been drastically reduced. A reduction in runoff will be perhaps the most serious impact of global warming on the water environment. This is why climate change is "amplified" in the water cycle. But the other things, which include the types of vegetation as well as the timing and intensity of rainfall, are unlikely to remain equal. Vegetation will change as a result of changes in temperature, rainfall and CO_2 concentrations. The intensity and timing of rainfall will change as a consequence of the changing circulation patterns inherent in generalized atmospheric warming (Sadoff and Muller 2009a).

Only 10% of the sewage is treated before being discharged into the sea. Silting and uncontrolled solid waste dumping cause blockages and stagnate water channels/wastewater runoff. Consequently, drains choke and overflow into neighbouring areas. Water bodies around the city, which acted as flood moderators, have also witnessed silting. Sometimes the stormwater and domestic sewage overloads

Table 2 Groundwater utilization (in MCM/year). Source: central groundwater board

Total replenishable groundwater resources	Provision for domestic, industrial & other uses	Available for irrigation	Projected net draft	Balance for future use	Level of groundwater development (%)
174.6	26.2	148.4	115.5	32.9	77.85

the open drainage system and inundates the low-lying areas of the town in some parts of Oulgaret municipality.

The case of the Kaveri (otherwise called Cauvery) river in South India has been the focus of intense controversy since the early 1990s. The river's drainage basin provides most of the water for the states of Karnataka, Tamil Nadu, Kerala, and the Union Territory of Puducherry. Unsurprisingly, the sum of each of the water claims of the different states is far greater than the actual availability of water.

Temperature, Evaporation and Aridity

Changes in temperature and precipitation will have a direct impact on the demand and supply of energy, and the quantity and quality of water available for irrigation, hydropower, and domestic and industrial use (Mall et al. 2006). Figure 4 shows the humidity percentage in Puducherry from 1984 to 1996.

Even slight changes in temperature and precipitation will affect the quantity and quality of water available for irrigation, hydropower, domestic and industrial uses. Evaporation rates increase as a result of an increase in temperature. As a result, aridity, which is the ratio of rainfall to potential evaporation, also increases, since the rising temperatures are not matched by rising rainfalls. Changes in aridity will have a substantial impact on both surface water runoff and groundwater recharge (Mujumdar 2008).

In the present condition, the potential evaporation exceeds the amount of rainfall and actual evaporation depends on the amount of water available to be evaporated. Thus only in periods in which there is sufficient rainfall to "swamp" the evaporation does runoff infiltrate the groundwater aquifers.

Changing Groundwater Recharge Patterns

Puducherry region is situated on the Coramandal coast, covering an area of 293 km^2. The main source of water supply for all purposes such as agriculture, drinking and domestic use is groundwater, which is extracted from three major aquifers: Alluvium, Tertiary, and Cretaceous. The irrigation is supplemented by 84 system and non-system tanks. Owing to population growth, intensive agriculture and rapid phase of industrial development, the demand for water has increased

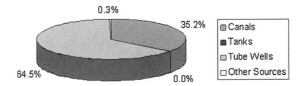

Fig. 5 Percentage area irrigated from various sources

considerably, which has resulted in the overexploitation of groundwater, which ultimately leads to drastic depletion in groundwater levels and deterioration of water quality due to seawater intrusion and also upward movement of chemical constituents present in deep-seated aquifer.

In the Puducherry region, the farmers started constructing shallow tube wells tapping alluvial aquifers up to a depth of 30–40 metres. Initially, the shallow tube wells were found free-flowing. But as the number of shallow tube wells started increasing, the water level started declining due to overload. This has been greatly felt in parts of the Mannadipet and Nettapakkam communes.

Due to the decline in the water level, the farmers had to dig out pits to lower centrifugal motors. At one stage, the centrifugal pumps were also not working as the water level declined below the lifting capacity of these pumps. The wealthy farmers switched over to submersible pumps, whereas the small and marginal farmers could not construct deep tube wells tapping older formations. They were forced to keep their land fallow or cultivate their land by getting water on a hire basis from the nearby sources at the rate of eight bags of paddy per acre per season. Whatever the farmers earned went towards hire charges for water. So the farmers could not sustain themselves and many were even pushed below the poverty line.

Table 2 provides the details of the utilization of groundwater in the Union Territory of Puducherry.

Figure 5 indicates the percentage area irrigated from various sources. 64.5% of the irrigation is through the tube wells which directly contribute to the lowering of the water table, leading to saltwater intrusion. Currently around 70 MLD is drawn through bore wells maintained by the Public Works Department for public water supply. As a result, the ground is over-pumped in many places in order to satisfy the increasing demand driven by the economy and population growth.

The impact on the environment is severe, including soil degradation, erosion and increase in sedimentation, poorer infiltration of water, reduction of aquifer recharge, loss of wetland communities, and increased concentration of pollutants and salt in groundwater. The government has already proposed a desalination plant to meet the rising demand. Though this is an option, desalination is in no way a sustainable alternative to water security, as this greatly increases the carbon footprint.

One of the most difficult water resources management challenges is monitoring and managing underground water, which many rely on for their water supply (Geocenter Denmark 2008). Furthermore, the Union Territory is already facing a growing urban demand which has to be balanced against the declining agricultural use for irrigation. These trade-offs can be managed only if there is a proper regulation of groundwater extraction.

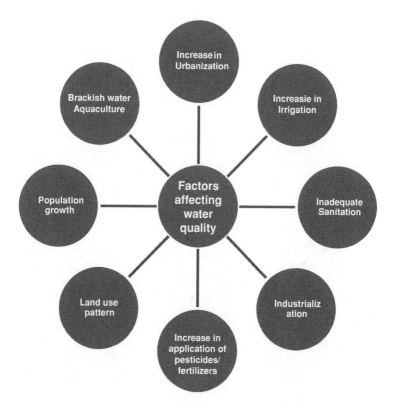

Fig. 6 Factors contributing to decline in water quality in Puducherry

Water Quality

The quality of the groundwater is greatly affected by the nitrates from agricultural fertilizers and chloride from saltwater intrusion in coastal aquifers. The seawater is affected by the disposal of untreated sewage. At present, only 10% of the sewage is treated and the rest is disposed as it is into the sea. Marine discharge from industry and heavy metal pollution also contributes to the pollution. Changing runoff patterns and increasing temperatures result in water quality effects that either render water unusable or make it require additional treatment. Figure 6 shows the factors contributing to the decline of water quality in the Puducherry region.

The poorly planned water usage patterns adopted by man, through urbanization, growth of population, growth of industries and employment of auxiliary means in agriculture, have disturbed the natural quality of water bodies in many regions, along with the water availability aspect. As a result, the water bodies have become unsuitable for potable uses at many locations. Another possible impact could be the intrusion of seawater into coastal freshwater systems. This might occur in the areas affected by sea level rise and where river flows are insufficient to prevent seawater

Table 3 Land use pattern in Puducherry region

Classification	2004–05 (%)	2008–09 (%)
Forest	0.00	0.00
Non-agricultural uses	35.22	37.58
Barren and uncultivable lands	0.14	0.15
Permanent pastures and other grazing lands	0.02	0.00
Land under miscellaneous tree crops not included in net area sown	2.38	2.49
Cultivable waste	8.44	9.02
Other fallow land	5.32	5.04
Current fallow land	5.25	6.24
Net area sown	43.23	39.48

from flowing upstream. The inland fishery which contributes to over 30% of fish/prawn production also contributes to the increase in salinity of the groundwater.

Land Use

Table 3 shows the changes in the land use pattern in the Union Territory over five years. In recent years, owing to economic growth and population increase, there has been a drastic change in land use. Puducherry is slowly moving closer to an industrialized economy. In some places, the farmers are forced to keep their land fallow owing to non-availability of groundwater for irrigation purposes (Nobi et al. 2009).

Floods, Droughts and Storms

The 2008 UN Human Development Report stated that from 2000 to 2004, some 262 million people were affected by climate related disasters annually. Over 98% of them lived in the developing world. Rising temperatures increase the evaporation from the water bodies and other sources, which in turn leads to more severe rainfall. This will lead to more floods and droughts and in a random pattern, affecting the urban and rural poor mostly. Recently the Karaikal region experienced severe drought coupled with water scarcity resulting in the reduced yield or failure of crops. Such scenarios might increase in frequency if adequate measures are not taken proactively to combat the impacts of climate change.

Social Dimension

The major impact of climate change in many areas may be to increase the cost of water services. As a result, the water required to sustain the ecosystem will be put under tremendous pressure. This will not only be the case for drinking water, but also for agriculture and power production, as well as for industry. The increased incidence of floods and droughts will impact lives, livelihoods, land values and investment incentives in vulnerable areas. Undoubtedly, it is the poor who are most vulnerable to floods, sea level rise, groundwater intrusion and loss of arable land.

The prevailing scenarios in the region that affect society are as follows:

- Over 40 villages are still to get access to potable water and over 80 villages have only partial access to it.
- The water price does not reflect the true costs of the production of potable water because of the huge and unsustainable subsidies.
- About 100% of the population are connected to the supply of potable water, but only 10% of households are connected to the wastewater system.
- The development of tourism has an impact on the water demand, so that in summer months where the demand is particularly increasing, there are problems of supply, affecting industry and tourism.
- The quality of water in rural areas is more affected because of the agricultural activities and the increasing use of fertilizers.
- The population must adapt the use of water to its availability.

Economic Dimension

The increased variability and changes in the availability and reliability of rainfall will directly impact the water-using sectors and indirectly impact the growth potential of the economy. These impacts, in turn, will affect the environmental and social needs (Report by the United Nations 2009). Recently, as a result of seawater intrusion, the water-intensive industries near the coastal zone were shut down or transferred to a different location. Climate change's impact on water resources has implications far beyond the Millennium Development Goals. The changes in distribution and timing of rainfall will change patterns of access to water, creating new surpluses in some areas and increased competition in others. Managing this evolving hydrology will impose significant demands on water management at regional, national and international level (Kolokytha 1994).

Water Security Through Integrated Water Resources Management

IWRM is a process which promotes the coordinated development and management of water, land and related resources, in order to maximize the resultant economic and social welfare in an equitable manner without compromising the sustainability of vital ecosystems. IWRM has been the accepted management paradigm for efficient, equitable and sustainable management of water resource since the early 1990s (Sadoff and Muller 2009b).

An IWRM perspective ensures that social, economic, environmental, and technical dimensions are taken into account in the management and development of water resources. It is not an answer to climate change, but an approach that reflects the integrated nature of the water cycle by addressing different users, uses, threats, and threatened resources. As such, IWRM has to deal with all natural resources, not only water, but also soils, ecosystem, surface water, groundwater, water quantity, quality, as well as other ecological aspects.

The multi-sectoral nature of IWRM should be recognized in the context of socioeconomic development (InfoResources Focus 2003). Serious water management problems are likely to occur in the Puducherry region if adequate remedial measures and policies are not taken up proactively by the Union Territory. In recent years, the Union Territory has been spending more on water resources management by implementing various projects. The objective is mostly to meet the demand with the supply. But such a traditional and fragmented approach will not lead to sustainable growth and therefore all the current and future water planning and strategies should be aligned and integrated with social, economic, and environmental goals (Slootweg 2009).

Policy Framework to Respond to Climate Change Through IWRM

Natural and Human System Interaction

Integration can be considered under two basic categories: the natural system, which focuses on the resource availability and quality, and the human system, which determines the resource use, waste production and pollution of the resource, and which must also set the development priorities (Agarwal et al. 2000).

The natural system integration focuses on the integration of land and water management, integration of freshwater and coastal water management, integration of soil water and runoff water management, integration of quantity and quality of water management, integration of surface and ground water management and integration of the upstream and downstream water related interests.

The human system integration deals with the mainstreaming of water resources, cross-sectoral integration in national policy development, macroeconomic effects of water developments, integrated policy-making, influencing economic sector decisions, integration of all stakeholders in the planning and decision process, and integrating of water supply and wastewater management.

The different stakeholders should ensure that integration has to happen both within and between these categories, taking into account variability in time and space.

Groundwater Regulation

Bahour commune in the Puducherry region accounts for one-third of the rice cultivation in the region. Recently, salinity in the groundwater has increased in the Bahour and southern part of Villianur communes, which has affected the soil and crop yield. The increase in salinity is mainly attributed to the over-extraction of groundwater using high-energy pump sets. If this trend continues, the Bahour commune will become unsuitable for cultivation in future (National Action Plan on Climate Change—Puducherry 2009).

It is encouraging that the government of the Union Territory has passed a Groundwater Regulation Act and set up the administrative machinery for monitoring the extraction and use of groundwater. It also promotes rainwater harvesting and maintains a hydrological data centre which aids in decision-making. Groundwater levels of many aquifers in the region show a decreasing trend. This is generally due to groundwater pumping exceeding groundwater recharge rate. Groundwater will be less directly and more slowly impacted by climate change, as compared to surface waters. As many groundwater systems both discharge into and are recharged from surface water, impacts on surface water flow regimes are expected to affect groundwater. Thus, neglecting the consideration of groundwater in the process of IWRM can result in the mismanagement of surface water, with severe effects on the population and the environment (Steenbergen and Tuinhof 2009).

Water Pricing

A major cause for the over-extraction of groundwater is the subsidies provided for water. IWRM should ensure that industrial users should be required to bear the full cost while some subsidy could be given to domestic users. The pricing model should be changed in such a way that everybody gets their fair share of water for a subsidized cost, and beyond which the user should be charged heavily. This will not only ensure that the basic need is taken care of but also discourage the water-users from wasting this precious resource. But for this to happen, first precise and uniform metering should be implemented in all places (Rakesh et al. 2005).

Water Quality Monitoring

The policy framework should ensure the safe discharge of industrial effluents, biomedical waste, electronic waste and domestic waste. Waste segregation, reuse and safe disposal of hazardous wastes should be done as close to the waste generation source as possible. A water quality monitoring network has to be established involving the NGOs, elite educational and research institutions to supplement the efforts of the government.

Tank Rehabilitation

The government is undertaking several tank rehabilitation projects with aid from the European Union. The tanks and ponds of the Puducherry region have to be rehabilitated not only for irrigation but also to recharge the groundwater and restore the ecosystems.

Conservation Measures

Apart from pricing and regulation, the government should also focus on creating awareness among the schools, colleges and households. Local communities who are most affected by the impacts of climate change should be educated about climate change and its impacts and ways to mitigate the risks.

Quality Drinking Water

At least 40 villages do not have any municipal water supply system and over 80 villages are only partially covered despite the abundance of water in rural areas compared to urban parts. The women in these villages are most affected as they are forced to travel a long way to collect water and thus lead a poor quality of life. The socioeconomic conditions should be given due importance as per the IWRM principles and thus should meet the basic need for the citizens. The government must also focus on tapping the excess water from the Ousteri and Bahour tanks to augment the urban water supply.

Better Drainage System

The water is supplied at the rate of 135 litres per capita per day. Of this, 80% of the water shall be wastewater. Hence, it is estimated that by the year 2026, about 94.5 MLD of sewage will be generated in the Puducherry urban area. The existing capacity of the STP can treat only 15.3 MLD of sewage. Hence, there is shortfall of 79.2 MLD. Presently, the sewage is treated through oxidation ponds, which require a large area but lower operation and maintenance costs. Due to land constraints in Puducherry, Up-flow Anaerobic Sludge Blanket (UASB) process of STP is designed for the uncovered areas. The UASB treatment will consume a smaller area of land and have higher operation and maintenance costs.

Monitoring and Change

Water resource managers need the ability to track changes and to devise and support the implementation of appropriate responses. This requires extensive data and the ability to analyse and interpret it in order to guide planning and inform the stakeholders of its implications. Data should be cleansed and integrated from various source systems, so that intelligent reporting can be implemented to enable to policy-makers to make decisions based on the facts alone.

Conclusion

Water is the primary medium through which climate change influences the Earth's ecosystems and therefore people's livelihoods and well-being. Adaptation to climate change is mainly about better water resources management. A traditional, fragmented approach to water management is not going to yield sustainable benefits in the long run. Water resources and how they are managed impact almost all aspects of society and the economy, in particular health, food production and security, domestic water supply and sanitation, energy, industry, and the functioning of ecosystems. Thus, adapting to the new challenge of climate change requires an integrated approach to water management. Integrated solutions coupled with innovative technologies are needed at the appropriate levels, for adaptation as well as mitigation. Appropriate adaptation measures should be built upon the existing water management practices to foster resilience to climate change, thereby enhancing water security.

References

Agarwal A, delos Angeles MS, Bhatia R, Chéret I, Davila-Poblete S, Falkenmark M, Gonzalez Villarreal F, Jønch-Clausen T, Aït Kadi M, Kindler J, Rees J, Roberts P, Rogers P, Solanes M, Wright A (2000) Integrated water resources management. Global Water Partnership, Denmark

Geocenter Denmark (2008) Groundwater and climate change: challenges and possibilities, pp 1–15

InfoResources Focus (2003) Integrated water resources management (IWRM): a way to sustainability, inforesources focus, no 1/03, pp 5–9

Jones JAA (1999) Climate change and sustainable water resources: placing the threat of global warming in perspective. Hydrol Sci 44(4):541–557

Kolokytha E (1994) European policies for confronting the challenges of climate change in water resources. Curr Sci 98(8):1069–1076

Mall RK, Gupta A, Singh R, Singh RS, Rathore LS (2006) Water resources and climate change: an Indian perspective. Curr Sci 90(12):1610–1626

Mujumdar PP (2008) Implications of climate change for sustainable water resources management in India. Phys Chem Earth, Elsevier 33(5):354–358

Nobi EP, Umamaheswari R, Stella C, Thangaradjou T (2009) Land use and land cover assessment along Pondicherry and its surroundings using Indian remote sensing satellite and GIS. Am-Eurasian J Sci Res 4(2):54–58

Rakesh K, Singh RD, Sharma KD (2005) Water resources of India. Curr Sci 89(5):794–811

Report by the Government of Puducherry (2009) National action plan on climate change, Department of Science Technology, Puducherry, India

Report by the United Nations (2009) The implications of climate change on water, UN World Water Assessment Programme, Perugia, Italy

Report by Wilbur Smith Associates Private Limited (2007) Puducherry city development plan, Town and Country Planning Department, Puducherry, India

Sadoff CW, Muller M (2009a) Better water resources management—greater resilience today, more effective adaptation tomorrow. Water and climate change adaptation, produced together with the World Water Council, IUCN and IWA, Paper 4, pp 1–14

Sadoff CW, Muller M (2009b) Water management, water security and climate change adaptation: early impacts and essential responses, Global Water Partnership, pp 1–85

Slootweg R (2009) Integrated water resources management and strategic environmental assessment–joining forces for climate proofing. Water and climate change adaptation, produced together with the World Water Council, IUCN and IWA, Paper 16, pp 1–13

Stakhiv EZ, Pietrowsky RA (2009) Adapting to climate change in water resources and water services. Water and climate change adaptation, produced together with the World Water Council, IUCN and IWA, Paper 15, pp 1–15

van Steenbergen F, Tuinhof A (2009) Managing the water buffer for development and climate change adaptation. UNESCO publication, Netherlands

Chapter 28
Problems of Water Resources Management in the Drainage Basin of Lake Balkhash with Respect to Political Development

Pavel Propastin

Abstract Lake Balkhash is Kazakhstan's second most important natural inland water reservoir, whose catchment basin is shared between the Republic of Kazakhstan (60% of the catchment's area) and the People's Republic of China (40% of the catchment). The lake is a source of fishery production and it serves as a means of transport, while the water of the catchment basin is extensively used for hydropower production, and irrigated agriculture and industry located in the Balkhash Lake basin. These socio-economic activities together with climate variability have strongly affected the lake's hydrological regime during the last 100 years. The centralized, master plan-oriented management system and use of water regulation during the late Soviet era have been a major drawback to sustainable management of the lake, leading to a significant decrease in its water level and degradation of the surrounding wetlands. After the collapse of the Soviet Union, the economic activity in Kazakhstan's part of the Balkhash Lake basin rapidly diminished, causing the current broad rehabilitation of the whole ecosystem. Unfortunately, despite the benefit for the Balkhash Lake basin's ecosystem accrued from the diminished anthropogenic impact in Kazakhstan, the basin's existing management strategy of post-Soviet Kazakhstan faces several challenges that limit their effectiveness in the realization of sustainable development for the region. Moreover, a new challenge for sustainable development in the Lake Balkhash basin appears in the form of rising water consumption in the upper parts of the basin in China. This paper gives an overview of the change in the Lake Balkhash basin along with the political change in the area. The paper also illustrates problems regarding water resources management in the drainage basin of Lake Balkhash.

P. Propastin (✉)
Department of Geography, Georg-August University Göttingen,
Goldschmidtstr. 5, 37077 Göttingen, Germany
e-mail: ppropas@uni-goettingen.de

Keywords Central Asia · Lake Balkhash · River Ili · Water table · Climate change · Political change · Water resources conflicts

Introduction

Lake Balkhash, Kazakhstan, is the world's fifth largest isolated water reservoir, with a volume of about 90 km^3 and a catchment area of more than 0.5 million km. Long-term periodical fluctuations in Lake Balkhash's water levels are suggested to be conditioned primarily climatically, and demonstrate their intimate connection with components of its water regime, especially runoff from its main contributor, the Ili river that flows in the Republic of China. The construction of the Kapchagay dam in the middle part of the Ili river in 1970 took place during the last downward fluctuation and led to the deepest drop in the lake's water level since the beginning of the measurements. This strongly affected the fragile ecosystems of Lake Balkhash and its surroundings, leading to strong diminishing of fish stocks and degradation of reed grass ecosystems in coastal areas and in the Ili river delta. After the collapse of the Soviet Union in 1991, the filling of the Kapchagay water reservoir was stopped and huge areas of irrigated cropland in the Ili river delta were abandoned. This led to a strong reduction in the anthropogenic impact on the ecosystem of Lake Balkhash. The reduction in the anthropogenic impact overlapped with a reduction of irrigated farming areas in Kazakhstan, and favourable climate conditions led to a significant rise in the water level in Lake Balkhash and rehabilitation of its ecosystems. However, new support for sustainable use of water resources of Lake Balkhash appeared during the last decade through the development of large irrigated areas in the upper part of the Ili river in the Republic of China.

This paper illustrates the problems of Lake Balkhash basin with respect to management of its resources under the impact of water losses for hydropower production, irrigated agriculture, and political change.

Summary of Basin Characteristics

Lake Balkhash lies in one of the most arid discharge-less watersheds of the world. The lake length is 605 km, and the lake width varies from 4 to 74 km. Lake Balkhash is situated within the Balkhash-Alakol depression which is enclosed by the Tien-Shan Mountain in the south, the Djungarsky Alatau in the east, the Chu-Ili Alatau in the west, and the Kazakh Low Hill Lands in the north (Fig. 1). The mean surface area of Lake Balkhash is about 17,000 km^2. Lake Balkhash has a catchment of 500,000 km^2 with about 85% of the drainage area located in the Republic of Kazakhstan and 15% in the People's Republic of China (Fig. 1).

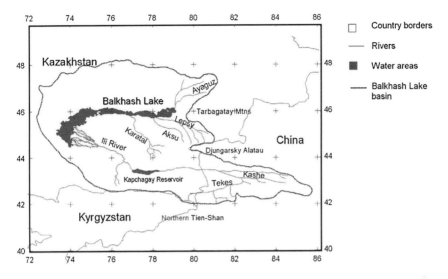

Fig. 1 Map of the Lake Balkhash basin in Central Asia

All inflow to Lake Balkhash is received from the Tien-Shan and the Djungarsky Alatau and the runoff from their ridges. The two largest rivers flowing into Lake Balkhash are the Ili (about 78% from the total inflow) and the Karatal (about 15% of the total inflow). Other contributors are the Karatal river from the western slopes of the Dzhungarskiy Alatau (15.1%); the Aksu river, also from the Dzhungarskiy Alatau (0.13%); and the Lepsy river from the same origin (5.4%). The Balkhash Lake basin is internally drained, the major loss of the inflow water is caused by evaporation from the lake surface, and a small part is infiltrated into the loosely detrital deposition of the drainage basin bottom (Shnitnikov 1973).

Variations of Water Level in Lake Balkhash

The level of Lake Balkhash is one of the major indicators for the whole basin ecosystem condition. Corresponding to analyses of sediments from Lake Balkhash, the water level has fluctuated during historical times by 10 m. The maximum water level of 346 m above sea level was reached during the first half of the eighteenth century, while the minimum, 338 m in 1840 (Petr 1992; Tursunov 2002). Gauge observations on water level have been made since the mid-1800s (Fig. 2). During the gauge observation period, with no runoff regulation, the lake level used to change, following the intersecular cycle, from 340.7 m (1987) to 344.4 m (1911), measured against the Baltic Sea level. The dynamics of water levels in Lake Balkhash during the gauge observation period reveal three phases of regression (1878–1890, 1910–1950 and 1970–1989), and three phases of transgression.

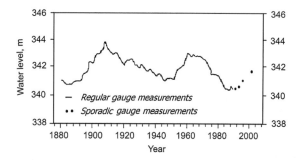

Fig. 2 Long-term variations of the water level of Lake Balkhash measured by gauges (1880–2001). The line is the year-to-year values from regular measurements while the *black circles* represent observed values obtained by sporadic measurements (data compiled from Petr 1992 and Tursunov 2002). Values are given in m above the Baltic Sea water level

The strongest transgression was observed during 1890–1911, with the maximum of 344.1 m in 1911. The second transgression phase was in the mid-1990s. Between 1946 and 1961, water level rose by 2.3 m, but in 1970 it started to decline again. The last regression of Lake Balkhash in the twentieth century was strengthened due to the regulation of the Ili river runoff. In 1987, the lake level reached the lowest values recorded during the period of gauge observations, i.e. 340.7 m (Petr 1992). The last transgression phase started at the end of 1980s and continued till the mid-2000s.

Biodiversity Values

The Balkhash Lake basin is rich with plentiful wetland landscapes in the deltas and valleys of its inflow rivers and has high biological diversity. The Balkhash Lake wetlands, particularly the largest of the Ili river delta, provide habitat for some 250 species of bird, including 22 internationally protected species. In the Ili river delta alone, some 1.5 million otters were harvested annually, i.e. 80% of all otters harvested in Kazakhstan. The deltas of the inflows provide rich pasture for livestock and wetlands function as natural filtering beds and represent the source of nutrients essential for high fish stocks in Lake Balkhash. The Balkhash Lake and adjacent wetlands are home to 22 species of fish, 50 species of mammals, more than 30 species of amphibians and around 20 species of reptiles.

Thickets of reed grass cover the floodplains of the wetlands and are dominated by reed grass species such as *Calamagrostis pseudophragmites* and *Phragmites communis*. The wetlands in the deltas include several hydrographic systems, which are interconnected during high water levels, but are otherwise separated by alluvial deposits. There are a number of lakes within the wetlands and these retain different volumes of water according to flood levels. Delta arms and channels are fringed by the tugay (a gallery forest association, mixed with shrubs and tall grasses).

Swamps and thickets of reed grass cover the floodplains. Plentiful plains between the tributaries and swamps are covered by grassland landscapes which serve as pastures (Nikolayev 1959).

Priority Issues for Management of the Balkhash Lake Basin

Kapchagay Reservoir and Its Ecological Impact on Lake Balkhash

Since 1970, the Kapchagay reservoir has been one of the most significant threats to the Balkhash Lake basin, distorting the natural hydrological regime of the lake (Kudrin and Rabinovich 1976). Before 1970, despite the development of irrigation and some other types of water use in the basin during this period, the fluctuations of the hydrological budget and level of Lake Balkhash were mainly determined by climatic factors (i.e. precipitation and temperature), as well as by losses of water in the river Ili delta associated with the development of its channel network. The retention of the Ili river water for the filling of the Kapchagay reservoir produced a drastic change in the natural hydrological regime of the Ili river and Balkhash Lake leading to a fall in the water level in Lake Balkhash (Petr 1992; Kudrin and Rabinovich 1976). This strongly affected fragile ecosystems of Lake Balkhash and its environments (Tlenbekov and Piven 1993; Petr and Mitrofanov 1998). The impacts included: (i) degradation of wetlands in the Lake Balkhash basin, (ii) rising salinity in the lake, (iii) a decline in the fish stocks and (iv) an alteration of natural hydrological patterns.

In the mid-1980s, through an awareness-raising campaign in the republican and federal media, society was alarmed by the approaching ecological disaster in the Balkhash Lake basin. The government of the Kazakh Soviet Republic undertook some efforts to change the ecological situation and reduce the consequences of the Balkhash disturbance. After stopping the filling of the Kapchagay reservoir in 1989, the situation began to improve. The collapse of the Soviet Union and proclamation of independence of Kazakhstan in 1991 caused an economic crisis in Kazakhstan. Most of the irrigated areas in the Balkhash Lake basin were abandoned during the period 1992–2000. This resulted in a colossal decrease in the total anthropogenic impact in the region. As a consequence, the Lake Balkhash water level rose by 100 cm in 10 years. Stopping the filling of the Kapchagay reservoir in the early 1990s, at the same time as the beginning of a new transgression phase, enforced a very rapid rise of the water level, and in 2001 Lake Balkhash reached a local maximum of 341.87 m.

This period was also characterized by a general increase in precipitation in the Balkhash catchment basin (Matsuyama and Kezer 2009). Balkhash Lake and its surrounding area have rapidly rehabilitated during the last decade of the twentieth century and the first decade of the twenty-first century.

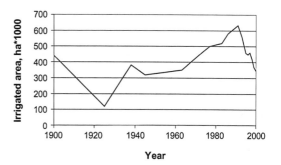

Fig. 3 Development of irrigation area in the Balkhash Lake basin during the twentieth century (after Tursunov 2002)

Development of Irrigated Lands

The use of water resources for irrigation in the Balkhash Lake basin has taken place for several hundreds of years. Under the existing climatic conditions and soil quality in the region, irrigation can significantly increase yields of rice, tobacco, cotton and other agricultural products. During the period of the Russian Empire (before 1917), extensive irrigation took up considerable areas in the Balkhash Lake basin (Fig. 3). In the Soviet era after World War II, a doctrine about intensification of irrigation in the Balkhash Lake basin was developed. This doctrine planned taking into irrigation an area of 1 million ha till the year 2000. Corresponding to this doctrine, an intensive extension of the irrigated land in the region began in the 1960s. The area of irrigated land increased gradually during the following three decades (Fig. 3). By 1991 (the year of Soviet Union collapse), the total area of irrigated land comprised 0.633 million ha. The extension of irrigated area and intensification of its production led to an increase in soil salinization, a decrease in vegetation area and desertification of the irrigated area and neighbouring lands (Petr 1992).

However, the collapse of the Soviet Union distorted the realization of this doctrine. The area of irrigated land decreased during the 1990s. The reduction of irrigated areas occurred very rapidly. By the year 2000, there were only 0.346 million ha under irrigation. Most of the remained irrigation areas are located in the Ili river valley in the Akdala region, comprising the Bakbakhty and Bakanas irrigation farms. Figure 4 shows development of the irrigated land in the Bakbakhty farm during the last decade of the twentieth century. The maps were produced by classification of two Landsat images acquired in May 1990 and May 2000. The maps indicate a large reduction in the rice area and a significant decrease in other agricultural land. However, traces of the former irrigation practice remain in the form of soil salinization, which take up considerable areas within the abandoned land.

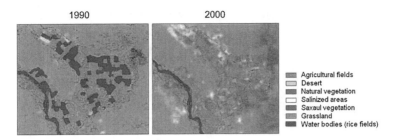

Fig. 4 Development of irrigated land in the Bakbakhty irrigation farm

Fig. 5 An abandoned rural settlement in the valley of the Ili river. In the centre of the photo is the ruin of the former middle school (photo taken by the author in 2008)

Social Development

The reduction in agricultural production during 1990–2000 resulted in a reduction in jobs and a high unemployment rate in the region. According to official statistics, up to 15% of the population do not have a job. Unofficial estimates talk of an unemployment rate of up to 50%. The high unemployment rate is the major reason for the enormous migration of the rural population, especially the youth, to towns during the period after the collapse of the Soviet Union. During the last decade of the twentieth century, the population of most rural settlements in the Balkhash Lake basin decreased by 30–50%. Many villages are abandoned and now convey a discouraging impression (Fig. 5). As a whole, the rural population of the Ili-Balkhash region dropped by 97,000 people.

Increasing Consumption of Water Resources in China

The political and economic development in China over the last two decades created a new challenge for water resources use in the Lake Balkhash basin. More than 75% of the total inflow into Lake Balkhash originates from the territory of China. While the irrigation area in Kazakhstan has rapidly reduced, water consumption in the Chinese part of the Balkhash Lake basin has progressively increased. Moreover, the People's Republic of China plans to increase its water intake from the Ili river in the nearest future. The Chinese government has plans to extend irrigation areas in the Ili river catchment by 450,000 ha. Such an extension of the irrigation production in this region would require additional water consumption of more than 5 km^3/year. In this case, the annual Ili river runoff to Lake Balkhash would be only 7.5 km^3/year. There is also a plan to build a channel connecting the Kashe river (the largest contributor to the Ili river in the Chinese territory) with the industrial centre Karamay. Another intended channel should bring the water of the Tekes (the second largest contributor to the Ili river in the China) to the Tarim depression, in order to increase agricultural production in that region (Tursunov 2002). The Chinese government also intends to build 15 water reservoirs in the upper flows of all three of the major Ili's contributories Tekes, Kashe und Kunes. If all these Chinese plans are realized, the future of the Balkhash Lake could be gloomy. A new massive disruption of the Balkhash hydrological regime would be accompanied by several environmental, economic and social consequences in the whole region. According to the specialists, the realization of these plans would cause a natural disaster similar to that of the Aral Lake. The Aral Lake disaster is known worldwide as a symbol for human mismanagement of water resources leading to the drying up of natural lakes and devastating environments of large regions.

Tursunov (2002) modelled consequences for the Balkhash Lake water balance with respect to different degrees of anthropogenic impact. The author modelled different scenarios. The best scenario implies that the Ili runoff would reduce from 15 to 10 km^3. In this case, the area of the Balkhash water surface would shrink to 11,720 km^2 and the water level would fall to 340 masl. According to the worst scenario, the runoff of the Ili river would be reduced from 15 to 6 km^3/year. In this case, the water level would fall below 338 m and the lake surface area would decrease to about 6,800 km^2. In this drastic scenario, the environmental disaster in the Balkhash Lake basin would be of a degree comparable to the Aral Lake.

Sustainable Development of the Ili-Balkhash Basin

Management Strategy During the Soviet Era

In the former Soviet Union, under the Ministry of Water Management (Minvodkhoz), distinct water management bodies (Bassejnovoe Vodnoje Ob'edinenie, BVO) were set up in each of the Central Asia republics. The BVOs coordinated and supervised the inter-republican utilization of waters, and administrated the water storage and diversion structures in the perspective river basins. In practice, these basin water management authorities worked out annual apportioning plans, which specified water-sharing between different agricultural and industrial water-users. The plans, in so doing, took into account the principles suggested in the annual master plans of the Federal State Planning Committee, the demands and perceptions of the regional administrations in the basin area, and periodic estimates on the basin water resources by the Central Asia Hydrometeorological Service. The water apportions indicated in the master plans of the BVOs would in turn receive the authorization of the federal Ministry. With a view to allowing readjustments in the apportioning schemes, which might be required by a variety of factors, the BVOs could modify the approved shares for different users by up to 10%. The BVOs, however, did not regulate the water quality in the basins, nor did they have the authority to impose the household sphere.

With respect to the Balkhash Lake basin, the system of unified water resources management has some advantages, one of which was good interaction between water-managing authorities and water-users. The major disadvantages were: (1) too strong centralization of the water-apportioning processes which often failed to take into account recommendations and demands of local administrations in the basin area, and (2) too strong focusing on fulfilment of the five-year master plans coming down from the Federal State Planning Committee. The Federal State Planning Committee provided five-year master plans for each of the economic sectors. This meant that each of the water-users in the Balkhash Lake basin received its individual five-year plans. Often, these plans did not consider the conflict of interests between different water-users. For example, hydropower production holds water over summer, when the irrigation agriculture in the lower reaches of the Ili river most requires it, releasing it in winter, with the resulting overflow that soaks the downstream irrigated areas in winter and spring. Hydropower appears to be in a greatly antipodean relationship with irrigation, menacing the agriculture in the lower reaches of the waters with drought in summer time, followed by floods in winter and spring. Being encouraged by five-year plans, both the hydropower and irrigated agriculture in the Balkhash Lake basin strived to increase their production. This often led to an uncontrolled increase in water consumption and to overexploitation of the Balkhash Lake resources.

Management Problems in Post-Soviet Kazakhstan

Unfortunately, the administrative system in the post-Soviet Republic of Kazakhstan remained highly ineffective for management of water resources in problematic regions such as the Balkhash Lake drainage basin. The system of administration in the region had a very complex structure with weak interaction between its elements. Realization of a programme for sustainable development of this region would require close coordination of efforts from many actors that are responsible for different compartments of the complex system. However, irrigated lands, pastures, the Kapchagay reservoir and the Kapchagay hydroelectric power station are managed by different ministries and responsible institutions. Sometimes, several difficult procedures are required to make any coordination between these authorities. Thus, the Kapchagay hydroelectric power station is managed by Almaty Power Consolidated Company (APCC), which does not have any relation to the management of the irrigation system in the Ili-Balkhash region. APCC is also not interested in sustainable modes of water release from the Kapchagay reservoir. Local farmers, being major consumers of the Ili's water, pay no fees for water consumption. In this way, farmers are not encouraged to introduce efficient technologies for water use. Water resources in the region are managed by several institutions: APCC manages the use of water for production of hydro-energy, agricultural producer cooperation groups manage irrigated areas in the region, etc. At least 15 acting institutions, central and local authorities are involved in administration and use of water resources in the Balkhash Lake basin. Most of the water resources users are only interested in obtaining short-term profit but not sustainable development of the Balkhash Lake basin. Corruption among responsible authorities also plays an important role in enabling mismanagement of water resources and disabling control of their use. Due to a lack of planning and management of multi-sector human activity, the resources of the Balkhash Lake basin are not used effectively and the region is characterized by economic stagnation, high levels of unemployment, poverty and population emigration.

Nonetheless, it is wrong to say that nothing has been done with respect to establishing strategies for sustainable development of the Balkhash Lake basin. In 2000, an international forum "Balkhash 2000" took place in Almaty. The forum, initiated by local authorities and supported by the government, adopted a resolution that acknowledged an extremely unstable ecological situation in the Balkhash Lake basin caused by irrational water use, decreased aridity of the region and other jeopardizing factors. The forum recommended developing a comprehensive programme of sustainable development of the Balkhash Lake basin (Ili-Balkhash Agenda 21) and resolved to create a Coordination Council on Sustainable Development (CCSD) for the region under the government of the Republic of Kazakhstan. The CCSD should involve local authorities, businesses, community, independent experts and international organizations. Agenda 21 developed a framework for analysing key problems of the Balkhash Lake basin, identifying the objectives of sustainable development and outlining the major trends. In the progress report of

2002, missing aspects in the existing management strategy and recommendations for improvement have been discussed (CA-REC 2002).

Over the past decade some large international projects and initiatives have been carried out with the aim of solving the Balkhash Lake basin problem. One of them, for example, was the project "Development of Ili-Balkhash basin integrated management plan" funded by the European Union and the government of the Republic of Kazakhstan. The project goal was "to develop an ecosystem-based management model in respect to the Ili-Balkhash basin for the purposes of ecosystem conservation, integrated water management and rational use of natural resources" (Kazakhstan TACIS/European Union Programme 2005). According to the reported project outcomes, "a management model and structure have been developed on the practices to ensure the sustainable regional development and expand the existing legislative and institutional frameworks of water resources management" (Kazakhstan TACIS/European Union Programme 2005).

In 2003, recognizing the need for administrative, public and water-user participation in water resources management decision-making, the government of Kazakhstan instituted River Basin Councils (RBCs). Prior to RBCs, no institution existed in Kazakhstan to coordinate stakeholders for water management. Establishing the RBCs was a totally new concept in Kazakhstan. Indeed, there are very few sectors of society in which there is significant public or stakeholder participation. However, the administrative system in post-Soviet Kazakhstan is known to be highly corrupt and ineffective for the fulfilment of all possible goodwill initiatives and purposes. Nowadays, moreover, sustainable development of the Balkhash Lake basin cannot be achieved without involving all related stakeholders. In the post-Soviet era, the major consumers are in the Chinese territory. Therefore, all strategies of sustainable development will fail if they do not consider transboundary issues of the Balkhash Lake basin problem.

Transboundary Issues of the Balkhash Lake Basin Management

Kazakhstan shares the drainage area of the major contributor to the Balkhash Lake, the Ili river, with China (Fig. 1). The Ili river, about 1,000 km in length, rises in the Chinese Tien-Shan and flows west, through Yining in Xinjiang, into Kazakhstan, where it terminates in Lake Balkhash. The Ili river basin includes several small streams, some shared by the two countries, such as the Tekes, which is an extension of the Ili. The total area of the Ili river basin is approximately 170,000 km^2, of which around 60% is Kazakh territory.

Since early 1999, some dissension has emerged between Kazakhstan and China over the utilization of transboundary water resources in the Ili basin. Planned measures by China on increased water intake from this river have caused alarm in Kazakhstan, which is anxious about consequences for its irrigated agriculture and environment in the lower riches of the Ili river. Reportedly, China is planning to reduce the outflow of the Ili into Kazakhstan by 15%. The diminished flow of the Ili is likely to upset the ecosystem in the Balkhash Lake basin, not to mention the undesirable effects on agriculture and energy production in the region.

The problem of shared water resources between the two countries is very complicated. The use of water resources in the Ili basin was regulated in some measure by a border protocol between the Russian Empire and China in 1915. This protocol established joint utilization in equal shares between the two countries. Unfortunately for Kazakhstan, the issue of the management of transboundary rivers was never addressed, and remained outstanding in the Sino-Soviet border relations until the end of the Soviet Union. Kazakhstan and China began talks on the status of the shared rivers in 1999. The talks lasted for two years and resulted in a bilateral agreement on transboundary waters. The Kazakh government did not present this agreement to the public community and did not respond to comments and recommendations from specialists in water resources management. However, according to information leaked from some government officials, the agreement is very unfavourable for Kazakhstan. Consequently, the threat remains to the Balkhash Lake from the territory of China.

Conclusions

A review of recent literature on the Balkhash Lake problem differentiates three phases in the development of Lake Balkhash during the last century. Each of these phases is characterized by a certain degree of anthropogenic impact. The degree of anthropogenic impact is strongly dependent on political development in the Balkhash Lake basin.

1. First phase (before 1969) could be characterized as almost natural. During this phase, the anthropogenic impact on the Balkhash Lake basin comprised moderate development of irrigation in the Kazakh part of the basin and practically no human activity on the part of the Chinese.
2. Second phase (1970–1991) was characterized by a strong increase in the anthropogenic impact associated with construction of the Kapchagay reservoir and the rapid extension of the irrigation area in the Kazakh part of the Balkhash Lake basin. By the end of the 1980s, the degradation of Lake Balkhash and its environments reached its highest degree.
3. Third phase (1991–early 2000s) started after the collapse of the Soviet Union and was associated with a rapid and strong decrease in the anthropogenic impact caused by the economical crisis in Kazakhstan. The water level of Lake Balkhash rose gradually during the third phase and reached 342 m. As a result, a wide rehabilitation of ecosystems in the whole region took place. However, parallel to positive change in the Balkhash Lake ecosystem, this period was characterized by a very depressing socio-economic situation in the region.
4. Recent economic and political development of the People's Republic of China has witnessed a great demand on the resources of the Balkhash Lake basin.

The threat of increase in water extraction from the Ili river by China is an increasing risk factor for the sustainable development of Lake Balkhash and the region as a whole.

The paper has illustrated that the strategies of water management in Lake Balkhash used during the Soviet era and after the collapse of the Soviet Union failed to stabilize the water level of the lake and to improve the ecological conditions in the region. In post-Soviet Kazakhstan, the development of market relations in all sectors of the economy led to the collapse of the centrally controlled water management system. The development of a revised version of water management system has started. This system should reflect the specifics of water use as that of the most important resource for human activity and existence. A number of international projects and initiatives that are focused on the development and testing of new principles of water resources management have been implemented. In the nearest future, preservation of the ecosystem in the basin and regulation of the Balkhash Lake hydrological regime will mostly depend on water consumption in the Chinese part of the Balkhash Lake basin. Taking into account this new reality, any successful management strategy for the Kazakh part of the Balkhash Lake basin can be established only in close cooperation with the Republic of China.

References

CA-REC (2002) 10-year progress report on Agenda 21 implementation in Kazakhstan, UNDP office in Kazakhstan, Almaty. Available at http://www.un.org/jsummit/html/prep_process/national_reports/kazakstan_natl_assess3008.pdf

Kazakhstan TACIS/European Union Programme (2005) Development of Ili-Balkhash basin integrated management plan: project results. Available at http://waterwiki.net/images/8/8e/CA-REC_ili-Balkhash_Result_Buklet_Eng.pdf (last assessed on 4 Oct 2010)

Kudrin RD, Rabinovich SA (1976) Forecast of change in the hydrological budget level and mineral content of Lake Balkhash in the next ten years. Soviet Hydrol 15:54–64

Matsuyama H, Kezer K (2009) Long-term variation of precipitation around Lake Balkhash in Central Asia from the end of the 19th century. SOLA 5:73–76

Nikolayev VA (1959) Delta of the Ili River and Bakanassy. Trudy Sektora Geographii AN KazSSR 4:35–54

Petr T (1992) Lake Balkhash. Kazakhstan Int J Salt Lake Res 1:21–46

Petr T, Mitrofanov VP (1998) The Impact on fish stocks of river regulation in central Asia and Kazakhstan. Lakes Reserv: Res Manag 3:143–164

Shnitnikov AV (1973) Water balance variability of lakes Aral, Balkhash, Issyk-Kul and Chany. In: Hydrology of lakes. Proceedings of the international symposium, IAHS-AISH Publication No. 109. Adlard & Son Ltd, Bartholomew Press, Dorking, Sudney, pp 130–140

Tlenbekov OK, Piven EN (1993) Anthropogenic change in runoff of the rivers in the Balkhash Lake basin. In: Tursunov AA (ed) Geographic problems of the Ili-Balkhash region. Almaty, Gylym, pp 54–62

Tursunov AA (2002) Ot Arala do Lobnor. Gidrologija besstotchnykh bassejnov Centralnoj Azii, Almaty

Chapter 29
Saline Irrigation Management for Sustainable Use

Medhat Mekhail Tawfik, M. A. Ahmed, Amany A. Bahr, M. F. El Karamany and M. S. Zeidan

Abstract The effect of climate change upon the distribution of plants and water is potentially severe, especially in arid and semi-arid regions of the world. Drought, attributable in significant part to climate change, is already causing acute water shortages in large parts of the world. Therefore, the utilization of marginal water and seawater is gaining considerable importance in such regions where freshwater is insufficient to develop all potential arable land. Currently, foliar-applied nutrients have limited direct use for the enhancement of stress-resistance mechanisms in plants. To achieve the aforementioned objectives, two pot experiments were conducted in the halophytic greenhouse of the National Research Centre, Dokki, Giza to study the effect of foliar application of potassium fertilizer (0, 20, 40, 60 ppm) on productivity, biochemical composition and some physiological aspects of *Sporobolus virginicus* (Dixi) plants grown under different levels of seawater irrigation (tap water, 12.5, 37.5, 25.0 and 50.0%). Increasing the saline irrigation level generally increased the content of soluble carbohydrates, proline, sodium, calcium and the value of succulence and osmotic potential (OP) as well as salinity tolerance index (STI), particularly under 50.0% seawater concentration. On the other hand, raising the level of seawater used for irrigation adversely affected the content of potassium as well as the K/Na and Ca/Na ratio. However, moderate concentration of seawater increased biomass production, crop growth rate (CGR) and the content of chlorophyll a + b and crude protein. Foliar application positively affected all the growth and physiological criteria, as well as the salinity tolerance of the tested plants. Foliar application with 60 ppm potassium surpasses the other treatments, especially at high levels of saline irrigation.

M. M. Tawfik (✉) · M. A. Ahmed · A. A. Bahr · M. F. El Karamany · M. S. Zeidan
Field Crop Research Department, National Research Centre,
El Bohooth Str., Dokki, Giza, Egypt
e-mail: medhatnrc@hotmail.com
URL: www.nrc.sci.eg

Keywords Seawater irrigation · *Sporobolus virginicus* (Dixi) · Foliar potassium fertilizer

Introduction

Salinity is currently one of the most severe abiotic factors limiting agricultural production. The high rates of population growth and global warming are expected to further exacerbate the threat of salinity, especially in areas with a semi-arid climate as in the Mediterranean region, where there is insufficient freshwater to develop all potential arable land. Therefore, the use of saline water in agriculture is a subject of vital importance for arid and semi-arid zones to meet the increasing food demand. The notion of using seawater to grow halophytic plants seems to be an ideal management practice when freshwater is not sufficient.

Saline agriculture is an alternative to allow the environment to select the crops, to match salt-tolerant plants with desirable characteristics to the available saline resources found in extensive areas of degraded and arid land in many developing countries.

Salt-tolerant plants (halophytes) are highly evolved and specialized organisms with well-adapted morphological, phonological and physiological characteristics allowing them to proliferate in high salinity conditions and offer a low-cost approach to reclaiming and rehabilitating saline habitats. This approach would lead to the domestication of wild, salt-tolerant plants for use as forage crops (David et al. 2007).

Sporobolus virginicus is a low-growing vigorous perennial grass. The only practical way to propagate it is by vegetative rhizomatous slips. It does very well as a dune stabilizer. It has potential for stream bank stabilization and also roadside slope stabilization. *S. virginicus* is well adapted to low rainfall and high salinity and is fairly widespread throughout the tropical regions and native to the Pacific.

A thorough understanding of the mechanisms conferring salt tolerance is therefore essential under the expected climatic change, as it will enable the selection of salt-tolerant genotypes and the adoption of appropriate practices to alleviate salinity impacts on agricultural production.

Currently, foliar-applied nutrients have limited direct use for the enhancement of stress-resistance mechanisms in field crops. Nevertheless, the interactions between plant nutrient levels and stress-repair mechanisms are now being studied (Lavon et al. 1999). They added that foliar application of potassium during vegetative growth is one of these precautions. Potassium is essential in the maintenance of osmotic potential and water uptake and has a positive impact on stomatal closure which increases tolerance to water stress. Moreover, it is involved in activating a wide range of enzyme systems which regulate photosynthesis, water use efficiency and movement, nitrogen uptake and protein building (Nguyen et al. 2002).

Therefore, this investigation was undertaken to evaluate the efficiency of foliar application of potassium to reduce the harmful effect of salt stress on biomass production, biochemical composition and salt tolerance of *S. virginicus* plants, and develop a management technique for the productive use of halophytes grown under high levels of seawater irrigation.

Materials and Methods

Two pot experiments were conducted in the halophytic greenhouse of the National Research Centre, Dokki, Giza during the two successive summer seasons of 2006 and 2007. Rhizomes of *S. virginicus* were transplanted on 9 and 13 May in the first and second seasons, respectively, in plastic pots 40 cm in diameter filled with a mixture of peat moss and sand (1:3). The mechanical analysis of the soil was 98.36% sand, 0.91% silt and 0.73% clay.

The experiment included 20 treatments which were the combination of four foliar treatments (tap water [control], 20, 40 and 60 ppm of potassium nitrate × five levels of seawater irrigation (tap water, 12.5, 25.0, 37.5 and 50.0% seawater concentration).

Foliar application was carried out 21 days after each cutting. Each pot was irrigated three times per week with the specified seawater concentration. The chemical analysis of the irrigation water is given in Table 1.

Three cuttings were taken at 42 days intervals to determine biomass production (g), total productivity of the three cuttings (g) and crop growth rate CGR = [(W2−W1)/(T2−T1) g/week], where W1 and W2 refer to dry weight of the whole plant at time T1 and T2 in week, respectively. Salt tolerance index was calculated as STI = [(TDW at Sx/TDW at S1) x100], where STI = salt tolerance index, TDW = total dry weight, S1 = control treatment, Sx = x treatment (Seydi et al. 2003).

The following physiochemical measurements were determined in the harvested shoot of the second cutting: chlorophyll a + b, proline, osmotic potential, succulence (ratio of fresh weight/dry weight), crude protein content, soluble carbohydrates content, sodium content, potassium content and calcium content according to AOAC (2000). K/Na and Ca/Na ratio were also calculated for each treatment. The obtained results were subjected to statistical analysis of variance according to Snedecor and Cochran (1982) and the combined analysis of the two seasons was calculated according to the method of Steel and Torrie (1980).

Results and Discussion

Effect of Diluted Seawater Irrigation on Dry Weight, Crop Growth Rate and Total Productivity

Data presented in Table 2 shows that increasing seawater concentration in the irrigation water to 12.5% increased dry weight, CGR and total productivity compared to tap water and there are no significant differences between them, while increasing seawater concentration in the irrigation water up to 25.00%

Table 1 Chemical analysis of diluted seawater irrigation (combined data of 2006 and 2007 seasons)

Characters	Tap water	12.50%	25.00%	37.50%	50.00%
pH	7.56	8.02	8.06	8.26	8.36
EC (ds/m)	0.88	10.36	12.36	18.25	23.58
Na (mg/l)	76.36	1,869.00	3,136.25	5,864.36	6,589.26
K (mg/l)	3.24	65.36	115.25	151.35	181.65
Cl (mg/l)	561.02	3,659.25	6,555.48	7,842.36	8,947.26
Ca (mg/l)	94.36	98.36	111.36	122.36	131.36
Mg (mg/l)	11.03	33.35	66.58	81.56	97.25

significantly increased dry weight and CGR in the three cuttings and consequently the total productivity. However, higher saline irrigation levels adversely affected the previous characters. Such stimulatory effect of moderate salinity on the growth of some halophytic plants may be attributed to improved shoot osmotic status as a result of increased ion uptake metabolism (Naidoo et al. 1995).

On the other hand, the reduction in growth and yield under high salinity levels could be due to a reduction in photosynthesis, disturbance in mineral uptake, protein synthesis or carbohydrate metabolism (Al-Garni 2006). He added that in most halophytic species, growth decreases gradually with the increase of salt rate in the culture medium above a critical threshold specific to each species. Similar results were obtained by Tawfik et al. (2009), who reported that low NaCl concentrations stimulate growth of some halophytic species.

Effect of Foliar Potassium Application on Dry Weight, Crop Growth Rate and Total Productivity

Data presented in Table 3 shows that all potassium foliar-spraying treatments significantly increased dry weight and CGR in the three cuttings and consequently the total productivity. Moreover, 60 ppm treatment surpassed the other treatments. In this concern, Michael et al. (2004) attributed such an enhancement effect of spraying plants with K on growth might be attributed to the favourable influence of this nutrient on metabolism and biological activity and its stimulating effect on photosynthetic pigments and enzyme activity which in turn encourage vegetative growth of plants. Our results are in agreement with those obtained by Thalooth et al. (2006).

Effect of Diluted Seawater Irrigation on Biochemical Composition and Some Physiological Aspects

Diluted seawater irrigation affects the studied parameters in different ways. Data presented in Table 4 shows that raising irrigation salinity levels up to 50%

Table 2 Effect of diluted seawater irrigation on some growth characters of *S. virginicus* (Dixi) (combined data of 2006 and 2007 seasons)

Seawater concentration	First cutting		Second cutting		Third cutting		Total productivity	
	Dry weight (g)	CGR	Dry weight (g)	CGR	Dry weight (g)	CGR	Dry weight (g)	CGR
Tap water	26.73	4.45	27.74	4.62	32.92	5.49	87.38	4.85
12.5%	30.50	5.08	31.05	5.18	37.09	6.18	98.63	5.48
25.00%	39.92	6.65	40.70	6.78	42.04	7.01	122.67	6.81
37.5%	23.60	3.93	24.49	4.08	29.07	4.84	77.16	4.29
50.00%	20.15	3.36	20.91	3.49	24.81	4.14	65.88	3.66
LSD 5%	1.95	0.39	1.71	0.43	1.78	0.44	4.42	0.45

Table 3 Effect of foliar potassium application on some growth characters of *S. virginicus* (Dixi) (combined data of 2006 and 2007 seasons)

Potassium concentration	First cutting		Second cutting		Third cutting		Total productivity	
	Dry weight (g)	CGR	Dry weight (g)	CGR	Dry weight (g)	CGR	Dry weight (g)	CGR
Tap water	26.87	4.48	28.28	4.71	31.78	5.30	86.93	4.83
20 ppm	28.07	4.68	28.50	4.75	33.49	5.58	90.05	5.00
40 ppm	28.44	4.74	28.80	4.80	32.86	5.48	90.10	5.01
60 ppm	29.34	4.89	30.34	5.06	34.61	5.77	94.29	5.24
LSD 5%	2.30	0.46	1.95	0.49	2.11	0.53	5.07	0.49

significantly increases the content of soluble carbohydrates, proline, calcium and sodium content, as well as succulence and osmotic potential values. On the other hand, the same treatment decreased the content of potassium and the values of STI as well as the ratio of K/Na and Ca/Na. However, moderate saline irrigation up to 25.0% generally increased chlorophyll a + b and crude protein content. Regarding the effect of diluted seawater irrigation on STI of *S. virginicus* plants, it is evident that irrigating the plants with high levels of seawater (37.5 and 50.0%) has the most deleterious effect on (STI). In other words, the deleterious effect of salinity on its productivity increases with the increase in the level of salinity. In this respect, Murphy et al. (2003) suggested that both proline and soluble carbohydrates act as compatible solutes under high salinity levels. Kusaka et al. (2005) added that the observed increase in the osmotic potential might be due to the accumulation of inorganic solutes, several organic components such as sucrose, glucose, quaternary ammonium compounds, and amino acids including proline. Furthermore, Flowers and Yeo (1986) stated that the most noticeable features of halophytes is the correlation between uptake of alkali ions and whole plant succulence with the aim of balancing out ion toxicity created in saline conditions by increasing the total plant water content. Bilquees et al. (2009) came to the same conclusion.

Table 4 Effect of diluted seawater irrigation on chemical constituents and some physiological aspects of *S. virginicus* (Dixi) (combined data of 2006 and 2007 seasons)

Seawater concentration	Chlorophyll a + b (mg/g dry weight)	Soluble carbohydrates (%)	Crude protein (%)	Proline (μg/g dry weight)	Potassium content (mg/g dry weight)	Sodium content (mg/g dry weight)	Calcium content (mg/g dry weight)	K/Na ratio	Ca/Na ratio	Succulence (fresh weight/dry weight)	Osmotic potential	Salinity tolerance index (STI)
Tap water	2.55	41.53	9.81	236.50	10.92	8.10	2.28	1.35	0.28	2.05	7.18	100.00
12.50%	2.67	43.71	10.62	271.42	10.31	9.85	3.00	1.05	0.30	2.27	8.23	112.88
25.00%	2.71	46.49	11.32	287.88	9.60	11.56	3.48	0.83	0.30	2.57	9.08	140.38
37.50%	2.42	47.66	9.62	330.80	9.13	12.12	3.87	0.76	0.32	2.74	10.38	88.30
50.00%	2.32	48.93	8.67	362.53	8.72	12.92	4.37	0.68	0.34	2.80	10.96	75.39
LSD 5%	0.14	2.44	0.59	13.53	0.52	0.62	0.31	0.04	0.01	0.12	0.62	4.92

The greatest accumulation of sodium by plants at high salt concentration may be attributed to the damage of the protoplasm of plant cells and as a result of the selective salt absorption being replaced by passive absorption, which causes abnormal accumulation of salts in plant organs (Kader and Lindberg 2005). They added that under saline conditions, sodium influx across the plasmalemma to the vacuole might play a major role in permitting turgor maintenance. He et al. (2005) added that the accumulation of sodium ions inside the vacuoles reduces the toxic levels of sodium in cytosol and increases the vacuolar osmotic potential with the concomitant generation of a more negative water potential that favours water uptake by the cell and better tissue water retention under high salinity levels. Similar results were obtained by Tawfik et al. (2009). On the other hand, the depressing effect of salinity on potassium could be attributed to the difficulty of its uptake due to competition with the high concentration of the sodium in the root medium.

Furthermore, Lacerda et al. (2005) reported that the greatest salinity tolerance observed in plants under saline conditions was associated a with lower Na/K ratio and greater capacity for osmotic adjustment. Lycoskoufis (2005) stated that the inhibition of photosynthesis under high salinity levels was predominantly due to reduced stomatal conductance.

The stimulating effect of moderate salinity on protein accumulation of some halophytic plants may be due to the increased synthesis of certain new sorts of proteins (Dubey and Rani 1989). On the other hand, the reduction in protein content under a high salinity level may be due to the disturbance in nitrogen metabolism, inhibition of nitrate absorption or the decrease of the availability of amino acids and denaturation of the enzymes involved in amino acid and protein synthesis (Mohamed et al. 1994).

Effect of Foliar Potassium Application on Biochemical Composition and Some Physiological Aspects

As for the foliar-spraying treatments, Table 5 shows that foliar-spraying with K generally increased the content of chlorophyll a + b, crude protein, potassium and calcium contents, as well as the ratio of K/Na and Ca/Na as compared with control plants, however 60 ppm treatment surpassed the other ones. In this concern, Rodriguez (2000) found that salt tolerance requires not only adaptation to sodium toxicity, but also the acquisition of potassium (an essential nutrient), whose uptake is affected by high external sodium concentration, due to the chemical similarity of the two ions. Therefore, potassium transport systems involving good selectivity of potassium over sodium can also be considered an important salt tolerance determinant.

On the other hand, the previous treatments decrease the content of soluble carbohydrates, proline and sodium content, as well as the values of succulence and osmotic potential. These results coincide with those obtained by Thalooth et al. (2006) and Youssef (2009). It is clear that all potassium foliar treatments improved

Table 5 Effect of foliar potassium application on chemical constituents and some physiological aspects of S. virginicus (Dixi) (combined data of 2006 and 2007 seasons)

Potassium foliar application	Chlorophyll a + b (mg/g dry weight)	Soluble carbohydrates (%)	Crude protein (%)	Proline (µg/g dry weight)	Potassium content (mg/g dry weight)	Sodium content (mg/g dry weight)	Calcium content (mg/g dry weight)	K/Na ratio	Ca/Na ratio	Succulence (fresh weight /dry weight)	Osmotic potential	Salinity tolerance index (STI)
Tap water	2.32	46.54	9.77	306.92	9.48	12.04	3.56	0.83	0.29	2.58	9.39	100.00
20 ppm	2.49	46.23	9.88	298.97	9.64	11.13	3.50	0.91	0.31	2.53	9.23	103.59
40 ppm	2.61	45.42	10.06	295.92	9.81	10.56	3.35	0.96	0.31	2.44	9.13	103.64
60 ppm	2.70	44.46	10.32	289.48	10.00	9.92	3.19	1.04	0.32	2.39	8.91	108.47
LSD 5%	0.16	2.87	0.69	15.95	0.61	0.73	0.36	0.04	0.01	0.15	0.73	5.80

all the tolerance features of *S. virginicus* plants and increased plant adaptation to saline irrigation. On the other hand, it can be noticed that all foliar-spraying treatments increase plant tolerance to salinity stress by increasing STI values.

Conclusions

Population growth and global warming will substantially impact the availability and quality of existing freshwater supplies. As a consequence, the risk of land salinization will further threaten agricultural production, particularly in areas with a semi-arid or arid climate. However, more detailed studies are needed to quantify the temporal and spatial effects of climate change on water resources. Such information is of paramount importance to adopt appropriate management practices to minimize the salinization of agricultural land and the impacts of salinity on crops productivity.

A study of salinity tolerance of plants irrigated with seawater in a quick-check system under greenhouse conditions is the first step for the development of cash crops. Increased research on the selection of halophytic species which have an economic utilization may enable the rehabilitation and re-vegetation of salt-affected lands provided that the appropriate soil and irrigation management is applied.

Foliar application with potassium positively affected all the growth and physiological criteria, as well as salt tolerance of the tested plants, especially at high levels of saline irrigation.

However, future investigations must be concentrated on the mobilization of halophyte world resources, creation of their gene pool, to multi-sided investigation of their ecology, biology, physiology, to perfect technologies of cultivation, harvesting and using halophytes in agriculture.

References

Al-Garni SMS (2006) Increasing NaCl-salt tolerance of a halophytic plant *Phragmites australis* by mycorrhizal symbiosis. American-Eurasian J Agric Environ Sci 1(2):119–126

AOAC (2000) Official method of analysis 12th association official analytical chemists. Washington

Bilquees G, Raziuddin A, Khan AM (2009) Salt tolerance of *Salicornia utahensis* from the great basin desert. Pak J Bot 41(6):2925–2932

David GM, Sharon EB, Hayley CN (2007) Biosaline agriculture for forage and livestock production. Agric Ecosyst Environ 119(3–4):234–248

Dubey RS, Rani M (1989) Influence of NaCl salinity on growth and metabolic status of proteins and amino acids in rice seedlings. J Agron Crop Sci 162:97

Flowers TJ, Yeo A (1986) Ion relations of plants under drought and salinity. Aus J Plant Physiol 13:75–91

He CX, Yan JQ, Shen GX, Fu LH, Holaday AS, Auld D, Blumwald E, Zhang H (2005) Expression of an arabidopsis vacuolar sodium/proton antiporter gene in cotton improves photosynthetic performance under salt conditions and increases fiber yield in the field. Plant Cell Physiol 46(11):1848–1854

Kader MA, Lindberg S (2005) Uptake of sodium in protoplasts of salt-sensitive and salt-tolerant cultivars of rice *Oryza sativa* L. determined by the fluorescent dye SBFI. J Exp Bot 56(422):3149–3158

Kusaka M, Ohta M, Fujimura T (2005) Contribution of inorganic components to osmotic adjustment and leaf folding for drought tolerance in pearl millet. Physiol Plant 125(4):474–489

Lacerda CF, Cambraia J, Oliva MA, Ruiz HA (2005) Changes in growth and in solute concentrations in sorghum leaves and roots during salt stress recovery. Environ Exp Bot 54(1):69–76

Lavon R, Salomon R, Goldschmidt EE (1999) Effect of potassium magnesium, and calcium deficiencies on nitrogen constituents and chloroplast components in citrus leaves. J Am Soc Hort Sci 124:158–162

Lycoskoufis IH, Savvas D, Mavrogianopoulos G (2005) Growth, gas exchange, and nutrient status in pepper (*Capsicum annuum* L.) grown in recirculating nutrient solution as affected by salinity imposed to half of the root system. Sci Hortic 106(2):147–161

Michael T, Walter T, Astrid W, Walter G, Dieter G, Maria SJ, Domingo M (2004) A survey of foliar mineral nutrient concentrations of *Pinus canariensis* at field plots in Tenerife. For Ecol Manage 189(1–3):49–55

Mohamed S, Sen DN, Mohamed S (1994) Seasonal variation in sugar and protein contents of halophytes in Indian desert. Ann Arid Zone 33(3):249–251

Murphy LR, Kinsey ST, Durako MJ (2003) Physiological effects of short-term salinity changes on *Ruppia maritima*. Aquat Bot 75(4):293–309

Naidoo Y, Jahnke J, von Willert DJ (1995) Gas exchange responses of the C4 grass *Sporobolus virginicus* (Poaceae) to salinity stress. Biology of Salt Tolerant Plants. University of Karachi, Karachi, pp 121–130

Nguyen HT, Nguyen AT, Lee BW, Schoenau J (2002) Effects of long-term fertilization for cassava production on soil nutrient availability as measured by ion exchange membrane probe and by corn and canola nutrient uptake. Korean J Crop Sci 47(2):108–115

Seydi AB, Hassan EK, Yilmaz ZA (2003) Determination of the salt tolerance of some barley genotypes and the characteristics affecting tolerance. Turk J Agric 27:253–260

Snedecor GW, Cochran WG (1982) Statistical methods, 7th edn. Iowa State Press, Iowa

Steel RGD, Torrie JH (1980) Principles and procedures of statistics. McCrow-Hill Book Co Inc., New York

Tawfik MM, Mohamed MH, El-Habbasha SF (2009) Optimizing management practices for increasing the efficiency of using seawater as alternating methods of irrigation bull. NRC (in press)

Thalooth AT, Tawfik MM, Mohamed MH (2006) A comparative study on the effect of foliar application of zinc potassium and magnesium on growth, yield and some chemical constituents of mungbean plants grown under water stress conditions. World J Agric Sci 2(1):37–46

Youssef MY (2009) Salt tolerance mechanisms in some halophytes from Saudi Arabia and Egypt. Res J Agric Biol Sci 5(3):191–206

Chapter 30
Modelling of Water Cycle Processes

Viktor Batyreu and Siarhei Zenchanka

Abstract World drinking-water stocks are limited and this means that water is a limiting factor that defines people's living conditions and ecosystems all over the world. It is clear that modelling the global water cycle is a complex task and can give an approximate estimation only. More exact estimations may be done for separate water cycle processes only when their coefficients and boundary conditions are well defined. The aim of this work is to develop the model of water use as a water cycle process and to define necessary and sufficient conditions of water-using regimes' stability. As a result of modelling, a common criterion of stability of water-using regimes is proposed. This criterion includes the necessary condition based on the radioactive balance criterion, and the sufficient condition that is defined by a pulse migration value. The considered approach permits to define the boundary of stable areas of water system processes and systems themselves. In the area of instability, a crash water issue arises that leads to the disturbance of an assimilated system capacity.

Keywords Modelling · Water cycle · Consumption · Pollution · Concentration · Necessary condition · Sufficient condition

---

V. Batyreu (✉) · S. Zenchanka
Minsk Branch of Moscow State University for Economics,
Statistics and Informatics, 127 Build. 2, Mayakovskogo Street,
220028 Minsk, Belarus
e-mail: batyrev1945@mail.ru

Introduction

The global water cycle includes water in all physical states (solid state—ice, liquid state—water, gaseous state—water vapour) and considers its spreading in different physical media (solid—soil, liquid—water streams with different physical properties, for example, the Gulf Stream, the El Niño phenomenon in the Pacific Ocean, gas—the atmosphere). Accordingly, different physical states and physical properties have different rates of circulations. The water residence time in different natural objects differs significantly and can change from 1 week for water in biosphere objects to 10,000 years in soil (Vital Water Graphics 2002).

The estimations show that clean water in its different states constitutes 2.5% of total water quantity and available drinking water equals 30% of this quantity.

The limit of drinking-water stocks makes water a limiting factor which defines the living conditions for people and the ecosystems on all the Earth's continents. Lately, the constantly increasing human impact on the environment has led to a significant change in water-use and significant human impact on water systems. Problems connected with the stability and sustainability of such systems are arising. It is clear that modelling of the global water cycle is a complicated problem and can give only approximate estimations. More exact estimations can be made for a separate process of a water cycle with well-defined coefficients.

Milićević et al. (2010) pointed to the fact that "the mathematical models become increasingly important for implementation of Water Framework Directive, particularly in terms of pollution control and management of water resources quality in river basin areas".

Different types of water modelling have been suggested. Among them are the hydrological model suggested by Chang et al. (2009) and the object-oriented model based on the concept of system dynamics suggested by Elshorbagy and Ormsbee (2005).

Saysel (2007) and Fiksel (2006) underlined the complexity, dynamics and nonlinear nature of environmental systems and showed that in systems with multiple feedback loops, it is complicated to predict further system development.

The aim of this paper is to describe the model of water consumption as a process of a water cycle and to define essential and sufficient conditions of water-using regimes.

Terms and Definitions

Any process occurring in nature should be considered in terms of its stability and sustainability.

Consider these terms:

- Stability—quality of being stable (Hornby 1974).

- Stable—having or showing an ability or tendency to maintain, or resist change in, position or form (Surjeet 1988).
- Sustain—keep from falling or sinking (Hornby 1974).

These terms are synonymous but they have some differences.

Fiksel (2006) pointed out that the commonly used notion of "sustainability" as a steady-state equilibrium is not realistic and that "achieving sustainability will arguably require the development of resilient, adaptive industrial and societal systems that mirror the dynamic attributes of ecological systems".

Resilience can be defined as the capacity of a system to tolerate disturbances while retaining its structure and function (Fiksel 2006).

In this work, *stability* is considered as the ability of the system to function without change to its intrinsic structure and to be in a state of equilibrium with the environment.

Sustainability is considered as the ability of the system to return to its initial state after exposure to external impacts. This definition is close to resilience.

For a more exact description of an ecosystem's response to external impacts, it is appropriate to use such characteristics as elasticity and plasticity.

Elasticity is the quality of being elastic, i.e. having the tendency to go back to the normal or previous size or shape after being pulled or pressed (Hornby 1974). At the same time, certain impacts that exceed the threshold value of such a system mean that it is typically destroyed or it transfers to a new quality.

Plasticity is the ability of the system to change its structure under external loading. When loading cuts off the system, it returns to a state close to the initial one.

Concepts such as "elasticity" and "plasticity" are closely connected with such concepts as "potential capacity of the system" ("assimilated system capacity") and "critical load".

Potential capacity of the system is the maximum allowable level of flow into the system during a certain period of time.

Critical load is a maximum allowable level of outflow during continuous time.

These definitions show the importance of the time of the impact on the system, hence all processes must be considered from the point of view of the own system time.

The intrinsic system time is the time that is considered within the period of the system existing or its processes. The intrinsic system time has a different scale, as well as a system dimension.

A Common Model of the Water Cycle Process

The model of water consumption is based on the dynamic model presented in Fig. 1.

Fig. 1 Dynamic model of estimation of environment quality

Model components:

1. The law of change of concentration with time:
$$C(t) = C0\exp(-kt); \quad C0 = \text{const}; \quad C0\exp(kt). \quad (1)$$

2. The impact of a pollutant on the system (dose) is defined as an integral of the change of concentration:
$$D(t) = D0(1 - \exp(-kt)); \quad D0(kt); \quad D0(\exp(kt) - 1); D0 = C0/k. \quad (2)$$

3. The probability of system lethal outcome (system destroying) as a function of the dose:
$$P(D) = P0\exp(-rD); \quad P0 = \text{const}; \quad P0\exp(rD). \quad (3)$$

4. The risk of lethal outcome (system destroying) that is defined as an integral from the law of variation of the probability of the system lethal outcome as a function dose value:
$$R(D) = R0(1 - \exp(-rD)); \quad R0\,rD; \quad R0(\exp(rD) - 1). \quad (4)$$

D0   initial value of concentration;
C0   initial dose;
P0   initial probability (corresponding initial concentration C0);
R0   initial values of the risk of the lethal outcome;
k   process constant (rate);
r   rate of dose absorption.

Scheme of Water Consumption

Water consumption is an essential part of a global water circle and affects the condition of water resources.

The typical scheme of water consumption is shown in Fig. 2.

The "Diluting" block may be both in the direct chain and be used as a feedback when dumping is realized in a water object near the source.

Consumption regimes can be described as follows:

1. Water economy in the process of production of food products, consumer goods, etc.:
$$C_1 = C_0 e^{-k_1 t};$$

Fig. 2 Scheme of water consumption

2. Water consumption by living organisms:

$$C_1 = C_0;$$

3. Emergency dumping of water from the reservoir, dam breaking, snow avalanche, etc.

$$C_1 = C_0 e^{k_1 t}.$$

C_i, k_i—water concentrations and process rates in the appropriate chain links of water consumption.

The generalized criteria of sustainability of the water consumption regime are suggested, including:

1. the essential condition that was obtained analogously to the criterion for radioactive equilibrium and
2. the sufficient condition determined by the value of the migration pulse $D = \int C(t)dt$.

Migration pulse is a common concept describing a process of transport of energy, matter, etc. from one object to another during a defined time interval.

Such an approach permits us to define the limits of sustainability of the process occurring in water systems and systems in general. In the area of unsustainability, emergency dumping takes place that leads to the disturbance of the potential capacity of the system.

It is possible to consider the system as consisting of N serial links (a finite Markov chain) (Howard 1971):

$$C_1 \rightarrow C_2; \; C_1 \rightarrow C_2 \rightarrow C_3; \; C_1 \rightarrow C_2 \rightarrow \cdots \rightarrow C_N.$$

The value of N for the ecosystems including water systems can be equal to 2–4, for physical and chemical processes—10 and greater, but N is finite (N—number of links in a chain, n—a link number in a chain).

Value C_n is easily calculated at any link of the chain and for any value of N (Kozlov 1991). The obtained formulas for N are very bulky and it is difficult to analyse them. Besides, it is necessary to know the exact ratio between the process rates in different links k_n.

The essential condition of sustainable equilibrium defined as $\frac{C_n}{C_{n-2}} = \text{const}$, when $t \rightarrow \infty$ actually defines what part of the concentration in the previous link of the chain passes to the next link and influences its existence.

The sufficient condition (migration pulse) when taking into account the value of the intrinsic time of the system combines the concepts of maximum loading and the potential capacity and defines actually the time of the system's existence in a sustainable state.

Let us consider several examples within the frame of the model shown in Fig. 2.

Linear Regime of Water Consumption

As a linear regime of water consumption, consider the regime when pollutant concentration and water volume are transferred along serial links of the chain and dumping occurs at large distance from the source.

The Two-Element Chain

Consider the two-element chain $C1 \rightarrow C2$ (Fig. 3):

The source C_1 is described by $C_1 = C_0 * e^{-kt}$. In this chain, three regimes are possible: (1) $k < 0$, (2) $k = 0$, (3) $k > 0$.

For the first regime:

$$\frac{C_2}{C_1} \rightarrow \frac{k_1}{k} * e^{kt}; \quad D_2 \rightarrow \frac{k_1}{k} * C_0 * t.$$

For the second regime:

$$\frac{C_2}{C_1} = k_1 * t; \quad D_2 = k_1 C_0 \frac{t^2}{2}.$$

For the third regime:

$$\frac{C_2}{C_1} = \frac{k_1}{k}\left(1 - e^{-kt}\right); \quad D_2 = C_0 * \frac{k_2}{k^2} * e^{kt}.$$

The Three-Element Chain

For such a chain (Fig. 4), the expressions for essential and sufficient conditions are:

When $k_2 < k$:

$$\frac{C_2}{C_1} \cong \frac{k_1}{k_2 - k}\left(1 - e^{-kt}\right); \quad D_2 \cong C_0 * \frac{k_1}{k_2 - k}.$$

Fig. 3 Model of the two-element chain

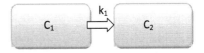

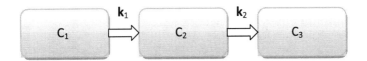

Fig. 4 Model of the three element chain

When $k_2 = k$:

$$\frac{C_2}{C_1} \cong kt = k_2 t, \quad D_2 \cong \frac{k_1}{k^2}.$$

When $k_2 > k$:

$$\frac{C_2}{C_1} \cong \frac{k_1}{k_2 + k}\left(e^{(k_2+k)t} - 1\right); \quad D_2 \sim e^{kt}.$$

The Four-Element Chain

The four-element chain is shown in Fig. 5.
The essential condition:

$$\frac{C_3}{C_2} \cong \frac{k_2}{k_3 - k}.$$

The sufficient condition:

$$D_3 \cong C_0 * \frac{k_2}{k_3 - k}.$$

The Analysis of the Results for Linear Chains

The analysis of the obtained expressions shows that the system consisting of two links (Fig. 3) is unsustainable for any regimes. Adding of the third link (Fig. 4) increases the sustainability of the system for several regimes. When $N \geq 3 \geq 3$ (Fig. 5), the dynamic regime is sustainable in accordance with the first and second conditions. For the chain consisting of N links we can obtain:

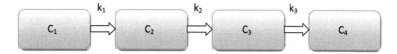

Fig. 5 Model of the four element chain

$$\frac{C_n}{C_{n-1}} \cong \frac{k_{n-1}}{k_n - k}. \quad D_3 \cong C_0 * \frac{k_{n-1}}{k_n - k}.$$

These expressions show that the increase in the number of links leads to system sustainability.

Regime of Water Consumption with Feedback

Consider the influence of feedback on the behaviour of the systems described above.

The Two-Element Chain with Feedback

Let us introduce feedback (k_0) (Fig. 6).
Such a chain is described by a set of equations:

$$\frac{dC_1}{dt} = -k_1 C_1 + k_0 C_2;$$

$$\frac{dC_2}{dt} = k_1 C_1 - k_0 C_2.$$

The essential condition for the system existence is:

$$\frac{C_2}{C_1} \sim \frac{A_3}{A_1}$$

and the sufficient condition is:

$$D_1 \sim A_1 t; \quad D_2 \sim A_3 t;$$

where A_1, A_3 are some numerical coefficients.

The Three-Element Chain with Feedback

Let us add the third element to the system described above and consider stability and sustainability conditions for the existence of such a system (Fig. 7).
Consider the case with feedback between elements C_2 and C_1. Such a chain is described by the set of equations:

Fig. 6 Model of the two element chain with feedback

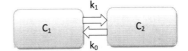

Fig. 7 Model of three-element chain with feedback

$$\frac{dC_1}{dt} = -k_1 C_1 + k_{01} C_2 - k_2 C_2;$$

$$\frac{dC_2}{dt} = k_1 C_1 - k_{01} C_2 - k_2 C_2;$$

$$\frac{dC_3}{dt} = k_2 C_2.$$

When $0 < \lambda_1 < \lambda_2$ (roots of the system determinant), the solution of this system gives the following conditions for stability and sustainability:

$$\frac{C_2}{C_1} \sim \frac{A_3}{A_1}; \quad D_1 \sim \frac{A_1}{\lambda_1} + \frac{A_2}{\lambda_2}; \quad D_2 \sim \frac{A_3}{\lambda_1} + \frac{A_2}{\lambda_2},$$

where A_1, A_2, A_3 are some numerical coefficients.

When $\lambda_1 = \lambda_2 = \lambda$:

$$\frac{C_2}{C_1} = \frac{A_3 + A_4}{A_1 + A_2}; \quad D_1 \sim \frac{A_1 + A_2}{\lambda}; \quad D_2 \sim \frac{A_3 + A_4}{\lambda}.$$

When λ_1 and λ_2 are imaginary, the solution is an oscillating process.

The Four-Element Chain with Feedback

For the four-element chain, consider the case with feedback between link C_2 and C_3 (Fig. 8).

The system behaviour is completely defined by the source regime, i.e. conditions $k < 0$, $k = 0$ or $k > 0$. The results are shown in Table 1. D is a determinant of the set of equations.

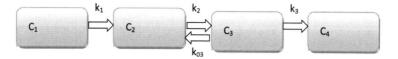

Fig. 8 Model of the four-element chain with feedback

Table 1 Solutions for the four-element chain

	>0D > 0		=0D = 0	
	$\dfrac{C_3}{C_2}$	D_i	$\dfrac{C_3}{C_2}$	D_i
k < 0	Constant	Constant	Constant	~t
k = 0	Constant	~t	Constant	~t
k > 0	Constant	e^{kt}	Constant	e^{kt}

The Analysis of the Results for Chains with Feedback

The consideration of the results obtained for the systems with feedback and without it shows that the implementation of feedback significantly influences the system behaviour. In many cases, stabilization of the system functioning takes place. At the same time, there are critical areas where the system collapses or its transition to another state occurs during the time.

Conclusion

An analysis of the functioning of multi-element systems without feedback (linear) and with feedback was made. It was shown that any system has its intrinsic lifetime. At the same time, there are regimes of the system functioning, the lifetimes of which are defined by the lifetime of the source only.

As the water cycle in nature and in industrial systems of water consumption includes elements with the direct link and feedback, the considered approach can be applied to such systems. In many cases, the optimization of the functioning of industrial systems can be reached by means of process rate optimization. In the cases where natural process rates cannot be optimized, it is necessary to take technical and preventive measures to limit them.

References

Chang G, Lowry K, Nemany R, Skiles JW, Schmidt C (2009) Modeling current and future water use in Utan with NASA's terrestrial observation and prediction system. ASPRS Annual Conference. www.asprs.org/publications/proceedings/baltimore09/0108.pdf. Accessed 30 Sept 2010

Elshorbagy A, Ormsbee L (2005) Object-oriented modeling approach to surface water quality management. Environ Model Softw 21:689–698. http://www.hydropyramids.com/files/Elshorbagy-Article.pdf. Accessed 30 Sept 2010

Fiksel J (2006) Sustainability and resilience: toward a systems approach. Sustain Sci Pract Policy (SSPP) 2(2):14–21. http://ejournal.nbii.org/about/about.html. Accessed 30 Sept 2010

Howard RA (1971) Dynamic probabilistic systems, volume 1: Markov models. Wiley, USA

Hornby AS (ed) (1974) Oxford advanced learner's dictionary of current English. Oxford University Press, Oxford, p 854

Kozlov VF (1991) Handbook on radiation safety. Energoatomizdat, Moscow (in Russian)

Milićević D, Milenković S, Potić O (2010) Water quality modeling role in implementation of the Water Frame Directive. Facta Universitatis Ser Archit Civ Eng 8(2):247–260

Saysel AK (2007) System dynamics: systemic feedback modeling for water resources management. International River Basin Management Congress Book, Chapter III. http://www.dsi.gov.tr/english/congress2007/index.htm. Accessed 30 Sept 2010

Surjeet (1988) New Webster dictionary of the English language. Surjeet Publications, Delhi, India, p 1482

Vital Water Graphics (2002) http:/unep.org/dewa/assessments/ecosystems/water/vitalwater/freshwater.htm. Accessed 30 Sept 2010

Chapter 31
Bridging the Communication Gap: An Exploration of the Climate Science–Water Management Interface

Katinka Lund Waagsaether and Gina Ziervogel

Abstract Exploration of the science–management interface in the Western Cape, South Africa, in terms of the communication of climate change-related data to water resource managers (WRMs), has revealed how continuous communication between WRMs and climate scientists is necessary in order to help narrow the gap between the two camps. This is firstly to meet the challenge of establishing an outline of the information that WRMs need, as well as developing a framework for how the climate change information can be incorporated into planning and procedure. Secondly, interaction between the two camps is necessary in order to establish comprehensible data presentations that account for model uncertainty and the uncertainty related to different future scenarios, while at the same time offering useful guidance for planning and adaptation strategies. Thirdly, continuous interaction is necessary in order to establish a new form of management required to deal with the new uncertainty and element of risk associated with climate change and increased climate variability: a proactive and resilient form of management based on "learning to manage by managing to learn". A Climate Change Forum can provide the required platform for communication, further allowing for co-production of knowledge, ensuring the production of comprehensible, relevant and applicable climate data.

Keywords Climate change · Water resources management · Climate science communication

---

K. L. Waagsaether (✉) · G. Ziervogel
Climate System Analysis Group, Department of Environmental and Geographical Science, University of Cape Town, Private Bag X3, Rondebosch, 7701, South Africa
e-mail: katinka@csag.uct.ac.za

Introduction

In responding to climate change projections, challenges face the water sector in terms of addressing the gap between science and resource management. The development of planning tools, management strategies and general capacity must build on sound science that allows for effective climate change adaptation on the ground. This means that climate knowledge must be made available and comprehensible to non-scientists, and that scientists and managers must collaborate to ensure that relevant information is communicated to stakeholders.

Based on current water use and population growth, South Africa is projected to run out of useable water by 2050. The pressure on the water resources management sector in South Africa is therefore large, and projections of rapid changes and increased variability in the global climate are raising further concern for the water sector. Increased variability in the climate tends to exacerbate and amplify the oscillations in the hydrological cycle, and is expected to increase vulnerability (Schulze 2000). An exploration of the interface between scientists and management—exemplified by the communication of climate change-related data to WRMs in the Western Cape province of South Africa—has revealed three challenges. The first is to establish an outline of the information that WRMs need, which requires climate scientists, together with WRMs, to build an understanding of how the field will be affected by climate change, as well as developing a framework for how the climate change information can be incorporated into planning and procedure. The second challenge is to improve the interaction between the two camps—water management and science—to establish easily comprehensible climate data presentations that account for model uncertainty and the uncertainty related to different future scenarios, while at the same time offering useful guidance for planning and adaptation strategies in the water sector. The third challenge is to ensure continuous interaction between WRMs and climate scientists in order to establish a new form of management that is required to deal with the uncertainty and element of risk associated with climate change and increased climate variability: a proactive and resilient form of management based on "learning to manage by managing to learn".

The Gaps in the Climate Science–Water Management interface

> The man who has the time, the discrimination, and the sagacity to collect and comprehend the principal facts and the man who must act upon them must draw near to one another and feel that they are engaged in a common enterprise (Woodrow Wilson, 1856–1924).

The need for collaboration and information sharing between the knowledge producers—the scientists—and the practitioners that manage human and ecological systems, who are the end-users of the knowledge, is just as urgent as it was 100 years ago. As argued by Jury and Vaux Jr (2005) in a study on science and water resources management, relevant and significant scientific information exists, but due to the lack of effective communication and outreach programmes, it is not being translated into practice. A range of literature has recognized the need for improving communication between WRMs and climate scientists in order to tackle the challenges faced in water management under projected climate change scenarios (Power et al. 2005; Kashyap 2004; Thomalla et al. 2006; Tompkins and Adger 2004; Roux et al. 2006; One World Sustainable Investments 2008).

In a study on the divide between science and management, Roux et al. (2006) explore how to improve the communication between "experts" and "users". They argue that in order to address the cultural disparities between resource managers, policy-makers and researchers, there needs to be a co-production of knowledge, implying that the users, in this context the WRMs, need to be involved in the knowledge production process. Hence there is a need for continuous interaction between the WRMs and the climate scientists, to ensure that the WRMs are in the end faced with a product which is applicable to their needs and of which they can feel ownership (Roux et al. 2006). This argument is supported by a study in Western Australia where continuous communication between knowledge producers and practitioners was found to bridge some of the cultural disparities between climate scientists and WRMs, and also to guide researchers to directly serve the needs of practitioners and the wider society (Power et al. 2005). But, as was pointed out by Roux et al. (2006), WRMs have to attend to day-to-day issues, and so taking the time to sit down and discuss issues that may be many years ahead is not necessarily a priority. This is especially the case in a less developed country like South Africa. As Schulze (2007: 274–275) points out: while developed countries can focus "on quality of life and environment as well as long-term issues", less developed countries have more urgent issues such as "'harnessing' the local environment (versus sustaining it)" and "providing basic potable water supplies to households (versus providing water of the highest quality)."

A scoping paper for International Dialog on Water and Climate, an international platform that bridges the information gaps between the water and climate sectors, suggests that one of the greatest barriers to the uptake of climate-related scientific data by water managers is its complexity (Kabat et al. 2003). Supporting this, Roux et al. (2006) argues that scientific information can only be useful if it is not too complex and if it is presented as unambiguous, hence indicating that presentation methods must be simple and provide a clear, unmistakable message. Uncertainty is an undeniable element of climate projections, both through the emission scenarios on which they are based and through inherent climate model uncertainty. This makes it challenging to communicate a sound and robust message that accounts for these uncertainties, while at the same time providing a simple, clear and usable message.

When investigating the potential impact of future climate change, it is more difficult to decide upon an adaptation strategy if drawing on several scenarios (New et al. 2007). Hence, while presenting climate change-related data using several future scenarios might give a good reflection of the range of possible futures, this could make it harder for a WRM to understand and to act upon it. It is further found to be difficult to make management decisions when the distribution, as well as the magnitude, of possible impacts is unclear due to the range of model predictions (Tompkins and Adger 2004). But as one study of the Sacramento–San Joaquin river basin argues, it is important to focus on quantifying uncertainties in predictions in order to ensure that these uncertainties are taken into account in decision-making (Vanrheenen et al. 2004). This reflects the challenge faced in providing comprehensible data presentations that account for model uncertainty and the uncertainty related to different future scenarios, while at the same time offering useful guidance for planning and adaptation strategies.

In order for scientific information to be useful it must not only be comprehensible, but also well matched with the planning models in use (Roux et al. 2006). This implies that practitioners such as WRMs need access to information that is relevant to their objectives and management paradigm. A water quality manager is likely to require different information from a groundwater manager, and so information must somehow be tailored to fit the specific user. While great attention has been given to the importance of climate science communication, little research has focused more specifically on issues such as outlining the data presentation methods best understood by practitioners or on customizing packages of data relevant to the water management sector.

The purpose of this research has been to address the science–management gap by exploring the communication of climate change-related data to WRMs in the Western Cape, South Africa. More explicitly, it has focused on finding comprehensible data presentation methods, on recognizing data relevant to WRMs and on assessing the challenges of communicating climate science to stakeholders in the water resources management sector.

Climate Science Communication in the Western Cape, South Africa

The Western Cape, South Africa, is located at the southwestern tip of the African continent. The province holds approximately 4.2 million inhabitations, out of which 3.4 million live in the City of Cape Town. Every year, the city alone consumes about 300 million m^3 of water, and if left unconstrained, water demand is projected to hit 1,000 million m^3 of water per annum by 2037 (Amanzi Obom Consulting cc 2008). Water resources management is hence an important sector in the province, and the Western Cape's Climate Change Strategy and Action Plan (2008: 59) recognizes that in the context of water supply "an ongoing platform is required that facilitates dialog between climate scientists and government." In September 2009, a climate change communication workshop was organized by the Climate System Analysis Group (CSAG) from the University of Cape Town, in collaboration with the Department of Water and Forestry (DWAF) in the Western Cape, bringing together 20 WRMs and a team of climate scientists from CSAG.

The workshop provided a starting point for communication between WRMs and CSAG scientists, while serving as a setting for the exploration of the climate science–water management interface. Interactive engagement between the scientists and WRMs was encouraged through facilitated discussion and group work, providing an insight into how climate change science and communication is perceived in the water resources management community. The CSAG winter school, a two-week intensive course on the interpretation and use of climate information, provided another setting for investigating practitioners' understanding of climate science. The 17 professionals attending the course were presented with a survey that assessed their understanding of climate change data displays, creating an understanding of some of the barriers to communicating climate science caused by current data presentation methods.

Contextualizing Challenges in the Communication of Climate Change Data

The climate change communication workshop brought about some understanding of how sectors of water resources management might become, and possibly already are, affected by climate change. There seemed to be a general consensus on an observed increase in intensity and frequency of heavy rainfall events, and a corresponding increase in the intensity and frequency of flooding. The Western Cape's Climate Change Strategy and Action Plan (2008) reports on projections with complex regional patterns, with the greatest concern being drying in the western parts of the province, away from the mountains. The plan also reports projections of small increases in rainfall in some areas, and further states that cut-off lows, which tend to cause intense rain and flooding, are likely to become more frequent. Hence, even if the perceived increase in extreme rainfall events can not be backed by current station data, it is likely to become a relevant issue. While several implications of increased intensity and frequency of heavy rainfall events brought up by the WRMs have been summarized broadly in Fig. 1, it is important to note that the implications of drying are likely to be greater and might thus require more extensive responses in the Western Cape. The WRMs' lack of attention to decreasing rainfall and droughts reflects the need for communication of climate change projections to WRMs, as strategies forming as responses to experiences on the ground may not address issues that are potentially more severe in the long-term.

Based on their experiences, WRMs emphasized how infrastructure, such as sewage works, dams and private and public property, tends to be constructed based on historical flood lines, and as one respondent stated, "80% of wastewater works in Western Cape lie below the 1 in 20 year flood line." Experiences among the

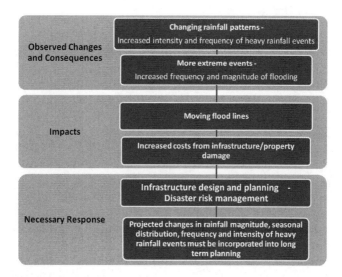

Fig. 1 Potential implications of changing rainfall patterns

WRMs showed that the 1/20 year flood lines are now being exceeded more often than every 20 years, with one sewage work experiencing flooding three years in a row. This led the WRMs to recognize the need to plan better, to analyze the capacity of infrastructure and to design better responses to extreme events. They further suggested that sewage works and other infrastructure might need to be constructed out of the 1 in 100 year flood line to avoid damage. While not aware of any specific action currently taking place in DWAF in response to changes in flood patterns, the WRMs highlighted action taken by farmers. In George, on the east coast of South Africa, the farmers were asking for permission to raise their dam walls as a response to increased frequency of flash floods, reflecting that an individual response to change is taking place faster than institutional response.

While farmers can act on experience, institutions may require more and better information in order to respond. In the case of changing rainfall patterns, climate change projections can provide a guideline for planning for a future of change, ensuring that changes that are not yet apparent on the ground are also taken into account in planning and procedures. Projected changes in rainfall magnitude, seasonal distribution, as well as frequency and intensity of heavy rainfall events could aid the long-term management of water resources. In order to assist WRMs in planning for infrastructure that can deal with changes in the frequency and intensity of heavy rainfall events, it might be necessary for climate scientists to look at whether it is possible to present the likelihood of reaching certain thresholds into the future, such as the current 1 in 100 year flood line. For magnitude and seasonal distribution, there should be a focus on presenting the range of projections from different models, indicating the direction of change. This can be related to one of the challenges with communicating climate change projections, namely model uncertainty. To provide a relatively robust message of change, it is important to present a number of model predictions that can indicate a direction of change rather than specific magnitudes of rainfall or exact seasonal shifts.

In order to incorporate rainfall projections into both short- and long-term planning, the WRMs indicated the need for projections for the next 6–10 years, as well as for the next 20, 30 and 40 years. While seasonal forecasts are available for three-month periods, projections of climatic change tend to be for the middle of this century and onwards. There is thus a gap between the information desired by the WRMs and that which is currently available from climate scientists.

The Search for Relevant Information

When asked to outline the three types of climate information they would like to see in an ideal world, WRMs at the climate change communication workshop came up with a relatively short list of variables that only included temperature, rainfall, evapotranspiration, evaporation, and combined evapotranspiration and rainfall. The list was even shorter when it came to the format and the temporal scale of the information, outlining only spatial coverage format and monthly scale. While giving an indication of the variables that the WRMs, with their current knowledge and understanding, recognize as relevant to their field of work, it further highlights the challenge of establishing a specific outline of climate change-related information relevant to WRMs. This can be related to the fact that WRMs are currently using historical information from instrumental records, and hence do not have frameworks for assessing future information from climate models. Water managers have been exposed to seasonal climate forecasts over the last decade and have used this seasonal information to support some decisions such as managing water restrictions and planning dam releases (Ziervogel et al. 2010). However, in the City of Cape Town, water managers have not had good access to detailed climate change scenarios yet, so have not explored ways to use this information. Furthermore, they are not yet sure of how exactly climate change will impact their sector and, partly because of this, they are unsure of how climate change-related data can be incorporated into planning and procedures. Therefore, it is not simply about the WRMs saying what they need and the climate scientists looking at whether they can provide it or not. The WRMs have to engage with climate change information in order to understand how climate change could impact their sector. To establish a specific outline of the information that WRMs require, one needs to build an understanding of how their specific field will be affected by climate change, as well as a framework for how the climate change information can be incorporated into planning and procedure. As climate scientists tend to be most concerned with understanding the climate system and providing climate change projections, they will in general have a limited understanding of how projected climate changes impact systems on the ground, such sewage works, dams and property.

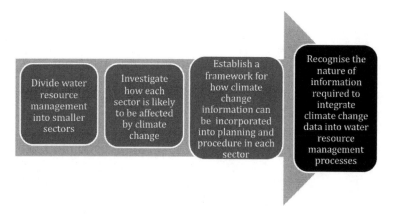

Fig. 2 Process of recognizing relevant data

Dialog between climate scientists and WRMs is thus required in order to outline how the different sectors of water resources management are likely to be affected. Various literature supports the argument of this research: that continuous communication between WRMs and climate scientists is necessary in order for climate change information to be incorporated into water resources management (Power et al. 2005; Kashyap 2004; Thomalla et al. 2006; Tompkins and Adger 2004; Roux et al. 2006; OneWorld Sustainable Investments 2008). This continuous dialog can allow for the establishment of a framework for how climate change information can be incorporated into planning and procedures, as well as the recognition of the nature of climate change information desired by WRMs. Figure 2 outlines a process by which WRMs and climate scientists together can establish the climate data relevant to the water management sector.

The Challenge of Providing Comprehensible Yet Robust Climate Data

The survey assessing the understanding of climate data displays, presented to the participants in the CSAG winter school, showed a number of graphs with related questions. The results of the survey highlighted some of the basic challenges experienced by non-climate scientists when faced with climate change data presentation methods. So did an exercise at the climate change communication workshop, where the WRMs were shown four posters with a number of presentation methods which they were asked to observe, discuss and comment on. From the poster exercise at the climate change communications workshop it was found that the WRMs best understand:

- model average over model range
- spatial display over column chart

- monthly over annual display
- actual values over difference or percentage change

By analysing the responses from the survey and the poster exercise, it was found that clearly labeled and well-explained graphs with relatively few lines and a clear message were best understood. It also became clear that respondents struggle with interpreting color ranges and working with different units within one display. What is more, the methods that were best understood by the respondents are not necessarily the displays that provide the most robust message. For example, in the poster exercise it was found that the respondents understood the model average better than the model range. A presentation of the model range reflects the range of different model outputs and so accounts for the uncertainty inherent in all climate models. The model average, on the other hand, only shows the mean of the model outputs, and so does not account for the different model outputs. Hence, this research supports the predicament that was highlighted in the literature review, namely the challenge of providing data presentation methods that are well understood and provide a useful guidance for planning and adaptation strategies, but at the same time account for uncertainties related to models and different future scenarios (Tompkins and Adger 2004; New et al. 2007).

Both the poster exercise and the survey also highlighted the importance of providing sufficient context and information to ensure that the data is well understood. The respondents kept requesting more information for the data displays. While usually understanding the initial message of the display, it was found that the respondents were at times unsure about what more the display could actually tell them. In some cases this led the respondents to see information in the graph that was not actually there. This shows how users, without sufficient information of what a display can provide, might not only miss the actual message, but furthermore search for information that is not there.

Approaching a New Form of Management

It was highlighted in a paper concerned with the influence of climate science on water management in Western Australia (Power et al. 2005) that the use of climate change projections is not necessarily accepted throughout the water management community, and in the Australian example there had been disputes about whether the projections were too uncertain to use for planning purposes. At the climate change communication workshop organized by DWAF and CSAG in the Western Cape, South Africa, participants were asked how reliable climate models are, and answers ranged from 30 to 80%. Hence the WRMs' perception of projection reliability can be found to vary, potentially influencing the willingness to incorporate them into planning and strategies. As pointed out by Hampel in "Different concepts of risk—A challenge for risk communication" (Hampel 2006), the fundamental comprehension of risk differs both among scientists and within societies.

Media has been found to play a major role in shaping the perception of the public, portraying climate change as an issue of great uncertainty, with disputing arguments within the scientific community (Moser and Dilling 2004). By giving equal coverage to unequal sides of the debate, the media creates a confusion and lack of faith in the scientific community (Moser and Dilling 2004). When communicating climate change data to practitioners, such as the WRMs, it is therefore important to overcome this barrier of mistrust. Scientists must create a rigid understanding of anthropogenic climate change, while at the same time explaining the uncertainties related to climate modeling and climate change projections. The new uncertainty related to climate change projections and the correspondingly different perceptions of risk can be seen as an important element of the paradigm shift that needs to take place within water resources management. As previously mentioned, WRMs are only familiar with using historical information, which relates to what Pahl-Wostl (2007: 49) characterizes as "prediction and control management". The perception that natural systems are stationary in the sense that they oscillate within a definable range of variability has long been the understanding of WRMs (O'Brien et al. 2008). But the new uncertainty and element of risk associated with climate change and increased climate variability should therefore lead to a new form of management, one characterized by "learning to manage by managing to learn" (Pahl-Wostl 2007: 49). The uncertainty associated with climate change predictions should not be seen as a reason not to act, but a reason to change some of the fundamentals found in water resources management. A proactive and resilient form of management can better deal with what can be defined as a somewhat ambiguous future of change. Hence, the different perceptions of uncertainty and the related climate change risk, though a great challenge, should be seen as an important and integral part of water resources management.

Bridging the Communication Gap: Establishing a Climate Change Forum

Working together we can do more (Election campaign slogan, African National Congress, 2008)

It is clear that in order to improve communication between WRMs and climate scientists, further interaction between the two groups is necessary. This is obvious both in the challenges of providing comprehensible data presentation methods that at the same time provide a robust message, and in trying to outline the climate change-related information that WRMs find relevant to their work. Communication allows for the incorporation of relevant and comprehensible, yet robust, climate data into water resources management, and will further aid the move towards a more proactive and resilient management paradigm.

A Climate Change Forum can provide a platform for continuous engagement, bringing climate scientists and WRMs together with regular intervals. In Australia, such a forum was established in 1998 through the Indian Ocean Climate Initiative (IOCI) (Power et al. 2005), a panel made up of scientists and partner agencies, including water managers, that set out to aid the communication between climate scientists and decision-makers. Research has shown that the initiative has provided important assistance to decision-makers (Power et al. 2005). For the establishment of a Climate Change Forum in South Africa, a few recommendations can be suggested, based on experience from the water management sector in the Western Cape. Firstly, in order to ensure that the gap between experts and users is narrowed at the Forum, they should be set out as interactive rather than one-way, presentation-based workshops. This will enable both the WRMs and climate scientists to feel as equal parts of the process, and result in co-produced knowledge that is of interest to both parties. Secondly, in order to address the different fields that can be found within water resources management, the Forum should look at providing group sessions that focus on specific sectors separately. Thirdly, it should be expected that this engagement will be time-consuming and at times frustrating, as there is likely to be some friction and disagreement along the way. Furthermore, with climate science continuously advancing, as technology and skills develop, and as practitioners on the ground gain experience in a changing climate, it is important that the Climate Change Forum should not be seen as temporary structure for the initial introduction of climate change science into water resources management. It must rather become a permanent and intrinsic part of a new water resources management paradigm, one based on resilience and continuous learning.

As mentioned earlier, WRMs in developing countries like South Africa face more immediate issues, such as providing basic water supply for all (Schulze 2007), and thus tend to lack the necessary human resources to deal with more long-term issues such as climate change. The country faces the constant challenge of balancing priorities in the social, economic and environmental spheres, while also tackling the threats of climate impacts (Ziervogel et al. 2010). The integration of climate change information into water resources management is therefore not only about bridging the communication between WRMs and climate scientists, but also about making climate change adaptation an integral part of general development objectives.

Conclusion

In trying to outline comprehensible data presentation methods and the data relevant to WRMs, it was found that these are some of the great challenges of communicating climate science to stakeholders in the water resources management sector. When interpreting different presentation methods, only the very simple, well-explained graphs with relatively few lines and a clear message were well understood—a predicament considering the fact that climate change projections are generally not simple and entirely clear. In trying to establish the nature of climate change data relevant to the water resources management sector it was discovered that WRMs do not know what data they need. This is partly because they do not know how climate change will impact their sector and partly because they are only used to working with historical data from instrumental records, and hence do not have frameworks for assessing future information from climate models. There was also found to be a gap between the time frames along which the WRMs work and the climate projection time frames that the climate scientists can provide. A Climate Change Forum can provide the required platform for communication, further allowing for co-production of knowledge, ensuring the production of comprehensible, relevant and applicable climate data.

Climate scientists should take note of the need to develop graphical communication methods that can provide a robust message with maximum information transfer and minimum confusion. They should further recognize the importance of engaging with stakeholders to ensure that their research is designed with the end-users—practitioners and policy-makers—in mind.

References

Amanzi Obom Consulting cc (2008) Water services development plan for City of Cape Town. Report commissioned by The City of Cape Town, Cape Town

Hampel J (2006) Different concepts of risk–a challenge for risk communication. Int J Med Microbiol 296:5–10

Jury W, Vaux H Jr (2005) The role of science in solving the world's emerging water problems. PNAS 102(44):15715–15720

Kabat P, Schulze RE, Hellmuth ME, Veraart JA (eds) (2003) Coping with impacts of climate variability and climate change in water management: a scoping paper. DWC-Report No.DWCSSO-01, International Secretariat of the Dialogue on water and climate, Wageningen

Kashyap A (2004) Water Governance: learning by developing adaptive capacity to incorporate climate variability and change. Water Sci Technol 49(7):141–146

Moss SC, Dilling L (2004) Making climate hot. Communicating the urgency and challenge of global climate change. Environment 46(10):32–46

New M, Lopez A, Dessai S, Wilby R (2007) Challenges in using probabilistic climate change information for impact assessments: an example from the water sector. Philos Trans Royal Soc 365:2117–2131

O'Brien K, Sygna L, Leichenko R, Adger WN, Barnett J, Mitchell T, Schipper L, Tanner T, Vogel C, Mortreux C (2008) Disaster risk reduction, climate change adaptation and human security, a commissioned report for the Norwegian Ministry of Foreign Affairs. GECHS Report 2008:3

Oneworld Sustainable Investments (2008) A climate change strategy and action plan for the Western Cape, Report commissioned by the Provincial Government of the Western Cape. Department of Environmental Affairs and Development Planning, Western Cape

Pahl-Whost C (2007) Transition towards adaptive management of water facing climate and global change. Water Resour Manag 21:49–62

Power S, Sadler B, Nicholls N (2005) The influence of climate science on water management in Western Australia: lessons for climate scientists. Bull Am Meteorol Soc 87(2):839–844

Roux DJ, Rogers KH, Biggs HC, Ashton PJ, Sergeant A (2006) Bridging the science-management divide: moving from unidirectional knowledge transfer to knowledge interfacing and sharing. Ecol Soc 11(1):4

Schulze RE (2000) Modeling hydrological responses to land use and climate change: a Southern African perspective. Ambio 29:12–22

Schulze R (2007) Some foci of integrated water resource management in the 'South' which are oft-forgotten by the 'North': a perspective from Southern Africa. Water Resour Manag 21:269–294

Thomalla F, Downing T, Spanger E, Han G, Rockstrom J (2006) Reducing hazard vulnerability: towards a common approach between disaster risk reduction and climate adaptation. Disaster 30(1):39–48

Thompkins E, Adger N (2004) Does adaptive management of natural resources enhance resilience to climate change? Ecol Soc 9(2):10

Vanrheenen N, Wood A, Palmer R, Lettenmaier D (2004) Potential implications of PCM climate change scenarios for Sacramento–San Joaquin River basin hydrology and water resources. Clim Chang 62:257–281

Ziervogel G, Shale M, Du M (2010) Climate change adaptations in a developing country context: the case of urban water supply in Cape Town. Clim Dev 2:94–110

Chapter 32
Influences of Public Ecological Awareness and Price on Potable Water Consumption in Geneva

Giuseppe Catenazzo, Jennifer D'Urso, Emmanuel Fragnière and Jean Tuberosa

Abstract The specific attitudes and behaviour of individuals in a given society must be properly understood in order to develop adequate and relevant public policies regarding water consumption. For this reason, we conducted a survey to measure the perception of the Geneva population regarding the main social aspects of water consumption. Our sample, collected between November 2006 and February 2007, numbers 907 valid questionnaires. We observed that most respondents indicate that they have an ecological awareness regarding their daily consumption of freshwater. However, we also discovered a lack of information, e.g. people are neither able to quantify water price, nor their individual consumption (for drinking, cleaning, washing, etc.). Non-parametrical statistical tests have highlighted a weak correlation between water price increase and consumption. Also, people who directly pay their water bills seem to be more concerned by water consumption than others. This and other elements of perception regarding public water consumption have been analysed with the aim to provide recommendations for better demand management policies.

Keywords Water demand · Water management · Survey · Water consumption · Public policies · Services management

---

G. Catenazzo (✉) · J. D'Urso · E. Fragnière · J. Tuberosa
Haute École de Gestion de Genève, Geneva School of Business Administration, Campus Battelle—Bât. F 7, Route de Drize, 1227 Carouge, Geneva, Switzerland
e-mail: giuseppe.catenazzo@unige.ch

Introduction

Geneva is the second largest Swiss city, located close to the Swiss and French Alps, at the bottom of a major European lake, Lake Geneva. Thanks to this position, the city benefits from several possible water supplies, i.e. Lake Geneva, Rhône and Arve rivers, and underground water. The city and its surrounding urbanized area, the Canton of Geneva, take advantage of an efficient functioning network that delivers high quality freshwater every day to all private and public buildings (D'Urso et al. 2006a).

In the Canton of Geneva, freshwater is supplied by SIG (Services Industriels de Genève) a state-owned monopoly that provides households and industries with a top-quality resource and efficient delivery service. This facility is provided at an affordable price for the locals: the average individual monthly cost for water consumption is CHF 6.27, while the standardized Canton median individual monthly salary is estimated at CHF 6,350.

In this context, water is a cheap and abundant commodity, thus individuals risk overusing it. We learn from Service Science that clients are co-producers of a service (Dubosson and Fragnière 2009): if most inhabitants behave properly (e.g. pay the service, do not flush away toxic liquids, consume it wisely), the service production can easily meet high-quality standards.

To better explore these issues, we conducted a survey among the Geneva population regarding their perception, attitudes and behaviours related to individual water consumption. We collected a sample of 907 valid questionnaires between November 2006 and February 2007. More precisely, we focused on individuals' awareness of the service they provided and we attempted to pattern the local demand-price elasticity. To investigate more in depth the latter, we included in our questionnaire hypothetical scenarios (Hoevenagel 1994) to assess the consumption sensitivity relative to price. Relationships between variables were analysed and then research hypotheses were verified on the basis of non-parametrical statistical tests.

Therefore, this empirical research aims to provide policy-makers and water managers with some elements of perception regarding individuals' water consumption leading to recommendations for better demand management.

This short paper is organized as follows. In the next section, we present some of the existing literature connected with our research. Then, we present the main descriptive statistics obtained from the survey. A few hypotheses related to the theme retained for this paper follow: the influence of public ecological awareness and price on potable water consumption in the Geneva area. In conclusion, we indicate limitations of this study and directions for future research.

Literature Review

The analysis of household water demand is a subject of high concern among researchers. From the Americas to Europe and Eastern countries, several empirical researches (Monteiro 2005) have been conducted to identify relevant factors to design appropriate freshwater management policies. To tackle these objectives, i.e. to provide households with high-quality and fair and affordable water prices, authors (Cavanagh et al. 2002; Monteiro 2005; Pashardes et al. 2002) have designed and applied econometrical models that take into account both operational delivery costs and some social aspects related to water consumption (Corral-Verdugo et al. 2003; Keshavarzi et al. 2006; Kolokytha et al. 2002; Nauges and Reynaud 2001; Wong and Mui 2006; Zhang and Brown 2005).

Analysis of the advantages and drawbacks of current pricing schemes is the theme of a careful literature review by Monteiro (2005). The paper reviews in depth pricing schemes based on seasonal/temporal variations, capacity constraints, multistage marginal cost pricing and revenue requirements: factors currently used to appraise retail water price. According to this author, cost pricing is the most efficient water pricing policy.

Several surveys have been administered both in Canada and in the US to determine household water demand sensitivity related to different pricing schemes. Authors such as Cavanagh (2002) and Nieswiadomy (1989) have explored water demand under increasing block prices. Similarly, other pricing systems have been tested in Cyprus (Pashardes et al. 2002) and in Iran (Keshavarzi et al. 2006) to point out the perceived advantages and drawbacks of each of them.

The latter study, a survey on the Iranian Fars province (Keshavarzi et al. 2006), highlights an inverse relationship between household size and water demand: the smaller the household, the higher its water consumption. Oppositely, in a different context (Hong Kong), a survey administered at 60 typical apartments situated in a crowded urbanized area shows that the annual domestic consumption is positively correlated with the occupant load of a flat (Wong and Mui 2006). Therefore, household size is a factor relevant to water demand with an outcome that is likely to differ between locations.

A survey conducted in France (Nauges and Reynaud 2001) points out that age, number of household components, educational level of the family head and climate are relevant. Furthermore, household water demand seems to be partly inelastic to price. In Sri Lanka, price elasticity for public water delivery is estimated at −0.34 and income elasticity is +0.08 (Gunatilake et al. 2001).

In Sonora, Mexico, further empirical research (Corral-Verdugo et al. 2003) confirms these findings: water consumption depends on some demographic characteristics such as gender, age and socioeconomic class. They also claim that information campaigns could positively influence adults and rich people, who seem to be the main water waste class.

In addition to socioeconomic and climate factors, housing environment and billing systems are two additional variables that are likely to influence residential water use (Zhang and Brown 2005). A further factor likely to influence water demand is individuals' trust in household freshwater supply. In a context where the level of the water supply infrastructure is low, people don't trust the quality of their tap water, the individual consumption level is not correctly acknowledged and price has a limited effect on residential water demand. This is the main finding of a survey conducted in Tessaloniki, Greece (Kolokytha et al. 2002) and confirmed by the Mexican survey (Corral-Verdugo et al. 2003). In the Greek region, authors (Kolokytha et al. 2002) also highlight low willingness to pay (WTP) for improvements in water supply services and a high level of public awareness concerning current and future water issues.

In western Europe, a survey conducted in 2000 across EU-15 countries highlighted that 94.53% of the population had easy access to a water supply. Quality standard and water price were considered to be fairly good by 92.8% of the population. These highly perceived supply levels seem to keep steady in the following years according to Fiorio et al. (2007).

Finally, in western Switzerland, about 90% of the population drink tap water and 94.9% are aware that drinking it is a more ecological behaviour than buying bottled water. Also, the author (D'Urso et al. 2006a, b) reviews freshwater household pricing in western Switzerland.

Far from being exhaustive, this literature review indicates that sociological aspects of water consumption are relevant but not sufficiently studied. This is a key issue necessary to device models comprehensive of both pricing and behavioural issues. Furthermore, attitudes and behaviours towards water consumption in Switzerland have not been analysed yet. Thus, in this study, we intend to discover social patterns associated with water consumption in a wealthy and international city, Geneva, enjoying an efficient public water service.

Methodology

This empirical study attempts to investigate some sociological clues connected with water consumption in Geneva. To discover the main patterns associated with this theme, we designed a survey administered to a representative sample of the population living in Geneva and its surrounding area (Canton). This research encompassed the following steps: qualitative exploratory phase, survey design, data collection and analysis.

The first phase of our research attempted to identify the underlying themes to be developed. In this exploratory phase, we conducted in-depth interviews to explore the overall perception among Geneva population concerning this topic. Around 40 volunteers agreed to talk freely to our interviewers about water-supply and consumption issues. Among other subjects explored with our interviewees, we observed that, in general, people don't feel concerned about water supply and consumption. In fact, most people are unaware of their consumption, its price and possible conservation methods.

On the basis of these indications, we designed a questionnaire administered to a sample of Geneva and its surrounding area (Canton of Geneva). The questionnaire is made up of 22 close-ended multiple-choice questions. The first part of the questionnaire (8 questions) attempted to investigate individuals' attitudes and behaviours towards water consumption. More precisely, we attempted to explore individuals' consumption as well as their sensitivity towards ecological issues.

This was followed by five questions more precisely related to water price. In the last section of the questionnaire (2 further questions), we presented our sample two hypothetical markets with the aim to predict possible individuals' behaviours within given scenarios. We included this contingent valuation (Bateman and Turner 1992; Garrods and Willis 1999; Hoevenagel 1994) to collect further elements of analysis leading to public policy and organizational suggestions. Thanks to this experimental method borrowed from psychology and the environmental sciences (Catenazzo and Fragnière 2009, 2010b; Debély et al. 2008; Hansla et al. 2008; Higgins et al. 2002; Nomura and Akai 2004), we attempt to assess individuals' typical attitudes and behaviours within hypothetical settings.

The respondents for this survey were selected on a random basis in the streets, open spaces and other public places in Geneva and surrounding areas. Our interviewers submitted the questionnaire in different parts of the city, different hours and days of the week according to a plan to make the collected sample as representative as possible of Geneva population.

907 from the Canton of Geneva's 447,584 inhabitants (OCSTAT 2008) responded to the questionnaire. Data collection was held between November 2006 and February 2007. The sample socio-demographic characteristics are given in Table 1.

Results

Descriptive Statistics

In this section, we present the analysis of the main descriptive results obtained by the analysis of the 907 questionnaires collected. The submitted questionnaire was structured in the following way: first, we asked our respondents about their attitudes towards water consumption followed by behavioural queries and the presentation of hypothetical scenarios referring to pricing issues.

First of all, we asked our sample whether they pay attention to not wasting water. 69% of our sample confirmed that they pay attention while 24.7% don't, and 6.3% don't know. Furthermore, 28.4% confirm that they often pay attention to potentially toxic liquids flushed away through pipelines. 25.7% confirm they are sometimes attentive to this issue, 19.9% feel concerned all the time, 15.6% are rarely concerned and 10.4% are never attentive to the potentially toxic liquids they pour through their pipelines (Fig. 1).

If we gather together respondents who say "concerned", "always", "often" and "sometimes", we can affirm that 74% feel concerned about this topic. However, 50.9% of the whole sample are not sufficiently informed about their current water consumption and available conservation methods; 41.7% are and 7.1% don't know.

The lack of information about water consumption is also confirmed by the answers provided to a further query of our questionnaire by which we discover that 83.2% are unable to state their average monthly water consumption, while 16.8% can.

Table 1 Sampling

Variable	Sample	Official data
Sex	54.7% men	48.2% men
	45.3% women	51.8% women
Nationality	68.4% Swiss citizens	61.6% Swiss
	31.6% foreigners	38.4% foreigners
Age range	18–89 years old	N/A
Household size	Compulsory education: 7%	N/A
	Practical post-compulsory education: 17.5%	
	High school: 29.9%	
	College or University: 28.3%	
	Other: 17.2%	
Education	18–89 years old	N/A
Household monthly income	- CHF 2,500: 10.1%	Single household me an income: CHF 3,899
	CHF 2,500–5,000: 22.8%	Married household mean income (with and without children): CHF 9,548
	CHF 5,000–7,500: 29%	
	CHF 7,500–10,000: 18.7%	
	CHF 10,000–12,000: 83%	
	CHF 12,000–15,000: 5.1%	
	+ CHF 15,000: 6%	
	Mean and median fall between CHF 5,000 and 7,000	
Flat—house property	58.9% rent the flat or house where they live with water charge included	83% are renters
	13.5% are renters with water charges not included	
	23.6% are owners and directly pay water bills	17% are owners
	3.9% don't know	

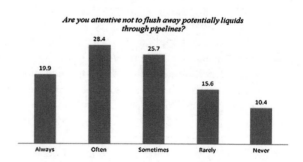

Fig. 1 Are you attentive not to flush away potentially liquids through pipelines?

We therefore informed them that, on average, each Swiss inhabitant consumes 162 litres of freshwater per day. Interviewees judge this consumption level as "high" (43.7%), "too high" (24.3%), "normal" (21.1%); 3.2% found it "low" and 7.9% don't know. Again, the rate of people showing sensitiveness towards this

Fig. 2 Would you accept a water price increase for ecological reasons?

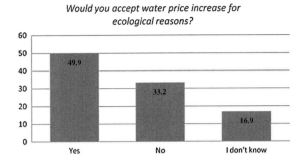

issue (i.e. people feeling this consumption is "too high" and "very high") represents 68% of the sample.

Afterwards, we asked our sample to rank some of the proposed features that mostly influence their purchase of a dishwasher or a washing machine. We chose these questioning devices since in-depth interviews (qualitative phase) highlighted that both devices are widely known by the population in use and are perceived as highly water-consuming. 56.7% ranked "price" as the first driver of their purchase of either or both devices; water and energy consumption is solely quoted only as the fourth criterion.

We also discover that 80.2% of the sample can't estimate the water price in Geneva for domestic use (1,000 litres), while 19.8% can. When this information is given to the respondents, i.e. CHF 1.60 m^3, which corresponds roughly to CHF 8 per month per person, 52.8% indicate that it is a normal or fair price while 21.8% estimate it is cheap, 12.7% say it is rather expensive, 5.9% say it is expensive, 5.3% don't know and 1.4% claim it is too expensive. Then, 49.9% say they would accept a price increase for ecological reasons, 33.2% would not, and 16.9% don't know (Fig. 2).

Although most Geneva inhabitants don't know water price, 74.7% are aware that water supply tariffs are paid to SIG (Services Industriels de Genève), the Cantonal public company in charge of water and energy supplies (gas and electricity) that operates as the local public monopoly, 12.4% don't know, 9.1% say to the city (or town) council, and 3.8% to a private company.

Next, we presented our sample two hypothetical scenarios with the aim to predict possible individuals' attitudes and behaviours if the water supply pricing changed in Geneva.

In the first scenario, we enquired as to individuals' possible reactions, in terms of water consumption, if water price shifted from CHF 0.5 to 2 per month and per person: 13.9% of the respondents affirm they would lower their consumption. An extra 21.6% would reduce their water consumption if the price increased between CHF 2 and 4. 11.2% would change their habits if water price increased by CHF 4–6. 7.7% would for an increase of CHF 6–8. 9.3% of the sample would be likely to change habits only for a water price increase higher than CHF 8. On the contrary, 19.9% of the sample affirm that price does not influence their water consumption; 16.4% cannot answer.

Fig. 3 If household water delivery were totally free, how would you change your consumption?

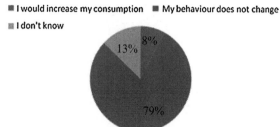

Finally, respondents were faced with an opposite scenario, with no market for water and free supply to individuals' consumption. We discover that only 7.7% of our sample would increase their individual consumption, while 79.1% affirm not to be willing to change their behaviour. 13.1% don't know (Fig. 3).

These statistics show that Geneva's population generally feels a deep lack of information towards household water consumption. Also, it seems that water demand is slightly elastic to price; and a portion of the population ranging between 10.4 and 33.2% is not concerned at all by water consumption issues and not willing to change their behaviour to improve water delivery services.

These attitudes suggest that people do not value the quality of the water supply service any more. The lack of interest from the concerned parties causes a problem whenever authorities take into account opportunities of investment in this sector: the resulting policies would risk being unpopular and inefficient.

In the following part of this paper, we look at the main relationships existing among variables under study and attempt to identify key levers to improve global awareness of this topic.

Hypotheses Testing

Descriptive statistics highlighted the main attitudes and behaviours connected with household water consumption demand. To provide policy-makers and water services managers with relevant information for decision-making, we explored in depth possible relationships between the different variables (questions) submitted to our sample. Collected questionnaires (907) were coded through SPSS 15 software and tests were made according to methods described by Bryman and Cramer (2006).

In the literature section we have presented some relevant papers identifying factors to be taken into account in the design and analysis of public policies, e.g. pricing (Monteiro 2005), socioeconomic characteristics and local climate (Nauges and Reynaud 2001); household (Keshavarzi et al. 2006) and flat size (Wong and Mui 2006).

In this paper, household water supply service pricing is analysed in terms of WTP for ecological reasons and individuals' action of paying water bills. The latter is an extra factor of interest: in Geneva, 83% of household housing is occupied by renters (year 2000 data, OCSTAT 2006) who either directly pay their water bills to SIG (a minority, 13.5% of our sample) or the water bills are included in their rent (58.9% in our sample). Alternatively, all flat owners (23.6% of our sample) pay their water charges in full.

As presented in the descriptive statistics section, to determine the relevancy of household water pricing in individuals' consumption, we designed two hypothetical scenarios (CVM; Catenazzo et al. 2010; Mourato 1998) to predict possible behaviours following water price changes.

According to studies conducted in other geographical contexts, water demand seems to be slightly correlated with price (Gunatilake et al. 2001; Nauges and Reynaud 2001); therefore, we could expect that households are sensitive to water price increases and consequently reduce their consumption.

To test this hypothesis, we rely upon the analysis of two queries (variables) of the survey. In the first question, respondents were informed that their average freshwater consumption is about CHF 8 per month. Knowing this figure, they were asked the minimum price increase that would force them to reduce their consumption. Seven possible choices were given: "between CHF 0.50 and 2", "from CHF 2 to 4", "between CHF 4 and 6", "from CHF 6 to 8", "more than CHF 8", "price does not influence my consumption" and "I don't know".

The second variable we used to test this hypothesis is a socio-economic question asked at the end of the questionnaire. We asked the sample to state their average monthly household income according to a corresponding scale divided into: "less than CHF 2,500", "between CHF 2,500 and 5,000", "between CHF 5,000 and 7,500", "between CHF 7,500 and 10,000", "between CHF 10,000 and 12,500", "between CHF 12,500 and 15,000" and "more than CHF 15,000".

Starting from these variables, we have designed our hypothesis scheme as follows:

Ho: There is a correlation between water consumption reduction in case of price increase and monthly household income.

Ha: There is no correlation between water consumption reduction in case of price increase and monthly household income.

Both variables under study here are ordinal variables. For this reason, the statistical test we have employed is a non-parametric test called rank correlation. More precisely, we are employing two statistics: Spearman's rho and Kendall's tau. We have retained a significance level of 1% in which we would fall in a Type-I error.

The p-value of the rank correlation test highlights a statistically significant relationship between variables since both correlation coefficients are positive (0.077; 0.099). This is confirmed by both tests (Spearman's rho and Kendall's tau). As a consequence, we fail to reject the alternative hypothesis and affirm that there is a correlation between household water consumption reduction in case of price increase and the monthly household income. Therefore, we can argue that the more the household income, the more the willingness to pay for household water delivery service.

However, Spearman's rho (0.099) and Kendall's tau (0.077) correlations are extremely weak. Thus, since these correlations are very low, we conclude that the price, despite being a factor in water consumption, does not represent a sufficient driver to influence individuals' behaviours. Indeed, people perceive that their water consumption would not be significantly affected by a change in its price.

This confirms findings of other researchers, e.g. in France (Nauges and Reynaud 2001) and in Sri Lanka (Gunatilake et al. 2001), who argue that household water consumption seems to be partly inelastic to price. Because of its low loading in influencing consumption, we can then conclude that pricing is not a strong enough lever to be included in an efficient public water policy.

As mentioned earlier in this paper, Geneva Canton freshwater distribution is managed by SIG (Services Industriels de Genève), the state-owned company in charge of the main public utilities (water and energy). We learn from the services management theory that customers are co-producers of a service as they take part into its production process (Hsu 2009). Among other behaviours, people's behaviour can help keep the high quality of the service provided by being attentive not to flush away toxic liquids, by using water consumption wisely and by paying the water bills.

The payment action by itself might be an additional factor: studies in other services have pointed out that individuals who regularly and directly pay for a service are more likely to have a higher commitment and loyalty towards it than the others (Della Vigna and Malmendier 2006). In Geneva, not everybody pays water bills directly. Indeed, renters, who represent a large part of the population, often see this cost included in their housing fee. Our sample confirms this characteristic, as only one-quarter of the population own the flat or house where they live. If we add together owners and all those people whose water bills are not included with their rents, we can argue that only 37% of the population pay their water bills directly. Therefore, we could expect that individuals who pay their water bills feel more concerned about their service purchase than the others and attempt to maintain a high-quality water delivery service.

To better explore the existence of this relationship, i.e. the payment action being a relevant factor influencing household water demand, we refer to two variables of our questionnaire. Respondents were asked "are you attentive not to waste water?" and the possible given answers were "yes" and "no". The second variable under analysis refers to another question asked of our sample. Interviewees were asked whether they pay themselves their water bills; they could choose between "yes", "no" and "I don't know" as possible answers. Here we deal with two nominal choice variables to make our statistical tests. Therefore, we tested our hypothesis according to the following scheme:

Ho: There is no relationship between the fact that individuals pay or do not pay their water bills and their water waste awareness.

Ha: There is a relationship between the fact that individuals pay or do not pay their water bills and their water waste awareness.

Table 2 Rank correlation statistical test between average monthly household income and price increase likely to reduce individuals' water consumption

		Do you pay your water bills?		Total
		Yes	No	
Are you attentive not to waste water?				
Yes	N	407	212	619
	Water waste (%)	65.8	34.2	100.0
	Pay water bills (%)	75.8	59.6	69.3
	Total (%)	45.6	23.7	69.3
No	N	102	117	219
	Water waste (%)	46.6	53.4	100.0
	Pay water bills (%)	19.0	32.9	24.5
	Total (%)	11.4	13.1	24.5
I don't know	N	28	27	55
	Water waste (%)	50.9	49.1	100.0
	Pay water bills (%)	5.2	7.6	6.2
	Total (%)	3.1	3.0	6.2
Total	N	537	356	893
	Water waste (%)	60.1	39.9	100.0
	Pay water bills (%)	100.0	100.0	100.0
	Total (%)	60.1	39.9	100.0

We have used the Pearson's chi-square test to study two independent nominal variables. We have retained a significance level of 5% that is the first-type error (or the risk to reject the null hypothesis when it is actually correct). The p-value of 0 (Table 2) indicates that we can reject the null hypothesis at the significance level of 5%. Therefore, we can conclude that there is a relationship between the fact that individuals pay or do not pay their water bills and their water waste awareness. More precisely, the cross-table shows that individuals who pay their bills themselves feel significantly more concerned with the quality of the service provided than the others. Thus, we can say that the payment action seems to influence individual behaviour regarding water consumption (Tables 3 and 4).

Discussion and Conclusions

In this paper, we present the main results of a survey conducted between autumn and winter 2006–2007 among 907 individuals living in Geneva, Switzerland, both in the city and surrounding area (Canton).

This work attempts to measure some elements of perception related to individuals' sensibility over individual freshwater consumption. As highlighted in the literature review, water consumption depends on several factors: water pricing seems to be relevant, ecological sensibility, information availability, socioeconomic characteristics and further elements appear to be worthwhile including in

Table 3 Chi-square test between individuals paying their water bills directly and their water waste awareness

	Value	df	Asymp. Sig. (2-sides)
Pearson's Chi-Square	26.894[a]	2	0.000
Likelihood Ratio	26.593	2	0.000
Line ar-by-Linear	20.813	1	0.000
X of Valid Cases	893		

[a] 0 cells (0.0%) have expected count less than 5. The minimum expected count is 21.93

Table 4 Cross-table between individuals paying their water bills directly and their water waste awareness

		Do you pay your water bills?		Total
		Yes	No	
Are you attentive not to waste water?				
Yes	N	407	212	619
	Water waste (%)	65.8	34.2	100.0
	Pay water bills (%)	75.8	59.6	69.3
	Total (%)	45.6	23.7	69.3
No	N	102	117	219
	Water waste (%)	46.6	53.4	100.0
	Pay water bills (%)	19.0	32.9	24.5
	Total (%)	11.4	13.1	24.5
I don't know	N	28	27	55
	Water waste (%)	50.9	49.1	100.0
	Pay water bills (%)	5.2	7.6	6.2
	Total (%)	3.1	3.0	6.2
Total	N	537	356	893
	Water waste (%)	60.1	39.9	100.0
	Pay water bills (%)	100.0	100.0	100.0
	Total (%)	60.1	39.9	100.0

the analysis. For this reason, we have endeavoured to identify some elements that are likely to influence public ecological awareness on water consumption in the Canton of Geneva.

In the context under study, the Canton of Geneva, people are used to getting high-quality tap water at a very affordable price. This has made people less and less aware of the benefits they are provided with. In fact, although most people affirm to be attentive not to waste water and not to flush away toxic liquid through the pipelines, they don't consider water consumption relevant when purchasing high water consumption home devices such as a dishwasher or washing machine.

Individuals are poorly informed about water consumption. They neither know its extent nor current tariffs. Indeed, whenever people are properly informed about consumption, they mostly feel it to be excessive. Public information campaigns should take this into account in their design.

Moreover, it should be noted that water pricing has a limited impact on individual behaviour. As evidenced in this paper, water price shifts designed to reduce consumption risk being unsuccessful.

We also observed that people who directly pay their water bills seem to be more concerned by water savings than others. Therefore, the payment action seems to have some influence on individual commitment. We can then argue that this is an additional factor leading to the improvement of the service provided.

This research encounters several limitations and possible directions for further studies. First of all, sampling and methodology weaknesses exist: we stressed raising a representative sample of the local population. However, sampling errors are possible; slight weaknesses and differences between our sample and official data are already described in the "Methodology" section of this paper. Also, as in all surveys based on questionnaires, ours suffers further tool limitations and heuristic biases (Catenazzo et al. 2010; Tversky and Kahneman 1974; OECD 1995).

Furthermore, we have employed Contingent Valuation Methods in our questionnaire that might lead to design biases, information availability and other biases. Strategic biases and differences between stated behaviours in the hypothetical market presented and the real market might affect the content and predictive validity of our measures (Catenazzo and Fragnière 2008; Garrods and Willis 1999; OECD 1995). Finally, further biases are associated with this type of empirical research (Catenazzo and Fragnière 2010a).

Further research should follow two main directions: the first should attempt to replicate this work in another context, both in Switzerland and in other European cities with similar features in terms of water supply and pricing. Secondly, the payment action factor developed in this paper should be explored to better understand its impact on individual behaviours within Geneva or other contexts.

Acknowledgments We would like to thank the students of Laboratoire d'Études de Marché (LEM) of Geneva Haute École de Gestion—Geneva School of Business Administration who participated in the construction of the survey, the data collection, and the transcription of answers. Without them, this research would not have been possible. An earlier version of this paper was presented at Climate–Klima 2010 online conference organized by the Hamburg University of Applied Sciences, Hamburg, Germany, 1–7 November 2010. Also, a previous version was presented at the International World Water Association's XIIIth World Water Congress held in Montpellier, France, 1–4 September 2008. Thus, we would like to thank the participants of both events as well as an anonymous reviewer for their valuable feedback.

References

Bateman IJ, Turner K (1992) Evaluation of the environment: the contingent valuation method, CSERGE Working Paper GEC 92-18
Bryman A, Cramer D (2006) Quantitative data analysis for the social scientist with SPSS 15 and 16. Routledge, London
Catenazzo G, Fragnière E (2008) La gestion des services. Economica, Paris

Catenazzo G, Fragnière E (2009) Identifying bank run signals through sociological factors: an empirical research in the Geneva area. In: Gregoriou G (ed) The handbook of banking crises. CRC Press, Chapman and Hall, London, pp 489–502, Chapter 26

Catenazzo G, Fragnière E (2010a) Implementation of compulsory internal control system within swiss organisations: perceptions among accountants and workers in Geneva. Int J Econ Account 1(1/2):53–75

Catenazzo G, Fragnière E (2010b) Pricing traditional travel agency services: a theatre-based experimental study. J Serv Sci Manag 3(2):272–280

Catenazzo G, Fragnière E, Ribordy B, Tuberosa J (2010) Is the 2008 financial turmoil increasing the risk of a bank run? An empirical research in Geneva. Int J Mod Account Audit 6(1):29–45

Cavanagh MS, Hanemann WM, Stavins RN (2002) Muffled price signals: household water demand under increasing-block prices, FEEM–Fondazione Enrico Mattei, Working Paper No. 40

Corral-Verdugo V, Bechtel RB, Fraijo-Sing B (2003) Environmental beliefs and water conservation: an empirical study. J Environ Psychol 23:247–257

Debély J, Dubosson M, Fragnière E (2008) The pricing of knowledge-based services: insights from the environmental sciences. J Serv Res 8:167–181 [Special Issue (February 2008)]

Della Vigna S, Malmendier U (2006) Paying not to go to the gym. Am Econ Rev 96(3):694–719

Dubosson M, Fragnière E (2009) The consequences of information overload in knowledge based service economies: an empirical research conducted in Geneva. Serv Sci 1(1):56–62

D'Urso J, Dubosson M, Fragnière E (2006a) The public management of water in Switzerland. http://ssrn.com/abstract=931617

D'Urso J, Dubosson M, Fragnière E (2006b) Drinkable water: logistics and attitude of the consumer. http://ssrn.com/abstract=931615

Fiorio CV, Florio M, Salini S, Ferrari P (2007) Consumers' attitudes on services of general interest in the EU: accessibility, price and quality 2000–2004, FEEM–Fondazione Eni Enrico Mattei, Working Paper No 2

Garrods G, Willis KG (1999) Economic valuation of the environment, methods and case studies. Edward Elgar, Cheltenham

Gunatilake MH, Gopalakrishnan C, Chandrasena I (2001) The economics of household demand for water: the case of Kandy Municipality, Sri Lanka. Water Resour Dev 17(3):277–288

Hansla A, Gamble A, Juliusson A, Gärling T (2008) Psychological determinants of attitude towards and willingness to pay for green electricity. Energy Policy 36:768–774

Higgins J, Warnken J, Sherman PP, Teasdale PR (2002) Survey of users and providers of recycled water: quality concerns and directions for applied research. Water Res 36:5045–5056

Hoevenagel R (1994) An assessment of the contingent valuation method. In: Pethig R (ed) Valuing the environment: methodological and measurement issues. Kluwer Academic Publishers, Dodrecht, pp 195–227

Hsu C (2009) Service science: design for scaling and transformation. World Scientific, Singapore

Keshavarzi AR, Sharifzadeh M, Kamgar AA, Amin S, Keshtkar SH, Bamdad A (2006) Rural domestic water consumption behaviour: a case study in Ramjerd area Fars province, I R Iran. Water Res 40:1173–1178

Kolokytha EG, Mylopoulos YA, Mentes AK (2002) Evaluating demand management aspects of urban water policy—a field survey in the city of Thessaloniki Greece. Urban Water 4:391–400

Monteiro HPC (2005) Water Pricing Models: a survey, DINAMIA–Research Centre on Socioeconomic Change Working Paper No. 2005/45

Mourato S (1998) Economic valuation in transition economies: an application of contingent valuation methods to Lake Balaton in Hungary. In: Acutt M, Mason P (eds) Environmental valuation, economic policy and sustainability: recent advances in environmental economics. Edward Edgar, Cheltenham, pp 15–34

Nauges C, Reynaud A (2001) Estimation de la demande domestique d'eau potable en France. Revue économique 52(1):167–185

Nieswiadomy ML, Molina DJ (1989) Comparing residential water demand estimates under decreasing and increasing block rates using household data. Land Econ 65(3):280–289

Nomura N, Akai M (2004) Willingness to pay for green electricity in Japan as estimated through contingent valuation method. Appl Energy 78:453–463

OCSTAT–Geneva Cantonal Office of Statistics (2006) Communication Statistiques. Surface et loyers des logements dans le canton de Genève, Geneva, Switzerland

OCSTAT–Geneva Cantonal Office of Statistics (2008) Mémento statistique du Canton de Genève, Geneva, Switzerland

OECD (1995) The economic appraisal of environmental projects and policies—a practical guide. OECD Publications, Paris

Pashardes P, Koundouri P, Hajispyrou S (2002) Household demand and welfare implications for water pricing in Cyprus. Environ Dev Econ 7(4):659–685

Tversky A, Kahnemann D (1974) Judgement under uncertainty: heuristics and biases. Science 185:1124–1131

Wong LT, Mui KW (2006) Epistemic consumption benchmarks for residential building. Build Environ 43:1031–1035

Zhang HH, Brown DF (2005) Understanding urban residential water use in Beijing and Tianjin China. Habitat Int 29:469–491

Chapter 33
Sustainable Development Through Pathways of Mitigation and Adaptation to Offset Adverse Climate Change Impacts

A. N. Sarkar

Abstract Climate change poses a multidimensional international challenge with no straightforward solution. There is a growing international consensus that the Earth's climate is changing by anthropogenic emissions of greenhouse gases primarily resulting from extensive industrial, commercial, transport and domestic use of fossil fuels. Adverse impacts of global warming that influence climate change include: melting glaciers, floods, earthquakes, drought, colossal loss of biodiversity, famine and demographic movement across continents. This warrants very careful planning and execution of mitigation and adaptation actions in a system-specific and sustainable manner to offset and minimize adverse impacts of anthropogenic interventions. Sustainable development involves a comprehensive and integrated approach to economic, social, and environmental processes. Cli mate change is influenced not only by the climate-specific policies but also by the mix of development choices and preparing for the resulting development trajectories. Therefore, mainstreaming the sustainability concept as an integral part/component of climate change-related policy options requires that non-climate policies, programmes, and/or individual actions take climate change mitigation and adaptation into consideration, in both developing and developed countries. The paper attempts to integrate the principles and modus operandi of mitigation and adaptation with sustainable development with a focus on low-carbon economy in order to help formulate suitable strategies to offset adverse impacts of climate change.

Keywords Global climate · Climate change impacts · Sustainable development · Mitigation · Adaptation

---

A. N. Sarkar (✉)
International Business, Asia–Pacific Institute of Management,
3&4, Sarita Vihar, New Delhi, 110025, India
e-mail: ansarkar1@gmail.com

Introduction

Climate change has become the most significant environment and development challenge to human society in the twenty first century. Responding to climate change is the core task to achieve global sustainable development. International negotiations on the prevention of global warming and related actions not only concern the human living environment, but also directly impact the modernization process of developing countries set on a planned growth path. Although the process of global climate protection depends on the consensus of our scientific awareness, political wills, economic interests, society's level of acceptance, as well as measures adopted, a low-carbon development path should be the critical choice of future human development. Climate change is not only an environmental problem; it is also clearly a development problem since its adverse effects will disproportionately affect poorer countries with economies predominantly based on natural resources and related economic sectors (e.g. agriculture, forestry and fisheries). Yet, even countries with more diversified economies are vulnerable to climate change. It is therefore in the general interest of humanity at large to promote sustainable trends in mitigating greenhouse gas emissions (GHG)—the main culprit of climate change. Adaptation actions should run parallel to mitigation actions to achieve synergy and long-term sustainability.

In 1992, the United Nations Rio de Janeiro Earth Summit focused international attention on the growing environmental and development problems facing our planet. This was a landmark event that put the issues of sustainable development on the international agenda for the first time. Spurred on by a real sense of urgency, the 178 governments attending the Earth Summit signed up to Agenda 21—an ambitious global action plan for achieving sustainable development. This document set out a long-term vision for balancing economic and social needs within the carrying capacity of the Earth's natural resources. In the World Summit on Sustainable Development (WSSD) (2002), which took place in Johannesburg at the end of August 2002, five priorities which are connected to climate change were identified as water, energy, health, agriculture, and biodiversity. The United Nations Framework Convention on Climate Change (UNFCCC), which came into force in 1994, and the Kyoto Protocol in 1997, had established the link between environment and development. The Monterrey Conference in 2002 stressed the need to mobilize and increase the effective use of financial resources to eliminate poverty, improve social conditions, raise living standards and protect the environment (Berman 2008). The eleventh Conference of the Parties (COP-11) at Montreal in 2005, inter alia, highlighted the importance of adaptation measures, in addition to mitigation, as a positive and constructive response to climate change.

With the increasing pace of globalization, industrialization and rapid change of lifestyle, the demand and consumption of these hydrocarbon-based fossil fuels to stimulate economic growth is steadily rising—both in developed and developing economies. In the process, the emissions level is also rising phenomenally, and fast becoming alarming in its impact on the global climate. The Kyoto Protocol signed in 1997 (COP-3) and the Bali Roadmap designed by COP-14, a decade later in December 2007, inter alia, made a concerted effort to reduce emissions substantially in a legally binding manner for the Annex-I developed/industrialized nations by 2012 and beyond. The Action Plan provides a roadmap to the post-2012 climate regime to be agreed upon by the end of 2009. The key objective of the Copenhagen Climate Summit in December 2009 was to continue and improve the existing Kyoto agreement from 1997, which ends in 2012. The new Copenhagen Protocol and the amended Kyoto Protocol (i.e. two tracks) would form the core of the Agreement in December 2010 in Mexico.

At the international level, global climate change therefore has its place on the agenda for sustainable development. However, it is widely recognized now that the Sustainable Development Agenda will be grossly incomplete without the "Human Development" and "Democratic Option" components to choose and decide for any imperative measures linked to mitigation or adaptation actions to curb adverse impacts of climate change. The UNDP's Human Development Index measures reflect and mirror performance, and the concept of "human development" represents, together with "human rights", a basis for democracy in the thinking of O'Donnell (2004). This performance emphasis of the Democracy Ranking earns additional credibility by a spreading general understanding on how important good governance is. Societies (politics, the economy) certainly have a responsibility for the environment. The environment represents a crucial context, and should the environment collapse, then societies would be severely impacted. Therefore, societies cannot ignore features of environmental sustainability (UNDP 2007). Campbell (2008) in his paper on "The Basic Concept for the Democracy Ranking of the Quality of Democracy" gives a broad framework on how the various parameters of Democracy Ranking can pave the way to the quality of democracy to attribute to good governance, among others, for evolving sustainable environmental strategy. Furthermore, both mitigation and adaptation actions should a priori take cognizance of "equitable and differentiated responsibilities", as per the "Copenhagen Accord Principles", to ensure greater participation and involvement of the government, private (including joint partnership) and the local community at large for a holistic and sustainable development of the vulnerable areas affected by adverse impacts of climate change (Sarker and Banerjee 2010). The paper attempts to integrate the principles and modus operandi of mitigation and adaptation with sustainable development within the framework of people's participation with democratic option, with a focus on low-carbon economy, in order to help formulate suitable development strategies to offset adverse impacts of climate change.

Objective

Sustainable development involves a comprehensive and integrated approach to economic, social and environmental processes. Climate change is influenced not only by the climate-specific policies but also by the mix of development choices and the formulation of resulting development trajectories. Therefore, mainstreaming the sustainability concept as an integral part/component of climate change-related policy options requires that non-climate policies, programmes, and/ or individual or community-led democratic actions take climate change mitigation and adaptation into consideration and logical conclusions, in both developing and developed countries. The paper reviews the various adverse impacts of climate change and to what extent various policies, programmes and projects designed to address mitigation and adaptive actions on a planned and selective basis can minimize the adverse impacts of climate change. The paper also attempts to integrate the principles and modus operandi of mitigation and adaptation with sustainable development giving a sharper focus on the significance of promoting a low-carbon economy in order to help formulate suitable climate response strategies to offset adverse impacts of climate change with the following specific objectives in mind:

- To review the extent of environmental, economic and social impacts of climate change, mainly caused by the process of global warming, triggered by greenhouse gases.
- Secondly, how far the mitigation and adaptation policies, programmes and proactive actions help minimize the adverse impact of climate change.
- Thirdly, the economics and cost implications and funding modalities of mitigation as well as adaptation measures.
- Fourthly, to review the operational linkages and broad framework for integration of climate change with sustainable development.
- Lastly, addressing the energy and environmental security issues in the context of formulating a global climate change response.

Detailed discussions and critical analysis on each of the above-mentioned objectives follow in the succeeding sections of the paper.

Environmental Impact of Climate Change

Addressing global climate change is a paramount challenge of the twenty first century. Since the beginning of the industrial revolution, atmospheric concentrations of carbon dioxide (CO_2), the main heat-trapping GHG, have risen by 35%. This increase is primarily from the burning of fossil fuels and from deforestation. If current trends in GHG emission growth are not altered, global temperatures are expected to rise between 1.4 and 5.8°C (2.5–10.4°F) by 2100, according to the Intergovernmental Panel on Climate Change (IPCC 2007). Scientists predict the rise in global average surface temperature by another 1.4–5.8°C over the next 100 years. This projected rate of warming is the highest in 10,000 years. It is predicted to have strong adverse effects of global warming to rising sea levels, more irregular precipitation and an increase in extreme weather events such as droughts, floods, famine, earthquakes and storms. Such temperature changes are likely to have detrimental impacts on agricultural production, water supply, forests and overall human development.

The IPCC integrated assessment shows (IPCC 2007) that since 1750, human activities have been a major cause of global warming, while in the last 50 years, most of the global warming is the consequence of human activities with a probability of more than 90%, in particular from the GHG emissions due to the human use of fossil fuels. According the Third Working Group Report of the IPCC 4th Assessment (IPCC 2007), human actions to mitigate climate change are feasible, both economically and technologically. Actions to deploy key mitigation technologies in various sectors, adopt policy and administrative interference and shift the development pathway could all contribute greatly to mitigation of climate change. Thus, the IPCC assessment report has become the most important scientific foundation for global climate political decisions. The IPCC have produced a report (SRES) where a number of different future scenarios are described. The scenarios differ in whether the future will be "more environmental" (scenarios with the "B" prefix) or "market-oriented" (scenarios with the "A" prefix) and "more global" (scenarios with the "1" suffix) or "more regional" (scenarios with the "2" suffix) (Stern 2006). Forecasts of changes in global mean temperature for these scenarios are given in the IPCC report (Godard 2008). A 2009 study by the MIT Joint Program on the Science and Policy of Global Change looked at future predictions of climate change (Peston 2006). According to this study, there is a 57% likelihood of exceeding a 5°C increase in global mean temperature by 2100 (relative to the base years of 1981–2000).

The Third Assessment Report of the IPCC (2001) estimated that, globally, the average land and sea surface temperature has increased by $0.6 \pm 0.20°C$ since the mid- nineteenth century with much of the change occurring since 1976. Warming has been observed in all continents, with the greatest temperature changes occurring at middle and high latitudes in the northern hemisphere. Patterns of precipitation have also changed: arid and semi-arid regions are apparently becoming drier, while other areas, especially mid-to-high latitudes, are becoming wetter. Where precipitation has increased, there has also been a disproportionate increase in the frequency of the heaviest precipitation events (Karl and Knight 1998; Mason et al. 1999). The small amount of climatic change that has occurred so far has already had demonstrable effects on a wide variety of natural ecosystems (Walther et al. 2002). Unsustainable consumption patterns of the rich industrialized nations are responsible for the threat of climate change. Only 25% of the global population lives in these countries, but they emit more than 70% of the total global CO_2 emissions and consume 75–80% of many of the other resources of the world (Parikh et al. 1991). By 2050, the global population will reach 9 billion, barring substantial changes in demographic trends, with 2.5 billion more people in today's developing countries. Larger populations put more pressure on ecosystems and natural resources, intensify the competition for land and water, and increase the demand for energy. Most of the population increase will be in cities, which could help limit resource degradation and individual energy consumption.

But both could increase, along with human vulnerability, if urbanization is poorly managed.

There is international consensus that climate change is one of the most serious threats to sustainable development, both currently and in the future. Adverse impacts of climate change have already been observed on natural resources, food security, human health, the environment, economic activity and physical infrastructures (IISD 2007). Projections of future impacts are sobering: for example, by 2020, almost 50 million additional people may be at greater risk of hunger due to climate change. If a global temperature increase of 3–4°C is reached, changed runoff patterns and glacial melt could force an additional 1.8 billion people to live in a water-scarce environment by 2080 (UNDP 2007). Climate change imposes an added burden on development. Its impacts are already visible, and the most recent scientific evidence shows the problem is worsening fast, with current trajectories of GHG emissions and sea level rise outpacing previous projections. Climate change will affect numerous sectors and productive environments, including agriculture, forestry, energy, and coastal zones, in developed and developing countries. Developing economies will be more affected by climate change, in part because of their greater exposure to climate shocks and in part because of their low adaptive capacity.

Economic Impact Assessment of Global Warming and Climate Change

Economic Impact of Climate Change

The IPCC Fourth Assessment Report, published in 2007, looked at the aggregate economic impacts of climate change as follows: impacts of climate change are very likely (greater than 90% probability) to impose net annual costs, which will increase over time as global temperatures increase. Peer-reviewed estimates of the social cost of carbon (Net economic costs of damages from climate change aggregated across the globe and discounted to the specified year) in 2005 average US \$12 per tonne of CO_2, but the range from 100 estimates is large (-\$3–\$95/ tCO_2). This is due in large part to differences in assumptions regarding climate sensitivity, response lags, the treatment of risk and equity, economic and non-economic impacts, the inclusion of potentially catastrophic losses and discount rates. Aggregate estimates of costs mask significant differences in impacts across sectors, regions and populations and very likely underestimate damage costs because they cannot include many non-quantifiable impacts (Godard 2008).

Economic diversification is becoming increasingly important as an adaptation strategy to increase economic resilience and decrease reliance on climate-vulnerable economic sectors. At the sectoral level, increased economic resilience is often achieved by adapting existing practices to reduce exposure to risk (UNFCCC 2007a).

In the agricultural context, for example, increased resilience may be achieved by improved water management practices. Or it may involve diversifying the mix of crops planted and adding value to the primary products—which is the focus here. Reducing reliance on vulnerable sectors involves diversification of livelihood strategies. More diversified livelihood strategies can lead both to enhanced incomes and to spreading the risk for poor people, whose livelihoods are largely based on natural resources. Diversification strategies are thus important in managing current climate risks, particularly for subsistence agricultural communities, and will be critical for future adaptation to climate change.

According to a comprehensive IFAD study (2007), diversification strategies for managing current climate risks are in use in settings as varied as subsistence agricultural communities in northern Nigeria, the Sudan, Mexico, the Lao People's Democratic Republic and the Philippines; smallholder commercial farms in Argentina, Mexico and Thailand; and pastoral systems in Mongolia. Recommendations from these and other studies urge further economic diversification, as important for future adaptation to climate change. However, there are significant constraints on diversification that are identified in several Assessments of Impacts and Adaptations to Climate Change (AIACC) studies. Success will require integration of development strategies with adaptation planning. In Latin America and the Caribbean, projects with ecotourism activities (actual or planned) are to be found in Bolivia, Guatemala, Mexico, Panama and Peru [IFAD (2007) IFAD's Latin America and the Caribbean division: desk review of the portfolio related to indigenous peoples (unpublished)]. There are a number of successful examples from the Andean region, linked to cultural assets, and in the Amazon, projects have showcased environmental and cultural assets of indigenous groups. For example, the Mapajo Indigenous Ecotourist Project (Beni, Bolivia), funded by the IFAD grant PRAIA, constitutes an indigenous initiative to create and diversify employment, increase income-generating opportunities for indigenous communities, preserve biodiversity, and promotes intercultural understanding. More diversified livelihood strategies can lead to enhanced incomes and can diminish the risk for poor people whose livelihoods are largely based on natural resources. The IFAD study also revealed that the richer countries have more resources to cope with climate impacts, and better educated and healthier populations are inherently more resilient. But the process of growth may exacerbate vulnerability to climate change, as in the ever-increasing extraction of water for farming, industry, and consumption in the drought-prone provinces around Beijing, and as in Indonesia, Madagascar, Thailand, and US Gulf Coast, where protective mangroves have been cleared for tourism and shrimp farms. Growth is not likely to be fast enough for low-income countries to afford the kind of protection that the rich can afford. Bangladesh and the Netherlands are among the countries most exposed to rising sea levels. Bangladesh is already doing a lot to reduce the vulnerability of its population, with a highly effective community-based early warning system for cyclones and a flood forecasting and response programme drawing on local and international expertise.

Climate change impacts can be measured as an economic cost (Smith et al. 2001). This is particularly well-suited to market impact, which is an impact that is linked to market transactions and directly affects GDP. Monetary measures of non-market impacts, e.g. impacts on human health and ecosystems, are more difficult to calculate. Evaluating the economic consequences of the climate change, as well as assessing the environmental economic policies associated with it, requires a good understanding of both natural and socioeconomic processes. Specific models, named Integrated Assessment Models (IAM), are used to this purpose. The idea behind the IAM models is relatively straightforward (in theory): climatologic sub-models get information about human-induced GHG emissions from economic sub-models, simulating levels of economic activity, whereas information about climate and temperature changes are used as an input in the determination of economic scenarios. The economic valuation of the various impacts of the climate change is a difficult task. It requires interdisciplinary knowledge from many fields, other than economics: agronomy, geology, meteorology, demography, medicine, political science, biology, engineering and physics.

Information on the impacts of global warming is now available for several regions and countries. The best-studied regions are developed countries, in particular the United States, where climate change impacts have been analysed in a series of studies, following up on initial work by Tirpak (2008). In the context of an Asian Development Bank (ADB 2007) project on climate change in Asia, global warming impacts have also been analysed for a number of Asian countries. The economic cost of climate change will be the cost of the adaptation measures implemented plus the value of the residual damage. The available information is not sufficient to support estimates of the cost-effective use of various adaptation measures and hence of the total cost of adaptation and the residual damages. The opportunities to adapt to climate change and the potential to reduce climate change damage vary by country. It is clear that adaptation will cost less than the potential damages for at least some of the impacts of climate change in all countries. Thus, the economic cost should be substantially lower than the high estimate of the potential economic impact in the absence of adaptation actions. Insofar as the analysis of economic impact of climate change is concerned, the most crucial research topics for the future investigation may include: (a) damage estimates for less developed countries; (b) improved estimates for non-market losses, particularly human morbidity and unmanaged ecosystems; (c) assessment of the importance of variability and extreme events; (d) models of the process of adaptation and the dynamics of vulnerability; and (e) formal uncertainty assessments and analyses on the outcomes.

Linking climate Change and Disaster Risk Reduction in Vulnerable Ecosystems

An important area for adaptation understands the impacts of and vulnerability to current and future climate variability and extreme events—and the implications for sustainable development. There are many possibilities for synergies between disaster risk reduction and climate adaptation, with both approaches aiming to develop resilient communities with reduced levels of vulnerability. Both approaches are ways of managing uncertainty. This includes tackling risks as well as considering how to prepare for events that cannot be predicted (LCA Network 2006). One difference is that, apart from extreme events, adaptation to climate change also involves adjusting to changes in mean climate variability, which has not been a customary focus of disaster risk reduction. Hundreds of thousands of poor people in developing countries live with constant threats to their lives and livelihoods due to weather-related natural disasters—drought, flooding, landslides, soil erosion and desertification (Practical Action 2007). These threats include homelessness, loss of possessions and livelihoods, and risks of injury and death. Recent years have seen an increasing focus on disaster risk reduction that highlights many synergies with climate change adaptation. Scientific evidence points to human-induced climate change as the underlying cause of the rise in hydrometeorological events over the past decade. Natural disasters are increasing in number and intensity, with extremely severe costs in human lives and social and economic infrastructure. In this context, important components of adaptation strategies are a better understanding of the risks associated with environmental change and climate change, and development of mechanisms to manage these, such as early warning systems. In the context of climate-related risks and extreme weather events, the level of risk depends on exposure to the hazard, the vulnerability both of people's livelihoods and of human settlements, and the status of disaster mitigation activities. A further key factor is the ability of people to protect themselves and to cope with hazards (Uitto and Shaw 2006).

In many places, previously uncommon risks are becoming more widespread. Consider floods, once rare but now increasingly common, in Africa and the first hurricane ever recorded in the South Atlantic, which hit Brazil in 2004. Reducing disaster risk—through community-based early warning systems, climate monitoring, safer infrastructure, and strengthened and enforced zoning and building codes, along with other measures—becomes more important in a changing climate. Financial and institutional innovations can also limit risks to health and livelihoods. This requires domestic action—but domestic action will be greatly enhanced if it is supported by international finance and sharing of best practice. A substantial part of the potential economic impact of climate change is due to loss of facilities due to sea level rise and hurricanes. Such losses would occur over decades. Adaptation will occur in response to, or in anticipation of, such damage. Adaptation incurs costs, but planned adaptation can reduce the costs. The adverse effects of climate change on agriculture can be mitigated by switching to drought- and heat-resistant varieties of the existing crops, switching to different crops, and by irrigating crops where sufficient freshwater supplies are available. Damage to natural ecosystems can also be reduced by protecting the ecosystems, helping species migrate in response to climate change and habitat shifts, and restoring damaged ecosystems.

Linking Sustainable Development with Climate Change Responses

The Concept of Sustainable Development

The concept of sustainable development had its roots in the idea of a sustainable society and in the management of renewable and non-renewable resources. The concept was introduced in the World Conservation Strategy by the International Union for the Conservation of Nature (IUCN). The World Commission on Environment and Development adopted the concept and launched sustainability into political, public and academic discourses. The concept was defined as "development that meets the needs of the present without compromising the ability of future generations to meet their own needs" (Bojo et al. 1992). While this definition is commonly cited, there are divergent views in academic and policy circles on the concept and how to apply it in practice (Banuri et al. 2001; Cocklin 1995; Pezzoli 1997; Robinson and Herbert 2001).

The concept of sustainable development was adopted by the World Commission on Environment and Development (WCED), and there is agreement that sustainable development involves a comprehensive and integrated approach to economic, social and environmental processes. Discourses on sustainable development, however, have focused primarily on the environmental and economic dimensions. The importance of social, political, and cultural factors is only now getting more recognition. Integration is essential in order to articulate development trajectories that are sustainable, including addressing the climate change problem. There is growing emphasis in the literature on the two-way relationship between climate change mitigation and sustainable development. The relationship may not always be mutually beneficial. In most instances, mitigation can have ancillary benefits or co-benefits that contribute to other sustainable development goals (climate first). Development that is sustainable in many other respects can create conditions in which mitigation can be effectively pursued.

There is a growing understanding of the possibilities to choose mitigation options and their implementation such that there is no conflict with other dimensions of sustainable development; or, where trade-offs are inevitable, to allow a rational choice to be made. The sustainable development benefits of mitigation options vary within a sector and over regions in the following manner:

- Generally, mitigation options that improve productivity of resource use, whether energy, water or land, yield positive benefits across all three dimensions of sustainable development. Other categories of mitigation options have a more uncertain impact and depend on the wider socioeconomic context within which the option is implemented.
- Climate-related policies, such as energy efficiency, are often economically beneficial, improve energy security, and reduce local pollutant emissions. Many energy supply mitigation options can also be designed to achieve other

sustainable development benefits, such as avoided displacement of local populations, job creation, and rationalized human settlement design.
- Reducing deforestation can have significant biodiversity, soil and water conservation benefits, but may result in loss of economic welfare for some stakeholders. Appropriately designed forestation and bioenergy plantations can lead to reclamation of degraded land, manage water runoff, retain soil carbon and benefit rural economies, but could compete with land for agriculture and may be negative for biodiversity.
- There are good possibilities for reinforcing sustainable development though mitigation actions in most sectors, but particularly in waste management, transportation, and building sectors, notably through decreased energy use and reduced pollution.

Mitigation and Adaptation in the Context of Social Dynamics and Sustainable Development

The UNFCCC highlights two fundamental response strategies: mitigation and adaptation. While mitigation seeks to limit climate change by reducing the emissions of GHGs and by enhancing "sink" opportunities, adaptation aims to alleviate the adverse impacts through a wide range of system-specific actions (Fussel and Klein 2002). Although both mitigation and adaptation measures must be pursued to tackle the climate change problem and to create an effective and inclusive international climate change regime, more attention has been devoted to mitigation in the past, both in scientific research and policy debate. Sensitivity to the issue of adaptation has grown over the last couple of years, particularly after the Intergovernmental Panel on Climate Change (IPCC) Third Assessment Report (TAR). Adaptation has now emerged as an urgent policy priority, prompting action both within and outside the climate change negotiations (Parry et al. 2005).

In the larger context of achieving long-term sustainability, understanding of mitigation and adaptation actions assumes prime importance to cope with climate change-related vulnerability. Hence, we need to know the implications of mitigation and adaptation in relation to climate change parameters in more clear terms. Climate mitigation is any action taken to permanently eliminate or reduce the long-term risk and hazards of climate change to human life and property. The IPCC defines mitigation as: "An anthropogenic intervention to reduce the sources or enhance the sinks of greenhouse gases." Climate adaptation refers to the ability of a system to adjust to climate change (including climate variability and extremes) to moderate potential damage, to take advantage of opportunities, or to cope with the consequences. The IPCC defines adaptation as the "adjustment in natural or human systems to a new or changing environment". Adaptation depends greatly on the adaptive capacity or adaptability of an affected system, region, or community to cope with the impacts and risks of climate change. The adaptive capacity of communities is determined by their socioeconomic characteristics. Enhancement of adaptive capacity represents a practical means of coping with changes and uncertainties in climate, including variability and extremes. The key features of climate change for vulnerability and adaptation are those related to variability and extremes, not simply changed average conditions. Most sectors and regions and communities are reasonably adaptable to changes in average conditions, particularly if they are gradual. However, these communities are more vulnerable and less adaptable to changes in the frequency and/or magnitude of conditions other than average, especially extremes. Sectors and regions will tend to adapt autonomously to changes in climate conditions.

The capacity to adapt varies considerably among regions, countries and socioeconomic groups and will vary over time. The most vulnerable regions and communities are those that are highly exposed to hazardous climate change effects and have limited adaptive capacity. Adaptation to climate change presents dilemmas of justice to the international community, including those around the responsibility of developed countries to assist developing countries in adapting to changing climate. Adaptation is comprised of inaction and proactive and reactive responses at the international, national, local and individual levels. Inaction at higher levels delegates the responsibility for adaptation to lower levels, and higher-level responses influence alternatives that are available at lower levels. Justice is thus always implicit in the choice of adaptive responses (Paavola and Adger 2002). The framework proposed by Paavola and Adger for analysing justice issues in the context of adaptation is broadly pluralist and pragmatic. First, we do not distinguish political from moral choices: all choices are moral and have justice implications in the sense that they need to be informed by some values that guide the comparison of alternatives and choice between them. Second, they recognized that different choices are informed by different values and that often compromises have to balance between multiple standpoints. Distributive justice in adaptation probably also needs to acknowledge other concerns in addition to distribution of utility or welfare (Table 1). For example, security and the absence of dangerous climate change impacts can be understood as hallmarks of justice that are valuable independently of their welfare consequences. Alternatively, security and absence of danger can be framed as rights. The Framework Convention's Article 2 indeed treats the integrity of climate system and the absence of dangerous climate change either as a valuable thing in itself or its preservation as a guiding principle.

Different countries and people and different activities have different mitigation potentials and costs, different vulnerabilities, and different adaptive capacities. It is unfortunate that the greatest mitigation potentials are normally associated with rich economies and people with lower vulnerabilities and higher adaptive capacities, whereas poorer people and economies tend to have lower emissions (and hence current mitigation potential), higher vulnerabilities and lower adaptive capacities. This has important implications for development pathways, for climate negotiations, and for the welfare of poor people.

Table 1 Justice issues in adaptation to climate change

Justice issues	Examples of justice rules	Questions and problems
Distributive justice	*Utilitarian rules*	
	–Welfare maximisation	Commensurability of types of benefit
	–Maximax; most able adapt	Who defines adaptive capacity and how?
	–Maximin; most vulnerable prioritised	Who defines vulnerability and how?
	Other consequentialist rules	
	–Avoidance of climate danger	
	Deontoiogical rules	Who defines danger and how?
	–Equality of burden-sharing	Equality of impacts, exposure, or adaptation?
	–Existence right of non-human species	Who defines rights?
Procedural justice	*Utilitarian rules*	
	–Rules that give effect to preferences and ability and willingness to pay	Why not willingness to accept compensation?
	Other consequentialist rules	
	–Self-determination: affected parties only	Who is internal and who is external?
	Deontoiogical rules	
	–Equality	Why not recognise the intensity of interests of e.g. most vulnerable?

Source: Paavola and Adger 2002

The rationale for integrating adaptation into development strategies and practices is underlined by the fact that interventions required to increase resilience to climate variability and change generally further development objectives. Adaptation calls for natural resource management, buttressing food security, development of social and human capital and strengthening of institutional systems (Adger et al. 2003). Such processes, besides building the resilience of communities, regions and countries to all shocks and stresses, including climate variability and change, are good development practice in themselves. Hence, the inclusion of climatic risks in the design and implementation of development initiatives is vital to reduce vulnerability and enhance sustainability.

Since vulnerability and its causes play essential roles in determining impacts, comprehending the dynamics of vulnerability is as important as understanding climate itself (Handmer et al. 1999). Definitions of vulnerability vary widely among different scholars. Researchers in social geography and political ecology regard vulnerability as an a priori condition of a household or community that is determined by socioeconomic and political factors (Blaikie et al. 1994; Bohle et al. 1994). Contiguously, Kelly and Adger (2000) declare that vulnerability is the "ability or inability of individuals or social groupings to respond to, in the sense of cope with, recover from, or adapt to, any external stress placed on their livelihoods and well-being". A common theme in the climate change impacts and vulnerability literature is the idea that communities, social groups, sectors, regions and nations differ in the degree of vulnerability to climate change,

i.e. there exist differential vulnerabilities (Bohle et al. 1994). Developing countries, SIDS (Small Island Developing States), people living in arid and semi-arid lands, water-limited or flood-prone areas, as well as countries and sectors heavily dependent on climate sensitive sectors—agriculture, water resources, forestry, fisheries, etc. are particularly at risk. This is partly due to the fact that climate-induced changes in temperature and precipitation will occur unevenly and hence climate change impacts will be unevenly distributed across the
globe.

Adaptation to climate change can be based on uncoordinated choices and actions of individuals, firms and organizations or on collective action and choice at local, national, international, as well as intermediate and multiple levels (Table 2). The distinction between individual and collective responses can be based on different theoretical approaches to these varieties of choice. It is also important because collective choices bring up issues such as representation, participation, procedure and assent that do not characterize individual choices. To date, research on adaptation to climate change has focused on empirical differences in adaptive responses. However, for the analysis of justice, it is more important to understand who decides on adaptive responses and how adaptive responses are timed with respect to climate change impacts (Paavola and Adger 2002).

The effects of climate change have increasingly come to be seen as a threat to human security. Human beings are causing environmental degradation. At the same time, they are among the victims, as their lives and livelihoods are distorted along with the distortion of whole ecosystems. However, those who are most vulnerable to the effects of climate change are often not those who have contributed to global warming the most (Zipprich et al. 2010). The poor and marginalized, as well as future generations, have to endure the consequences of the actions of the wealthy in the present and the past. The situational and structural effects of environmental pollution, such as natural hazards, extreme weather events and climate change are a threat to human security. Considering the need to curb GHG emissions so as to mitigate the effects of climate change, energy consumption is best considered a limited resource in the sense that we cannot dramatically increase the supply of energy for reasons of achieving human security without regard for the consequences of additional energy use on human security through the affects of climate change as depicted in Fig. 1.

The problematic of energy equity and human security can hence be framed as an issue of distributive justice between populations and between generations: ensuring that no population or generation threatens the security (welfare) of other populations or generations through over-consumption. Putting these two aspects of the distributional question together we can see that energy use is both a condition of human security as well as a threat to human security and there does exist an equilibrium between the two (Fig. 2).

Because of the broad nature of the energy equity and human security issue, there are many stakeholders. Every individual has some kind of involvement with

Table 2 A typology of adaptive responses to climate change impacts with examples of responses in the context of agricultural productivity and food security

Response	Proactive	Reactive	Inaction
International	Guidelines for national adaptation strategies, support for development of new crop varieties	Food aid measures	No responses are taken to instigate context-specific behavioural responses
National	Grain storage, investments and changes in agricultural policies to adapt crop mix and agricultural practices to changing climate	Changes in tariffs and fiscal policy to augment food imports; disaster relief and food aid	No small-scale proactive investments in infrastructure that confer only local adaptive benefits
Local	Small-scale infrastructure investments for groundwater recharge, irrigation and flood protection, local seed banks, and coordination of adaptive responses	Collective action and reciprocity in overcoming obstacles in agricultural production and mitigating the effects of shortages of food and water	Migration ignored as an adaptive response
Individual	Diversification of livelihood, investment in human capital, physical capital, and alteration of agricultural practices	Migration	Adjustment of increased vulnerability and/or reduced welfare

Source: Paavola and Adger 2002

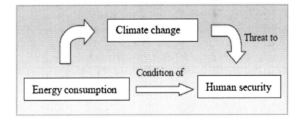

Fig. 1 Relationship between climate change, energy consumption and human security. Source: Zipprich et al. (ECCAP Project Report 2010)

energy, be it through making use of various forms of energy directly or being deprived of the use of energy in decent quality. The extent to which we are able to access energy is a factor which influences our way and quality of life. At the same time, the amount of energy used for our own convenience is also having an effect on others and the environment in the present and in the future. It is not only the interest in the benefits of sustained energy availability in private and public surroundings, but also the interest in the minimization of externalities that shape the attitude of individuals towards energy equity and human security.

Fig. 2 Equilibrium between energy consumption and human security. Source: Zipprich et al. (ECCAP Project Report 2010)

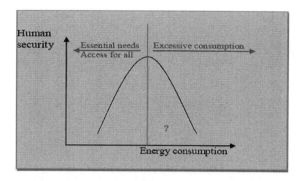

Economic Consideration for Mitigation and Adaptation Actions

Economic benefits and costs are important criteria but are not sufficient to adequately determine the appropriateness of adaptation measures; there also has been little research to date on the roles and responsibilities in adaptation of individuals, communities, corporations, private and public institutions, governments, and international organizations. Given the scope and variety of specific adaptation options across sectors, individuals, communities, and locations, as well as the variety of participants—private and public—involved in most adaptation initiatives, it is probably infeasible to systematically evaluate lists of particular adaptation measures; improving and applying knowledge on the constraints and opportunities for enhancing adaptive capacity is necessary to reduce vulnerabilities associated with climate change. The capacity to adapt to climate change is determined by factors such as economic resources and other assets, technology and information, infrastructure and stable and effective institutions. Since many partner countries are poorly endowed with these attributes and are consequently highly vulnerable to climate change, the enhancement of their adaptive capacity is therefore likely both to reduce vulnerability to climate change and promote sustainable development. Adverse effects on ecosystems, natural resources and related economic sectors will affect poor people hardest. Climate change could cause many complex alterations: a shift in temperature zones caused by climate change could lead to the extinction of species in many locations while many areas where the warm seasons will become drier will be at risk of land degradation, drought and desertification.

The Stern review (Stern 2007) proposes stabilizing the concentration of GHG emissions in the atmosphere at a maximum of 550 ppm CO_2e by 2050. The Review estimates that this would mean cutting total GHG emissions to three-quarters of 2007 levels (Stern 2007). The Review further estimates that the cost of these cuts would be in the range of −1.0 to +3.5% of GDP, with an average estimate of approximately 1%. Stern has since revised his estimate to 2% of GDP.

The Review emphasizes that these costs are contingent on steady reductions in the cost of low-carbon technologies. Mitigation costs will also vary according to how and when emissions are cut: early, well-planned action will minimize the costs. One way of estimating the cost of reducing emissions is by considering the likely costs of potential technological and output changes. Policy-makers can compare the marginal abatement costs of different methods to assess the cost and amount of possible abatement over time. The marginal abatement costs of the various measures will differ by country, by sector, and over time.

Mitigation and Adaptation Strategies for Global Change addresses a wide range of timely environment, economic and energy topics including global climate change, stratospheric ozone depletion, acid deposition, eutrophication of terrestrial and aquatic ecosystems, species extinction and loss of biological diversity, deforestation and forest degradation, desertification, soil resource degradation, land use change, sea level rise, destruction of coastal zones, depletion of freshwater and marine fisheries, loss of wetlands and riparian zones and hazardous waste management. The world bank has identified critical linkages between climate impacts and poverty reduction, pointing out that if atmospheric CO_2 concentrations were to double from pre-industrial levels, "developing countries would suffer economic costs of 5–9% of GDP, several times higher than industrialized countries, and the poor in the Bank's borrowing countries would be at the greatest disadvantage" (IPCC 2001). Industrialized countries are largely responsible for the build-up of GHGs in the atmosphere thus far, and must bear the brunt of the mitigation effort. But developing countries can play an important role in reducing emissions growth within the context of their continued economic development. Nearly 80% of the world's population lives in developing countries, which already account for over 40% of current world emissions and given present trends, this share will rise to 56% by 2025 (Shah 2008).

Financing Mitigation and Adaptation for Meeting the Climate Challenge

Investment and Financial Flows to Address Climate Change

In 2007, the UNFCCC Secretariat prepared a report on "Investment and Financial Flows to Address Climate Change" (UNFCCC 2007a, b). The report covers mitigation and adaptation in various sectors over the period to 2030. The report defines an investment as the initial (capital) cost of a new physical asset with a life of more than one year, such as the capital cost of a gas-fired generating unit or a water supply system. A financial flow is an ongoing expenditure related to climate change mitigation or adaptation that does not involve physical assets, such as research or healthcare. These investment and financial flows are not the same as the cost of addressing climate change; changes to the operating costs of

Table 3 Change to the annual investment and financial flows in 2030 for climate change mitigation

Sectors	Global (billions of 2005 US dollars)	Share of developing parties (%)
Fossil fuel supply	(−) 59	50–55
Electricity supply	(−) 7	50–55
Fossil-fired generation, transmission and distribution	(−) 156	50–55
"Renewables", nuclear and CCS	148	50–55
Industry	36	50–55
Building	51	25–30
Waste	0.9	66–70
Transport	88	40–4
Forestry	21	Almost 100
Agriculture	35	35–40
Energy research, development and demonstration (RD&D)	35–45	–
Net change	200–210	35–40

Source: UNFCCC 2007a, Tables IX-61, IX-62 and IX-63, pp. 173–174

investments are not considered, nor are damages due to climate change estimated. The changes to the investment and financial flows in 2030 for climate change mitigation are shown in Table 3. Global additional investment and financial flows of US $200–210 billion will be needed in 2030 to return global GHG emissions to 2005 levels with about US $75 billion of this funding needed in developing countries (Tirpak 2008). The net increase involves reduced investment for fossil fuel supply and large shifts in investment for electricity generation. Annual investment in fossil fuel supply and associated infrastructure in 2030 is almost US $60 billion lower due to increased energy efficiency. However, global fossil fuel equipment in industry and buildings consumption is still about 30% higher than in 2000. Improved vehicle efficiency, including hybrid vehicles, increases energy efficiency in the transportation sector. Annual spending on energy research and development (R&D), US $10 billion, and on demonstration, US $25–35 billion, was projected to double by 2030. Currently, most research is undertaken in a few developed countries; what share of the research will be conducted in developing countries in 2030 is difficult to predict.

The IEA (2008) also undertook a bottom-up assessment of the investment needs out to 2050 for a wide range of power generation, infrastructure, transport and energy demand technologies in 2008. Importantly, the IEA study examined the implications of meeting long-term abatement targets on near- to mid-term investments. A comparison of current and additional finance from this analysis for achieving a 500 parts per million scenarios is presented in Table 4. The estimated additional finance needed increases with each stage of the innovation cycle, from US $10 to 100 billion per year for R&D to US $1,000 billion annually for diffusion. These estimates are the total annual investment needed at each stage.

Table 4 Estimated additional financing needs by stage of technology innovation pathway

	R&D (global)	Demonstration (global)	Deployment (global)	Deployment (developing)	Diffusion (global)	Diffusion (developing)
Existing finance	US $20 billion per annum 1	Not available	Not available	US $45 billion per annum 1	US $71 billion in 2006 1, 2	US $14.2 billion per annum 1, 2
Additional finance required	US $10–100 billion per annum	US $27–36 billion per annum until 2030	US $18.25–40.75 billion per annum 1	US $73–163 billion per annum 1	US $1,000 billion per annum from 2010–2050 1, 3	US $370 billion per annum from 2010–2050 4

Source: derived from: IEA 2006, 2008; Greenwood et al. 2007; Doornbosch et al. 2008

Table 5 Change to the annual investment and financial flows in 2030 for climate change adaptation

	Global (billions of 2005 US dollars)	Developing countries (%)
Agriculture	14	50
Water supply	11	85
Human health	5	100
Coastal protection	11	45
Infrastructure	8–130	25–35
Total	49–171	35–60

Source: UNFCCC 2007a, Table IX-65, p. 177

The support needed for deployment and diffusion will be only the cost in excess of the cost of the incumbent technology.

The UNFCCC estimates the incremental investment and financial flows needed to adapt to climate change in selected sectors total US $49–171 billion globally in 2030 with US $28–67 billion of this total being needed in developing countries. Other recent estimates of adaptation costs for developing countries include World Bank (US $9–41 billion), (World Bank 2007, 2009a), Oxford Institute for Energy Studies (US $2–17 billion) (Müller and Hepburn 2006), Oxfam (greater than US $50 billion) (Oxfam 2007) and UNDP (US $86 billion) (Watkins 2007). While these estimates differ in terms of their scope and approach, and hence are not directly comparable, they all show that tens of billions US dollars annually will be needed by developing countries to adapt to climate change. The estimated additional investment and financial flows needed for climate change adaptation in 2030 are shown in Table 5. The agriculture, forestry and fisheries sector is estimated to need an additional investment of US $11 billion annually in new capital, such as irrigation systems, equipment for new crops and fishing practices, and relocation and modification of processing facilities. An additional US $3 billion will be needed annually for research and extension activities to facilitate adaptation. About half of the total requirement will be for developing countries.

The capital cost of the water supply infrastructure needed to meet the projected population and economic growth to 2030 given the projected climate in 2050 is about US $800 billion with about 28% of this—US $225 billion—estimated to be due to climate change (UNFCCC 2007b). Spreading the capital cost over the 20-year life of the facilities leads to an annual adaptation cost of US $11 billion. About 85% of the additional investment would be needed in developing countries (UNFCCC 2007b). The UNFCCC report (2007b) on investment and financial flows concluded that meeting the additional investment and financial flows would require a combination of:

- commitments by developed countries to provide additional financial assistance to developing countries under the Convention;
- appropriate national policies to encourage private investment and domestic government investment in mitigation and adaptation measures;
- optimal use of the funds available under the Convention and from other sources to spread the risk across public and private sources;

Table 6 Change to the annual investment and financial flows in 2030 for climate change mitigation

Sectors	Global (billions of 2005 US dollars)	Share of developing parties (%)
Fossil fuel supply	(−) 59	50–55
Electricity supply	(−) 7	50–55
Fossil-fired generation, transmission and distribution	(−) 156	50–55
"Renewables", nuclear and CCS	148	50–55
Industry	36	50–55
Building	51	25–30
Waste	0.9	66–70
Transport	88	40–4
Forestry	21	Almost 100
Agriculture	35	35–40
Energy research, development and demonstration (RD&D)	35–45	–
Net change	200–210	35–40

- expansion of the carbon market through more stringent commitments by Annex I Parties to increase demand and possible additional mechanisms to increase supply; and
- new sources of predictable funds to provide additional external financial flows to developing countries for adaptation and mitigation.

The funding required for mitigation, adaptation, and technology is massive. In developing countries, mitigation could cost $140–175 billion a year over the next 20 years (with associated financing needs of $265–565 billion); over the period 2010–2050, adaptation investments could average $30–100 billion a year (in round numbers). The private sector will have a key role in financing mitigation through carbon markets and related instruments. But official flows or other international funding will be an important complement to build capacity, correct market imperfections, and target areas overlooked by the market. Private finance will also be important for adaptation, because private agents—households and firms—will carry much of the adaptation burden. But good adaptation is very closely linked to good development, and those most in need of adaptation assistance are the poor and disadvantaged in the developing world. This means public finance will have a key role (The World Development Report 2010).

The International Institute for Sustainable Development (IISD) has prepared three papers to explore how major developing economies might become effectively engaged in a post-2012 global climate change regime (Sadler 1996). The goal of this first background paper, "Financing Mitigation and Adaptation in Developing Countries: New options and mechanisms", is to review financing issues relating to mitigation and adaptation under the UNFCCC. The changes to the investment and financial flows in 2030 for climate change mitigation are shown in Table 6. Global additional investment and financial flows of US $200–210 billion will be needed in 2030 to return GHG emissions to 2005 levels with about US $75 billion of this funding needed in developing countries. The net increase involves reduced investment for fossil fuel supply and large shifts in investment for electricity generation. Annual investment in fossil fuel supply and associated infrastructure in 2030 is almost US $60 billion lower due to increased energy efficiency. However, global fossil fuel consumption is still about 30% higher than in 2000.

Substantial shifts in investment for electricity supply will be needed. Mitigation is projected to reduce investment for fossil-fired generation, transmission and distribution of the power supply by US $156 billion in 2030. Almost all, about US $148 billion, needs to be shifted to renewables, nuclear, and carbon capture and storage (CCS). Increased energy efficiency requires additional investment for electrical and fossil fuel equipment in industry and buildings. The UNFCCC estimates the incremental investment and financial flows needed to adapt to climate change in selected sectors total US $49–171 billion globally in 2030 with US $28–67 billion of this total being needed in developing countries.

Financial Needs for Adaptation

Adaptation to climate change will bring with it additional costs for both the public and the private sector. However, assessing the costs and, especially, the benefits of adaptation is considerably more complicated than it is for mitigation. The Intergovernmental Panel on Climate Change in its Fourth Assessment Report observed that the current literature on adaptation costs and benefits is quite limited and fragmented, and that equity considerations (i.e. the distribution of costs and benefits) are hardly addressed at all. A recent review by the Organization for Economic Cooperation and Development (OECD) on the same subject (Parry et al. 2008) found that there is very little quantified information on the costs of adaptation in developing countries, and most studies are constrained to a few sectors within countries (mostly coastal zones; to a lesser extent water, agriculture and health).

The financing needs of climate change mitigation and adaptation are uncertain and large. In response, an extensive architecture of financing for climate change mitigation and adaptation has started to develop around the world. Is there a role for a regional approach in this developing financing architecture? The paper on Financing Climate Change Mitigation and Adaptation Role of Regional Financing Arrangements by ADB (Sharan 2008) examines the issue in the context of the Asia and Pacific region. The paper argues that, while climate change is a global public good, it has strong regional features, and, consequently, regional institutions and financing arrangements play an important role. In particular, regional financing arrangements have a special role in climate change adaptation. Regional financing arrangements complement and strengthen the global collective action on climate change. Another important priority is to create specialized regional funds to address key region-specific climate change adaptation needs. Finally, opportunities for private sector financing in a regional setting should be promoted through appropriate policy and regulatory support.

Adaptive Capacity Building

Potential Link Between Climate Change and Sustainable Development

Development—that is socially, economically, and environmentally sustainable—is a challenge, even without global warming. Economic growth is needed, but growth alone is not enough if it does not reduce poverty and increase the equality of opportunity. And failing to safeguard the environment eventually threatens economic and social achievements. These points are not new. They only echo what still is, after more than 20 years, perhaps the most widely used definition of sustainable development: "development that meets the needs of the present without compromising the ability of future generations to meet their own needs." By definition, then, unmitigated climate change is incompatible with sustainable development. Adaptation is adjustment in ecological, social, or economic systems in response to actual or expected climatic stimuli and their effects or impacts. This term refers to changes in processes, practices or structures to moderate or offset potential damages or to take advantage of opportunities associated with changes in climate. It involves adjustments to reduce the vulnerability of communities, regions or activities to climatic change and variability. Adaptation is important in the climate change issue in two ways—one relating to the assessment of impacts and vulnerabilities, the other to the development and evaluation of response options. Understanding expected adaptations is essential to impact and vulnerability assessment and hence is fundamental to estimating the costs or risks of climate change (Fankhauser 1996; Yohe et al. 1996; Tol et al. 1998; UNEP 1998; Smit et al. 1996; Pittock and Jones 2000). Article 2 of the UNFCCC (2007b) refers to "dangerous" human influences on climate in terms of whether they would "allow ecosystems to adapt, ensure food production is not threatened, and enable economic development to proceed in a sustainable manner." The extent to which ecosystems, food supplies and sustainable development are vulnerable or "in danger" depends on their exposure to climate change effects and on the ability of impacted systems to adapt. Thus, to assess the dangerousness of climate change, impact and vulnerability assessments must address the likelihood of autonomous adaptations. Adaptation also is considered an important response option or strategy, along with mitigation (Smith et al. 1996). Even with reductions in GHG emissions, global temperatures are expected to increase, other changes in climate—including extremes—are likely, and sea level will continue to rise (Raper et al. 1996). Hence, development of planned adaptation strategies to deal with

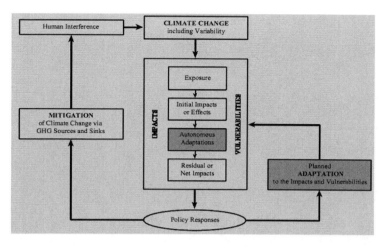

Fig. 3 Places of adaptation in climate change issues (Smit et al. 1999)

these risks is regarded as a necessary complement to mitigation actions (Smith et al. 1996; Smit et al. 1999—Fig. 3). Article 4.1 of the UNFCCC commits parties to formulating, cooperating on, and implementing "measures to facilitate adequate adaptation to climate change."

The first empirical studies of climate adaptation (reviewed and assessed in the IPCC AR4) showed that the success of adaptation in developing countries relies strongly on broader development progress. When adaptation is limited to responses specific to climate change, it neglects the fact that vulnerability to climate change does not emerge in isolation. For example, it may help to provide a rural household that grows a particular subsistence crop with a more drought-resistant variety, but a more robust and comprehensive adaptation strategy would seek to improve food security through a set of coordinated measures that include agricultural extension, crop diversification, integrated pest management and rainwater harvesting. In addition, a poor rural household is more likely to use these options if it has a literate family member, if it has access to investment capital through local financial institutions, if it enjoys relatively intact social networks, and if it can hold policy-makers accountable. In other words, it takes more than narrow, climate-focused measures to build adaptive capacity. A recent study by McGray et al. (2007) confirmed this view. It reviewed more than 100 initiatives labelled as adaptation in developing countries and found that in practice there is little difference between these adaptation initiatives and what can be considered good development. The difference lies more in the definition of the problem and the setting of priorities than in the implementation of solutions. The study presented adaptation as a continuum, ranging from more narrowly defined activities aimed specifically at addressing impacts of climate change, to building response capacity and addressing the drivers of vulnerability (Fig. 4).

Many developing countries have already begun to integrate climate risks into their mainstream sectoral and national development planning. The benefit of this

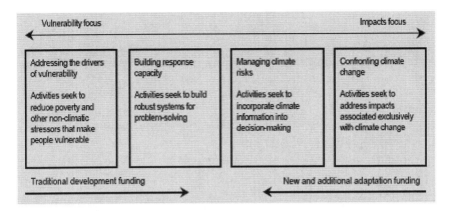

Fig. 4 Adaptation as a continuum from addressing the drivers of vulnerability to confronting the impacts of climate change. Source: adapted from McGray et al. (2007)

integration effort is that it often refers to today's and tomorrow's climate, thus ensuring the effectiveness and sustainability of investments. India, for example, has adopted policies to reduce risks and enhance the adaptive capacity of its most vulnerable sectors and groups. The policies are primarily driven by the objective of ensuring sustainable livelihoods and alleviating poverty. For example, adaptation in the agricultural sector includes the development of drought-resistant crop varieties, the promotion of crop diversification and the extension of the National Agricultural Insurance Scheme. Overall, India reports to be spending 2% of its gross domestic product on adaptation activities in the areas of agriculture, water resources, health and sanitation, coastal zones, forests and disaster risk reduction (Ray 2007).

Successful human societies are characterized by their adaptability, evidenced throughout human existence. However, climate change introduces a new challenge, not only because of the expected rise in temperature and sea levels, but also due to the current context of failure to address the causes of poverty adequately. As a result, policy supporting adaptation has been cast as a necessary strategy for responding to climate change and supporting development, making adaptation the focus of much recent scholarly and policy research. Research conducted by Schipper (2007) at the Tyndall Centre for Climate Change Research School of Environmental Sciences, University of East Anglia, addresses this new adaptation discourse, arguing that work on adaptation so far has focused on responding to the impacts of climate change, rather than sufficiently addressing the underlying factors that cause vulnerability. What is particular about climate change is that adaptive processes will have to take into account not entirely understood, but certainly extensive, impacts on hydrology and water resources, agriculture and food security, terrestrial and freshwater ecosystems, coastal zones and marine ecosystems, human health, human settlements, energy and industry, and insurance and other financial services (Adger et al. 2003). These impacts will have profound economic, social, demographic, technological and political implications.

Such impacts will be unprecedented in their scope and range, going beyond national boundaries and penetrating a broad array of policy and theory discourses because of the global nature of climate change.

From a policy perspective, the adaptation approach to development is simpler: it centres on mainstreaming adaptation, which comes down to taking into account climate change in social, institutional and infrastructural development planning. Schipper (2007) has identified two tracks toward adaptation: an "adaptation approach" to development and a "vulnerability reduction approach" to development (see Box 1).

Box 1 Different approaches to linking adaptation and development

Adaptation approach

Adaptation to climate change impacts → Vulnerability reduction → Development

In this view, adaptation is carried out in response to the observed and experienced impacts of climate change on society (including ecosystems). These responses ensure that the vulnerability to the impacts is reduced. This in turn ensures that less is lost each time a climate-related hazard takes place, which means risk is reduced. With reduced risk, development can be more sustainable.

Vulnerability reduction approach

Development → Vulnerability reduction → Impact reduction → Adaptation

In this view, development processes help reduce vulnerability to climate change. By reducing the vulnerability, impacts of climate hazards are also reduced, as there is less sensitivity and exposure to the hazards. This translates into a process of adaptation to climate change.

Even the Adaptation Policy Framework, developed by the UN Development Programme/Global Environmental Facility (Lim et al. 2005) emphasizes the mainstreaming approach, as do the National Adaptation Programmes of Action guidelines for least developed countries. But mainstreaming will not be effective if existing development trajectories are inconsistent with the objectives of adaptation, i.e. if they explicitly contribute to vulnerability. This is also relevant as adaptation needs to confront the same constraints as those faced by development, and therefore an adaptation process is only possible if there is successful sustainable development to support it. However, for this to be successful, an awareness of climate change impacts and the needs for successful vulnerability reduction are imperative within the development process.

Cost Implications of Mitigation Measures

Mitigation measures to reduce GHG emissions have a certain cost. However, they also constitute an economic benefit by reducing the impacts of climate change, and the costs associated with them. In addition, they can bring economic benefits by reducing local air pollution and energy resource depletion. If the benefits of avoided climate change are taken into account and a "carbon price" is established for each unit of GHG emissions, this could create incentives for producers and consumers to significantly invest in products, technologies and processes which emit less GHGs. The resulting mitigation potential is substantial and could offset the projected growth of global emissions over the coming decades or reduce emissions below current levels (Garg et al. 2003). Mitigation measures could contribute to stabilizing the concentration of GHGs in the atmosphere by 2100 or later. To achieve low stabilization levels, stringent mitigation efforts are needed in the coming decades. This could reduce global GDP by up to a few per cent. Changes in lifestyle and behaviour that favour resource conservation can contribute to climate change mitigation. Mitigation measures can also have other benefits for society, such as health cost savings resulting from reduced air pollution. However, mitigation in one country or group of countries could lead to higher emissions elsewhere or effects on the global economy. Mitigation potential for different sectors is a function of carbon price (Schaffer et al. 2002).

No one sector or technology can address the entire mitigation challenge. All sectors, including construction, industry, energy production, agriculture, transport, forestry and waste management could contribute to the overall mitigation efforts, for instance through greater energy efficiency. Many technologies and processes which emit less GHGs are already commercially available or will be in the coming decades. In order to stabilize the concentration of GHGs in the atmosphere, emissions would have to stop increasing and then decline. The lower the stabilization level aimed for, the more quickly this decline would need to occur. Worldwide investments in mitigation technologies, as well as research into new energy sources, will be necessary to achieve stabilization. Delaying emission reduction measures limits the opportunities to achieve low stabilization levels and increases the risk of severe climate change impacts (Nair et al. 2003).

Energy Security and Sustainable Development Nexus

There is a trade-off between stability and sustainability within energy equity. Should more emphasis been placed on the stability and equity of present energy supply or should intergenerational equity be prioritized? The former option includes the demand for equitable geographical distribution of energy supply. The World Energy Council (2000) stated that "though we expect no major problem in providing the energy that will be necessary to sustain the rates of economic growth that one might realistically expect in the coming decades, the current situation is characterized by distributional inequalities and detrimental impacts on the global, regional and local environment and on human health". Additionally, the protection of the commons and avoidance of energy imperialism is a precondition for energy equity. There are also negative effects of increasing energy access including cycles of ever higher demand, higher energy production, and higher pollution, threatening environmental security. For the sustainability criterion to be met the idea of a "limits-to-growth scenario" has to be accepted. While making energy accessible and affordable to all to fulfil their basic needs, energy use for luxurious purposes can be reduced without infringing basic human rights. Thus, the ethical demands to meet concerns of equity also mean restrictions for those who make excessive use of energy. The challenge of the twenty first century will be for human society to regain a healthy relationship with our "living" Earth. The signs are clear that right now we are failing in that task. Unless we change our ways, it is likely that comfortable civilization as we know it will disappear. Climate change is not just an environmental problem, it is a societal problem. The simple truth is that humanity is changing the atmosphere—every day we add around 70 million tonnes of invisible carbon dioxide, the very gas that we know controls the temperature of the planet. The pathway to integrate energy security and sustainable development is through achieving "carbon neutrality". Carbon neutrality offers individuals, businesses and other institutions the opportunity to take personal responsibility for the global warming implications of their lifestyle (Carbon Neutral 2008).

Leading energy-generating companies have the ability to break the cycle by helping raise consumer awareness, challenging government and looking for business opportunities in a carbon-constrained world. There is clearly a need for more research to be done in understanding what exactly encourages people to reduce their emissions, on the extent to which the practice of offsetting has an effect on such behaviour, and on how much it can play a role in educating people about climate change (Carbon Literacy 2008).

The biggest spending consumers in the voluntary offset market are businesses. As with individuals, the motivation for carbon emissions offsetting varies: some companies argue that they recognize the threat of climate change and try to act responsibly; but also some of the motivations for offsetting here will be strategic—they might be to meet Corporate Social Responsibility (CSR) obligations, to generate goodwill, or to attract the growing number of customers attracted by environmental action. Future efforts should be directed to bridge the gap between the carbon-literate vis-à-vis carbon-illiterate society. In this regard, sustained efforts will also be required to address concerns on developing sustainable energy portfolios in the context of environment management, as also simultaneously meeting the growing needs of society for enhancing Developmental Matrices and improving Human Developmental Indices (Carbon Sense 2006).

According to a recent World Bank Study reported by Patricia da Camara titled "Green Growth is Possible in Mexico (2009)", Mexico could reduce its CO_2 emissions by at least 42% (or 477 million tonnes) per year by 2030 without sacrificing economic development. The report was presented to 120 legislators from major economies at the GLOBE Copenhagen Legislators Forum organized by the Global Legislators Organization (GLOBE) and the COM+ Alliance of Communicators for Sustainable Development. The Forum took place at a critical scenario ahead of the formal UNFCCC negotiations in December in this city. Low-Carbon Development for Mexico identifies a series of 40 actions for reducing CO_2 emissions that could be implemented in the short term at low costs. Significant opportunities for CO_2 reduction were identified in the areas of transport, power generation, oil and gas, agriculture and forestry, and energy efficiency. Implementing these initiatives on a larger scale over the next 20 years, however, will require changes to the country's financial, regulatory and institutional frameworks.

The IEA (2007) has estimated that $26 trillion in investment in the energy sector is needed between 2005 and 2030 under the business as usual scenario, but another $4.1 trillion would be required to limit temperature increase to 3°C (0.2% of global GDP), or an additional $9.3 trillion (0.6% of global GDP) to attain the 2°C target. It is widely agreed that international trade will play a crucial role in making these huge investments possible. At the same time, there are various issues that potentially will have a negative impact on trade and investments in clean technologies, such as: (i) lack of clear guidance on future energy policy and lack of fiscal incentives for clean energy production; (ii) weak environmental regulation and enforcement; (iii) tariffs on environmental goods and services (EGS); (iv) non-tariff barriers on trade in EGS, such as standards and labels; (iv) weak and fragmented protection of investments; (v) subsidies for conventional energy sources, and a domestic financial sector that has little experience with new technologies. The Organization for Economic Cooperation and Development (OECD) and International Energy Agency (IEA), and recent reports on World Energy Outlook 2004 (IEA 2004) project that over the next 30 years global primary energy demand will grow by 1.7% per annum from 9.20 to 15.30 billion tonnes of oil equivalent, and that this demand will be met primarily by conventional fossil energy such as oil, natural gas and coal in the near term. Energy from renewable resources is also expected to grow in the mid-century term, but will remain in the small percentages of the total energy mix in the near term. Based on a number of statistics, it is also projected that many communities across the globe (1.40 billion people according to IEA) living at or below the poverty line will remain without access to modern energy systems such as electricity, which is an essential requirement for social and economic development.

The overall value of the global carbon market was estimated at over $10 billion in 2005. In the first quarter of 2006 alone, the value of carbon market transactions reached $7.5 billion. This growth is driven by high prices in the EU-ETS market for allowances. EUAs worth $8.2 billion were traded in 2005. The World Bank has been a pioneer in the carbon market, mainly through the establishment of carbon procurement funds to secure carbon credits on behalf of investors. These funds typically enter into pay-on-delivery contracts and contribute to the positive cash flow of projects after the start of operations and the delivery of emission reductions. Recently, some funds have secured limited insurance against non-delivery of CERs and are able to offer partial upfront payments. The funds in the World Bank portfolio were not solely intended to procure carbon credits, but also to help create demand and spur the global carbon market (The World Bank 2009a).

Through increasing carbon sink and using CCS technology, the GHG emissions from fossil fuels that are hard to reduce can be offset. Meanwhile, through the establishment of reasonable and fair technology transfer and financial support mechanisms, developing countries can undertake the costs to shift towards low-carbon patterns while being at the lowest end of the value chain in the international trade structure. The perspectives of development value need to be changed in order to promote the transition of consumption towards a sustainable and low-carbon future, especially for developing economies (Wang 2009). The United Kingdom is the first country that put forward the concept of a "low-carbon economy". The UK had its own historic and realistic reason for doing so. Its major purpose is to guarantee energy security, mitigate the impact of climate change, utilize the opportunities from retrofitting its energy facility infrastructure and its advantage in low-carbon technology, increase its own economic efficiency and vitality, take a potentially larger piece of the future low-carbon technology and products market, and obtain its leadership role in international politics, as well as increase its international influence. Other EU countries, Japan and other major developed countries have also gained progress around a "low-carbon economy" on the basis of their own advantages in energy, environment, economy and politics, as well as their own social and economic contexts and global strategy. They are beginning to lead the trend of global revolution towards a low-carbon future through various strategies and models.

Carbon taxes can be an efficient instrument for controlling carbon emissions—but changes in the tax system to incorporate environmental costs (green taxes) could be regressive, depending on the country's economic structure, the quality of targeting, and the distribution of burden sharing. In the UK, a carbon tax imposed equally on all households would be very regressive, consistent with findings from other OECD countries. Green taxes in developing countries could even be progressive, as suggested by a recent study for China. Most poor households in China reside in rural areas and consume products that are much less carbon-intensive than those consumed by generally better-off urban households. If revenues from a carbon tax were recycled into the economy on an equal per-capita basis, the progressive effect would be larger still. Developing countries, with less existing infrastructure than developed countries, have a flexibility advantage and could potentially leapfrog to cleaner technologies. Developed countries must provide leadership in bringing new technologies to market and sharing knowledge from their experiences of deployment. The ability to change emissions trajectories depends on the availability of appropriate and affordable technology, which will not be in place at some future date without research and development (R&D) investment, dissemination, and learning-by-doing starting today.

Addressing Climate Change and Sustainable Development Issues

The most commonly accepted understanding of environmentally sustainable development (ESD) is encapsulated by the Brundtland definition: "meeting the needs of present generations without compromising the ability of future generations to meet their own needs" (WSSD 2002). It is recognized that meeting essential needs requires economic growth and equity facilitated by "political systems that secure effective citizen participation in decision-making" (ibid.).

The report of the Brundtland Commission (1983) was central to the process leading to a wide recognition of the "sustainability" concept as the guiding rule for future planning. IUCN (1980) defines sustainable development as "the management of human use of the biospheres so that it may yield the greatest potential to present generations while maintaining its potential to meet the needs and aspirations of future generations". Two key elements in planning for sustainable development are the precautionary principle and the internalization of external effects. The precautionary principle states that any development action should take into account future risks, hazards and adverse impacts. In the absence of clear evidence, the principle emphasizes safety considerations (Gilpin 1996).

Sustainable development has become part of all climate change policy discussions at the global level, particularly due to adoption of Agenda 21 and the various Conventions resulting from the UNCED (1992). The generally accepted and used definition as given by the Brundtland Commission is "development that meets the needs of the present without compromising the ability of future generations to meet their own needs". Climate change and other sustainable development policies are often but not always synergistic. There is growing evidence that decisions about macroeconomic policy, agricultural policy, multilateral development bank lending, insurance practices, electricity market reform, energy security and forest conservation, for example, which are often treated as being apart from climate policy, can significantly reduce emissions. On the other hand, decisions about improving rural access to modern energy sources, for example, may not have much influence on global GHG emissions. Climate change policies related to energy efficiency and renewable energy are often economically beneficial, improve energy security and reduce local pollutant emissions. Other energy supply mitigation options can be designed to also achieve sustainable development benefits such as avoided displacement of local populations, job creation and health benefits.

Sustainable development has become an integrating concept embracing economic, social and environmental issues (Lélé 1991). Sustainable development does not preclude the use of exhaustible natural resources but requires that any use be appropriately offset. This concept is not acceptable to many developing countries since it seems to disregard their aspirations for growth and development. Further, sustainable development cannot be achieved without significant economic growth in the developing countries. Three critical components in promoting sustainable development are economic growth, social equity and environmental sustainability. There is a growing understanding of the possibilities to choose and implement mitigation options in several sectors to create synergies and avoid conflicts with other aspects of sustainable development. Making development more sustainable by changing development paths can make a major contribution to climate change mitigation, but implementation may require resources to overcome multiple barriers.

The Ministry of Power (BEE 2007), Government of India, brought out a detailed status position paper on "Addressing Energy Security and Climate Change" in October 2007, which will bring about integration and various initiatives institutionalized through specialized institutions, such as the Ministry of New and Renewable Energy, the Bureau of Energy Efficiency, and the Technology Information, Forecasting and Assessment Council, with specific mandates to promote climate-friendly technologies. The National Environment Policy (2006) provides the basis for the integration of environmental considerations in the policies of various sectors. The Policy Statement for Abatement of Pollution (1992) stressed the prevention of pollution at the source based on the "polluter pays" principle. The Forest Policy (1988) highlights environmental protection through preservation and restoration of the ecological balance. Sustainable development has become part of all climate change policy discussions at the global level, particularly due to the adoption of Agenda 21 and the various Conventions resulting from the UNCED (1992).

The generally accepted and used definition as given by the Brundtland Commission in 1983 is "development that meets the needs of the present without compromising the ability of future generations to meet their own needs". In this regard, the statement of the former Secretary General of the United Nations, General Ban Ki-moon, is relevant: "Eradicating extreme poverty continues to be one of the main challenges of our time, and is a major concern of the international community. Ending this scourge will require the combined efforts of all, governments, civil society organizations and the private sector, in the context of a stronger and more effective global partnership for development."

Developing countries now contribute about half of annual GHG emissions but have nearly 85% of the world's population; the energy-related carbon footprint of the average citizen of a low- or middle-income country is 1.3 or 4.5 metric tonnes of carbon dioxide equivalent (CO_2e), respectively, compared with 15.3 in high-income countries. Moreover, the bulk of past emissions—and thus the bulk of the existing stock of GHGs in the atmosphere—is the responsibility of developed countries. Adaptation and mitigation need to be integrated into a climate-smart development strategy that increases resilience, reduces the threat of further warming, and improves development outcomes. Adaptation and mitigation measures can advance development, and prosperity can raise incomes and foster better institutions. A healthier population living in better-built houses and with access to bank loans and social security is better equipped to deal with a changing climate and its consequences. Advancing robust, resilient development policies that promote adaptation is needed today because changes in the climate, already begun, will increase even in the short term. Development, sustainability and equity (DES) are key elements of any sustainable development strategy. In many countries, response to climate change cannot be seen independently from more general sustainable development strategies.

Emerging insights from adaptive ecosystem management and new institutional economics suggest that building resilience into both human and ecological systems is the optimal way to deal with future surprises, or unknowable risks. But do these emerging insights have implications for policies and strategies for responding to anthropogenic climate change? Herein, we ought to review our perspectives on collective action for natural resource management and use insights from this area to inform our understanding of climate response capacity and to demonstrate the importance of social acceptance of strategies that build social and ecological resilience. All societies need to enhance their response capacity to face future climate impacts that could lie outside their experienced coping range. The challenge, posed at both the scale of local natural resource management and at the scale of international agreements and actions, is to promote adaptive capacity in the context of competing sustainable development objectives.

Conclusions

Given the backdrop of accumulated knowledge on the scientific reasons for global climate change and research findings emanating therefrom on the various adverse impacts of these changes on ecology, human and animal life, etc., it necessitates a proper documentation and analysis of these events to guide the future options, strategies and long-term climate policies leading to formulating goal-oriented abatement of the declining climate change trend and their stabilization in order to save the planet from further ecological devastation. There is international consensus that climate change is one of the most serious threats to sustainable development, both currently and in the future. Adverse impacts of climate change have already been observed on natural resources, food security, human health, the environment, economic activity and physical infrastructures. Climate change, therefore, should not be seen in isolation merely as an environmental problem; it should be perceived as a development problem, since its adverse effects will disproportionately affect poorer countries with economies predominantly based on natural resources and related economic sectors. The rationale for integrating adaptation into development strategies and practices is underlined by the fact that interventions required to increase resilience to climate variability and change generally further development objectives and the underlying processes. Emerging insights from adaptive ecosystem management and new institutional economics would suggest that building resilience into both human and ecological systems is the optimal way to deal with future surprises, or unknown risks. Herein, we ought to review our perspectives on collective action for natural resource management and use our insights from this area to inform our understanding of climate response capacity to demonstrate the importance of social acceptance of strategies that build social and ecological resilience. All societies need to enhance their response capacity to face future climate impacts that could lie outside their experienced coping range. The challenge, posed at both the scale of local natural resource management and at the scale of international agreements and actions, will there be to promote adaptive capacity in the context of competing sustainable development objectives.

At the international level, global climate change will have to have its deserved place on the agenda for sustainable development. Sustainable development involves a comprehensive and integrated approach to economic, social and environmental processes. Climate change is influenced not only by the climate-specific policies but also by the mix of development choices. Therefore, the future Sustainable Development Agenda will be grossly incomplete without inclusion of the "Human Development" and "Democratic Option" components to decide on any imperative measures linked to mitigation or adaptation actions to curb adverse impacts of climate change. By implication, this is also to be linked a priori with the quality of democracy of the climate response actors to attribute to good governance, among others, for evolving sustainable environmental strategy. Among other things, important components of adaptation strategies should be a better understanding of the risks associated with environmental change and climate change, and development of mechanisms to manage these, such as early warning systems. The capacity to adapt varies considerably among regions, countries and socioeconomic groups, and will vary over time. The most vulnerable regions and communities will be those that are highly exposed to hazardous climate change effects and have limited adaptive capacity. Nevertheless, adaptation to climate change can confront in certain given situations dilemmas of rendering justice to the international community, including those around the responsibility of developed countries to assist developing countries in adapting to changing climate. Justice to community and the stakeholders among the climate response actors (viz. government, private, joint partnership actors) should thus always have to be implicit in the choice of adaptive responses.

The effects of climate change have increasingly come to be seen as a threat to human security. Human beings are causing environmental degradation. At the same time, they are among the victims, as their lives and livelihoods are distorted along with the distortion of whole ecosystems. However, those who are most vulnerable to the effects of climate change are often not those who have contributed to global warming the most. The problematic of energy equity and human security can hence be framed as an issue of distributive justice between populations and between generations: ensuring that no population or generation threatens the security (welfare) of other populations or generations through over consumption. The extent to which we are able to access energy is a factor which influences our way and quality of life. At the same time, the amount of energy used for our own convenience is also having an effect on others and the environment in the present and in the future. The pathway to integrate energy security and sustainable development is through achieving "carbon neutrality". Carbon neutrality offers individuals, businesses and other institutions the opportunity to take personal responsibility for the global warming implications of their lifestyle. Therefore, for effective mitigation, emissions reduction and corresponding strategic options would need to be aligned with the expanding energy demands on the one hand, sustainable economic development with increasing environmental concerns on the other hand. In an effort to do so, there has to be parallel development of renewable forms of energy at a competitive cost to gradually replace conventional non-renewable sources of energy that are predominantly dependent on fossil fuels. Efficient carbon management by way of energy conservation, recycling of renewables, carbon sequestration, geo-sequestration, eco-technology, etc. should be the key drivers for mitigation and adaptation actions to help build a carbon–neutral global society. Carbon taxes can be an efficient instrument for controlling carbon emissions—but changes in the tax system to incorporate environmental costs (green taxes) could be regressive, depending on the country's economic structure, the quality of targeting, and the distribution of burden sharing. Developing countries, with less existing infrastructure than developed countries, have a flexibility advantage and could potentially develop cleaner technologies. Developed countries, in turn, must also provide much-needed leadership in bringing new technologies to market and sharing knowledge from their experiences of deployment.

Economic diversification will become increasingly important as an adaptation strategy to increase economic resilience and decrease reliance on climate-vulnerable economic sectors. In the larger context of achieving long-term sustainability, clearer understanding of mitigation and adaptation action actions assume prime importance to cope with climate change-related vulnerability. Hence, we need to know the implications of mitigation and adaptation in relation to climate change parameters in more clear terms. At the sectoral level, increased economic resilience is to be achieved by adapting existing practices to reduce exposure to risk. Evaluating the economic consequences of the climate change, as well as assessing the environmental economic policies associated with it, will require a good understanding of both natural and socioeconomic processes and finding new business opportunities while addressing climate change impact assessment and its planned mitigation action thereof. Economic implications of climate change, according to various studies made in recent times borders around funding for mitigation and adaptation and reversing the adverse impacts of global warming, largely caused by greenhouse gases and various anthropogenic factors. Integration of carbon finance (largely the projects aimed at reduction of carbon intensity and carbon neutrality/offsetting) with implicit cost-economics along with regulatory provisions contained in it would go a long way in making carbon financing a cost-effective and socioeconomically viable proposition. The new Copenhagen Protocol and the amended Kyoto Protocol would form the core of the Agreement in December 2010 in Mexico with the main elements agreed and a process decided to finalize the details through decisions in the year or so following, in order to ensure ratification by 2011.

A new Institutional mechanism—namely the Copenhagen Climate Facility as well as "Carbon Fund"—will bring together developing country action plans on mitigation, deforestation, technology and adaptation with the needed support, with transparent and equitable governance as a core starting point. It is clear that the new Protocol must include a mechanism or mechanisms to deliver the new and additional finance for mitigation and adaptation actions in a predictable fashion scale. The Mitigation and Adaptation Action Framework must be robust and include not only new funding but also an insurance mechanism and a compensation and restitution mechanism. Technology cooperation should occur quickly on both mitigation and adaptation but be aided by a longer-term vision and a set of action programmes that ensure that the world is delivering technology at an adequate scale and speed and within a "protect and share" framework of intellectual property rights. The new agreement must also build trust through transparency and rigorous data collection and verification in a manner that reflects the different capabilities of countries. The new Agreement should also try to minimize cross-border asymmetry in natural assets sharing such as river/irrigation water, etc. for meeting development goals and minimize inter-regional imbalance in economic growth.

References

Adger N, Brooks N (2003) Does global environmental change cause vulnerability to disaster? In: Pelling M (ed) Natural disasters and development in a globalizing world. Routledge, London

Asian development bank (ADB) (2007) Supporting provision of regional public goods in the Asia and Pacific Region, Manila

Banuri T, Weyant J, Akumu G, Najam A, Pinguelli Rosa L, Rayner S, Sachs W, Sharma R, Yohe G (2001) Setting the stage: climate change and sustainable development. In: Metz B, Davidson O, Swart R, Pan J (eds) Climate change 2001: mitigation report of working group III intergovernmental panel on climate change (IPCC). Cambridge University Press, Cambridge

Berman T (2008) Report on the on the follow-up to the Monterrey conference of 2002 on financing for development (2008/2050(INI)). http://www.europarl.europa.eu/sides/getDoc.do?type=REPORT&reference=A6-2008-0310&language=EN

Blaikie P, Cannon T, Davis I, Wisner W (1994) At risk: natural hazards, people's vulnerability and disasters. Routledge, London

Bohle HG, Downing TE, Watts MJ (1994) Climate change and social vulnerability: toward a sociology and geography of food insecurity. Glob Environ Chang 4:37–48

Bojo J, Mäler KG, Unemo L (1992) Environment and development: an economic approach, Second revised edition edn. Kluwer Academic Publishers, Dordrecht, p 211

Brundtland Commission (1983) http://www.simple.wikipedia.org/wiki/Brundtland_Commission#cite_note-0

Bureau of Energy Efficiency (BEE) (2007) Ministry of Power, The Government of India, New Delhi

Campbell FJ (2008) The basic concept for the democracy ranking of the quality of democracy, 29 Sept 2008, Vienna: Democracy Ranking. http://www.democracyranking.org/downloads/basic_concept_democracy_ranking_2008_A4.pdf

Carbon Literacy (2008) a global challenge; http://www.carbonsense.com/documents/Developing-CarbonLiteracy.pdf

Carbon Neutral (2008) http://www.windows2universe.org/bio/travis.html, http://www.noco2.com.au/, http://www.windows2universe.org/earth/climate/neutral.html, http://www.ictandclimatechange.com/USERIMAGES/Beyond_carbon_neutral.pdf

Carbon Sense (2006) http://www.carbonsense.com/documents/CarbonSense_whatwouldagenuinelycneutralBTlooklike_000.pdf

Cocklin C (1995) Agriculture society, and environment: discourses on sustainability. Int J Sustain Dev World Ecol 2:240–256

Doornbosch R, Gielen D, Koutstaal P (2008) Mobilising investments in low-emission energy technologies on the scale needed to reduce the risks of climate change. OECD, Paris

Fankhauser S (1996) The potential costs of climate change adaptation. In: Smith J, Bhatti N, Menzhulin G, Benioff R, Budyko MI, Campos M, Jallow B, Rijsberman F (eds) Adapting to climate change: an international perspective. Springer, New York, pp 80–96

Fussel HM, Klein RJT (2002) Assessing the vulnerability and adaptation to climate change: an evolution of conceptual thinking. Paper presented at the UNDP expert group meeting on integrating disaster reduction and adaptation to climate change, Havana, Cuba, 17–19 June 2002

Garg A, Ghosh D, Shukla PR (2003) Energy sector policies and mitigation of GHG emissions from India. In: Toman M (ed) Climate change economics and policy: Indian perspectives. Resources for the Future Publication, Washington DC

Gilpin ME, Soulé ME (1996) Minimum viable populations: Processes of species extinction. In: Soulé ME (ed.) Conservation Biology: The Science of Scarcity and Diversity, Sinauer Associates, Sunderland, pp 19–34

Godard O (2008) The stern review on the economics of climate change: contents, insights and assessment of the critical debate. S.A.P.I.EN.S 1:1

Greenwood C, Hohler A, Hunt G, Liebreich M, Sonntag-O'Brien V, Usher E (2007) Global trends in sustainable energy investment 2007: analysis of trends and issues in the financing of renewable energy and energy efficiency. UNEP SEFI and New Energy Finance, Paris

Green Growth in Mexico: Patricia da Camara (2009) http://beta.worldbank.org/climatechange/node/5038

Handmer J, Dovers S, Downing TE (1999) Societal vulnerability to climate change and variability. Mitig Adapt Strateg Glob Chang 4(3 & 4):267–281

IEA (2004) International Energy Agency World Energy Outlook, Paris

IEA (2007) International Energy Agency, World Energy Outlook, 2006, pp 428, 430

IFAD (2007) IFAD's response to climate change through support to adaptation and related actions, Comprehensive report: Final version, New York

International Energy Agency (2006) World Energy Outlook 2006. IEA, Paris

International Energy Agency (2008) World Energy Outlook 2008. IEA, Paris

International Institute of Sustainable Development (IISD) (2007) A special report on selected side events at the twenty-sixth sessions of the subsidiary bodies (SB 26) of the United Nations framework convention on climate change (UNFCCC), Bonn, Germany, 7–8 May. http://www.iisd.ca/climate/sb26/enbots/

IPCC (2001) Intergovernmental panel on climate change Secretariat, Geneva, Switzerland. http://www.ipcc.ch/

IPCC (2007) Fourth assessment report, intergovernmental panel on climate change Secretariat, Geneva, Switzerland. http://www.ipcc.ch/

IUCN (1980) The World Conservation Strategy: Living Resource Conservation for Sustainable Development, prepared by IUCN in consultation with UNEP, WWF, FAO & UNESCO, 1980

Karl TR, Knight RW (1998) Secular trends of precipitation amount, frequency, and intensity in the United States. Bull Am Meteorol Soc 79:231–241

Kelly PM, Adger WN (2000) Theory and practice in assessing vulnerability to climate change and facilitating adaptation. Clim Change 47:325–352

LCA Network (2006) LCA discussion background paper 2: reducing disaster risk while adapting to climate change. Paper prepared for the LCA's web-based discussion on 'climate change and disasters', 27 February to 9 March. Linking Climate Adaptation Network, Eldis Community, Institute of Development Studies (IDS), University of Sussex, Brighton, UK. http://community.eldis.org/.59b65000/bp2.pdf

Lélé SM (1991) Sustainable development: a critical review. World Dev 19(6):606–607

Lim B, Burton I, Huq S, Doherty B, Spanger-Siegfried E, Adger N, Aguilar Y, Boer R, Bosch H, Brooks N, Conde C, Downing T, Ebi K, Jones R, Khan SR, Lonsdale K, Malone E, Mearns L, Niang-Diop I, Patwardhan A, Perez RT, La Rovere E, Smith J, Yohe G (2005) Adaptation policy frameworks for climate change. Cambridge University Press, Cambridge

Mason SJ, Waylen PR, Mimmack GM, Rajaratnam B, Harrison JM (1999) Changes in extreme rainfall events in South Africa. Clim Change 41:249–257

McGray H, Hammill A, Bradley R (2007) Weathering the storm: options for framing adaptation and development. World Resources Institute, Washington DC

Müller B, Hepburn C (2006) Current needs based on extrapolations of LDC national adaptation programmes of action (NAPAs), p 14

Nair R, Shukla PR, Kapshe M, Garg A, Rana A (2003) Analysis of long-term energy and carbon emission scenarios for India. Mitigation and Adaptation Strategies for Global Change, vol 8. Kluwer Academic Publishers, Dordrecht, pp 53–69

O'Donnell G (2004) Human development, human rights, and democracy. In: O'Donnell G, Vargas Cullell J, Iazzetta OM (eds) The quality of democracy. Theory and applications. University of Notre Dame Press, Notre Dame, pp 9–92

Oxfam (2007) Financing adaptation: why the UN's Bali climate conference must mandate the search for new funds. Oxfam Briefing Note, Oxford

Paavola J, Adger WN (2002) Justice and adaptation to climate change. Tyndall Centre for Climate Change Research Working Paper 23

Parikh J, Parikh K, Gokarn S, Painuly JP, Saha B, Shukla V (1991) Consumption patterns: the driving force of environmental stress, IGIDR prepared for the United Nations Conference on Environment and Development (UNCED), IGIDR Monograph

Parry JE, Hammill A, Drexhage J (2005) Climate change and adaptation. IISD, Canada

Parry ML, Canziani OF, Palutikof JP, van der Linden PJ, Agrawala CE, Fankhauser S (eds) (2008) Economic aspects of adaptation to climate change: costs, benefits and policy instruments, Organisation for Economic Cooperation and Development. Paris

Peston R (2006) Report's stark warning on climate, BBC, 29 Oct 2006

Pezzoli K (1997) Sustainable development: a trans-disciplinary overview of the literature. J Environ Plan Manag 40(5):549–574

Pittock B, Jones RN (2000) Adaptation to what and why? Environ Monit Assess 61(1):9–35

Practical action (2007) Increasing the resilience of poor communities to cope with the impact of climate change: project summary. Bourton on Dunsmore, Rugby, UK. http://practicalactionconsulting.org/?id=climatechange_resilience. Accessed 29 Oct 2008

Raper SCB, Wigley TML, Warrick RA (1996) Global sea level rise: past and future. In: Milliman JD, Haq BU (eds) Sea level rise and coastal subsidence: causes, consequences and strategies. Kluwer Academic Publishers, Dordrecht, pp 11–45

Ray R (2007) India: adaptation approaches and strategies, presentation at the workshop on addressing action on adaptation, UNFCCC Dialogue on long-term cooperative action to address climate change by enhancing implementation of the convention, Bonn, 17 May

Robinson J, Herbert D (2001) Integrating climate change and sustainable development. Int J Global Environ Issues 1(2):130–148

Sadler B (1996) Environmental assessment in a changing world: evaluating practice to improve performance, international study of the effectiveness of environmental assessment, Final report, June 1996, Canadian Environmental Assessment Agency, International Association for Impact Assessment

Sarkar AN, Banerjee DK (2010) Global climate change: a strategic and operational framework for future negotiations beyond Copenhagen. Asia-Pacific Bus Rev 6(1):139–169

Schaffer R, Chandler W, Dadi R, Shukla PR, Tudela F, Davidson O, Alpan-Atamar S (2002) Climate changes mitigation in developing countries. Report, The Pew Center on Global Climate Change, Washington DC, Oct 2002

Schipper E, Lisa F (2007) Climate change adaptation and development: exploring the linkages, Tyndall centre for climate change research school of environmental sciences, University of East Anglia, Norwich, UK, Tyndall Centre Working Paper No. 107, July 2007

Shah A (2008) Climate Change and Global Warming, Global Issues

Sharan D (2008) Financing climate change mitigation and adaptation role of regional financing arrangements, ADB Sustainable Development Working Paper Series (No. 4), December 2008

Smit B, Burton I, Klein RJT, Street R (1999) The science of adaptation: a framework for assessment. Mitig Adapt Strateg Glob Chang 4:199–213

Smith JB, Bhatti N, Menzhulin G, Benioff R, Budyko MI, Campos M, Jallow B, Rijsberman F (eds) (1996) Adapting to climate change: an international perspective. Springer, New York, p 475

Smith JB et al. (2001) Vulnerability to climate change and reasons for concern: a synthesis. In: McCarthy JJ et al. (eds) Climate change 2001: impacts, adaptation and vulnerability, contribution of working group II to the third assessment report of the intergovernmental panel on climate change, Cambridge University Press, Cambridge and New York. http://www.ipcc.ch/publications_and_data/publications_and_data_reports.htm. Accessed 10 Jan 2010

Stern N (2006) Stern review on the economics of climate change, Final report. http://webarchive.nationalarchives.gov.uk/+/http:/www.hm-treasury.gov.uk/independent_reviews/stern_review_economics_climate_change/stern_review_report.cfm

Stern N (2007) The economics of climate change: the stern review. Cambridge University Press, Cambridge

The World Bank (2007) State and trends of carbon market, Washington DC

The World Bank (2009) Economics to adaptation of climate change. http://www.iisd.ca/publications_resources/trade_investment.htm

Tirpak D (2008) National policies and their linkages to negotiations over a future international climate change agreement. Report prepared for UNDP. www.undp.org/climatechange/docs/UNDP_NationalPolicy.pdf

Tol RSJ, Fankhauser S, Smith JB (1998) The scope for adaptation to climate change: what can we learn from the impact literature? Glob Environ Change 8(2):109–123

Uitto J, Shaw R (2006) Adaptation to changing climate: promoting community-based approaches to developing countries. Sansai 1:93–108

UNCED (1992) http://www.un.org/geninfo/bp/enviro.html

UNDP (2007) Human Development Report 2007/2008, Fighting climate change: Human solidarity in a divided world, New York

UNEP (1998) Handbook on methods for climate impact assessment and adaptation strategies. In: Feenstra J, Burton I, Smith J, Tol R (eds) United Nations Environment Program. Institute for Environmental Studies, Amsterdam, 359 pp

UNFCCC (2007a) Vulnerability and adaptation to climate change in small island developing states, Background paper for the expert meeting on adaptation for Small Island developing States, UNFCCC Secretariat, Bonn, Germany. http://unfccc.int/files/adaptation/adverse_effects_and_response_measures_art_48/application/pdf/200702_sids_adaptation_bg.pdf

UNFCCC (2007b) Background paper—impacts, vulnerability and adaptation to climate change in Asia, UNFCCC Secretariat, Bonn, Germany. http://unfccc.int/files/adaptation/methodologies_for/vulnerability_and_adaptation/application/pdf/unfccc_asian_workshop_

Walther GR, Post E, Convey P et al (2002) Ecological responses to recent climate change. Nature 416:389–395

Wang Y (2009) http://www.world-governance.org/IMG/pdf_China_Sustainable_Development_Strategy_Report_2009_WANG_Yi_english.pdf

Watkins K (2007) Human Development Report 2007/2008: fighting climate change: human solidarity in a divided world. UNDP, New York

WEC (2000) Energy for Tomorrow's World – Acting Now! p 2, 175 pp, Atalink Projects Ltd, London
World Development Report (2010) Development and Climate Change. The World Bank, Washington
World Summit on Sustainable Development (WSSD) (2002) Development imperatives. http://www.gdrc.org/uem/energy/renewable-energy-agenda-2004.html
Yohe G, Neumann J, Marshall P, Ameden A (1996) The economic costs of sea-level rise on developed property in the United States. Clim Change 32:387–410
Zipprich A, Moss J, Manohar N, Wolbring G (2010) Energy Equity and Environmental Security, Preliminary Draft 3 of the report of WG7 of the ECCAP project, 12 Jan 2010

Part IV

Chapter 34
Activity-Based Water Resources and Climate Change Education Among School Students in Puducherry

R. Alexandar and G. Poyyamoli

Abstract Water resources and climate change education is an urgent need for encouraging students to understand the water issues related to climate change at their local level and make the changes in their attitude, behaviour for protection and conservation, and protection of local precious natural resources and capability to sustain their life under changing climatic conditions in future. The main objectives of this study were to bring changes to students' behaviour and attitudes to protect and conserve local drinking water resources. In order to address local water quality issues, students were provided with the scientific skills to monitor water quality, essential water quality parameters to protect against/prevent water pollution and were motivated and trained to practise various climatic change adaptations in order to conserve their local water resources and to create awareness among the local community about climatic change adaptations and the protection/conservation of local water resources through service-learning education. In this study, a pre- and post-test with control group design was used to measure students' efficiency of knowledge, attitude, and behaviour pre- and post-test. Activity-based environmental education methods were used to address water quality issues and climate change at local level, including classroom sessions such as lectures, PowerPoint presentations, films on various water issues and climate change, debates, essay/painting competitions, and quizzes and tests. During the outdoor activities, students were facilitated with hands-on activities, field trips, community visits, games and group projects. The post-test results revealed that there was an increase in students' knowledge, understanding, attitudes and skills to solve local water resources and climate change problems. This study therefore stresses the need to extend teaching and learning activities into the immediate environment

R. Alexandar (✉) · G. Poyyamoli
Department of Ecology and Environmental Sciences,
Pondicherry University, Puducherry, 605014, India
e-mail: enviroalexandar@gmail.com

of the pupils beyond the classroom so that they are made aware of and understand water resources and climate change studies.

Keywords Climate change · Water resources · Environmental education · Active participation · Local natural resources · Attitude and knowledge

Introduction

Climate change is recognized as a significant man-made global environmental challenge. It is also treated as a threat. Consequently, there is now growing recognition of the vulnerability of key sectors of economy and development to climate change. The various assessment reports brought out since the 1990s by the Intergovernmental Panel on Climate Change (IPCC) have progressively tracked the development and build-up of knowledge and understanding of the science, and the impacts and mitigation of climate change at global and regional levels. The IPCC concluded that the impact of human activities on climate is unequivocal (IPCC 2007). The debate at this point is over the extent and magnitude of climate change.

The quality of water is central to all the roles that water plays in our lives. From the beauty of natural waterways teeming with wildlife, to the vital livelihoods that clean rivers and streams support, to the essential role that safe water plays in drinking water and health—good water quality is fundamental to the network of life and livelihood that water supports. Water is the source of life on earth, and human civilizations blossomed where there was reliable and clean freshwater. Use of water by humans—for drinking, washing and recreation—requires water free from biological, chemical and physical sources of contamination. Plants, animals and the habitats that support biological diversity also need clean water. Water of a certain quality is needed to grow food, to power cities and to run industries. Water quality is as important as water quantity for satisfying basic human and environmental needs, yet it has received far less investment, scientific support, and public attention in recent decades than water quantity, even though the two issues are closely linked. As part of the effort to improve water quality, the United Nations Environment Programme (UNEP) is supporting educational efforts around the world to call attention to water quality challenges and solutions.

Over one billion people lack safe water; 80% of infectious diseases are waterborne, killing millions of children each year. Several parts of India are facing an immense challenge to meet the basic needs of water (Tambekar and Banginwar 2005). There arises an urgent need for understanding the status of drinking water quality, the related problems and also the reasons for the problems (CLEAN-India Project 2005). This understanding will help those people most affected by water quality problems to evaluate and change the situation. Water has no taste, no colour, no smell, yet it is life itself. Nothing could be more fundamental to life than water. Yet a third of our people do not have access to safe drinking water. If the water we use is not of good quality, it causes a number of health-related problems. In India, almost 44 million people suffer from water-related problems. The availability of water of good quality has become a major concern today.

During the past 150 years, the global average surface temperature has increased by about 0.76°C. In addition to the warming up of the Earth's surface, there have been increased incidences of heat waves; accelerated melting of continental glaciers and polar ice caps; rise in sea level of up to 20 cm; heavy rainfall in some regions, resulting in frequent floods; and reduced rains in other regions of the world, resulting in severe drought. Climate change may alter the distribution and quality of India's natural resources and adversely affect the livelihoods of people. With an economy closely linked to its natural resource base and climatically sensitive sectors such as agriculture, water and forestry, India may face a major threat because of the projected change in climate. Rising temperature, changes in rainfall patterns, and increased frequency of floods and droughts are likely to have serious effects on rural populations. The major human contribution to climate change in India is from carbon dioxide, methane, nitrous oxide and halocarbons. Their impact on the world is not therefore determined by the source from which they are emitted. Particulate emissions constitute the second group: they do not disperse as quickly from their region of origin. India's poor people remain highly dependent on agriculture and, in the absence of creative policies that insure rural incomes and provide alternative livelihoods, increased poverty may well accompany changes in physical geography.

Water Quality Challenges

A wide range of human and natural processes affect the biological, chemical and physical characteristics of water, and thus impact water quality. Contamination by pathogenic organisms, trace metals, and human-produced and toxic chemicals; the introduction of non-native species; and changes in the acidity, temperature and salinity of water can all harm aquatic ecosystems and make water unsuitable for human use. Numerous human activities impact water quality, including agriculture, industry and mining, disposal of human waste, population growth, urbanization and climate change. Agriculture can cause nutrient and pesticide contamination and increased salinity. Nutrient enrichment has become one of the planet's most widespread water quality problems (UN WWAP 2009) and worldwide, pesticide application is estimated to be over 2 million metric tonnes per year. Industrial activity releases about 300–400 million tons of heavy metals, solvents, toxic sludge and other waste into the world's waters each year (UN WWAP, Water and Industry). About 700 new chemicals are introduced into commerce each year in the United States alone. Mining and drilling create large quantities of waste materials and by-products and large-scale waste-disposal challenges. Widespread lack of adequate disposal of human waste leads to contamination of water—

worldwide, 2.5 billion people live without improved sanitation (UNICEF and WHO 2008), and over 80% of the sewage in developing countries is discharged untreated in receiving water bodies (UN WWAP 2009). Meanwhile, growing populations will potentially magnify these impacts, while climate change will create new water quality challenges.

Water Quality Impacts

Water contamination weakens or destroys natural ecosystems that support human health, food production and biodiversity. Studies have estimated that the value of ecosystem services is double the gross national product of the global economy, and the role of freshwater ecosystems in purifying water and assimilating wastes has been valued at US $400 billion (2008 $) (Costanza et al. 1997). Freshwater ecosystems are among the most degraded on the planet, and have suffered proportionately greater species and habitat losses than terrestrial or marine ecosystems (Revenga et al. 2000). Most polluted freshwater ends up in the oceans, damaging coastal areas and fisheries.

Domestic water consumption represents only 2% of global abstraction (agriculture requires a much higher proportion), the demand for water in cities is steadily rising due to changing consumption patterns and increasing wealth (Wilbanks et al. 2007, p 370). Water supply in urban areas must be considered as a key for human development, but it is a bottleneck factor in many respects: water supply is critical for economic development; massive groundwater withdrawal can lead to land subsidence and increasing flood risks in coastal and delta areas; the issue of water supply is related to wastewater management and the associated implications for health (sanitation, waterborne and vector-borne diseases).

Communicable and Other Diseases

Not only most water-related diseases, but also vector-borne and rodent-borne diseases are highly influenced by climate conditions, and several are very common within cities. The most important vector-borne viral infection is dengue, which has increased dramatically in tropical developing regions in the past few decades due to the weakening of vertical control programmes in many regions and the rapid unplanned urbanization that produces breeding sites for Aedes mosquitoes. The global spreading of the four different types of dengue is favoured by increased travelling, on the one hand, and higher temperatures and rainfall, on the other. However, there are also studies showing that drought can lead to an increase in dengue if household water storage increases the number of mosquito breeding sites (Campbell-Lendrum and Corvalán 2007, p i112; Confalonieri et al. 2007, p 403). Climate change and variability will be likely to impose additional pressure on water availability, water accessibility and water demand (Boko et al. 2007, p 444), and an increased conflict potential between rural and urban water supply, agriculture and nonfarm activities is to be expected.

Glacier and Snow Pack Decline

Glaciers the world over are thinning and shrinking as the planet warms, and glaciers in the Himalayas are thinning faster than anywhere else. If the Earth keeps warming at the current rate, Himalayan glaciers are likely to disappear altogether in 25 years (Cruz et al. 2007). In the absence of glaciers, rivers in the Indo-Gangatic Plain will become much more seasonal, threatening the Rabi crop (spring harvest), as well as domestic and industrial water supplies in the non-monsoon months. In addition, more precipitation will fall as rain rather than snow and the greater water runoffs will increase flooding. Up to half of the glacier decline is thought to have occurred as a result of upper atmospheric heating from the black carbon particles in the south Asian brown cloud (Ramanathan and Carmichael 2008). In addition, deposits of these soot particles in snow and ice accelerates melting. This implies that India has the ability to slow the melting of glaciers and snow by reducing particulate emissions.

Sea Level Rise

Global warming has been raising the sea level because warm water has greater volume. In addition, there is a melting of Greenland and west Antarctic ice packs. A recent study that takes into account both thermal expansion of the ocean and ice-pack melt suggests that the likely range of the rise by 2100 is 0.8–2 m (Pfeffer et al. 2008). This will lead to the permanent displacement of millions of people in coastal areas in India (Dasgupta et al. 2007). The IPCC expects extreme weather and weather-related events to become more frequent and/or intense, with serious consequences for human health and well-being. Scientists expect heat waves, droughts, wildfires, floods, severe storms and dust transported between continents to cause locally severe economic damage and substantial social and cultural disruption.

Rapid temperature changes would affect the seasons, causing variations in season length. Changes such as shorter winters could lead to mismatches between key elements in an ecosystem, such as feeding periods for young birds and availability of worms or insects for food. Increases in the frequency of droughts and floods would negatively affect local food production, and communities in mountain regions would face an increased risk of floods caused by melting glaciers. In addition, the risk of flood-induced illness and death from diarrhoeal diseases could rise in south and southeast Asia. A region's vulnerability to such extreme events depends both on how much the climate changes and whether or not nations develop effective responses to potential threats.

With the increase in the concentration of greenhouse gases in the atmosphere since the pre-industrial period, the planet has warmed by about 0.76°C. The eleven hottest years in the past 650,000 years occurred between 1991 and 2006. The hottest year on record was 2004 and the second hottest year on record was 2005. It is projected that the warming will continue to occur and by the end of this century, the planet will warm by a further 1.4–5.8°C.

The estimated amount of freshwater potentially available for human use is as follows: groundwater, freshwater lakes and rivers comprise 0.629%; inland seas and salt lakes 97.28%; polar ice caps 2%; lakes and rivers 0.00091%. India is fortunate to have a lot of freshwater, which is replenished by annual monsoons. India receives an annual rainfall equivalent to about 4,000 km^3. This is unevenly distributed across different parts of the country (Engelman and Roy 1993). In India, water is used with abandon and available supplies of water have become increasingly contaminated with waste and pollution from industry, agriculture and households. Water is essential for achieving sustainable development and the Millennium Development Goals. Properly managing water resources is an essential component of growth, social and economic development, poverty reduction and equity: all essential for achieving the Millennium Development Goals. Water is linked to the crises of climate change, energy and food supplies and prices, and troubled financial markets. Unless their links with water are addressed and water crises around the world are resolved, these other crises may intensify and local water crises may worsen, converging into a global water crisis and leading to political insecurity and conflict at various levels (UNESCO 2009).

Active Participation in Environmental Education

Research conducted in the past several decades suggests that participation in environmental education helps students develop knowledge about the environment, positive environmental attitudes and environmentally friendly behaviours (Barnett et al. 2006). In addition to these outcomes, some forms of environmental education have also been shown to be associated with learning advantages, such as improvements in achievement, critical thinking, attitudes toward learning, and motivation to learn and achieve in school (Wheeler et al. 2007). Environmental education is not simply a strategy for educating young people about environmental issues, but rather a system for using the environment as an organizing theme around which other content areas are organized (Lieberman and Hoody 1998). Environmental education uses environmental issues and topics to integrate various subject area concepts which students explore through hands-on, community-based and self-directed inquiry.

Children's involvement in environmental education can provide a foundation for environmental stewardship behaviour (Chawla 2007; Fisman 2005; Wells and Lekies 2006). A study that presented data from several schools across New England, including Boston, showed that students with more environmental education were more likely to report taking action to help the environment (Place-based Education Evaluation Collaborative, 2008). An evaluation of a Boston-based programme emphasized the importance of providing young people with initial and repeat outdoor experiences (Duffin et al. 2007). Early adolescents in another study reported more confidence in their own capacity to make a difference and displayed more optimism for the future as they worked with other students on environmental issues (Blanchet-Cohen 2008). Yet education is only part of the story of getting kids outside. Adults (in or out of school) can be a significant influence in nurturing children's interest in the environment (Blanchet-Cohen 2008). Feelings of safety and security have been identified as being important to a child's ability to bond to a place (Falco 2004). Childhood experiences of special natural places, often with the companionship of a caring adult, have been reported by adults as top reasons for their continued involvement in environmental engagement and activism (Blanchet-Cohen 2008).

Environmental education programmes immerse students in their physical surroundings (Fisman 2005; Lieberman and Hoody 1998; PEEC 2008). A field-based urban ecology programme in Boston helped students (especially girls) to connect to their community through scientific investigation. Students in research from across the country have shown more interest in learning when their programmes used the environment as a context for integrating maths, language arts, social studies, science and/or the arts (Athman and Monroe 2004; Barnett et al. 2006; Lieberman and Hoody 1998; PEEC 2008). Extensive research exists documenting the link between enthusiasm for learning and subsequent academic achievement. It is little surprise that children learn better when their hands and minds are actively engaged.

Research by the Sustainable Development Commission (2009) and by Thomas and Thomson (2004) shows that environmental quality and young people's well-being are inextricably linked, and that young people's everyday experience of living and learning in the environment, and the health of the environment itself, are critical to overall well-being. Broadhurst et al. (2008) show that the quality of the physical environment surrounding the school affects behaviour within schools, as do attendance, academic achievement and parental support. Thomas and Thomson (2004) found that the worse a local environment looks, the less children are able to play freely, and develop the habits and commitments that will enable them to address environmental problems in the future. This research base also shows that schools have an important contribution to make in improving the quality of young people's environments.

Gayford (2009) found clear evidence from young people that telling them what to think and do about environmental issues is not effective. A participatory and collaborative learning approach helps young people to enjoy and achieve everyday life. Rickinson (2001) bring learners and their experiences to the fore, and show that skilled teaching can build on young people's own views and interests in order to enhance understanding. Percy-Smith et al. (2009) reported that by using action-based forms of learning, young people were able to examine issues more effectively, understand the complexities of their real world contexts, and consequently develop effective intervention strategies.

McClaren (1992), in his description of elements of environmental literacy, states that children need to develop the ability to think about systems and move from awareness to knowledge to positive action. Ecological principles and concepts are important organizers for experience in the environment and provide insights for critically thinking about environmental issues.

An important consideration for environmental educators in the development of attitudes and knowledge among children is the importance of direct experience (Orion and Hofstein 1994). According to Kellert (1985), children who have direct contact with animals rather than just learning about them are more appreciative and knowledgeable, and show concern towards the animals. Moreover, Jaus (1994) discovered in his study that even minimal amounts of instruction in environmental education is effective in producing highly positive attitudes and that it is possible for these attitudes to be retained over time.

Teaching in the field brings students into the outdoors to a natural place where they are given the opportunity to see things in their wholeness (Thomashow 2001; Burkholder 2003), where they can merge landscape (the natural world) with mindscape (the mental world) (Orr 1992), and where they can deepen their connection with nature (Cuthbertson et al. 2003). This way of teaching provides students with a way to unify theoretical book knowledge with personal direct experience (Orr 1992; Wilson 1998). It is a way to move students towards a middle ground where the self, nature and theory overlap and are integrated (Burkholder 2003). Overall, this process elevates the role of nature (or place) from something to observe to the role of teacher. This way of learning encourages students to see nature as a primary source of information and inspiration, as motivator and classroom, and, primarily, as home—a place that provides comfort and protection, and gives meaning to life—all of which supports the process of learning and creates a meaningful and memorable learning experience (Cuthbertson et al. 2003).

Environmental education creates opportunities where students can have a direct experience with what is, for many of them, a separate reality. It is implied that through the process of environmental education students will develop knowledge about and concern for the natural world. The task of the instructor is to craft appropriate experiences in reflection of both "place" and "learner" (Haskin 1999).

Effective environmental education brings about an increased awareness and appreciation of nature. It also reveals, in context, the wounds or separations that exist between the human and nonhuman worlds. Therefore, the problem every environmental educator must address is how to motivate students into action without terrifying them into inaction (Haskin 1999).

Research shows that service-learning participation was associated with higher scores in the state test of basic skills (Anderson et al. 1991) and higher grades (Shumer 1994; Shaffer 1993; Dean and Murdock 1992; O'Bannon et al. 1999). Students in over half of the high-quality service-learning schools studied showed moderate to strong positive gains in student achievement tests in language arts and/or reading, engagement in school, sense of educational accomplishment and homework completion (Weiler et al. 1998).

Service-learning provides a hands-on, collaborative approach to learning. It also engages students in real world problem-solving and in exercising their own initiative and opportunities that are rare in most school curricula (McPherson 1991). Service-learning is a method of teaching and learning that engages students with meaningful hands-on service and immerses them in addressing community issues gains in valuable knowledge and skills that connect classroom studies with community improvements (McPherson 1991).

Education and Awareness

Improving water quality, preventing point and non-point source pollution, treating wastewater before disposal and restoring the quality of waterways all require political will. Education and awareness-building campaigns play a critical role in building community knowledge and support for the importance of protecting and improving water quality. This support for water quality improvements can lead to increased pressure on policy-makers and elected officials to implement legislation and regulations to protect water quality improve enforcement of these regulations, and increase the willingness on the part of the policy-makers and elected officials to act (UNEP 2010).

Changing Individual Behaviour

Individual behaviours impact water quality. For example, residents living, working or playing near waterways may dispose of solid waste, human waste, or chemicals in water bodies, which impairs water quality. Solid waste can clog water movement and leach toxins into the water. Inappropriate disposal of human waste in or near waterways can cause serious health problems. Education and awareness-building efforts can help change behaviour among individuals. An example is the numerous awareness-building campaigns in municipalities urging people to dispose of oil and grease properly to avoid letting it drain into stormwater drains, eventually compromising water quality. There are also education efforts to help people using open defecation to understand its impacts on water quality and disease (UNEP 2010).

Education on the connections between water quality, habitats and biodiversity can help increase the interest of conservationists in promoting and protecting water quality. Conservationists, anglers, birders and others who enjoy fishing or birdwatching need to understand the connections between water quality and their experiences of wildlife. Making these connections can help increase the constituency that is advocating for water quality improvements (UNEP 2010).

Water resources and climate change is one of the most important environmental topics that these active teaching methods aim to engage students about. They learn about water quality, water conservation and water's connections to many other environmental concerns, such as climate change, communicable diseases and health, and acquire skills to analyse biological and chemical testing of local drinking water sources, ranging from groundwater sources to municipal water sources supplied by the government.

The increasing population in the Puducherry region has placed greater demands on water resources due to climate change. Even though water is available at shallow depths in the coastal areas, high salinity reduces the availability of potable water. Excessive withdrawal of groundwater in coastal areas also creates pressure in the form of saltwater intrusion. The increasing population also increases wastewater generation and the need for more infrastructure facilities such as water supply and a sewerage system. Lack of sanitation facilities and improper disposal of human and animal waste contribute to poor surface water quality and thus lead to the spread of waterborne diseases. In urban areas, lack of adequate sewage collection, treatment and disposal facilities negatively impacts water quality. This often leads to high levels of biological pollution and resulting eutrophication. Slum growth is another problem. Improper disposal of plastics often leads to blockage of sewer pipes, etc. leading to unhygienic conditions. Leachate from the disposal site may contaminate groundwater. Puducherry is facing environmental threats such as overpopulation, industrial growth, vehicular increasing, improper solid waste management, open drainage, decreasing agricultural lands into real estates, lack of forest cover posing serious air pollution, water pollution, solid waste issues, domestic sewage, and industrial pollution. Considering these urban environmental problems, the service-learning environmental education programme is considered for urban school students.

Water resources and climate change method is an important educational tool to understand the local environmental issues and creates awareness and protect against/prevent local environmental problems with special reference to water quality issues and climate change among the local community. In considering the essentials of activity-based environmental education, the programme's aim was to give students knowledge and the necessary skills to change their attitudes towards local drinking water resources such as municipal supplied water, groundwater and open water sources such as lakes and ponds, to provide them with scientific skills such as analysing, interpreting the drinking water quality, and to put into practice various protection and conservation steps towards local drinking water sources, and to create mass awareness among the communities to conserve and protect this natural resource.

Water resources and climate change education is a pathway to quality education through active teaching and learning methods in schools, which is intended to support the local community, teachers and parents to grasp the challenges and opportunities posed by climate change and water pollution. It is a rights-based tool for skills and empowerment, which is designed to fill a perceptual gap between a child's physical environment and its daily life.

This environmental education for water resources and climate change programme aims to provide students, the future leaders of our country and beneficiaries of natural resources, with opportunities to understand, and improve their local drinking water resources. The programme empowers young people to investigate the drinking water quality in their neighbourhoods and use their findings to implement solutions. To understand the prevalent climate change effects on water quality in students' residential places involved seasonally assessing the drinking water quality, and protecting water quality from various pollution sources such as domestic and industrial.

Objectives of the Programme

The main objectives of this study were to change students' behaviour and attitudes to encourage them to protect and conserve local drinking water resources. In order to address the local water quality issues, students were provided with the scientific skills to monitor water quality, essential water quality parameters to protect against/prevent water pollution and were motivated to practise various climatic change adaptations in order to conserve their local water resources and to create awareness among the local community about climatic change adaptations and protect/conserve local water resources through service-learning education.

Study Area

Puducherry comes under the semi-arid type of climate, with a mean annual temperature of around 30°C and 70–85% relative humidity is prevalent in the Union Territory of Puducherry, which receives most of its rain from the northeast monsoon and southwest monsoon. As per the 2001 census, the population of the UT is 974,345 which is about 0.09% of India's total population. With a geographical area of just 480 km^2, the UT accounts for 0.000154% of that of India. The population density is 2029 per km^2, one of the highest in India. The community which was selected for the water quality study is located adjacent to the selected schools. This residential area comprises around 2,000 households with various economic statuses and varied cultural backgrounds. Half of the community is made up of fishermen. This community was affected by the 2004 tsunami—some research conducted in the area shows that the area has been highly affected by saltwater intrusion and faces industrial pollution in the groundwater sources area. Both the study schools, Chewalier Sellane Government Higher Secondary School (CSS) and Javagar Navodya Vidyalaya School (JNV) are located on the bank of the east coast of Puducherry. The residential places Kalapet and Chinna Kalapet were chosen from the area near the schools.

Population and Sample

School students are the target group because it is believed that one of the best ways to provide for a healthy environment in the future is to produce an environmentally literate and active citizenry as well as decision-makers. The study was conducted from July 2008 to May 2009. The schools are located in a suburban area of Puducherry town and have a diverse student population with varied cultural backgrounds, urban/suburban and rural settings with various economic groups. The schools are run by the government of Puducherry. The participants in the study were chosen from the 13- to 15-year-old age group at middle-school level. The rationale behind selecting this age group of students was that they have a lower curriculum load than those at high-school level. The total sample size was 140 students. They were divided into two groups: 70 students from each school, matched with the control group.

Study Design

A pre- and post-test design was used to measure the effectiveness of the environmental service-learning programme among students, as well as to assess the students' knowledge, interest and skills related to water quality monitoring study. Students participating in both the experimental and control group in the environmental service-learning education programme were measured as to various aspects in a questionnaire related to foundations of water quality study in the beginning (pre-test) and then they were given service-learning environmental education, and after education (post-test), they were again measured with the same questionnaire (this included both experimental and control group students).

The first phase of this education programme was held in the classroom. During this session, students were explained the basic concept of water resources, various threats to drinking water sources, water pollution control and conservation methods, and factors related to climate change, various threats of climate change and climate change adaptations. These were presented through active classroom sessions such as lectures, PowerPoint presentations, films on various water issues and climate change, debates, essay/painting competitions, quizzes and content knowledge tests.

The programme takes a student-centred, skills-based approach to empowered learning and it was designed to support the involvement and engagement of students. It aims to increase students' understanding of the interdependent relationship between the environment and their life and community. The programme is designed to address water quality issues at community level, vulnerabilities to chronic and sudden impacts of climate change, and environmental degradation. It includes formal and non-formal activities and resources for local decision-makers, school administrators and teachers and is unique in that it is designed to integrate practical guidelines for "facilities-based solutions" such as school gardens, tree planting and rainwater harvesting, with a participatory, child-centred skills-based curriculum.

In the second phase of this programme, the experimental group students were involved in service-learning education in monitoring drinking water quality (Figs. 1–3) in about four different localities in the nearby residential area, ranging from low-income colonies, market places, religious places, residential colonies, as well as their own school taps, and the impacts of climate change on these drinking water sources in the local community. The aim was to find out the quality of water consumed by the community at large. The water is supplied by the government or groundwater extraction by hand pumps or bore wells. During the water quality monitoring, students collected water samples and analysed the essential parameters in physical, chemical and biological tests. These include simple tests such as temperature, pH, chlorine, chloride, ammonia, and dissolved oxygen (DO) at the spot, since these water quality parameters change according to geographic and environmental conditions. Some of the essential water quality parameters were collected and taken to the school laboratory and analysed with regard to nitrate, fluoride, e-coli form, hardness and iron using the portable water-testing kit.

The third phase of this education programme was to bring awareness to the community, with students being provided with basic information about water resources and its importance, and prevention and control methods based on the students' water quality tests of various essential physical, chemical and biological parameters of drinking water sources. While doing water quality analyses, students interacted with the community about their water quality analysis and gave information to the community. If they found that any water quality exceeded permissible limits then they gave simple control methods to prevent pollution and asked the community to lodge complaints with the local panchayats (village councils) and with the government agencies.

Assessment Framework

We developed a comprehensive outcomes framework to assess the efficacy of water resources and climate change education modules in enhancing teaching and training in water resources and climate change. The framework measured changes in conceptual understanding, improvements in self-perceptions of process skills, confidence in water resources and climate change knowledge, interest in water resources and climate change topics, and changes in environmental orientation and worldview. The methodology adapted and integrated three types of evaluation instruments in a pre-module exposure test/post-module exposure test format.

Fig. 1 Students collecting water sample from slum area

Fig. 2 Students analysing water quality

Fig. 3 Students interacting with community

Water Resources and Climate Change

What is water resources and climate change? Why is the topic of water resources and climate change important? And what about threats to water resources? Each module (Table 1) includes an interactive PowerPoint lecture of slides with notes and discussion questions, a detailed topical synthesis paper, and a series of hands-on exercises in which students collect, analyse and synthesize water resources and climate change data from multiple sources. We have used presentations to introduce and discuss topics and applied the exercises as complements to lectures. Students were introduced to the activity and answered questions at the end of the lecture; they were allowed to work on the problems and then discussed the results in the following class; they used an entire activity during one class session and another group modified an exercise to last the whole semester. Variability in use and adaptation was allowed in this study since we were testing the proposed assessment framework rather than applying a quasi-experimental design.

Table 1 Teaching learning theme of water quality and climate change study through activity-based environmental education

Knowledge
1. How much water is there for us and how it distributed on earth?
2. What is the definition of water pollution?
3. How does water help us?
4. Why should we be concerned about water pollution?
5. Types/sources of water pollutants?
6. Impacts of water pollution?

Understanding
7. Understand local water pollution problems through analysis of drinking water quality due to climate change as well as anthropogenic sources in the local community
8. Provide scientific skills to monitor/analyse drinking water sources using a field portable water-testing kit in the local community and provide climate change adaptations at local level

Attitudes
9. Bring an attitudinal change protecting drinking water quality from contamination of water sources and its protection in terms of rainwater harvesting, water and sanitation methods

Action
10. Put into practice various simple water pollution control/prevention and conservation methods at school/household and local level

Content Knowledge Tests

Content knowledge assessments measure students' learning from the module component used (Table 2). These assessments include questions, multiple choice, matching, short answer, problem sets, and short essays. In addition to measuring knowledge recall, assessments focus on higher-order learning, including comprehension and application of material and problem-solving in new situations. The water resources and climate change module (Table 1) used a written content knowledge test, consisting of multiple choice and matching questions that were selected from the modules, to measure changes in students' knowledge of water resources and climate change. Pre-tests were given prior to classroom use of the modules. The post-test was administered either immediately after teaching the modules, or was embedded in the final exam.

Questions Grouped into Categories

Water Resources Knowledge Confidence: Assessed the Students' Confidence

1. Defining climate change and its related water resources issues at global, regional, local level.
2. Identifying threats to local water resources ranging from seawater to freshwater due to climate change.

Table 2 Questions grouped into categories

An assessment of student's knowledge and understanding of water resources
1. Defining climate change and its related water resources/issues global/regional/local
2. Identifying threats to local water resources ranging from sea water and fresh water due to climate change
3. Providing examples of the importance of fresh water sources for plants, animals and human survival
4. Describing methods such as analysing local drinking water quality used in water protection and conservation
5. Identifying water issues-anthropogenic as well as natural in a protection controversy
6. Analysing/synthesizing information on water issue in order to make awareness among local community
7. Critically reviewing the content quality of researched material

An assessment of student's interest on water resources
1. Understanding the relevance of water resources to real world issues
2. Taking additional courses related to climate change and mitigation
3. Majoring in a related subject of water resources and climate change
4. Exploring career opportunities in related protection and conservation of water resources and climate change
5. Considering changes in lifestyle choices

An assessment of student's process skills on climate change and water resources
1. Oral communication
2. Written communication
3. Identifying underlying climate change problems
4. Gathering credible information to support a small group projects on water resources and climate change
5. Sorting and filtering diverse sources of information
6. Predicting potential outcomes
7. Applying critical thinking about water issues and climate change
8. Collecting data and managing information
9. Analysing and interpreting data
10. Creating and interpreting graphs and tables
11. Working collaboratively with and in a group
12. Coordinating and managing a class or real world project

3. Providing examples of the importance of freshwater sources for plants, animals and human survival.
4. Describing methods such as analysing local drinking water quality used in water protection and conservation.
5. Identifying water issues—anthropogenic as well as natural in protection controversy.
6. Analysing/synthesizing information on water issues in order to bring awareness to the local community.
7. Critically reviewing the content quality of researched material.

Water Resources Interest: Assessed the Students' Interest

1. Understanding the relevance of water resources to real world issues.
2. Taking additional courses related to climate change and mitigation.
3. Studying at university a related subject of water resources and climate change.
4. Exploring career opportunities in related protection and conservation of water resources and climate change.
5. Considering changes in lifestyle choices.

Climate Change and Water Resources Process Skills: Assessed the Students' Confidence

1. Oral communication.
2. Written communication.
3. Identifying underlying climate change problems.
4. Gathering credible information to support small group projects on water resources and climate change.
5. Sorting and filtering diverse sources of information.
6. Predicting potential outcomes.
7. Applying critical thinking about water issues and climate change.
8. Collecting data and managing information.
9. Analysing and interpreting data.
10. Creating and interpreting graphs and tables.
11. Working collaboratively with and in a group.
12. Coordinating and managing a class or real world project.

Statistical Analysis

Paired-sample two-tailed t-tests compared pre-test and post-test means for each question on the content knowledge test for all respondents. Questions were grouped into three categories to compare overall reported changes in learning: knowledge in water quality study, interest in service-learning environmental education, and skills in analysis of various physic, chemical and biological water quality parameters. Paired-sample t-tests assessed differences across pre- and post-tests for each of these measures. The test questionnaire ensured with test–retest reliability that the coefficient of inventory was 80. The content of the questionnaire, an adequate number of questions with all aspects of water quality study issues and protection and conservation, was ensured with subject experts.

Results and Discussions

Changes in Content Knowledge

Students significantly increased their content knowledge of climate change and water resources programme differences in pre- and post-testing, while in the control group there was no change in pre-test and post-test. Students gained significantly ($t = 0.32254$, $p < 0.005$) in their confidence in climate change and water resources knowledge, showing a significant increase in interest in water pollution control. Student confidence in water resources and climate change knowledge increased significantly in questions including those relating to defining climate change, identifying principal threats, providing examples of how climate change is effecting water resources, describing methods and strategies used in mitigation and adaptation, identifying underlying issues in protection, analysing/synthesizing information on an issue, and critically reviewing the content quality of researched material.

While students reported increased interest in taking additional courses, exploring a degree or career in conservation, and considering lifestyle changes to support local water sources, somewhat surprisingly, they indicated increased interest in understanding the relevance of climate change to real world issues. Since the pre-test indicated a little less than average interest in the relevance of water resources, the change in post-test was increased and insignificant relative to the high starting value. In terms of climate change and water resources process skills, students reported significant ($t = 2.48126$, $p < 0.005$) gains in confidence in their skills in identifying conservation issues and evaluating diverse sources of information. However, students showed a significant increase in overall confidence in climate change and water resources-related process skills.

Students showed their significant increases ($t = 6.76841$, $p < 0.005$) in their post-test attitudes questionnaire and behaviours (4.67794, $p < 0.005$) towards steps to conservation and protection of local water resources in terms of rainwater harvesting at household level, community level and at school level and tree plantation programmes, continuous monitoring of drinking water quality at household as well as community level, communicating regarding climate change and water resources with school students and with the local community.

Finding

Students were trained in scientific skills for monitoring of water quality that enable them to put into practice, in real life conditions, the theoretical aspects of water pollution that were learnt in the classroom. Students also realized the value and importance of clean water in personal and at school level, by checking water quality parameters and water conservation. Regular interactions with the community helped the students to understand other related factors which affect water quality. Recommendations and a plan of action for clean water were also made by the students. These vary as per the persisting problem. It may be as simple as lowering vehicular emissions and maintaining a green environment. From the analysis it is apparent through this hands-on-training, that an informal environmental education programme increases the students' knowledge, understanding, skills, attitudes and participation in order to protect and abate air pollution in their local community. The students were led to understand the various physical and chemical problems in air quality by monitoring their surrounding environment. This study therefore stresses the need to extend teaching and learning activities into the immediate environment of the pupils beyond the classroom.

This activity-based informal learning environmental education method for climate change and water resources education for sustainable development encompasses comprehensive aspects of students' cognitive, affective and behavioural development, related to the perception and understanding of environment. These types of experiments can make learning about their local environmental issues practical and meaningful, potentially having long-term impacts on students' attitudes towards conserving local water resources and also protecting their future environment. As a result of these experiments, the students reflected on their experiences and actions in their homes, school and community, which will get them pondering about everyday habits and occurrences in the water resources and climate changes dimension. These programmes will help them to acquaint themselves with the local water resources problems, and arouse interest in them to identify more environmental problems at local level.

Children with the skills and knowledge to address the challenges faced by their local community are empowered to take action and better prepared to respond to threats, evacuations and restoration activities, thereby reducing impacts of climate change. Further, as the success of facilities-based solutions is demonstrated to local communities and decision-makers, they implement similar programmes which improve local environments and offer an opportunity for children to engage in action- and skills-based learning, thereby acknowledging their rights, needs and capacities. Partnerships with governments and voluntary organizations offer specialized tools and skills to deliver a quality education for children, while reducing the impact of climate change on water resources and associated ecosystems.

Acknowledgments The author would like to thank the Department of Education and Policy Perspectives (DERPP) and NCERT, New Delhi, India for their financial assistance for this study.

References

Anderson V, Kinsley C, Negroni P, Price C (1991) Community service-learning and school improvement in Springfield, Massachusetts. Phi Delta Kappan 72:761–764

Athman J, Monroe M (2004) The effects of environment-based education on students' achievement motivation. J Interpret Res 9(1):9–25

Barnett M, Lord C, Strauss E, Rosca C, Langford H, Chavez D et al (2006) Using the urban environment to engage youths in urban ecology field studies. J Environ Educ 37(2):3–11

Blanchet-Cohen N (2008) Taking a stance: child agency across the dimensions of early adolescents' environmental involvement. Environ Educ Res 14(3):257–272

Boko M et al (2007) Africa: climate change 2007. In: Parry M et al (eds) Climate change 2007: impacts, adaptation and vulnerability: contribution of working group II to the fourth assessment report of the IPCC. Cambridge University Press, Cambridge, pp 433–467

Broadhurst K, Owens K, Keats G, Taylor E (2008) One more broken window: the impact of the physic a environment on schools. NAS/UWT, Birmingham

Burkholder RE (2003) To see things in their wholeness: consilience, natural history, and teaching literature outdoors. In: Crimmel H (ed) Teaching in the field: working with students in the outdoor classroom. The University of Utah Press, Salt Lake City, pp 17–32

Campbell-Lendrum D, Corvalán C (2007) Climate change and developing-country cities: implications for environmental health and equity. J Urban Health: Bull New York Acad Med 84(1):109–117

Chawla L (2007) Childhood experiences associated with care for the natural world: a theoretical framework for empirical results. Child Youth Environ 17(4):145–170

CLEAN India Project (2005) Community Led Environment Action Network

Confalonieri U et al (2007) Human health. In: Parry M et al (eds) Climate change 2007: impacts, adaptation and vulnerability: contribution of working group II to the fourth assessment report of the IPCC. Cambridge University Press, Cambridge, pp 391–431

Costanza R, d'Arge R, de Groot R, Farberk S, Grasso M, Hannon B, Limburg K, Naeem S, O'Neill RV, Paruelo J, Raskin RG, Sutton P, van den Belt M (1997) The value of the world's ecosystem services and natural capital. Nature 387:353–360

Cruz RV, Harasawa H et al (2007) Asia. In: Parry ML, Canziani OF, Palutikof JP, van der Linden PJ, Hansen CE (eds) Climate change 2007: impacts, adaptation and vulnerability. Contribution of working group II to the fourth assessment report of the intergovernmental panel on climate change. Cambridge University Press, Cambridge, pp 469–506

Cuthbertson B, Dyment J, Curtnoys LP, Potter TG, O'Connell T (2003) Engaging nature: a Canadian case study of learning in the outdoors. In: Crimmel H (ed) Teaching in the field. The University of Utah Press, Salt Lake City, pp 77–98

Dasgupta S, Laplante B et al (2007) The impact of sea level rise on developing countries: a comparative analysis. World Bank, Washington DC

Dean L, Murdock S (1992) Effect of voluntary service on adolescent attitudes toward learning. J Volunt Adm

Duffin M, Becker-Klein R, Plumb S, PEER Associates (2007) Summary report: an evaluation of the Youth Opportunities Program and the A Mountain Classroom Program. http://www.peecworks.org/PEEC/PEEC_Research/S0238128B?Close=-1. Accessed 15 Aug 2008

Engelman R, Roy P (1993) Sustaining water: population and the future of renewable water supplies, Population and Environment Program, Population Action International, Washington DC, p 56

Falco E (2004) Environment-based education: improving attitudes and academics for adolescents. http://www.myscschools.com/Offices/CSO/enved/docments/EducationUsingtheEnvironment FINAL2004_000.doc. Accessed 3 Nov 2005

Fisman L (2005) The effects of local learning on environmental awareness in children: an empirical investigation. J Environ Educ 36(3):39–50

Gayford C (2009) Learning for sustainability: from the pupils' perspective. World Wide Fund for Nature, Goodling, Surrey

Haskin J (1999) Place-based learning: the technology frontier in environmental education. Educational Technology. Indian Network for climate change assessment November 2010 Ministry of Environment and Forest Government of India

IPCC (Intergovernmental Panel on Climate Change) (2007) Summary for policymakers. In: Parry M et al (eds), Climate change 2007: impacts, adaptation and vulnerability: contribution of working group II to IPCC, Cambridge University Press, Cambridge, pp 433–467

Jaus HH (1994) Activity-oriented science: Is it really that good Science and Children, 14(7):26–27

Kellert, Stephen R (1985) Attitudes toward animals: Age-related development among children. J Environ Educ 16(3):29–39

Lieberman GA, Hoody L (1998) Closing the Achievement Gap. State Education and Environment Roundtable, San Diego, CA

McClaren M (1992) Environmental literacy. A critical element of a liberal education for the 21st century. Alces 25:168–171

McPherson K (1991) Learning through service: by caring for others, students learn how to care for themselves—and their world, Winter 27:53

O'Bannon BW, Hammond R, Williams E, McKinley S (1999) Preparing for the 21st century: professional development efforts for university faculty and K-12 teachers. Paper presented at the 16th international conference on technology and education. Edinburgh, Scotland

Orion N, Hofstein A (1994) Factors that influence learning during a scientific field trip in a natural environment. J Res Sci Teach 31:1097–1119

Orr D(1992) Ecological Literacy: Education and the Transition to a Postmodern World. SUNY Press, Albany

Percy-Smith B et al. (2009) Exploring the Role of Schools in the Development of Sustainable Communities. Full Research Report ESRC End of Award Report, RES-182-25-0038. Swindon: ESRC

Pfeffer WT, Harper JT et al (2008) Kinematic constraints on glacier contributions to 21st century Sea-level rise. Science 321(5894):1340–1343

Place-based Education Evaluation Collaborative (PEEC) (2008) The benefits of place-based education: a report by the Place-based Education Evaluation Collaborative. http://www.peecworks.org/PEEC/Benefits_of_PBE-PEEC_2007_web.pdf. Accessed 1 Sept 2008

Ramanathan V, Carmichael G (2008) Global and regional climate changes due to black carbon. Nature Geosci 1(4):221–227

Revenga C, Brunner J, Henninger N, Kassem K, Payne R (2000) Pilot analysis of global ecosystems: freshwater systems. World Resources Institute, Washington DC. http://www.wri.org/wr2000. Accessed 28 Jan 2010

Rickinson M (2001) Learners and learning in environmental education: a critical review of the research. Environ Educ Res 7(3):207–320

SDC (2009) Every child's future matters, 3rd edn. Sustainable Development Commission, London

Shaffer B (1993) Service-learning: an academic methodology: Stanford. Stanford, CA University Department of Education. Cited in R. Bhaerman, K. Cordell, B. Gomez (1998). The role of service-learning in educational reform. National Society for Experiential Education and Needham, Raleigh, NC: MA: Simon and Shuster, Inc State of Environment Report—2005 Puducherry Pollution Control Committee Government of Puducherry

Shumer R (1994) Community-based learning: humanizing education. J Adolesc 17(4):357–367

Tambekar DH, Banginwar YS (2005) Studies on potential intervention for control of water borne diseases: promotion of storage: handling and service practices of drinking water in hotel/restaurants. Pollut Res 24:171–176

Thomas G, Thomson G (2004) A child's place: why environment matters to children. Green Alliance/DEMOS, London

Thomashow M (2001) A biospheric natural history. ORION 20(4):24–37

United Nations World Water Assessment Programme (UNWWAP) Water and Industry. http://www.unesco.org/water/wwap/facts_figures/water_industry.shtml. Accessed 16 Dec 2009

UNESCO (2009) UN World Water Development Report. UNESCO, Paris

UNEP (2010) Environment for development. Annual report. UNEP, Nairobi

Weiler D, LaGoy A, Crane E, Rovner A (1998) An evaluation of K-12 service-learning in California: Phase II Final Report, RPP International with the Search Institute, Emeryville, CA

Wells NM, Kristi SL (2006) Nature and the Life Course: Pathways from Childhood Nature Experiences to Adult Environmentalism. Child Youth Environ 16(1):1–24

Wilbanks T et al (2007) Industry, settlement and society: climate change 2007. In: Parry M et al (eds) Climate change 2007: impacts, adaptation and vulnerability, contribution of working group II to the fourth assessment report of the IPCC. Cambridge University Press, Cambridge, pp 357–390

Wilson EO (1998) Consilience: the unity of knowledge. Alfred A. Knopf, New York

Wheeler G, Thumlert C, Glaser L, Schoellhamer M, Bartosh O (2007) Environmental education report: empirical evidence, exemplary models, and recommendations on the impact of environmental education on K-12 students, Office of Superintendent of Public Instruction, Olympia, WA

World Health Organization (WHO) (2008) Guidelines for drinking-water quality: incorporating the first and second addenda, vol 1, recommendations, 3rd edn

Chapter 35
Water Resources Management in the Peruvian Andes: Participatory Adaptive Measures to Climate Change

Rebecca Clements and Juan Torres

Abstract Glacial retreat caused by global climate change is creating uncertain conditions for the rural farming communities of Yungay province in the Peruvian Andes. Peru's glacial mass has been reduced by an average of 22% over the last 30 years, leading to a reduction in water supplies vital for local farming and agricultural activities. Local water management practices are inefficient and are increasing pressure on already strained supplies. In 2006, Practical Action Peru initiated a four-year capacity-building project in sustainable water management in 12 rural communities. Participatory approaches shaped project design and implementation, including experimental plots and community training programmes. Consequently, five adaptation measures were prioritized for implementation: improved irrigation techniques, plague management, crop rotation, recovery of local knowledge about water management, improved dissemination of water management knowledge and technology, and strengthening the organization of Water User Boards. The introduction of improved technologies saved up to one-fifth of the water being used for farming. Community awareness of links between climate change impacts, improved technologies and reduced vulnerability were improved. By providing a focal point for community decision-making on water distribution and use, User Boards reduced the potential for social conflict and led to increased preventative measures implementation.

Keywords Climate change · Glacial retreat · Adaptation measures · Rural vulnerability · Water use · Improved technologies · Participatory methods

---

R. Clements (✉) · J. Torres
Soluciones Practicas–ITDG, Av. Jorge Chavez 275,
Miraflores, Lima 18, Peru
e-mail: rclement@solucionespracticas.org.pe

Introduction

Global warming is central to climate change: one of the world's most worrying and pressing topics on the current agenda. This phenomenon is apparent from observations of increased atmospheric and ocean temperatures, widespread melting of snow and ice, and rising sea levels Intergovernmental Panel on Climate Change: fourth Assessment Report (AR4) (2010a). The scientific community and leading environmental protection agencies collectively accept that anthropogenic greenhouse gas (GHG) emissions play a fundamental role in global warming Intergovernmental Panel on Climate Change: fourth Assessment Report (AR4) (2010b). Consequently, the reduction of GHG emissions is considered to be a key strategy in slowing global warming, and is the driving force behind a series of international agreements, most notably the Kyoto Protocol.

Despite its globalized effect, the chain of reactions linked to climate change also affects diverse ecosystems on a local scale. Populations most directly impacted are those which are vulnerable to increased climatic variability, such as the occurrence of extremes events and desertification, often suffered by rural dwelling, farming/agriculture communities already enduring poverty-related hardships. Therefore, in addition to the development of global strategies and solutions to climate change, local answers are necessary in response to the uncertainties caused by microclimate change. In other words, local agenda must focus on investigating and generating situation-appropriate adaptation and mitigation measures. Under this conceptual framework, the Practical Action's Peruvian office designed and implemented the project Water Management and Climate Change in 2006, as part of a nationwide initiative entitled Climate Change Adaptation and Mitigation Technologies.

Climate change is producing both negative and positive effects at a local level, and adaptation methods should seek to simultaneously reduce the negative effects and enhance the positive. On the basis of this premise, the objective of this project was to reduce the vulnerability of local populations to climate change through participatory capacity building in watershed management practices amongst rural communities living in the Yungay province, specifically in the San Toribio sub-basin and upper reaches of the Santa river. Ultimately, participatory strategies and results obtained in this project have great potential for implementation in similar contexts, namely rural farming communities based in tropical Andean ecosystems.

Yungay Province and Climate Change

The province of Yungay extends over the Cordillera Negra and the western flank of the Cordillera Blanca, within Ancash, a department located on the western side of the central Andes. Areas in this region of Peru have been noticeably affected by climate change over the past 30 years, with the strongest repercussions being felt by the rural farming majority of approximately 8 million inhabitants. Ancash is the

Table 1 Vulnerability factors in Yungay

Component	Vulnerability
Geomorphic	Surrounding snow-capped mountains and high, steep slopes lead to detached ice-flows, lake overflows, flooding and avalanches
Environmental	Contaminated and depleted water reserves
	Contaminated and unproductive soils
	Deforestation
	Pollution
Physical–Infrastructural	Precarious and dangerous construction: canals, highways, housing and medical clinics
	Insufficient service infrastructure: reservoirs, canals, etc.
	Dangerous infrastructure location
	Poor technical knowledge of safe housing construction
Technological	Inadequate crop production technologies (fertilization and irrigation), yet high cost
	Low-quality fertilizers
	Little or no unbiased technical information regarding fertilizing techniques (merchants consulted)
	Undernourished and unvaccinated livestock
	Water loss during irrigation
	Inadequate disease and pest treatment
	Weak crops susceptible to plagues and disease
	Unplanned and unorganized crop cultivation, based on habit or highest immediate profit
	Ignorance of soil quality classification for housing construction
Sociocultural	Weak organization
	Limited sense of control over risks
	Poor access to information about risks
	Food insecurity (malnutrition) and poor sanitation
Economic–Financial	Low income and poverty
	Lack of participation and consultation mechanisms
	High cost of fertilizers
	Produce sold for very little
Political–Institutional	Little support from authorities
	Authorities unaware of climate change issues and the impacts suffered by the general population

most vulnerable of Peru's 24 departments, and Yungay is the province with the most vulnerability factors, as described in Table 1.

Yungay has seen climate change manifested in rainfall pattern changes: advance or delay in the onset of rainy seasons, increased rain intensity characterized by shorter duration and prolonged absence; as well as the presence of increasingly damaging frosts yet higher daily temperatures. Agriculture represents at least 80% of the region's economy, and therefore climate change directly affects local economic activity as well as daily life itself.

Other impacts attributed to local climate change commonly reported by regional farming communities are the increased presence of crop plagues, pests

Table 2 Glacial retreat in the Cordillera Blanca until 2006

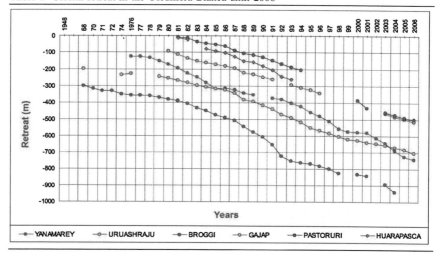

The table illustrates the accumulative annual retreat (in metres) of six Cordillera Blanca glaciers until 2006.

and associated diseases carried by these vectors. But while there is consensus about the problems experienced as a result of climate change, there is a distinct lack of management strategies organized by communities in order to better face them, with the exception of a series of spontaneous adaptive measures such as the construction of irrigation canals and adjustments to the agricultural calendar. This characteristic lack of organized management within and between communities is worsened by the prevalence of misinformation with respect to the real causes of climate change-related problems.

Yungay is an important province in Peru's seismic history, most notably for the immensely destructive earthquake suffered in 1970. The devastation caused by the quake and the resultant landslide was such that much of the population now attributes many of the regional-wide climatic changes to the event itself.

Climate Change and Glacier Retreat in the Cordillera

While glaciers have been absent from the Cordillera Negra for several centuries, they are still present in the Cordillera Blanca. This is illustrated in translation of their Spanish names, the *Black Ranges* and *White Ranges*, respectively. However, the accelerated retreat of these glaciers is well known throughout the region, and perhaps one of the best documented, as demonstrated in Table 2. Of the 723 km^2 of Cordillera Blanca present at the end of the 1970s, only 536 km^2 remained in 2003 (Zapata 2008).

From this data it is evident that glacial mass has been in steady decline since the 1970s. Approximately 30% of the glacial mass in the Cordillera Blanca region has been lost over the past 30 years, which implies a very significant loss to additional water reservoirs that are so vital to local agricultural needs during the summer months.

Agricultural water stress is already experienced regularly in areas within the Cordillera Negra, and while adaptation efforts consisting of water storage and regulation do exist, these are spontaneous and isolated, and therefore there is a heavy dependence on rainfall.

Problems in Yungay

The five districts within Yungay that were chosen for the Practical Action capacity building project represent the range of socioeconomic conditions and variety of ecosystems found within the province. Perpetual poverty, as indicated by a low human development index of 0.487 UNDP 2010, and strong economic dependence on small-scale agricultural activities is characteristic of this region. Staple foods such as maize, potatoes, wheat, barley, garden vegetables and other grains make up the bulk of crops grown. Upon consultation throughout the project, farmers commonly expressed that their principle concern was inefficient water use, rather than a severe shortage of water.

A number of factors have led to the ever-increasing vulnerability faced by the population of Yungay today: poor water management, unsustainable and outdated farming and crop cultivation practices; and detrimental climate changes and associated effects. Increased migration and cross-culturalism amongst younger generations has led to the progressive collective loss of traditional agricultural know-how, and this is exacerbated by a general lack of modern crop management technologies. The various components of this problem are summarized in Fig. 1.

Project Methodology

Project implementation consisted of a series of stages: social approach, introduction of locally appropriate technologies, definition and validation of adaptation measures, and influence on social, institutional and political aspects.

Social Approach

Potential intervention areas were identified through a workshop attended by local government representatives and community leaders from all Yungay districts, who each proposed intervention areas based on the water availability problems their

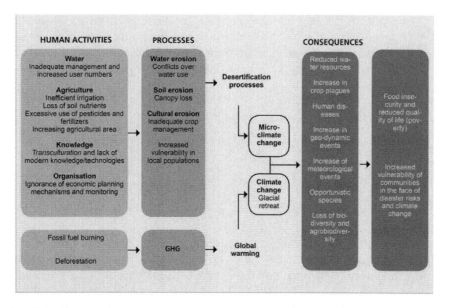

Fig. 1 Problem components

communities were experiencing. Subsequent one-day workshops were held within each of the selected districts–Yungay, Mancos, Ranrahirca, Shupluy-Cascapara and Yanama–in order to gain a first-hand understanding of community perceptions of climate change, and to identify the most vulnerable elements in local social organizations. Eleven project sites were selected from the original five districts, and a participative Adaptation Action Plan was formulated.

Introduction of Adaptation Technologies

The results obtained during the first stage of the project were instrumental in identifying the demands and technological deficiencies in current farming practices, as well as water usage concerns in the Yungay province. This phase called for the implementation of test plots in order to introduce good practice techniques such as pressure irrigation and disease monitoring. Plot areas and crops were selected in consultation with community leaders and farmers themselves.

Key steps taken during this phase were the participative formulation of an agricultural Adaptation Action Plan, via two planning and leadership workshops; the cataloguing of current local knowledge and techniques through a planning session between agricultural specialists and local farmers; the establishment of test plots sown with potato, maize, avocado, peach and alfalfa, where pressure irrigation and disease monitoring was applied; implementation of a knowledge transfer course from specialists to farmers during field sessions, emphasizing proper water and crop management; an evaluation of current potato production processes via interviews with farmers; and studies into the impacts of climate change on crop diversity and water resources.

Adaptation Measures

Four priority components to be addressed during the project were defined and validated during the planning phase: water, farming, knowledge and social organization.

Water

Water is a key element in agricultural development, and it is the resource that is most directly affected by climate change in the region. Water supply to the province is declining due to the gradual disappearance of the glacial mass in the Cordillera Blanca, and the general misuse of water through inappropriate irrigation techniques in particular. It is estimated that irrigation in common highland farming practice is less than 15% efficient, and that approximately 80–85% of current water resources is being consumed by agricultural activity, according to local information.

Diminished water supply is a continual source of conflict within and between communities, and therefore improved water management is critical to climate change adaptation efforts.

Farming

Temperature increases leading to more frequent incidences of crop disease and plagues have resulted in indiscriminate use of chemical fertilizers, insecticides and pesticides by farmers ill-informed about the detrimental effects these practices have. Often crop plague and disease management are heavily influenced by the biased recommendations of commercial merchants motivated purely by short-term economic gain. Local farmers mentioned the following commonly witnessed diseases and plagues during the initial workshops:

- Increased presence of the so-called white fly (Aleiroridos) with the El Niño phenomenon.
- Increased citrus rust mite presence during periods of high humidity.
- The development of scale insects, favoured by drought.

- High occurrence of water mould, or blight, during periods of high humidity.
- Increased fruit-fly presence.
- Presence of sand flies (Phlebotomus) in cold areas (Huaraz).
- Increased plague presence in grains and stored produce.
- Increased rodent presence.

In collaboration with the National Service of Agrarian Health (SENASA), the Centre for Development and Participation Studies (CEDEP) and the National University of Ancash, Practical Action held workshops and field sessions to train farmers in ecological pest management techniques.

Knowledge

Although over recent years farmers have indeed been adapting their practices to changing weather conditions, there is a clear need to improve and adapt their basic farming techniques, including water flow measurement, pest management technology and adequate housing and infrastructure construction in the face of climate change. Small meteorological stations were set up at project sites, and community members were then able to quantitatively monitor daily parameters such as rainfall and temperature. The information collected from these community-managed stations allowed for a more comprehensive understanding of local climate variability.

Social Organization

Community water management is currently either inefficient or unprepared for increasing climate variability in both the Cordillera Negra and Cordillera Blanca. The organization of communities into committees is a key strategy to inclusive water management, as it enables water users to make unified decisions and exercise greater influence on local government with regard to demand for and use of water supplies. While communities in the Cordillera Negra are already organized into water boards, given their common water shortage, this level of organization and planning is not found in the Cordillera Blanca. However, in both cases, water management is simply adapted to current circumstances and is ill-prepared for the water shortages projected for the future.

Influence on Social, Institutional and Political Aspects

In order for project results and experiences to have an effective and long-term impact, local governments were lobbied to incorporate the adaptation strategies into local development plans.

Table 3 Significant climatic events

Gender	Event	Year	Perceived effect	Areas Affected
Men	Torrential rain and hailstones	1965	Sowing season adversely affected	Poncos, Ocshpachán
	Freezing weather conditions	1997	Crops burnt	Bellavista, Primorpampa
	Heavy rains from January to March	1997	Roads, schools and farmland destroyed	Casacapara, Shupluy
	Rain shortage	2004–2005	Sowing season adversely affected	Entire sub-basin
	Plagues of rats and mosquitoes	2004	Crops destroyed	Entire sub-basin
Women	Prolonged drought	Pre-1970	Water shortage	Entire Cordillera Negra
	Drought from November to March	1989	Poor harvest and no sowing	Entire Cordillera Negra
	Heavy rains and hailstones	1995	Transport made impossible	Steep slopes
	Severe drought	2004	Disappearance of springs and streams	Spring-fed farmland

Project Outcomes

Popular Perception of Climate Change-Associated Risks

In order to gauge local population perceptions of climate change impacts and associated risks, information was compiled from interviews, testimonies and workshops that took place in the eleven project sites. Testimonial information on significant climatic events in the past (summarized in Table 3) reflects the variability of weather conditions and patterns in this area of the Peruvian Andes, and confirms that the people are aware of the increasingly unpredictable and often destructive nature of the region's climate, with some anecdotes dating back to the 1960s.

Rainfall seasonality and intensity is the most common weather indicator used by locals. Surveys revealed that the population has perceived a gradual change in rainfall characteristics. Whereas previously the rainy season was well defined and regulated the crop cultivation calendar, now seasonality is far less defined and varies greatly from year to year. Similarly, rain intensity is no longer predictable given the month, and rains are said to be shorter yet more intense, occur out of season–interrupting the growth cycle of important crops–and are often punctuated by atypical droughts or summer-like spells.

Perceptions of temperature related variations in local weather were also investigated. Daily temperature extremes were commonly reported: hotter by day yet colder by night. Increased daily temperatures have a positive effect on evapotranspiration, which requires more frequent crop irrigation. However, increasingly limited water availability means farmers have diminished irrigation capacity, and crops may go up to 28 days between watering. While frosts previously occurred between July and December once every 2–4 years, now they are endured annually between June and January. Frost occurrence in atypical periods of the year has serious consequences for agriculture, most commonly resulting in great losses in production.

The two sectors most vulnerable to climate change impacts were identified as agriculture and livestock. Severe climatic events such as those previously mentioned have direct negative effects on the health of crops and livestock, whether as a result of disease or physical damage. As a consequence, the population suffers impacts such as insufficient production of food for household consumption, diminished income, increased cultivation costs, land contamination due to increased use of agrochemicals and plague proliferation.

Significant and widespread impacts on infrastructure, housing and health were also recognized. Heavy rains primarily damage roads and highways, irrigation systems and canals, and sewerage systems, and weaken already unstable housing construction. Elevated lead content in stored drinking water is attributed to drought, and locals perceive frosts as influential in the development or exacerbation of respiratory illness and rheumatism.

Findings and Recommendations

Local adaptation efforts in response to climate change were largely limited to alterations to the farming calendar and the gradual construction of irrigation canals with institutional support. More technical adaptation responses, such as drip irrigation and specific plague management, were isolated efforts carried out at the level of families or groups of families. As a result of the surveys, Practical Action proposed a number of adaptation measures to assist communities in climate change adaptation and help them to reduce their vulnerability to associated risks.

Water and Improved Irrigation

Through the various training workshops with existing water boards, community members acquired basic knowledge of efficient water use practices, including the construction of irrigation canals, improvement of existing systems and the development of sustainable technology alternatives such as overhead irrigation. Six test plots featuring pressurized irrigation systems were implemented, in order to illustrate the benefits of water-saving practices, where between one-fifth and one-tenth of the volume of water usually used was saved. To date, potatoes, maize and avocado are successfully being grown on these test plots, demonstrating to farmers that with less water than traditionally used they can obtain equal or greater harvests.

Given the inevitable decline in water supply in the Yungay province, future projects should prioritize water management practices by minimizing leakage and evaporation of water resources. To this end, it is recommended that water from the reservoirs be channelled through pipes rather than canals. In order to maximize benefits to local populations, priority should be placed on installing irrigation systems on communal land, systems should be affordable, and concerted action research efforts are required in order to achieve the technological innovations still necessary.

Adaptive Agricultural Practices

The community training programme offered in collaboration with the National Service of Agrarian Health (SENASA), the Centre for Development and Participation Studies (CEDEP), and the National University of Ancash offered local farmers the chance to learn on-site improved agricultural practices grounded in environmental conservation principles. Emphasis was placed on basic soil and water analysis before initiating sowing, the importance of investigating their markets so as to better select which crops to cultivate, the use of fertilizers that are appropriate to soil characteristics, integral pest management, and basic techniques in the installation, operation and maintenance of irrigation systems. A key recommendation of the project is that a territorial farming system be established for the entire Santa river basin.

Encouraged Knowledge Sharing

One of the key objectives of the project was to salvage traditional adaptation methods in response to microclimate irregularity, and to disseminate this information through community workshops and field training sessions.

Innovative agricultural initiatives, noteworthy for their creative simplicity, include the use of ground *rocoto* (native chilli pepper) as a natural pest deterrent, and the nightly removal of egg-bearing weevils from crops. Together with novel techniques demonstrated by project specialists, improved methodological farming practices were proposed consensually within communities, simultaneously reinforcing successful, locally produced solutions, and introducing new techniques as a way of adapting to ever-changing conditions.

While the population already senses the gradual reduction of water resources in the region, there is a general lack of organized local initiatives and responses to the problem. The project's capacity-building process for the introduction and acceptance of improved irrigation systems consisted of a series of logical stages: raise awareness of the need to save water, stimulate ideas for improved water use, jointly produce water-saving methods and the implementation of test plots featuring improved irrigation systems.

Water management techniques were introduced and implemented in all five intervention districts. Nevertheless, promotion of research efforts applied to agriculture and water management in Ancash, as well as the rest of the region, is necessary if local communities are to continue improving their understanding of climate change impacts and viable adaptation methods.

Social Organization

Ancestral water usage was based on the colloquially termed "rule of the third", whereby each user has the right to a third of a waterway's flow, without regard to farming area or crop type. Currently, in water usage schemes in place throughout the Cordillera Negra, where water resources are scarce, this traditional scheme has been replaced by one based on categories, dependent on territory size and crop type. Water usage schemes in the still water-abundant region of the Cordillera Blanca, in contrast, are much more incipient, and are very wasteful of water.

As a result of the numerous workshops and training sessions, consensus of climate change and impact perceptions was reached between local authorities, farmers and other community members. Subsequently, a participative climate change Adaptation Plan that linked into local and regional development plans was put forth to local authorities.

Reference

Zapata M (2008) Desglaciación en la cordillera Blanca y cambio climático, Instituto Nacional de Recursos Naturales, Ministerio de Agricultura

Website References

Intergovernmental Panel on Climate Change: fourth Assessment Report (AR4) (2010a), p 30, http://www.ipcc.ch/pdf/assessment-report/ar4/syr/ar4_syr.pdf. Accessed March 2010

Intergovernmental Panel on Climate Change: fourth Assessment Report (AR4) (2010b), p 39, http://www.ipcc.ch/pdf/assessment-report/ar4/syr/ar4_syr.pdf. Accessed March 2010

UNDP (2010) National Human Development Index: United Nations Development Programme, http://www.pnud.org.pe/frmDatosIDH.aspx (data retrieved from the 2005 National Census). Accessed March 2010

Chapter 36
Community-Level Environmental and Climate Change Adaptation Initiatives in Nawalparasi, Nepal

Eloise M. Biggs, Gary R. Watmough and Craig W. Hutton

Abstract Nepal is one of the poorest countries in the world and much of its rural population is at, or near, subsistence level. In recent years the timing and intensity of the monsoon in Nepal, as well as temperature extremities, have changed and this is severely impacting upon agriculture, the mainstay for over 80% of the population. Flash flooding and drought has led to landslides, water shortages and irrigation problems, which have adversely affected subsistence farming. This research conducted social surveys in rural locations to ascertain which adaptation initiatives have been implemented at the community level and determine how indigenous populations have adapted to climate-induced environmental change, with a focus on water resources. The principle research aim was to qualitatively understand how rural inhabitants have adapted/are adapting to changes in climate, the environment and water from a bottom-up perspective. Water is an essential resource for sustaining community livelihoods in rural Nepal, providing an indispensable resource for irrigation, consumption and sanitation. Research conducted in communities within the Nawalparasi district found disparities in living standards relative to resource availability. Results indicated that water stress is impacting on food security and there is a need to better adapt crop production and irrigation systems to ensure viable future sustainability. In addition, illiteracy, education facilities and accessibility were found to be strongly linked to community adaptability.

Keywords Climate change · Adaptation · Water resources · Food security · Rural · Nepal · Communities · Delphi

---

E. M. Biggs (✉) · G. R. Watmough · C. W. Hutton
School of Geography, University of Southampton,
Southampton, SO17 1BJ, UK
e-mail: Eloise.Biggs@soton.ac.uk

Introduction

Listed as one of the poorest countries in the world, Nepal has 82.5% of the population living below the international poverty line of US$2 day^{-1} and it is ranked as the thirtieth most vulnerable country in terms of water-induced disasters (World Bank 2003). These disasters cause nearly 29% of the annual deaths and 43% of the annual property losses from all disasters (Khanal 2005). Much of its rural population is at, or near, subsistence level and agriculture is the mainstay for over 80% of the population (Pariyar 2003). Poverty is prevalent, particularly in rural regions where there are inadequate resources to aid sustainable development. Food insecurity and inadequate water resources are a major inducer of poverty, and generally, impoverishment is directly related to the remoteness of a community (Pariyar 2003). The disparity between communities is increasing as those lacking development opportunities are descending into an increased state of impoverishment. Climate change is exacerbating socio-environmental issues, especially water-related problems, and livelihoods are subsequently being adversely affected (Lohani 2010).

Water Resources

The national abundance of water resources in Nepal holds great potential for the development of hydropower, irrigation facilities, navigation, and domestic and industrial uses (Pariyar 2003; Shrestha 2005). In terms of community livelihoods, water provides an indispensable resource for irrigation, drinking and sanitation. It is an essential commodity, yet water is one of the country's greatest natural hazard risks with a high vulnerability from water-related disasters. Flood frequency is high during the monsoon season (June–September), with intense precipitation, catchment saturation and steepness of watersheds and river channels contributing to a varying flood magnitude. Resultant flood damage is particularly aggravated by rapid flows in upland areas of altered land use and poor land management practice, accelerating populations and the accumulation and concentration of assets and population in hazardous areas (Shrestha 2005). Common tangible impacts of flood-induced hazard are sediment-related disasters, causing direct damage to communities through environmental deterioration and thereby impacting on the social and economic development of the nation (Tuladhar 2005).

Despite water being Nepal's largest natural resource, with abundant supplies in the summer season, there is a significant shortage of water reserves during the winter period, particularly in many rural locations. Isolated rural communities often lack the development and resources to adequately store or access water resources during the dry season, and to cope with associated natural hazards of flooding in the rainy season (Pariyar 2003). In a rain-fed subsistence culture this poses serious social threats to livelihoods from a lack of potable water, sanitation and food security. Nepal's vulnerability is highly associated with the change of water regime (Alum and Regmi 2004). Rural impoverishment and water resources are being further aggravated by significant alterations in the country's hydrological system instigated by changing climatic conditions.

Climate Change

Research shows that the timing and intensity of the monsoon in Nepal, as well as temperature extremities, have changed in recent years. Temperatures have been rising at an average rate of 0.02°C annum^{-1} (1975–2005; HICODEF 2009), which is 0.6°C decade^{-1} higher than the global average (IPCC 2007; Webersik and Thapa 2008). Precipitation has increased at an average of 4 mm annum^{-1} (1975–2005), with observations indicating that winter rainfall has decreased to a point of near non-existence and the delayed onset of the monsoon has resulted in intensified downpours leading to flash flooding and sediment-related disasters (HICODEF 2009). Subsistence farming has benefited from climatic changes in some locations through increased crop production (Gurung and Bhandari 2009). Elsewhere, in the climate-sensitive sector of rain-fed agriculture, flooding, water shortages and irrigation problems have adversely affected yields. Climate model predictions for Nepal indicate changes in precipitation variability and intensity throughout the twenty first century, yet uncertainty about the future is exacerbated by climatic influence on the hydrological cycle, making droughts and floods harder to predict (Dixit 2006). Nonetheless, water resources are predicted to change in temporal availability, quality and quantity. It is probable that these climate-driven environmental changes will adversely impact on livelihoods of many communities through (i) inadequate access to clean water for drinking and sanitation, and (ii) increased pressure on irrigation systems to ensure food security. Nepal is heavily reliant on natural resources, and the depletion of natural resources by climate-induced change coupled with a poorly developed technological infrastructure and financial sector, make adaptation to climate change a substantial challenge (Agrawala 2003; HICODEF 2009). But, as Wilenius and Tirkkonen (1997) stated, with a reduction in greenhouse gas concentrations unlikely in the foreseeable future, climate change is often regarded in terms of adaptation to its impacts.

Aim of Research

To determine what factors affect rural community livelihoods in Nepal and understand how inhabitants have adapted to climate-induced environmental changes in recent years, with a particular focus on water resources.

The Delphi Technique

The Delphi technique is used to allow a group of individuals as a whole to deal with a complex problem and obtain an overall group consensus. It is a social survey method first devised in the 1950s (Dalkey and Helmer 1963) which allows the diversity of views to be expressed and refined, offering participants a better chance to stand by their opinions (Wilenius and Tirkkonen 1997). The method avoids direct confrontation with experts, which reduces the risk of influencing participants through preconceived notions (Okoli and Pawlowski 2004). Linstone and Turoff (1975) detected that the Delphi technique is especially suitable where the problem being investigated does not lend itself to precise analytical techniques and subjective judgements can contribute to a collective solution.

In practice, the Delphi technique is always adapted to the requirements and goals of the research in question (Wilenius and Tirkkonen 1997; Okoli and Pawlowski 2004). A variant of Delphi was used in this research to develop group consensus about the relative importance of issues using a "ranking type" based on the approach developed by Schmidt (1997). Within each community visited, inhabitants were asked to participate in the survey. The technique was divided into four rounds as follows:

1. The participants were asked, "Make a list of all the main issues which affect your livelihoods in this community"
2. Once a list had been generated, the participants were asked to formulate a group consensus of the six most important issues; these were written on to individual cards
3. The scenario of having RP200,000 of capital to invest (indicated with markers of RP1,000 amounts) was explained and participants were asked to distribute these "funds" as they wished amongst the six selected important issues; this effectively created an indicator of weighted importance to each of the issues
4. Any issues not raised by the Delphi survey that the researchers wished to ascertain importance to were posed as direct questions following survey completion

Although the researchers were present to direct the survey, the need for a translator effectively acted as a barrier for any direct contact with the participants. Klabber's (1994) interaction model to search for climate policy solutions is particularly relevant to this study, where problem-solving capacity is increased between researchers and decision-makers through a learning process. The Delphi technique was selected as a productive method to generate public debate and through this there is hope to successfully aid the construction of proposals for social change (Wilenius and Tirkkonen 1997). By employing this social survey method within each community visited, the key issues affecting livelihoods could be identified using a "bottom-up" approach.

Communities in Nawalparasi District

Social survey research was conducted in rural communities located in the subtropical Siwalik Hills of Nawalparasi, a relatively undeveloped district located in the Lumbini zone of the Western Development Region. The majority of flood-affected communities in Nepal inhabit land alongside marginalized rivers, which are rain-fed systems and originate in the southern faces of the Siwalik range (Shrestha 2005). These hills are comprised of steep, weakly consolidated mudstone, siltstone and sandstone bedrock slopes with a high surface erosion rate (Agrawala et al. 2003). In the Terai (lowland located downstream of the hilly regions), extensive soil erosion, formation of alluvial fans and raising of river beds due to debris deposition are major hazards which trigger flooding, and large flow volumes intensely erode river banks (Tuladhar 2005). Annual rainfall averages around 2,500 mm in this region. A report published by the Himalayan Community Development Forum (HICODEF 2009), a local non-governmental organization (NGO), indicated that precipitation patterns are changing in the region with depletion in winter rainfall events and an intensification of monsoon downpours replacing steady continuous rainfall. Drought conditions are also now commonly occurring.

A total of six communities were visited (Fig. 1): Divyapuri II (Divyapuri VDC); Girubari and Namjakot (Deurali VDC); Bhandare, Jagadev, Pelega and Damar (Hupsekot VDC). Elevation of the communities ranged from 150 to1,200 m, with those located in the upstream reaches situated on steep, unconsolidated land where widespread deforestation practices since the 1960s have greatly decreased the stability of the area. The communities located at higher elevations were lacking accessibility and were noticeably more isolated from development, unlike the communities at lower elevations. A range of communities was selected to ensure that the relationship between the geographical location of rural livelihoods and environmental resources, development and an ability to adapt to climate change could be determined.

Principle Livelihood Issues

Conducting the Delphi social surveys, and any subsequent questioning, resulted in a "bottom-up" approach in determining how livelihoods have been affected by social and environmental change. In addition, the influence of changing climate conditions was either directly communicated or could be inferred through participant responses. An idea of which resources capital should be invested into sustain, or improve, livelihood conditions was indicated through the weighting of the Delphi process. Through the combination of these results obtained, social survey findings have allowed inferences to be made regarding the social adaptation to climate-driven environmental change at the community level.

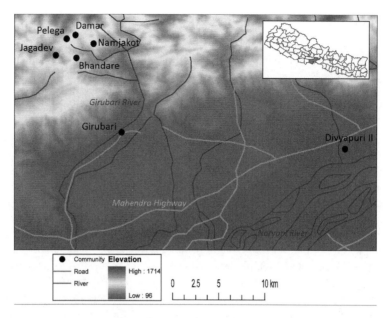

Fig. 1 Location of communities visited within the Nawalparasi district (district indicated as highlighted region in the Nepal inset map)

Results from the Delphi surveys conducted in each of the communities indicated a wide range of social and environmental factors which have influenced, and are continuing to influence, inhabitant's livelihoods. The results of the Delphi weighting exercise are indicated in Table 1. Several common issues were raised within the visited communities. The main factors highlighted were water-related issues, education, health and access, and these were generally weighted the most important in responding to hypothetical investment requirements. Within most communities, environmental issues (e.g. drought, untimely rainfall, landslides, floods, etc.) were raised as impacting upon livelihoods but were not ranked as the most important. Potential proxies for identifying national-level vulnerability to climate change recognized by Brooks et al. (2005) included economy, health and nutrition, education, infrastructure, governance, geography and demography, agriculture, ecology and technology. These socio-environmental indicators will be discussed further for the communities in Nawalparasi.

Socio-Environmental Observations

Education and Awareness

Illiteracy and a lack of education facilities in the majority of communities highlighted a limited knowledge on climate change issues, with those having heard of the

Table 1 The six major issues which affect livelihoods in each of the communities visited; with the percentage weighting indicated (bold indicates that the factor was raised as an issue at more than one community)

Divyapuri II		Girubari		Bhandare		Jegadev		Pelaga		Damar		Namjakot	
Irrigation	35	**Education—school**	35	**Education—school**	35	**Drinking water**	25	**Irrigation**	35	**Irrigation**	30	**Roads**	30
Agricultural tools	25	**Roads**	25	**Drinking water**	25	**Education—school**	25	**Education—school**	25	**Education—school**	20	**Education—school**	25
Unemployment	15	**Health**	15	**Sanitation**	15	**Health**	20	Agricultural productivity	20	**Food availability**	20	**Electricity**	15
Roads	10	Agricultural facilities	10	**Irrigation**	10	Agricultural training	10	Sanitation	10	**Health**	10	**Food availability**	15
Training	10	**Drinking water**	10	**Electricity**	10	**Roads**	15	**Roads**	5	Govt. facilities	10	**Irrigation**	10
Desertification	5	Bank erosion	5	**Roads**	5	Agricultural knowledge	5	Landslides	5	**Awareness**	10	**Awareness**	5

concept not understanding its meaning. In the remote rural communities, particularly Jagadev and Namjakot, it was evident that education was a major hindrance to successful climate change adaptation. Inhabitants did not highlight the need to ensure sustainable subsistence agriculture into the future and this is likely to be attributable to the stated lack of awareness. Contrastingly, in Divyapuri II, literacy levels were higher, with investment being placed into adult education to develop agricultural awareness and training. It was noted that although schools were present in some of the communities visited, they lacked resources to teach and there was a shortage of qualified teachers. Wealthier inhabitants of some communities could afford to send their children from isolated mountainside communities to schools in more developed areas lower down the valley.

Infrastructure, Accessibility and Technology

The inhabitants of the high-altitude, remote communities of Bhandere, Jagadev, Pelega, Damar and Namjakot, with no current road infrastructure, stated that roads were important to sustain future livelihoods through access to markets and alternative sources of income (labouring employment). With further questioning in Namjakot (one of the most impoverished communities visited) as to why they wanted road infrastructure given the unsustainability of their subsistence crops, let alone enough surplus to generate income to fund vehicle transportation to access markets, they stated that they wanted to develop like their neighbouring community, which had received investment for road construction. This reiterates the lack of adult awareness and education regarding adaptation and development; as Blankie et al. (2002) stated: "roads do not deliver the benefits of increased agricultural production, increased commercialisation, and trace as forecast in the economic appraisal documents […]. [Rather roads] essentially serve to deepen dependency and underdevelopment rather than alleviate it".

Although the former part of this quotation is somewhat reflective of reality in Nawalparasi, the latter part is not. The communities visited with road access were substantially more developed that those without, albeit in different ways. The degree to which communities operated independently was observed through their commitment to improving their livelihoods despite an inevitable increased dependency on infrastructure. Divyapuri II, located close to the national highway, also listed roads as important for livelihoods, and in particular to sustain communication links. Being located at a lower altitude on level ground, this community has a geographical advantage in making successful use of infrastructural development. This was also the case in Girubari, with a road linking to the national highway used for market access and public transportation.

Indirect road access has also assisted in sustaining livelihoods. In Jagadev, access to the highway has reduced from a 15-h walk to a 1-h walk plus 3-h jeep drive, following the creation of a connecting road with vehicle provisions. Road investment has certainly resulted in development for communities but there are many environmental conditions which may affect an increase in accessibility. For example, to utilize roads for market access, produce is required for export. Within a rain-fed monoculture environment with one crop produced each year, and assuming no monsoon failure or insect infestation, there is little surplus available for selling. In this respect, drought-resistant crops, pest control and irrigation systems are concurrently required to ensure successful development and self-sufficiency. Livestock are generally of substandard quality to sell at market and remain within the communities. In lowland regions, infrastructure investment is more sustainable compared to areas of steep terrain where slope failure makes road maintenance very difficult, especially with climate-induced increases in soil desiccation during winter and flash flooding during summer.

A change in technological conditions towards more commercial agricultural production is not really a viable option in these communities. The environmental conditions of land in Nawalparasi is a state of poor soil fertility and a reduction in land plot size from family inheritance and increased river bank erosion, as found in Girubari. Dixit (2006) stated that the value of a farmer's use of new technology depends on access, capacity to use and product marketability. The hardest part of technological advancement is sourcing capital or attracting investment to kick-start a project. In this part of Nepal, large-scale agriculture ventures are not generally seen as worthwhile. However, small-scale technological improvements were observed to be making a great deal of difference to some communities, for example, the installation of bridges for access, bamboo channels for irrigation and solar panels for electricity. The greatest technology investments were observed in hydropower and irrigation systems.

The hydropower industry has grown extensively in Nepal in recent years, but little of the total investment is filtered back into local economies (Agrawala et al. 2003). The only hydropower investment away from the developed national highway route was observed in Damar. No external capital had been obtained or given to the community; following observations of a hydropower scheme in a different community, the inhabitants of Damar had formed their own hydropower committee and each household involved had invested an equal share into the construction of a small-scale hydropower scheme. The whole enterprise was driven and constructed entirely by the community, with households now receiving electricity as a result. Impoverishment in this community is noticeably less than the surrounding communities, as electricity has provided a resource to increase their wealth through various means.

Downstream in Girubari, wealth is also greater, as there has been substantial investment and this has been aided by a discernible difference in accessibility. A programme for disaster mitigation was implemented by the government which funded gabion construction on river banks to reduce the impact of landslides. An even greater scheme impacting Girubari residents' livelihoods was the construction of a dam by a Japanese NGO to siphon water off from the Girubari river into an extensive irrigation network flowing into Girubari and communities located downstream. The dam on the Girubari river is a spur/levee structure partially covering the width of the river and is one of the most popular river control methods in Nepal (Tuladhar 2005). Not only has it provided an irrigation network, the dam also helps to control the increasing flashiness of summer flows, and reduces consequential flooding, bank instability, landslides and erosion. This was highlighted in the Girubari Delphi survey results, with landslides, drought (and associated crop disease incidence) and flooding all stated as major livelihood impacts pre-dam construction, with problems being alleviated at present. Dixit (2006) stated that over the last five decades, US$3 billion, financed by governments and international development banks, has been invested into expanding the physical and institutional infrastructure for irrigation, but this investment has failed to contribute to Nepal's agricultural development and the irrigation sector is in a state of stagnation. This may be true for commercial agricultural production but it is unquestionably enhancing the livelihoods of Nawalparasi's subsistence communities, especially under the increased environmental pressures associated with climate change.

Food Security and Potable Water

The need for increased awareness, as highlighted by the social survey results in many of the communities, has risen from an inability to irrigate the land. With depleted winter rainfall and untimely summer rainfall, food security from crops has become a major issue due to uncertainty in rain-fed agriculture and an increase in crop diseases associated with increasing air temperatures. This issue was not found to be constrained to higher elevations where water is less accessible; in Divyapuri II, irrigation was raised as a major issue despite being located within 2 km of the Naryani river (one of Nepal's largest rivers). Food security issues were attributed to unpredictable rainfall resulting in reduced crop production. The cost of cultivation in this community was stated as inadequate to cover expenses, and crop failure, due to a lack of rain and drought conditions, has become increasingly common.

Crop diversification could be a suitable option for improving local food security in this vicinity through supporting a longer growing season and increasing diet diversity (Dahal and Khanal 2010). This could also strengthen the local economy. However, diversification is largely dependent on irrigation and driven by market incentives and labour availability. In Divyapuri II this may be viable given infrastructure and accessibility, but in the remote hilly communities this will only be sustainable for ensuring subsistence survival. That said, crop diversification would reduce the risk of crop failure for some communities, e.g. in Bhandare they annually reuse crop seeds and crop loss is common. Nonetheless, crop diversification needs to be used in conjunction with drought resistant crop varieties and education of agricultural awareness. In Bhandare, increasing temperatures have caused a decline in soil moisture and it has become difficult to sow crops, meaning land is left fallow and reduced produce is insufficient to meet subsistence needs. In Jagadev, Pelega, Damar and Namjakot agricultural awareness is also poor. Despite inhabitants' observed changes in climate and environmental conditions, no adjustments in the types of crops being produced, or the dates they are being sown, have been made. If crops fail due to insufficient rain or destruction by disease, the land remains fallow. Contrastingly, food security in Girubari, a community with an extensive irrigation network, is more stable, as several crops can now be grown and harvested throughout the year.

Alterations in hydrological regimes are also adversely affecting drinking-water resources as spring sources and river flows have decreased. The time taken to collect water has an influence on water uses and family health, with women and girls usually suffering the most when water supply, sanitation and hygiene facilities are inadequate (Dixit 2006). Fortunately, problems with obtaining adequate potable water provisions have attracted investment, some sources from government schemes, but primarily from NGOs. Although some communities still have unsustainable provisions with water not constantly available, there has been a marked improvement in accessibility. Despite increases in drinking water access over recent years, this was still listed by every community as a significant issue affecting their livelihoods, with Bhandare and Jagadev ranking this as a highly important factor. Details of community water provisions are as follows:

- Divyapuri: provided through a pipe connected to a cement rainwater storage tank located within a forested area of the mountains; water is available for two hours in the morning and two hours in the evening
- Girubari: several taps installed throughout the community as part of the NGO irrigation network scheme; also available from irrigation channels
- Bhandare: tap in community which is fed from a spring and water also available at spring 5-min walk away (both for approximately 9 months of the year; further spring is 1-h walk away with all-year availability)
- Jagadev: tap in community but the water supply is insufficient during the dry season
- Pelega: tap in community with water availability all year
- Damar: tap in community with water fed from stream with all-year availability
- Namjakot: tap in community fed from stream but long pipeline with no maintenance means it suffers from consistent leaks

Drinking water still needs constant attention. Changes in the climate, combined with population growth, are causing springs, pools, rivers and groundwater stores to be depleted. Water springs in the mid-hills of Nepal have been drying up in recent years, with up to a 45% reduction observed (Gurung and Bhandari 2009). The monsoon and winter rainfall do not provide enough water to replenish groundwater reservoirs and recharge. In addition to potable supplies and irrigation, this is having substantive consequences for health and sanitation.

Healthcare and Hygiene

Health and sanitation were identified by most communities in the Delphi surveys as being important issues in their livelihoods. Rather than a lack of accessibility to clean water for sanitation, it is the insufficient amount of water and the risk of wasting potable water on hygiene which is an issue. Healthcare was not available in any of the villages and there is a need to journey to the highway should medicine or medical attention be required. In addition to inadequate amounts of water for sanitation, unawareness regarding health education was an evident issue. In the noticeably poorer communities of Bhandare and Jagadev, skin diseases were observable. This is likely to be attributable to a combination of water availability, accessibility and education. In Pelega and Damar, where water is available all year, and inhabitants of Damar are wealthier from their hydropower scheme and can afford to send their children to be educated, there were less noticeable health problems. The majority of communities did not have toilet facilities. Those which did—Pelega/Damar shared toilets in the school and Namjakot toilets in the school—had had them installed through NGO investment but they were not being used or maintained. This is likely to be due to traditional cultural values and a lack of educational awareness regarding good hygiene practice.

Employment, Income and Investment

Research by Blankie et al. (2002) concluded that although there has been no transition into a state of elevated poverty for the majority of the Nepali population, there have been no noticeable developments in commercialized agriculture either, and very minimal investment. These findings are reinforced by research produced by Dixit (2006), as discussed earlier. Household surveys conducted by Blankie et al. (2002) in west-central Nepal implied that agricultural commodity production has retreated from the market, with households providing adequate means of subsistence but with low income productivity per hour worked. The study found that a lack of labour, land, processing facilities, transportation, agricultural products (chemical fertilizer and insecticides) and seed varieties has constrained a large proportion of producers. With noticeable increases in temperatures causing drought conditions, and an increased prevalence and incidence of crop disease, supplementary income is important given the increased likelihood of crop failure.

Additional sources of income were noted as being important in sustaining community livelihoods. With insufficient income, and even subsistence, generated from farming, it was found that in nearly all the communities, the male population (approximately aged 12 and above) were more commonly emigrating to find additional sources of income and send earnings back to their families. Employment destinations include the Middle East, India, Southeast Asia, the United Kingdom, as well as other locations in Nepal, predominantly to work in construction or join the armed forces. Blankie et al. (2002) stated that releasing a member of the household to migrate to earn income is easier and less of a risk to prioritize food security than to enter the market as a producer. Alternatives like this are much less uncertain than investing in agriculture; particularly in areas where irrigation is not available. This has resulted in a biased demographic structure of rural communities, with rural poverty becoming largely feminized (Dixit 2006). Pariyar (2003) stated that women are much more affected by poverty and in these impoverished communities women are disproportionately vulnerable to climate change, with the limited access to resources and decision-making processes preventing women from using their knowledge and experience to successfully adapt (Dissanaike 2009). This was evident in Divyapuri II where Sohamati, a local NGO, implemented a skills project programme but since funding has ceased, the women cannot implement their new skills. Women in this community have to make extra income from daily wages and they are contemplating taking loans from wealthier community inhabitants to ensure there is an adequate annual food supply for their family. Similarly, in Pelega, inhabitants have to earn additional wages to ensure food and they sometimes need to take credit from neighbouring Damar inhabitants, who are contrastingly wealthier due to their hydropower scheme. Road construction in the valley adjacent to Jagadev is providing waged employment for inhabitants, where income is being used to fund the construction of a school, as the government is not providing any support; school construction is also the case in Pelega.

Some NGO and government investment has already been discussed. There have been intermittent projects throughout the communities visited, except Pelega, where the inhabitants stated that they have never received any form of external investment. HICODEF has a presence in Bhandare, Jagadev and Damar and is continuing to promote capacity building within the communities. Other projects mentioned include Jeiger, an international NGO (iNGO), who implemented an afforestation scheme in Damar to reduce the number of landslides and their impact. In Namjakot, a Finnish iNGO installed a water pipe from the stream (2 h' walk away) and built a school. However, the level of illiteracy and lack of awareness within this community has rendered the investment of limited usefulness. The community formed a construction committee to assemble the water system, but following the withdrawal of the iNGO-investors post-project completion, no maintenance committee was formed and the pipe is in a state of disrepair. The school also doesn't seem to be in use due to a lack of resources. Investment provides a major issue in terms of maintenance.

Climate, Biodiversity and Disease

Several issues regarding weather patterns and the physical environment were raised during the Delphi surveys by the participants. These are summarised in Table 2. General trends indicate a shorter, erratic and more intense monsoon. Rainfall has reduced overall, especially in winter with precipitation being unpredictable. Temperatures are mainly increasing concurrent with drought. Drought has increased the incidence of crop disease and prevalence of insects such as aphids. This was discussed by all communities during Delphi surveys and, as well as noting increases, they have observed new types of disease. Gurung et al. (2010) conducted a community perception study in Nawalparasi and noted an increase in potato blight, fungal infections and tick occurrence in livestock. In Bhandare,

Table 2 Observed changes in climate and water resources

	Monsoon	Temperature	Drought	River flow	Rainfall	Fog
Divyapuri II		↑	↑	Flash floods	↓ Unpredictable	↑
Girubari	Shorter	↑Summer hot days ↑Winter	Stable	Stable	↓ Unpredictable	↓
Bhandare	Unpredictable Non-constant	↑Summer ↑Winter	↑		↓	↑
Jagadev		↑Summer ↑Winter warm days	↑		Unpredictable ↓ Frequency	↓
Pelega	↑Intensity Non-constant	↑Summer ↓Winter		Flash floods ↓Winter	↓ Winter Unpredictable	↑
Damar	↑Intensity	↓Winter ↓Summer hot days		Flash floods ↓Winter	↓ Hailstorms Unpredictable	
Namjakot	↑Intensity			Flash floods ↓Winter	Unpredictable	↑

inhabitants have noticed new types of weeds and insects in rice plantations. In Damar and Namjakot, there has been an increase in tree disease, which has subsequently rendered the trees useless for the construction of housing, irrigation infrastructure and bridges.

Large-scale deforestation in the upland area has decreased timber resources, reduced soil stability, increased the frequency of landslides, decreased biodiversity and habitats, increased the likelihood of flash flooding, and most probably contributed to increased desiccation and reduced rainfall through a decline in evapotranspiration. Desertification was mentioned as affecting livelihoods in Divyapuri II, Girbuari and Pelega. River flows have become flashier in nature, with winter streamflows so low that river beds are drying out in Pelega, Damar and Namjakot. The number of foggy days has mostly increased, with the exception of Girubari and Jagadev. Fog increases the reflectance of sunlight and reduces the rate of photosynthesis, causing a reduction in crop growth. Gurung et al. (2010) observed that there has been a reduction in winter dew, previously a soil moisture source, which has become a food security hazard as crops are drying out.

Vulnerability and Adaptation

Successful climate change adaptation is most often best evaluated by those adapting or affected by the adaptation measures, with experts present to assist this process by ensuring that adaptation is well understood and that all the relevant factors are considered (Doria et al. 2009). Wilenius and Tirkkonen (1997) stated that the production of knowledge is a discursive practice forming an essential part of social control, where scientific knowledge is a socially generated discourse both reflecting and shaping reality. The Delphi technique adopted within this research aimed to ascertain physical geographical changes from a social perspective. Through this approach, adaptation measures could be better understood. Survey results indicated a noticeable increase in vulnerability by communities to climate-induced change and natural hazards over the last 10 years or so, with the principle environmental hazard being water.

Brooks et al. (2005) identified Nepal as a country of moderate to high vulnerability within the upper quintile of their multiple vulnerability indices. Results from their Delphi survey emphasized the importance of governance, sanitation and life expectancy as indicators. The analysis indicated that a reduction in mortality outcomes could be achieved through increasing government effectiveness and accountability, civil and political rights, and literacy. The most important indicator found to influence the level of vulnerability in the Nawalparasi communities is that of adult education and awareness (usually associated with literacy), as this was observed to be the most influential in terms of successful adaptability. Adaptation to environmental change is strongly influenced by the level of poverty and scope for opportunity within these communities. According to Adger et al. (2005) successful adaptation is that which takes into account cost-effectiveness, efficiency, the distribution of benefits and the legitimacy of the adaptation. In these rural communities of Nepal, climate change adaptation needs to be gender-responsive to address water scarcity and food security issues, as women are largely responsible for water collection and providing edible produce. Women's expertise in adaptation, mitigation and disaster risk reduction strategies needs to be recognized (Dissanaike 2009).

Sustainability for the Future

Issues investigated in these rural Nepali communities have highlighted key development categories where future research and investment should be directed. The research outcomes emphasize the essentiality in ensuring adaptation initiatives remain practicable and sustainable. A definition provide by Van Aalst et al. (2008) states that

> Adaptation at the community level means being able to maintain (and preferably improve) the current living standard in the face of expected changes in climate trends and the intensity and frequency of severe events that may affect people's livelihoods.

The principle issues in communities visited in Nawalparasi affecting livelihoods, as identified by the inhabitants themselves, were:

1. Education and awareness
2. Water resources
3. Food security
4. Health
5. Infrastructure

Water resource issues are certainly of great importance at the community level, but this research has highlighted the importance of the need to implement education and awareness alongside water resource projects to ensure future sustainability. Long-term adaptation strategies are lacking due to limited understanding of climate change and its impact. Inadequate awareness is prevalent in communities but this is also evident within the professional government and NGO communities, at a grassroots and national level (Gurung and Bhandari 2009). With climate change localized, it is essential that adaptation and mitigation strategies remain as local enterprises (Gurung and Bhandari 2009). The livelihood benefits of water and sanitation projects tend to be under-represented despite growing evidence that they are widespread and despite considerable irrigation investment; the approach to water development and management has remained top–down, hierarchic and technocratic (Dixit 2006).

A document produced by Practical Action, an iNGO (Dissanaike 2009), identified that a new concept is needed to address disaster risk reduction in South Asia, to transform development into an opportunity to reduce disaster risk. This is illustrated well in Fig. 2. Top–down governance approaches treat disasters separately from development, despite a significant association between the two, and many South Asians have become poorer as a result of development, with climate change becoming the greatest source of maldevelopment-induced disasters (Dissanaike 2009). Globally, there has been a shift away from top-down adaptation strategies to community-centred disaster risk reduction and adaptation initiatives (Van Aalst et al. 2008). However, this has yet to be successfully integrated within Nepali governance, with the selection of funded adaptation schemes tending to remain focused on those living in the most accessible areas and where they are best able to express their demands (Dixit 2006). Alum and Regmi (2004) identify numerous soft and hard adaptation measures which could be implemented to secure Nepali development and livelihoods, and the government's National Development Council is going some way towards pushing policy in the right direction of livelihood sustainability. Blankie et al. (2002) identified in the 1970s that the future of Nepal was seen as slow yet inexorable economic decline, with increasing poverty and systematic failure of policies that assumed an intrinsic propensity toward capitalist development. Despite this pessimistic outlook, there has been progress towards development and the sustainability of Nepal's growing population over the last 30–40 years. Climate change has impacted, and will continue to impact, on the socio-environmental conditions of community-level livelihoods. But with a current state of increased political stability and a somewhat "NGO-ization of life" (quoted in Dixit 2006) in Nepal, there is hope that adaptation can be implemented from a bottom–up approach, through community-level capacity building and ensuring the implementation of long-term sustainable initiatives.

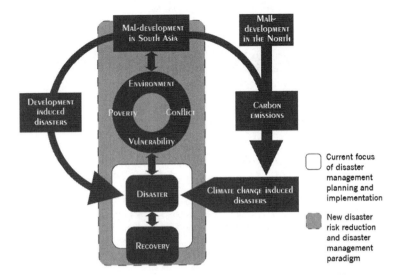

Fig. 2 A suggested conceptual shift in disaster risk reduction by Practical Action (Source: Dissanaike 2009)

Summary

Nepal lacks adequate coping capacity to deal with the challenges of climate-induced environmental change due to underdeveloped economies and limited scientific and technical capabilities (Lohani 2010). However, Pariyar (2003) concluded that the potential for water resource development in Nepal, termed as the "blue revolution", could bring economic revival and play a vital role in alleviating poverty. Successful adaptation involves the promotion of water storage and conservation under changing climate conditions (Bandyyopadhyay 2009). This research has highlighted the need for developing sustainable adaptation strategies to preserve water resources and ensure food security, particularly in the remotest locations where communities lack sufficient support and opportunity for development. Within the communities surveyed in the Nawalparasi district it was found that education, awareness, infrastructure, accessibility, technology, food security, potable water, healthcare, hygiene, employment, income and investment all played a part in sustaining livelihoods, with climate change found to be impacting on community vulnerability, and all factors interacting to determine a community's capacity to successfully and sustainably adapt to environmental change.

Acknowledgments This research was funded by The Glacier Trust and the University of Southampton. The authors would like to thank Mohan Khatiwadi and Dolraj Gaire for providing invaluable assistance during data collection as translator and local field guide.

References

Adger WM, Arnell NW, Tompkins EL (2005) Successful adaptation to climate change across scales. Glob Environ Chang 15:77–86

Agrawala A, Raksakulthai V, van Aalst M, Larsen P, Smith J, Reynolds J (2003) Development and climate change in Nepal: focus on water resources and hydropower. Environment directorate development co-operation directorate, Organisation for Economic Co-operation and Development, Paris, p 64

Alum M, Regmi BR (2004) Adverse impacts of climate change on development in Nepal: integrating adaptation into policies and activities. Working paper No. 3: capacity strengthening in the least developed countries (LDCs) for adaptation to climate change (CLACC), Bangladesh Centre for Advanced Studies, Bangladesh, p 40

Bandyyopadhyay J (2009) Climate change and Hindu Kush-Himalayan waters—knowledge gaps and priorities in adaptation. Sustain Mt Dev 56:17–20

Blankie P, Cameron J, Seddon D (2002) Understanding 20 years of change in west-central Nepal: continuity and change in lives and ideas. World Dev 30:1255–1270

Brooks N, Adger WN, Kelly PM (2005) The determinants of vulnerability and adaptive capacity at the national level and the implications for adaptation. Glob Environ Chang 15:151–163

Dahal H, Khanal DJ (2010) Food security and climate change adaptation framework: issues and challenges. Available at http://moac.gov.np/uploads/pdf/wnPoverty%20and%20Food%20Security%20Updated.pdf. Accessed Sep 2010

Dalkey N, Helmer O (1963) An experimental application of the Delphi method to the use of experts. Manag Sci 9:458–467

Dissanaike T (2009) South Asia disaster report special Copenhagen issue. Practical action, p 19. Available at http://practicalaction.org/print/docs/region_south_asia/south-asia-disaster-report-copenhagen.pdf. Accessed March 2010

Dixit A (2006) Water research in Nepal: creating a contested terrain. SaciWaters, Hyderabad, p 50

Doria MF, Boyd E, Tompkins EL, Adger WN (2009) Using expert elicitation to define successful adaptation to climate change. Environ Sci Policy 12:810–819

Gurung GB, Bhandari D (2009) Integrated approach to climate change adaptation. J For Livelihood 8:90–98

Gurung G, Pradhanananga D, Karmacharya J, Subedi A, Gurung K, Shrestha S (2010) Impacts of climate change: voices of the people. Practical Action Nepal Office, Kathmandu, p 40

HICODEF (2009) Baseline survey of integrated water resources management project for climate change adaptation in Nawalparasi District. Produced by HICODEF for Practical Action, p 33

IPCC (2007) Climate change 2007: the physical sciences basis. Summary for policy makers. IPCC, Geneva

Khanal NR (2005) Water induced disasters: case studies from the Nepal Himalayas. In: Herrmann A (ed) Landschaftsökologie und Umweltforschung 48 Proceedings of international conference on hydrology of mountain environments Berchtesgaden, Germany, pp 179–188

Klabbers J (1994) Climate policy: management of organised complexity through gaming. In: Proceedings of international conference on international simulation and gaming association. Sage Publications, London

Linestone H, Turoff M (1975) The Delphi method, techniques and applications. Addison-Wesley, Reading

Lohani SN (2010) Climate change in Nepal—shall we wait until bitter consequences? Available at http://www.nepjol.info/index.php/AEJ/article/viewFile/725/45. Accessed Sep 2010

Okoli C, Pawlowski SD (2004) The Delphi method as a research tool: an example, design considerations and applications. Inf Manag 42:15–29

Pariyar MP (2003) Water and poverty linkages in mountainous areas: a case study from Nepal. In: Hussain I, Giordano M (eds) Water and poverty linkages: case studies from Nepal, Pakistan and Sri Lanka, International Water Management Institute, Sri Lanka, pp 5–50

Schmidt RC (1997) Managing Delphi surveys using nonparametric statistical techniques. Decis Sci 28:763–774

Shrestha PM (2005) Water-induced disaster mortality of Nepal. In: Pradhan LC, Sthapit SM, Strestha PM (eds) DWIDP Bulletin, Department of Water Induced Disaster Prevention, Kathmandu, Nepal, pp 14–17

Tuladhar RM (2005) Water-induced disaster mortality of Nepal. In: Pradhan LC, Sthapit SM, Shrestha PM (eds) DWIDP Bulletin. Department of Water Induced Disaster Prevention, Kathmandu, pp 14–17

Van Aalst MK, Cannon T, Burton I (2008) Community level adaptation to climate change: the potential of participatory community risk assessment. Glob Environ Chang 18:165–179

Webersik C, Thapa M (2008) Nepal climate change and security factsheet. United Nations University, Institute of Advanced Studies. Available at http://www.ias.unu.edu/resource_centre/Nepal_Climate%20Change%20Facts%20Sheets%20Series_2008_1_lowres.pdf. Accessed Feb 2010

Wilenius M, Tirkkonen J (1997) Climate in the making: using Delphi for Finnish climate policy. Futures 29:845–862

World Bank (2003) Poverty and climate change: reducing the vulnerability of the poor through adaptation. World Bank, New York, p 43

Chapter 37
Multifunctional Land Use in Urban Spaces to Adapt Urban Infrastructure

Marko Siekmann, Nina Vomberg, Michael Mirgartz, Johannes Pinnekamp and Sarah Mühle

Abstract The research project presented here is an evaluation of problems arising from global warming, increasing stormwater intensities, demographic changes and migration in highly industrialized regions. 1D/2D simulations are used to show the advantages of water sensitive urban design (WSUD) strategies using the MIKE URBAN Flood Software (Danish Hydraulic Institute, DHI, Copenhagen, Denmark). Two regions in the Ruhr area, Germany, were analysed in order to illustrate the advantages of WSUD. More and more frequently, rainwater after an extreme rainfall event cannot be drained by the existing sewer system [DWA (2010) DWA-Themenheft, Klimawandel–Herausforderungen und Lösungsansätze für die deutsche Wasserwirtschaft", Deutsche Vereinigung für Wasserwirtschaft, Abwasser und Abfall e. V., DWA, Hennef] and flooding of urban city centres occurs. One reason why the central sewer system fails to absorb the rainwater is the increased load in the sewers due to a higher discharge of wastewater. More frequent extreme rainfall events due to climate change will worsen the situation [Gerstengarbe F (2009) Klimawandel in Ballungsräumen: Das Beispiel Ruhrgebiet, Bildungszentrum für die Entsorgungs- und Wasserwirtschaft GmbH]. That rainfall characteristics have already started to change is shown by numerous flash flood events all over Europe [Castro D, Einfalt T, Frerichs S, Friedeheim K, Hatzfeld F, Kubik A, Mittelstädt R, Müller M, Seltmann J, Wagner A (2008) Prediction and management of flash floods in urban areas in Germany (URBAS). Final Report, Federal Ministry of Education and Research, Bonn]. Although such events will occur more frequently in Germany in the future, the extension of sewage systems is not financeable. Therefore, decentralized infiltration and storage devices should be introduced to reduce runoff.

M. Siekmann (✉) · N. Vomberg · M. Mirgartz · J. Pinnekamp · S. Mühle
Institute of Environmental Engineering, RWTH Aachen University,
Mies-van-der-Rohe-Straße 1, 52074, Aachen, Germany
e-mail: siekmann@isa.rwth-aachen.de

Additionally, the directed discharge of stormwater runoff across the surface in an urban catchment is required. Within a city, streets and squares could be flooded in order to protect buildings and urban infrastructure. Additionally, the directed discharge into decentralized storage structures is useful to limit the negative effects of flooding. The required storage capacity needs to be integrated in the environment. It has been shown that the directed flooding of public parks or public squares is a useful measure to adapt urban infrastructure to the consequences of climate change.

Keywords Adaptation · Climate change · Water sensitive urban design · Multifunctional land use · Participation and acceptance

Introduction

The management of rainwater runoff in settlements is of great importance for environmental engineering. Considering the potential consequences of climate change, it is advised to re-evaluate infrastructural facilities soon because they can only be adapted slowly. Due to the high damage potential in urban areas, environmental engineering and urban management authorities are pressed to act. Based on a holistic approach, flexible solution strategies have to be developed. Accordingly, a continuous adaptation of settlements and their infrastructure to future climatic changes can be realized.

The project was located in the Ruhr area (North Rhine-Westphalia, Germany) as a representative example for a highly industrialized and developed region in central Europe. The results are based on case studies for the local authorities of Bochum and Herne. Figure 1 shows the location of the investigated area in Germany.

Climate Model Results

In order to gather precise information about the extent of climate change with regard to urban drainage, it is necessary to determine important parameters such as precipitation and rainfall intensity more accurately than is done in global trends (IPCC 2007). Based on the IPCC SRES scenarios, different regional climate models (statistical: STAR II, Wettreg; dynamical: CLM, REMO) have been built in Germany. But so far, all published future climate scenarios, though using those different climate models, agree in the trend on the impact of climate change to the changes in precipitation patterns (EXUS 2010). There is still, however, a great lack of knowledge and great uncertainty in the results of the climate models, so future research is needed.

In the presented study, results of the regional statistical climate model STARII (Gerstengarbe 2009) are used to show the consequences of climate change to the

37 Multifunctional Land Use in Urban Spaces

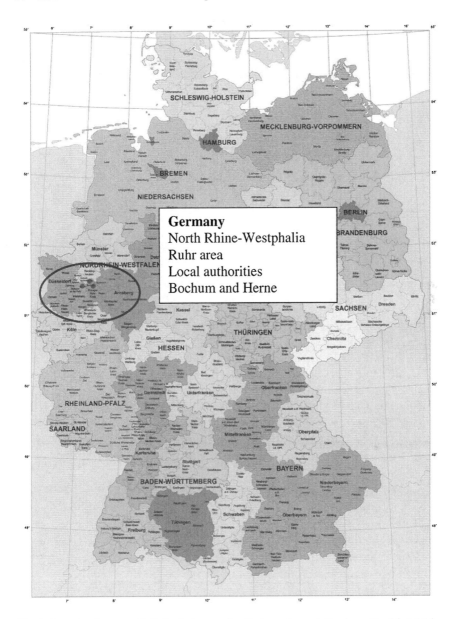

Fig. 1 Investigation area © Federal Agency for Cartography and Geodesy, Frankfurt Main, Germany

changes in the precipitation patterns for the selected study area North Rhine-Westphalia. The demonstrated regional climate scenarios calculated with STAR II are based on the IPCC emission scenario A1B.

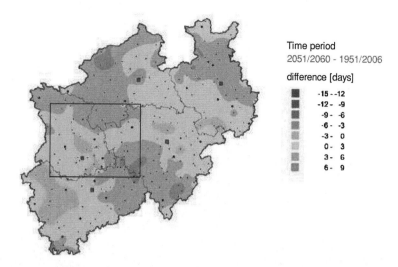

Fig. 2 Difference in days with heavy rainfalls (>10 mm) comparing time periods of 2051/2060 and 1951/2001 (Gerstengarbe 2009)

Gerstengarbe compared the number of days with more than 10 mm rainfall at present (1951–2001) and in the near future (2051–2060). The results are shown in Fig. 2. It can be seen that in most parts of North Rhine-Westphalia, an increase of days with more than 10 mm precipitation is expected. As a result of changing precipitation patterns, the frequency, intensity and duration of precipitation events will change, too. In the investigated Ruhr area in North Rhine-Westphalia, the following effects of changing precipitation patterns are expected (DWA 2010):

- Higher runoff due to precipitation and higher risk of urban flooding
- More frequent and/or higher overflow from combined sewers into receiving waters
- Strongly fluctuating water levels of rivers
- Reduced recharge rates of aquifers, which cause a decline of the groundwater level during summer
- Risk of infiltration into buildings due to an increasing level of groundwater, caused by changes in the hydrological cycle (displacement of precipitation towards winter)

Methods and Objectives

As part of the R&D project "water sensitive urban design" funded by the Federal Ministry of Education and Research, the potential of a water sensitive urban design in the Ruhr region has been analysed (www.wassersensible-stadtentwicklung.de). The Ruhr region has been chosen because of the current changes in economy and population.

In Germany, as in the rest of Europe, population and demographic trends vary regionally or even locally, but overall population is decreasing. On a regional level, in the investigated Ruhr area it can be observed that urban population is decreasing significantly whereas in the suburban areas, population is increasing (Beckmann 2005).

In areas with growing population, an increase in paved areas is to be expected, which will cause an acceleration of hydrological processes, a higher risk of flooding, increasing discharge loads and decreasing recharge rates of groundwater. Furthermore, densification of areas and sealing of as yet unused urban areas occupy space, which could otherwise be used for water management purposes. In addition, the drinking water demand will rise in these areas, which will cause an increasing strain on drinking water resources.

In areas with decreasing population, the deconstruction of buildings allows the use of the regained open space for water management purposes. In the long run, a decreasing drinking water demand caused by the decline in population and hence in industry will cause problems with the operation of drinking water supply facilities (Hiessl et al. 2002) since a maximum residence time in the sewers must not be exceeded. The same applies to the sewage system, as deposits in combined wastewater systems can cause problems. The financial burden for each resident will rise, because fixed costs which cover the bigger part of the fees have to be raised by fewer residents.

The aim of the project is to show the effects of climate change on existing settlement areas and to develop adaptation strategies by using the method of water sensitive urban design. These strategies are intended to neutralize or at least minimize the negative impacts of climate change on urban living space.

In order to enable a comprehensive approach, a cooperation network of experts from different disciplines and professionals in the Ruhr region was founded (Siekmann 2009). Apart from representatives of the water management department, this network includes urban planners, landscape architects, social scientists and representatives of the local city council and water association. The discussion and evaluation of the results of the simulations, as well as the conclusions drawn by the cooperation network, have been very helpful in order to identify the advantages and problems of water sensitive urban design.

In general, the principle of water sensitive urban design, as it has already been applied in Australia (Wong 2005), is based on a combination of environmental engineering planning approaches and urban and landscape planning requirements under consideration of the interests of all parties involved. It represents a holistic approach to stormwater management that aims at a sustainable urban development. The frequently discussed measures concern infiltration, temporary storage, delayed runoff and rainwater harvesting. Additionally, multifunctional land use, such as temporary storage of runoff in open spaces or parks, is considered during the planning process of urban areas. If water sensitive urban design is applied on a large scale, there will be several positive effects such as increasing groundwater recharge rates and decreasing flooding frequency. Additionally, costs may be reduced by avoiding building measures in the centralized drainage systems and synergy effects in the planning and operation of multifunctional facilities.

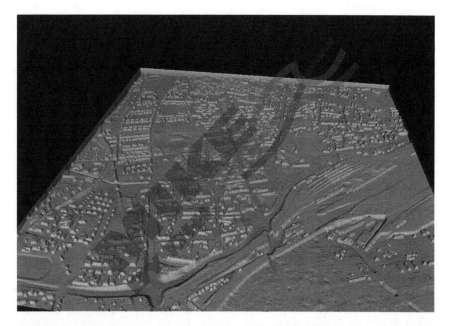

Fig. 3 Surface model, Herne

In this research project, the effects of water sensitive urban design were simulated for two project areas in the Ruhr region, Bochum and Herne. 2D models of the surface area were developed (e.g. Fig. 3) and connected to 1D sewer network models to build up an integrated model of the urban system (Ellis at al. 2009). To that end, the MIKE URBAN Flood software was chosen (Danish Hydraulic Institute, DHI, Copenhagen Denmark). The results of this project will be summarized in a guideline for water sensitive urban development in autumn 2010. The intended purpose of this guideline is to demonstrate to professionals concerned with urban design (e.g. water management, urban development and other involved institutions), how water sensitive urban design can be applied successfully. Furthermore, it shows which basic, local conditions have to be created to allow effective adaptation and maximum acceptance by the affected people.

Results and Discussion

Estimation of Future Precipitation under the Influence of Climate Change

With regard to the anticipated changes of the precipitation characteristics, two different categories of rainfall events have to be clearly distinguished (Staufer 2008):

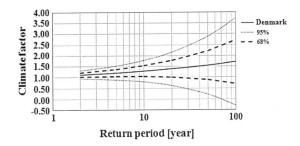

Fig. 4 Climate factors, example Denmark (Larsen et al. 2008)

- Design precipitation events (return period <10 a)
- Extreme precipitation events (return period >30 a)

The effects of rainfall events of both categories will be analysed for the two model regions.

The examination of the effects of climate change on precipitation patterns have been the object of many research projects. For example, Arnbjerg-Nielsen (2005) discovered that the frequency as well as the intensity of a ten-minute rainfall has increased, by analysing rainfall data of the last 20 years in Denmark. Although this tendency does not apply to the entire country, the authors suggest that existing intensity-duration hydrographs have become obsolete for the dimensioning of rainwater collection systems.

Based on these results, Larsen et al. (2008) determined climate factors for Denmark that allow one to dimension drainage systems adapted to the new precipitation patterns. In this context, the climate factor is used as a multiplier to enhance the calculated discharge in the sewer system.

Figure 4 shows the climate factor in relation to the return period of the precipitation event. For commonly used design events (return periods of 1–5 years), the climate factor is between 1.1 and 1.25. Extremely rare events (return period of 100 years) are linked to a climate factor of up to 1.5. It has to be considered that the inaccuracy of the climate factor increases with the return period. The conducted study led to the integration of the climate factor into design practice in Denmark and hence to an extension of the cross-section of sewer pipes. In Germany, the application of climate factors is not bound by law. Conventional dimensioning standards are still in use, particularly because of the insufficient reliability of climate prognoses at local levels.

However, technical guideline No. 4.3/3 published by the Bavarian Environment Agency (LFU 2009) suggests to increase design parameters by 10–40%, depending on the region, the rainfall duration and the return period in Bavaria (cf. Table 1). The results are in the same range as the results of Larsen et al. (2008) discussed earlier.

Table 1 illustrates how the assumed design precipitation event for a sewer system has to be adjusted according to technical guideline No. 4.3/3. Today, the dimensioning of a sewer system in a residential area is usually based on a

Table 1 Changed design precipitation intensities as a potential precaution of the effects of climate change (LFU 2009)

Design frequencies according to DWA-A 118 rule (1-time in n years)	Recommended reduced frequencies (1-time in n years)	Increase in design rainfall intensity [acc. to KOSTRA-DWD (1997)] (%)
1 in 1	1 in 2	22–40
1 in 2	1 in 3	10–19
1 in 3	1 in 5	12–21
1 in 5	1 in 10	14–23
1 in 10	1 in 20	12–19

rainfall event with a return period of 3 years. In the future, it will have to be replaced by what is currently a five-year rainfall event.

Therefore, a five year precipitation event is chosen to demonstrate the effects of climate change with and without the application of water sensitive urban design measures in the study area. The five year precipitation event with a duration of 60 min is shown in Fig. 5.

Additionally, strategies to manage extreme rainfall events have to be included in future planning processes. To show the effects of extreme rainfall events in this study a 30 year precipitation event is chosen. Figure 6 shows the precipitation event with a return period of 30 years and a duration of 60 min.

Increase of Design Precipitation Events, Adjustment of Urban Drainage with Decentralized Measures of Rainwater Management

Water sensitive urban design aims to reduce the catchment area of a sewage system by disconnecting different parts of the system from each other or by including infiltration systems, retention tanks for combined storage or stormwater retention facilities into the system. Additionally, it should be checked if runoff producing surfaces can be converted into runoff absorbing surfaces (e.g. green roofs) and if sealed areas (e.g. car parks) can be used for temporary storage. All these measures will result in a reduced discharge during design precipitation events, and an expensive extension of the cross-sections of the sewers can be avoided.

Constructional methods which aim to reduce the risk of flooding have to comply with the guidelines for sustainable development. Hence, water balance methods such as infiltration and disconnection of impervious areas have to be prioritized over integrated water management methods with similar effects. Thus, if water sensitive urban design is applied, the first step is to reduce the catchment area of the centralized drainage system. However, the adoption of water sensitive urban design methods requires the participation of property owners. Public acceptance is crucial for the implementation of water sensitive urban design methods.

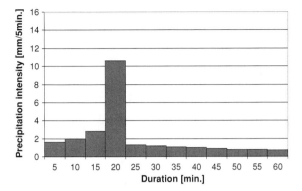

Fig. 5 Design precipitation event, return period five years, 25 l/m² in 60 min, (Kostra-DWD 1997)

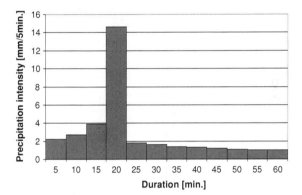

Fig. 6 Extreme precipitation event, return period 30 years, 34 l/m² in 60 min, (Kostra-DWD 1997)

Consideration of Extreme Events, Multifunctional Land Use in Order to Reduce the Negative Effects of Flooding

Extreme events, such as one in August 2008 in the Ruhr region (Dortmund, rainfall intensity of 200 l/m² in 3 h), that have a statistical return period of over 100 years (Grünewald et al. 2009) can certainly not be soundly managed by expanding the drainage systems. Even if such extreme events become more frequent in the future (EXUS 2010), adequate subsurface drainage systems would neither be neither financially viable nor operationally feasible during dry weather periods (DWA 2010).

Experiences from monsoon regions where extreme rainfalls events with up to 1,000 mm occur show that the most practical option to manage extreme rainfall events is to allow a controlled runoff and flooding of pervious areas (Staufer 2008). Although flooding is not avoided completely, negative impacts can be minimized.

The design of "emergency flood paths" or intra-urban retention areas shows that multifunctional land use is possible within an urban area. For example, roads with low traffic volume could be used as an "emergency flood path" or areas such

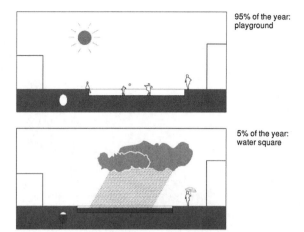

Fig. 7 Water square, illustration according to Boer, de Urbanisten (Vallée and Benden 2010)

as marketplaces or playgrounds could be converted into temporary retention areas (Fig. 7).

Figures 8 and 9 show the results of a coupled 1D/2D simulation of the sewage system and surface area in the project area of Bochum-Harpen after stressing it with the 30-year return period rainfall discussed above (see results and discussion). In the illustrated map free space, streets, buildings and flooded areas (blue) are easy to identify.

Figure 8 illustrates the results for the existing drainage system, where wide areas are flooded as an consequence of the rainfall event. In Fig. 9, additional flood paths have been included in the model. In this case, roads with modified kerbstones were used as flood paths. It can be seen that the storm runoff is directed along the roads. As is shown, the adverse effects of flooding in the city centre are reduced drastically by routing the discharge in the streets.

Modified Drainage Systems in Urban Centres, Water Squares and Open Drainage Channels to Decentralize Combined Drained Areas

In the past, newly subdued areas of a city have usually been attached to the central sewage system which led to an increased discharge of wastewater and hence an increased saturation of the central sewers. Today, the sewage systems are often loaded to full saturation or they are even oversaturated. As urban areas and especially city centres are usually drained by a combined sewer network, this development can lead to serious problems during rainfall events. The runoff cannot be absorbed by the sewer system and the city centre is flooded.

Water sensitive urban design is able to provide safe urban drainage for current conditions and even for future precipitation characteristics. Figure 10 shows the

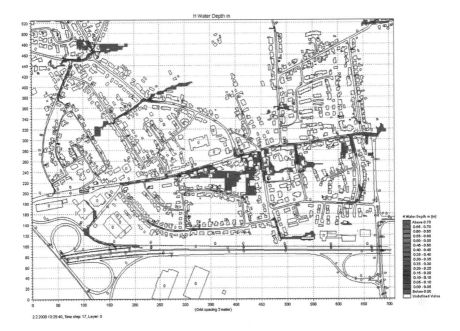

Fig. 8 Flooding following an extreme event—Bochum Harpen, current situation

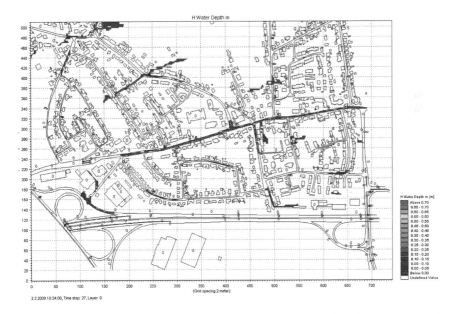

Fig. 9 Flooding following an extreme event—Bochum Harpen, multifunctional land use and emergency flood paths in the streets

Fig. 10 Flooding after a rainfall event with a return period of five years—Herne-Wanne

drainage system of the second project area in Herne-Wanne under the effect of the design rainfall event with a five year return period.

It can be seen that the runoff, which is highlighted in blue, cannot be drained and floods the pedestrian precinct of Herne-Wanne. Infiltration into adjacent buildings cannot be prevented. Due to the intra-urban situation, disconnection of the sewer system and creation of infiltration areas is not possible. Instead, the surface areas of the pedestrian precinct and the adjoining streets, as well as the street-facing roof areas, are converted into an open channel system. Consequently, 50% of the catchment area is independent from the combined sewer system, whose capacity is almost exhausted during dry periods. Additionally, a retention area is integrated in the west of Herne-Wanne to temporarily store runoffs. After passing the retention area, the discharge is conducted into subsequent drainage elements (sewage system, if necessary a directed discharge to the receiving water).

The results of the application of water sensitive urban design methods in Herne-Wanne (retention areas, open channels) are displayed in Fig. 11. It is shown that spilling of combined sewer systems can be avoided without increasing the cross-section of the pipes. This is achieved by conducting part of the rainfall runoff in open channels to a retention area. Additionally, the modification of the surface of

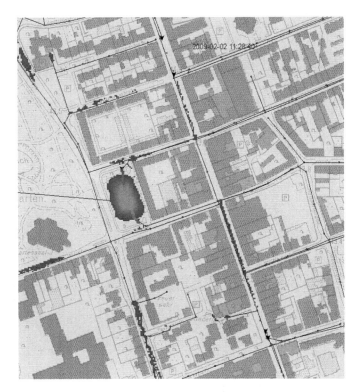

Fig. 11 Channel system and water square after an event with a return period of five years—Herne-Wanne

the pedestrian precinct prevents the infiltration of runoff into adjacent buildings. In fact, the whole surface of the pedestrian precinct would serve as a channel for discharging rainwater in the case of an extreme rainfall event.

Conclusion and Perspectives

The presented R&D project tackles the challenges of climate change in urban areas by applying water sensitive urban design methods, taking the limited local funds and the uncertainty of the existing climate scenarios into account.

Rainfall events are expected to become more frequent and more intensive due to climate change. Therefore, the existing urban sewer systems increasingly often fail to drain the runoff. Instead of difficult and expensive modification of the sewers, decentralized water management schemes (e.g. infiltration facilities) are recommended. Furthermore, the risk of flooding during extreme rainfall events has to be reduced. For example, flooding in densely settled areas can be prevented by multifunctional land use. The 1D/2D models implemented in MIKE URBAN

Flood software of the two investigated urban areas clearly show the positive effects of water sensitive urban design.

Multifunctional land use has not yet been established as a flood prevention concept in Germany. The R&D project "water sensitive urban development" aims to examine the practicability of multifunctional land use in Germany.

To allow an adaptation of the drainage systems to the effects of climate change in time, legal regulations are required. For example, a "climate aspect" in wastewater disposal concepts could be requested. Wastewater disposal concepts would be continuously updated and adapted to the current climatic conditions. Based on this, a legal background could be defined.

Acknowledgment The results presented here originate from the project "Wassersensible Stadtentwicklung, Maßnahmen für nachhaltige Anpassung der regionalen Siedlungswasserwirtschaft an Klimatrends und Extremwetter" ("Water sensitive urban development with regard to sustained adaptation of environmental engineering concerning climate change and extreme weather events"). Special thanks go to the Federal Ministry of Education and Research for funding the named project (reference number 01 LS 05017 A - C).*Project partners*. RWTH Aachen University—Institute for Environmental Engineering (project coordination), Institute for Urban Planning and Traffic. University of Duisburg-Essen—Faculty of Engineering, Department of Landscape Architecture and Rural Planning. University of Bochum Faculty of Psychology—Workgroup for Environmental Psychology and Cognition.

References

Arnbjerg-Nielsen K (2005) Significant climate change of extreme rainfall in Denmark. 10th international conference on urban drainage, Copenhagen

Beckmann KJ, Holz-Rau, C, Scheiner J (2005) Mobilität älterer Menschen – Analysen und verkehrsplanerische Konsequenzen. In: Echerhoff W (ed) Strategien zur Sicherung der Mobilität älterer Menschen, Schriftenreihe der Eugen-Otto-Butz-Stiftung, vol 1, pp 43–71

Castro D, Einfalt T, Frerichs S, Friedeheim K, Hatzfeld F, Kubik A, Mittelstädt R, Müller M, Seltmann J, Wagner A (2008) Prediction and management of flash floods in urban areas in Germany (URBAS). Final Report, Federal Ministry of Education and Research, Bonn

DWA (2010) DWA-Themenheft, Klimawandel–Herausforderungen und Lösungsansätze für die deutsche Wasserwirtschaft, Deutsche Vereinigung für Wasserwirtschaft, Abwasser und Abfall e. V., DWA, Hennef

Ellis B, Viavattene C, Revitt M, Peters C, Sieker H (2009) A modeling approach to support the management of flood and pollution risks for extreme events in urban stormwater drainage systems. 4th switch scientific meeting, 4–7 October 2009, Delft

EXUS (2010) Extremwertstatistische Untersuchung von Starkniederschlägen in NRW (ExUS)–Veränderung in Dauer, Intensität und Raum auf Basis beobachteter Ereignisse und Auswirkungen auf Eintretenswahrscheinlichkeit, Landesamt für Natur, Umwelt und Verbraucherschutz Nordrhein-Westfalen (LANUV NRW), Essen

Gerstengarbe F (2009) Klimawandel in Ballungsräumen: Das Beispiel Ruhrgebiet, Bildungszentrum für die Entsorgungs- und Wasserwirtschaft GmbH

Grünewald U, Schümberg S, Wöllecke B, Graf-van Riesenbeck G, Pieroth K (2009) Zu Entstehung und Verlauf des extremen Niederschlag-Abfluss-Ereignisses am 26.07.2008 im Stadtgebiet von Dortmund–einschließlich der Untersuchung der Funktionsfähigkeit von wasserwirtschaftlichen Anlagen und Einrichtungen der Stadt, Emschergenossenschaft und

Dritter in den Gebieten Dortmund-Marten, -Dorstfeld und –Schönau. Cottbus, Karlsruhe, Köln, January 2009, Germany

Hiessl H, Toussaint D, Becker M, Dyrbusch A, Geisler S, Herbst H, Prager J (2002) AKWA 2100—Alternative Kommunaler Wasserver- und Abwasserentsorgung". Abschlussbericht des Verbundvorhabens gefördert durch die WestLB-Stiftung Zukunft NRW, Karlsruhe

IPCC (2007) Climate change 2007: the physical science basis—summary for policymakers. www.ipcc.ch. Accessed 12 Dec 2007

KOSTRA-DWD (1997) Koordinierte Starkregen Regionalisierungs Auswertung. Deutscher Wetterdienst (DWD), ITWH, Hannover

Larsen AN, Gregersen IB, Christensen OB, Linde JJ, Mikkelsen PS (2008) Future development in extreme one-hour precipitation over Europe due to climate change, 11th International Conference on Urban Drainage, Edinburgh, Scotland

LFU (2009) Merkblatt Nr. 4.3/3. Bemessung von Misch- und Regenwasserkanälen. Teil 1: Klimawandel und möglicher Anpassungsbedarf. Referat 66 des Bayerischen Landesamtes für Umwelt, July 2009

Siekmann M, Staufer P, Roder S, Hellbach C, Pinekamp J (2009) A trans-disciplinary approach to confronting climate trends and extreme weather in urban areas. In: Feyen J, Shannon K, Neville M (eds). Water and urban development paradigms, pp 289–295

Staufer P (2008) Zwingt der Klimawandel zu dezentralen Lösungen in der Niederschlagswasserbewirtschaftung, 2. Aachener Kongress: Dezentrale Infrastruktur Aachen

Vallée D, Benden J (2010) Städtebauliche Anpassung an Starkregenereignisse durch multifunktionale Flächennutzung. Beispiele aus den Niederlanden, 43. Essener Tagung für Wassere und Abfallwirtschaft, Gewässerschutz – Wasser – Abwasser, GWA, vol 220, pp 6/1–6/15

Wong THF (2005) An overview of water sensitive urban design practices in Australia. In: Proceedings on CD, 10th international conference on urban drainage, Copenhagen

Chapter 38
Integrated Industrial Water Management as a Tool for Mitigating Climate Change

Rasmi Patnaik and Gopalsamy Poyyamoli

Abstract Observational evidence from all continents and various spheres shows that many natural systems are being affected by anthropogenic climate changes. One of those systems affected by the climate change scenario is the hydrological cycle, which encompasses water availability and water quality, as well as water services. In any industrial plant, the discharge of by-products and waste materials in various forms may pollute receiving waters rendering them unsuitable as a water supply. Excessive nutrients can lead to algal blooms, oxygen deficits, the release of toxic gases and increased pollution load, which in turn contribute either directly or indirectly to the climate change. In return, climate change can also negatively affect the industrial water sector in its own unique way. Adaptation to climate change is, consequently, of urgent importance in today's world. While end-of-pipe treatment of industrial waste was a popular management approach only 20 years ago, advanced treatment/management concepts such as Industrial Ecology, Clean Development Mechanism, Industrial Symbiosis and Design for Sustainability are more common today. This paper discusses the theoretical background of industrial water management as an important tool in putting water adaptation to climate change into practice. It focuses on the impacts of climate change on sustainable water management, the additional new challenges for water management deriving from climate change, and how the planning should be modified to adapt to climate change from an industrial ecology point of view. Although water management seems to be an operational issue, the choice of appropriate site location, provision of services, facilities design, best operational management practices along with supported institutional structures such as site development, risk mitigation measures, quality control measures, awareness creation, emergency planning and

R. Patnaik (✉) · G. Poyyamoli
Department of Ecology and Environmental Sciences,
Pondicherry University, Puducherry, 605014, India
e-mail: rasenvs@gmail.com

monitoring are needed to ensure sustainable industrial water management for mitigating climate change.

Keywords Climate change · Industrial wastewater · Adaptation · Mitigation · Industrial ecology · Policy · Integrated water management · System engineering

Introduction

Observational evidence from all continents and various spheres shows that many natural systems are being affected by anthropogenic climate changes. One of those systems affected by the climate change scenario is the hydrological cycle, which encompasses water availability and water quality, as well as water services. Water is vital to all companies, from those that use water in production processes to those that build, operate and finance water infrastructure (World Business Council for Sustainable Development 2005b). But, unfortunately, the discharge of by-products and waste materials from inefficient use, improper maintenance and excessive consumption by the industrial processes may pollute receiving waters, rendering them unsuitable as a water supply. Excessive nutrients can lead to algal blooms, oxygen deficits, the release of toxic gases and increased pollution load, which in turn contribute either directly or indirectly to the climate change. In return, climate change can also negatively affect the industrial water sector in its own unique way. If timely efforts are not being made for stringent actions against the severe industrial pollution now, the industrial sector will have to pay a huge unbearable price in the changing scenario of climate change. Mitigation as well as adaptation to climate change is, consequently, of urgent importance in today's world.

Climate Change and Industrial Water Resources

Industrial water pollution is reaching high levels of non-sustainability day by day. Wastewater is becoming a heavy burden to industries and the natural environment. Industrial wastewater consists of liquid discharges generated by raw material extraction, transformation processes and manufacturing of industrial products or consumer goods, along with a broad range of chemical pollutants such as solid or dissolved compounds, organic and mineral materials, metals, hydrocarbons, solvents, polymers, oil, grease, salts, etc., with various toxicity levels. The release of toxic gases from these pollutants acts as an amplifier for creating the suitable conditions for global climate change. Not only extraction or manufacturing processes, but also the later phases across the life cycle of industrial production generate huge amount of pollution load. During the water treatment process, wastewater treatment plants produce greenhouse gases (GHGs) that include carbon dioxide (CO_2), methane (CH_4) and nitrogen oxides (NO_x), thus contributing to global warming. CH_4 and CO_2 are produced when residual sludge is handled under anaerobic conditions, thus contributing to the global warming potential or the greenhouse effect (Shahabadi et al. 2009). Also, industrial wastewater reclamation plants produce GHGs as they contain more suspended solids and have a higher demand for biochemical oxygen (Tucker 2010).

On the other hand, climate change will affect water resources through its impact on the quantity, variability, timing, form and intensity of precipitation (Global Water Partnership 2009). The continuity and future success of any business are impacted by the availability, cost and quality of water along the "value chain", including "upstream" (in the production and supply of raw materials), "midstream" (in what businesses make from raw materials or other pre-processed inputs), and "downstream" (in treatment and recycling phase) (Deliverable D2-2 Report 2005). But alongside climate change, the existing and projected scarcity of clean water is likely to be one of the key challenges facing the world in the twenty first century. Influential voices in the global economy are increasingly talking about water-related risk as an emerging threat to businesses (Matthews and Quesne 2009). In this scenario, the huge demand of water consumption by the industrial sector will bring problems in areas where the water shortage and depletion of water level due to excessive pollution load is a common feature.

Under this view, the need for an integrated industrial water management system covering the cross-cutting themes of these two sectors will be of fundamental importance.

The Challenges

Industrial wastewater management through modern and efficient technologies is becoming an increasingly important strategy for individual industries. However, managing industrial waste is a difficult task and the challenges are many.

The major challenge to industrial wastewater management is its diverse nature. As mentioned earlier, industrial waste is combination of a broad range of pollutants of chemical and biological nature, which makes the released water extremely heterogeneous. Moreover, its quantity and quality vary depending on the process implemented and industrial domain. This large diversity requires a specific approach for each type of wastewater, which appears a primary constraint for the successful implementation of an efficient management system (Scientific Chronicles Supplement 2007).

Considering the geographic distribution of a certain area and the systematic absence of impact studies, some industrial plants are built on fragile ecosystems such as arable lands and costal belt and require overexploitation of the water resources. This situation presents a major challenge for the wastewater management, as it requires special consideration with finer details towards sensitiveness of the area.

The situation is complicated by the absence of treatment facilities of the industrial effluents. A significant part of the industries were not equipped with antipollution equipment. Improved and cleaner technologies guaranteeing a sustainable quality of wastewater are expensive. It makes the situation worse for small- and medium-scale enterprises (SMEs), which generate considerable amounts of waste discharge and have to face significant negative effects in the changing scenario of climate change.

Current water management practice at the industrial park level is still fragmented and non-systematic. The most pressing issue seems to be that there are many government departments dealing with water problems, but their activities are not well coordinated. There are many agencies or administrations whose jurisdiction overlaps on the same issues, sometimes resulting in scarce or no coordination and conflicting decisions. In some cases, high centralization of decision is complained about, together with minimal involvement of the stakeholders in the decision-making process. Also, the water law system is old-fashioned and widely scattered. The law tends to be deficient in the case of pollution issues, where quality standards for water bodies and/or effluents of complex characterization have not been clearly established. A general lack of impact assessment and monitoring studies is reported in most of the industries. Furthermore, the sporadic consideration of water quality from a policy point of view and the absence of systematic, uniform and enforceable pollution charges have compounded problems of integrated water management. The legislation is inefficiently or unsatisfactorily implemented, because of various factors, among which the most common are the lack of enforcement measures in order to guarantee the correct law application, the insufficient level of resources (human, structural, financial), and the inefficiency or slowness of application procedures (Geng and Yi 2006). Under these conditions, it is quite natural that the cross-cutting themes involving water-climate change issues are considerably neglected.

In some cases, unsolved questions of territorial sovereignty/transboundary issues related to water resources emerge, creating problems for proper management of industrial water resources.

Business as Part of Solution

To cope with the changing scenario of climate change, industry has a much larger role to play than just protecting its access to water. It can bring the technological capability to treat water and manage water supplies. Businesses from both the demand side and the supply side of the water equation must see themselves as part of the solution to issues related to water. To ensure this equitable balance, companies must gain an improved understanding of the dynamics of climate change and its broader impacts on all water-using sub-sectors. This will guide better water resources management inside an industry, which, in turn, will build resilience to current climate variability, while strengthening systems and capacity for longer-term climate risk management. Achieving and sustaining water security—broadly defined as harnessing water's social and productive potential and limiting its destructive potential—provides a focus for adaptation strategies and a framework for action. A coherent approach to promote this drive for water security that will address both current climate variability as well as the challenges of climate change can provide a solution for integrated water resources management (Global Water Partnership 2009). In the early 1980s and 1990s, end-of-pipe treatment of industrial wastes was a popular management strategy to deal with this approach. Gradually, advanced treatment/management strategies were developed and concepts such as Industrial Ecology, Clean Development Mechanism, Industrial Symbiosis, Design for Sustainability are more common today (Brenner 1999).

Industrial Ecology (IE) is defined as the system-oriented study of the physical, chemical and biological interactions and interrelationships both within industrial systems and between industrial and natural ecological systems. It deals with the problems faced by the sector to which it is applied and creates a framework that allows business to contribute effectively to sustainable management. It develops new instruments and technologies for planning, design and development of eco-industry and search for ecologically sound and economically effective strategies and policies. Application of IE offers the conceptual tools to reduce the ecological footprint of industrial development through changing production mode, consumption behaviour and decision instruments based on ecological economics and system engineering (Deliverable D2-2 Report 2005).

The following section will present a framework for integrated wastewater management within an industrial sector through application of industrial ecology approach. The framework is composed of various elements such as management of information system, policies and regulations, economic instruments and capacity building. We will discuss below each one of these elements in greater detail.

Data and Analysis

Business needs reliable meteorological and hydrological data at regional, national and sub-national levels in order to assess risk and make informed decisions or plans. The tools and systems used to collect and analyse this data need to be consistent. A globally accepted measurement tool that quantifies water efficiency throughout the life cycle would enable society to make more informed decisions. Such a tool would need to incorporate complex variables such as type and sustainability of the water withdrawal, as well as an understanding of the cost and benefit of different management options (World Business Council for Sustainable Development 2009a). Understanding the water footprint of the business, both inside and outside the corporate fence line, acts as structural guidelines for management planning (World Business Council for Sustainable Development 2009b). Medium- to long-term impacts of climate change due to potential changing hydrological patters need to be better understood, valued and integrated in current economic planning frameworks. The sustainability of existing and future water supply schemes should be analysed using hydrogeological analysis and models. Applying routine performance measurement and establishing performance improvement programmes with respect to the use of water and the production of wastewater can reduce significant environmental impacts arising from water use (World Business Council for Sustainable Development 2009a). Taking into consideration all kinds of related uncertainties, there is a need to develop practical tools to improve our knowledge of vulnerability to climate change, to adopt approaches that address climate change in a risk management context in order to better assess the risks and benefits posed by a changing climate and the ways that existing policies, programmes and socioeconomic circumstances serve to reduce (or exacerbate) vulnerability to climate change; and, ultimately, to build the foundation upon which appropriate decisions about adaptation can be made. Applications of GIS technologies and remote sensing systems to strengthen knowledge capabilities and assessment of risk and vulnerability will be quite useful in this context (Kassem et al. 2005). Advances in nanotechnology and information technology open the doors to new standards of reliability and security for such systems. Access to climate information and developing stronger linkages with climate scientists, in order to gain better understanding of the significant recent improvements in today's scientific world with some degree of accuracy, climate variability at seasonal and inter-annual scales, will help for better planning and decision-making. Incorporating this information effectively as part of water resources management could be a crucial tool for coping more effectively with climate variability and building capacity for adapting and mitigating climate change (Global Water Partnership 2009).

Partnerships

Business cannot act alone and in order to succeed, we need support and collaboration from legislators, policy-makers, civil society and academics in winning partnerships. Governments in particular have a responsibility to provide the necessary framework conditions to encourage companies to take effective action (World Business Council for Sustainable Development 2005a; ICLEI—Local Governments for Sustainability 2007). Business should engage with the communities they share water or water services with to promote sustainable water management. Private sector involvement in water resources management functions should be encouraged, under the condition that it serves public interests and is sustainable. Non-governmental groups of local regions will be quite helpful to encourage water conservation and improved water management system (World Business Council for Sustainable Development 2009a). There are significant opportunities for public–private sector cooperation; in this case, the World Bank provided incentives to foster such cooperation (World Business Council for Sustainable Development 1998). In order to achieve efficient deployments of climate-mitigating measures, there need to be partnerships and favourable inclusions of scientific institutions with capacity to enable the development of those functions in the targeted setting. Support to applied research in partnership between the public and private sector should be encouraged to build and strengthen capacity to plan for climate change. A productive relationship with partner country governments and local institutions is an important step in order to win support for and implement new ideas of physical incursions in existing infrastructural schemes. It also helps in solving the transboundary issues/territorial sovereignty problems (Kassem et al. 2005).

Policy

Given the impending challenges, it is crucial for policy-makers to recognize the role of water as a primary medium through which climate change will have an impact on development and to incorporate these considerations in overall development planning and management. Water policies and practices must aim to build institutions, information and capacity to predict, plan for and cope with seasonal and inter-annual climate variability, as a strategy to adapt to long-term climate change (Global Water Partnership 2009). Integrated policies for changes towards environmentally friendly economics, institutions, production and consumption patterns, and lifestyles should be promoted. Policy needs to be long term and flexible to allow for the use of the most appropriate approach, depending on local conditions (World Business Council for Sustainable Development 2005a). Policies to support accountability and transparency in all water-related processes should be encouraged in order to make them more inclusive. All current policies and programmes will need to be re-evaluated to introduce new approaches which can enhance our understanding of the impacts that climate change will have on water demands, water supplies, water quality and the whole water system inside an industry (World Business Council for Sustainable Development 2005b). Other favoured strategies involve further advancements of water storage technologies and policies aiming to enhance multiple uses of water resources to mitigate impacts of changing hydrological patterns, in order to sustain many different production activities (Kassem et al. 2005).

Effective Water Governance Framework

Providing effective water governance, including basic environmental, water and commercial law and a national water resources plan based on Integrated Water Resources Management, is the responsibility of governments, and business should be supportive of such a framework. Climate change impacts need to be strongly integrated in the water governance framework, including in planning, design and implementation by both public and private sector actors (Kassem et al. 2005). The management development aspects must include accounting, management information, customer billing and collection, organizational restructuring, operations and maintenance management, environmental management, public information and consumer relations, risk management and change management (World Business Council for Sustainable Development 2009a). The development of strong institutional capacity, adaptive and effective governance, and the ability to successfully implement sound adaptation policies should be regarded as the most important task in facilitating successful adaptation to climate change in freshwater. In all of these cases, these institutions need to discharge their functions independently and in the absence of undue interference, corruption, or local capture (Matthews and Quesne 2009). By dealing with the ecological footprint at the parcel level, semi-autonomous "demand management" developments can be created that will deal with their own infrastructure needs on site, including water supply, stormwater control, sewage treatment, thermal demand for heating and cooling, and electrical demands. Creating these nested systems will buffer the demand on centralized infrastructure and add system robustness and resilience—all necessary in a world with increased uncertainty in climate effects on infrastructure (World Business Council for Sustainable Development 2010).

Effective Regulation

Businesses should encourage appropriate and efficient regulation and conform to its requirements (World Business Council for Sustainable Development 2009a). All the relevant regulations must be flexible enough to cope with the uncertainties and local problems while at the same time bringing proper justice to any kind of violation.

Unlocking Finance

As integrated water resources management uses a combination of both conventional and modern approaches, successful implementation of it needs a firm financial investment strategy. Investment needs can only be met when effective cost recovery systems are adopted, based on the use of tariffs and taxes that are appropriate for the local social, economic and environmental context. Where local authorities are willing and able to implement sustainable economic management, the business sector can explore with them alternative financing, investment and service delivery options tailored to local conditions and capacity to meet capital and operational costs covering the climate mitigation measures (World Business Council for Sustainable Development 2005b).

Awareness-Raising and Capacity-Building

Awareness-raising is a crucial driving factor in the process and this should be promoted through both general and targeted campaigns, through educational systems, and by enabling easy access to relevant data and essential information

(Kassem et al. 2005). The community should be educated to improve understanding of water resources and their management and to develop technologies to get the most value out of the water cycle (World Business Council for Sustainable Development 2009b). Strategies to raise general awareness, with particular focus on water and climate change and its underlying drivers and measures needed to be taken, and to mitigate impacts should be developed. Thus, wide-ranging information campaigns targeting a broad spectrum of societies should be promoted (Kassem et al. 2005). Building the capacity of local businesses and skilled workers is essential to undertake operational, maintenance and renewal works for the long-term management of water systems. This avoids dependence on imported skills and ensures that water facilities continue to deliver their social, economic and environmental benefits in the longer term. Assisting small- and medium-sized enterprises to improve water management needs special attention. Successfully coordinating all the development programmes through technology transfer, culture change and training can contribute to significant improvements to bridge the existing gap among various management approaches. This kind of capacity-building effort plays a central role in the development of basic skills and contributes to wider objectives by opening avenues of development to the people who benefit from it. It also has a beneficial impact on involving the people and building the trust in the wider community (World Business Council for Sustainable Development 2009a). Creating a sense of shared responsibility among all the stakeholders will help in bringing out new innovations from different sectors. Supporting programmes to change mindsets, behaviour and practices should be promoted.

System Engineering

Although water management seems to be an operational issue, the choice of appropriate site location, provision of services, facilities design and best operational management practices, along with supported institutional structures such as risk mitigation measures, quality control measures, emergency planning and monitoring, are needed to ensure sustainable industrial water management for mitigating climate change. Intelligent institutions that can go beyond managing water on a day-to-day basis to identify water-use trends, areas vulnerable to climate change and opportunities to respond as well as possible to the emerging challenges are needed at all levels (Global Water Partnership 2009).

Site Location

While locating a site, water has to be considered one important parameter for calculating the suitability maps for industrial areas and industrial parks. Available water resources and competing demands for those resources should be considered when choosing a site for new facilities (World Business Council for Sustainable Development 2009b). A decision-support system can be used to choose or rank the industrial alternatives suitable for the socioeconomic development of one area (Water Quality Protection Note 2009).

Design and Development

The hierarchical approach in water management plays an important in its successful implementation. Once choices are made as to where the industries are settled, strategies have to be applied to reduce the ecological footprint of the industries by establishing programmes of cleaner production and water-saving. By better designing their products, companies can minimize their impact on the water environment in terms of the water they use or the pollution they cause. They range from acquiring raw materials to producing finished products. Existing facilities can be redesigned to minimize water use and improve resilience through water-saving technologies and services (World Business Council for Sustainable Development 2009a). The availability of adequate service infrastructure, such as emergency management, waste recycling and disposal services, sewerage services, reticulated water supply, electricity, gas, communications and transport access should be assessed when selecting a potential site. Water-sensitive land use planning must incorporate water issues in spatial planning and construction standards, especially for new-build areas (World Business Council for Sustainable Development 2010).

Operation and Maintenance

The water management to be adopted in industry is depending on the choice of the type of industrial development. Industrial wastewaters are subject to acceptance criteria for discharge to sewer (set by local government, the water corporation or other water providers' systems) and may require on-site treatment. Sensitive water resources require a range of management techniques to ensure their adequate protection such as definition and community awareness, separation buffers from intensive land usage, effective containment of potentially mobile contaminants, waste characterization and segregation, regulation of land-use activities, appropriate waste management decisions by land-use operators, recycling and reusing water, lowering toxic and other contaminants in all operations involving water, catchment surveillance/monitoring, control measures, emergency planning and remedial action to address historical contamination problems. Renewable energy investment strategies in the water sector including hydropower (artificial storage and watershed management) and biofuel production where major potential exists is a good mitigation strategy which can bring substantial economic benefits at the small, medium and large scale (Kassem et al. 2005). A combination of system engineering techniques using seawater for industry, diverting water from water-rich to water-scarce areas, using the heat content of wastewater as a source for regenerative energy supply, using wastewater irrigation to increase production, new groundwater exploration techniques, microbiological wastewater treatment techniques combined with novel membrane separation techniques, nanotechnology, innovative desalination techniques, crystallization techniques, membrane development, cheap point-of-use treatment systems and consumer products to remove bacteria, viruses, parasites, and heavy metal, etc. plays pivotal role in integrated management of wastewater (World Business Council for Sustainable Development 1998).

Business should also consider some new/alternative procedures such as increased rainwater absorption through aggressive tree planting and green roof initiatives, rainwater harvesting, stormwater infiltration to recharge groundwater, using membrane bioreactor systems for wastewater treatment and reuse, along with other advanced techniques. Such developments dramatically reduce on-site demand for potable water and use of sewers and storm drains, thus deferring the need to expand infrastructure (World Business Council for Sustainable Development 2010). Both runoff and industrial emissions can be best tackled and prevented at source through better planning of water treatment and resource-efficient practices. This is particularly relevant to dealing with emissions from industry, which may not be biodegradable. The use of treatment processes to remove pollution after it has occurred should be a last resort because they are usually energy-intensive and can become unreliable if the pollution being treated is severe (World Business Council for Sustainable Development 2009a).

An alternative approach to end-of-pipe treatment of complex and toxic wastes may be the introduction of an in-plant control programme, including waste segregation and process-specific pretreatments. Separation technologies such as sedimentation, adsorption, extraction, etc., change the physical status of the contaminants and usually make them more concentrated in their new environment. However, it is sometimes economically feasible to recover valuable products such as solvents to be recycled and reused. Any policy at designing for environment should attempt to allow for reuse at all possible levels, but should encourage the scenarios of reuse over reprocessing or recycling, as reuse will generally require less energy input than others and as such will have lower overall negative environmental impacts. Policy needs to encourage full development of resource recovery systems and phase out continued dependence on landfills and incinerators as the primary means of handling waste.

First, chemical and toxicological characterization of each waste stream should be done. This should include basic measures such as pH, total and dissolved solids, total organic content, nutrients, main minerals, and toxicity. The second stage is the screening of removability potential. It can be based on simplified procedures which include aerobic biodegradation screening for the selection of waste sources that are suited to a conventional biological treatment, volatilization screening to identify VOC-contaminated streams that should be pretreated before undergoing biological treatment and carbon adsorption screening for the evaluation of carbon requirements and adsorption efficiency, to be considered as a pretreatment means of reducing toxicity. A simplified laboratory procedure combining three parallel assays—biodegradation, volatilization and adsorption—can be used successfully for the evaluation of removability potential of waste sources forming a complex chemical industrial wastewater. The biodegradation procedure may serve as an efficient tool for the evaluation of complex streams' suitability for biological treatment, in place of the traditional BOD test, which is of limited value for such wastes. The volatilization procedure may add a valuable index regarding VOC air pollution, on the one hand, and possible recovery of solvents, on the other hand. The adsorption procedure, based on toxicity balances, may enable simultaneous determination of both adsorption capacity and the impact of the treated water on biota in receiving waters. Put together, the proposed approach may lead to an optimal in-plant management programme, based upon stream-specific solutions such as biological treatment of readily degradable organic streams, carbon adsorption or chemical oxidation of refractory organics, controlled stripping of VOC-containing streams, and evaporation of brines (Hertle 2008).

Depending on the amount of land available for an industry, wastewater can be used to irrigate plantations to provide Kyoto-compliant carbon sinks (Scientific Chronicles Supplement 2007). The best and most efficient treatment method for an industrial wastewater reclamation plant is a hybrid of aerobic rectors and anaerobic solid digestion with the use and recovery of biogas to operate the plant and reduce the demand for external electricity. The hybrid system offers a higher rate of nutrient and contaminant removal compared to anaerobic and aerobic processes alone and produces enough biogas from the decomposition of solid waste to power the whole plant. Furthermore, by lowering the temperature of anaerobic solid digestion, GHG emissions can further be reduced. In addition, producing all the materials needed for treatment on site instead of transporting them from farther away can reduce GHG emissions produced from transportation (Tucker 2010).

The combination of pro-efficient government regulations, effective implementing mechanisms for water governance, alignment of state and provincial relationships, increasing investment in research and development, support of venture capital for the new water businesses, along with a strong base for capacity building, will open new opportunities for industries towards achieving a win–win situation for climate change mitigation.

References

Brenner A (1999) New concepts in industrial wastewater management. Environ Eng Policy 1:217–222

Deliverable D2-2 Report (2005) Network on governance, science and technology for sustainable water resource management in the Mediterranean. The role of DSS tools. Deliverable D2-2 report on industrial water use in the Mediterranean countries. http://www.feem-web.it/nostrum/doc/d2-2_rev.pdf

Geng Y, Yi J (2006) Integrated water resource management at the industrial park level: a case of the Tianjin economic development area. Int J Sustain Dev World Ecol 13(1):37–50

Global Water Partnership (2009) Paper series on water and climate change adaptation, "Perspectives on water and climate change adaptation—better water resources management—greater resilience today, more effective adaptation tomorrow"

Hertle C (2008) Excellence in industrial water management—challenges in a climate change world, GHD. http://www.resourcefulevents.com/userfiles/file/eiw08-Chris-Hertle-GHD.pdf

ICLEI—Local Governments for Sustainability (2007) Paper series on water and climate change adaptation, "Perspectives on water and climate change adaptation—local government perspective on adapting water management to climate change"

Kassem A, McRae T, Sydor M (2005) Integrated Water Resources Management. OECD workshop on agriculture and water: sustainability, markets and policies. Adelaide, Australia

Matthews JH, Quesne TL (2009) Adapting water management a primer on coping with climate change. WWF Water Security Series 3

Scientific Chronicles Supplement (2007) Why is industrial wastewater difficult to treat? Ind Wastewater 10:1–5

Shahabadi B, Yerushalmi L, Haghighat F (2009) Impact of process design on greenhouse gas (GHG) generation by wastewater treatment plants. Water Res 43:2679–2687

Tucker A (2010) The impact of greenhouse gas emissions generated by industrial wastewater treatment plants. http://www.claremontclimatereport.com/2010/02/impact-of-greenhousegasemissions.html

UNFCCC (undated) Mitigation of greenhouse gas emissions from treatment of industrial wastewater. Approved consolidated baseline and monitoring methodology, ACM0014/Version 03.1, Sectoral Scope: 13, EB 47

World Business Council for Sustainable Development and UNEP (1998) Industry, fresh water and sustainable development. http://www.wbcsd.org/web/publications/freshwater.pdf

Water Quality Protection Note (2009) Government of Western Australia, Industrial wastewater management and disposal. http://www.water.wa.gov.au

World Business Council for Sustainable Development (2005a) Collaborative actions for sustainable water management. http://www.wbcsd.org/DocRoot/6b28PhsvW6E7LaUKKOjj/Water%20collaborative%20actions.pdf

World Business Council for Sustainable Development (2005b) Water and sustainable development—a business perspective. http://www.wbcsd.org/DocRoot/TN6PV20XbQJr06oKAOYe/water_and_SD_bizpersp_march05.pdf

World Business Council for Sustainable Development (2009a) Water, energy and climate change—a contribution from the business community. http://www.wbcsd.org/DocRoot/Dg6GYWJq7xuaLO0OwZOi/WaterEnergyandClimateChange.pdf

World Business Council for Sustainable Development (2009b) Water facts and trends, version 2. http://www.wbcsd.org/DocRoot/lD1tMGiLZ7NL9mBOL2aQ/WaterFactsAndTrends-Update.pdf

World Business Council for Sustainable Development (2010) Water for business—initiatives guiding sustainable water management in the private sector, version 2. http://www.wbcsd.org/DocRoot/3wlfDj0SSDsKcJWIBbKu/WBCSD_Water_for_Business_WEB.pdf

Chapter 39
Unjust Water: An Investigation of the Drinking Water Situation in Argentina

Sonja Maria Geiger and Natalia Salvático

Abstract With their new campaign "I am made of water", the work group on water of Amigos de la Tierra, Argentina (Friends of the Earth, Argentina) wants to call attention to the unjust and increasingly precarious drinking water situation in Argentina. This situation is characterized by great geographical, climatic and socio-economic differences between the dry and comparatively poor north and west of the country, on the one hand, and the humid, fertile Pampas and Buenos Aires region in the east, on the other. The basic climatic differences are intensified by differential effects of climate change on these regions. Additionally to this disparity in natural drinking water supplies, the national government employs a pricing and supply policy that favours the relatively rich, humid urban areas, while disadvantaging rural areas and the poor city fringes with low drinking water connectivity. The authors will follow up on how this unjust pricing system and lack of education have strengthened environment-damaging habits among the general public in those areas where water supply, at first glance, does not seem to be a problem. The role of cognitive factors, such as beliefs and (lack of) knowledge about water-related facts in creating and maintaining these unsustainable behaviours is discussed. Finally a threefold initiative for change is suggested, including personal education, strengthening social awareness and creating structural foundations as constitutional law warranty for access to water for fair prices.

Keywords Drinking water · Mental model · Behaviour · Sustainability · Cost structure · Supply · Injustice

---

S. M. Geiger (✉) · N. Salvático
Cordoba 5051, 1414 BAI, Buenos Aires (Cap Fed), Argentina
e-mail: sonja.m.geiger@gmail.com

Introduction

"A lot have lived without love, but no one without water". This is the opening line of the movie "Flow: For Love of Water"[1] that was shown in an open air cinema in Buenos Aires as part of a chain of events in "Water Week 2010" organized by Amigos de la Tierra, Argentina (ATA from now on[2]). Water Week was created to celebrate the most valuable natural common good during International World Water Day on 22 March and National Water Day on 31 March. With different activities, which will later be described, ATA intends to highlight and explain different focal points of the drinking water situation in Argentina.

This situation is characterized by great geographical, climatic and socioeconomic differences within the country's regions. In the highly populated and humid region of Buenos Aires, availability of drinking water seems, at first glance, to be unproblematic compared to the dry and hot northwestern and western regions. Combined with highly subsidized and therefore extremely low water prices, this basic situation has led to extraordinarily wasteful behaviour patterns in the population of Buenos Aires and its surroundings (Gran Buenos Aires). Low ecological consciousness and lack of information about water-related facts perpetuate the unsustainable consumption patterns of this sub-population of Argentina, which makes up close to a third of the country's population. The supply of clean drinking water to the complete population seems to be a growing challenge for the publicly owned water service suppliers in Argentina, which face low connectivity figures in remote areas, unsteady natural water supply due to changing weather patterns and lack of compliance by consumers.

With its new campaign, ATA claims first of all that access to water should be constitutionally acknowledged as a basic human right, in order to ascertain its significance legally for all following discussions about this valuable common good. Regarding its management in Argentina, ATA tries to point out injustices in the existing water supply and pricing system. Some potential explanations for the resulting individual behavioural patterns are given from a psychological point of view. For a sustainable management of water, a change in attitude and the actual use of water is absolutely indispensable, even more so, as will be explained, with continuing climate change the situation will worsen rather than improve. How this change might be achieved will be outlined last. ATA suggests a threefold approach consisting of personal education, public information and structural revisions in legal status, supply and cost schemes.

---

[1] Flow: For Love of Water (May 2010): http://www.flowthefilm.com/.
[2] Website of ATA (May 2010): http://www.amigosdelatierrra.com.ar.

The Water Situation in Argentina

To give an overview of the drinking water situation in Argentina, the authors will roughly sketch the geographic situation and related climatic factors, followed by a description of socioeconomic aspects, including water connectivity. The situation of drinking water in Argentina is as diverse as the country itself, one that stretches from desert regions bordering Bolivia, rainforests bordering Brazil, to the very tip of the glacier-covered Andes in the south. Thus, it is rendered imperative to explain the two factors separately.

Geographic and Climatic Factors

When it comes to the situation of drinking water, the territory of Argentina has to be considered in at least two parts.

The dry, hot region of the north and west generally experiences little rainfall over the year. Thus the provinces of Salta, Jujuy, Catamarca, La Rioja, San Juan and Mendoza are arid or semi-arid zones with <800 mm of annual rainfall (Ministerio de Educación 2010). In recent years, due to less precipitation in winter and the retraction of glaciers and ice mountain tops throughout the year, the river outflow in the Andean regions has been greatly diminished. The Andean foot region suffers more frequent drought, and the drinking water supply is severely threatened (Leiva 2006).

On the other hand, in the Pampas (area comprised of the provinces of Buenos Aires, Santa Fe, La Pampa and parts of Cordoba and San Luis) and the northeast (Missiones, Corrientes, Entre Rios), precipitation is higher than 800 mm a year (Ministerio de Educación 2010), thus giving the area the fertile grounds for the cattle breeding it is famous for. The rising temperatures of recent years have led to strong precipitation from the Atlantic coast, which in turns leads to significant increase in river flows in the Paraná Delta and water abundance in the catchment area of the river Plate (Magrin et al. 2007; Barros 2005). As Buenos Aires borders the widest river on the planet (up to 220 km), which carries 30,000 m^3 of freshwater into the Atlantic every second, water shortages have never been such a problem in the area of Gran Buenos Aires.

In the south of Argentina, the climate is arid to semi-arid, coupled with high dependency on the river outflows of glaciers and ice shields. Nevertheless, due to low population density, the drinking water supply has not (yet) been a problem in this area, as explained below.

Socio-Economic Factors and Drinking Water

As detailed above, natural water purveyance is greatly unequal in the different regions of Argentina. Superimposed on these geographic and climatic realities, there is a steep socioeconomic disparity between the relatively rich centre of Buenos Aires (with the exception of its slum fringe cities) and the Pampa region to the poorest regions in the north and northeast. In the provinces of Corrientes, Formosa, Chaco and Missiones, up to 25% of the population lives below the Argentine poverty line, followed by the northern provinces of Catamarca, Tucuman, Salta and Jujuy, where this applies to approximately 17% of the population (INDEC 2009). Infant mortality (up to 2.2%), another index for extreme poverty, is also highest in the provinces of Chaco, Formosa, Salta and Jujuy (formerly also Tucuman and Catamarca) (INDEC 2007).

On the other hand, the area-wide water supply of a country is considered an important index of its development. Consequently, the socioeconomic inequality is mirrored in the percentage of Argentine people who do not have access to drinking water or wastewater sanitation respectively. Figure 1 depicts the percentage of Argentines lacking these vital services according to provinces (Paris et al. 2009).

Whereas in the humid and sparsely populated south, coverage of drinking water supply is almost complete (<5% of the population lack freshwater in the provinces of Chubut, Santa Cruz and Fireland), the situation worsens to the north, where in the provinces of Buenos Aires, Formosa, Chaco, Santiago del Estero and Missiones, up to 40% lack a supply of drinking water to their homes.

Regarding the situation of sanitation coverage, the overall pattern is similar but on a more extreme level. Whereas in the south, close to a third of all people lack sanitary installation, this figure rises to 86% in the aforementioned provinces of the north, with Buenos Aires at a worrying level of 61.3% of all homes without connections to cloacae. The relatively high figures in both aspects for the political and economical centre of the country, the province of Buenos Aires, are only understood in the urbanization context of a South American metropolis.

Almost 30% of the country's inhabitants, 11.5 million according to the 2001 census (INDEC 2001), then lived in the 24 districts of Gran Buenos Aires (as for the age of the census probably an underestimation as annual positive urbanization rates show; CIA 2010). The southern suburbs of the metropolis Buenos Aires, despite their relative closeness to the prosperous economic centre of Argentina, are characterized by high poverty rates, precarious housing and close to zero connectivity of basic sanitation installations.

The main problem of this area with regard to drinking water is another one: due to poor legislation and historical paralysis, the waters of the river Plate are severely contaminated by the industrial wastewater of its two major affluent rivers, Reconquista and Riachuelo (Wurgaft 2009; Greenpeace 2010). Among the most common contaminants are heavy metals such as lead, chrome and mercury, but also the organic waste of 350,000 m^3 of unfiltered sewer waste daily, which aggravates the situation (Lobos et al. 2004). The two main water outtakes for the

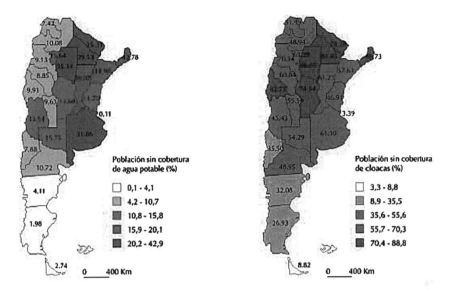

Fig. 1 Percentage of Argentine population without drinking water (*left*) or sanitation (*right*) coverage in Argentina

drinking water plants of San Martin (in the suburb of Palermo) and General Belgrano (in the suburb of Bernal), which are responsible for the drinking water supply of 7.8 million people, are located within only 500 m of this main industrial pollution belt (Borthagaray et al. 2002; AySA 2008).

Cost Structure of Water Supply in Argentina

Water supply in Argentina was privatized in 1993 to allegedly secure competitive prices and advancement in the nationwide water and wastewater sanitation supply. The privatized Obras Sanitarias de la Nación (the major shareholder of French company Suez) was dissolved in 2006 after the general collapse of the poor sanitation services and inflationary prices and the national water supply contract was retracted.

After the dissolution of Obras Sanitarias de la Nación, water and sanitation services were devolved out to the provinces, where to this day every province manages its drinking water supply with different mix forms of (semi-)public organizations and enterprises (MECON 2010).

With the area of Buenos Aires and 17 surrounding suburbs of Gran Buenos Aires, Agua y Saneamientos Argentinos (AySA), is responsible for the water supply to 7.8 million people (AySA 2010a). For the main purpose of the article to follow (and the difficulty to obtain data for the rest of the country), we will focus on the drinking water supply of this area characterized by high (alleged) availability, high decontamination costs and lack of appreciation of the value of water.

In the area of provision under AySA there are two different systems of charging the end-user. According to the enterprise's website (AySA 2010b) these are:

a. per flat rate (applied in ca. 12% of accounts);
b. per meter (applied in ca. 88% of accounts).

The first system is based on a complicated calculation that involves the size, age and type (house or apartment) of the property. In the case of a 50-year-old, medium-sized apartment (60 m^2), this flat rate amounts to Arg$17.14 for the period of two months, equalling a monthly charge of Arg$8.57 plus value added tax resulting in Arg$10 for unrestricted access to drinking water (water bill to the first author of 20 February 2010).

The second system involves a base rate at half of the above flat rate, namely Arg$4.29 monthly plus a surcharge of Arg$0.33/m^3 (drinking water only, without sewage treatment) that exceeds the free use of 10 m^3/house or living unit/month (AySA 2010a). At the same estimate of the daily use of 500 l/day and capita plus value added tax (22%), we arrive at a monthly cost of Arg$5 for the base rate plus ca. Arg$20 for excess use (50 m^3), resulting in Arg$25 of monthly costs for a family of 4 with an estimated (extremely high) average use of 60 m^3.

On the other hand, in areas without drinking water coverage (in the area of AySA 19% of the population, and up to 32% in the whole province of Buenos Aires), people rely on buying bottled water from the supermarket. The basic price for bottled water in Buenos Aires in the same period was about Arg$1.35/l (packaged in big water containers of 6 l). Setting an absolute minimum of 2 l a day/person, this results in a monthly cost of $2 \times 4 \times 30 \times 1.35 = $ Arg$324. Table 1 shows the costs for this minimum of safe drinking water under different pricing systems in the area of Gran Buenos Aires.

As is clear from Table 1, household costs for drinking water vary considerably, depending on which arbitrarily applied price scheme a household falls under. The differences in prices according to whether or not a meter is used to measure water expenditure in a household vary extremely. There have been attempts to equip all households with water meters, but due to great resistance from the urban population (which comes as no surprise looking at the price scheme) and lack of funds on the side of the enterprise AySA, meters have not been installed area-wide.

This inequality exhibited in the area of Gran Buenos Aires is only exemplary for the rest of the country, where the main difference between connected and unconnected households prevails, regardless of the exact pricing systems applied.

Above and beyond these internal contradictions and historically grown injustices of the drinking water supply in this area, the costs of purification lie according to an estimation of AySA (oral communication from the head of the library) around $0.80/m^3. This estimate, five times as high as the average flat rate price that the consumer is charged, includes only pure operational costs of the purification plants. Thus, it ignores administrative costs or the rest of the infrastructure, making it a very conservative estimate.

Table 1 Estimated average costs (in Argentine pesos) for drinking water for a family of four in the area of Gran Buenos Aires

Price system	Flat rate	Per meter	Without coverage
Inhabitants affected	71%	10%	19%
Average monthly cost	$10	$25	$324
Projected cost[a]/m^3	$0.16	$0.42	$1,350

[a] Where applicable, based on the average use of 500 l/daily = 15 m^3/monthly/capita = 60 m^3 for a family of four

As AySA is 90% publicly owned, the annual losses of the enterprise such as this year's Arg$1.7 billion is carried by the state (Rossi 2010). This means that the people of all of Argentina, including those who have to buy their water at the supermarket, subsidize the extremely low water prices of the capital and the consequent unprecedented waste of it.

To summarize, all over Argentina, people face an extreme disparity in their drinking water supply, socioeconomic circumstances and the resulting opposing costs for safe drinking water. For the purpose of visualization, we outlined the projected costs of cubic metres (m^3 = 1,000 l) of water, although we are clear that no one, either in Argentina or elsewhere, would buy their daily supply of water at the supermarket. Nevertheless, we ask ourselves how exactly people in the sparsely connected northern rural regions or the slum regions around Buenos Aires do acquire their indispensable share of water, albeit a lot less than the average estimated use of 15 m^3/person/month.

As detailed above, it is precisely in those regions where water supply connectivity is the lowest. Thus, understandably but ironically, the poorest of the poor in Argentina face the highest costs for their water supply, if only for the minimum of 2 l/person/day.

We would like to summarize the most outstanding injustices:

1. *The swimming pool irony*: under the subsidized flat-rate scheme, a person pays the same amount to quench his/her thirst with a minimum of 2 l a day as filling up a private swimming pool as often as desired.
2. *The 10,000 factor irony*: in unconnected areas the price per cubic metre (as calculated by supermarket prices) is roughly 10,000 times higher than in connected areas under the flat-rate system.
3. *The meter trap*: although water suppliers depend on voluntary installations of meters, average metered water prices are still around 2.5 times higher compared to flat-rate ones.
4. *The price trap*: water has to be purified at five times the cost than charged to its consumers.
5. *The subsidy irony*: national tax money is used to finance the swimming pool irony instead of providing minimum water quantities to the poor.

The Psychology of Water Use

According to AySA, the estimated average use per day per person in Gran Buenos Aires is around 500 l, which rises to 620 l on hot summer days (AySA 2010b). With this record use of drinking water, the 7.8 million inhabitants of Buenos Aires use more than 10 times as much as the WHO suggests as the absolute daily minimum of 50 l and four times as much as e.g. the ca. 130 l of an average German (Bundesministerium für Umwelt und Reaktorsicherheit, BMU 2010). According to the Water Footprint Organization, which published worldwide data for the year 2001, Argentina's national average lies at about 384 l/day/person. This is still 2.5 times above the worldwide average of 157 l/day and capita (Water Footprint Organization 2010).

Given that the usual distribution of water use in private household is as follows: approximately two-thirds (66%) is used for personal hygiene (body washing, showers, bath, toilet flushing) whereas only around 1% (up to 2 l) is used for actual drinking (the rest is shared between household hygiene, cars, gardening and washing of clothes) (BMU 2010), the question arises: for exactly *what* do the inhabitants of Buenos Aires use their daily 500 l?

(Un)sustainable Customs

According to a recent study (and the first of its kind to our knowledge), Cattaneo and Lopez (2010) found some of the most common water-wasting habits of 1,162 inhabitants of Gran Buenos Aires:

- Leaving the tap running the whole time while:
  - showering (up to 30 min),
  - washing dishes,
  - washing fruits,
  - brushing teeth.

- Daily hosing down of pavements (a very common habit in Buenos Aires).
- Washing the car with running water.
- Always using the longest washing machine programme.
- High tolerance for dripping or running taps and toilet equipment.

All these habits differ greatly in their potential share of the wasting of water: a hose with a fully open tap for example, as commonly seen in Buenos Aires as its habitants hose down their walkways, loses around 1,140 l of water an h, whereas brushing teeth with the tap running only adds around 30 l more to this balance. Washing dishes with an open tap uses 100 l more than necessary for an average load of dishes, while showering for 20 min can add up to another 200 l. All the above listed habits add to a consumption way above the recommended and necessary use of water.

Interestingly, they did not find a difference between users charged by flat rate (45% of their sample) and users charged by meter (55%). They assume this is partly because water prices are so low that stopping the wasting of water is still not seen as a way of saving money.

Mental Models

One of the reasons we believe that this extensive amount of water is used by the inhabitants of Buenos Aires is their internal representation of the water cycle in which they place their own water use. Internal representations, or "mental models", is a concept of cognitive psychology that explains the way we organize and represent knowledge. Mental models are like miniature representations of an object or situation and they serve to depict and structure the information we have about a domain. In the example for this article, it would be about the whereabouts of water before it arrives at the tap in my house and after I let it run down the drain.

Mental models are constructed actively to represent the basic features of an object, system or situation. They are spontaneously constructed to understand a given situation or make a prediction about it. Because more often than not they are based on intuitive theories of the world rather than direct experience, they are prone to erroneous presentation of details or interrelations of different parts (Johnson-Laird 1983; Gentner and Stevens 1983).

Explicitly or not, everybody has some kind of an inner representation of the water cycle surrounding our domestic use of it. A very simple mental model of an Argentine in Buenos Aires might be greatly impacted by the closeness of one of the widest freshwater rivers in the world and might look like this in Fig. 2a (extracted from a personal communication):

On the left (Fig. 2a), a simple mental model of a water cycle including the river, a house and some pipes is shown. The simple assumption is made that water is taken from the river and part of it is redirected to my house where I use it whenever I turn on a tap. Meanwhile, "the rest" of the water keeps flowing back to the river (as well as the water I keep using and adding substances to, such as dirt, soap, domestic chemicals or kitchen oils). This simple model makes it self-evident why saving water does not make sense to this person. The less water used domestically, the more flows back in the parallel pipe or canal (or is used by others). In this model, water is a lot like air: it flows around incessantly and although the use might alter its composition (by adding soap or domestic waste), it won't change the overall availability.

A somewhat more elaborate model (Fig. 2b) might have added the representation of a drinking water abstraction plant, where water has to be pumped out of the river, is cleaned and chemically treated before it reaches my house. After I have used the water, it is taken to a sewage treatment plant where the biggest

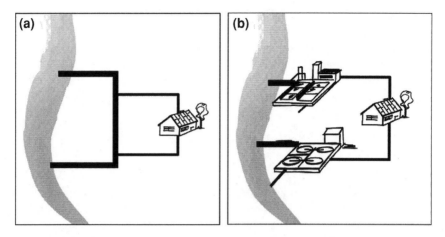

Fig. 2 a and b simple mental model of the water cycle (*left*); more elaborate model (*right*)

particles and objects are extracted and the water is filtered (ideally) before it is let back into the river.

In this somewhat more elaborate model we see some core elements that are lacking in the most simple of possible models:

- Water has to be treated before it reaches my house (somebody makes it drinkable).
- Once I have used it, it has to be treated and cleaned again.
- Clean (running tap) and dirty (a lot of added substances) sewage water alike end up in the sewage treatment plant.
- Energy is needed for the pre- and post-treatment of water.

If we extend the model to numerous users of the same river (namely our neighbours), some of the social aspects of a shared common "commodity" become evident as, for example:

- The water I use is no longer available for other people.
- The pollution I add, everybody else has to bear.

Almost all ecological dilemmas have a social side to them as we share our common goods between people and groups, an aspect that complicates sustainable management and fair distribution (Hardins 1968; Ernst 1997).

According to whichever model of a water cycle we may use, we might answer questions regarding some aspects of it differently. For example, the question: "What happens to water in the pipes when the tap is turned off?" can be answered in different ways:

a. "It keeps running (back to the river)"—answer from somebody with a simple model; turning off the tap does not make a difference.

b. "It stands still, it stops running"—answer from a person with a more complex model, turning off the tap saves energy in the treatment pants; only what goes down the drain has to be filtered again.
c. "It stands still, there's more for others"—answer from a person with a more complex model, including the social factor of a natural common commodity, the characteristic of (obligatory) sharing.

Mental models are based on what we (believe to) know about the world. In this, they are susceptible to new information and enhanced knowledge. For example, in the above example, it is relatively easy to inform the holder of a simple mental model on water, that neither the entering nor the exiting of water in an urban area function without the intervention of water treatment plants, so they can include them in their mental models.

(Climate-Relevant) Knowledge about Water

For an adequate interaction with the environment, we render the knowledge of some basic facts essential. In the following, we give a list of topics related to water that are necessary to transport a meaningful and comprehensive picture of sustainable behavioural patterns.

The Water Cycle

The natural water cycle comprises evaporation over the oceans and transpiration of vegetation over land, the formation of clouds, which in turn carry the atmospherically stored water in-land inwards on to the mountainous regions, where the precipitation in form of snow or rain fills the snow and ice storages of the world. Snow and ice melt charge the runoffs of streams that in turn form rivers that run back into the ocean, while infiltration into groundwater feeds subterranean water storages, springs and freshwater lakes that in their part lead back to the ocean. This very short description of the basic water cycle is taught in secondary schools around the world and should be considered basic biological knowledge of our environment.

More complicated are the aspects of human intervention in this biological equilibrium. Humans intervene in the water cycle in multiple ways. The majority of worldwide settlements (small and big) are placed at streams and rivers; people take the surface river water for personal and industrial use and leave their pollution trace. Agriculture heavily impacts on groundwater quality through fertilizers and pesticides. In dry regions, water is pumped out of the ground, lowering the groundwater level, thus intervening with vegetation.

Oceans are polluted indirectly through the river effluents carrying urban pollution and directly through heavy ship traffic, plus many more examples.

Human Dependencies and Uses of Water

Humans use water for many different purposes, among them playing, washing, cleaning, moving, travelling, cooking and ultimately drinking. In its multiple functions for entertainment, transport, hygiene and nourishment, different aspects are touched: there is a social component of sharing it, an ecological component of (not) contaminating it, a political component of administering and protecting it and, lastly, an economic component of supplying it. A human body consists on average of 70% water. What it all comes down to is that unlike no other consumption good, water is indispensable for life and in this only resembles one other element: air. This is the main reason why water is so vulnerable to political and economical exploitation.

Water Exploitation: Drinking Water Industry

One big aspect of economic exploitation of water is the bottled water industry. The consumption of bottled purified or mineral water has a long list of negative consequences and only few positives. The water-bottling industry exploits local natural spring water, bottles it mostly into one-way plastic bottles, and ships them around the globe to sell at extremely high prices, as the short documentary "The Story of Bottled Water"[3] poignantly shows.

This approach adds in a lot of ways to a negative environmental balance, among the main ones: adding to the local industrial pollution, increasing the volume of plastic refuse and monetary and energy costs to transport the bottles. The water industry prospers in spite of an existing water conduit network in most countries of the world, lesser legal quality regulations for bottled water compared to tap water and prices up to 10,000-fold (UN 2009). For an in-depth film documentation of problems of bottled water see the movie "Flow: For Love of Water".[1]

Cost of Water Purification and Contamination

In almost all densely populated regions where people do not take drinking water directly from a pristine natural fountain, water has to be purified for human consumption. The related costs can be expressed in terms of money and energy.

---

[3] The story of bottled water (May 2010): http://storyofstuff.org/bottledwater/.

In Buenos Aires, the purification costs are at a minimum Arg$0.80/m^3, although the water is supplied to people in most cases for less than a quarter of that.

Water prices vary hugely worldwide (National Geographic 2010) and are not always direct reflections of the real monetary costs of its provision. Because of public subsidization schemes, often the real monetary costs of a given water supply are not reflected in the charges made to private end-users.

Above and beyond the financial aspects, the energy costs directly relate to climate change, as the energy expenditure to make water drinkable gives off greenhouse gases that close the vicious cycle of boosting climate change (which in turn threatens drinking water storages).

In areas of high contamination, such as the river Plate around Buenos Aires, the costs of water purification are considerably higher (as contamination grows) than in natural areas. The price of poor purification consists in rising environmentally related illnesses such as sneaking metal intoxications, cancer and respiratory problems (Greenpeace 2010).

A Person's (or Enterprise's or Nation's) Water Footprint

As a relatively new concept, and analogous to a person's carbon footprint, a personal or collective "water footprint" has been suggested (Allan 2003). According to the Water Footprint Organization, the footprint of a person consists of the volume of water that is used for all the goods and services this person makes use of. This comprises the "virtual" use of water for the manufacturing of all the products we consume (e.g. 2,000 l for a cotton shirt; Hoekstra and Chapagain 2007) as well as the direct domestic water use of an individual. Domestic water use only comprises the smallest part of an individual's footprint (348 average litres of domestic consumption compared to a general water footprint of 3,854 l daily for an average Argentine).

This virtual water trace can be calculated on a geographical level of nations or continents, as well as on a smaller level such as that of enterprises, public buildings or individuals. A person's or nation's water footprint is mainly determined by their consumption patterns of food and other agricultural products, as the agricultural sector is responsible for the biggest share of water consumption worldwide. For a comprehensive overview see the website of the Water Footprint Organization[4] and studies by the WWF.[5]

---

[4] Water foot print organsiation (May 2010): http://www.waterfootprint.org/?page=files/home.

[5] WWF on virtual water (May 2010): http://www.wwf.org.uk/what_we_do/safeguarding_the_natural_world/rivers_and_lakes/water_footprint/.

Climate Change Vulnerability

The availability of safe drinking water is intricately linked to climate change and, depending on the region, has many different faces. It should be part of people's basic knowledge how exactly climate change affects their drinking water situation. In Argentina, climate change mainly has two different consequences, again according to the greater region, as alluded to in the first section. As for the complexity of the interplay and the importance of the topic, we dedicate the next section to it.

Climate Change and Drinking Water in Argentina

In the successive occurrences of climate change, one thing will be sure: the drinking water situation will worsen—not only in Argentina. On the one hand, as detailed above, lesser precipitation in the west led to greater droughts and lower water level in rivers. On top of that, Argentina is experiencing a retraction of glaciers due to rising temperatures. As detailed in the last report of the IPCC (Magrin et al. 2007), there are a number of studies that indicate that most of the South American glaciers, including those in Argentina (up to 25° S), are drastically reducing their volume at an accelerated rate (Leiva 2006). As ice shields do not grow any more over the heavier precipitation periods in winter, the outflow in low precipitation periods is greatly diminished, leaving the inhabitants of low-lying pre-mountain regions with severe shortages of drinking water in summer moths. To keep on securing water access for the inhabitants of these provinces and amplify the water net up to 100% coverage will have already high costs.

On the other hand, there is an over-abundance of water in the northeast and east. As already mentioned in the opening section, strong precipitation occurs in this part of the country. As our worldwide climate keeps changing, this precipitation comes increasingly in the form of extreme weather events. For central and eastern Argentina, occurrences of more than 100 mm in <2 days have roughly trebled over the last 50 years (Berbery et al. 2006). For the area of Gran Buenos Aires, five out of only 12 days in weather-recording history (first one in 1915) with more than 80 mm of rain have occurred after 1999. Two of those five were the 15 and 19 February 2010, when the usual monthly average of 112 mm of rainfall occurred in both occasions in <2 h (92 and 80 mm respectively; UBA 2010).

As we see, the over-abundance of water comes at a cost: the risk and actual occurrences of heavy flooding have increased immensely over the last few years, as, for example, flooding of the Pampa region in 2000–2002 and the great flooding of Santa Fe and Entre Rios in 2007, (Magrin et al. 2007) and flooding of major streets in Buenos Aires become the new norm on a rainy day, as a bystander's video impressively shows.[6]

---

[6] Tsunami en Palermo (May 2010): http://www.youtube.com/watch?v=JgdWvIUKFII.

When these heavy rainfalls coincide with strong landward winds ("Sudestadas") the coastline of Gran Buenos Aires is already severely affected, making flooding events even more violent (Bischoff 2005). Above and beyond the problem of heavy rainfall and "Sudestadas", the general sea level pushes back freshwater boundaries of the river Plate. The resulting higher risk of seawater intrusion is threatening the safe drinking water supply of approximately 13 million people (Barros 2005).

Initiatives for Change

The campaign launched by ATA, "I am made of water", pursues multiple goals, which can only be attained through addressing necessary changes on a structural, social and personal level. The main goals toward just and sustainable management consist of:

1. Acknowledgement of access to water as a basic human right.
2. Supply of water and sewage purification services at fair and affordable prices.
3. Extension of water connectivity up to 100% coverage.
4. Cut of industrial pollution of ground and surface water to 0%.
5. Levelling to reasonable mean usage per capita per day.

We will outline at which of the three levels we see necessary measures to reach each of the goals and outline the activities carried out by ATA towards these ends.

On a Structural Level: Laws, Policies and Regulations

Acknowledgement of Access to Water as a Human Right

One of the main goals of ATA is that access to water be recognized as a human right and incorporated into the constitution of Argentina, as for example it is in its neighbouring countries Bolivia and Uruguay.

Towards this end, ATA has written an official statement (Amigos de la Tierra Argentina 2010) on the occasion of the national day of human rights in Argentina that commemorates the last military coup on 24 March 1976. The statement has been sent to the organizers of the day's big March and the most important human rights organization in Argentina, the Mothers of the Plaza de Mayo.[7]

---

[7] Madres de la plaza del Mayo (May 2010): http://www.madres.org/.

Revision of Price Policies

As outlined in this article, for the area of Buenos Aires and its outskirts of 17 suburbs, the existing supply and price system shows some major flaws. Different pricing systems are applied without any underlying logic saving historical developments and actual incapacities of the public supplier to change them. Especially the low flat-rate scheme leads, in our opinion, to unsustainable behaviour patterns of extreme water waste and overall inefficient use of existing water reservoirs.

ATA suggests revising the outdated pricing system applied in the AySA region and put to discussion (at least) four different variants of pricing systems:

a. *Per meter, low general price*

Analogous to carbon emissions, the world's average use of 150 l (Water Footprint Organization 2010) could be considered as a reasonable quantity of domestic water use for the sake of argumentation. This would sum up to 4.5 m^3/month/capita.

An appropriate price could be calculated now by assigning a reasonable percentage (e.g. no more than 2%) of a basic income in the country at a given time. For a family of four and a basic income of Arg$1,600, the resulting price/m^3 would be Arg$1.80 (about ten times as much as the average price under the flat rate scheme, but with an overall justifiably higher bill of Arg$32 for those who manage with the quantity of 200 l/day).

b. *Free minimum, highly priced exceeding use*

Another approach could be guaranteeing a free socket quantity. This could be calculated e.g. by the recommendation of the WHO (World Health Organization), which holds 50 l/day as dignified. This would lead to a free-of-charge socket quantity of 1.5 m^3/person/month. The exact price for exceeding water use would be subject to discussion: one approach could be a significantly higher price (e.g. double the aforementioned price, namely Arg$3.60/$m^3$) to discourage excessive use. Obviously, lower free quantity would go hand in hand with lower exceeding prices, and higher free quantity with higher exceeding prices.

c. *Block system*

A combination of a and b could be the introduction of a block system, where according to used quantities of e.g. 1.5 m^3 (as indispensable minimum), prices rise linearly or, to penal excessive above-average use, even exponentially. This is to say, whereas the first block of 1.5 m^3 would be free, the second might have a cost of Arg$2.70 (according to price under system (a)), the third Arg$5.40, and so on. The block system carries the highest social aspect with it, as usually water-intense consumption patterns come with objects of relative luxury, such as washing cars and filling swimming pools.

d. *Real value directive plus social subsidy*

A "real value directive", as applied in Chile, for example, implies that water should be sold at the actual costs reflected in its generation (or purification, respectively). As this can lead to considerably high prices depending on regions, in Chile it is accompanied by a social subsidy system that pays up to 100% for the first 15 m^3 consumed in low-income households (Gomez-Lobo 2001).

All these different pricing systems rest on the area-wide installation of meters to homes and enterprises. For a transition period, there are two options to achieve the installation of meters: through enforcement by law or though a preliminary change in prices that favours a meter-based system by raising the flat-rate price, and lowering the exceeding cubic-metre costs for a meter-based scheme.

All of the discussed systems have advantages and disadvantages, but all rest on the aspect of incentive to low consumption that, together with the educational goals promoted by ATA, are considered essential for the necessary change in excessive usage.

National Law and Contamination

"Contamination is criminal" says one of our thematic banners. For years, ATA has been working together with established base organizations (e.g. Asamblea Delta Tigre; Espacio Intercuenca[8,9]) that care for the most emergent environmental problems in the territory of the river Plate, which is undoubtedly the severe contamination by chemical and beef-processing industry. The most worrying case of industrial contamination in South America has recently been acknowledged by the Supreme Court of Argentina. With an unprecedented law case (Causa Mendoza), the CSJN (Corte Suprema de Justicia de la Nación) ruled in 2007 that the heavy industrial contamination of the river Riachuelo has to be decontaminated and major polluters have to be closed down. In 2009, ATA was asked to join "Espacio Riachuelo",[10] which is following up the fulfilment of this court ruling.

Public Level: Awareness-Raising

To raise public awareness of the different pressing problems regarding water (in all its appearances), ATA organizes events of all different kinds; in the following, we will highlight our main activities in this sector.

---

[8] Asamblea Tigre (May 2010): http://asambleaurbanoambiental.blogspot.com.

[9] Espacio Intercuencas (May 2010): http://www.espaciointercuencas.org.

[10] Espacio Riachuelo (May 2010): http://www.espacioriachuelo.org.ar.

Water Week in March and Blue October

For four years now, Water Week has been celebrated over 14 March (Day Against Dams), 22 March (World Water Day) and sometimes extending up to 31 March (Argentine Water Day). As well as in March for the world and national Water Days, ATA celebrates water in the "Blue Month of October", an international movement fighting the commercialization of water. ATA joined "Blue October" for the first time in 2008.

In both periods, March and October, different organizations join their efforts to spread information about water-related problems to the public. The events held include festivals, public manifestations, expert talks in community centres, artistic activities such as theatre productions and concerts, among other things.

Cultural Events: Hydro Festival, Open Cinema, Thematic Concerts

Throughout the year, ATA hosts cultural events with different organizations. Among the most popular ones are the Hydro festival Tigre in cooperation with "La Asamblea Tigre",[8] which took place within Water Week in March, an environmental cinema cycle in an alternative open air cinema in Buenos Aires and concerts with cooperating Buenos Aires musicians, which led to the release of the CD "Musica contra el sed" in 2008 together with "Radio la Tribu, 88.7 FM".

Thematic Exhibitions

There are two current exhibitions hosted by ATA. One, "Reflections on Water" is an itinerant exhibition that consists of 11 poster-sized presentations of case studies of different problems concerning water. The poster topics were gathered in a nationwide call for examples of environmental problems around water and range from a description of the industrial pollution belt of Buenos Aires to the health problems of people of Cordoba related to contaminated waters. With this nationwide approach, it becomes evident that similar problems repeat themselves in different areas of Argentina.

The other exhibition, part of the territorial campaign "The River for the People" is a collection of historical pictures of the river Plate, showing how the waterfront along Buenos Aires was used for riverside bathing before it was all closed off to public use in the 1970s, due to rising contamination. At the present moment, there are regulations at community level to prohibit bathing in the river Plate between the north of the city and La Plata (60 km south of Buenos Aires). The exhibition invites contemplation of the huge differences between the waterfronts of now and then and makes it very obvious how far from natural use the Buenos Aires waterfront has strayed from.

Support of Local Neighbourhood Assemblies

Members of ATA visit and testify local hot spots of environmental problems, such as the immense colony construction site and related contamination in the Paraná Delta, approximately 50 km north of Buenos Aires. The Paraná Delta is one of the biggest deltas worldwide, comprising countless islands and marshland with UNESCO biosphere status[11] before flowing into the river Plate. Interviews and short film documentaries have been made in order to broadcast environmental problems in different media such as websites and the radio.

Individual Level: Personal Education

In the end, real change in whatever sector of societal or even international proportions begins with oneself. Although ATA believes adequate basic structures and an according social atmosphere are beneficial to breed certain behaviour in people, the ultimate ruling instances are personal convictions and beliefs. To this end, ATA is convinced that the only way to provoke lasting pro-environmental behaviour patterns is profound education of individuals. The negative effects of simplistic mental models and the basic water-related units of knowledge to overcome them have already been outlined. Here we will briefly describe the measures with which we convey this knowledge.

Seminar of Environmental Education: "From My School into My Suburb"

The education team of ATA gives a 2-h workshop to schools of the Gran Buenos Aires region. The workshop treats consequences of the unsustainable use of water and gives proposals for ways out of it. The various parts contain a suggestion of a holistic water-cycle view, an analysis of our relation to water on a macro (universal) and micro (personal) level, and inspections of examples for mismanagement on personal, social and political levels.

The form of the workshop is participative; participants work in small groups, constructing a network of knowledge including their personal experience instead

---

[11] UNESCO Biosphere list (May 2010): http://www.unesco.org/mab/doc/brs/Lac.pdf.

of presenting mere facts to be learned. So far, it has been presented to more than 800 students between 10 and 17 years of age and more than 100 teachers of primary and secondary schools. As suggested to all the groups at the end of the workshops, some have produced a poster dealing with their most emergent environmental problem included in our itinerant exhibition "Reflections on Water".

Educational Material: "From My School into My Suburb"

Topics of the water workshop with accompanying educational material as basic information, group exercise assignments, tools for the analysis of water, (mis)management in schools or public buildings of the suburb, and much more, were collated on a CD-ROM for the teachers to keep. With this first step, ATA tries to initiate a scholar environmental education programme in Argentina. Extensions to community centres are in planning.

Public Education Material

ATA presented its first edited book "Sobre el Agua" ("About Water") during this year's Water Week in March at the National Library of Buenos Aires. The first edition of 500 copies was made available to schools, interested organizations and libraries for free. Other material produced includes informative brochures and flyers about different water-related topics, e.g. the current state of affairs of the Mendoza law case or the situation of the local freshwater basins.

In Closing

As we have seen, water-related problems in Argentina are as manifold and diverse as the country itself. The majority of Argentines experience a (drinking) water-related problem of some form, be it scarcity, destructive over-abundance, contamination or high commercial prices. Amigos de la Tierra Argentina wants to raise awareness of these problems and related injustices with activities at the three different levels of society: personal, social and political. Although ATA believes that aiming at all three levels is essential to produce the necessary change for a balanced blue future of the country, it also gives the highest priority to the personal level as far as this agrees with Margaret Mead in her often-cited affirmation that it has always been a small handful of committed citizens who have ever changed the world. We interpret this slogan as a motivation to awaken the future-committed core in every citizen.

References

Allan JA (2003) Virtual water—the water, food and trade nexus: useful concept or misleading metaphor? Water Int 28:4–11

Amigos de la Tierra Argentina (2010) El derecho humano al agua. Informe para el día de derechos humanos, 24 March 2010

AySA (2008) Expansión y desarrollo, comprometida con la inclusión. Informes al usuario, Facts of December 2008

AySA (2010a) Area de acción de AySA. Available at http://www.aysa.com.ar/index.php?id_seccion=194. Accessed May 2010

AySA (2010b) Nuestro trabajo. Available at http://www.aysa.com.ar/index.php?id_seccion=196. Accessed May 2010

Barros A (2005) Inundación y cambio climático: costa argentina del Rio de la Plata. In: Barros V, Menéndez A, Nagy GJ (eds) El Cambio Climático en el Río de la Plata. Consejo Nacional de Investigaciones Científicas y técnicas (CONICET), Buenos Aires, pp 41–52

Berbery EH, Doyle M, Barros V (2006) Tendencias regionales en la precipitación. In: Barros V, Clarke R, Silva Dias P (eds) El Cambio Climático en la cuenca del plata. Consejo Nacional de Investigaciones Científicas y técnicas (CONICET), Buenos Aires, pp 67–79

Bischoff S (2005) Sudestadas. In: Barros V, Menéndez A, Nagy GJ (eds) El Cambio Climático en el Río de la Plata. Consejo Nacional de Investigaciones Científicas y técnicas (CONICET), Buenos Aires, pp 53–67

BMU (2010) Bundesministerium für Umwelt, Naturschutz und Reaktorsicherheit. Trinkwasser in privaten Haushalten. Available at http://www.bmu.de/gewaesserschutz/fb/trinkwasser_priv_haushalte/doc/3652.php. Accessed May 2010

Borthagaray JM, Igarzabal M, Pereyra FX (2002) Buenos Aires y el agua. El Rio de la plata, fuente y cloaca. In: Borthagaray JM (ed) El rio de la Plata como territorio. Ediciones infinito, Buenos Aires, pp 361–393

Cattaneo M, López Sardi M (2010) Los ciudadanos y su relación con el agua, Paper to be presented at the Congress of Sustainable Engineering and Urban Ecology, Universidad de Palermo, Buenos Aires, Argentina

CIA (2010) The World Fact book, South America, Argentina, People. Available at https://www.cia.gov/library/publications/the-world-factbook/geos/ar.html. Accessed May 2010

Ernst A (1997) Ökologisch-soziale Dilemmata, Psychologische Wirkmechanismen des Umweltverhaltens. Beltz, PVU, Weinheim

Gentner D, Stevens AL (1983) Mental models. Lawrence Earlbaum Associates, Hillsdale NJ

Gomez-Lobo A (2001) Making water affordable. Output-based consumption subsidies in Chile, World Bank, Public Policy for the Private Sector, Note Number 232, June 2001, pp 23–29

Greenpeace (2010) Riachuleo. 200 años de contaminación, Greenpeace Argentina, Campaña toxicos. Available at http://www.greenpeace.org/raw/content/argentina/contaminaci-n/agua/riachuelo/riachuelo-doscientos-contaminacion-informe.pdf. Accessed May 2010

Hardin G (1968) The tragedy of the commons. Science 162(3859):1243–1248

Hoekstra AY, Chapagain AK (2007) Water footprints of nations: water use by people as a function of their consumption pattern. Water Resour Manag 21(1):35–48

INDEC (2001) Censo Nacional de Población, Hogares y Viviendas. Available at http://www.indec.gov.ar/webcenso/index.asp. Accessed May 2010

INDEC (2007) Evolución de las tasas de mortalidad infantil por 1000 nacidos vivos según provincia de residencia de la madre. Total del país. Años 2004–2007. Available at http://www.indec.mecon.ar/nuevaweb/cuadros/65/q030205.xls. Accessed May 2010

INDEC (2009) Incidencia de la pobreza e indigencia—Segundo semestre de 200. Available at http://www.indec.mecon.ar/nuevaweb/cuadros/74/grafpobreza1_ephcontinua.xls. Accessed May 2010

Johnson-Laird PN (1983) Mental models: towards a cognitive science of Language, inference, and consciousness. Harvard University Press, Cambridge

Leiva JC (2006) Assessment of climate change impacts on the water resources at the northern oases of Mendoza province, Argentina. In: Price MF (ed) Global change in mountain regions. Sapiens Publishing, Kirkmahoe, Dumfriesshire, pp 81–83
Lobos JE, Duran J Roizen RM (2004) Evaluación de la calidad ambiental de la cuenca Matanza-Riachuelo. Informe Final, Instituto Nacional del Agua, Buenos Aires. Available at http://www.ina.gov.ar/boletines. Accessed May 2010
Magrin G, García D, Choque C, Giménez JC, Moreno JC, Nagy GJ, Nobre C, Villamizar A (2007) Latin America. In: Parry ML, Canziani OF, Palutikof JP, van der Linden PJ, Hanson CE (eds) Impacts, adaptation and vulnerability contribution of working group II to the fourth assessment report of the intergovernmental panel on climate change. Cambridge University Press, Cambridge, pp 581–615
Ministerio de Educación (2010) Las condiciones naturales del territorio argentino. Available at http://www.encuentro.gov.ar/Gallery/5980.pdf. Accessed May 2010
National Geographic (2010) Water: our thirsty world, Special Print Issue for World Water Day, March 2010
Paris MdC, Zucarelli GV, Pagura MF (2009) Las Miradas del Agua. Universidad Nacional de Litoral, Santa Fe
Rossi A (2010) El Estado pone $1.685 millones para tapar el "rojo" de AySA, Clarin. Available at http://www.clarin.com/diario/2010/05/08/elpais/p-02192777.htm. Accessed May 2010
UBA (2010) Rainfall statistics of the Department of Ocean and Atmosphere Science, University of Buenos Aires. Available at http://www-atmo.at.fcen.uba.ar/tiempo/150210/ev_15022010.htm. Accessed May 2010
UN (2009) World water development report, 3rd edn. Available at http://www.unesco.org/water/wwap/wwdr/wwdr3/pdf/WWDR3_Water_in_a_Changing_World.pdf. Accessed may 2010
Water Footprint Organization (2010) Water footprint of nations, 1997–2001. Available at http://www.waterfootprint.org/?page=files/NationalStatistics. Accessed May 2010
Wurgaft R (2009) Navegando por el río más contaminado de Sudamérica, El Mundo. Available at http://www.elmundo.es/elmundo/2009/02/28/internacional/1235834943.html. Accessed May 2010

Chapter 40
Socio-Environmental Evaluation of Drip Irrigation System Implementation as a Climate Change Adaptation Measure Within the N'hambita Community Carbon Project Area, Mozambique

Jelena Barbir and Walter Leal

Abstract According to a recent document produced by the Intergovernmental Panel on Climate Change (Le Treut et al., Climate change 2007: the physical science basis, contribution of working group I to the fourth assessment report of the intergovernmental panel on climate change, Cambridge University Press, Cambridge, 2007), climate change is expected to increase the impact of droughts and water shortages in sub-Saharan Africa, which is already significant, intensifying the need to acquaint local people with adaptation measures in the near future. One of the tools that can be used to help rural areas adapt to climate change can be Drip Irrigation Systems (DIS). The research described on this paper was undertaken in rural Mozambique, as part of the N'hambita Community Carbon Project (NCCP), in order to evaluate the efficiency of DIS and the success of its implementation in the area. The methodology of the research consisted of on interviews, questionnaires, GPS mapping of water sources and observation of the water use within agricultural activities. The efficiency of DIS irrigation, investigated among the local farmers, is estimated as higher than manual irrigation, but less efficient than its presumed theoretical potential. In addition, the results also showed that the investigated approach to DIS implementation in the NCCP area did not appear as successful as expected and has yet to be improved by increasingly involving social aspects in the process. Nevertheless, it was concluded that DIS have a high potential for improving water management, productivity of farms and households' food security within the NCCP area.

J. Barbir (✉) · W. Leal
Department of Crop Protection, Institute of Agricultural Sciences,
Centre for Environmental Sciences, CSIC, Calle Serrano 115,
28006 Madrid, Spain
e-mail: jelenabarbir@ccma.csic.es

Keywords Mozambique · Socio-environmental indicators · Water management · Drip irrigation systems · Climate change adaptation

Introduction

Climate change is one of the biggest global problems and threats the world is facing today (Le Treut et al. 2007; Barbir et al. 2009). There is thus a worldwide need to estimate its impact on the environment (Sanchez 2000; Christensen et al. 2007; Bates et al. 2008) and to human populations. According to the Intergovernmental Panel on Climate Change (IPCC) prediction, the global annual median surface air temperature may increase by 3–4°C (Christensen et al. 2007) by the end of the twenty-first century, leading to changes to both climate and environmental conditions. The extent to which climate change will differ among the continents and regions depends on geophysical conditions and the vulnerability of ecosystems.

Africa, as the second largest continent in the world with the highest rate of population growth (Grid Arendal 2002), is described as one of the regions that is the most vulnerable to climate change (Callaway 2004). This is especially so in respect of water availability (Christensen et al. 2007). Due to the fact that this research has been conducted in the rural area of Mozambique, some estimations regarding climate change impacts on the water resources of sub-Saharan Africa are of significant importance for creating adequate future water management decisions for the region.

According to Shah et al. (2008), arid and semi-arid areas in Africa will expand by about 5–8% as a result of climate change. In addition, Bates et al. (2008) estimate that average runoff and water availability will decline in northern and southern Africa, impacting on freshwater ecosystems. According to Batisani and Yarnal (2009), both the drying trend and decrease in rainy days are in accordance with climate change projections for southern Africa. All these, and many other provisions, point out that northern and southern Africa are among the most vulnerable regions of Africa to water scarcity in the future as a result of climate change.

The most widely considered responses to climate change are mitigation (Reduction of Greenhouse Gas (GHG) emissions and enhancing sinks) and adaptation to the impacts of climate change (UNFCCC 2003). Coping with climate change in developing countries (which are highly vulnerable and often economically fragile) may be facilitated by creating synergies between adaptation and mitigation measures, which consequently increases the cost-effectiveness of the actions, making them more attractive to stakeholders (Barbir et al. 2009). Considering that mitigation efforts may not fully prevent climate change in the next few decades (Christensen et al. 2007), the reduction in vulnerability of ecosystems, sectors and communities, by the means of adapting to climate change impacts, is urgently necessary (Stringer et al. 2009).

The Need for Study

Africa is in great need of proper adaptation strategies (Adger and Barnett 2009) as climate change is already underway in many sub-Saharan countries. The climate change projections results (Batisani and Yarnal 2009; Solomon et al. 2007) have important policy implications for the government, emphasizing the significance of helping dryland farmers to adapt to climate change. Therefore, it is important to look at ways in which adaptation projects may be implemented, especially those which maximize the use of scarce resources such as water.

As presented by the World Resources Institute (WRI) (WRI 2000), Africa is the continent most highly reliant on the agricultural sector due to the fact that the agricultural sector contributes on average 21% of total African Gross Domestic Product (GDP), ranging from 10 to 70% between African countries (Christensen et al. 2007). Although this range is relatively broad, when it is compared with the agricultural contribution to GDP within developed countries (which varies from 0.5 to 3% (UNDP 2008)), the degree of importance attached to agriculture in Africa becomes apparent.

Furthermore, as stated by Usman and Reason (2004), the agricultural sector is particularly sensitive to periods of climate variability, especially in the region of sub-Saharan Africa where, according to Rockstrom et al. (2002), only 5% of agricultural land is irrigated, while 95% is rain-fed. In accordance with that, the yields in this part of the world are estimated as rather low, oscillating on average around 1 tonnes/ha (Rockstrom et al. 2001).

Based on the above perceived need, this research was undertaken in rural sub-Saharan Africa, within the N'hambita Community Carbon Project (NCCP) area, located in the buffer zone of National Park Gorongosa, Sofala province, Mozambique (Fig. 1). Regarding land use, NCCP farmers distinguish between *machambas* and gardens. *Machambas* are rain-fed agricultural fields, whereas the local gardens are smaller and irrigated by farmers (field observation 2009).

As the main soil type within the study area is generally poor, extremely weathered, and freely draining sandy loams with a low fertility (Williams et al. 2008), the equilibrium between cultivated and semi-natural or natural vegetated areas in the dry forests and woodlands is shifting to cultivated areas (Jansen et al. 2008). This upward trend demands urgent action and radical changes regarding land use management in the area (Jansen et al. 2008).

An additional factor which affects agricultural production and food security is the tropical and subtropical climate in that region. This climate means that it is noticeable that during the wet season, communities and vegetation are well provided with water, but then during the dry season, based on the fact that no reliable rainwater harvesting or water storage systems are in place, the water deficit becomes obvious, especially affecting the communities reliant on the seasonal river streams (Van Zyl and Powell 2009—personal interview).

There is a great challenge to maintain successful agricultural production under these climate conditions, and according to Tompkins's research (2005), it will be

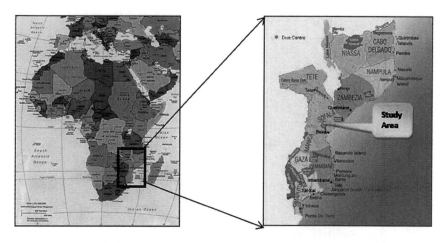

Fig. 40.1 Location of Mozambique, Sofala province and the study area

even more difficult as the frequencies of extreme events (droughts and floods) are likely to increase rapidly with climate change in sub-Saharan Africa. Based on the conditions of the site, characterized by a tropical climate and occurrences of extremely dry periods, there is a pressing need to implement reliable measures towards climate change adaptation (Andersson et al. 2006).

Improving food security and agricultural production in developing countries is in accordance with the Millennium Development Goals (UNFCCC 2007) and high on the agenda when adapting to climate change in poor, developing countries. Even though the policy-makers are responsive about the effects of climate change in Mozambique, the results of Patt's research (Patt and Schroter 2008) showed that local people are not linking the current increase in flooding and droughts with the predicted ones. Many case studies confirmed that local people's opinions and attitudes are highly significant when implementing new policies and technologies, as their disapproval can easily lead to the implementation's failure.

Even though researchers and policy-makers alike are aware of the importance of considering social aspects when creating a plan for agricultural changes, they often struggle to find the right way of implementing it (Doss 2001; Osbahr et al. 2008), bearing in mind that there is always a possibility of failure in the adoption of successful technology, as reported in many case studies (Marra et al. 2003). The implementation of new technologies, without a significant impact on the local traditions and the local environment, remains a great challenge for the management of climate change (Allen et al. 2001).

Due to the above-mentioned reasons, it is immensely important to develop appropriate indicators to be investigated before implementing a new technology in agriculture, which will also be used further on in the process of monitoring and evaluation. The choice of indicators to be investigated is always closely related to the scope of the research (Shields et al. 2002) and, considering the socio-environmental aspect of this research, social and environmental indicators have been developed.

Environmental indicators are chosen with an aim to analyse the efficiency of Drip Irrigation Systems (DIS) in agricultural irrigation and to understand the irrigation and agricultural practices within the NCCP area. The social indicators, on the other hand, are developed in order to understand social networks, relations, customs and habits of the local people.

The N'hambita Community Carbon Project's Activities

The NCCP's activities started in 2003, with the aim of developing sustainable rural communities and alleviating poverty through the process of carbon sequestration. Although the NCCP's objectives are mainly focused on forestry and land management within the area, water management has recently been incorporated into their activities as well (Van Zyl and Powell 2009—personal interview).

According to Van Zyl and Powell (personal interview, 2009), in 2007, the main connection between forestry management, water management, food production and local communities was found in the implementation of DIS, organized in order to improve local farmer's production, to encourage them to stop slash-and-burn practices, motivate them to practise long-term usage of the limited agricultural area and to reduce their vulnerability to water scarcity (the main objective). This initiative was conceived as a pilot project which is to be further investigated.

The group of 38 farmers agreed to sign a contract with NCCP, which obliged them to purchase the DIS within a three-year period, paying annual instalments of 2,500 MTZ (€57). For that price, the farmers were offered a drum kit DIS, consisting of a 250-l volume water tank, filter, and the pipes. This drip irrigation system covers and irrigates the garden area of 250 m^2.

The pilot project of DIS implementation has been in operation for two years, which is a suitable amount of time to evaluate its success and relevance for the area throughout an external evaluation, which has been done in order to assist considerably in a more objective estimation of the DIS efficiency initiated by NCCP.

The Study Area

The boundaries of the NCCP area are the Pungwe river to the south, the Vanduzi river to the west and Gorongosa National Park to the east and north (Fig. 2). The biggest river passing through the research area is the Pungwe river, meaning that all the sources of the surface water used by the local people are part of the Pungwe river basin.

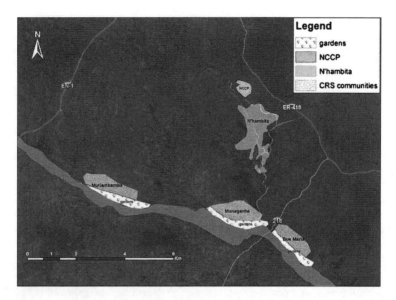

Fig. 40.2 The study area (field data processed by ArcGIS)

The NCCP area is bisected by the national road (EN-1) that runs between the towns of Gorongosa and Inchope. The larger part of the area lies east of the EN-1 road, where the NCCP establishment and the villages of Bue Maria, Munaganha, N'hambita and Mutiambamba are situated (Fig. 2). The rural road (ER-418) runs east to west from the EN-1 road, providing access to these villages. The villages to the west of the EN-1 road, Mbulawa, Povua and Pungwe, lie outside the designated buffer zone.

The study was conducted among the drip irrigation users from the three villages (Mutiambamba, Munaganha and Bue Maria) with gardens located next to the Pungwe river (Fig. 2). The seasonal floods occur quite frequently within the Pungwe river basin, destroying garden yields, and on the other hand, during the dry season, strong droughts affect agricultural production to a high extent, as irrigation practices are not developed in the area. This current situation in rural Mozambique calls for urgent agricultural and irrigational improvements in local and regional management.

Data Collection and Analysis

During the four-week fieldwork period in the NCCP area, qualitative and quantitative data was collected through semi-structured interviews and questionnaires conducted among NCCP managers, N'hambita Community leaders and some of the NCCP's technicians, with the aim of collecting information about the project activities and future plans.

Table 1 Social and environmental indicators

Environmental indicators	Social indicators
1. The irrigation water sources and water use	1. Awareness of water significance
2. Land use	2. Social networking
3. Usage of fertilizers and pesticides	3. Socioeconomic potential to purchase the DIS
4. Plant losses	
5. "D" and "T" parameters	

The 38 drip irrigation farmers (DIFs) from the three villages (Mutiambamba, Munhanganha and Bue Maria), who relied on the constant stream of Pungwe river (Fig. 2), were offered the DIS as a new technology in their agricultural activities. The group of 35 farmers (92% of total number) was interviewed respectively about the agricultural practices of the last three years (2006, 2007 and 2008). Furthermore, of the 35 farmers interviewed, 9 stated that they were not using the DIS, reducing the DIF group to 26 farmers interviewed through the questionnaire.

The data from the survey questionnaires was analysed using descriptive statistics and graphical representations, while some of the gathered data were further processed with a statistical programme called Statistical Package for the Social Sciences (SPSS), version 16.0. In order to estimate if there is any significant correlation between two parameters, the Pearson parametric test was used.

In order to cluster the collected data and undertake the socio-environmental research, the following social and environmental indicators have been developed (Table 1):

The numerical values of all the indicators, except "D" and "T" parameters, were gathered directly from the farmers. The "D" parameter (average daily walking distance, garden to water source) and the "T" parameter (average daily time spent in collecting the irrigation water) were calculated by using two equations created in accordance with the field data (Eqs. 1 and 2).

$$D = \frac{\sum [Dp \times 2 \times Wp]}{20n} \quad (1)$$

$$T = \frac{\sum [Tp \times 2 \times Wp]}{20n} \quad (2)$$

(D—average daily walking distance [m/day]; Dp—distance, garden to water source (m); Wp—irrigation water withdrawal (l/day); T—average daily time spent collecting the irrigation water [min/day]; Tp—time spent in walking, garden to water source (min); n—number of farmers)

The values of Dp and Tp are multiplied by two in order to calculate the return distance/time from the garden to water source. As all the farmers use the 20-l water containers to fetch the water, the value of Wp is divided by 20 to estimate how many times a farmer walks the distance (Dp).

Additionally, the map of the NCCP area was created by the Arc Geographical Information System (GIS) program to present the locations of communities and gardens of the study area.

Results

Environmental Indicators

The Irrigation Water Sources and Water Use

The only sources of irrigation water for DIS supply are the river streams, as irrigated gardens are located next to the river's banks (Fig. 2). Furthermore, of 26 farmers, only 15% irrigate their gardens using DIS alone, while the other 85% of the farmers own the additional part of the garden that is irrigated manually. Those 85% of the farmers have been asked to compare their irrigation practices between DIS and manually irrigated parts of the garden (half of those farmers were able to distinguish the amount of irrigated water spent manually and by using DIS, and that data is used).

The results, gathered from those farmers for 2008, show that the average amount of irrigation water used daily per square metre by the DIS (2.03 l/m^2/day) is significantly higher than the manual consumption of irrigated water (0.47 l/m^2/day). Although higher water consumption is noticeable in the case of DIS usage, the farmers argued that they achieved better productivity, better quality of vegetables, faster growth of the vegetables and reduced plant losses in the part of the garden which was irrigated by the DIS.

Even though the results for 2008 indicate that there is significantly higher water consumption when using the DIS in comparison with using manual irrigation, the average amount of total irrigated water (manual and drip) used by the DIF farmers shows an overall decrease in water consumption over a three-year period (Fig. 3).

From observing the water consumption data over the three-year period (2006–2008), it is noticeable that the amount of water (l/m^2/day) had slightly decreased when DIS was implemented (in 2007), and remained unchanged in the following year (in 2008).

Land Use

As already mentioned, the local farmers in the NCCP area distinguish between two types of agricultural areas: *machambas* and gardens. The results gathered from the DIFs, while comparing the average areas of machambas and gardens over the period of three years, show that the average area of *machambas* increases yearly (Fig. 4).

Considering the gardens (Fig. 5) in the period between 2006 and 2007, the average area of the gardens expanded from 0.54 to 0.57 ha, but after implementing the DIS in 2007, the average area of the gardens remained unchanged. This result is in accordance with the main NCCP's pilot project objective to use DIS as a technology to increase agricultural efficiency and motivate farmers not to expand their agricultural area.

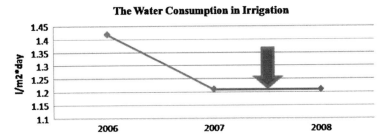

Fig. 3 Irrigated water consumption (field data)

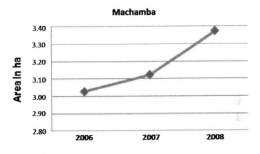

Fig. 4 Average area of *machamba* (DIF group) (field data)

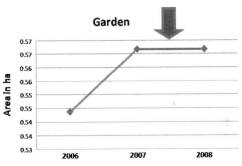

Fig. 5 Average area of garden (DIF group) (field data)

Additionally, 52% of the DIFs explained their plans to expand the area of their *Machambas* and gardens in the future. The reason for these plans was not to increase the production of essential food for the household's needs, but to increase the production of food to sell on the local market. This partly shows that the DIFs produce enough food for their household's needs, but the production is not efficient enough for earning additional income. The remaining 48% of the farmers, who do not plan to expand their gardens, justified their decisions mainly by reporting sufficient food production. Only a few of the farmers mentioned the lack of additional agricultural land next to the riverbank as a reason.

Pearson's parametric test was run in order to estimate if there was any significant correlation between the garden area and the amount of irrigated water in 2008. The results of the test show a significant negative correlation between those two parameters, when using a statistical significance of $p = 0.01$. This means that when the garden area increases, the farmers use less irrigated water per square metre of the garden.

Usage of Fertilizers and Pesticides

The results conveying the use of the pesticides show that the DIFs do not use any pesticides or any other alternative method to fight insects. Regarding fertilizers, only one DIF uses fertilizers as part of the DIS. Additionally, the farmers were asked if they prepare compost or if they use any other natural fertilizer. The answers were always negative, with the explanation that if they notice a decrease in soil fertility, they will switch to a new piece of land.

Plant Losses

Plant loss is defined as the percentage of plants lost out of the total number of plants planted. The only reason for the plant losses described by the farmers is insect pests. An unexpected finding is related to the fact that the farmers did not report any difference in plant losses over the three-year period, although they did describe 2008 as very hot and dry. Nevertheless, they all were eager to explain the differences in the plant losses between the DIS and the part of the garden irrigated manually. The losses in the DIS part of the garden were lower (23%) than in the manual part of the garden (35%).

The "D" and "T" Parameters

The "D" and "T" parameters were analysed in order to estimate if the "garden to water source" distance plays a significant role in agricultural irrigation practice for the farmers. The "D" parameter is the average distance that farmer walks daily to fetch the irrigated water, and the "T" parameter is the average time that a farmer spends daily to fetch the irrigated water.

Table 2 presents the average "garden to water source" distances, the average amount of irrigated water, and D and T parameters. Besides the average values, minimums and maximums are presented in the table in order to clarify the numerical oscillations. As can be seen from the table (Table 2), some maximum values (56 or 49 km per day) are related with the fact that some African families consist of many family members who collect irrigated water.

While all the farmers were able to recall the data for the last two years (2007 and 2008), many of them had difficulties providing the data from 2006. For that purpose, the comparison between the parameters has been calculated considering data only from the ten farmers who were able to give data for the last three years (2006–2008).

Table 2 Calculations of DIF's total walking distance (D) and total time consumption (T) in collecting irrigation water daily

DIU group: 2006		No data	n = 10	MIN	MAX
AVERAGE distance "garden-water source" (m)			160	10	700
AVERAGE amount of irrigated water (l/day)			446	100	800
D(m)			8,284	100	56,000
T(min)			166	2	1,120

DIU group: 2007	n = 25	MIN	MAX	n = 10	MIN	MAX
AVERAGE distance "garden-water source" (m)	97	5	700	160	10	700
AVERAGE amount of irrigated water (l/day)	514	100	900	488	100	700
D(m)	4,638	340	49,000	7,588	340	49,000
T(min)	93	7	980	152	7	980

DIU group: 2008	n = 25	MIN	MAX	n = 10	MIN	MAX
AVERAGE distance "garden-water source" (m)	100	5	700	160	10	700
AVERAGE amount of irrigated water (l/day)	529	100	900	488	100	700
D(m)	4,814	340	49,000	7,588	340	49,000
T(min)	96	7	980	152	7	980

(n—number of the farmers; MIN—minimum; MAX—maximum) (field data)

The parameter which proves the reliability of the farmers' answers has been found in the average "garden to water source" distance, maintaining the same average value of 160 m from year to year, as well as unchanged minimum and maximum distances (Table 2). Also, the negative correlation has been found between the average amount of irrigated water (l/day) and the "garden to water source" distance comparing the average data in the second column ($n = 25$) and the fifth column ($n = 10$) presented in the table (Table 2). When the distance is longer, the farmers fetch less water daily to irrigate their gardens.

The reason why the "D" and "T" parameters always increase in combination with a reduced amount in fetched water is due to the fact that the farmers who walked longer distances to fetch water reduced their water consumption, while the farmers whose gardens are located closer to the water source increased their consumption after DIS implementation in 2007.

Social Indicators

Awareness of the Significance of Water

The environmental awareness of the NCCP management and the people living in the NCCP area is of significant importance for the successful development of the project. Considering that this research is focused on irrigation practices, the environmental awareness of these groups is investigated within these boundaries.

Regarding the NCCP managers (Van Zyl and Powell 2009—personal interview), awareness is high regarding water scarcity within the area, climate change vulnerability, the need to adapt to the climate change and to alleviate poverty by ensuring better food production. Despite this high awareness and motivation, the NCCP managers still did not include a water management plan in their policy, and very little improvement and investigation has been done so far regarding this topic.

During the interviews, the NCCP managers recognized that water management should be investigated and implemented to a higher extent because numerous activities within the NCCP were closely related to water use. Although there was high awareness of the importance of water management among the NCCP members, improvements in that field were not prioritized compared to other objectives of the project and, as a result, progress in agricultural irrigation were consequently slightly underestimated (Van Zyl and Powell 2009—personal interview).

The opinions regarding water conservation were gathered from 71% of the DIFs. The farmers stated that they had enough water for their household needs, although the use of, on average, 70 l/day/household, is far above the average water consumption in developed countries. In the dry period, farmers proved to be very practical when coping with water scarcity during the dry periods of the year, mostly digging traditional wells in the dry river bank or walking longer distances than usual to fetch water. Awareness of the importance of water originates from the difficulties they face when supplying water to their households.

Furthermore, the farmers demonstrated good understanding of the correlation between water streams and forests, stating that in order to keep the water streams running, they have to protect the forests around them.

Social Networking

The NCCP managers focused their concerns around the two main points regarding social networking with DIFs. One was related to the farmers' constant expectations of the NCCP management to solve all their problems, while another one was related to the farmers' low comprehension of the contracts' significance.

Concerning the first issue, the NCCP managers stated that they offered the farmers help regarding the drip irrigation systems, but that they are not in a position to be able to solve all their doubts and issues. Furthermore, considering the fact that the NCCP will not stay in the area forever, the local farmers are expected to learn how to become less dependent on the NCCP. Regarding the second issue, the NCCP managers argued that for local people it is hard to understand their obligations once the contract is signed.

Upon investigating the social networking within the DIF group, it was discovered that cooperation between farmers from the same village exists, but the level of interaction between the farmers from different villages (e.g. DIFs from Munaghanha and Mutiambamba) is not very well developed.

Table 3 The farmer's economic potential to purchase the instalments for the DIS (field data)

	Ideal state	Data from DIFs
Payment for DIS	100% paid 2 of 3 rates for DIS	0% paid 1 of 3 rates for DIS 0% paid 2 of 3 rates for DIS
Earning enough to pay the rate/year (€57/year)	2007: 100% 2008: 100%	2007: 27% 2008: 19%

Another limitation for the farmers is the lack of transportation to access the markets. All the farmers remarked that NCCP should assure them transportation of the vegetables to the market, in order to pay their instalments for the DIS.

Socio-Economic Potential to Purchase the DIS

In order to estimate the farmers' ability to purchase the DIS, data collection has been focused on the average earnings from the garden production and their approach to the significance of the contract.

Looking at the results (Table 3), the percentage of farmers who earned enough to pay the instalments was lower than was expected by the NCCP's administration (for 2007 and 2008).

In 2007, only 27% of the farmers were earning enough to purchase the DIS instalment, but none of the farmers paid. In 2008, the percentage of the farmers who were earning an income high enough to pay for the DIS instalments was reduced to 19%. As in 2007, the farmers did not pay any instalments for DIS in 2008 (Table 3).

According to the contract, farmers were obliged to pay off the DIS within a three-year period or if not, to give the irrigation system back. 30% of DIFs stated that they are planning to return the DIS because they cannot earn enough to pay for the systems. The two reasons noted by the farmers were the low production (the lack of fertilizers and pesticides caused the production of low-quality vegetables) and the lack of access to the market (enabling farmers to sell the vegetable they produce).

Discussion

Efficiency of the DIS in Agricultural Irrigation

In theory (Maisiri et al. 2005; Sijali 2001), for successful drip irrigation, it is necessary for farmers to be supplied with a permanent water source, agricultural land (a garden to grow the vegetables), adequate fertilizers (to ensure the soil fertility) and pesticides (to be successful against insect pests). In practice, among

DIF farmers, 100% of them have access to water streams, 100% of them own a garden, but only 4% of them (1 out of 26 farmers) use fertilizers, and none of them have the ability to supply, buy and use pesticides.

According to Maisiri's DIS efficiency investigation in Zimbabwe, it is concluded that DIS cannot be implemented successfully without proper water and nutrient management, and that the yield of the vegetables depends mostly on the fertilizer application method (Maisiri et al. 2005). In addition, Stoorvogel and Smaling (1990) emphasize in their assessment of soil nutrient depletion that fertilizers affect water productivity in sub-Saharan Africa, describing how the irrigation is more effective when soil is fertile.

As mentioned, the soil within the NCCP area is not particularly fertile, and usage of fertilizers could improve production significantly (Graham et al. 2002). However, NCCP managers have already suggested the production and usage of compost, instead introducing the local people to chemical fertilizers, pointing out that overusing fertilizers can cause environmental problems.

With all this in mind, it can be assumed that, in practice, DIS efficiency is lower than it is in theory. In order to further investigate this assumption, the efficiency of drip irrigation was compared with the efficiency of manual irrigation, by using two comparisons.

One comparison evaluates the farmers' agricultural production before implementing DIS (2006) and after DIS implementation (2007 and 2008), and another comparison presents the difference in irrigational efficiency between the two parts of the gardens (manual and drip-irrigated). In order to estimate the efficiency of the DIS operated by the farmers, water use, land use and plant losses are discussed as environmental indicators, including the social indicators as a contribution to create a complete socio-environmental outlook.

Access to water sources is a crucial factor in agricultural irrigation (Sijali 2001). The local farmers' irrigation merely relies on the surface water sources, which is caused by the gardens' location next to the riverbanks (Fig. 2). Additionally, this is also the reason why the farmers' irrigational practice and water use are closely affected by the seasonal changes in the streamflow (Forster et al. 2007).

The results (Fig. 3) demonstrate that the overall water withdrawal was reduced by DIS implementation, but not that farmers use less water for drip irrigation than for manual irrigation. Although many case studies (Hanson et al. 1996; Shrivastava et al. 1993) demonstrated significant decreases in irrigated water withdrawal when using DIS, the results of this research showed that the farmers used four times as much water to irrigate the DIS part of the garden then the manually irrigated part of the garden. This fact confirms that the farmers base their agricultural production mainly on DIS, supplying the systems with enough water and neglecting the manual irrigation. The research's results are in accordance with Skaggs's research in New Mexico (Skaggs 2001), where it is argued that DIS might increase water consumption, because farmers never allow their plants to go into water deficit.

Taking into account the common farmers' practice to own a garden next to the river banks, there is a possibility that, if the agricultural practice within the NCCP area is not controlled, and if the usage of fertilizers and pesticides, on the one hand, and water withdrawal, on the other hand, starts to increase, it could lead to eutrophication and pollution of the river streams (Andersson et al. 2006). Considering this, the irrigation management and the drip irrigation implementation have to be controlled and monitored.

Although Fig. 4 demonstrates that the average area of a *machamba* increases yearly, the fact that the area of the garden did not change in the period between 2007 and 2008 reveals the influence of the DIS on the agricultural area expansion.

The initial reason for NCCP management to begin with the DIS implementation was to motivate farmers to stop their slash-and-burn practice by assuring efficient food production on an unchanged agricultural field. Regarding the results from 2007 to 2008 (Fig. 5), this NCCP aim was achieved, but the fact that 52% of the farmers still plan to expand their agricultural area in the future (interview data) creates a potential threat to the NCCP's objective. In conclusion, this stagnation of the gardens' expansion can be understood as a temporary occurrence, as a result of which it has to be monitored as an environmental indicator in the future.

The negative correlation between the garden area and the amount of irrigated water used per square metre daily is not fixed in the way that even if the farmer has a big garden, he may not be able to supply enough water to irrigate it. Additionally, this parameter partly confirms the NCCP's initial hypothesis that, if the farmers use water in a more productive way, they do not need to expand their agricultural area.

The explanation that the farmers still do not earn enough to pay for the DIS instalments can be obtained by examining social indicators and farmers' observations in the field. The farmers argued that low production and the lack of access to the market were the main reasons for low earnings, but that they want to continue using the DIS because it helps them to survive dry periods similar to the one in 2008 (interview data).

Another indicator for the DIS efficiency estimation is the percentage of the plant losses. The farmers argued that the reason for the plant failure was only insects, and that the losses under the drip irrigation are smaller than those under the manual irrigation because drip-irrigated plants are bigger, greener and more resistant. These results are also in accordance with Skaggs's findings about DIS efficiency in improving garden productivity (Skaggs 2001). Although the DIS did improve the yield of vegetables in the farmers' gardens, the fact that farmers provide DIS with a better water supply than manual irrigation must not be underestimated.

In summary, DIS is proposed as a good technology for adaptation to climate change by many studies (Karlberg et al. 2007), but for successful adoption it is important to ensure that farmers are able to use it in a correct way. The level of DIS efficiency achieved in NCCP area is illustrated in the pyramid figure (Fig. 6).

The base of the pyramid represents manual irrigation as the most inefficient, and also illustrates that this type of irrigation is the most common within the NCCP area. At the top of the pyramid is DIS efficiency in theory, which represents the level of efficiency achievable when DIS is adopted and used successfully in

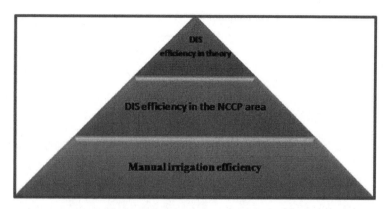

Fig. 6 The drip irrigation system's efficiency in the NCCP area

practice. As the pyramid demonstrates, the theoretical DIS efficiency is hard to achieve in the modest agricultural conditions of rural sub-Saharan Africa, and as such it is located at the top of the pyramid illustration.

Using DIS without fertilizers and pesticides (or any alternative), using more water for DIS than when irrigating manually, not producing enough food in 100% of the households, and not earning enough to pay for the DIS instalments—all these point out that DIS' efficiency is lower than it is in theory. On the other hand, DIS show a higher level of efficiency than manual irrigation does; the plants are bigger, greener and more resistant to insects, DIS implementation caused stagnation in the garden area expansion, the average overall irrigated water consumption has decreased since DIS was implemented, the farmers reported a lower percentage of plant losses in the part of the garden irrigated with DIS than in the manually irrigated part, and farmers are willing to continue to use it.

Including a description of all the previously mentioned indicators in the DIS efficiency analysis, the DIS efficiency of NCCP area is to be found in the middle of the pyramid (Fig. 6).

Success of DIS Implementation in the NCCP Area

Many studies have proven that the theoretical efficiency of new technologies is not frequently achieved, mentioning as the main reason for this a poorly managed implementation process which often disregards the social aspect (FAO 2000; Shaw 1987). The fact that 9 out of 35 farmers gave up on the system before even trying to adopt it shows that those farmers were not ready to take the risk of adopting the new technology. In accordance with this, Marra et al. (2003) also points out that adoption processes are strongly affected by risk-related issues. In addition to this, NCCP managers' experiences in implementing carbon sequestration activities in the area (Van Zyl and Powell 2009—personal interview), confirmed that community members are suspicious and careful when accepting something innovative and foreign, and only after several cases of local success does the tide of acceptance and desire for implementation become apparent.

The overall results regarding the social indicators (e.g. lack of access to the markets, lack of inter-farmer cooperation, and the low socio-economic potential to purchase DIS instalments), emphasize that additional effort has to be dedicated to social work and active farmers' participation in order to inform and improve decision-making processes related to the project. The key factor in the success of DIS adoption among rural farmers and their ability to pay for the yearly instalments of DIS is their capacity to compete in the market (Markelova et al. 2009). In addition to this, Markelova et al. (2009) investigated the ability of rural farmers to access the markets and concluded that collective actions, organized by the farmers themselves, can contribute to their market access to a great extent.

In order to demonstrate the present state and overall success achieved so far in the DIS adoption process, the Adaptive Management Framework (AMF) scheme is used.

Adaptive Management Framework

Although the adaptive management framework was not initially used by the NCCP management when the process of DIS adoption started, it can be discussed now as the appropriate and useful framework to adopt new technology and also as a useful tool to estimate success of the DIS adoption (Lockie and Rockloff 2005).

Normally, AMF consists of five steps, starting the cycle with the *information collation* step, but in this case, there was no basis on which the process of pre-DIS implementation *information collation* was established and the AMF cycle's initial step is the *system analysis and vision* step while presenting the actual process of DIS adoption (by using AMF). In this step, the NCCP managers developed the idea and vision of DIS adoption, while in the following step (*planning*) they developed methods regarding how to distribute the DIS among the local farmers. In the third step (*implementation*), they offered the DIS to the farmers/volunteers (Fig. 7).

Of the total of five steps included in this framework, the process of DIS adoption is currently the third step—*implementation* (Fig. 7a)—and with this socio-environmental research it reaches the fourth step (*monitoring and reviewing*). In the figure (Fig. 7a), the finalized steps of the DIS adoption process are coloured red while the steps which have been initiated with this research are coloured orange.

Taking into account mostly environmental indicators to estimate the efficiency of DIS, and social indicators to estimate how local people think and cooperate among themselves, this points out the necessity for evaluating and improving the strategy of DIS adoption by involving local farmers to a greater extent. The need for the inclusion of social aspects is in accordance with Kusel's findings, where he

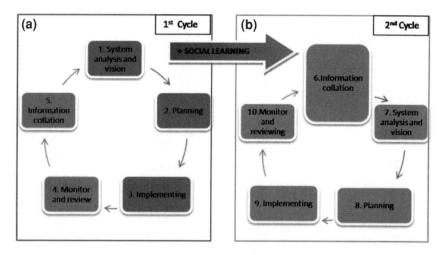

Fig. 7 Adaptive management framework: **a** finished steps in red, steps in process in orange; **b** recommended steps in green

suggests that successful adaptive management requires a public role to be taken up the same level as scientific (Kusel et al. 1996).

The AMF's first cycle (Fig. 7a) is a method of social learning for both parties (NCCP managers and the local farmers) which encourages NCCP managers to collate the information and lessons learned from this process and to decide on further steps in the DIS adoption process. The second part of Fig. 7b, demonstrates the suggested initiation of the new AMF cycle (coloured green), with integration of social learning from the first cycle.

As this NCCP pilot project evaluation emphasizes the call for the inclusion of social aspects, the AMF seems to be a good tool to accomplish successful implementation, due to its potential to integrate social components (Lockie and Rockloff 2005). Although the DIS has already been implemented, it has not been entirely adopted. The adoption strategy has not been previously developed and socio-environmental indicators have not been clearly stated. As a result, it is recommended to start with the second cycle of AMF (Fig. 7b), where all the steps will be included and explained in advance. In order to define the indicators and to achieve more successful decision-making, it is recommended to include the economic indicators to fulfil all three aspects of sustainability when implementing DIS.

Conclusions and Recommendations

It is widely known that suitable practices and techniques for managing climate change should incorporate stakeholder participation and facilitate long-term thinking and a meaningful comparison and integration of environmental, economic and social costs and benefits. In order to address the multifaceted climate challenges the world faces, there is a perceived need for tangible, locally oriented adaptation initiatives with a constant regard of the fact that an integrated agenda for adaptation research is needed (Patwardhan et al. 2009).

The N'hambita Community Carbon Project's initiative to introduce and implement the drip irrigation system (DIS) in the NCCP area is viewed as being appropriate for this area. Due to the high risk of feeling the impact of climate change in the near future (Christensen et al. 2007), DIS adoption in that area is a good measure of climate change adaptation. Nevertheless, the investigated approach of DIS adoption in the NCCP area did not appear completely successful, and has yet to be improved by involving local people's active participation to a greater extent.

As already discussed, the socio-environmental potential exists to implement drip irrigation systems as a climate change adaptation measure within the NCCP area but DIS efficiency did not reach the expected level. The efficiency of DIS irrigation that was investigated among the local farmers is estimated as higher than manual irrigation, but lower than it is in theory. Although the farmers produce more food when using the DIS, they still do not earn enough money to purchase the instalments for the system.

In order to improve local DIS usage, two types of recommendations are offered as an output of this research:

Recommendations for the NCCP management:

- To develop a plan for the DIS adoption process within the NCCP area, by following the AMF
- To include an appropriate methodology when evaluating DIS efficiency (considering for example: soil quality analysis, crops productivity, water availability, usage of irrigated water, etc.)
- To improve the developed AMF by including social learning and learning from previous mistakes
- To consider the social aspect to a higher extent, investigating the local people's needs carefully and their desire to adopt new technology before starting with implementation
- Furthermore, to share the experience from the fieldwork regarding DIS implementation with other carbon sequestration projects

Recommendations for the local farmers in the NCCP area:

- To improve their social networking in order to become less dependent on the NCCP
- To cooperate together, organizing transportation to ensure themselves access to local markets

The overall conclusion of this research points out the high potential of DIS for improving water management within the NCCP area as well as for improving local farmers' production and households' food security. Additionally, if successful in implementing the DIS, NCCP could encourage the same in further carbon sequestration projects, pointing out the difficulties experienced by local farmers in the adoption of the DIS.

Acknowledgments Above all, thanks to the local farmers who were willing to share their experience and information regarding their agricultural activities. Special thanks to NCCP managers Piet van Zyl, Antonio Serra and Philip Powell; local employees Gary Goss and Nhamo Gochococho, and the local forest researcher Jose Carlos Monteiro. A grateful acknowledgement goes to the European Commission for funding this research.

References

Adger WN, Barnett J (2009) Four reasons for concern about adaptation to climate change. Environ Plan A 41:2800–2805

Allen W, Bosch O, Kilvington M, Brown I, Harley D (2001) Monitoring and adaptive management: addressing social and organisational issues to improve information sharing. Nat Resour Forum 25(3):225–233

Andersson L, Hellström SS, Kjellström E, Losjö K, Rummukainen M, Samuelsson P, Wilk J (2006) Modeling report. Climate change impact on water resources in the Pungwe drainage basin, SMHI Reports. Norrköping, Sweden, p 92

Arendal G (2002) Africa environment outlook past, present and future perspectives. http://www.grida.no/publications/other/aeo. Accessed 14 Mar 2009

Barbir J, Leal Filho W, Spiric J (2009) An appraisal of measures to cope with climate change in the Baltic Sea. Manag Environ Qual 20:82–91

Bates BC, Kundzewicz ZW, Wu S, Palutikof JP (2008) Climate change and water, Technical paper of the intergovernmental panel on climate change. vol 210, no 43, IPCC Secretariat, Geneva, pp 34–77

Batisani N, Yarnal B (2009) Rainfall variability and trends in semi-arid Botswana: implications for climate change adaptation policy. Appl Geogr 30(4):483–489

Callaway JM (2004) Adaptation benefits and costs: how important are they in the global policy picture and how can we estimate them? Glob Environ Chang 14:273–284

Christensen JH, Hewitson B, Busuioc A, Chen A, Gao X, Held I, Jones R, Kolli RK, Kwon WT, Laprise R, Magaña Rueda V, Mearns L, Menéndez CG, Räisänen J, Rinke A, Sarr A, Whetton P (2007) Regional climate projections. In: Solomon S, Qin D, Manning M, Chen Z, Marquis M, Averyt KB, Tignor M, Miller HL (eds) Climate change 2007: the physical science basis, contribution of working group I to the fourth assessment report of the intergovernmental panel on climate change, vol 11, no 93. Cambridge University Press, Cambridge, pp 847–940

Doss CR (2001) Designing agricultural technology for African women farmers: lessons from 25 years of experience. World Dev 29(12):2075–2092

FAO (2000) Socio-economic impact of smallholder irrigation development in Zimbabwe: case studies of ten irrigation schemes, FAO-SAFR, Appendix 3. http://www.fao.org/docrep/x5594e/x5594e00.htm. Accessed 1 June 2010

Forster P, Ramaswamy V, Artaxo P, Berntsen T, Betts R, Fahey DW, Haywood J, Lean J, Lowe DC, Myhre G, Nganga J, Prinn R, Raga G, Schulz M, van Dorland R (2007) Changes in atmospheric constituents and in radiative forcing. In: Solomon S, Qin D, Manning M, Chen Z, Marquis M, Averyt KB, Tignor M, Miller HL (eds) Climate change 2007: the physical science basis, contribution of working group I to the fourth assessment report of the intergovernmental panel on climate change, vol 2. Cambridge University Press, Cambridge, pp 129–234

Graham MH, Haynes RJ, Meyer JH (2002) Soil organic matter content and quality: effects of fertilizer applications, burning and trash retention on a long-term sugarcane experiment in South Africa. Soil Biol Biochem 34(1):93–102

Hanson BR, Schwankl LJ, Schulbach KF, Pettygrove GS (1996) A comparison of furrow, surface drip and subsurface drip irrigation on lettuce yield and applied water. Agric Water Manag 33:139–157

Jansen LJM, Bagnoli M, Focacci M (2008) Analysis of land-cover/use change dynamics in Manica province in Mozambique in a period of transition (1990–2004). For Ecol Manag 254: 308–326

Karlberg L, Rockstrom J, Annandale J, Steyn M (2007) Low-cost drip irrigation–a suitable technology for Southern Africa? An example with tomatoes using saline irrigation water. Agric Water Manag 89(1–2):59–70

Kusel J, Doak S, Carpenter S, Sturtevant VE (1996) The role of the public in adaptive ecosystem management. Assessments and scientific basis for management options, Sierra Nevada ecosystem project, Final Report to Congress, Centers for Water and Wildland Resources, vol 2. University of California, Davis, CA, pp 611–624

Le Treut H, Somerville R, Cubasch U, Ding Y, Mauritzen C, Mokssit A, Peterson T, Prather M (2007) Historical overview of climate change. In: Solomon S, Qin D, Manning M, Chen Z, Marquis M, Averyt KB, Tignor M, Miller HL (eds) Climate change 2007: the physical science basis, contribution of working group I to the fourth assessment report of the intergovernmental panel on climate change. Cambridge University Press, Cambridge

Lockie S, Rockloff S (2005) Decision frameworks: assessment of the social aspects of decision frameworks & development of a conceptual model, Technical Report 20, pp 17–28

Maisiri N, Sezanje A, Rockstrom J, Twomlow SJ (2005) On farm evaluation of the effect of low cost drip irrigation on water and crop productivity compared to conventional surface irrigation system. Phys Chem Earth Parts A/B/C 30(11–16):783–791

Markelova H, Meinzen-Dick R, Hellin J, Dohrn S (2009) Collective action for smallholder market access. Food Policy 34:1–7

Marra M, Pannell DJ, Ghadim AA (2003) The economics of risk, uncertainty and learning in the adoption of new agricultural technologies: where are we on the learning curve? Agric Syst 75(2–3):215–234

Osbahr H, Twyman C, Adger WN, Thomas DSG (2008) Effective livelihood adaptation to climate change disturbance: scale dimensions of practice in Mozambique. Geoforum 39(13): 1951–1964

Patt AG, Schroter D (2008) Perception of climate risk in Mozambique: implications for the success of adaptation strategies. Glob Environ Change 18:458–467

Patwardhan A, Downing T, Leary N, Wilbanks T (2009) Towards an integrated agenda for adaptation research: theory, practice and policy: Strategy paper. Curr Opin Environ Sustain 1:219–225

Rockstrom J, Kaumbutho P, Mwalley P, Temesgen M (2001) Conservation farming among smallholder farmers in E. Africa: adapting and adopting innovative land management options. First World Congress on Conservation Agriculture, pp 363–373

Rockstrom J, Barron J, Fox P (2002) Rainwater management for increased productivity among small-holder farmers in drought prone environments. Phys Chem Earth 27:949–959

Sanchez PA (2000) Linking climate change research with food security and poverty reduction in the tropics. Agric Ecosyst Environ 82:371–383

Shah M, Fisher G, van Velthuizen H (2008) Rainfed agriculture, climate change and food security. Report submitted at the FAO expert meeting on climate change, water and food security, Rome, pp 26–28

Shaw AB (1987) Approaches to agricultural technology adoption and consequences of adoption in the third world: a critical review. Geoforum 18(1):1–19

Shields DJ, Solar SV, Martin WE (2002) The role of values and objectives in communicating indicators of sustainability. Ecol Indic 2(1–2):149–160

Shrivastava PK, Parikh MM, Sawani NG, Raman S (1993) Effect of drip irrigation and mulching on tomato yield. Agric Water Manag 25:179–184

Sijali IV (2001) Drip Irrigation: options for smallholder farmers in Eastern and Southern Africa. RELMA Tech Handb 24:16–19

Skaggs RK (2001) Predicting drip irrigation use and adoption in a desert region. Agric Water Manag 51(2):125–142

Solomon S et al. (eds) (2007) Climate change 2007: the physical science basis, contribution of working group I to the fourth assessment report of the intergovernmental panel on climate change. Cambridge University Press, Cambridge, p 996

Stoorvogel JJ, Smaling EMA (1990) Assessment of soil nutrient depletion in sub-Saharan Africa: 1983–2000, nutrient balances per crop and per land use systems, vol 2, Report 28

Stringer LC, Dyer JC, Reed MS, Dougill AJ, Twyman C, Mkwambisi D (2009) Adaptations to climate change, drought and desertification: local insights to enhance policy in Southern Africa. Environ Sci Policy 12:748–765

Tompkins EL (2005) Planning for climate change in small islands: insights from national hurricane preparedness in the Cayman Islands. Glob Environ Change 15:139–149

UNDP (2008) Inter-regional project document: Africa and Arab States regions. http://www.undpadaptation.org/africaprogramme/docs/Climate_Change_Adaptation_in_Africa_Project_Document_EN.pdf. Accessed 7 Sept 2009

UNFCCC (2003) Kyoto protocol. http://unfccc.int/kyoto_protocol/mechanisms/emissions_trading/items/2731.php. Accessed 15 Mar 2009

UNFCCC (2007) Climate change: impacts, vulnerabilities and adaptation in developing countries. UNFCCC Secretariat, Bonn

Usman MT, Reason CJC (2004) Dry spell frequencies and their variability over Southern Africa. Clim Res 26:199–211

Van Zyl P, Powell P (N'hambita Community Carbon Project Managers) (2009) Personal Interview, 30 April 2009

Williams M, Ryan CM, Rees RM, Sambane E, Fernando J, Grace J (2008) Carbon sequestration and biodiversity of re-growing Moimbo Woodlands in Mozambique. For Ecol Manag 254:145–155

World Resource Institute (2000) World resources 2000–2001, people and ecosystems: the fraying web of life. Washington DC

Chapter 41
Social Learning Through Local Water Governance Institutions

Margot Hurlbert

Abstract This paper attempts to answer the question, has the creation of local water governance groups improved water governance through social learning? The study reviews and assesses a sampling of local water governance groups in the three Canadian Prairie provinces Alberta, Saskatchewan and Manitoba, based on their legislation and policy mandate, planning and activities to date in respect to social learning. Important to adaptiveness to climate change are processes of social learning which go beyond production of new and additional knowledge but include reflecting and re-evaluating, building capacity, creating trust, incorporating local knowledge, and internalizing externalities into water governance decisions. The three local water governance groups studied in each of the three Prairie Provinces show positive developments in social learning and improved water governance. There is evidence of new data accumulation, application of existing data and local knowledge to decisions, capacity and trust building, as well as ongoing planning and re-evaluation. This review and comparative analysis provides important insights into the development of policy surrounding the mandates and activities of local watershed advisory groups to improve their effectiveness, increase social learning surrounding their activities and reduce vulnerability to climate change.

Keywords Sustainable development · Water governance · Adaptive water governance · Social learning · Deliberative decision-making

---

M. Hurlbert (✉)
Department of Justice Studies and Department of Sociology
and Social Studies, CL 235, University of Regina,
3737 Wascana Parkway, Regina, SK S4S 0A2, Canada
e-mail: Margot.hurlbert@uregina.ca

Introduction

Issues of water governance have increasingly emerged in the Canadian context because of issues of water quality and contentious quantity decisions arising in the last two decades. Government inquiries into deaths and illnesses from water-quality issues confirmed that the water crisis in Canada is often a crisis of governance, not of natural conditions (Laing 2003). As a result of the reports prepared from these inquiries, changes have been made in water governance and engagement of people in water planning has occurred through the formation of local watershed advisory groups (Diaz et al. 2009).

This governance development is consistent with literature based on building capacity, creating trust, incorporating local knowledge and internalizing externalities into water governance decisions (Brooks 2002; WWCWAU 2003). By encouraging the participation of local people in the decision-making surrounding water, social learning can occur.

Social learning is a process of including the active engagement of various stakeholder groups in decision-making for more than just production of new and additional knowledge but promoting collective perceptions and paradigms about understanding of well-being, happiness and development goals (Pahl-Wostl et al. 2007). In this way governance can be designed as a learning process or a reflexive governance mode (Voss et al. 2006). Two important changes to traditional top-down governance are important. Firstly, reflexive governance takes place on different scales as individuals, organizations and social groups participate; secondly, these different actors take on different and new roles and relationships. Science and government bureaucrats are no longer the revered "experts"; science is central, but both knowledge-generating mechanisms and related action patterns are formed, diffused and institutionalized by different actor groups, including non-governmental organizations, political agencies, the media, etc. (Haas 1992, 2001).

This chapter reviews three local water governance groups in the Canadian Prairie provinces of Saskatchewan, Manitoba and Alberta for evidence of social learning. The legislation, policy mandate, planning and activities to date are examined for activities supporting reflection and re-evaluation, capacity- and trust-building, the incorporation of local knowledge and the internalizing of externalities.

Legislation and Policy Foundation

Several years ago, amendments to water legislation in the three Prairie provinces of Alberta, Saskatchewan and Manitoba were enacted, incorporating the participation of civil society into the water governance framework. In the study region, both Alberta and Saskatchewan have provided for the formation of local water advisory committees on a watershed basis involved in water source protection issues.

The legislation in Alberta specifically provides that water management plans be developed with public consultation (ss. 7, 9 and 11, Water Act). As a result, many river councils have been formed. Membership in these groups is spread among such groups as municipal governments, conservation organizations, environmental groups and other community volunteers. Prior to the establishment of these specific institutions incorporating civil society into ongoing decision-making, Alberta had also produced a water strategy, "Water for Life", through a comprehensive consultation with the public and water stakeholders, which has since been updated or reformulated (Alberta Environment 2003, 2005b).

Saskatchewan has not amended its legislation any further than allowing the pre-existing ability of the Crown Corporation responsible for water to form advisory committees in respect of any issue (s. 19, Saskatchewan Watershed Authority Act). However, the Crown corporation has established many watershed advisory committees in the past several years throughout the province tasked with writing water source protection plans. Manitoba water stewardship initiated local water planning by assigning conservation districts with the task of preparing water plans pursuant to the Water Protection Act. Conservation districts were created in the 1970s with the main objective of soil and water conservation pursuant to the Conservation Districts Act. Utilizing these pre-existing institutions already involved in water conservation, several watershed plans have been initiated and completed (Government of Manitoba 2009). Other Canadian provinces which were not part of this study have gone further and not only provided for these types of civil society participating organizations, but also provided these types of organizations with the ability to have their plans enforced by regulation (s. 65, Water Act, R.S.B.C., s. 23). This study focuses on three groups involved in local watershed planning in each of the three Prairie provinces of Manitoba, Saskatchewan and Alberta.

Because local watershed advisory groups have been in existence for several years, an assessment can be made whether a transformation of water governance has occurred embracing "reflexive governance" in which social learning has happened and exigencies of climate are recorded and acted upon. After a review and comparative analysis of selected groups in each province, based on their legislation and policy mandate, planning and activities to date, this paper makes recommendations for improving social learning in respect of climate change and water stress.

Methodology

This paper is based on information gathered from public documents, legislation, policy and planning documents surrounding water and watershed planning in Alberta, Manitoba and Saskatchewan. The area encompassing southern Alberta and southwestern Saskatchewan is an ideal area to study in relation to water because not only are significant climate change effects anticipated (Sauchyn and Kulshreshtha 2007, p. 291), but the area has experienced significant water shortages historically with the drought of the 1930s and more recently in 1961,

1988 and 2001–2002 (Marchildon et al. 2009). Water from this study region drains into Manitoba and Lake Winnipeg. The watershed advisory groups selected include the Oldman Watershed Council in Alberta, the swift current creek watershed stewards in Saskatchewan, and the Turtle Mountain Conservation District in Manitoba.

Based on a review of the legislation, policy and public documents surrounding these groups and their water plan, content was identified relating to the issue of social learning and specifically the themes of the application and development of local knowledge, the building of local capacity, and finally reflection and re-evaluation in water decision-making. This content was analysed and the findings reported.

This documentary analysis is limited to the selected watershed group in each of the three provinces and so may not be reflective of other watersheds in the province. Further, the study would be improved with qualitative data supporting or expanding on the information contained in the public documents and would be research in sequel.

Research Findings

The local watershed advisory groups reviewed showed high levels of planning, with some evidence of reflection and re-evaluation. Continued support from government as well as development of independent funding through local business and rural municipality participation will be needed in this effort. Further, these groups demonstrated the application of local knowledge to decision-making. For example, there was evidence of the utilization of historical conservation and water-related information, as well as involvement of these groups in the ongoing economic development decisions of the community. These watershed groups were also active in developing and building further local knowledge with data gathering and generation activities. The leveraging of existing civil society organizations built capacity through the activities of the local watershed advisory groups and through their contacts and networks within which they operated. Each of these findings will be discussed in turn.

Local Knowledge

Local knowledge is a key factor in the working and performance of local watershed advisory groups in each of the three Prairie Provinces. Not only was local knowledge amassed and compiled in the plans, new knowledge and data-gathering initiatives were planned and effected, and local knowledge was applied to ongoing water decisions.

Considerable local knowledge was amassed and then applied by each of the watershed advisory groups in each of the three provinces through the preparation of their respective source water and water protection plans (Swift Current 2008a; Turtle Mountain 2006; Alberta Environment 2005a). These plans detail actions required to protect watersheds from sealing wells, monitoring water quality, advancing better agriculture practices for improving water quality, and education. Alberta's watershed group specifically dealt with water quantity issues. The similarities between these plans are that they embody local knowledge about the water resource reflected in ongoing plans. Very specific issues that people living next to and depending on a water resource have vested and contextual knowledge of were captured in this manner.

In addition to this compilation of local knowledge, all the three plans in each respective province include measures to monitor and collect more local knowledge about water quality and trends. In the East Souris river watershed, information respecting surface and groundwater quality in each of the four sub-watersheds is to be collected and to be presented in an Annual Watershed Report Card and five-year State of the Watershed Report Card (Turtle Mountain 2006, at p. 16). In Alberta, in the South Saskatchewan river basin, a strong recommendation is made that performance monitoring of the aquatic environment should occur, including flow monitoring, the tracking and reporting of actual water diversions, water quality (especially in relation to dissolved oxygen and temperature), and assessments of biological communities (Alberta Environment 2005a, p. 16). In Swift Current, Saskatchewan, recommendations for improved reporting of data and use of this data were also made (Swift Current 2008a, p. 35).

Watershed plans also reflected and encouraged specific data important to the specific watershed. For instance, in Saskatchewan, the Swift Current Creek Watershed Monitoring Project was designed to assess the health of the watershed based on a water quality index and riparian health through fish surveys to assess the impact of current management practices of water, agricultural management practices and the impact of the City of Swift Current on water downstream (Swift Current 2008b). In Alberta, a survey of stormwater practices was made with over 900 residents responding (Urban Beneficial Management Practices Team 2004).

In addition to the compilation and documentation of local knowledge and expansion of local water and environmental knowledge, another important theme emerges from the analysis of the local watershed group documentation. This is the incorporation of local community values and knowledge into decisions respecting the use of water, or the internalization of economic externalities (Brooks 2002; Hickey and Mohan 2004). Although models of integrated water management in the literature vary, one common theme in many is the involvement of local people in water decisions affecting them for this very reason (Brooks 2002; Rahaman 2005).

In Alberta, the South Saskatchewan river basin advisory committees were tasked with determining how much water should be left in the rivers for the aquatic environment and complementarily, how much water should be available for consumption (Alberta Environment 2005a, p. 1). This required a determination of the best balance between protection of the aquatic environment and consumption of water for economic growth. From the documentation records, this was no easy task. Regarding the Bow River, a moratorium was recommended on accepting applications for water licences until further information concluded such allocations are reasonable given the intention to protect the aquatic environment. This determination passed by a majority of 73% (Alberta Environment 2005a, p. 6). In other rivers, it appears such decisions were more contentious. For instance in the Oldman river, only a small majority (not consensus) of the basin advisory committee favoured a moratorium on allocation until studies show further allocation is possible. As a result, the committee agreed that irrigation water allocations should be limited (Alberta Environment 2005a, p. 3). This data confirms previous literature that there is not always consensus in community decision-making (Smith 2008; Hillman and Howitt 2008; Few et al. 2006; Gleick 2006), and also confirms that the watershed advisory groups can be an important social institution to make difficult determinations balancing economic growth with environmental limitations.

In Manitoba, the same theme appears. When addressing "Why is a watershed plan important?" it is acknowledged that there are pressures on a watershed due to a growing economy and that the plan ensures the resources in the watershed are managed in a sustainable fashion; further, the plan helps define what is important to the community (Turtle Mountain 2006, p. 4). Thereafter, the plan details activities of maintaining ecological and hydrological functions of the watershed through providing incentives for the conservation of wetlands and promoting the concept of ecological goods and services payments (Turtle Mountain 2006, pp. 29, 31). There is further evidence of this important application of local knowledge in later documentation that reports under the action item "improve drainage licensing system" that the Turtle Mountain Conservation District is commenting on and reviewing applications for drainage licensing requirements for the Manitoba Watershed Stewardship (Government of Manitoba 2009, p. 1). In the future, it is expected that the district will have a greater ability to reform licensing conditions once a surface water management plan is completed (ibid.).

In Saskatchewan, this idea of local community input contributing to delicate decisions of balancing economic development and environmental quality issues appears in the planning documents in a preliminary manner. A goal is to "explore options to develop a project to determine current valuation of ecological goods and services" and to provide extension activities to provide information on the value of ecological goods and services (Swift Current 2008a, p. 19). Notwithstanding the efforts towards valuation of ecological goods and services and the associated tax and grant incentives, ecological endowments are invaluable or rather absolute and efforts are necessary towards eco-upgradation and preservation.

Building Capacity

The documentation of the process of completing a water plan and the end product of these plans confirm the capacity of the watershed advisory groups studied to participate in local water planning. In addition, the partnerships, networking and linkages set out in the documentation allow the conclusion that community capacity was created. Lastly, the documentary evidence provides confirmation that important decisions relating to drainage licences (Manitoba) and water quantity restrictions (Alberta) have been delegated to and made by these water advisory groups.

In the three case studies selected for each of the three provinces, existing groups concerned with water and the environment were utilized for the purpose of preparing the water protection plans. In Alberta, the Bow River Basin Council, which had 140 members at the time, was an active participant in the recommendations prepared along with three other basin advisory councils created to represent the adjoining basins in making recommendations for the plan (Basin Advisory Committees 2004). As a result of this process of consultation, one other basin advisory committee was forming (Oldman River), and another was being recommended (Red Deer). In this way, the initiative of watershed planning created two additional groups in two additional sub-basins to continue with future planning. This shows increasing local capacity for watershed planning.

In Saskatchewan and Manitoba, the groups participating in developing the water plans pre-existed the watershed planning initiative. In Manitoba, the Conservation Districts have existed since the 1970s for the purpose of both land and water conservation. The exercise of watershed planning built on this decadal experience of the Turtle Mountain conservation district and focused the current planning and activities on sustaining a healthy watershed. In addition, the planning process allowed the Conservation District to hold extensive consultation with others in the watershed to elicit support and inputs into the various watershed initiatives (Turtle Mountain 2006, p. 37). Similarly, in Saskatchewan, the pre-existing Swift Current Creek Watershed Stewards brought together other environmental, urban and rural groups in the development of the Swift Current plan. The broad consultation with rural and urban representatives brought the community together such that the chair stated, "Our planning process has indirectly represented each and every citizen in our watershed" (Swift Current 2008a, p. 3). Indeed a case of an inclusivist local and ministration; a lesson to the entire world.

Watershed groups in all the three provinces are very active in networking with their respective communities to build awareness and educate about watershed issues. The goals of each group focus significantly on this. For instance, goals one and three of the Swift Current Creek Watershed are to "educate users on a continuous basis about issues affecting water quality", and "foster an attitude of individual responsibility toward watershed stewardship" (Swift Current Creek Watershed Stewards 2010, p. 1). In Manitoba, Turtle Mountain has a goal to "increase awareness of the community about the benefits of improved watershed management" (Turtle Mountain 2006, p. 18). The Oldman Watershed Council Goals include integrating its activities "through knowledge, research, partnerships and education" (Oldman Watershed Council 2010, p. 1).

Activities listed by each of these organizations in this regard are substantial and referred to in their reporting documents (Swift Current 2008a; Turtle Mountain 2006; Alberta Environment 2005a). In Saskatchewan, the watershed advisory group organizes and holds youth stewardship workshops in order to teach students why the creek is important by covering topics of geography, riparian zones, animal, fish and waterbug habitat, and nitrate and pH in the water. In addition, publications of "Know Your Watershed" are regularly distributed (Swift Current Creek Watershed Stewards 2010). Similar activities occur in Alberta; also, science forums at the university and stakeholder forums with the community entitled "Holding the Reins" are in place (Oldman Watershed Council 2010). Manitoba combines water education with its conservation education in a variety of festivals, school-based learning, interpretative areas, programmes and activities (Government of Manitoba 2009).

The Alberta and Manitoba groups show a great deal of capacity in demonstrating relevance, scope and effect of their respective plans. In Alberta, the water management recommendations specific to flow were implemented in the Alberta Environment Water Management Plan. Recommendations as specific as imposing a moratorium of allocation in the Oldman river, the minimum flows in the South Saskatchewan and the Blindman river were adopted by the government of Alberta in the plan and have been implemented (Basin Advisory Committees 2004, p. 3; Alberta Environment 2005a, p. 2). By effecting such recommendation in policy, the government of Alberta has built trust with this water advisory committee, its members and the community that had provided inputs into the recommendations that their input is both relevant and valued. Eliciting support and participation in future water planning and management should not be impeded but encouraged based on this significant event. This demonstrates capacity to have a highly relevant ongoing presence.

In Manitoba, the Turtle Mountain Conservation district is in a constant process of reviewing drainage licence requests in coordination with the Manitoba Watershed Stewardship (Turtle Mountain Conservation District 2009, p. 1). In addition, the Turtle Mountain conservation district is preparing an East Souris River Surface Water Management Plan and participating in a Pembina River Watershed Management Plan (Government of Manitoba 2009, p. 46). These activities show a high degree of relevance, trust and capacity in the future work of this organization.

Planning, Reflection, and Re-evaluation

All the three groups in each province show a high degree of planning. This is most probably due to the fact that these groups in recent years have been specifically tasked with watershed planning. In Saskatchewan, this initiative's genesis is from water quality issues and illness (Laing 2003); Alberta from water quantity issues (Oldman Watershed Council 2010); and Manitoba from a 2002 initiative to integrate water planning (Turtle Mountain 2006, p. 6). The focus of planning for each watershed group differs. In Saskatchewan, the focus is on source and watershed protection, given the genesis in an outbreak of illness due to water quality. In Alberta, the focus was on quantity of water available to new licence applicants because of water shortages and environmental concerns. In Manitoba, the focus was on integrated watershed planning. The similarity across all of these planning initiatives is the utilization of local people and organizations to participate and plan in respect of water.

As discussed in reviewing local knowledge above, these plans reflect the planners, the local people and organizations, and their knowledge and values. A further question arises as to whether these plans are being monitored, revised and changed as new information, priorities or circumstances arise. Observations can be made based on documentation including the plan and subsequent updates and reports.

In Manitoba, the Turtle Mountain conservation district is tasked with the ongoing monitoring and reporting on the Watershed Plan for the East Souris river watershed. The plan dates back to January 2006. The plan is related to the Conservation Districts of Manitoba annual reports (Government of Manitoba 2009) and budget updates and plan updates (Turtle Mountain Conservation District 2009). These documents detail specific actions which will be taken in future. Information is provided based on each objective and then further detailed with respect to each action.

It is evident some of the initial actions have been reviewed. For instance, one action was to "design and implement a Wetland Rehabilitation Incentive Program" (Turtle Mountain 2006, p. 29). This action is reported as "currently offered by Ducks Unlimited" (Turtle Mountain Conservation District 2009). Subsequently, it is listed that this may be potentially accomplished with Manitoba Watershed Stewardship and as a programme through a government department (Turtle Mountain Conservation District 2009). Clearly, the method of achieving the overarching goal of wetland conservation is being revised over time based on new information and social learning.

Further, the objective of "promoting drinking water source protection" was detailed as accomplished through a conservation agreement programme targeting certain reservoirs, and also incentive programmes for farmyard runoff control, relocation of livestock farms, riparian area management and land management for soils at risk. A future action is listed as requesting assistance from Manitoba Watershed Stewardship to complete a East Souris River Watershed Source Water Protection Plan (Turtle Mountain Conservation District 2009). This supports the conclusion that when initial actions are taken and completed, often further actions are deemed necessary to achieve the goal. This is evidence of social learning—planning based on experience.

In Alberta, the Water Management Plan (Alberta Environment 2005a) was commissioned to assess the issue of water quantity. However, pursuant to the governing legislation, comments could be made on matters and factors to be considered in making decisions about issuing licensing. The Plan details that existing, potential and cumulative effects on the aquatic environment and any applicable interim objective or water conservation must be considered (Alberta Environment 2005a, p. 15). In addition, the issue of water quality was recommended as a topic that merited a more detailed study along the South

Saskatchewan river basin to support further decision-making as to sustaining growth and managing rivers to enhance aquatic life (ibid., p. 16). Although no specific schedule was in place to review the plan and no evidence of having done so could be found, the Oldman Watershed Council is actively pursuing both the creation of a "State of the Watershed Report" and an integrated watershed management plan (Oldman Watershed Council 2010). It is apparent that both planning and implementation are occurring.

In Saskatchewan, there is no evidence of further review and revision of the water plan; however, the plan is dated 2008 but didnt receive an official government announcement until October of 2009 (Government of Saskatchewan 2009). It would appear too premature for a conclusion of reassessment and learning based on this watershed advisory group and plan.

None of the plans of the watershed advisory groups have documented evidence regarding the exploration of climate change, considered and incorporated into their respective watershed plan. Manitoba's documentation is silent as to climate change. Saskatchewan specifically concludes that climate change is still being debated and one recommendation is to explore the effect of climate change on quantity (Swift Current 2008a, pp. 11, 13). The Alberta plan specifically excluded the effects of climate change from consideration due to the "absence of 'definiteness' regarding research conclusions on future water supply and demand that may result from climate change." It was noted that a major study was underway with results anticipated in 2006. However, as mentioned, no further revisions or reassessments of the Alberta plan have occurred (Alberta Environment 2005a, p. 35).

Recommendations

Based upon a review of the documents, the issues of scale or linkages, membership and the role of government should be explored and considered in the future to improve social learning. Because these conclusions are based only on the available public record, it may be that significant initiatives are occurring or being planned in this regard. Further exploration and research of these issues are needed. Further detail of these recommendations follows.

In Alberta, input from four watershed advisory groups contributed to the recommendations which ultimately formed the plan. The relationship of the water management plan to other planning initiatives is clearly outlined in an Appendix (Alberta Environment 2005a, p. 28). However, the need for further work on water quality and additional planning and research, not all of which has been noticed, appears in the recommendations (Basin Advisory Committees 2004, pp. 8–9). The geographical and political aspect of tackling these issues and the actors involved requires careful consideration and action in the future. Further, in Manitoba, the geographical boundaries of Conservation Districts do not necessarily coincide with watershed boundaries. Conservation Districts are the actors tasked with participating and developing the Manitoba plans. Again, the geographical and political issues are pertinent. As these have been identified as a potential hurdle in governance; a careful consideration is advisable (Hurlbert et al. 2009)

Although all three groups constituted a wide variety of people, groups and organizations participating in the development of their respective plans, despite their disparate variety, these stakeholders should continue to cooperate. Utilizing and applying local knowledge in decision-making is only as good as the quality of the local knowledge being accessed. Ensuring all knowledges are accessed, including those of First Nations, Inuit and Métis, in future plans and executions is important. Often this important connection is missed in relation to community decision-making (Hurlbert 2009).

Although the seeking and documentation of data is an important achievement of all the three watershed advisory groups, the role of government in assisting and coordinating this process cannot be ignored. Data collection and data management systems are not only an indicator of a healthy institutional system; they are also the fundamental components of informational capital (which is an important determinant of adaptive capacity). As relevant as other forms of capital—economic, social and human—informational capital contributes to a better knowledge of the existing resources, facilitating their management in situations of uncertainty and surprise. The existence of a large accumulation and good use of information capital is a must in ensuring the social and economic sustainability of livelihoods of productive communities. This data will inform the local water groups while re-evaluating and planning for the future. The role of government in managing this information should not be ignored.

Assisting the local water groups with collecting and processing information, and sharing it with other actors, is crucial. This assistance in data management improves social learning about the impact of climate events and adaptation measures, including how particular policies and programmes are being applied in response to these climate events in order to reduce vulnerability. This review should include not just the plans and policy of the local water groups, but also the plans and policy of the government.

The development of flexible or adaptive policies, those "that can foresee and respond to an array of conditions that lie ahead, and can navigate towards successful outcomes when surprised by the unforeseen", is an important requirement in reducing vulnerability to climate change (Venema and Drexhage 2009, p. 1). Rigid policies that are unable to cope efficiently with the uncertainties and dynamics of new climate conditions are obviously a serious obstacle that has to be overcome. Adaptive policies, to be effective, must be supported by a public institutional system that is able to learn constantly from those other systems with which it interacts. Institutional learning increases the capacity of public organizations to learn from experience, and change trajectories and practices as required. Implicit in institutional learning is a preparedness to experiment, preventing rigid persistence and purposefulness of practice (Goodin 1996). It is because of this that the informational flows of local knowledge, data collections and communication to government and other actors within the system of local water advisory groups is important for social learning and for reducing vulnerability to climate change. Functionaries at all levels must attitudinally get to cope with this requirement.

Conclusions

Based on the assessment of local watershed groups in the three Prairie Provinces, experiences of social learning occurring in these three groups is satisfactory. All the three groups show a marked tendency to incorporate local knowledge into their plans and activities. Measures important to their respective watersheds and their protection are reflected in each plan. Further, the local watershed groups are able to identify important data gaps and initiate steps to collect and monitor this data. In addition, there is evidence that the groups are participating in important ways in the application of local knowledge to decisions determining the delicate balance between strain on environmental water resources and economic development.

Capturing these data or information flows and processes has only occurred through the application of watershed advisory groups in the past several years (although some of these activities occurred in other fora) so the historical record is only approximately five years old. Because of this reassessment or revisiting of decisions, this data or information is just emerging. There are important provisions in the Alberta plan for revisiting some of the decisions made based on new or updated information; similarly, the documentation in Manitoba shows that action items are continuously monitored and revised.

All the three provinces' watershed groups manifest a wide participation of community organizations in the preparation and development of the plan and in the educational activities of the group every year. This shows that learning is occurring within each watershed advisory community. Especially important is that the activities of each group focus on youth, the contributing members to future watershed plans, decisions and programmes. Further research on the effectiveness of these programmes and sharing of research findings specific to these programmes and their optimization would be beneficial.

It is evident from the documentation that a repetitive review of decisions and action items is occurring. In Manitoba, the Turtle Mountain Conservation District is commenting on the issuance of drainage licences and in the future, it is anticipated they will have a greater ability to comment when the East Souris River Surface Water Management Plan is completed (Turtle Mountain Conservation District 2009, p. 1). This shows a disciplined and informed manner of considering and reviewing an important watershed decision, all at the community level.

In Alberta, the specific water management recommendations made by the water advisory committees with respect to streamflow were implemented in the water management plan of Alberta Environment. The significance of this is the trust which is built with the water advisory groups that their voluntarism, time and commitment is valued and worthwhile. Continued research of this process combined with additional methods of data-gathering such as qualitative interviews with water advisory group members and government personnel to explore this interaction and social learning would be beneficial. The Alberta documentation reflected a desire to have more inputs towards managing water quantity; it is unclear whether this has occurred.

Lastly, all the three groups in the three Prairie provinces make provision for uncertainties in climate. Excess water and resultant drainage issues (in Manitoba) and lack of water and scarcity resulting in quantity issues (in Alberta) are predominant themes. However, the lack of evidence of planning and adapting for climate change is indeed a concern. Research into whether the plans and activities are taking into account and preparing for the exigent events of climate change would be beneficial. Again, other methods of data-gathering would be required to explore this fully.

This review of social learning occurring in relation to the watershed advisory committees in the three Prairie provinces confirms that social learning is occurring. Data is being gathered, collated and applied by these groups in relation to important water issues and decisions. Capacity is being developed within each watershed advisory group and its community in relation to making decisions or providing input into decisions respecting water. Lastly, there is evidence that these groups are making decisions and reviewing decision-making factors and information in an ongoing fashion; watershed plans are not stagnating from lack of relevance and usage.

References

Alberta Environment (2005a) DRAFT (Approved) water management plan for the South Saskatchewan River Basin in Alberta, 18 October 2005. http://www3.gov.ab.ca/env/water/regions/ssrb/pdf/Draft_SSRB_Plan.pdf

Alberta Environment (2005b) Enabling partnerships: a framework in support of water for life. Alberta Environment, Edmonton

Alberta Environment, Water for Life (2003) Alberta's strategy for sustainability, Edmonton, Alberta. http://www.waterforlife.gov.ab.ca/html/background2.html, p 32

Basin Advisory Committees for the Oldman River, Red Deer River, Bow River and South Saskatchewan (sub-basin) River (2004) South Saskatchewan River Basin, Water Management Recommendations. A Report to Alberta Environment, Alberta

Brooks DB (2002) Water, Local-Level Management, International Development Research Centre, Ottawa

Conservation Districts Act, The, C.C.S.M., c. 175

Diaz H, Hurlbert M, Warren J, Corkal D (2009) Saskatchewan Water Governance Assessment Final Report, Unit 1E Institutional Adaptation to Climate Change Project (cited September 4). www.parc.ca/mcri/pdfs/papers

Few R, Brown K, Tompkins EL (2006) Public participation and climate change adaptation. Tyndall Centre for Climate Change Research Working Paper 95. University of East Anglia, Norwich

Gleick P (2006) The world's water, 2006–2007, biannual report on the world's freshwater resources. Island Press, Washington

Goodin RE (1996) Institutions and their design. In: Goodin R (ed) Theory of institutional design. Cambridge University Press, Cambridge

Government of Manitoba (2009) Conservation Districts of Manitoba 2008–2009 Annual Report, Manitoba. http://www.gov.mb.ca/waterstewardship/agencies/cd/cd_annual_report2008-2009.pdf

Government of Saskatchewan (2009) News release, 23 October 2009. www.swa.ca and www.gov.sk.ca/news?newsId=00b17ebf-4e02-4b2d-albb-b272d5681770

Haas PM (1992) Banning chlorofluorocarbons: epistemic community efforts to protect stratospheric ozone. Int Organ 46:187–224

Haas PM (2001) Policy knowledge: epistemic communities. In: Smelser NJ, Baltes PB (eds) International encyclopedia of the social and behavioural sciences. Pergamon, Amsterdam, pp 11578–11586

Hickey S, Mohan G (2004) Towards participation as transformation: critical themes and challenges. In: Hickey S, Mohan G (eds) Participation: from tyranny to transformation? Exploring new approaches to participation in development. Zed Books, London

Hillman M, Howitt R (2008) Institutional change in natural resource management in New South Wales, Australia, sustaining capacity and justice. Local Environ 13(1):55–66

Hurlbert M (2009) Integrating climate change adaptation into the law. Retfaerd Argang 32(3):23–39

Hurlbert M, Corkal D, Diaz H, Warren J (2009) Climate change and water governance in Saskatchewan, Canada. Int J Clim Chang Strateg Manag 1(2):118–132

Laing R (2003) Report of the commission of inquiry into matters related to the safety of the public drinking water in the city of North Battleford, Saskatchewan, Regina: office of the Queen's Printer

Marchildon G, Pittman J, Sauchyn D (2009) The dry belt and changing aridity in the Palliser triangle, 1895–2000. Prairie Forum 23(1):31–44

Oldman Watershed Council (2010) Goals and Integrated water plan. http://www.oldmanbasin.org/our_goals.html and http://www.oldmanbasin.org/integrated_team.html

Pahl-Wostl C, Craps M, Dewulf A, Mostert E, Tabara D, Taillieu T (2007) Social learning and water resources management. Ecol Soc 12(2), Article 5. www.ecologyandsociety.org/vol12/iss2/art5/

Rahaman MM, Varis O (2005) Integrated water resources management: evolution, prospects and future challenges. Sustain Sci Pract Policy 1:15–21

Saskatchewan Watershed Authority Act (2005) S. S., c. S-35.03

Sauchyn D, Kulshreshtha S (2007) Chapter 7—Prairies. In: Lemmen DS, Warren FJ, Lacroix J, Bush E (eds) From impacts to adaptation: Canada in a changing climate. Natural Resources Canada, Ottawa, pp 275–328

Smith JL (2008) A critical appreciation of the 'bottom-up' approach to sustainable water management: embracing complexity rather than desirability. Local Environ 13(4):353–366

Swift Current (2008a) Swift current creek watershed protection plan. http://www.sccws.com/sccws2008/Welcome/Swift%20Current%20Creek%20Watershed%20Protection%20Plan%20%28FINAL%29.pdf

Swift Current (2008b) Swift current creek watershed monitoring project final report. http://www.sccws.com/sccws2008/Monitoring_Project.html

Swift Current Creek Watershed Stewards (2010) Goals. http://www.sccws.com/sccws2008/index.html

Turtle Mountain (2006) Taking care of our watershed, a watershed plan for the East Souris River watershed, Manitoba, Canada, January 2006

Turtle Mountain Conservation District (2009) Budget presentation 2009–2010. http://www.tmcd.ca/documents/ESR_Update.pdf

Urban Beneficial Management Practices Team (2004) Report. http://www.oldmanbasin.org/stormwater.html

Venema H, Drexhage J (2009) The need for adaptive policies. In: Swanson D, Bhadwal S (eds) Creating adaptive policies. A guide for policy-making in an uncertain world. Sage, New Delhi. www.idrc.ca/openbooks

Voss P-J, Bauknecht D, Kemp R (eds) (2006) Reflexive governance for sustainable development. Edward Elgar, Cheltenham

Water Act (1996) The, R.S.B.C., c. 483

Water Act (2000) The, R.S.A., c. W-3

Water Protection Act, The C.C.S.M. c. W-65

(WWCWAU) World Water Council Water Action Unit (2003) World water actions, making water flow for all. Earthscan Publications Ltd, London

Chapter 42
These are Our Water Pipes—Sand Dams, Women and Donkeys: Dealing with Water Scarcity in Kenya's Arid and Semi-Arid Lands

Abigail Cruickshank and Velma I. Grover

Abstract The uncertainty and intensity of rainfall, and the high evaporation rates in Africa's arid and semi-arid lands are making traditional water sources undependable. Sand Dams represent a development and adaptation strategy that is increasingly being used to combat this growing water insecurity. When operationalized, these small-scale structures increase the size of the riverbed reservoir, thereby allowing more water to be captured and stored during the rainy season. Additionally, a developed Sand Dam is argued to improve the quality of water and build resilience through enabling ecological restoration, diversified livelihoods and the creation of socioeconomic capital. The first section of this paper focuses on how the Sand Dam strategy brings added advantages by reviewing the literature and evaluating the effectiveness of Sand Dams as a development and adaptation strategy. The second section presents findings from five Sand Dam projects in Kenya and explores which social factors and processes are contributing to the success of projects. The case studies emphasize the importance of community cohesion and drive, capacity building and follow-up from the implementing institutions. It has been noticed that these social factors are equal in importance to the proper construction and technical considerations of the concrete structure and are in fact the key factors ensuring that a Dam matures into a proper functioning Sand Dam.

A. Cruickshank (✉) · V. I. Grover
Abigail Cruickshank, 24048 109th Ave, Maple Ridge,
BC V2W 1Z4, Canada
e-mail: abcruicks@gmail.com

Introduction

The fourth Intergovernmental Panel on Climate Change (IPCC) report (2007) states that 25% of Africa's population (200 million people) are currently experiencing high water stress and that by 2020, between 75 and 250 million people are projected to be exposed to an increased water stress due to climate change. Among the areas most affected will be the arid and semi-arid lands (ASALs), which account for more than two-thirds of the African landmass. These zones are extremely susceptible and vulnerable to water shortages and drought, receive an annual rainfall of below 700 mm, and support a population of over 400 million Africans (Darkoh 1993; IISD 2007). Climate change aggravates the existing problems of poverty, water and food insecurity.

Coping mechanisms for dealing with drought and uncertain rainfall are failing, as rains have become more unpredictable and drought more frequent. The Government of Kenya (GoK) (2009) explains that in the past, droughts used to occur every 10 years; now the frequency is every five years. Traditionally used water sources such as scoopholes in dry riverbeds and hand-dug shallow wells are drying. Excellent Development (2007), an NGO dealing with water security in Kenya, states: "during drought women can spend 5–8 h a day to collect 20 l of dirty water, leaving little time for other activities".

Guided by the premise that water security and water conservation needs to be a priority in any successful development initiative in ASALs, this study investigates the potential of Sand Dams as a strategy to increase the adaptive capacity of rural communities. This is accomplished by assessing the effectiveness of Sand Dams as an adaptation strategy and by evaluating five Sand Dam projects in Kenya to determine which factors have contributed to successful Sand Dam projects.

Sand Dams as Effective Adaptation Strategies in Africa's ASALs

The lack of perennial rivers, the combination of brief intensive rainfall events, excessive runoff and high evaporation in Africa's ASALSs culminate in water insecurity. Typically, during the rainy season, rivers have a surface flow for only short periods after intense rainfall (Mutiso n.d.). The flow appears as a flash-flood event, leaving local communities to access the remaining river baseflow by digging deep scoopholes in the riverbed (Nissen-Petersen 2006). Under climate change, even riverbed aquifers are drying up during prolonged drought (Mutiso n.d.). A Sand Dam is an impervious structure built across an ephemeral stream, which blocks the loss of water downstream, providing more retention time for water to infiltrate (Fig. 1). A mature Sand Dam is one in

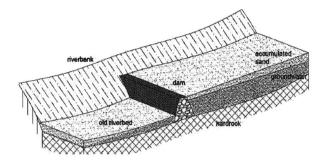

Fig. 1 Schematic cross-section of a typical Sand Dam (Borst and De Haas 2006)

which the Dam has filled with Sand, thereby increasing the storage capacity of the riverbed.

Sand Dams are being advocated as effective climate change adaptation strategies for arid and semi-arid environments in developing countries. The low cost and simple technology is promoted as being easily transferable and accessible to rural communities which have limited financial resources but have access to labour and local materials (Sand, rocks and water). The technology of harvesting and storing water below ground is deemed appropriate for an environment that brings heavy rainfall that quickly disappears through runoff and with the high evaporation. In addition, the strategy's participatory nature is viewed as a building block, facilitating collective action and subsequent development activities.

Case Study Area: Kitui and Makueni Districts, Kenya

The study focussed on five Sand Dams in Kenya's Eastern province, Kenya's second largest province. It borders Ethiopia to the north, Northeastern and Coastal provinces to the east, and Central and Rift Valley provinces to the west. The principal focus of this research is on the arid and semi-arid districts of Makueni and Kitui, located approximately 150 km east of Nairobi (Fig. 2). Eastern province is home to some of the poorest areas in Kenya, and has more than 58% of its population living below the poverty line. Significantly higher than the province's average, Makueni and Kitui districts have 70 and 69% of their populations living in poverty. In addition, in both districts the majority of the population live in rural areas (90 and 82%, respectively), which have a higher percentage of people living below the poverty line than in urban areas. Rural dwellers are dependent on small-scale farming and pastoralism for livelihood, which combined with low and erratic rainfall creates a population vulnerable to water scarcity (Few et al. 2006; Ministry of Planning, National

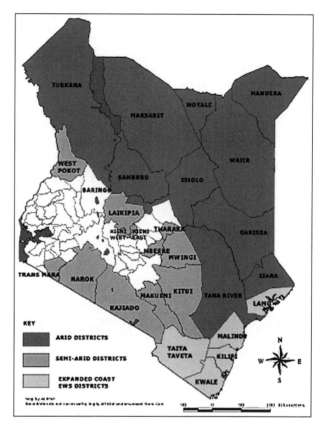

Fig. 2 Location of Makueni and Kitui districts within Kenya's arid and semi-arid lands (source: www.aridland.go.ke)

Development 2002a, b). These two districts also represent areas that lack reliable potable water sources and experience high levels of food insecurity, creating vulnerability that would benefit from the effective implementation of adaptation strategies. In fact, both areas have frequent emergencies due to crop failures and food shortages.

The climate in all five case study sites is semi-arid, characterized by high temperatures and low erratic rainfall. On average, the temperature in Makueni district is 22.1°C, while Kitui's is slightly higher at 24°C (Borst and De Haas 2006; Ministry of Planning, National Development 2002a, b). The high temperatures result in high potential evaporation rates, which exceed rainfall and cause scarce surface water to dry up quickly. Rainfall in both districts varies with altitude and topography and is bimodal. In the upland areas, rainfall can range from 800–1200 mm, whereas the lowland areas receive less than 730 mm of rain per year (Ministry of Planning, National Development 2002a, b).

The rains fall in October to January and March to May. In all the study sites, precipitation occurs as intense storms causing excessive runoff, little infiltration and massive losses of water downstream (Mutiso n.d.). Drought is a recurring phenomenon in semi-arid climates, as it is common for rains to fail. On average, both districts experience drought once every 4–5 years (MDK 2005 in Lasage et al. 2008). Kenya is considered a water-scarce country with surface water covering only 2% of the landmass and an annual availability of 647 m^3 of renewable freshwater per capita (UN-Water, Water/Africa 2003; UNEP 2006). By 2020, this availability is expected to drop to 359 m^3 per capita due to population growth (UNEP 2006). While 68% of the urban population have access to safe water, only 49% have access in rural settlements (UNEP 2006). The percentages are even lower in Makueni and Kitui districts, with 10 and 6% of households, respectively, having access to potable water (Ministry of State for Planning, National Development and Vision 2030 2008; District Commissioner of Kitui, 2002 in Lasage et al. 2008).

In both districts, the climate and vegetation have been altered by the clearing of land for agriculture and settlement and by the felling of trees for firewood and charcoal production. Fuelwood and charcoal account for over 66% of Kenya's energy production. Kitui district produces over 300,000 bags of charcoal annually (FAO Kenya 2007). Gazetted evergreen and mist forests currently account of only 2.2 and 1% of Makueni and Kitui districts' respective total area (Ministry of Development and Planning 2002a).

Methodology

The investigation was a case study approach, focusing on five Sand Dams in Kenya's Eastern province. Primary data was collected through household questionnaires, focus group discussions, participant observations and formal semi-structured interviews with project leaders, community development workers, local and district administration, farmers and community groups. Four of the five Sand Dams had already been constructed and have been operating for various lengths of time. The fifth Sand Dam was still in the construction phase of the project (winter 2008). Two different institutions located in two of Kenya's semi-arid districts completed the five Sand Dams. The two institutions were selected as they were addressing water scarcity through the construction of Sand Dams.

The two institutions implementing the Sand Dam projects in the case studies were (1) the Arid Lands Resource Management Project (ALRMP) and the Centre for Science Technology Innovations implementing the "Increasing community resilience to drought in Makueni district" (Resilience) pilot project[1]; and (2) Sahelian Solutions (SASOL),[2] a Kenyan NGO based in Kitui district, Kenya. The Resilience pilot project sought to address community vulnerability to drought through prioritizing several goals simultaneously. While Sand Dams were a priority for improving water security, it formed only one of the main goals of the project. In contrast, SASOL, which overall also seeks to reduce vulnerability to drought, does so through using water as the main platform. The belief is that the Sand Dam projects provide the necessary access to water for other development

---

[1] "Increasing Community Resilience to Drought in Makueni District" Adaptation Project, Makueni District, Kenya

The regional project "Integrating Vulnerability and Adaptation to Climate Change into Sustainable Development Policy Planning and Implementation in Southern and Eastern Africa" was initiated by the United Nations Environment Programme (UNEP) in 2006. Through pilot projects undertaken in Kenya, Mozambique and Rwanda, which focused on priority areas such as water and agriculture, the regional project expected to build the capacity of each country to generate and use climate information and to implement adaptation strategies in the field as well as to integrate climate knowledge into policies and plans. In Kenya, the "Increasing community resilience to drought in Makueni District" pilot was a 3-year project focusing on increasing household food security through increased livelihood resilience and reduced vulnerability to drought, on reducing poverty through improved livelihoods, and facilitating the integration of climate change and adaptation into policy development and planning.

Sakai sub-location in Makueni District was selected for the pilot project. Sakai is situated 20 km from one of Makueni's major centres, Wote. It is home to approximately 750 households and a population of 4,800 (Resilience project 2008). The current study focused on the three villages in Sakia where the pilot project was to build Sand Dams. The first Sand Dam constructed, Kwa Dison, is located in Muiu village, along Lumu stream (GoK 2008). The second one, Kwa Mutingu, is located in Nthongoni village, along Mutingu stream. Kwa Mutingu was not completed and poorly constructed. The construction of the third Sand Dam, Kwa Ndeto, in Kathamba village, along Mutomo stream, has not begun. The Sand Dam committee is active and has prepared the area three times, as the cement did not arrive in time after the first and second attempts at digging the trench, resulting in it being washed away by the rains.

[2] SASOL's projects, Kitui District, Kenya

Sahelian Solutions, or SASOL Foundation, is a local Kenyan NGO that has been operating in ASALs since 1990. The NGO focuses on dealing with drought mitigation and capacity building and seeks to address household and production water scarcity through Sand Dam technology. To date, SASOL has built over 500 Sand Dams in Kitui District, Kenya, which is the highest concentration of Sand Dams around the world. Today, they operate under three principles. The first principle outlines their belief that although the technology is simple, it has enormous social organization potential. The second principle is to focus on "green" water instead of the predominant focus on "blue" water. Sand Dams act to enhance the system as opposed to wells, which are purely extractive. The third principle expresses the need for the communities to be involved in the whole process. SASOL provides the community with the material and knowledge but requires that the demand, drive and labour all come from the community. Two randomly selected SASOL Sand Dam projects were visited. The projects were the Musingu Sand Dam along the Muvuko river, in Mutha division, Kanziku location, Keutunda sub-location and Kwa Ndunda Sand Dam along the Kiindu River, Kitui Central division, Mulango location, Kyangunga sub-location, Kitunda village. The second location is approximately 15 km from Kitui town, which is the district's capital, while the first is located approximately 25 km from Mutomo.

activities and they also help to build social capital, which can then be utilized to accomplish other initiatives (Mutiso and Mutiso n.d.).

Evaluation of Sand Dams as an Adaptation Strategy

This section will evaluate the use of Sand Dams as an adaptation strategy. Combining characteristics outlined by De Loë et al. (2001) with those from the literature led to six evaluation criteria selected as exemplifying appropriate adaptation strategies in the water sector (Burton et al. 1998; Smit et al. 2000; Smit and Pilifosova 2001). As such, effective adaptations should be: no regret measures; improve equitable access to freshwater; reduce socioeconomic vulnerabilities; minimize environmental impacts; be cost-effective; and feasible, both in the sense of feasibility to implement and long-term sustainability.

No Regrets Measures

De Loe et al. (2001) used the criterion "no regret measures" to illustrate that planned adaptations should offer additional benefits and be effective strategies under current climate conditions. Many water supply strategies, such as shallow wells and boreholes, are extractive. These methods rely on the continual withdrawal of groundwater, depleting the aquifer and causing the source to become dry. In contrast, rainwater harvesting (RWH) methods, such as Sand Dams, seek to enhance the amount of water available. These methods catch and store rainwater, preventing the waste of water downstream.

A typical mature Sand Dam, with a height of 1–4 m above the surface, can have a stored volume of 100,000–50,000,000 l and can be fully recharged from a single flash flood (Ertsen et al. 2005). The Sand also acts to protect the water from evaporation. Water that is 60 cm or deeper below the surface experiences virtually zero evaporation (Nissen-Petersen 2006). Lasage et al. (2008) estimated that the potential evaporation in Kitui district is 1,500–1,600 mm/year, far exceeding the average precipitation for the same area of 1,000 mm/year. The recharge rate and protected storage capacity of Sand Dams make this strategy suitable for dealing with current climate typical in ASALs, as well as future scenarios under climate change.

Improve Equitable Access to Freshwater

With SASOL's high concentration of Sand Dams in Kitui district, over 120,000 people are reported to have improved access to water (Borst and De Haas 2006). Sand Dams do increase the amount of available water and perform optimally if the

Dam is properly managed and it reaches maturity. A Sand Dam can take 5–7 years to mature, i.e. for the reservoir to fill with Sand (Mutiso and Mutiso n.d.). Within this time period, catchment degradation and Sand harvesting in the riverbed needs to be properly managed. Many constructed Sand Dams have not realized their potential, as they are experiencing siltation caused by catchment degradation and erosion (J. Nakiyu, personal communication, 15 April 2009). However, to ensure equity of access, Sand Dam projects need to consider which households are the intended beneficiaries, how far the households are from the source, and what is the means of transporting water to the household. The strategy's ability to influence equitable access to freshwater is ultimately dependent on project planning, implementation and management.

For both institutions the point of entry was through the local administration. As a result, the group of households designated to be involved in the project was characterized by political boundaries. Unfortunately, Sand Dam placement is not always amenable or equitable within this predetermined community in a specific political boundary. For many members in the case study projects, the Sand Dam proved too far to be a useful source of water. In one community, nearly a third of households interviewed felt that the Sand Dam was located too far away to be beneficial.

Sand Dam projects are cost-effective because they rely on community participation in the form of labour, but households that cannot view the benefits are unlikely to participate. Participation further impacts equity as the majority of Sand Dam committees interviewed intended to restrict access to only those who participated in the construction.

In reality, the committees may or may not be able to put this into practice. To ensure that communities benefit from the natural percolation provided by the Sand, water should be accessed from the Dam by either a shallow well on the bank or a pipe protruding on the downstream side of the Dam (Fig.3). If collection was restricted to these methods, then committees could exercise some control on access. However, there are no sure measures to stop members from accessing the stored water, as anyone could dig a scoophole in the riverbed. Equitable access is also determined by the means with which households transport the water to their homesteads. Those that have other means of transporting water back to their homesteads, such as with a donkey or cart, can collect more water (Kitundu respondent, personal communication, 17 February 2009). Only one committee had capped the amount of water each household could withdraw and to avoid congestion implemented a schedule outlining when each household could access the Dam.

On a larger scale, equitable access needs to take into consideration the entire catchment. Each Sand Dam project should evaluate the potential of negatively impacting downstream users. A study completed by Borst and Haas (2006) indicated that a single Sand Dam may reduce the flow downstream by 2–3%. Ersten and Hut (2009) cautioned that these findings corresponded to Kitui district's upland areas, which experience higher rainfall than Kitui's lowlands areas. Sand Dams in these areas might collect a higher percentage of the total annual runoff.

Fig. 3 Option for accessing water from a Sand Dam. A shallow well can be placed on the bank (*top*) or a pipe can be fitted through the Sand Dam (*bottom*) (Protos 2006)

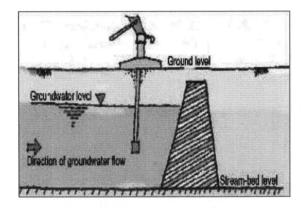
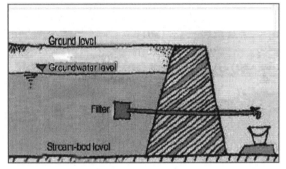

Ersten and Hut (2009) found that analysing the entire watershed findings might indicate that the water being lost to downstream users was not of beneficial use, because the short fast runoff is not harvested or because the runoff causes flat areas to be inundated.

Reduce Socio-Economic Vulnerabilities

The increased availability and the improved proximity of water through the construction of Sand Dams have had a positive effect on community social and economic vulnerabilities. SASOL found that "households on land adjacent to the regenerated rivers are now earning more than 100,000 Ksh from bucket irrigation during the three dry months of August, September and October" (Mutiso and Mutiso n.d.). More available water contributes to increased agricultural output, allowing farmers to grow more vegetables and build larger herds of healthier livestock. A study comparing changes in income over 10 years in Kiindu and Koma catchments in Kenya showed that income increased by 9,000 Ksh for households in Kiindu catchment, which had three Sand Dams (Lasage et al. 2008). The same study found that the percentage of irrigated crops (bucket irrigation) had risen from 37 to 68% in the catchment with the Sand Dams (Lasage et al. 2008). Other research has noted that communities were able to diversify crop production thus decreasing their susceptibility to a total crop failure (Manzi 2005). Results from the present study indicated that 41% of households in the community with the oldest Sand Dam were able to grow more vegetables as a direct result of the Sand Dam and 19% believed there had been an increase in crop production. In the four case study communities where the Sand Dams were complete, households noted livestock were healthier and grew faster as they were not walking long distances in search of water. As a result of Sand Dams, farmers in Kenya have been able to collect more milk and sell livestock for higher prices (F. Ndeto, personal communication, 31 March 2009). The benefits garnered by livestock can quickly result in community vulnerabilities if access of livestock to Sand Dams is not adequately managed. Livestock directly accessing the baseflow through traditional scoopholes will also defecate and urinate, thereby contaminating water collected downstream. The large numbers of cattle entering the riverbed might also exacerbate problems of erosion. It is therefore necessary that communities devise measures to reduce contamination and erosion such as by building watering troughs adjacent to the Dam.

The shortened distance and increased supply enables women to save a considerable amount of time normally reserved for fetching water. Remple et al. (2005) found that, on average, Sand Dams with water available saved households 2.4 h per trip. As a result, women are able to spend more time on household chores, agricultural activities, and other income-generating enterprises. Lasage et al. (2008) found that households in the Kiindu catchment had increased their domestic and agricultural water consumption by 50% within ten years after their Sand Dam had been constructed. The increased use of domestic water allows for improved hygiene practices such as more frequent washing of hands, body and clothes. A focus group participant in the current study explained that before their Sand Dam, her children used to go for days without washing or laundering their clothes but now she can clean them every day (Kitunda focus group participant, personal communication, 16 March 2009). This same participant claimed that there has been a reduction in head and bed lice since the Sand Dam, probably due to more frequent washing. The higher washing frequency decreases the incidences of skin infections and diarrhoea. Remple et al. (2005) noted that the medical director of a clinic in Kitui district where Sand Dams have been built attributed the decrease in diarrhoea and skin ailments to increased water availability. The Sand also acts as a natural filtration system, disinfecting the water people will use for domestic purposes (RAIN Foundation and Acacia Waterstop 2007). Vegetable production, often cited as a benefit of Sand Dams, will also boost household well-being and nutrition through better access to produce in local markets and omitting the cost of travelling to larger urban markets.

Household economic standing increases as time saved allows women to implement better agricultural practices and to focus on other income-generating activities. Manzi (2005) found that a production increase of approximately 20% for maize, beans and legumes could be directly attributed to the time saved and water availability from the local Sand Dam. Diversification of livelihoods is facilitated not only by time, but also through the Sand Dam itself. The improved water tables and necessary erosion control measures lead the community to grow grasses along the riverbanks. The grasses can be sustainably harvested and sold or used to feed livestock. Excess Sand can also be collected and used in construction or the making of bricks. One focus group participant described that before the Sand Dam there were few permanent buildings in the village but now there are many (Kitunda focus group participant, personal communication, 16 March 2009).

Immature Sand Dams can create community vulnerabilities. If Sand hasn't yet filled the reservoir, surface water can collect behind the Dam, creating a pool of stagnant water. This presents a hazard to children who are susceptible to drowning and to the community as exposed surface water is predisposed to contamination and provides of a suitable breeding ground for mosquitoes, resulting in more frequent outbreaks of malaria (ALRMP n.d.). This can be avoided by constructing the Dam in stages, thereby reducing the depth of surface water.

Socioeconomic vulnerabilities can also be reduced through the creation of social capital. The participation involved in Sand Dam projects creates social capital, which can be harnessed to complete other development projects. The extension of group activities directly attributed to the construction of the Sand Dam was difficult to assess. Three of the five Sand Dam committees evaluated stated that other initiatives were started because the project showed the community the power of uniting. The community with the oldest Sand Dam had definitively carried out other ventures such road maintenance and started other livelihood activities such as tree nurseries and beekeeping. These developments can, however, at least partially be attributed to training sessions offered by SASOL and the Kitui Development Centre (KDC). Remple et al. (2005), in their review of 30 of SASOL's Sand Dam projects, found limited evidence of collective action, but noted that 22% of households interviewed did report an increase in group activities post Dam construction.

Impact on the Surrounding Environment

The construction of a single Sand Dam can influence water levels to a distance of 350 m upstream and downstream (Quilis et al. 2009). The improved water table allows ecosystems to regenerate. Sand Dams decelerate catchment degradation by enabling vegetative growth. In order for Sand Dams to reach maturity and function at optimum efficiency, effective erosion control needs to be in place. In each Sand Dam location, communities have been encouraged to develop terraces in their homesteads and to plant grasses, shrubs and trees along the natural river line to stabilize the banks and decrease erosion. The time saved from fetching water enables communities to put these practices in place and increases the amount of stored water, facilitating vegetative growth. Substantial regeneration of natural vegetation is evident where Sand Dams have been constructed. Sand Dams themselves have also been constructed to reduce erosion and functionalize gullies. Nissen-Petersen (2006) noted that Sand Dams were built in gullies to harvest Sand and stop it from silting up Lake Victoria.

Cost-Effective

Sand Dam projects use a cost-sharing approach. Communities are required to provide the labour and locally available materials such as Sand, stones and water, while the project proponent supplies the technical support, supervision and necessary materials such as cement, ballasts and pipes. SASOL additionally asks the community to provide housing and food to the artisans. The cost-sharing approach used improves the project viability by making it cost-effective and therefore accessible for rural communities, NGOs and governments with limited resources. The cost of an average SASOL Sand Dam is US$5,000 plus community contributions (Ersten and Hut 2009; Lasage et al. 2008), which equates to an average cost of US$5–10/consumer. Mati (2005 in Maddrell and Bown n.d.) noted that Sand and subsurface Dams are the cheapest water-harvesting technique to implement, as compared to water tanks, which are ten times the cost.

Feasibility

The practicality of Sand Dams as an option for decreasing vulnerability to water scarcity in rural ASAL communities in Africa depends primarily on the technology's affordability and achievability. As illustrated above, the low cost and cost-sharing approach enables Sand Dam projects to be readily affordable to NGOs or governments and the local communities. The achievability of the strategy is enhanced because it is culturally appropriate as Kenyans have been collecting water from ephemeral streams through scoopholes for centuries. There are, however, impediments and areas of concern regarding Sand Dam construction and maturation. In addition to problems of siltation and contamination listed above, each community must take into consideration labour requirements and land ownership. Sand Dams are labour-intensive strategies. All of the Sand Dam committees interviewed agreed that the organizing and maintaining adequate participation was challenging. The communities are required to collect heavy stones, to search for water in order to mix cement, and to dig the trench for the Sand Dam. The lack of food impacted community participation both in terms of members lacking energy and as fields needed tending. This participation tended to be not readily forthcoming due to their heavy preoccupations.

As an adaptation to climate change, the long-term feasibility of the project is determined by the ability of the technology to operate under expected changes in climate. Unlike many water supply infrastructures, Sand Dams are not extractive technologies. They contribute to water conservation and groundwater recharge, which will lessen the impacts of climate change. In Kitui district, the annual runoff is expected to decrease by 1 and 34% in the November to March and April to October seasons, respectively, due to climate change (Aerts et al. 2007). While precipitation is expected to increase overall, the decrease in runoff is associated with the higher rate of potential evaporation due to the increase in average temperature of 3°C. As water held within a Sand Dam is less susceptible to evaporation, these techniques and the water stored become more important and lastingly productive. Under climate change scenarios, Aerts et al. (2007) found that there remained enough water to fill the current 500 Sand Dams located in Kitui district and the percentage of the annual catchment runoff collected by the Dam increased. During the April to October season, the percentage of annual runoff collected behind the Dams changed from 3% in the 1950s to 20% in 2100 (Aerts et al. 2007). While Sand Dams are still effective at collecting water, the higher percentage of runoff may cause conflict with downstream users.

Factors and Processes Contributing to or Constraining Successful Implementation of Sand Dam Development and Adaptation Strategies

Several studies have evaluated the physical and socioeconomic feasibility of these structures (Borst and De Haas 2006; Lasage et al. 2008; Quilis et al. 2009). They noted that the strategy increases the amount of stored water, reduces poverty and is cost-effective, robust and simple to implement. However, the ultimate success of the strategy within an area has been linked to community organization and commitment. Lasage et al. (2008) reflect on this in the following:

> more research needs to be done before Sand Dams can be up-scaled to other areas as adaptation to droughts. Besides the relative simplicity of the technique the commitment and organization of the community is a very important factor for the success. This must be taken into account when the feasibility of implementing the technique in other regions is explored.

For Sand Dams to be effectively implemented in other communities and by other institutions, it is important to identify the necessary conditions that ensure success and how to develop and nurture these conditions. The case studies highlighted the importance of community motivation, capacity building and follow-up from the implementing institutions. It was noticed that these social factors are equal in importance to the proper construction and technical considerations of the concrete structure and are in fact the key factors in a Dam maturing into a proper functioning Sand Dam.

Community Motivation

Community-based projects can either be initiated by a community group or by an implementing body. Motivation is implicit in projects initiated by the communities and is evident by the fact that the community has organized, identified problems, decided on possible solutions and has begun looking for resources. In this scenario, the community is the driving force. In contrast, in all five case studies, the implementing body initiated the project. It was the responsibility of the project proponents to mobilize the communities and ensure that the communities were equipped with the necessary know-how. From the case studies there emerged different factors and processes that both enabled and constrained the development of community motivation such as past experiences, perceptions, beliefs and ownership, community cohesion and spirit, and participation.

Past Experiences, Perceptions, Beliefs and Ownership

Each community joins a new project with past experiences, beliefs and perceptions, all of which will influence their motivation to complete the Sand Dam project. In Keutunda village, the community had previously constructed a Sand Dam with another organization, which according to SASOL never functioned (F. Kimwilu, personal communication, 19 February 2009). This past project impacted participation from the community, as many felt disillusioned and no longer believed in the technology, while others misunderstood the technology, believing simply that Sand Dams take a very long time to mature and collect water.

In order to form a composite of their beliefs and perceptions, households and Sand Dam committee members were asked to identify the main problems in their communities. Water shortages, lack of food and health were mentioned most frequently, but were not always linked as being addressed by the Sand Dam project. Sand Dam benefits most frequently listed were improving water availability and permitting vegetable gardens. Seldom mentioned were health benefits associated with cleaner water, the ability to wash more often, and time saved by having nearby water sources. Both past experiences and potential benefits need to be carefully integrated into capacity-building processes.

Community beliefs and religion also represent an important influence. For all the case study communities, religion impacted their worldview. Over 40% of households interviewed were convinced that God was the main cause of the failed rains. This trust in an omnipotent God also influenced the people's faith or lack thereof in their own solutions to the identified problems. In every community with the exception of one, there were individuals who felt that only God could fix their problems. Respect for and sensitivity of religious beliefs were crucial in gaining credibility and trust within the community.

Dependence on governments and non-governmental institutions was also evident in the case study communities. Many households expressed that solving the problems of water scarcity was the responsibility of the government. This impacted the community's motivation to complete their own projects. Institutions, especially government organizations, implementing Sand Dam projects need to ensure that it is not simply perceived as a government or non-governmental intervention but rather as a joint venture with the community. During the focus group discussions, the Sand Dam committees were asked to describe the driving force and community vision behind their project; most responses simply indicated that they provided labour to get water and did not understand linkages to a broader vested goal of community development. In fact, the Kitunda village Sand Dam committee clearly stated that it was SASOL's vision and not their own. Sand Dam project proponents claim that community empowerment is created through their participation in the project. However, in all but one case study community, the committees became inactive once the Dam was built. It was evident, strictly from the Sand Dam project, that, with the exception of Kitunda village, the communities had not realized their power or the potential of the project to act as a catalyst. In other words, development efforts were a participatory effort and could open the gateway for democratic progress.

Buiding ownership of the Sand Dam increases community motivation to complete and care for the project. Ownership needs to be addressed on several levels including legal ownership, care and financial ownership, and sense of ownership. Sand Dams in most of the case study locations appeared to be built adjacent to private land. In Kenya, permits to water resources run with the land; therefore, it is necessary that at the onset of the project the implementing institutions ensure clear definition of ownership and rights of access. Failure to ensure definiteness in these rules and regulations might result in the community being denied access. The committee should operationalize a signed memorandum of understanding between the landowner, the implementing institutions and the community. In addition, Remple et al. (2005) noted the importance of ensuring that the community gains ownership of the riverbanks surrounding the Dam. Control of this exterior boundary supports the Dams' reaching maturity, as the community can legally plant grasses and control erosion. Feelings of ownership and the assumption of responsibility are harder to attain and are not simply achieved through the signing of legal documents. Most community members interviewed believed that the community owned the Dam; however, responses were far less convincing when the majority felt that it was the responsibility of the implementing institution to fix/maintain the Dam. The Sand Dam committees, by contrast, knew it was their duty to take care of the Dam, but none of them had established any, or thought of any plans as to how this would be accomplished. Additionally, all of the Sand Dam committees indicated that their first step would be to contact the project proponents. It is clear that proper training in project sustainability and management is funDamental to the long-term success of Sand Dam projects. A sense of belonging and participation in the beneficiary community needs to be fostered from the beginning.

Findings from the case study communities, additionally, indicated that both organizations missed the opportunity to build ownership through emphasizing the cost contribution of labour and local materials. A key enticement for Sand Dam development projects is the notion that the community contributes half or more to the cost of the project. However, only two households interviewed (both from SASOL's projects) believed that the implementing institutions and the local community jointly funded the project. Involving the communities in cost analysis and budget planning could help the participants realize the value of their work and build ownership through participating in more levels of planning. This will have to be done by the (government) mandated village committee or administration since it is a considerably technical process.

Community Cohesion and Spirit

Trust, cohesion and proactive spirit are all the idyllic characteristics presumed to be inherent in many rural African communities. However, in each case study community, there was evidence of tensions and conflicts within the project team and community. In one community, the assistant chief felt that the Sand Dam committee had acted dishonestly and had promised community members that they would be paid for working on the project (D. Mutisya, personal communication, 23 February 2009). The committee explained that the only exchange of money was from households that did not participate, who were fined either 50 Ksh or had to bring some food for the participants. The assistant chief also indicated that there was a lack of communication from both the Sand Dam committee and the implementing institutions. Project proponents should recognize that the assistant chief is a significant bridge between the community and the district government and wields a certain amount of influence. Another project built their Sand Dam with only 18 members, in a community with 128 households. In this community, there were again conflicts over payments and the disappearance of materials. Despite the fact that the Sand Dam was built, there are ill feelings throughout the community.

Tensions may also originate outside of the project but need to be addressed. For example, limestone deposits found within one community impacted motivation to complete the project. Two cement companies were vying for mining rights and as a result, residents speculated that the local administration and others would benefit, while the rest would simply be displaced from their homes. Another indicator of this community's lack of cohesion was the fact that more than half of the households interviewed were not part of a single self-help group. Implementing institutions need to work beforehand to build social capital and address community problems through processes of social learning.

Initiating procedures can facilitate or hinder the development of a cooperative group spirit. Both implementing institutions in this study requested that the communities find potential sites for the Sand Dam and select a Sand Dam committee. The communities were advised to select sites that were accessible and would serve the largest number of people. Being involved in the process of placement contributes to community cohesion, as members feel as though they had had some input in the planning stage. All of the committee members from each case study claimed to be democratically elected by their communities. While SASOL would advise communities as to the composition of the committees, suggesting that there should be an equal representation from people of different age groups, genders, education levels, religious beliefs and political party affiliations, they do not dictate the process (Ertsen et al. 2005). The households interviewed came forth with multiple responses when asked of the existence of the Sand Dam committee and how the members were selected. From both the Kitui district locations, almost half of the respondents indicated that the committees were democratically elected; the remaining respondents were unsure of the process, the committees' existence, or believed that the village elders had chosen the members. The responses were more distributed for the projects in Makueni district. In general, the largest percentage of people indicated that they were unsure how the committee was selected. Organizations implementing Sand Dam projects should include as many participatory processes as possible to help build cohesion and make an effort towards inclusive and participatory development.

Participation

Participation is the cornerstone of Sand Dam development projects. Irrespective of the quantum of community benefits brought about by participation, the involvement of the people is necessary in order to keep the cost of the project low and viable to institutions. As there is a considerable amount of hard work, requiring physical strength and stamina, it is essential that communities gain as many participants as possible in order to share the work. One Sand Dam constructed by SASOL, required, on average, 604 man/woman working days. Balanced between 30 community members, this still amounts to 20 full days of labour. Excellent Development found that younger communities are less affected by the level of work and are therefore more often successful with Sand Dam projects (J. Mukusya, personal communication, 22 January 2009). Institutions are thus encouraged to ensure that younger community members are motivated to participate.

The Sand Dam committees promoted participation in several ways. People were mobilized to take part in the project through community meetings where the benefits of Sand Dams were explained and by the village elders, who went door to door, persuading members to participate. Groups also commonly used fines and restrictions. All of the committees described that households in the community that did not participate were fined and individuals who were late were also fined. Nevertheless, little evidence was found that fines in any form were ever collected.

Enforcing and monitoring participation proved to be complicated. In many households, the able-bodied were needed in the farms or had migrated in search of employment, leaving the sick or elderly to look after the children. For the committees, it was difficult to determine which reasons were justifiable for not participating. Enforcement agreed work norms were indeed difficult.

Several solutions were suggested during the focus group discussion to address the impediments to participation, such as distance to project sites, and household duties. Many households interviewed indicated that they did not participate because the Dam was too far to benefit them. One Sand Dam committee suggested creating a community vision. The vision would outline all the goals they had for their community and how these goals were to be accomplished. In this way, the households that were more remote would feel as though the community was working towards a common objective. All households struggled to participate throughout the dry season, as food was scarce. This scarcity influenced energy levels to complete the work in time as households needed to search for food and water. In two projects, participants brought food and cooked together. Another group suggested that institutions factor in food-for-work programmes, as then the community would be able to build faster and more members would participate. Food-for-work has been a successful part of Excellent Development's projects.

It general, the case studies revealed that incentives and effective mobilization were better motivators than restrictions or fines, which were perceived negatively and found infeasible to enforce. Project proponents need to invest the time and other resources to ensure that communities are genuinely keen and able to complete the work necessary for the project to be successful.

The idea behind the cost-sharing strategy is that an actively engaged community that expends their time and energy to the construction of the Sand Dam will develop de facto ownership of the Dam and be vested in its success. The contribution of free labour for projects is not a new phenomenon in rural African communities; despite this, projects across the continent have not been duly maintained. Building ownership and ensuring that Sand Dams reach maturity and work effectively will require community participation in addition to simply providing labour for the construction. In this regard, both the Resilience pilot project and SASOL had limitations in that the organizations were only offering the initial technical strategy. Both institutions initiated the project through participatory approaches; where the communities identified their problems, however, neither organization could enable the communities to develop their own plans to accomplish what they wanted to achieve. According to an Excellent Development co-founder, "success depends on freedom given to communities to make decisions, when freedom is given communities are more committed" (J. Mukusya, personal communication, 22 January 2009). During both the household questionnaires and the focus group discussions, it was apparent that community members saw the implementing bodies as leading and directing the project and not themselves. The majority of households interviewed in Makueni district stated that the community had very little to do with the initiating or planning stage of the project. In contrast, the majority of respondents in Kitui district believed the community played a large role in the planning, which included scheduling working days, selecting the site, and electing the committee. The more stages that the community is allowed and enabled to participate in, the greater the encouragement to build capacity and sustain it collectively.

Capacity Building and Training

Building a community's capacity to effectively maintain, manage and ensure their Sand Dam reaches maturity is essential for a successful and sustainable project. In each community there was evidence of past projects that were no longer functioning. In part, this can be related to lack of training and understanding of the technologies on the part of neighbourhood beneficiaries.

With the exception of basic knowledge, such as that a Sand Dam would bring more water and that with more water they would be able to grow more vegetables, all the case study communities but one lacked a true understanding of the function of a mature Sand Dam and how to ensure a Sand Dam reaches maturity. A common illustration of this was the belief that if communities continued to collect water from scoopholes adjacent to the Sand Dam, as opposed to the installed pipes or shallow wells, then the Sand Dam wasn't working. One community where the outlet pipe had never been fitted believed that even if rains came and filled the Dam, they would not be able to access the water for domestic consumption. Members from another project indicated that the Sand Dam was not working because there was no water in the shallow well built adjacent to the Dam. However, even after the second failed rainy season, the community was able to collect water from deep scoopholes in the vicinity of the Sand Dam, which previously would have been dry. The communities did not comprehend that the Sand Dam causes the water table in the local area to rise. Only one committee approached this level of understanding by explaining that one of the intended benefits of the Sand Dam is that the surrounding area would become greener.

As noted earlier, drowning was mentioned as a potential hazard associated with Sand Dams. Upon further inquiry this suggested a misunderstanding of the technology. Several members in each community believed that the purpose of the Dam was to block surface water, resulting in a collected body of open water. They did not realize that this open body would only be associated with an immature Sand Dam. A mature Sand Dam should be filled to the height of the Dam with Sand, minimizing the potential for drowning. This misconception will negatively impact Sand Dam maturation, as community members will be more likely to practise Sand harvesting close to the Dam. The tendency of this misunderstanding was also noted by Gaff (2005), who stated "communities did not recognize that the accumulation of Sand in a streambed was just as important a piece of their realization of water security as their construction of a cement and stone barrier—that a streambed without Sand was also a streambed without water".

Furthermore, only one case study village linked erosion control, promoted within the projects, as a requirement for a properly functioning and mature Sand Dam. The other communities recognized practices that would control erosion, such as building terraces and planting grasses. However, they felt that these strategies were implemented only as a means of improving agricultural production. This was especially evident with the three communities in Makueni district forming part of the Resilience pilot project. The structure of the project was such that all objectives and associated activities ran concurrently. The participants who received training on proper agricultural practices such as erosion control were not necessarily the same as those participating in the Sand Dam construction, hence there was little opportunity to link the natural resources management practices to the Sand Dam. This emphasizes the fact that the Sand Dam user community needs persistent long term advice and persuasion to utilize the intended full and diverse potentialities emanating from the project.

All of the Sand Dam committees claimed that they were promised training sessions regarding Dam maintenance; however, only the communities in Kitui district received the promised sessions. SASOL offers each community that builds a Dam a series of three training courses. These training sessions are fundamental to empowering the community and providing them with the tools to guarantee that optimal results are realized from their Sand Dam development project. The first training session taught the communities about proper natural resources management. The groups were told about the interactions between air, soil, water, and vegetation and the necessity to take care of each element. They were also shown how to harvest water and manage runoff, for example by building terraces and capturing road runoff. The second session outlined proper project management. The goals of this course were to ensure project sustainability by training the community to use water wisely and to generate income through horticulture, beekeeping and pottery (F. Ndeto, personal communication, 31 March 2009). The third and final session focused on participatory hygiene and sanitation. This course taught the community to be proactive about health and well-being by being aware of and mitigating disease transmission. They are educated on the prevention of malaria, the proper disposal of waste and how to implement personal hygiene. Each session occurred over the course of a week and was offered to ten members of the community. Those who are trained by SASOL are then expected to return to the community and train ten other members. Interaction and communication with the two communities under SASOL's guidance provided evidence that the delivery of the training sessions were less than ideal. In both communities, cattle could be seen accessing water in the riverbed in places upstream from where residents collected their domestic water. The connection that livestock were polluting the water flowing downstream to the community scoopholes was not realized. Gaff (2005) also noted that the sessions were dense and packed with information and that the participants, most of whom had had little formal education, were "left to pick and choose among the varying lessons that had been handed to them over the course of training, latched onto one or two that seemed most interesting and carried that, and that alone, back to their communities". Ideally, the lessons should be spread out over time, enabling the students to absorb the material and have opportunities to ask questions.

While the Resilience project disseminated similar information to the communities through multiple strategies, this training was not linked to the Sand Dam. Instead, the Sand Dam was simply viewed as a concrete structure which when built would supply more water. It has been clearly established that in order for a Sand Dam to be successful, the community must be engaged in much more than the construction. Jucinda Nakiyu from the Kenyan Water Resources Management Authority stated in an interview: "training is crucial; it is the key issue. It is this aspect that takes the most time. We have to sit down and decide how they want their catchment to be managed so that their water sources remain viable and healthy. If the Sand Dams are to have an effect the community needs to learn how to manage and have to be effective managers. How the Sand Dam will be managed needs to be written in from the very initial proposal, don't build the Dam without the community being involved and the management being written" (J. Nakiyu, personal communication, 15 April 2009).

Communication, Promises and Follow-Up

Organizations implementing development projects need to recognize that their role extends beyond the provision of material resources. Although delivering on the resource promises is essential, the manner, timeliness, and level of communication and follow-up also have significant impacts on project success.

Both the Resilience project and SASOL missed key opportunities to encourage and facilitate community ownership through their lack of follow-up. SASOL's mandate is to build as many Dams as possible in Kitui district and it therefore directs its resources to construction as opposed to follow-up and the ensuing project maintenance (Gaff 2005). However, contrasting the two SASOL case studies can show the importance of maintaining the connection with the community. Kitunda village is highly organized and has completed several other development projects, which they claim were catalysed by SASOL and the Sand Dam. The Sand Dam committee in Kitunda explained that they now come together as a community and look at their problems and discuss solutions. Currently, the community is divided into four subgroups, with each group tackling a particular problem or livelihood activity such as tree planting and the creation of a nursery, poultry keeping, goat keeping, and fruit tree planting. In addition, the community has also extended its enterprises into beekeeping and sunflower production, which provides them with edible oil. They have also decided to reunite as a group in order to rehabilitate the road network in their village (Kwa Ndunda focus group discussion, personal communication, 16 March 2009). In the 14 years since Kitunda village's Sand Dam was completed, SASOL has never left the community. As the village is close to Kitui town and SASOL's headquarters, it has often been used as a demonstration Sand Dam for funders and other interested proponents. SASOL has also formed key partnerships, which influence the communities close to Kitui town. Through an agreement with a university in the Netherlands, students are able to come to Kenya and complete research projects at this site. This often leads SASOL back to Kitunda village, where the community is asked to support the students and, in turn, the community benefits through encouragement and new ideas. Another fruitful alliance for Kitunda village is the partnership between SASOL and the KDC. SASOL engaged the KDC to run their three training programmes in Kitunda village, which consequently led to other development projects with the community. In contrast, Keuntunda sub-location has not benefited from any of these partnerships or the strong presence of SASOL.

The level of rapport and communication that communities have with project proponents also influences their pride, confidence and ability to strengthen their projects. The majority of households interviewed in Kitui district knew that it was SASOL that had completed the project and neither Sand Dam committee stated that there were any communication problems. In addition, both committees could easily identify ways to contact SASOL if necessary. Unfortunately, the same cannot be said for the projects in Makeuni district. The majority of households interviewed could not identify any of the proponents involved in the project. The Resilience pilot project was a complex, multi-stakeholder project involving researchers from the Centre for Science and Technology Innovations (CSTI), the Arid Lands Resource Management Project (ALRMP) and several other line ministries, such as the Ministry of Water and Irrigation, Ministry of Gender and Social Development and the Ministry of Agriculture. The many partners, as well as the several programmes being implemented concurrently and involving different groups of community members, created confusion. While the Sand Dam committees recognized that the Sand Dam projects were connected to the larger Resilience project and could identify several contacts, few households made the connection. The three committees also expressed frustration indicating that communication with the project proponents was lacking. One Sand Dam committee indicated that they had not heard from the implementing institution in almost a year. Officials from the pilot project were still operating within the community but with another subsection of the project, thereby demonstrating the disconnect between the various areas of the pilot project. This multi-faceted project design requires diligent communication to prevent missing of opportunities and detrimental impacts. The Sand Dam committees in all three communities in Makueni district felt as though the project had broken promises made to them and disappeared. In Muiu, the committee was waiting as they were told the Sand Dam would be raised over time. In Nthongoni, the Sand Dam was never fully completed, and in Kathamba, the community dug the trench for their Sand Dam three times and still had not had cement delivered. There were also communication breakdowns and coordination problems between all the different sectors implementing the project. The project was being overseen and coordinated by the ALRMP from the district headquarters in Wote; however, the line ministries implementing the activities were located in another town. With regard to the Sand Dam project, there were misunderstandings in terms of who was responsible for ordering the materials, arranging for it to be brought to the community and knowing who was keeping it (F. Ndeto, personal communication, 31 March 2009). There was also significant confusion regarding who was in charge of supervision, training, and signing off on the completed Dam. Staff resources and transportation were other limiting factors in maintaining contact with the communities (M. Kamwanza, personal communication, 26 March 2009; F. Ndeto, personal communication, 31 March 2009).

Even while projects are constrained by funding schedules and resources, short follow-up sessions where training is provided and the community is encouraged in their progress are essential. In addition, return visits force the Sand Dam committees to reconnect. Jucinda Nakiyu, Water Resources Management Authority, explains that when they work with Water Resource User Associations (WRUA), they plan on being involved with them for 10 years. She argues that it is only after approximately 10 years that the groups will be stable and self-sufficient enough to be left alone (J. Nakiyu, personal communication, 15 April 2009). The Resilience project did take this into consideration and while the pilot project is 3 years, the partnership with the ALRMP was meant to facilitate an ongoing resource for the communities.

Conclusion

As an adaptation strategy with the potential of lessening community vulnerability to water scarcity, Sand Dam development projects have been shown to be appropriate. In contrast to other water supply infrastructures, a Sand Dam adds value to the system by harvesting water that otherwise would have been lost downstream. In addition, the water is protected from high evaporation and cleaned through a natural filtration process. The cost-sharing approach ensures community involvement and provides an implementable project for many governmental and non-governmental organizations working in rural Africa. The benefits of a Sand Dam also reach beyond the provision of water. The reduced distance to a water source that remains yielding throughout the dry season provides households with more time to spend on other income-generating activities. The economic standing of the community also increases as livestock are healthier, vegetables can be grown, and alternative livelihoods can be established.

While these benefits can be realized, it is evident from the case study communities that a Sand Dam development project requires much more than simply building a concrete structure across an ephemeral stream. The Dam itself is only the beginning of the project; however, too often the project proponents fail to take into consideration the factors necessary to ensure that a "Dam" becomes a "Sand Dam". Participation and capacity building are the keys to success. The experiences from the case study communities illustrate that there are many underlying reasons affecting a community's willingness to participate. Effective projects need to take the time to develop community cohesion and drive in order to successfully complete the Sand Dam. The construction phase, although requiring hard work, is the easy part for the communities who are used to following directives. Ensuring that a Sand Dam reaches maturity and remains viable is highly unlikely without proponents investing in training and building capacity. These efforts require a large amount of time and resource investment, which is often beyond the scope of an organization's funding cycle. A solution, which was illustrated by SASOL's experience in the Kitunda community, is partnership building. If organizations building Sand Dams partner with another that would facilitate training in natural resources management, hygiene, project management, and building capacity for alternate livelihoods, communities would experience maximum benefit. The length of time that communities are encouraged and supported would be doubled. In addition, in times of need, the contacts and social resources that communities can draw on would have increased through the multiple partnerships.

References

Aerts J, Lasage R, Beets W, de Moel H, Mutiso G, Mutiso S, de Vries A (2007) Robustness of Sand storage Dams under climate change. Vadose Zone J 6(3):572–580
Borst L, de Haas SA (2006) Hydrology of Sand storage Dams: a case study in the Kiindu catchment Kitui District, Kenya. Vrije Universiteit, AmsterDam
Burton I, Smith JB, Lenhart S (1998) Adaptation to climate change theory and assessment. In: Feenstra JF, Burton I, Smith JB, Tol RSJ (eds) Handbook on methods for climate change impact assessment and adaptation strategies. Institute for environmental studies Vrije University and Nairobi, Atmosphere Unit, UNEP, AmsterDam
Darkoh MBK (1993) Desertification: the scourge of Africa. Tiempo 8:320–331. Available at http://www.cru.uea.ac.uk/cru/tiempo/issue08/desert.htm. Accessed July 2009
de Loë R, Kreutzwiser R, Moraru L (2001) Adaptation options for the near term: climate change and the Canadian water sector. Glob Environ Chang 11(1):231–245
Ersten M, Hut R (2009) Two waterfalls do not hear each other Sand-storage Dams, science and sustainable development in Kenya. Phys Chem Earth 34:14–23
Ertsen M, Biesbrouck B, Postma L, van Westerop M (2005) Community organisation and participatory design of Sand-storage Dams in Kenya. In: Goessling T, Jansen RJG, Oerlemans LAG (eds) Coalitions and collisions. Wolf Publishers, Nijmegen, pp 175–185
Excellent Development (2007) Excellent development 2006–07 annual review, London. Available at http://www.excellentdevelopment.com. Accessed April 2009
FAO (2007) Food security district profiles. Nairobi, Kenya. Available at http://www.fews.net/docs/Publications/Kenya_FSU_May_2009_final.pdf. Accessed April 2009
Few R, Osbahr H, Bouwer LM, Viner D, Sperling F(2006) Linking climate change adaptation and disaster risk management for sustainable poverty reduction, Synthesis report, study carried out for the vulnerability and adaptation resource group (VARG)
Gaff P (2005) An assessment of the strengths and weakness of the organizational and environmental aspects of the Sand Dam paradigm. American University, Washington

Government of Kenya (GoK) (2009) Draft national policy for disaster management in Kenya. Nairobi, Kenya

Government of Kenya (GoK) (2008) Adaptation to climate change project—Sakai pilot project: environmental management plans for the proposed construction of Sand Dams and shallow wells in Sakai community of Kisau division, Makueni district, ALRMP, Wote

International Institute for Sustainable Development (IISD) (2007) Integrating vulnerability and adaptation to climate change into sustainable development policy planning and implementation in eastern and Southern Africa (Regional Team Meeting Proceedings), Nairobi, Kenya

IPCC (2007) Climate change 2007: impacts, adaptation and vulnerability. contributions of working group ii to the fourth assessment report of the intergovernmental panel on climate change. In: Parry ML, Canziani OF, Palutikof JP, van der Linden PJ, Hanson CE (eds). Cambridge University Press, Cambridge

Lasage R, Aerts J, Mutiso G-CM, de Vries A (2008) Potential for community based adaptation to drought: Sand Dams in Kitui, Kenya. Phys Chem Earth 33(1):67–73

Manzi HK (2005) Kitui Sand Dam socio-economic impacts and outcomes from changes in agriculture practices and production, Egerton University, Faculty of Agriculture, Nakuru, Kenya

Mati B (2005) Overview of water and soil nutrient management under smallholder rain-fed agriculture in East Africa. Working paper 105. International Water Management Institute (IWMI), Colombo, 94 pp

Maddrell S, Bown S (n.d.) Sand Dams: Africa's answer to climate change? Excellent Development, UK. Available at http://www.excellentdevelopment.com/uploads/Excellent%20Downloads/sand_dams_africas_answer_to_climate_change.pdf

Meteorological Department of Kenya (MDK) (2005) Yearly precipitation in Kitui district, Government of Kenya, Kitui

Ministry of Planning and National Development (2002a) Makueni district development plan 2002–2008, Government of Kenya, Wote

Ministry of Planning and National Development (2002b) Kitui district development plan 2002–2008, Government of Kenya, Kitui

Ministry of State for Planning, National Development and Vision 2030 (2008) Makueni district development plan 2008–2012, Government of Kenya, Wote

Mutiso SM (n.d.) Sand Dams: a catchment groundwater storage approach in arid and semi-arid lands, SASOL foundation, Nairobi

Mutiso SM, Mutiso G-CM (n.d.) Kitui Sand Dams: a development paradigm, SASOL foundation, Nairobi

Nissen-Petersen E (2006) Water from dry riverbeds: how dry and Sandy riverbeds can be turned into water sources by hand-dug wells, subsurface Dams, weirs and Sand Dams. ASAL consultants ltd for the danish international development agency (DANIDA), Nairobi. Available at http://www.spate-irrigation.org/librar/documents/DrinkingWaterFromDryRivers.pdf. Accessed May 2010

Protos (2006) A manual of subsurface Dams construction based on an experience of Veterinaires Sans Frontieres in Turkana District (Kenya), Protos, Belgium. Available at http://www.protos.be/protosh2o/copy_of_Links-met-bestanden/Sub%20surface%20Dams%20defvolledig.pdf. Accessed May 2010

Quilis R, Hoogmoed M, Ertsen M, Foppen JW, Hut R, de Vries A (2009) Measuring and modeling hydrological processes of Sand-storage Dams on different spatial scales. Phys Chem Earth 34:289–298

RAIN Foundation & Acacia Water (2007) A practical guide to Sand Dam implementation: water supply through local structures as adaptation to climate change. RAIN Foundation, AmsterDam. Available at http://www.rainfoundation.org/fileadmin/PublicSite/Manuals/Sand_Dam_manual_FINAL.pdf. Accessed May 2010

Remple H, Nyaga CW, Manzi HK, Gaff P (2005) Water in the Sand: an evaluation of SASOL's Kitui Sand Dams project. SASOL Foundation, Nairobi December

Resilience project (Increasing community resilience to drought in Makueni District Pilot project) (2008) Overview of key issues and challenges. The Project Area, ALRMP Office, Wote. Available at http://www.csti.or.ke/publications/Kenya%20Pilot%20Brochure%20Dec%2008%20version%202.pdf

Smit B, Pilifosova O (2001) Adaptation to climate change in the context of sustainable development and equity. In: McCarthy JJ, Canziani OF, Leary NA, Dokken DJ, White KS (eds) Climate change 2001: impacts, adaptation and vulnerability (contributions of working group II to the third assessment report of the intergovernmental panel on climate change). Cambridge University Press, Cambridge

Smit B, Burton I, Klein RJT, Wandel J (2000) An anatomy of adaptation to climate change and variability. Clim Chang 45:223–251

United Nations Environment Programme (UNEP) (2006) The 2nd world water development report: water, a shared responsibility. United Nations, Geneva

United Nations—Water/Africa (2003) (UNWAfrica) (2003) The Africa water vision for 2025: equitable and sustainable use of water for socio-economic development. Economic Commission for Africa, Addi Ababa

Chapter 43
Urban Rainwater Management Using the Aquifer for Storage as One Measure to Confront Changed Water Availability

Reinhard Marth and Leonel González González

Abstract This paper presents an adapted set-up for urban rainwater management since the upscaling of the classical set-up tends to be problematic. It is found that this is mainly due to the high investment costs required for the construction of storage elements such as tanks or cisterns. The general idea of the proposed set-up, therefore, is to combine the decentralized catchment and infiltration of rainwater at household level with a centralized water supply system. The water supply of the households is provided through a high connection rate to the water supply system, exploiting the local groundwater resources. The rainwater is infiltrated on-site into the subsoil and stored in the local aquifer allowing for indirect reuse in the households through the water supply system. The proposed set-up requires the existence of a competent water management institution and the implementation of an appropriate management model. Possible benefits and important challenges of the proposed set-up are discussed herein, as well as a future extension including on-site treatment of greywater. It is concluded that the proposed set-up could be very beneficial to achieve sustainable usage of groundwater resources in the urban context. However, there is no experience, since the proposal has not been implemented yet.

Keywords Rainwater harvesting · Urban rainwater management · Decentralized infiltration · Adaptation · Storage in aquifer · Management model

---

R. Marth (✉) · L. G. González
Comisión Nacional del Agua (CONAGUA), Centre for International Migration and Development (CIM), Dirección Local, Estado de México, Conjunto SEDAGRO, Edificio C-1, 52148 Metepec, Estado de México, Mexico
e-mail: marth.cim@gmail.com

Introduction

Changed precipitation patterns and extended dry periods are among the most important challenges resulting from the climate change. Communities confronted with this situation need to increase the resilience of their water resources management in order to secure the water supply of the population. While surface water bodies such as rivers and lakes show more variability with changed hydrological conditions, the importance of groundwater resources will increase in future. Rainwater harvesting and recharge of groundwater resources are key elements for sustainable water management in arid or semi-arid regions.

Since rainwater harvesting and artificial recharge can play a major role in achieving sustainability of drinking water sources, local communities need to be encouraged to take them up on a large scale (Ray 2002). In Mexico, however, it is found that the upscaling of successful pilot projects is difficult. This paper presents an adapted set-up approach for urban rainwater harvesting, including a management model which could encourage the application.

Rainwater Harvesting in Mexico

The Situation in General

In Mexico, water availability per capita is decreasing. Figure 1 presents the annual freshwater availability per capita in cubic metres showing a clear tendency to critical values, particularly in central and northern Mexico (national average: 4427 m^3/cap/year in 2005).

The main reasons for this situation are:

- Growth of population.
- Increase of water demand for agricultural, industrial and commercial activities.
- Deficiency of the water networks.
- Lack of awareness in the population of the situation of the environment.

The situation is especially critical in central Mexico, a very important part of the country in terms of politics and economy, with more than 30 million inhabitants mainly living in urban areas. The decreasing water availability has been aggravated by changing precipitation patterns in recent years. Heavy rainfalls during the rainy season seem to be intensifying, while precipitation during the drought season is decreasing. The consequences are more frequent inundations during the rainy season and water shortage during the dry season. This observation is in line with the findings of Bates et al. (2008).

The growing interest in rainwater harvesting systems in Mexico led to some pilot projects. Figure 2 shows two examples in central Mexico which combine the classic design of the system with some interesting features, lowering the operation

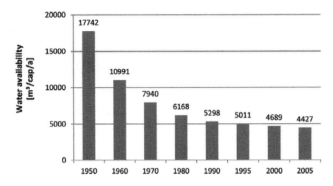

Fig. 1 Availability of freshwater per capita (m³/cap/year) in Mexico from 1950 to 2005 (with data from CONAGUA 2008)

Fig. 2 Examples of rainwater harvesting tanks in Mexico. On the *left*: catchment and storage of rainwater for a small community (129 inhabitants) in Aguascalientes (Source: Aguascalientes 2006). On the *right*: rainwater harvesting tank in a state school, Estado de México (Source: Marth)

costs. Figure 2 on the left shows a system which supplies a small rural community. A special roof system was built to catch the water on an elevated point outside the community allowing for the supply by gravity. Figure 2 on the right shows a storage tank of a system for a secondary school. The water is pumped by means of a solar power system to small rooftop tanks which assure the pressure for the supply. For both projects, high investments had to be made for the roof construction, the tanks and the solar power system.

In general, there have been good experiences made with rainwater harvesting pilot projects in Mexico. However, the upscaling of these projects remains very difficult.

This situation is mainly caused by two important factors:

1. Lack of adequate institutions which have the capacity to implement rainwater harvesting projects: only large urban municipalities can count on professional water utilities. The small urban municipalities mostly lack capacitated institutions for water management. The water systems, run directly by local governments, work inefficiently and without any long-term planning.

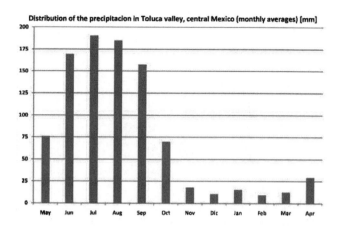

Fig. 3 Distribution of precipitation in Toluca valley in central Mexico [monthly averages (mm)]. The precipitation is shown from May to April to demonstrate the marked differences of the rainy and dry seasons (based on unpublished data from CONAGUA)

2. High investment costs: the implementation of a classic rooftop rainwater harvesting system requires rather high investment costs (compare example below). The implemented pilot projects usually rely on national or international funding. For a broad dissemination in private households, very little funding is available.

The result of this situation is that there is little incentive for private households (particularly in urban areas) to implement rainwater harvesting systems.

Example for a Draft of a Typical Rainwater Harvesting System

To demonstrate the difficulties of implementing rainwater harvesting systems in Mexico, the rough draft of a typical domestic system is given below. Figure 3 shows the monthly precipitation averages in the Toluca valley, situated in central Mexico close to Mexico City. Figure 3 clearly marks the differences between the rainy and the dry season, each of them lasting about six months. A rainwater harvesting system at household level should be designed to store the exceeding rainwater from May to October, in order to cover the water demand in the other six months, which have very occasional precipitation.

Table 1 shows the volumes of rainwater which could be harvested in a site in the Toluca valley applying different catchment areas and a long-term collection efficiency of 85% (recommended by Ree 1976).

In the official statistics of the national water authority, it is estimated that the average water demand of a family in Mexico is about 30 m^3, which is far more

Table. 1 Potential rainwater harvesting for a location in Toluca valley (central Mexico) collected with different rooftop catchment areas applying a collection efficiency of 85%

Month	Precipitacion (mm)	Volume collected (m^3) with a catchment area of			
		50 m^2	75 m^2	125 m^2	300 m^2
January	15.35	0.65	0.98	1.63	3.91
February	9.15	0.39	0.58	0.97	2.33
March	12.50	0.53	0.80	1.33	3.19
April	29.16	1.24	1.86	3.10	7.44
May	75.78	3.22	4.83	8.05	19.32
June	169.23	7.19	10.79	17.98	43.15
July	190.24	8.09	12.13	20.21	48.51
August	184.81	7.85	11.78	19.64	47.13
September	157.40	6.69	10.03	16.72	40.14
October	69.70	2.96	4.44	7.41	17.77
November	17.82	0.76	1.14	1.89	4.54
December	10.33	0.44	0.66	1.10	2.63
Total	941.45	40.01	60.02	100.03	240.07
Monthly average	78.45	3.33	5.00	8.34	20.01

than 200 l per person daily (CONAGUA: Comisión Nacional del Agua 2008). This high value is mainly caused by the bad state of the infrastructure. High water losses in the water supply systems are very common. For the design of a rainwater harvesting system, daily consumption per person of 150 l is realistic. Applying the average size of families, this amounts to consumption of about 20 m^3 in a month in a household.

The values obtained in Table 1 exhibit that this demand requires high catchment areas. Even during the rainy season, a rooftop area of more than 125 m^2 is necessary. In order to cover the water demand throughout the whole year with rainwater, a rooftop area of more than 300 m^2 would be required. The required storage capacity of the tank or cistern would be about 100 m^3, a magnitude far beyond the resources of a private household.

A study by the National Institute for Ecology of Mexico about the characteristics of households in central Mexico reveals that the average construction size in the urban context is about 105 m^2 (INE 2002). Modern constructions are usually two-storey-buildings, a consequence of increasing prices of land. The large areas required for rooftop catchment in most cases are not available. Even a reduced monthly water demand of 10 m^3 for a household would require a rooftop area of at least 150 m^2 and a storage capacity of 50 m^3 for coverage throughout all the year.

The most expensive part of a rainwater harvesting system is the tank or cistern for storage. In NWP: Netherlands Water Partnership (2007), it is estimated that the construction of a 60 m^3 ferrocement tank in a developing country could cost about US$1,900. Worm and van Hattum (2006) analysed the data of different NGOs which implemented rainwater harvesting systems in

Nepal and worked out that the construction cost per m^3 of storage volume are between US$78 and 142.

In Mexico, as a transition county, costs tend to be comparatively high: an estimated US$4,000 for a 60 m^3 tank, not including excavation costs for construction below the soil surface. A pilot project in the Toluca valley shows that above-ground plastic-lined tanks could be an economic alternative. The four tanks with a total volume of 320 m^3 cost about US$10,000. Unfortunately, the low physical life of such plastic-lined tanks of 5–10 years does not provide a long-term solution. Also, the space constraints of the urban context usually do not allow for above-ground solutions.

The conclusion from the hydrological boundary conditions is that very large catchment areas and storage volumes are required to cover the water demand completely by means of a rainwater harvesting system. The high investment costs, particularly due to the need for considerable storage volumes, are a main impediment to the upscaling of existing rainwater harvesting projects in the domestic sector.

Situation of Water Supply, Drainage and Treatment Systems

The coverage with connections to the public water supply is comparatively high in Mexico: 89.2% in total and 95.0% for urban areas (CONAGUA 2008). The most important source for the water supply is groundwater. The statistics of CONAGUA (2010) show that nationwide 15% of the aquifers are overexploited. The situation is even more dramatic for central and northern Mexico.

Whereas the population is well served with water supply, only 67.6% of the population was connected to public sewers in 2005 (CONAGUA 2008). It is difficult to find reliable data about the coverage with treatment. From the official statistics it can be deduced that less than 40% of the wastewater receives some kind of treatment. The observation of the authors is that many wastewater treatment plants work insufficiently and/or are operated irregularly.

It is very common that the sewerage systems are combined sewers which in the rainy season are frequently overloaded due to high stormwater volumes. Usually there is no infrastructure for the separation and retaining of stormwater. Also, the wastewater treatment plants suffer from high variable volumes. The consequence of the low coverage with wastewater treatment and overloading of the existing plants is the frequently observed heavy pollution of surface water close to urban areas.

The implementation of rainwater harvesting systems on a large scale offers not only opportunities for the water supply side. Since the rainwater is infiltrated on-site, lower volumes of stormwater will enter the drainage and water treatment systems.

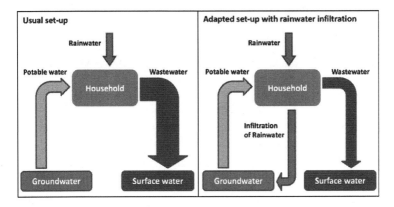

Fig. 4 Scheme of an adapted water management system including rainwater infiltration at domestic level

Proposal for Urban Rainwater Management

General Ideas

As mentioned above, rainwater harvesting at domestic level in Mexico is economically unviable due to the high investment costs required for large catchment areas and storage volumes. As the tank or cistern is the most expensive element of the system, a proposal was developed for a system operating without a storage element on-site.

The main ideas of the proposal are:

- To use the aquifer for storage of rainwater.
- Combination of decentralized rainwater harvesting systems with a centralized water supply system.

Figure 4 compares the usual set-up (on the left) with the proposed set-up (on the right). The household is served through the water supply system with potable water, originating from groundwater, and is connected to the drainage system. As mentioned above, in most urban municipalities in Mexico, the coverage with water supply is very high. The widths of the arrows used in the diagrams indicate the corresponding volumes. The management of the rainwater makes the difference between the two schemes. In the usual set-up the rainwater which falls on the impermeable areas of the household enters the drainage system and, in the case of combined sewers, mixes with the wastewater, resulting in large volumes to be treated in the treatment plant. In the proposed set-up, the rainwater captured on the rooftops is infiltrated on-site into the subsoil. Therefore, the volume of wastewater generated is considerably decreased. The rainwater infiltrated into the subsoil recharges the aquifer and is indirectly reused in the household by means of the public water supply.

The idea of using the aquifer for storage is nothing new. Sadoff and Muller (2009) state that discussions of how to provide sufficient storage tend to focus on large-scale, man-made storage dams, but there is a range of storage options, which include natural storage, such as groundwater (naturally or artificially recharged), wetlands and lakes. Garduño et al. (2006) indicate that the natural storage of groundwater bodies can be used to buffer the climatic variability of river flow and together provide a more reliable water supply at lower cost and higher efficiency than would otherwise be possible. In NWP (2007) it is noted that the use of sub-surface storage in aquifers offers advantages with regard to storage volumes and natural filtering but is relatively unknown. Consequently, examples of the application of sub-surface storage in aquifers are very hard to find in literature. Some successful implemented projects can be found in van Steenbergen and Tuinhof (2009).

Focusing only on the water supply side, the approach of storing rainwater in the aquifer has the important shortcoming that the reuse of the water requires pumping from the aquifer. But considering also the costs for drainage and treatment systems in an integrated scheme, the approach can lead to an economic solution.

Technical Issues

Figure 5 shows a possible implementation of the adapted set-up. The rainwater is captured from appropriate rooftops and then transported through gutters and channels to an artificial hollow or ditch, where it infiltrates into the subsoil. One indispensable element of the system is the water meter, measuring the withdrawal of water from the public water supply. The measurement of the water volume consumed in the household is a basis for the management system.

To aid broad application of the set-up, emphasis should be taken to choose elements which are economical and easy to maintain. The local market offers different materials for gutters and channels. There is also a whole range of different infiltration systems allowing the subsoil infiltration for different boundary conditions (soil characteristics, rain intensities, space constraints, land use, local climate, etc.), ranging from infiltration wells to simple infiltration ponds or trenches.

The infiltration basins shown in Figure 6 are very economical solutions. They are easy to construct and to maintain. They aim to copy the natural infiltration process. The upper natural soil layer with its rich content of biosphere helps to clean the infiltrated rainwater.

The volume of the infiltration basin needs to be designed according to the rain intensities, the rainwater volumes captured and the soil characteristics. In the case of intense rainwater events, the rainwater will be stored in the basin temporarily but should be infiltrated into the ground within a reasonable time to reduce evaporation losses and to prevent insect breeding. As a rule of thumb, the space requirement of such a basin can be calculated at 10–15% of the catchment area.

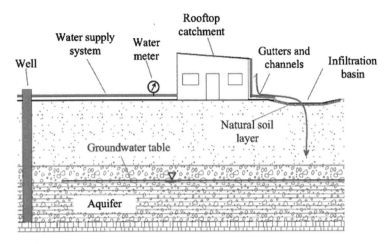

Fig. 5 Elements of the adapted set-up with rainwater infiltration

Fig. 6 Examples for the implementation of infiltration ponds in Germany. Photo on the left: infiltration basin with gravel filling for a multiple dwelling (AxelHH, Wikimedia Commons). Photo on the right: rectangular infiltration basin for a detached house, allowing for use as garden area (Pleil et al. 2005)

The form of the infiltration basin can be adapted to the local space constraints and demands for the utilization of the area. A wide but shallow basin, as shown in Fig. 6 on the right, allows for a wide range of utilization of the area when the basin is not filled with water. In areas with multiple dwellings, the captured rainwater can be directed to areas where space and proper soil is available for the infiltration (Fig. 6, on the left).

Management Model

Since the proposed rainwater harvesting systems operate without any on-site storage element, an appropriate management model is necessary. A prerequisite is the existence of a water management institution interested in better management of the hydraulic recourses. This water management institution—a municipal or private water utility or a community-led organization—secures the water supply of the households by means of a high connection rate to a centralized water supply system.

The general idea is that the household uses predominantly that amount of water which was infiltrated into the subsoil by the same household. The amount of rainwater infiltrated can be estimated easily by means of the local precipitation data and the catchment area connected to the infiltration system. This value needs to be reduced due to collection losses depending on the roof design and the gutter system, as well as to evaporation losses from the infiltration system. Calculating the infiltrated rainwater throughout a year gives a specific basic water volume for the corresponding household.

As part of the management model, an adapted tariff system with increasing block rates should be applied. In this tariff system, the first block corresponds to the specific basic water volume of each household. If the consumption exceeds the specific basic water volume, increased tariffs are applied. These household-specific tariffs offer a good basis to establish a balance between the exploitation of the aquifer and the recharge gained by the decentralized infiltration. The proposed tariff system would require a rather high level of administration capacity and public awareness-raising activities. An alternative is the classification of the households corresponding to the amount of rainwater infiltrated on-site. Households with high infiltration rates could be offered discounts on their water bills.

The management model proposes that the households would directly be responsible for the construction, operation and maintenance of the rainwater harvesting systems. This ownership fosters house-owners' understanding of functioning of the system and the importance of sustainable water resources management. The house-owners should be in close communication with the water utilities, which would have the following specific tasks within the management model:

- Providing water supply to the households.
- Analysing and monitoring the hydrological conditions, including the monitoring of the aquifer and the soil conditions.
- Identifying areas within the operation area where the implementation of decentralized rainwater harvesting systems is possible.
- Promoting the proposed decentralized rainwater harvesting systems.
- Technical advice for planning, layout, construction, operation and maintenance of the rainwater harvesting systems.
- Monitoring the implemented rainwater harvesting systems and the quality of the infiltrated water.
- Designing and implementing an appropriate tariff system.
- Financial support (loans, grants, temporarily reduced tariffs).
- Specific capacity building within the institution.

An alternative to the household-driven approach proposed here could be the introduction of user groups or neighbourhood organizations which could organize the operation and the maintenance of the systems in a better way than single households.

Benefits and Challenges of the Adapted Set-up

The main benefits of the proposed set-up are the following:

- The managed rainwater infiltration on-site prevents the pollution of stormwater due to superficial runoff, runoff in open drainage channels and/or mixing with wastewater.
- Considerably reduced volumes of wastewater, which allow for better operation and maintenance of the wastewater treatment.
- New housing projects can be designed without any stormwater systems or with considerably reduced diameters.
- Better and more secure performance of the groundwater resources due to decentralized recharge with every rain event.
- High storage volume in the aquifer, hygienically safe and long-lasting storage.
- Less danger of inundations due to reduced stormwater runoff and overloading of sewage systems.
- Every household is infiltrating "its" specific basic water volume. This, together with an appropriate tariff system, could help to prevent the overexploitation of aquifers.
- The implementation of the rainwater harvesting systems fosters the awareness about water resources management issues.
- Benefits for the users: low water price for a basic water volume and comparatively low investment costs for the implementation of the system.

The main challenges of the proposed set-up are:

- The rainwater stored in the aquifer needs to be pumped for reuse. A remedy for shallow aquifers could be the application of windmills for pumping. A project in Paraguay, presented in van Steenbergen and Tuinhof (2009), shows the successful implementation of such a concept for a small rural community. However, the energy reductions resulting from reduced wastewater volumes and the reduced inundation damages can more than compensate the energy needed for pumping.
- The set-up requires certain conditions for the aquifer and the subsoil. Bates et al. (2008) mention that the unsaturated zone should be sandy and highly permeable and that the depth to the groundwater table should be at least 4 m to avoid evaporation. Ideally, the rainwater is infiltrated in the same aquifer from which the groundwater is extracted for the water supply. Thus, clear subsoil conditions and uniform aquifers are advantageous.
- Awareness raising and motivation of the users to participate in such a rainwater infiltration programme. As the proposed set-up implies a lot of benefits for the water utilities, the water utility should actively promote the implementation and upscaling. The water utilities need to offer incentives for the implementation, including awareness-raising activities and financial support or facilitation of materials such as gutters, channels and other construction materials. The user-driven systems also need to be monitored by the water utilities and

complemented by measures in public areas such as public buildings, streets and squares implemented by the authorities to set examples.
- The definition of the tariffs and the amount of the subsidized specific basic water volume need to be carried out carefully. As the rainwater captured by a household depends on the design of the house, but the water demand depends on the number and on the activities of the persons living in the household, there is the danger of unequal conditions. A study in central Mexico indicates that for every additional person in a household, the water demand grows by 18–26%, whereas for every additional m^2 of construction, the water demand increases by 3–6% (Mosqueira 2004). The tariff system also should consider that a household which implements a rainwater harvesting system pays lower fees than a similar household which does not participate in the programme.
- Also, ground catchment of rainwater may be included to gain more infiltration. Ground catchment increases the danger of contamination. Another risk is the possible motivation of sealing up of land (the more impervious the area, the less is paid for potable water). To avoid this situation, it would be necessary to define an upper limit of the specific basic water volume which is subsidized by the tariff system.

Adapted Set-up Including Greywater Treatment and Reuse

A possible extension of the proposed set-up is the inclusion of on-site greywater treatment into the scheme.

Figure 7 outlines the adapted set-up with decentralized rainwater infiltration and greywater treatment on-site. The greywater inside the house is separated from the blackwater and directed to an on-site treatment facility. The treated greywater thereafter is used as service water in the same household or alternatively infiltrated into the subsoil. Since a large proportion of the domestic wastewater is greywater (around 70%) and an important part of the potable water can be replaced by service water, the consumption of potable water in the household and the volume of wastewater generated could be reduced significantly.

The high potential of greywater recycling as well as the design of economic on-site greywater treatment systems are very well known (for example Tilley et al. 2008). The reuse of treated greywater is possible at different levels. The high-level reuse replaces potable water as much as possible. The design requires considerable investments in new installations inside the building (plumbing, pumps, storage reservoir, overhead tanks). The low-level reuse proposes to use the treated greywater only for irrigation and cleaning purposes and for infiltration into the subsoil for aquifer recharge. For this case only small devices for short-time storage of the treated greywater are necessary.

The main additional benefits of the adapted set-up including the on-site greywater treatment are:

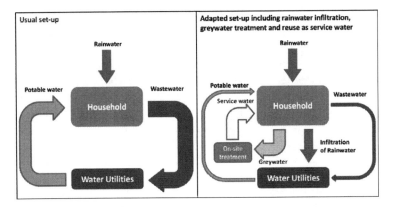

Fig. 7 Adapted set-up including greywater treatment and reuse in the household as service water applying "high-level reuse"

- For the case of high-level reuse: less potable water demand of the household.
- For the case of low-level reuse with infiltration into the subsoil: higher recharge rate of groundwater resources.
- Less wastewater generated from the household, reduced wastewater volume to be treated.

The most important challenges are:

- The household becomes responsible for greywater treatment. This is a paradigm shift which in most places is very difficult to overcome and requires intense awareness-raising activities and active communication between the water utility and the users.
- The infiltration of treated greywater into the subsoil under the responsibility of the households includes the risk that contaminated water is infiltrated due to bad operation or shortcuts.
- High investment costs for high-level reuse, lower investment costs for low-level reuse.
- The successful application of constructed wetlands for the treatment of the greywater would require the shift of washing habits (appropriate soaps and washing powders).

Conclusions

The observations made here show that even in a transition country like Mexico, which has far better economic conditions than most of the developing countries, the classical design of rainwater harvesting systems is too costly to be disseminated on a large scale. The high coverage with connections to the water supply system in urban communities in Mexico allows for a set-up which spares the storage element, the most expensive element of classic rainwater harvesting systems.

The proposed set-up combines decentralized rainwater harvesting systems at household level with a centralized water supply system, using the aquifer for storage, which, by nature, offers ideal conditions. The decentralized on-site infiltration on a large scale helps to prevent inundations and the pollution of stormwater. On the other hand, the treatment of wastewater is facilitated due to lower wastewater volumes and more steady conditions. It is important to accompany the rainwater harvesting systems with an appropriate management system, converting the households into the responsible units for operation and maintenance of their catchment and infiltration systems. The general idea of the management model is that the household uses the amount of water which corresponds to the infiltrated amount of rainwater on the site of the same household. A crucial condition for the successful implementation of the proposed set-up is the existence of a water management institution which has the technical and administrative capacity and the willingness to implement the management model in the long term. Water utilities meeting these conditions in Mexico are only found in major cities.

The discussion of the benefits and the challenges indicates that the proposed set-up can be an alternative for the urban context. The combination of decentralized rainwater harvesting systems with a centralized water supply system is very promising in order to achieve a better resilience of water resources management. A future extension of the proposed set-up, which includes the on-site treatment of greywater, offers even more benefits for water resources management. The proposed set-up is still lacking implementation. It is hoped that the basic ideas will be implemented soon in Mexico, or in another county with similar hydrological conditions.

References

Aguascalientes Gobierno del Estado de Aguascalientes (2006) Techo Cuenca (Rooftop catchment), Official webpage of the government of the state of Aguascalientes. http://www2.aguascalientes.gob.mx/inagua/EventosyParticipaciones/IVForo/TechoCuenca.aspx. Accessed June 2010

Bates BC, Kundzewicz ZW, Wu S, Palutikof JP (2008) Climate change and water. IPCC technical paper VI, intergovernmental panel on climate change, Geneva, 210 pp (ISBN: 978-92-9169-123-4)

CONAGUA: Comisión Nacional del Agua (2008) Estadísticas del Agua en México Water statistics for Mexico 2008, 1st edn. Secretaría de Medio Ambiente y Recursos Naturales, Mexico

CONAGUA: Comisión Nacional del Agua (2010) Estadísticas del Agua en México (Water statistics for Mexico 2010), 1st edn. Secretaría de Medio Ambiente y Recursos Naturales, Mexico

Garduño H, Foster S, Nanni M, Kemper K, Tuinhof A, Koundouri P (2006) Groundwater dimensions of national water resource and river basin planning, sustainable groundwater management, concepts and tools, The world bank. www.worldbank.org/gwmate. Accessed July 2010

INE (2002) Modelación econométrica de los precios en bloque para agua de uso urbano (Econometric modelling of the block prices for urban water use), Instituto Nacional de Ecología, Mexico, unpublished working report

Mosqueira LA (2004) Una Evaluación Econométrica de la Demanda de Agua de Uso Residencial en México (An econometric evaluation of the water demand for domestic usage in Mexico), Universidad Nacional Autónoma de México, Mexico, unpublished working report

NWP: Netherlands Water Partnership (2007) Smart water harvesting solutions examples of innovative low-cost technologies for rain, fog, runoff water and groundwater. In: Worm (ed.), Netherlands Water Partnership, p 65 (ISBN/EAN: 978-90-78986-01-0)

Pleil J, Wünsch M, Kaiser M (2005) Regenwasser versickern—Gebühren sparen (Infiltration of rainwater—saving fees). Informationen zu Planung und Bau von Versickerungsanlagen in München Water Utilities, Munich, p 48

Ray S (2002) Artificial recharge in small community water supplies. In: Smet J, van Wijk C (eds) Small community water supplies: technology people and partnership. International Water and Sanitation Centre IRC, Chapter 6, Den Haag, pp 101–128

Ree W (1976) Rooftop runoff for water supply, US Deparment. of Agriculture Report, ARS-s-133, Washington, DC, USA, Agricultural Research Service, US Department of Agriculture

Sadoff CW, Muller, M (2009) Better water resources management—greater resilience today, more effective adaptation tomorrow, perspectives on water and climate change adaptation, Global Water Partnership. www.gwpforum.org. Accessed June 2010

Tilley E, Lüthi C, Morel A, Zurbrügg C, Schertenleib R (2008) Compendium of sanitation systems and technologies, Swiss Federal Institute of Aquatic Science and Technology (Eawag), Dübendorf, Switzerland, p 158. (ISBN: 978-3-906484-44-0)

van Steenbergen F, Tuinhof A (2009) Managing the water buffer for development and climate change adaptation. Groundwater recharge, retention, reuse and rainwater storage, p 51 (ISBN: 978-90-79658-03-9)

Worm J, van Hattum T (2006) Rainwater harvesting for domestic use. Agrodok Agromisa Foundation and CTA, Wageningen (ISBN 90-8573-053-8)

Chapter 44
Educating for Sustainable Consumption as a Response to the Global Water Crisis: An Investigation of an Embedded Learning Approach

Daniel Fischer and Eva Freund

Abstract Patterns of consumption are considered as a main driver of climate change. Consequently, the promotion of sustainable consumption features prominently on the international agenda. Education and educational organizations are considered a key instrument to contribute to a more sustainable socialization of youth consumers. Surprisingly, in light of the importance accredited to education for the promotion of sustainable consumption, specifications of educational organizations' potential contributions remain vague and empirical research on school efforts to engage students with the notion of sustainable consumption incomprehensive and rare. The paper synthesizes conceptual contributions to the discussion of education for sustainable consumption and proposes an approach to an embedded consumer learning that blends formal and informal learning methods. In the empirical study presented, qualitative research was conducted to explore the effects of curricular and action-based education on vocational school home economics students' perceptions of and relations to sustainable consumption. The findings suggest that while the school's educational methods exert an influence on students' awareness, attitudes and behavioural intentions, these only have loose implications for self-reported consumer behaviour. Also, curricular methods that students are exposed to prefer certain issues such as health and environment over others such as social and cultural aspects. Drawing on the findings of the study, implications are drawn for a more holistic approach to

---

D. Fischer (✉)
Institute for Environmental and Sustainability Communication (INFU),
Leuphana Universität Lüneburg, Scharnhorststr. 1,
21335 Lüneburg, Germany
e-mail: daniel.fischer@uni.leuphana.de

E. Freund
Berufs- und Wirtschaftspädagogik, Didaktik der Wirtschaftslehre,
Leuphana Universität Lüneburg, Scharnhorststr. 1,
21335 Lüneburg, Germany

education for sustainable consumption in schools that accounts for both rich thematic contexts and formal and informal learning processes, seeking to better align consumption-related educational theory and practice.

Keywords Education for sustainable consumption • Consumer learning • Informal learning • Food and nutrition • Water • Student company • Vocational school

Water, Consumption and Sustainable Development

In the dawn of the twenty-first century, mankind is facing a situation in which human activities "are having impacts on a planetary scale" (WBGU 1996, p. 15) that have the "potential to make irreparable changes to the entire global fabric, including [...] climate, water distribution, land use and biodiversity" (Harris 2007, cited in Barth and Rieckmann 2009, p. 23). The exerted impacts have caused transformation processes covering a wide range of environmental, economic, social and cultural changes that can be referred to as global change. The German Scientific Advisory Council on Global Environmental Change has typified sixteen syndromes of global change (WBGU 1996) that cover utilization (e.g. overcultivation, overexploitation), development (e.g. uncontrolled growth, urban expansion) and sink (e.g. waste disposal, contamination) syndromes. The scarcity of water resources and the problems of distributing access to them are representative of overarching problems in the context of global change. Key issues in the freshwater crisis comprise potential conflicts ensuing from water resource problems, threats to human health and food provision, degradation of freshwater habitats and neighbouring biotopes, as well as the development of adequate water treatment technologies (WBGU 1997). The global water crisis marks an aggravating problem in the context of sustainable development and a priority field of action to ensure that humans' most basic and existential needs are sufficiently met.

> "Two-thirds of the global population live in areas receiving only one-quarter of the world's rainfall. And about 40 percent of the global population currently live in water stressed areas. Experts predict that this figure will rise to between 50 and 65 percent by 2025" (Zygmunt 2007, p. 9).

Given the urgency of the problems outlined, surprisingly little attention has been paid to it in the first decade after Rio. While the concepts of global warming potential (e.g. the recently popularized product carbon footprint) (Finkbeiner 2009), and particularly the ecological footprint (Wackernagel and Rees 1996), have established themselves in the broader discourse on product life cycle assessment approaches to sustainable consumption and production, the discussion of water resources management mainly remained focused on the production side and confined to expert circles. It was not until recent years that an elaborated consideration of water consumption in the entire life cycle of products has advanced. In 2002, the concept of the water footprint was developed "in order to have a consumption-based indicator of water use that could provide useful information in addition to the traditional production-sector-based indicators of water use" (Hoekstra and Chapagain 2007, p. 36). The definition of the concepts of water footprint and the related virtual water reveals close links to consumption:

> "The water footprint of a country is defined as the volume of water needed for the production of the goods and services consumed by the inhabitants of the country. Closely linked to the water footprint concept is the virtual water concept. Virtual water is defined as the volume of water required to produce a commodity or service. International trade of commodities implies flows of virtual water over large distances. The water footprint of a nation can be assessed by taking the use of domestic water resources, subtract the virtual water flow that leaves the country and add the virtual water flow that enters the country. The internal water footprint of a nation is the volume of water used from domestic water resources to produce the goods and services consumed by the inhabitants of the country. The external water footprint of a country is the volume of water used in other countries to produce goods and services imported and consumed by the inhabitants of the country." (Chapagain and Hoekstra, 2004: 9)

Estimations of different studies based on the water footprint approach show considerable differences in national water usage. The United States, for example, uses more than three times as much water per capita per year than China. How can these variations be explained? Besides climatic conditions and agricultural practices, the overall volume of consumption (related to the GDP) and the water intensity of consumption patterns (e.g. high versus low meat consumption) are direct factors determining the water footprint of a country (Hoekstra and Chapagain 2007).

From this cursory introduction it becomes clear that the water crisis is intertwined with and perfectly reflects many of the observed phenomena of global change. The identified drivers of unsustainable water consumption (overall consumption levels, particularly problematic patterns of consumption) suggest situating the demand-side problems of the global water crisis in the broader context of globalized patterns of consumption and production. Such an approach considers both the demand and supply side and accounts for its interconnectedness with (virtual) water consumption in other consumption domains (textiles, food). The escalating ideal of a material-intensive lifestyle, as is characteristic of the industrialized world, marks a central challenge to the preservation of water resources and the objective of fair and equal access to them, as expressed in the United Nations Millennium Development Goals. Hence, western-style consumer culture poses a key challenge to the concerns of sustainable development. Consequently, in Chap. 4, Agenda 21 acknowledges the need to change unsustainable patterns of consumption and production. For this change to take place, action is called for to develop a better understanding of the role of consumption and how to bring about more sustainable consumption patterns (UNCED 1993). Education is the most frequently used word in Agenda 21. Not surprisingly, in Chap. 36, education and educational organizations are unanimously ascribed to play a pivotal role in awareness raising and bringing about a change towards sustainable consumption. The following chapter explores the potential contributions of education to the promotion of sustainable consumption and concludes with the postulation of some propositions for the design of effective learning environments.

Learning to Consume Sustainably

Ten years after the adoption of Agenda 21, the Rio +10 summit in Johannesburg reaffirmed the importance given to changing unsustainable patterns of consumption and initiated the so-called Marrakech Process, which proposes a ten-year framework to progress globally towards more sustainable consumption and production patterns. One of the task forces established is concerned with "Education for Sustainable Consumption". To strengthen the role of education, the summit recommended the launching of a UN decade on "Education for Sustainable Development". The decade (2005–2014) sets out to "to integrate the principles, values, and practices of sustainable development into all aspects of education and learning" (UNESCO 2005, p. 6) and explicitly aims to develop "knowledgeable consumers who purchase goods with low lifecycle impacts and who use their purchasing power to support corporate social and environmental responsibility and sustainable business practices" (ibid.: 29). Education for sustainable consumption (ESC) has become a central theme in efforts related to education for sustainable development and the promotion of sustainable production and consumption. The following section provides a brief review of existing investigations and extracts some shared assumptions with regard to the design of effective learning environments.

Education for Sustainable Consumption: Some Propositions

The most common and widely used working definition of sustainable consumption was proposed by the Oslo Round Table of Sustainable Consumption and Production in 1994, defining sustainable consumption as the "use of goods and services that respond to basic needs and bring a better quality of life, while minimizing the use of natural resources, toxic materials and emissions of waste and pollutants over the life cycle, so as not to jeopardize the needs of future generations" (UNEP 2008, p. 19). Yet, beyond this seemingly consensual surface sustainable consumption remains a controversial concept that can best be perceived of as an "umbrella term" (UNEP 2001) accommodating a wide scope of competing concepts. Within the educational discourse, the Oslo definition of sustainable consumption is disputed and contested. Critics argue that it implies a simplistic relationship between supply and demand and reduces sustainable consumption to an unpolitical matter of individual choice (Fien 2000). Also, it is criticized for privileging ecological sustainability over social, cultural and economic aspects. This criticism is reaffirmed by a cursory investigation into contested themes in sustainable consumption and their representation in consumer education, which revealed a bias towards ecological themes (Fischer 2010).

As UNESCO states, the aspired transition needs to be primarily a cultural change: "To the extent that the global crisis facing humanity is a reflection of our collective values, behavior and lifestyles, it is, above all, a cultural crisis" (UNESCO 1998, p. 35). What is asked of education is nothing less than "to promote attitudes and behavior conductive to a culture of sustainability" (ibid.: 1). This requires a transformation of individual and institutional assumptions and practices and presupposes concerted efforts on a broad range of themes and issues. While it is widely consensual that sustainable consumption calls for a critical revision of highly routinized lifestyles and shared norms and values that can thus be perceived of as a fundamental challenge to the very essence of contemporary western culture (Reisch 2005), such transformative aspirations fall short behind the predominant reformative orientations as they are represented in teaching resources in the context of education for sustainable consumption (Fischer 2010). This is not surprising, given the fact that consumption has become a "non-natural matter of course" (Tully and Krug 2009, transl. by the authors) in everyday life. Yet the roles of consumption and material culture at least in large parts of the western world do not only pose a challenge, but also open up new opportunities as consumption becomes "a vehicle through which present and future solutions to the problems of unsustainability may be reached" (Fien 2000, p. 53).

According to a common distinction, consumer goods and services satisfy human needs through either their value-in-use or their symbolic value. While the former refers to goods and services in their "utilitarian nature" (Baba 2005, p. 237) and addresses their functional properties (e.g. performance, reliability, durability, number and type of features, price), the latter refers to their ability to carry and express meaning beyond their functional applicability (McCracken 1986). It is especially through their symbolic features that consumer goods and services fulfil several functions as satisfiers of basic needs. As symbolic meaning attached to material artefacts is "neither embodied in the artefacts themselves nor entirely open to personal interpretation"(Jackson 2008, p. 347), the reconstruction of shared symbolic meaning inevitably requires collective interpretative, reflective and deliberative processes in an interactive social setting. In light of this, social learning theory is proposed to reorient consumer education towards education for sustainable development (McGregor 2009).

The main contribution of education is seen in the provision of rich learning contexts that enable students to acquire competencies needed to participate in shaping a sustainable society (De Haan 2006). The orientation towards a competency-based approach to education in the context of sustainable development entails a fundamental shift from transmissive to transformative learning strategies (Sterling 2006; Wals and Corcoran 2006). In such understanding, educational interaction and communication is not about instruction to transport information, but about the exchange of meaning (Biesta and Miedema 2002). Consequently, to be transformative in nature, education for sustainable consumption must allow for meaning-making as "epistemological moves" (Rudsberg and Öhman 2010, p. 96). Against this background, competency-based approaches are underpinned by a participatory understanding of sustainable development that seeks to engage students in the societal discourse on and action towards a sustainable change. Explicit references in the debate about competency-based approaches to education for sustainable development are made to critical theory and the notion of education as *Bildung* in the understanding of educational theorists in the humanist tradition

(Mogensen and Schnack 2010). With respect to sustainable consumption, among the competencies discussed are skills to responsibly influence marketplace practices and structures, to engage in individual and collective consumer action, and to critically process and assess information (Barth and Fischer 2011). While an instrumental function of education (e.g. equipping the future workforce with skills required in a sustainable economy) is generally undisputed and regarded as indispensable (Bänninger et al. 2007), the current rhetoric of education as "one of the most powerful tools" (OECD 2008, p. 25) to achieving sustainable development objectives has been severely criticized for illegitimately attempting to instrumentalize education for political means (Jickling 1992). It is argued that in light of the normative nature of the sustainable consumption agenda and the emancipatory aspirations of transformative learning, a pedagogical engagement with sustainable consumption must refrain from imposing politically preconceived objectives upon students (De Haan et al. 2008). As Schnack (1996, p. 11) puts it:

> It is not and cannot be the task of the school to solve the political problems of society. It is not the task to improve the world with the help of the pupils' activities. These must be assessed on the basis of their formative value and, thus, according to educational criteria. A school, regarded as a school, does not become 'green' by conserving energy, collecting batteries or sorting waste. The crucial factor must be what the pupils are learning from participating in such activities.

Hence, education for sustainable consumption must consist of "nothing less than the acquisition of knowledge, attitudes and skills necessary for functioning in today's society" (Pierre and Thoresen 2009). For this, consumer education must widen its scope and connect "the responsibilities of individuals as consumers with their responsibilities as citizens" (O'Riordan 1994, cited in Fien 2000, p. 57) when addressing learners. As such, the concept of education for sustainable consumption is underpinned by the notion of civic action (as opposed to consumer behavioural change) (Fien 2000; Jensen and Schnack 2006).

An Embedded Approach to Consumer Learning in Schools

What implications can be drawn from the conceptual investigations on education for sustainable consumption for an elaborated approach to consumer learning and the design of respective learning settings?

Burandt and Barth (2010) have described specific requirements on learning settings that allow for the acquisition of such appropriate competencies and skills. They name self-directed learning, collaborative learning and problem-oriented learning as three distinctive principles characterizing such competency-based learning environments. This can be complemented by the notion of whole-school approaches that seek to align formal and informal learning environments and advance schools as models of sustainable practices (Henderson and Tilbury 2004; Mogensen and Schnack 2010). Table 1 provides an overview of some key conceptual components of education for sustainable consumption.

Table 1 Conceptual components of education for sustainable consumption

Level	Component	Exemplary sources
Individual objectives	• development of an understanding of a range of concepts (knowledge) • encouragement of reflection on the effects of personal values and lifestyle choice (valuing processes) • promotion of skills for personal reflection, decision making and civic action (action competence)	Fien 2000
Collective objectives	• a rational utilization of natural resources necessary to human life • an equitable economic and social development • a better quality of life for all	Pierre and Thoresen 2009
Principles	• comprehensive understanding of sustainable consumption • civic understanding of consumers • prohibition of overpowering (action vs. behavior) • whole-school approaches	Fien 2000 de Haan et al. 2008 Jensen and Schnack 2006 Henderson and Tilbury 2004
Methods	• self-directed learning • collaborative learning • problem-oriented learning • authentic and consistent formal and informal learning environments	Burandt and Barth 2010

A practical example of a whole school approach to sustainable consumer learning was developed in the context of the project BINK (Barth et al. 2010).[1] The three-year research and development project set out to examine the potential contributions of educational organizations to contribute to the promotion of sustainable consumption. Two overarching goals were pursued: on the organizational level, the project aimed to establish a culture of sustainable consumption in six affiliated educational organizations. On the individual level, the goal was to foster students' sustainable consumer competence, awareness and behaviour.

The BINK approach focuses on schools' twofold contributions to consumer learning. On the one hand, schools are understood to act as a facilitator of education related to sustainable consumption that makes students reflect and render their own consumption patterns more conscious. On the other, schools themselves are consumption households and as such places in which school members consume. This twofold character, being both a place of learning and a life world in itself, scaffolds a school's characteristic "culture of consumption" (Fischer 2011).

For a consistent learning about sustainable consumption, formal and informal learning occasions need to be equally addressed and systematically aligned.

---

[1] www.consumerculture.eu

Fig. 1 Embedded consumer learning for sustainable consumption

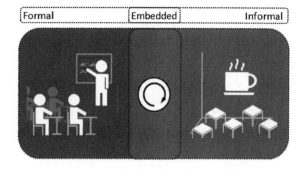

Adopting such an integrative perspective, an embedded learning approach seeks to integrate formal and informal learning activities on matters of sustainable consumption within the school environment (Fig. 1). Importantly, embedded learning must account for both sides of the coin and encompass the adoption of both supply and demand perspectives.

The following section discusses student companies as an exemplary method and format for embedded consumer learning on sustainable consumption.

Student Companies: An Action-Oriented Approach to Embedded Consumer Learning

Consumption can be understood as a hypernym or generic term that subsumes several domains of consumption activities. According to the analysis of the environmental impact of household consumption activities of the global affluent consumer class by Spangenberg and Lorek (2002), food and nutrition is one of three priority consumption clusters that (together with construction and housing, and transport and mobility) make up for nearly 70% of material extraction and energy consumption and more than 90% of land use (ibid.: 135). The food-producing sector accounts for approximately 20% of Germany's total primary energy demand and the national carbon dioxide emissions (Erdmann et al. 2003). It also plays a pivotal role in the genesis of water footprints, as about 85% of humanity's water footprint can be related to the consumption of agricultural products (Hoekstra 2010). Against this background, the consumption cluster of food and nutrition can be regarded as a crucial field of sustainable consumption that has potentials for significant stress release on energy and water resources.

How can educational approaches take up the consumption cluster of food and nutrition and provide embedded learning environments based on the tenets outlined above? Innovative learning formats that seek to enable practical learning and prepare students for working life in German vocational schools comprise specific learning opportunities such as student companies, project work or other forms of active learning. Student companies and related formats are an exceptional and exemplary suitable approach to sustainable consumer learning as they link activity-orientated and reality-based learning with embedded learning environments that blend formal and informal contexts. Students involved in such formats are offered insights to working life and the economic system in a safe setting. Valid data about the numbers of such projects is virtually unavailable. As an exploratory study estimates, student companies operate in about one in four secondary schools in Germany (De Haan et al. 2009, p. 16).

Student companies can be understood as "a pedagogical tool based on practical experience by means of running a complete enterprise project, and on interaction with the external environment" (European Commission 2005, p. 14). Originally, student companies were promoted because of their potential to help develop entrepreneurial attitudes and skills among the students through participative and creative activities. Only in the last decade has the format been applied in the context of education for sustainable development. Sustainable student companies were set up as a way to help students develop, implement and evaluate different strategies to cope with the requirements of the sustainable consumption and production agenda. Among the competencies fostered in student companies are the ability to motivate oneself and others or problem-solving skills for the handling of complex or incomplete information (De Haan 2003). With regard to work-related learning in vocational schools, such an action-learning approach was regarded as an opportunity to promote professional action competences. Action competences include skills which empower the individual to act in a problem-solving and independent manner in a practical sense (Ott 2007), but action competence can also be seen as an "educational ideal", as Mogensen and Schnack (2010, p. 71) put it. Central characteristics of the action-learning approach are planned acting, experience-oriented, self-organized, and exemplary and personality-developing learning, as well as a vital mix of methods and social forms (Arnold and Gonon 2006). The action-learning approach thus involves different methodical practices which allow an independent, multi-sensual exploration of learning contents (Gudjons 2008). An action approach to sustainable consumption refrains from imposing moral appeals and encourages experimenting with sustainable practices. Importantly, this requires the provision of alternative opportunities to act and consume sustainably in the personal and professional environments of the students (Arnold and Gonon 2006). Student companies interact with the school's internal and external environments. As students, teachers and parents are often customers of the student companies or otherwise recipients of their work. Student companies operating in the context of sustainable consumption also contribute to the integration of the sustainability discussion into school life and raise awareness of consumption matters.

Results of an Empirical Study

This section provides an insight into a practical approach in vocational school that reflects the principles of embedded consumer learning as outlined above. The study presented draws on the case of a student company operating at one school that participated in the BINK project.

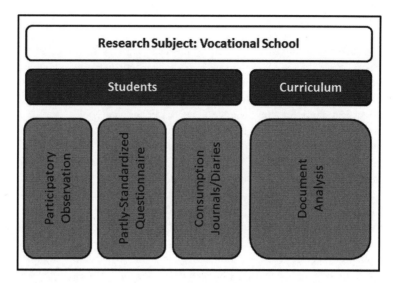

Fig. 2 Empirical methods used in the study

Study Design

In an empirical study in the context of a diploma thesis, qualitative research was conducted to explore the effects of curricular and action-based educational opportunities on vocational school students' perceptions and understandings of sustainable consumption. The goal of the study was to illustrate the relations and interactions between school activities and personal eating behaviour. The study sought to explore patterns of how the students deal with the themes of sustainable consumption and especially the theme of eating in their private and school life. In particular, exposure to and participation in student company activities was related to food and nutrition practices beyond the school premises. In addition to this, the potentials of sustainable consumption for class and school culture were analysed.

One of the research questions the study sought to answer was to what extent school and especially participation in student companies influence the eating behaviour of students. In particular, the effects on the students' sustainable consumption orientations were explored. Data was collected at a vocational school in Lower Saxony, Germany, during the 2008–2009 academic year. The research was limited to a class of full-time students receiving vocational training in home economics with theoretical and practical classes. The main topics during the school year were dietetics, planning and preparation of meals, food safety, cleaning, and others. As a part of the practical education, the students were in charge of operating the school cafeteria once a week. The study focused on the effects of both the experiences in the school cafeteria as well as curriculum-related inputs.

A comprehensive set of different qualitative methods was applied to collect the empiric material (Fig. 2). Participatory observation, a partly standardized questionnaire, consumption journals/diaries and document analysis were combined to cope with the complexity of the research subject (Flick 2006, 2008; Konrad 2001; Sichert et al. 1984). The sample comprised ten data sets, of which seven (all female) were complete and included in the final analysis. To analyse the patterns of eating behaviour of the students, data on attitudes towards eating, reflections of individual behaviour and curricular topics as well as self-reported behavioural changes was collected through a partly standardized questionnaire that was administered to the students. The students also kept an alimentary diary in which they recorded their daily meals and food consumption patterns for a period of 3 weeks. Further data was collected from participatory observation in the cafeteria in which the student company operated. Finally, a document analysis of schedules, resources and curricular documents was conducted to assess themes, topics and issues addressed in classroom teaching.

In the data collection, the term behaviour is understood as a generic term that comprises conscious and unconscious actions (Neef 2008, p. 123), whereby a broad understanding of the term behaviour was adopted that recognizes several types of behaviour (Stern 2000) and relates to what Mogensen and Schnack (2010) referred to in their conceptualization of action competence. Behaviours arise out of an interplay of knowledge, attitudes, values and situational facets such as offers and incentives (Fietkau and Kessel 1981). In the framework of the study presented, special attention is given to the interrelation of attitudes and actual behaviour, because awareness of and positive attitudes towards a given object (e.g. sustainable consumption) can be considered as a pre-condition to actual behaviour in a given way (e.g. to consume in a sustainable way).

The Learning Environment of the School Cafeteria

The principal activity in practical class in the school cafeteria is to sell different foods and drinks from breakfast to lunch to the other school members. Further activities include ordering goods from different suppliers and keeping the records for accounting. These tasks are supervised by the teachers. Time management and service quality control are important regular activities in the operation of the business. While operating the school cafeteria, the students encounter a variety of possibilities to get in touch with sustainable consumption. For example, the spectrum of aspects to be considered when composing the meals and selecting the ingredients ranges from the origin of the products (e.g. indicators for resource use such as the input of water or electric energy) via its way of production to its disposal (e.g. the output of waste and wastewater). Naturally, learning settings in practical class focus on the acquisition of practical skills. Only little time is allocated to more complex or abstract topics. However, these feature more prominently in theoretical classes. Although theory and practice are sought to be taught in an integrated approach, the full potential of inductive practical activities for learning about more abstract concepts in the context of sustainable consumption and global water usage is not yet fully tapped by the arrangements observed.

Attitudes Towards Eating and Changes in Eating Behaviour

The data retrieved shows that the students' alimentary styles are structured by similar external constraints. All students are still integrated in the household of their family of origin, where the main responsibility for food supply is up to the mothers. Besides that, similar patterns of food preferences and practices could be observed among the students. Almost all the students reported a high level of sweet consumption, showing a low intake of fruits and vegetables and taking meals at irregular times.

With respect to students' attitudes, changes towards more responsible eating practices can be observed. During the school year, the students were highly exposed to different curricular topics and activities in the field of food and nutrition. The data on self-reported increase in awareness about their own eating habits, routines and practices suggests that these curricular activities had an impact on the students. Practices of the students' family households and those taught in school were compared. Especially health aspects of nutrition were considered more important by the students since attending the vocational school. They show positive attitudes towards healthy nutrition and report intentions to change their own eating behaviour towards a more healthy style. The increased awareness also focuses on practical aspects of food preparation or the arrangement of mealtimes, which seem more important to the students now. Other aspects of sustainable nutrition such as ecological impacts of their food and food production, or social and cultural patterns of eating behaviours, only feature randomly in the students' reports. The domains of increased awareness reflect the range of themes covered in school: while health aspects are addressed and regularly taken up in different curricular contexts, other topics of sustainable consumption and nutrition only play a marginal role. The complex theme of water usage, for example, features only in the economic aspects of efficient use of water and is not discussed in detail for different areas of home economics. In this way, questions of what sustainable water usage means in a globalized context and practically, for example, in the context of cleaning or the preparation of meals, are dismissed. Consequently, the student reports did not account for any increase in awareness of the problems of global water usage.

Eating Behaviour in Comparison to Attitudes

The results of the study suggest two tendencies. One the one hand, the students' actual eating behaviour differs from their reported and intended behaviour in several ways. The data shows the discrepancy between the students' routine and habitualized eating behaviour that has been influenced and shaped by their family household and other social and cultural circumstances, and the knowledge of responsible alimentary practices based on facts and recommendations about nutrition presented and exemplified in school. Although good intentions about healthy nutrition are made, the practised eating behaviour is routinized and locked in (Sanne 2002). The consumption of fruit and vegetables, for example, is far below the five servings a day suggested by the German Association of Nutrition (DGE 2007). Water as a resource in household economics or as a distinct component of individual alimentary practices did not receive any particular attention by the study subjects. By contrast, individual changes in the students' habits can be observed with respect to the choice or combination of meals, as well as the development of practical skills in cooking.

Discussion and Outlook

From a long tradition of research on pro-environmental behaviour it is known that the relationship between knowledges, attitudes and behaviour are only loose and not of a monocausal nature (Kollmuss and Agyeman 2002). We also know that behavioural change is a complex endeavour that is particularly hard to sustain in the mid and long term. Against this background, changes in the eating behaviour of students can neither depend solely on knowledge and attitudinal factors, nor can they be expected to be realized from one day to another. However, if stable and routinized sustainable practices are considered as the final stage in a number of different stages an individual goes through, education has the potential to stimulate a transition from initial (pre-contemplation) to further advanced stages (contemplation, preparation, action) by critically reflecting one's own habits and those living conditions affecting and shaping these habits (Bamberg 2007). The results of the study presented suggest that approaches to action-oriented embedded consumer learning on sustainable food and nutrition in a student company in vocational schools offers a promising approach to awareness raising in different contexts, as it operates with real life situations and provides opportunities to experiment with different ways of problem-solving. Prospectively, the approach could also be a promising doorway to introducing topics of the global water crisis to the educational agenda and integrating them into embedded learning settings (e.g. student companies). A didactical proposition for this is an approach that connects abstract concepts such as virtual water to day-to-day issues that are personally relevant to the students and link these everyday experiences subsequently to local, national or global contexts. However, referring to the principles underpinning educational approaches to sustainable consumer learning, the objective of school activities must be directed at empowering students to be able to proceed through the different stages to maintained sustainable consumer practices. Whether the students eventually adopt such practices is beyond the scope of education (De Haan 2006).

The empirical analysis indicated the complexity that is in place regarding a reconstruction of human consumption patterns and styles. Knowledge, attitudes, and formal and informal learning in school are factors in this complex field, but they are neither the only nor the universally determining factors, given the varying intensity of other influences detected such as previous experiences in life or social influences such as peer groups, family or advertising. Given the limited scope and the exploratory nature of the study, a deeper understanding of these influences requires a more detailed and focused approach. Referring to the approach of embedded learning on sustainable consumption, the study identified some possible starting points for educational strategies seeking to promote sustainable consumption among students. Besides curricular offers that have the potential to stimulate contemplation on matters of sustainable consumption, it can also engage the students in specific actions and contribute to the acquisition of experimental knowledge and practical skills. The practical focus of the learning setting showed that changes are taking place when authentic practical action is taken and that this can be a way of achieving awareness of the global water crisis. With respect to the implementation of water issues into learning settings as described above, it must be made clear that the issue of water use does not mark an add-on to an already overloaded curriculum, but can be smoothly integrated in the day-to-day operations in different contexts of the practical work (e.g. cooking, cleaning, washing) to stimulate learning processes. Beyond that, schools can function as trustworthy models that understand themselves as learning organizations on the road to sustainable consumption. Hence, schools show various and individual starting points to put a culture of sustainable consumption into practice. There are various entry points for bringing the issue of the global water crises to school: from curricular-based theoretical inputs to structural changes in sanitation and practical experiments in the individual job-related classes. With respect to an integrative approach to these entry points, the study underlined the importance of linking theory and practice in vocational education for more comprehensive and richer student learning.

An implication that can be immediately derived from the study is the suggestion to widen the scope of curricular contents in the context of education for sustainable food consumption, pay more attention to the global water crises and explicate the links between these two agendas and between these abstract concepts and everyday household practices. The study found that a comprehensive approach to addressing the issue of sustainable eating is still missing. This is particularly true for water issues and water education. While water can be regarded as a cross-cutting theme in the consumption domain, it is so far widely neglected. Only recently have the concepts of virtual water and water footprint been incorporated in ESD, for example in the context of the annual theme water of the UN decade on ESD in Germany in 2008.[2] Still, carbon dioxide emissions and energy consumption remain predominant perspectives in education for sustainable consumption. Hence, the explicit integration of more and different food consumption aspects to the curriculum and into school life seems an important direction for the further advancement of sustainable food consumption (Fischer 2007). As one of the urgent global challenges of the 21st century, the global water crisis could serve as a focal point and lens to put forward such re-orientation and expansion of a more sustainable food education.

Acknowledgments The empirical part of the study stems from a diploma thesis written at the Institute for Environmental & Sustainability Communication (INFU) at Leuphana Universität Lüneburg that was supervised by Dr Horst Rode and Dr Matthias Barth. The authors would like to thank the supervisors for their valuable support and an anonymous reviewer for helpful comments.

References

Arnold R, Gonon P (2006) Einführung in die Berufspädagogik, Opladen
Baba ML (2005) Anthropological practice in business and industry. In: Kedia S, van Willigen J (eds) Applied anthropology, domains of application. Praeger, Westport, Connecticut, pp 221–262
Bamberg S (2007) Is a stage model a useful approach to explain car drivers' willingness to use public transportation? J Appl Soc Psychol 37(8):1757–1783
Bänninger C, di Giulio A, David CK (2007) Schule und nachhaltige Entwicklung. Gaia–Ecol Perspect Sci Soc 16(4):267–271
Barth M, Fischer D (2011) Key competencies for sustainable consumption (in press)
Barth M, Rieckmann M (2009) Experiencing the global dimension of sustainability: student dialogue in a European-Latin American virtual seminar. Int J Dev Educ Glob Learn 1(3):23–38
Barth M, Fischer D, Michelsen G (2010) Bildung für nachhaltigen Konsum. Gaia–Ecol Perspect Sci Soc 19(1):71
Biesta GJJ, Miedema S (2002) Instruction or pedagogy? The need for a transformative conception of education. Teach Teach Educ 18(2):173–181
Burandt S, Barth M (2010) Learning settings to face climate change. J Clean Prod 18(7):659–665
Chapagain AK, Hoekstra AY (2004) Water footprint of nations: appendices. Vol 2, http://www.waterfootprint.org/Reports/Report16Vol2.pdf. Accessed July 2010
de Haan G (2003) Bildungsstandards und Kompetenzen der Bildung für eine nachhaltige Entwicklung. DGU Nachrichten 27/28:24–32
de Haan G (2006) The BLK '21' programme in Germany: a 'Gestaltungskompetenz'-based model for education for sustainable development. Environ Educ Res 12(1):19–32
de Haan G, Kamp G, Lerch A, Martignon L, Müller-Christ G, Nutzinger HG, Wütscher F (2008) Nachhaltigkeit und Gerechtigkeit: Grundlagen und schulpraktische Konsequenzen, Berlin
de Haan G, Grundmann D, Plesse M (2009) Nachhaltige Schülerfirmen, Eine Explorationsstudie, Berlin
DGE (2007) Vollwertig Essen und Trinken nach den 10 Regeln der DGE. http://www.dge.de/pdf/10-Regeln-der-DGE.pdf Accessed July 2009

---

[2] http://bne-portal.de/coremedia/generator/unesco/en/04__The_20UN_20Decade_20in_20Germany/04__Annual_20Themes/Annual_20theme_3A_20Water.html

Erdmann L, Sohr S, Behrendt S, Kreibich R (2003) Nachhaltigkeit und Ernährung, Werkstattberichte, Bd 57. Institut für Zukunftsstudien und Technologiebewertung (IZT), Berlin

European Commission (2005) Mini-companies in secondary education, best procedure project. Final report of the expert group

Fien J (2000) Education for sustainable consumption: towards a framework for curriculum and pedagogy. In: Bruun Jensen B (ed) Critical environmental and health education. Research issues and challenges, vol 46. Research Centre for Environmental and Health Education of the Danish University of Education, Copenhagen, pp 45–66

Fietkau H-J, Kessel H (1981) Umweltlernen, Königstein im Taunus

Finkbeiner M (2009) Carbon footprinting—opportunities and threats. Int J Life Cycle Assess 14(2):91–94

Fischer D (2007) Ernährung, Bildung und Nachhaltigkeit: Konzeptionelle Überlegungen zu einer nachhaltigen Ernährungsbildung. Haushalt Bildung 84(2):13–21

Fischer D (2010) Sustainable consumption—mapping the terrain: an exploratory review of contested themes and their representation in consumer education. Int J Environ Cult Econ Soc Sustain 6(3):207–226

Fischer D (2011) Educational organisations as cultures of consumption: cultural contexts of consumer learning in schools. Eur Educational Res J 10(4), (accepted)

Flick U (2006) Qualitative Sozialforschung. Eine Einführung, 4. Auflage, Reinbeck

Flick U (2008) Triangulation. Eine Einführung, 2. Auflage, Wiesbaden

Gudjons H (2008) Handlungsorientiert lehren und lernen. Schüleraktivierung. Selbsttätigkeit. Projektarbeit, 7. Auflage, Bad Heilbrunn

Henderson K, Tilbury D (2004) Whole-school approaches to sustainability: an international review of sustainable school programs. Australian Government Department of the Environment and Heritage & Australian Research Institute in Education for Sustainability (ARIES), Canberra

Hoekstra AY (2010) The water footprint: water in the supply chain. The Environmentalist 93:12–13 http://www.waterfootprint.org/Reports/Hoekstra-2010-TheEnvironmentalist_01March_Issue93.pdf Accessed 11 May 2010

Hoekstra AY, Chapagain AK (2007) Water footprints of nations: water use by people as a function of their consumption pattern. Water Res Manag 21(1):35–48

Jackson T (2008) Sustainable consumption and lifestyle change. In: Lewis A (ed) The Cambridge handbook of psychology and economic behaviour. Cambridge University Press, Cambridge, pp 335–362

Jensen BB, Schnack K (2006) The action competence approach in environmental education. Environ Educ Res 12(3–4):471–486

Jickling B (1992) Why i don't want my children to be educated for sustainable development. J Environ Educ 23(4):5–8

Kollmuss A, Agyeman J (2002) Mind the gap: why do people act environmentally and what are the barriers to pro-environmental behavior? Environ Educ Res 8(3):239–260

Konrad K (2001) Mündliche und schriftliche Befragung. Ein Lehrbuch, 2. Auflage, Landau

McCracken GD (1986) Culture and consumption: a theoretical account of the structure and movement of the cultural meaning of consumer goods. J Consumer Res 13(1):71–84

McGregor SLT (2009) Reorienting consumer education using social learning theory: sustainable development via authentic consumer pedagogy. Int J Consumer Stud 33(3):258–266

Mogensen F, Schnack K (2010) The action competence approach and the 'new' discourses of education for sustainable development, competence and quality criteria. Environ Educ Res 16(1):59–74

Neef C (2008) Förderung beruflicher Handlungskompetenz. Ein experimenteller Vergleich zwischen handlungsorientiertem und traditionellem Unterricht, Stuttgart

Organisation for Economic Co-Operation Development (OECD) (2008) Promoting Sustainable Consumption: Good Practices in OECD Countries. OECD, Paris

Ott B (2007) Grundlagen des beruflichen Lernens und Lehrens. Ganzheitliches Lernen in der beruflichen Bildung, 3. Auflage, Berlin

Pierre F, Thoresen VW (2009) Advancing sustainable lifestyles and responsible consumption through ESD. In: Workshop 3, UNESCO World Conference on Education for Sustainable Development, 31 March—2 April 2009 in Bonn, Germany. http://www.esd-world-conference-2009.org/fileadmin/download/workshops/ESD2009WS3ConsumptionEN.pdf Accessed 27 May 2009

Reisch LA (2005) The cultivation of sustainability: the long way towards more sustainable consumption. Int J Environ Cult Econ Soc Sustain 1(3):165–172

Rudsberg K, Öhman J (2010) Pluralism in practice—experiences from Swedish evaluation, school development and research. Environ Educ Res 16(1):95–111

Sanne C (2002) Willing consumers—or locked-in? Policies for a sustainable consumption. Ecol Econ 42(1–2):273–287

Schnack K (1996) Internationalisation, democracy and environmental education. In: Breiting S, Nielsen K (eds) Environmental education research in the nordic countries. Royal Danish School of Educational Studies, Copenhagen, pp 7–21

Sichert W, Oltersdorf U, Winzen U, Leitzmann C (1984) Ernährungs-Erhebungs-Methoden. Methoden zur Charakterisierung der Nahrungsaufnahme des Menschen, Frankfurt am Main

Spangenberg JH, Lorek S (2002) Environmentally sustainable household consumption: from aggregate environmental pressures to priority fields of action. Ecol. Econ. 43(2–3):127–140

Sterling S (2006) Whole systems thinking as a basis for paradigm change in education explorations in the context of sustainability. University of Bath, Dissertation Bath

Stern P (2000) Toward a coherent theory of environmentally significant behavior. J Soc Issues 56(3):407–424

Tully CJ, Krug W (2009) Konsum und Umwelt im Jugendalter: Eine Sekundäranalyse (Wissenschaftliche Texte). Deutsches Jugendinstitut, München

UNCED—United Nations Conference on Environment, Development (1993) Agenda 21: programme of action for sustainable development, Rio declaration on environment and development; statement of forest principles. UN Department of Public Information, New York

UNEP—United Nations Environment Programme (2001) Consumption opportunities: strategies for change. A Report for decision-makers, Paris

UNEP—United Nations Environment Programme (2008) Planning for change: guidelines for national programmes on sustainable consumption and production. UNEP, Nairobi

UNESCO—United Nations Educational, Scientific, Cultural Organization (1998) Educating for a sustainable future: a transdiciplinary vision for concerted action. UNESCO, Paris

UNESCO—United Nations Educational, Scientific and Cultural Organization (2005) international implementation scheme: United Nations decade of education for sustainable development (2005–2014), Paris. http://unesdoc.unesco.org/images/0014/001403/140372e.pdf Accessed 1 February 2010

Wackernagel M, Rees WE (1996) Our ecological footprint: reducing human impact on the earth. The new catalyst bioregional series, vol 9. New Society Publishers, Gabriola Island

Wals A, Corcoran PB (2006) Sustainability as an outcome of transformative learning. In: Holmberg J, Samuelsson B (eds) Drivers and Barriers for Implementing Sustainable Development in Higher Education, Göteborg Workshop, 7–9 December 2005, education for sustainable development in action: Technical Paper, Vol 3, UNESCO Publishing, Paris, pp 103–110

WBGU—Wissenschaftlicher Beirat der Bundesregierung Globale Umweltveränderungen/German Advisory Council on Global Change (1996) World in transition: ways towards global environmental solutions. Annual Report, Springer, Berlin

WBGU—Wissenschaftlicher Beirat der Bundesregierung Globale Umweltveränderungen/German Advisory Council on Global Change (1997) World in transition: ways towards sustainable management of freshwater resources. Annual Report, Springer, Berlin

Zygmunt J (2007) Hidden waters. http://www.waterfootprint.org/Reports/Zygmunt_2007.pdf. Accessed July 2011

Chapter 45
Achieving Sustainable Wastewater and Organic Solid Waste Management Taking Advantage of the Clean Development Mechanism

Sören Rüd and Reinhard Marth

Abstract Combined management of municipal or industrial wastewater and organic solid waste offers various benefits for greenhouse gas reduction and sustainable environmental management. These can be achieved by a joint treatment step, thus reducing uncontrolled methane emissions and providing sanitary handling of waste streams. The clean development mechanism (CDM) offers the opportunity to co-finance the implementation and operation of the necessary infrastructure for integrated management of appropriate wastewater and organic solid waste. Due to the additional costs in the application procedure of a CDM project, a certain amount of greenhouse gas emission reduction is required in order to achieve a viable project. The minimum magnitude for a viable CDM project is calculated for various scenarios. The results show that CDM projects are generally not applicable for small municipalities and industries. However, the required project magnitude can be achieved by bundling different project activities into one CDM project, as is the case for the combined treatment of wastewater and organic solid waste.

Keywords Mexico · Wastewater · Organic waste · Integrated management · Biogas · Anaerobic digestion · CDM

S. Rüd (✉) · R. Marth
Deutsche Gesellschaft für Technische Zusammenarbeit (GTZ) GmbH,
Edif. Semarnat, Av. San Jerónimo 458 Col. Jardines del Pedregal,
01900 Ciudad de Mexico, México
e-mail: soeren.rued@gtz.de

Introduction

Water management and climate change are linked in two important fields: on the one hand, water supply, drainage and sanitation services have to be adapted to a changing environment and, on the other hand, appropriate water management can play an important role in mitigation efforts. Water supply and sanitation contributes to the emission of greenhouse gases (GHGs) via the operation of the necessary infrastructure, such as use of electricity for water pumps, or carbon dioxide (CO_2) and methane (CH_4) emissions in wastewater treatment plants, e.g. via aeration in activated sludge treatment. However, insufficient management of treatment infrastructure, and particularly a lack of wastewater treatment, also have a very high potential to produce GHGs.

The experience in many Latin American countries shows that the wastewater treatment coverage is generally very low. According to IRC (2005), only an average of 14% of the collected wastewater in Latin America receives proper treatment. Many municipalities or service operators do not have the sufficient financial resources to invest in the construction and operation of an appropriate infrastructure for wastewater treatment. Where available, treatment infrastructure is often not operated and maintained properly due to lacking technical capacities and liquid funds, leading to lower efficiencies in the treatment process and whole plant failures.

The situation in solid waste management is similar to the water sector: as financial resources are lacking, municipalities often struggle to maintain a simple recollection and disposal system for municipal and commercial solid waste and the federal governments fail to manage appropriately hazardous waste, which is often within their responsibilities. The lack of separate collection systems, treatment infrastructure for recyclable and biodegradable waste streams, and suitable sanitary landfills for all waste fractions leads to contamination of water bodies and soils. Especially waste dumps and landfills without a biogas collection system have a high impact on GHG emissions, as degrading organic matter emits harmful greenhouse gases for decades.

According to the Fourth Assessment Report of the Intergovernmental Panel on Climate Change (IPCC), the sector of waste and wastewater contributed 2.8% of global GHG emissions, including methane emissions from landfills and wastewater, as well as carbon dioxide from waste incineration (Barker et al. 2007). This figure leads to the wrong assumption that the GHG reduction potential in the waste sector is relatively low. By integrated management measures, such as the recycling of inorganic waste fractions and appropriate treatment of the organic matter contained in wastewater and organic solid waste, the waste sector has a high mitigation potential.

Uncontrolled anaerobic degradation processes of organic matter produce CO_2 and CH_4. Of special interest are the methane emissions as they have a climate-warming potential which is 21 times higher than CO_2. In order to simplify the comparison of emissions of different GHGs, mitigation is measured in form of tonnes of CO_2

equivalent. These can be transferred to certified emission reductions (CER), representing one CER unit; one tonne of CO_2 equivalent.

The Clean Development Mechanism and its Application

The Clean Development Mechanism (CDM) is a financing instrument for GHG-reduction projects in developing countries and emerging economies. It is part of the Kyoto Protocol and permits the generation of carbon credits which can be sold on international markets (Frondizi 2009; Martínez and Fernández 2004). CDM projects can be realized bilaterally when industrialized countries or enterprises from these countries (see Annex I countries of the Kyoto Protocol) invest in developing and emerging countries (see non Annex I countries of the Kyoto Protocol) to receive CERs. Projects can also be unilateral, permitting the project owner to sell the gained CERs on his own account.

The entity which decides the approval of a CDM project and issues carbon credits is the Executive Board of the United Nations Framework Convention on Climate Change. The project cycle until the generation of CERs includes:

- Elaboration of the Project Design Document (PDD) following standardized rules
- Validation of the PDD by an appointed operational entity
- Approval of the project by an appointed national authority
- Project registration by the Executive Board, start of the crediting period
- Monitoring of GHG reduction by project owners
- Regular verification by an appointed operational entity and emission of CERs by the Executive Board
- Sale of credits

CDM is thus a mechanism of post-financing, as CERs are received after demonstrating the reduction of GHG emissions. The basis for calculating mitigation is a baseline scenario which represents a projection of emissions occurring without implementation of the proposed project. CERs are generated during the crediting period whose term is eligible for either ten or 21 years; the second case requires a re-validation of the baseline after seven and 14 years. The most important parts of the PDD are the baseline definition which is needed for calculating GHG reductions, the proof of additionality (the project would not be feasible without the additional financing obtained by the sale of credits) and the monitoring plan. Furthermore, the discussion of environmental topics and the participation of local actors is of high importance in the project development.

For calculating the GHG emission reduction and the definition of the monitoring system, one has to use methodologies published by the executive board (UNFCCC 2009a). These methodologies are updated constantly on the basis of proposed projects. It is also possible to propose new methodologies for novel project types. For small-scale projects (emission reductions generally<60,000 CERs/year), simplified methodologies can be used (UNFCCC 2009a).

CER demand is basically generated by European enterprises which participate in emission trading as well as industrialized countries which aim to comply with their emission reduction objectives. At present, produced CERs have a value of approx. US$17 on the international carbon markets. In the last two years, prices varied between US$10 and US$25/CER. Generally, a contract between seller and buyer of CERs is signed which defines prices for the whole crediting period to give security to both parties. In CER sale contracts which have not yet been produced, prices tend to be lower depending on the risks of the future project.

One has to take into account that the realization of a CDM-Project implicates extensive and costly formalities. Transaction costs are estimated at approx. US$60,000 until registration of the project (contract with the operational entity, consultants for the elaboration of the PDD) and approx. US$15,000 for every verification of emission reductions. Due to these costs, only CDM projects with a CER production above 5,000 CERs/year are viable. To obtain sufficient CERs, various scenarios of baselines can be combined in the same project. It is also possible to combine various small-scale projects in a single CDM project.

There are two types of combining small-scale projects into one CDM project: bundling of activities and the programme of activities (PoA). For bundling, the exact project location of every activity is needed, every activity in the bundle represents a separate CDM project activity, every activity is represented by a project participant and project activities can be permitted based on several activities. PoA features include that the exact location does not need to be known in every case and that the CDM project activity represents the sum of all single activities, thus only defining activities as objectives and confirming the real activities with verification. Furthermore, only the legal person who is implementing the programme represents the project activities as CDM project participants (not every single person concerned by the project) and all project activities combined in a PoA must be permitted following the same methodology. In the waste sector, still no CDM projects have been registered in the programme of activities (BMU 2009).

A CDM project can obtain considerably higher incomes if an additional certification according the Gold Standard (Gold Standard 2009) is achieved. This implies higher requisites in environmental and social topics to assure project sustainability.

In the context of wastewater and solid waste management, a CDM project is especially suitable for the following baseline scenarios:

- Wastewater with high content of organic matter which decomposes in small artificial lagoons under clearly anaerobic conditions (e.g. animal stock breeding where liquid dung is stored in open lagoons or palm oil production).
- Anaerobic decomposition of stacked solid or semi-solid organic wastes (e.g. waste with high organic content in open dumps or landfills with no biogas collection systems, sludge from wastewater treatment plants).

- Renewable energy produced from organic waste replacing energy based on fossil fuels (e.g. electricity or heat from anaerobic digestion or biomass combustion).

The entire documentations of all approved CDM projects can be found on the website of the Executive Board of UNFCCC (UNFCCC 2009b). It reveals that financing organic waste and wastewater treatment infrastructure by CDM is already very common for private enterprises. Anaerobic reactors for a mescal factory in Mexico (UNFCCC 2009b, Ref. 2333) and for livestock farming in Brazil are just two of many examples (UNFCCC 2009b, Ref. 0047).

A project in Abidjan, Ivory Coast, includes an additional aspect (UNFCCC 2009b, Ref. 2250). The amount of 200,000 tonnes of solid waste is treated annually in an anaerobic process. The remaining process sludge undergoes an aerobic composting step and is used as fertilizer by local farmers while the biogas production from the anaerobic digester is used for electricity generation. The electricity sale (25 GWh/year) gives another benefit to the project, which reduces GHG emissions by approximately 500,000 tonnes of CO_2 equivalent in the first seven years.

Opportunities to Take Advantage of CDM for Financing Wastewater and Solid Waste Treatment Infrastructure

In order to demonstrate the dimensions of GHG mitigation for water management and sanitation, as well as for solid waste management, various scenarios are calculated in the following. The calculations refer to small-scale projects (generally fewer than 60,000 CERs/year) which permit the application of simplified tools and methodologies. The minimum dimensions needed for viable CDM projects (more than 5,000 CERs/year) are determined for the different scenarios.

For the calculation of GHG emissions in a municipality which does not have a wastewater treatment system or has insufficient treatment capacities (baseline), the methodology AMS-III.H "Methane recovery in wastewater treatment" (UNFCCC 2009a) is applied. Calculations are simplified if a municipality does not have a treatment plant. The most important factors in this case are the annual volume and the chemical oxygen demand (COD) of wastewater. Additionally, it is of high importance whether the decomposition occurs in an aerobic or anaerobic manner which produces methane emissions. Such uncontrolled anaerobic decomposition processes have much higher emissions according to CO_2 equivalents. Where wastewater is discharged into rivers, bigger lakes or the sea, the potential for emission reduction is much less, due to the prevalence of aerobic conditions.

GHG emissions resulting from inappropriate solid waste management are calculated for the scenarios applying an official UNFCCC tool to determine methane emissions avoided from disposal of waste at a solid waste disposal site (UNFCCC 2009a). The most important factors for using this tool are annual volume of the organic waste fraction, composition of organic wastes, local climatic conditions, and altitude or depth of the waste dump or landfill. The tool provides data on the annually generated GHG emissions during project duration. The reason for calculating emissions per year lies in the fact that GHG emissions are not constant but rise as the volume of disposed waste is increasing during project implementation.

Despite the availability of innovative project ideas, CDM projects in developing and emerging economies in the waste sector still concentrate mainly on the improvement of disposal technology such as the burning of biogas from landfills. Insufficient knowledge of modern waste treatment technologies and the lack of experience in the implementation of these projects are among the reasons for the concentration on waste disposal (BMU 2009).

Scenario 1

In scenario 1, the minimum dimension of a CDM project aiming at establishing a wastewater treatment system is calculated. Solid waste management is not considered in this scenario. For the calculation the following values are assumed:

- Average wastewater flow: 200 l per person per day.
- Average organic load: 500 mg COD/litre.

In the baseline it is assumed that wastewater is discharged into a river. It is assumed that with the implementation of the CDM project, wastewater is treated in an adequate way (anaerobic wastewater treatment). The methodology "Methane recovery in wastewater treatment" for selected small-scale CDM project activity categories is used for the calculation of the baseline scenario (UNFCCC 2010a). The methodology applies in this case to the introduction of an anaerobic wastewater treatment system with biogas recovery and combustion to an untreated wastewater stream (river discharge).

From the possible baseline emissions for this case, emissions on account of electricity or fossil fuel used, methane emissions from baseline wastewater or sludge treatment systems, methane emissions on account of inefficiencies in the baseline wastewater treatment systems, presence of degradable organic carbon in the treated wastewater discharged into river, lake or sea, and methane emissions from the decay of the final sludge generated by the baseline treatment systems, only the methane emissions for wastewater discharged into a river apply for this case.

Considering the assumed values of a wastewater flow of 200 litres per person per day and an average organic load of 500 mg COD/l and applying the given standard values for the global warming potential for methane (value of 21), for the methane-producing capacity of the wastewater (0.21 kg CH_4/kg COD), for the model correction factor to account for model uncertainties (value of 0.94) and the IPCC default value for the methane correction factor (value of 0.1), one gets a baseline emission for 10,000 inhabitants of approximately 151 tonnes of CO_2 equivalent.

Scenario 2

In contrast to scenario 1, the implementation of an adequate solid waste management such as composting of organic solid waste is considered without taking wastewater treatment into account. In order to simplify the estimation, the only calculation performed is the baseline scenario by help of the methodology "Tool to determine methane emissions avoided from disposal of waste at a solid waste disposal site" (UNFCCC 2010b). The following values for waste composition are used:

- Average solid waste generation: 1 kg per person per day.
- Organic fraction of total solid waste: 45%.

Food scraps are considered as the main part of organic waste and the altitude of waste at the landfill is below 5 m. As local climatic conditions are of high importance for GHG emissions, two different cases are investigated:

- Scenario 2a: A cold (annual average temperature<20°C) and dry (annual average precipitation<annual potential evapotranspiration) local climate.
- Scenario 2b: A warm (annual average temperature>20°C) and wet (annual average precipitation>annual potential evapotranspiration) local climate.

As GHG generation increases each year, in this case GHG emissions are calculated for the third year of project implementation. The methane emissions avoided from preventing waste disposal at the solid waste disposal site during the third year are calculated using the following parameters:

- Model correction factor to account for model uncertainties: 0.9
- Fraction of methane captured at the solid waste disposal site and flared, combusted or used in another manner: 0
- Global warming potential of methane: 21
- Oxidation factor (reflecting the amount of methane from the solid waste disposal site that is oxidized in the soil or other material covering the waste): 0
- Fraction of methane in the disposal site gas: 0.5
- Fraction of degradable organic carbon that can decompose: 0.5
- Methane correction factor: 0.4
- Fraction of degradable organic carbon in the waste type: 0.2 (mainly food waste, partly wood and garden waste)
- Decay rate for the waste type: 0.06/0.4

For Scenario 2a, one gets assuming 10,000 inhabitants and a decay rate for the waste type of 0.06 an amount of approximately 136 CO_2 equivalent; for Scenario 2b and a decay rate of 0.4, the CO_2 equivalent for the third year is much higher, namely 578.5.

Table 1 Minimum number of inhabitants to avoid GHG emissions of 5,000 tonnes of CO_2 equivalent for the described scenarios

Scenario		Local climate	Inhabitants
1	Wastewater	Not defined	331,000
2a	Solid waste	Cold, dry	368,000
2b		Warm, wet	87,000
3a	Wastewater and solid waste	Cold, dry	175,000
3b		Warm, wet	68,000

Scenario 3

In scenario 3, both the implementation of a wastewater treatment system and the introduction of an appropriate solid waste management are considered using the methodologies described above. The same values as in scenario 1 and 2 are applied and it also distinguishes between local climatic conditions as mentioned in scenario 2.

For each scenario, the minimum number of inhabitants necessary for avoiding GHG emissions of 5,000 tonnes of CO_2 equivalent is calculated. The calculation results are shown in Table 1.

As a consequence, a CDM project which only implements a wastewater treatment system (scenario 1) should include at least a service for 331,000 inhabitants. For small municipalities, it is not viable to start such a project. It becomes evident that the baseline for the implementation of an appropriate solid waste management system depends very much on the local climatic conditions. The combination of a project in which wastewater treatment and solid waste management are realized is advantageous, especially for cold and dry climatic conditions. In scenario 3a, the minimum number of inhabitants is reduced by half compared to scenario 2a.

It is worth mentioning that wrongly designed or badly operated aerobic wastewater treatment according to methodology III.H result in treble the emissions compared to the nonexistence of any type of treatment. In this case, emissions generated by treatment plant operation and resulting from sludge management should be subtracted.

Furthermore, other components of a CDM project can be added to improve viability. If the project considers anaerobic treatment, generated biogas can be used directly or for electricity generation. For the calculation of the baseline for replacing fossil energy by renewable energy, methodology AMS-I.D. "Grid-connected renewable electricity" generation is applied. The GHG emission reduction depends on the energy sources in the absence of the project. Utilizing energy generation in a solid waste treatment project can duplicate CERs resulting from GHG emission reduction. In a CDM project in Cali, Colombia (UNFCCC 2009b, Ref. 2285), generated biogas from anaerobic decomposition of sludge is used in a municipal treatment plant applying stated methodology.

Integrated Organic Solid Waste and Wastewater Management

The calculations for mitigation of GHG emissions (Table 1) show that the combination of wastewater treatment and solid waste management can improve the viability of a CDM project. The resulting integrated management of water, sanitation and solid waste provides several advantages for the operation of services (Marth et al. 2009).

As sustainable water and sanitation services are not possible without appropriate organic solid waste management, the service provider has a vital interest in avoiding environmental pollution by waste. Separation of different solid waste fractions such as paper, plastics and organic waste at the source eases recycling, which results in a reduction of waste volume to be treated and disposed of. The combined treatment of wastewater and/or sludge with organic waste gives the opportunity to sanitize the organic matter in a safe way and produce biogas. The most efficient biogas use is via a combined heat and power plant. In case no heat is needed (this is often the case in developing countries), the biogas might be used directly (e.g. for cooking or lighting) or for electricity production. The digestate should be dried, possibly by help of the excess process heat of the treatment and sanitized, e.g. by composting, before it may be used as fertilizers, which can bring even more benefits for the service provider. The concentration of services in a single provider also offers advantages by sharing human resources, capital, equipment and materials.

Treatment of organics can be done generally by aerobic or anaerobic technologies. Domestic or large-scale composting is an aerobic solution needing sufficient ventilation and appropriate management for obtaining the necessary humidity and temperature to generate a final product that can be used as soil conditioner or fertilizer.

In order to obtain biogas from wastewater treatment, it is necessary that sewage has a high organic load. A lot of types of industrial wastewater, sewage from livestock breeding or sludge from treatment plants have a sufficient organic load and a constant flow. Pavan et al. (2000) have investigated the effects of combined organic solid waste and sewage sludge in a pilot project and conclude that the mixture is advantageous for the degradation process and the generation of biogas. Likewise, Pavan et al. (2007) demonstrate that the stated mix brings economic benefits and represents a sustainable solution.

An example for aerobic co-treatment of solid waste with wastewater is implemented in a CDM project of a palm oil factory in Izabal, Guatemala (UNFCCC 2009b, Ref. 2527). In this plant, the solid palm residues are first accumulated and then mixed with wastewater originating from oil production, thus enhancing the composting and degradation process. The final product is then used as fertilizer on the palm plantations.

Potential of the Biogas Market in the Case of Mexico

The results of a market study on the potentials of the biogas market in Mexico (Schulze 2009) show a high potential for combined projects for treatment of wastewater and organic solid waste.

Mexico generated approximately 37.6 million tonnes of municipal solid waste in 2008 with an organic fraction of around 50%. Discounting the 14% of waste generated in rural and semi-urban regions where larger-scale organic waste treatment processes are not very likely, as organic waste is often composted on-site or collected waste amounts are too small, the energy potential is estimated around 39 PJ (petajoule)/year assuming a biogas generation potential of 60 m^3/tonne of waste with a methane concentration of 50% and an energy content of 20.5 MJ/m^3 (Stege 2003). Another 111.4 PJ are potentially available from the currently 128 landfills in operation which could fuel power plants with a capacity of 44 MW in the next 20 years. Around 48.6 PJ/year are already used in the existing landfill gas projects (Schulze 2009).

In the case of wastewater treatment, the biogas potential is estimated via the biological oxygen demand (BOD) parameter assuming a biogas generation of 0.3–0.5 m^3 for the removal of 1 kg BOD in an anaerobic treatment process considering an average BOD removal efficiency of 65% in this treatment step. Assuming anaerobic wastewater treatment of 60% of the municipal sewage gives a primary energy potential of 6.5 PJ, which is, however, very unlikely, as very few municipal wastewater treatment plants include an anaerobic treatment step and only mechanic treatment or aerobic biological treatment still prevail in new constructions of wastewater treatment plants. Industrial wastewaters offer a higher and better accessible biogas generation potential due to higher organic loads, e.g. in food and beverage-processing industries. Wastewater from these industries alone amount to an energetic potential of 21.9 PJ/year.

Finally, livestock breeding and slaughterhouses are also important points of interest when considering improved management of municipal waste. The almost 30 million animals kept in Mexico for cattle meat production could potentially contribute, with total manure production of 350 million tonnes per year, to an energy potential of 630 PJ. However, one has to consider that only about half of the cattle is kept in stables where anaerobic manure treatment is feasible and that many livestock farms are too small to allow an anaerobic treatment process (Schulze 2009).

Conclusions

The Clean Development Mechanism (CDM) offers financing to projects which help to mitigate emissions of greenhouse gases. The realization of a CDM project implicates extensive and costly procedures. That is why they require a certain amount of GHG emission reductions. It is estimated that a CDM project is viable if the project generates more than 5,000 CERs/year. In this paper, the most important official tools of the UNFCCC are used in the context of wastewater and solid waste management. In different scenarios, varying the implementation of a treatment and final disposal system for wastewater and/or solid waste, the required minimum dimension of a CDM project is calculated.

The results show that using CDM is not viable for small municipalities. In order to reach a sufficient amount of CERs, various baseline scenarios can be combined in the same CDM project. The calculations demonstrate that the implementation of a CDM project which includes both wastewater and/or sludge with high organic load and organic solid waste can have certain advantages for the service provider. Integrated water, wastewater and solid waste management represent an important step for the improvement of municipal services and in their contribution in reducing greenhouse gas emissions.

References

Barker T, Bashmakov I, Bernstein L, Bogner JE, Bosch PR, Dave R, Davidson OR, Fisher BS, Gupta S, Halsnaes K, Heij GJ, Kahn Ribeiro S, Kobayashi S, Levine MD, Martino DL, Masera O, Metz B, Meyer LA, Nabuurs GJ, Najam A, Nakicenovic N, Rogner HH, Roy J, Sathaye J, Schock R, Shukla P, Sims REH, Smith P, Tirpak DA, Urge-Vorsatz D, Zhou D (2007) Technical summary, climate change 2007: mitigation, contribution of working group III to the fourth assessment report of the intergovernmental panel on climate change. In: Metz B, Davidson OR, Bosch PR, Dave R, Meyer LA (eds) Cambridge University Press, Cambridge

Bundesministerium für Umwelt, Naturschutz und Reaktorsicherheit—BMU (2009) Umwelt—Sonderteil: Nutzung des CDM in der Abfallwirtschaft, Leitfaden für Investitionsprojekte im Ausland, Vol 10, pp 4, 6–7

Frondizi I (2009) El mecanismo de desarrollo limpio—Guía de orientación, Government of Brazil. Available at http://inter.bndes.gov.br/espanol/estudios.asp. Accessed September 2009

Gold Standard (2009) The gold standard—premium quality carbon credits. Available at http://www.cdmgoldstandard.org. Accessed September 2009

IRC International Water and Sanitation Centre (2005) América Latina y el Caribe y los Objetivos de Desarrollo del Milenio. Available at http://www.es.irc.nl/page/34056. Accessed August 2009

Marth R (2009) Including solid waste management as an option for small scale water and sanitation providers. Paper presented at 1st IWA development congress, IWA International Water Association,Mexico, November 2009

Martínez J (2004) Cambio climático: una visión desde México. INE Instituto Nacional de Ecología, Mexico City

Pavan P (2000) Integration of wastewater and OFMSW treatment cycles: from the pilot scale experiments to the industrial realisation–the new full scale plant of Treviso (Italy). Water Sci Technol 41(12):165–173

Pavan P (2007) Anaerobic co-digestion of sludge with other organic wastes in small wastewater treatment plants: an economic considerations evaluation. Water Sci Technol 56(10):45–53

Schulze P (2009) Potentials of the Biogas Market in Mexico commissioned by Deutsche Gesellschaft für Technische Zusammenarbeit (GTZ) GmbH, in cooperation with Federal Energy Ministry (SENER) of Mexico. Unedited

Stege G (2003) User's Manual Mexico Landfill Gas Model, Version 2.0, Landfill Methane Outreach Program, US Environmental Protection Agency

UNFCCC (2009a) CDM: methodologies, baseline and monitoring methodologies. Available at http://cdm.unfccc.int/methodologies/index.html. Accessed September 2009

UNFCCC (2009b) CDM: project activities, project search, search of a specific project via the reference number indicated in the text. Available at http://cdm.unfccc.int/Projects/projsearch.html. Accessed September 2009

UNFCCC (2010a) III.H. Methane recovery in wastewater treatment. Available at http://cdm.unfccc.int/UserManagement/FileStorage/ANF0MTK4BHZC9O7IEY68P5DJ2VRQ3X. Accessed July 2010

UNFCCC (2010b) Tool to determine methane emissions avoided from disposal of waste at a solid waste disposal site. Available at http://cdm.unfccc.int/methodologies/PAmethodologies/tools/am-tool-04-v4.pdf. Accessed July 2010

Chapter 46
Local Wisdom of Danowudu Community in Preserving Forest as a Water Source for the City of Bitung

Trina E. Tallei and Saroyo Sumarto

Abstract Bitung is a city located in North Sulawesi province in Indonesia. It is a unique city due to having three conserved forests. Covering an area of 304 km^2 with a population of about 175,000 inhabitants, the need for clean water for Bitung is high. In the midst of the difficulties of having clean water available, there are people who have an important role in maintaining clean water resources for the community needs in the city. These are people from Danowudu sub-district who still retain 21.5 ha of forest to maintain the availability of clean water for Bitung. Local wisdom is the community's local knowledge used by the community in carrying out environmental conservation for the sake of harmonious interaction between man and environment. To retain the sustainability of the forest and a clean water source, the indigenous community implements customary rules that entail customary sanctions to offenders. These include prohibition of the destruction and taking of forest products, as well as the obligation to maintain the forest's sustainability. Local wisdom of Danowudu community has been proved to be very effective in preserving the city forests so that the source of clean water for the city of Bitung is well maintained.

Keywords Local wisdom · Customary rule · City forest · Water source · Danowudu · Bitung

---

T. E. Tallei (✉) · S. Sumarto
Dept. of Biology, Faculty of Mathematics and Natural Sciences,
University of Sam Ratulangi, Jl. Kampus Unsrat, Manado, 95115, Indonesia
e-mail: trinatallei@yahoo.com; trina@daad-alumni.de

Introduction

Local wisdom is the local knowledge that is unique to a culture or society. This knowledge is passed on from generation to generation orally or through cultural rituals, and is the basis for agricultural practice, livestock production, food preparation, healthcare, education, environment conservation and a wide range of traditional activities that sustain a society and its environment in many parts of the word (Priosoeryanto and Risa 2007). Nowadays, local wisdom has been disappearing gradually due to the introduction of modern science and technologies. For a long time, local wisdom has significantly contributed to global knowledge. One striking example is drugs derived from medicinal plants. Therefore, sharing knowledge via local wisdom is important for the betterment of society.

Water is the essence of life and issues of water conservation need to be addressed since clean water is vital to the survival of the community. One of the main sources of clean water is the forest. Forest destruction in North Sulawesi, however, is occurring on a massive scale. Various efforts have been made to protect the forest as a heritage for the future. One example is the forest conservation for the availability of clean water for the city of Bitung, conducted by the local community called Danowudu.

The name Danowudu is literally derived from two words, namely Rano (water) and Wulu (bamboo), due to the abundance of bamboo plants in the vicinity of the river. Danowudu sub-district was formed on 6 May 1908. One striking phenomenon of Danowudu sub-district is the bond and strength of its customs and traditions. Danowudu is the only sub-district in Bitung that still retains and holds the values of its customs and traditions in high esteem. To preserve these, the people have formed an organization called Pemangku Adat Negeri Danowudu (PAND) or Custom Functionary of Danowudu Land. This organization has drawn the attention of Indonesia Supreme Court by having the honour to become a research site on customary/traditional law. The role of PAND and the sub-district administration is deeply felt in natural resource management. The local people particularly have benefited greatly from the water conservation due to the preservation of natural water bodies in this area. The water bodies, in turn, are preserved due to the preservation of city forest.

Danowudu City Forest and its Water Source Management

Danowudu land is situated in North Minahasa district, North Sulawesi province, Indonesia. Its geographical coordinates are 1° 27′ 59″ North, 125° 7′ 34″ East and its original name was Ranowulu (Fig. 1). Danowudu sub-district provides clean water accessibility to the city of Bitung. It has protected natural and artificial forests that function to retain ecosystem equilibrium, especially the rate of water flow. The size of the city forest is approximately 21.5 ha (Fig. 2). The city forest

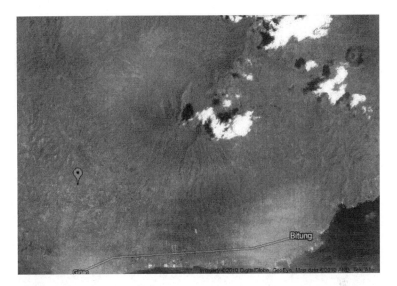

Fig. 1 Location of Danowudu land (blue marker) relative to the city of Bitung (*Source* Google Earth, 2010)

Fig. 2 City forest

consists of three custom forests, namely 12 ha of water forest, 4 ha of teak forest, and the rest is rainwater forest.

Sub-district management and PAND, together hand in hand, maintain the sustainability of the city forest. The place is only open for research purposes or nature-lover activities. Camping ground is available close to the forest. Perusahaan

Fig. 3 Danowudu reservoir managed by PDAM and PAND

Daerah Air Minum (PDAM) or Drinking Water Local Company, owned by the Indonesian government, cooperates with local people to manage the clean water source so that the water can be distributed to Bitung city community (Fig. 3). The agreement made by this company and the local people is that people of Danowudu sub-district receive clean water free of charge.

Drinking water for the Bitung city community comes from seven sources with the following capacities: danowudu I (120 l/s), Danowudu II and Air Hujan (Rainwater) Forest (both 20 l/s), Kumersot I and II (both 40 l/s), Tendeki (10 l/s), and Girian river (100 l/s). Production of clean water in 2005 was 8,021,537.8 m^3, compared to the production of 8,453,617.7 m^3 in 2004. It means that average distribution of clean water per capita in 2005 was 58.63 m^3 (Bitung City Government 2010). The water source at Danowudu I is at 194 m above sea level. Half of the water is distributed to the closest community and the rest goes by gravity to Danowudu reservoir. From this, the water is distributed to the city of Bitung. Part of the water from Danowudu reservoir goes to Madidir reservoir (44 m above sea level). Water from this reservoir is distributed to the port joins water from Aer Tembaga reservoir.

Customary Rules of Danowudu Community

North Sulawesi has been experiencing heavy deforestation. Protecting natural forest must remain as one of the fundamental conservation strategies in North Sulawesi. Forest destruction is primarily driven by anthropogenic developments such as agricultural activities and human settlement to cope with burgeoning populations (Lee et al. 2007). The credit for the existence of Bitung city forest in the midst of a populated city should be given to the Danowudu people. Inside the city forest, even fallen trees may not be taken. There is a belief that the spirit of their ancestors lives in the forest so there is a need to protect it. This is a traditional value of culture conservation that is implemented in nature conservation. This so-called community-based conservation is aimed at achieving dual goals of conserving nature and improving people's livelihoods by way of involving local communities as active partners in conservation efforts, and using traditional knowledge and local values in the management of resources (Adams and Hulme 2001). Culture conservation strengthens individual and collective commitment for a set of values, beliefs and attitudes towards nature conservation to maintain the sustainability of quality of life.

Concluding Remarks

Respecting local wisdom can give an unexpected bonus at times. Many indigenous groups have learned to manipulate natural forests to meet their demands. Local wisdom can help to develop sensitive and caring values and attitudes, thereby promoting a vision of a sustainable future. In view of its potential value for sustainable development, it is necessary to preserve indigenous knowledge for the benefit of future generations. One way to preserve indigenous knowledge is to integrate it into the school curriculum by encouraging students to learn from their community. This strategy would sufficiently decrease the gap of traditional knowledge between generations. Schools would act as agencies for transferring the culture of society from one generation to the next (Unesco 2010).

Another strategy of dissemination of local knowledge is the capacity-building strategy. This would empower national and regional government and all stakeholders to sustainably manage natural resources. Steps towards capacity building are identification of people involved, interested and competent in natural resource management. In the end, all stakeholders are encouraged to independently conduct the monitoring and evaluation of this effort.

References

Adams W, Hulme D (2001) If community conservation is the answer in africa, what is the question? Oryx 35(3):193–200
Bitung City Government (2010) Listrik dan air minum. Available at http://www.bitung.go.id
Lee TM, Sodhi NS, Prawiradilaga DM (2007) The importance of protected areas for the forest and endemic avifauna of sulawesi (Indonesia). Ecol Appl 17(6):1727–1741

Priosoeryanto BP, Risa T (eds) (2007) Empowering society through the animal health and production. Activities with the appreciation to the indigenous knowledge. J Agric Rural Develop Tropics Subtropics, vol 90. Kassel University Press, GmbH, 196 pp

Unesco (2010) Activity 5: enhancing the curriculum through indigenous knowledge. Available at http://www.unesco.org/education/tlsf/TLSF/theme_c/mod11/uncom11t05.htm

Chapter 47
Education in Global Climate Change at a Botanical Garden: Students' Perceptions and Inquiry-Based Learning

Daniela Sellmann and Franz X. Bogner

Abstract The current global climate change has increased the importance of environmental education substantially. Adolescents are known to have common concepts about global change that are often fragmentary or incorrect. Reinfried et al. (Geographie heute 265:24–33, 2008) suggest special learning materials by facing students with their own conceptions and confronting them with the correct ones. In our two-day environmental education programme, combining indoor and outdoor activities in an ecological botanical garden, we implemented specially designed materials to confront students with a common misconception about the relationship of ozone layer depletion with the greenhouse effect. Other common misconceptions about global climate change and the greenhouse effect were not explicitly mentioned in the learning material. We used concept mapping as a method to reveal conceptions and potential conceptual changes. 25 high-achieving high-school students have taken part in our study so far. Initial results show that conceptions of our participants were quite in line with the literature. We also showed that presentation of information does indeed play an important role in changing students' conceptions. Therefore, not only providing the necessary information but also confronting misconceptions should be part of any educational material.

Keywords Global climate change · Greenhouse effect · Education · Out-of-school learning · Botanical garden · Inquiry-based learning · Concept mapping · Conceptual change

---

D. Sellmann (✉) · F. X. Bogner
Department of Biology Education, University of Bayreuth, NW I,
Universitaetsstrasse 30, 95447 Bayreuth, Germany
e-mail: Daniela.sellmann@uni-bayreuth.de

Introduction

Most recent surveys describe adolescents as often feeling helpless in facing global problems such as climate change, although they are explicitly willing to take action against global warming (Forsa 2009; TNS Emnid 2009). Informing students about global climate change, its causes, mechanisms and impacts, therefore, is an important issue of teaching. This should include realistic options for climate-protecting actions (Bord et al. 2000), as well as provide sufficient knowledge about the efficacies of such actions. Environmental education can meet these demands.

Several studies with high-school students (e.g. Rebich and Gautier 2005; Sterman 2002) often showed just fragmentary knowledge levels or even misconceptions with regard to climate change, its mechanisms, causes and impacts. In a common conception, holes in the ozone layer cause an enhanced greenhouse effect or, in general, cause global warming as they would allow more radiation to reach the Earth (e.g. Andersson and Wallin 2000; Dove 1996; Ekborg and Areskoug 2006; Mason and Santi 1998). Reinfried et al. (2008) stated as a reason for such conceptions similar graphic presentations of the phenomena of "ozone layer depletion" and of the "greenhouse effect" in conventional learning materials (textbooks, etc.): specifically designed learning material might respond to individual conceptions by comparing misconceptions with correct concepts.

Students often believe that environmental problems (such as environmental pollution, air pollution or acid rain) enhance the greenhouse effect and lead to global warming (Andersson and Wallin 2000; Niebert 2007). Consequently, many students think that reducing pollution would be an appropriate solution for stopping global climate change (Shepardson et al. 2010).

Concept mapping was first developed in the 1970s (Novak and Gowin 1984) and can be used as an appropriate tool for many purposes in instructional research (Gerstner and Bogner 2009; Ruiz-Primo and Shavelson 1996; Schaal et al. 2010; Slotte and Lonka 1999). It also offers the opportunity to reveal students' conceptions and a potential conceptual change "since it allows an exploration of student knowledge at a sufficient level of complexity" (Rebich and Gautier 2005).

In our study, we implemented a two-day environmental education programme on global climate change. Our research questions were

1. Which conceptions do students have about the enhanced greenhouse effect and global warming?
2. Is it possible to change those conceptions if they are fragmentary or even incorrect by implementing our educational programme?
3. How should the implemented educational material be structured to reach this conceptual change?

Our hypothesis was that, according to Reinfried et al. (2008), the concept of ozone layer depletion causing global climate change would change the most because we implemented specially structured learning material to meet this common misconception.

Materials and Methods

25 high-achieving year 10 students (highest stratification level = *Gymnasium*) participated in a two-day environmental education programme on global climate change (the ongoing study is still producing a bigger sample size). The programme consisted of student-centred learning units based on learning at workstations, group work and inquiry-based lab experiments. Apart from imparting knowledge, the programme units were designed to provide students with climate protection actions.

The first day of our programme was conducted in the classroom and consisted of learning at workstations. The unit imparted knowledge on

1. mechanisms of the greenhouse effect and global climate change,
2. greenhouse gases and their sources,
3. impacts of global climate change on resources (water, nutrition),
4. impacts of global climate change on ecosystems,
5. climate research and scenarios,
6. climate policy,
7. climate protection actions,
8. climate change in the Earth's history.

In the learning material dealing with mechanisms of the greenhouse effect, the common misconceptions of students dealing with the relation between ozone layer depletion and the enhanced greenhouse effect were explicitly mentioned (see above). The students were forced to compare both the misconception and the actual facts about the greenhouse effect.

The second programme day took place at an ecological botanical garden featuring plants from all dominant climate zones in nature-oriented environments. Indoor and outdoor units were combined by using the public greenhouses as well as the outdoor area of the botanical garden. The unit presented different ecosystems to the students and dealt with the following topics concerning the impact of global climate change on vegetation:

1. impacts on the ecosystem of the tropical rainforests,
2. impacts on the fire ecosystems of Australia,
3. impacts on the alpine regions and their vegetation,
4. impacts on German forestry and potential actions to take,
5. fossil fuels versus energy from renewable primary products.

During the third unit of the programme, students conducted inquiry-based lab experiments dealing with simple climatic and physical phenomena related to global climate change.

To reveal students' conceptions on global climate change and the greenhouse effect students completed individual concept maps before participating in our programme (Fig. 1).

Fig. 1 Study design

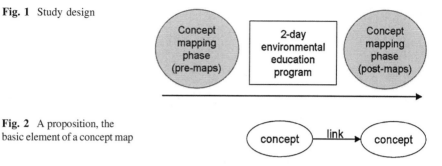

Fig. 2 A proposition, the basic element of a concept map

All participants were novices in concept mapping, which is why we implemented a 15-min introductory phase before the first concept mapping phase. To complete individual concept maps (in dyads), 30 min were available. Concept maps consist of concepts that are linked to show their relationship. Links are usually verbs. The basic element of a concept map, a proposition, is formed by two concepts linked to each other (Fig. 2). The students were provided with 20 pre-defined concepts essential for explaining the global greenhouse effect. Additionally, five concepts referring to common misconceptions reported in several studies (e.g. Dove 1996; Rebich and Gautier 2005; Reinfried et al. 2008) were also provided but not declared as such. While the concept "ozone hole" was explicitly outlined during the educational programme (see above), the other four concepts (acid rain, environmental pollution, air pollution, recycling) were meant to show whether students could logically link the information imparted within the programme to their own conceptions.

Students could use all of the provided concepts and were free to add new ones. All links had to be added by the students, none were pre-defined. In this concept mapping phase, up to now 12 concept maps (pre-maps) were completed (example in Fig. 3).

Subsequent to the two-day educational programme on global climate change, students inspected their individual concept map. They could actively revise their concept map and complete a new one within a 30-min working phase: in this way, we received 11 new concept maps (post-maps) which we compared with pre-maps. Therefore, a potential conceptual change individually could be analysed (by using CMap Tools, version 5.03): in the first step, all concept maps were analysed with regard to misconceptual portions and to the shifting rate of misconceptions/conceptions. The propositions of each concept map were extracted and analysed. Propositions that matched the common misconceptions were identified and listed. Further analysis is still in progress (enlarging the sample size).

Results

Analyses revealed that of five pre-defined concepts referring to misconceptions, "ozone hole" was the most applied in the pre-maps (92% of all pre-maps; Fig. 4). The concepts "air pollution", "environmental pollution" and "recycling" scored

47 Education in Global Climate Change at a Botanical Garden

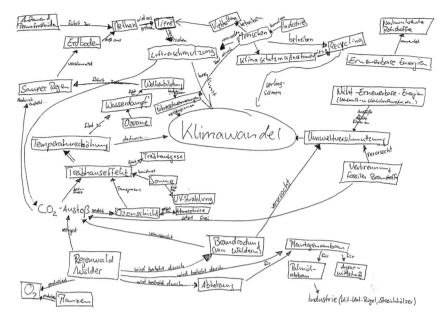

Fig. 3 Students' concept map (pre-map)

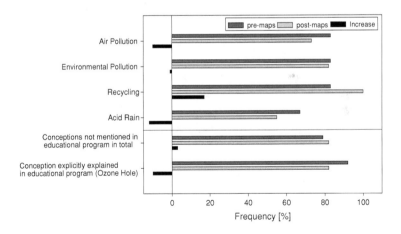

Fig. 4 Proportion of applied concepts referring to common misconceptions in pre- and post-maps

83% of the pre-maps, respectively. "Acid rain" was the least used, but still a frequent concept (67%).

Except for the concept "recycling", all post-maps' concept frequencies dropped (Fig. 4), at most in the case of "acid rain" (−12%), closely followed by "air pollution" and "ozone hole" (−10% each).

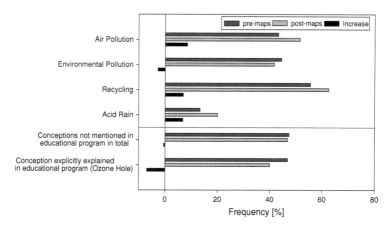

Fig. 5 Proportion of propositions which match common misconceptions in pre- and post-maps

Comparing the concept "ozone hole", which was explicitly explained in the educational programme, to the conceptions not integrated into special learning material, it can be seen that while the use of the latter ones increased slightly (3%), the former decreased by 10%.

In the following, the results of the content analysis of propositions in which the concepts referring to common misconceptions were used are presented (Fig. 5).

In the pre-maps, about 47% of propositions in which the concept "ozone hole" was used matched the common misconception. A typical example of such a proposition used by students is "The ozone hole causes global climate change". Within the post-maps, the proportion of such propositions decreased to 40%.

Propositions using "acid rain" had a proportion of 13.3% in the pre-maps, which increased to 20.0% in the post-maps. In this case, the actual number of propositions remained unchanged. The increase of proportion was caused by fewer post-maps (11) compared to the pre-maps (12). Students used only one proposition matching the referring misconception: "Acid rain is caused by carbon dioxide".

While the usage of propositions representing misconceptions increased for the concepts "recycling" and "air pollution", respectively, it decreased slightly for the concept "environmental pollution" (cf. Fig. 5). Students' typical propositions for these misconceptions are "Recycling is a climate protection action", "Air pollution causes the greenhouse effect", and "Environmental pollution causes global climate change".

A comparison of the usage of the concept "ozone hole" with the conceptions not mentioned in the educational material shows a decrease for both but with a much higher decrease for the conception explicitly explained in the learning material (6.8%).

Discussion

For all participants, similar conceptions could be found compared to related studies. Common misconceptions are present, with the conception that ozone layer depletion contributes to or causes global climate change as the most frequent conception. Our results, therefore, are quite in line with previous studies. Therefore, the reported misconceptions are widespread in different age cohorts such as children (Mason and Santi 1998), high-school students (Shepardson et al. 2009), university students (Dove 1996; Rebich and Gautier 2005) and adults (Bord et al. 2000).

One aim of our study was to investigate whether fragmentary conceptions and misconceptions concerning global climate change can be changed by implementing an environmental education programme. In fact, the maps' proportion with concepts referring to misconceptions decreased in four of five cases after the programme participation. In other words, fewer students used these concepts in their maps. This is a quite satisfactory result as it indicates that the number of students relating these concepts to global climate change and global warming (being the phenomena students were meant to describe with their maps) decreased.

Yet the proportion of propositions matching the common misconceptions increased in three of five cases. This fact could be explained by the way in which the information was presented to the students. In the case of the concept "ozone hole", the referring misconception was presented to students and was opposed to the actual mechanisms of the greenhouse effect. As the other concepts referring to misconceptions were not explicitly mentioned during the programme, we assume that the confrontation of students with the misconception seems to be helpful in causing a conceptual change in students' minds. Comparing the usage of the concept "ozone hole" with the other concepts in total also stresses this assumption. Our results therefore support the theory presented by Reinfried et al. (2008). Although students had the necessary information available to fully understand a presented topic, they were obviously not able to relate this new information to their existing conceptions. This implies that environmental education programmes or learning units desperately need a design which confronts them with common conceptions and provide them with a plausible, new conception (Reinfried et al. 2008). Further studies, therefore, need to concentrate on appropriate educational material.

References

Andersson B, Wallin A (2000) Students' understanding of the greenhouse effect, societal consequences of reducing CO_2 emissions and why ozone layer depletion is a problem. J Res Sci Teach 37(10):1096–1111

Bord RJ, O'Connor RE, Fisher A (2000) In what sense does the public need to understand global climate change? Public Underst Sci 9:205–218

Dove J (1996) Student teacher understanding of the greenhouse effect, ozone layer depletion and acid rain. Environ Educ Res 2(1):89–100

Ekborg M, Areskoug M (2006) How student teachers' understanding of the greenhouse effect develops during a teacher education programme. Nord Stud Sci Educ 5:17–29

Forsa. Gesellschaft für Sozialforschung und statistische Analysen mbH (2009) Klimawandel und Klimaschutz. http://www.bmu.de/publikationen/bildungsservice/aktuell/doc/45011.php. Accessed Dec 2009

Gerstner S, Bogner FX (2009) Concept map structure, gender and teaching methods: an investigation of students' science learning. Educ Res 51(4):425–438

Mason L, Santi M (1998) Discussing the greenhouse effect: children's collaborative discourse reasoning and conceptual change. Environ Educ Res 4(1):67–85

Niebert K (2007) Den Klimawandel verstehen—Eine theoriegeleitete und evidenzbasierte Entwicklung von Interventionen. Erkenntnisweg Biologiedidaktik 6:37–52

Novak JD, Gowin DB (1984) New strategies for evaluation: concept mapping. In: Novak JD, Gowin DB (eds) Learning how to learn. Cambridge University Press, Cambridge, pp 93–108

Rebich S, Gautier C (2005) Concept mapping to reveal prior knowledge and conceptual change in a mock summit course on global climate change. J Geosci Educ 53(4):355–365

Reinfried S, Schuler S, Aeschbacher U, Huber E (2008) Der Treibhauseffekt—Folge eines Lochs in der Atmosphäre? Geographie heute 265:24–33

Ruiz-Primo MA, Shavelson RJ (1996) Problems and issues in the use of concept maps in science assessment. J Res Sci Teach 33(6):569–600

Schaal S, Bogner FX, Girwidz R (2010) Concept mapping assessment of media assisted learning in interdisciplinary science education. J Res Sci Educ 40(3):339–352

Shepardson DP, Niyogi D, Choi S, Charusombat U (2009) Seventh grade students' conceptions of global warming and climate change. Environ Educ Res 15(5):549–570

Shepardson DP, Niyogi D, Choi S, Charusombat U (2010) Students' conceptions about the greenhouse effect, global warming, and climate change. Clim Change 104(3–4):481–507

Slotte V, Lonka K (1999) Spontaneous concept maps aiding the understanding of scientific concepts. Int J Sci Educ 21(5):515–531

Sterman JD, Sweeney LB (2002) Cloudy skies: assessing public understanding of global warming. Syst Dyn Rev 18(2):207–240

TNS Emnid & Bertelsmann Stiftung (2009) Jugend und die Zukunft der Welt. http://www.bertelsmann-stiftung.de/bst/de/media/xcms_bst_dms_29232_29233_2.pdf

Chapter 48
The Climate Change Challenge for the Urban Environment: The Use of an Integrated Management System

Esther Kreutz and Kirsi-Marja Lonkila

Abstract European citizens are experiencing directly the impacts of the rapidly changing climate. In the whole of Europe, extreme situations such as heat waves, floods, storms and forest fires, as well as sea level rise, droughts and scarcity of water in the southern parts of Europe, are occurring more frequently. Both European and national targets concerning the climate change challenge need to be implemented at the local and regional level. Local and regional authorities have huge potential to fight climate change by developing and implementing climate mitigation and adaptation strategies within their scope of responsibility and through stakeholder involvement. Local and regional authorities are responsible for a variety of planning and management issues and have the possibility to implement concrete measures. They are also directly involved in awareness raising, incentive setting and providing further support to their communities to tackle climate change. The focus of the CHAMP project is to support capacity development for local authorities to combat climate change through an integrated management system. The response to climate change calls for full integration of policies and structures that few local and regional authorities in Europe have in place. It also requires involvement of stakeholders to implement successful measures. The project initiates the establishment of national support centres in four European countries for training local authorities in implementing integrated management for climate change response.

Keywords Climate change · Integrated management system · Local authorities · Urban area

---

E. Kreutz (✉) · K.-M. Lonkila
Union of the Baltic Cities, Environment and Sustainable Development
Secretariat, Vanha Suurtori 7, 20500, Turku, Finland
e-mail: esther.kreutz@ubc.net

The Role of Local Governments in Responding to Climate Change

With a changing climate, extreme events such as floods, droughts, storms, etc. are predicted to occur more frequently all over the world. Large cuts in greenhouse gas emissions must be made to stop a temperature rise of 2°C above pre-industrial levels. Above this threshold, there is a risk that the impacts will be extremely difficult for mankind to cope with. However, even if this 2°C goal is reached, adaptive measures will be crucial for many cities and regions.

Cities have a key role in taking action. The majority of Europeans live in an urban environment. By 2020, 75–80% of the European population will live in urban areas (European Environment Agency 2009b). The demand for land is increasing; urban sprawl is re-shaping landscapes and influencing land use planning and people's quality of life. Urban planning and management face new challenges; climate change is high on all agendas, and local and regional authorities are expected to react in an adequate manner to secure the quality of life and security of their citizens. Energy production and use, as well as transportation and building, are the most important sectors for climate change response at a local level.

Urban areas are particularly vulnerable due to the dense population, high-density areas and sealed surfaces. The cities' structure and design has a significant impact on the effects caused by climate change—coastal cities face a serious risk of flooding and sea level rise; heat and droughts are another threat, especially for cities in southern Europe. Also, the "urban heat island" effect cannot be neglected. Effects from climate change will hit not only the city as a structure, but firstly the people and the environment, and will affect also social and economic aspects severely (European Environment Agency 2009b, pp. 67–70).

Multilevel Governance for Climate Change

Climate change is the biggest challenge humankind is facing and it is high on all political agendas, from global to local level. In its nature, the climate and also climate change is of a global nature, and this means that climate change is not limited to a certain area or to a certain government level.

On a global level, environmental governance is mainly conducted through so-called international regimes around an international agreement or treaty, e.g. the UN Framework Convention on Climate Change and the Kyoto Protocol (Fig. 1). International regimes are defined as social institutions which agree upon principles, norms, rules and decision-making procedures. International regimes are the framework for interaction of actors on a special topic (Betsill and Bulkeley 2006).

National governments are committing to targets concerning the climate change challenge through the UN and EU, for example. To reach these targets, emission

Fig. 1 Multi-level governance for climate change

cuts need to be implemented at the local and regional level. Local and regional authorities have huge potential to fight climate change by developing and implementing climate mitigation and adaptation strategies both within their scope of responsibility and through stakeholder involvement.

Local and regional authorities are responsible for a variety of planning and management issues and have the possibility to implement concrete measures. They are also directly involved in awareness raising, incentive setting and providing further support to their communities to tackle climate change. It is argued that local governments have a bigger potential influence over people's day-to-day lives and thus they can be more effective than nation states in controlling greenhouse gas emissions (Bulkeley and Betsill 2003, p. 49).

The local level is an increasingly important actor when it comes to climate change response—in taking concrete actions in mitigation and adaptation, since many of the decisions and concrete measures relevant for climate change are taken on that level. Local governments around the world often have considerable authority over land-use planning and waste management and also can have an important role in transportation issues and energy consumption. Moreover, they have significant experience in these areas of action (Bulkeley and Betsill 2003, pp. 48–49). Some regional and local governments have already developed action plans and strategies related to climate change. However, many cities still have no climate strategy or action plan and even the existing ones often problematically lack an integrated cross-sectoral approach.

A comprehensive and integrated cross-sectoral approach with concrete measures for climate change response is necessary. Climate change response measures are closely connected with the topics of energy, transport and buildings because the supply and use of energy—including transport—makes up to 80% of emissions in the EU-27 (European Environment Agency 2009a). The next chapter will outline and stress the need for an integrated management approach for climate change response.

Responding to Climate Change Through an Integrated Approach

Climate change is a complex and challenging issue for local governments. It is particularly challenging because the decisions that have the biggest effect on climate change mitigation and adaptation are usually made in sectors other than the environmental sector. Even more than most other environmental issues, responding to climate change demands for an integrated approach that covers the whole city organization.

A sustainable integrated management system therefore needs to concentrate on an integrated approach in different ways. Climate change strategies should be developed by people from different administrative departments and be integrated into the economic situation of the city. Naturally, they also need the necessary political commitment from the city government.

In the EEA report (2009b, p. 90), major integration gaps in local policy-making have been identified. They include gaps between sectoral policies, sectoral plan-making and implementation, between resources needed and those available, as well as between administrations and functional urban regions. These integration gaps definitely exist in the climate change work done at the local level. At least part of them can be grasped through an integrated management system introduced later in this paper.

Cities in different countries have various structures, management approaches, duties and functions, but they also have many elements in common, and they all function as a framework for local democracy (Union of the Baltic Cities Commission on Environment 2008a, p. 18). The integrated management approach is not a static, rigid system but can be adapted to the local situation to improve the public management and make it more efficient.

An Integrated Management System for Sustainability

Background and Development

The integrated management system (IMS) for sustainability, which is the topic of this paper and is applied to climate change challenges in the CHAMP project, has been developed by the consortium of the Managing Urban Europe 25 project (MUE-25) in 2005–2008.

Managing Urban Europe 25 included 25 European local and regional authorities working to improve their environmental quality and sustainability performance. The project developed a framework for better implementation of already existing environmental management systems such as EMAS, ISO 14000 and *eco*BUDGET. It provides a method for how cities and regions can practically work with integrated management, an approach recommended in many EU policies and strategies of today.

MUE-25 was developed based on recommendations from the Thematic Strategy on the Urban Environment within the 6th Environmental Action Plan of the European Union and the Leipzig Charter, among other EU-wide recommendations and policy frameworks. The integrated approach was strongly requested by the EU Commission, as stated in the EU Leipzig Charter on Sustainable European Cities (2007), where the ministers declare "to use the tool of integrated urban development".

The IMS cycle approach follows the well-known and widely used management cycle Plan–Do–Check–Act, also known as the Deming cycle. The IMS uses elements from existing environmental management systems and complements them.

Many cities are using different kinds of tools and environmental management systems (EMS) and have used them for a long time. In her study, Emilsson (2005) describes the importance of these tools for local authorities. In fact, they see the benefits in using an EMS not only for environmental issues but greatly for the organizations' overall general management: topics such as communication, involvement, commitment, continuity and competence are clear benefits that a management system can bring to the daily work (Emilsson 2005).

In the MUE-25 final research report, existing environmental management systems were examined as they lay the basis for the development of the IMS. The key question was "how to combine (existing) tools and systems to make them more holistic and coordinated" (Union of the Baltic Cities Commission on Environment 2008).

When addressing sustainability issues with a management system, it was clear that the environmental aspect alone is insufficient. Social and economic aspects also needed to be included. Nevertheless, existing environmental management systems are to be included and have to interact in the integrated management system. The IMS approach is not a completely new management system that replaces existing EMS, but it aims at combining them and benefit from synergies and tries to avoid duplication of work (Union of the Baltic Cities Commission on Environment 2008) (Fig. 2).

A Short Introduction to IMS

The integrated management system for sustainability aims to better serve the needs of local governments when it comes to managing sustainability issues. It is a tool that helps to improve the sustainability work and make public management more

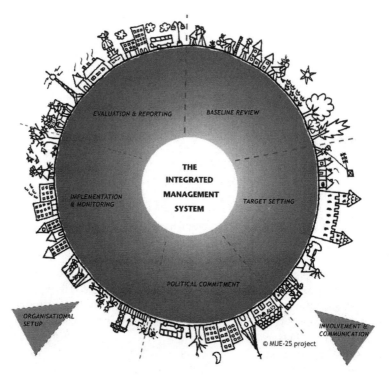

Fig. 2 The integrated management system cycle (Union of the Baltic Cities Commission on Environment 2008a, p. 7)

efficient. The system is based on the best parts of existing environmental management systems (such as EMAS, ISO 14000 and *eco*BUDGET) but broadens its application to all sustainability dimensions—environmental, social and economical.

The system consists of five steps that follow an annual cycle; full revision of the system is required after 3–5 years. The steps are:

- Baseline review
- Target setting
- Political commitment
- Implementation and monitoring
- Evaluation and reporting.

Integrated in those five steps are two cross-cutting elements: *communication and involvement* and *organizational set-up*. In this paper, we will briefly introduce all five steps and two cross-cutting elements of the IMS (Fig. 2).

The Integrated Management Process

The aim of the integrated management system for sustainability is to be flexible and adaptable and to avoid duplication of existing environmental or other management systems in a city. Existing EMAS or ISO 14000 processes could be the basis of the integrated management system for the city. All existing management instruments used in the city should be integrated to conduct the individual steps and elements of the system. These tools comprise, for example, state of the environment reports, local environmental action plans, indicators, local pollution registers, etc. The integrated management system approach will help to better coordinate existing instruments, strategies and operational plans by assembling the mosaic into one picture.

Environmental management systems are often the responsibility of the environmental department in a city. This should not be the case in an integrated management approach. It has to be organized centrally in the city management and needs to have cross-sectoral and cross-departmental cooperation (Union of the Baltic Cities Commission on Environment 2008a, p. 15; b, p. 11).

To avoid overwhelming the city administration, the integrated management system allows for the gradual expansion of the scope and content over time by using a modular approach. *Territorial expansion* will allow the implementation of the IMS to the whole urban area, also beyond the city's administrative boundaries. Environmental management systems are mostly used for the departments but it is also important to expand the system to the local authorities' core targets, i.e. strategic and operational planning and programmes.

The second way of *expansion is related to actors* and will include stakeholders in the city and cooperation with neighbours and other stakeholders. *Topic-related expansion* will allow the integration of further dimensions of sustainability to the system and develop it into a true integrated management system for sustainability, focusing not just on environmental issues but also economic, social and cultural dimensions (Union of the Baltic Cities Commission on Environment 2008a, p. 15; b, p. 11).

Practically, the system is not limited to a certain issue, such as environmental issues, but is explicitly useful when working with complex issues, such as the climate change challenge.

The following section will introduce the steps of the IMS in respect to climate change work in a city.

Step 1: The baseline review

During the first step, the baseline review—the local situation with regard to climate issues—is analysed. This is necessary to assess the starting point—only knowing the baseline makes the improvements visible. The baseline review will form the basis of all following steps in the management cycle and allow the setting of targets and the drafting of a climate strategy.

The baseline review consists of planning the process of drafting the baseline review, mapping the local situation, evaluating the results and finally presenting the results to the city council and the public. Mapping the local situation includes the following aspects:

- *Mapping the policy framework.* The existing legal requirements, emerging issues, trends and forthcoming policies and existing political priorities, decisions, commitments and strategies
- *Mapping the organizational set-up.* Climate relevant responsibilities and organizational set-up, existing relevant management instruments and procedures and stakeholder landscape including all relevant actors
- *Mapping the local climate change impacts.* Energy production and consumption patterns; greenhouse gas emissions; significant climate aspects; climate change impacts and vulnerabilities, risks and opportunities

(Union of the Baltic Cities Commission on Environment 2008c, pp. 11–12)

The baseline review process has to be carefully planned beforehand and possibilities for involvement and participation of various stakeholders should also be considered. Practically, the baseline review is a process of identifying, assessing and restructuring all existing information and actors in different administrative departments and agencies.

Step 2: Target setting

Based on the baseline review, a *strategic programme* and an *action plan* are prepared. First, it has to be decided which aspects of climate change mitigation and/or adaptation are given priority. The chosen priorities can then be expressed by using indicators. The *strategic programme* should be the document that sets long-term targets and measures for the agreed priorities. The targets should be measurable and quantified and they should be expressed in relation to a certain priority area. Concerning climate change mitigation and adaptation, the targets can, for example, contain a target for greenhouse gas emission cuts and share of energy used from renewable sources. In the strategic programme, a common vision for the future development of the city is also defined. The vision should be inspiring and attainable at the same time, and should originate in the chosen priorities (Union of the Baltic Cities Commission on Environment 2008d).

In addition to the strategic programme, an *action plan* results from target setting. It should include short-term targets that derive from long-term targets and define responsibilities and time frames to reach both long- and short-term targets. In the action plan, the allocation of human and financial resources must be clearly defined. If there were information gaps in the baseline review, one of the measures in the action plan can be filling them in (Union of the Baltic Cities Commission on Environment 2008a, pp. 17–18).

The process of drawing up a strategic programme for a city cannot succeed without participatory measures, and thus involvement and communication are vital for successful target setting. Setting up an appropriate organizational structure to run the system is also necessary.

Step 3: Political commitment

Political commitment should be sought from the very beginning of the process and needs to be maintained throughout the whole cycle. It is most crucial when the strategic programme or other outcome of the target setting process is approved by the highest decision-making body in the local government (i.e. the city council). For the legitimization of the IMS process, this is extremely important and can even be seen as the main success factor for the implementation of an IMS. In addition, many cities also approve the action plan and the entire organizational set-up designed to run the IMS. In addition, the involvement of decision-makers in the target setting and evaluation in line with the annual budget cycles guarantees political commitment, legitimacy and helps to maximize the positive impacts (Union of the Baltic Cities Commission on Environment 2008a, pp. 10, 15; e, pp. 4–5).

Without the approval and support of the politicians and the city top management, the implementation of the IMS process may be put off or it may never take place. Generated through debate, political commitment serves as a driving force of the whole system (Union of the Baltic Cities Commission on Environment 2008a, pp. 10, 15).

Step 4: Implementation and monitoring

During *implementation* and *monitoring*, the action plan is transferred into action and the IMS reaches its core. Usually the problem of local authorities is that they produce several strategies or plans that are never implemented. One of the most important aspects of an IMS is to integrate plan making and plan delivery, that is transferring strategies into action.

The implementation phase is demanding from the aspect of coordination and organization, as several parallel actions are taking place. Well-prepared involvement and communication, as well as organizational set-up, are the keys for a successful implementation.

During the whole process, actions and their effects need to be *monitored*. It should be noted that monitoring is only possible when actions refer to the targets that are based on indicators defined in the strategic programme (Union of the Baltic Cities Commission on Environment 2008a, p. 11, f, pp. 4–5).

Step 5: Evaluation and reporting

When actions have been implemented and monitoring data is available, it is possible to assess the achievements. As the last step of the cycle, evaluation and reporting gives a basis for starting a new cycle. The IMS is an expanding and continuous process and for this end it is extremely valuable to evaluate the cycle that has just ended in order to improve the system. In this phase, successes and failures should be analysed. Both the results obtained by the use of the cycle and the process of the IMS cycle should be assessed (Union of the Baltic Cities Commission on Environment 2008a, p. 12, e, p. 5).

The Organizational Set-Up

The organizational set-up of the integrated management system is the basis for a successful process. Clear responsibilities and tasks need to be assigned to the right people, working together for a common goal defined in the strategic programme. The organizational structure for implementing the IMS should be incorporated with the existing structures in municipal administration. Many local governments have already implemented some elements of an environmental management system or procedures that allow them to meet legal requirements. However, usually all climate-related direct responsibilities and indirect aspects of a local authority are not connected in an organizational set-up (Union of the Baltic Cities Commission on Environment 2008a, p. 13, g).

Within the IMS process, the most crucial part is the centrally arranged coordination team, which manages the coordination of the IMS. In addition, a cross-departmental coordination board ensures the integration of different departments and expertise. An important part of the organizational set-up is developing the capacities of the involved staff. Climate change is a challenging and complex issue and its impacts are not easy to predict and special and up-to-date information is, for that reason, necessary (Union of the Baltic Cities Commission on Environment 2008a, p. 13, g).

Some characteristics for an ideal organizational structure have been defined. They include flat hierarchy with a focus on informal organization, transparency and active information policy towards the public (this implies that the emphasis should be on the involvement and participation rather than on fulfilling the bureaucratic rules). The management style should be cooperative and employees should be encouraged and given room for their own initiatives (Union of the Baltic Cities Commission on Environment 2008g, p. 5).

The integrated sustainability management process will have a clear added value of having a dedicated organizational unit in the city administration, which coordinates the process and is the focus point for sustainability related issues. The SUSTAINMENT project developed a tool for local authorities to support them building up an organizational unit for sustainability (Union of the Baltic Cities Commission on Environment 2007).

Involvement and Communication

An IMS can only be sustainable if it is implemented in a legitimate way. This means that involvement and participation of different stakeholders has to be present in all phases of the IMS cycle (Union of the Baltic Cities Commission on Environment 2009, p. 8). A wide involvement of stakeholders is an integral part of the integrated management system for sustainability. Enabling participation and communicating the benefits that an IMS has will increase acceptance of and generate interest in the system. A successful IMS requires a communication and involvement strategy in which the relevant stakeholders and ways to involve them are defined (Union of the Baltic Cities Commission on Environment 2008a, p. 14; h, pp. 4–6).

A relevant stakeholder is anyone who

(1) is affected by the issue or affects the issue
(2) possesses information, resources and expertise needed or
(3) controls the implementation instruments.

The communication and involvement strategy should give the city a clear picture of who should be involved and at what stage—involvement and communication are part of every step of the IMS (Union of the Baltic Cities Commission on Environment 2008h, p. 7).

Although it can be argued that participatory processes demand a lot of time and might slow down the policy process, the advantages of a thorough involvement of various stakeholders are clear. Participatory processes are a way to increase trust between local authorities and various stakeholders. Participation also helps in making the policy process more transparent. Involvement and communication can also lead to a better outcome as different actors bring their own expertise and knowledge to the process (Union of the Baltic Cities Commission on Environment 2008h, p. 6).

Repeating and Expanding the Cycle

Working with IMS is a continuing cycle of assessing, reviewing and improving. The model suggests a journey approach, as it is not possible to take in all the steps at once. When implementing an IMS, local governments can start small and expand the whole time on their journey through the cycle. It should also be noted that setting up an IMS takes a certain amount of time (Union of the Baltic Cities Commission on Environment 2008a, p. 15).

Some of the steps (i.e. implementation and monitoring, evaluation and reporting) are done on an annual basis, but a full revision of the cycle is only needed every 3–5 years. This of course only applies when no changes in the surroundings defining the city's strategic orientation occur (Union of the Baltic Cities Commission on Environment 2008a, p. 15; b, p. 8).

Experiences With the IMS

During the MUE-25 project, 25 municipalities and cities all over Europe pilot-implemented the IMS approach. The challenge definitely lies in the complexity of the system. But as mentioned before, the aim is not to implement the system in its entirety at once but to gradually expand the system over the years. All participating cities have started with issues that were most critical to them and most of the cities are continuing to do so. Nevertheless, there is no city that has implemented the IMS for the whole urban area and all issues, since the idea of the system is that it can be continuously improved and expanded.

The IMS approach is also more a way of working, which means, with IMS in the background, the city is prepared to respond to upcoming sustainability challenges as they face them in a systematic and integrated way.

CHAMP: Local Response to Climate Change

CHAMP is a 3 year (2009–2011) project partly funded by the EU Life + instrument. The project is coordinated by the Union of the Baltic Cities and has six partners in four EU countries. CHAMP follows up the "Managing Urban Europe-25" project (2005–2008) where the consortium with 25 cities developed together the integrated management approach for sustainability and pilot-implemented the system in several cities.

In CHAMP this approach is applied on the topic of climate change response from the local level. In the CHAMP project, the focus is to support capacity development for local authorities to combat climate change in Europe. The response to climate change calls for full integration of policies and structures that few European local and regional authorities have in place. It also requires involvement of stakeholders to implement successful measures.

During the project's lifetime, four national training hubs have been established in Finland, Germany, Italy and Hungary. Training hubs support local authorities in implementing an integrated management for climate change response. Five training workshops will be organized in each country and the local authorities receive additional support from the national hubs by phone and email. Moreover, peer reviews will be conducted between the cities and municipalities, where they are able to assess their own situation in a certain field.

A main outcome of the project will also be the online version of a capacity development package. This will comprise materials, methods and good practices for local authorities, training Materials and methods for trainers and information materials for auditors and other multipliers.

More information about the project can be found at http://www.localmanagement.eu/index.php/champ:home

References

Betsill MM, Bulkeley H (2006) Cities and the multilevel governance of global climate change. Glob Gov 12(2):141–159 Apr–Jun 2006, Research Library

Bulkeley H, Betsill MM (2003) Cities and climate change. Urban sustainability and global environmental governance, Routledge studies in physical geography and environment. Routledge, London

Emilsson S (2005) Local authorities approaches to standardized environmental systems, Dissertation No. 939. Environmental Technology and Management, Department of Mechanical Engineering University of Linköping

European Environment Agency (2009a) Greenhouse gas profiles (Greenhouse gas trends and projections in Europe 2009) Nov 2009. http://www.eea.europa.eu/themes/climate/ghg-country-profiles.Accessed May 2010

European Environment Agency (2009b) Ensuring quality of life in Europe's cities and towns. Tackling the environmental challenges driven by European and global change. EEA Report No 5/2009

Leipzig Charter on Sustainable European Cities (24 May 2007). http://www.rfsustainablecities.eu/IMG/pdf/LeipzigCharte_EN_cle1d4c19.pdf

Union of the Baltic Cities Commission on Environment (2007) URBAN works. Integrated Solutions for Sustainability Management in the Baltic Cities

Union of the Baltic Cities Commission on Environment (2008) Research report Managing Urban Europe-25. http://www.localmanagement.eu/index.php/mue25:downloads

Union of the Baltic Cities Commission on Environment (2008a) Integrated management. Towards local and regional sustainability

Union of the Baltic Cities Commission on Environment: MUE-25 Guidelines, including (2008b): Overview of the Integrated Management System. http://www.localmanagement.eu/index.php/mue25:downloads

Union of the Baltic Cities Commission on Environment: MUE-25 Guidelines, including (2008c): The guidance paper on Baseline Review. http://www.localmanagement.eu/index.php/mue25:downloads

Union of the Baltic Cities Commission on Environment: MUE-25 Guidelines, including (2008d): The guidance paper on Target Setting. http://www.localmanagement.eu/index.php/mue25:downloads

Union of the Baltic Cities Commission on Environment: MUE-25 Guidelines, including (2008e): The guidance paper on Political Commitment. http://www.localmanagement.eu/index.php/mue25:downloads

Union of the Baltic Cities Commission on Environment: MUE-25 Guidelines, including (2008f): The guidance paper on Implementation and Monitoring. http://www.localmanagement.eu/index.php/mue25:downloads

Union of the Baltic Cities Commission on Environment: MUE-25 Guidelines, including (2008g): The guidance paper on Organisational Set-up. http://www.localmanagement.eu/index.php/mue25:downloads

Union of the Baltic Cities Commission on Environment: MUE-25 Guidelines, including (2008h): The guidance paper on Involvement & Communication. http://www.localmanagement.eu/index.php/mue25:downloads

Union of the Baltic Cities Commission on Environment (2009) Engage Your Stakeholders. Stakeholder Involvement Toolkit for Local Authorities

Thematic Strategy on the Urban Environment. 6th Environmental Action Plan of the European Union. http://ec.europa.eu/environment/urban/thematic_strategy.htm

About the Authors

Adesina Yinka is a Professor of Geography at Obafemi Awolowo University, Ile Ife, Nigeria. He undertakes a number of projects on climate change adaptation and technology transfer. He has also participated in several scientific and technical reports, at both national and international levels. Such reports include the International Panel on Climate Change, second edition of National Communication and Climate Change Policy for Nigeria. He is also the Dean, Faculty of Social Sciences at the same institution.
Contact: Department of Geography, Obafemi Awolowo University, Ile Ife, Osun State, Nigeria. Email: faadesin@yahoo.com

Ahmed Abd El Monem Mohamed Professor of Agronomy, National Research Center. He was the PI in many projects for increasing productivity of field crops under Egyptian conditions.
Contact: El Bohooth Str., Dokki, Giza, Egypt, P.O. Box 12311.
Email: tawfiknrc@hotmail.com

Alemayehu Fikadu Reta is among the founding members of RDFS professionals association and he is on the academic staff at Hawassa University, Ethiopia. He studied for an M.Sc. degree in Agricultural Development at the University of Copenhagen. His research priorities include: rural livelihood, agriculture and food security in sub-Saharan Africa, human nutrition, climate change and natural resource management.
Contact: Permanent address: Hawassa University, College of Agriculture, P.O. Box 5, Hawassa, Ethiopia. Current address: Terrasserne 18 St 93, 2700 Copenhagen, Denmark. Email: fikadureta@gmail.com

Alexandar R Junior Research Fellow (NCERT) "Ecology & Environmental Sciences" at Pondicherry University in India, where he is doing a Ph.D. on Environmental Education for Sustainable Development (EESD).

Contact: No 3/21, Dr Ambedkar Street, Kadaperikuppam & Post, Vanur Taluk, Villupuram District, Tamil Nadu, India. Email: enviroalexandar@gmail.com.

Aloj Eugenia is a Professor in Ecology and Professor in Hygiene at the Sannio University of Benevento. She has also been head of the team for "Biology and Ecology for the Protection of the Environment" at the Regional Centre of Competence BENECON endorsed by Regione Campania. She is the author of more than 150 publications (handbooks, chapters in books, papers in journals) at national and international level on the topic of sustainable development, focusing either on global environmental problems (such as climate change, pollution) or environmental resources (water, air, etc.).
Contact: Viale Maria Cristina di Savoia 18/d, 81100 Naples, Italy.
Email: galoj@email.it

Azad Al Mamun is a senior disaster management officer with Oxfam Bangladesh. Prior to his present engagement, he served in different positions at the World Food Programme. He has a bachelor degree in Urban and Rural Planning from Khulna University.
Email: mamun_azad@yahoo.com

Bahr Abbas Amany Professor of Agronomy, National Research Center. She has worked on many projects for increasing the productivity of field crops under Egyptian conditions.
Contact: El Bohooth Str., Dokki, Giza, Egypt, P.O. Box: 12311.
Email: Amany_nrc@yahoo.com

Bai Ramani Ph.D. (Environmental & Water Resources Engineering). Associate Professor in the Department of Civil Engineering at the University of Nottingham Malaysia Campus since May 2008. She has 17 years of significant service in environmental and water resources systems analysis, hydrology, water quality and water treatment and optimization techniques, computer modelling, artificial intelligence methods such as fuzzy logic, artificial neural networks and neuro-fuzzy logic systems and geographical information systems. She has been involved in various activities as academician, project leader, researcher, thesis examiner, international project reviewer, sole resource person of one-day, two-day seminars, etc. in the above-mentioned fields at Indian and Malaysian universities and organizations.

Barbir Jelena has a degree in Biology, joint European Master's degree in Environmental Studies, and is doing a Ph.D. in Biological Control in Agriculture. Her research focus is in the areas of sustainable agriculture and climate change adaptation.
Contact: Jelena Barbir, Department of Crop Protection, Institute of Agricultural Sciences, Centre for Environmental Sciences, CSIC, Calle Serrano 115, 28006 Madrid, Spain. Email: jelenabarbir@ccma.csic.es

About the Authors

Batyreu Victor is a part time senior scientist at the Minsk branch of Moscow State University for Economics, Statistics and Informatics. He has a Ph.D. in Physics. His research focus is in the area of modelling environmental processes.
Contact: Minsk Branch of Moscow State University for Economics, Statistics and Informatics, 127, Building 2, Room 301, Mayakovskogo Str., Minsk, 220028, Belarus. Telephone/Fax: +375-17-2914497. Email: batyrev1945@mail.ru

Biggs Eloise is a lecturer at the University of Southampton. She lectures in Geography within the field of Geographic Information Systems (GIS) and environmental applications. Ellie completed her Bachelor degrees in Environmental Sciences and her Ph.D. in Geography, both at the University of Southampton. Her postgraduate research investigated hydrological extremes and climate change in the Severn Uplands, UK. Ellie's current research is centred on climate change and water resources across rural Nepal, with particular focus on environmental risk.
Contact: School of Geography, University of Southampton, Southampton, SO17 1BJ, UK. Email: Eloise.Biggs@soton.ac.uk

Boer Rizaldi is a head of the Climatology Laboratory at the Department of Geophysics and Meteorology, Bogor Agricultural University. He has been involved in numerous national and international research activities dealing with climatic risk analysis, climate change mitigation and adaptation. He is also a director of the Centre for Climate Risk and Opportunity Management at Southeast Asia Pacific Bogor Agriculture University.
Contact: Head of Laboratory of Climatology, Department of Geophysics and Meteorology, Faculty of Mathematics and Natural Sciences, Bogor Agricultural University, Gedung FMIPA Level 4 Wing 19 Kampus IPB Darmaga, Bogor 16680, Indonesia. Telephone: +62-251-8313709 and +62-251-8623850 (work), Fax: +62-251-8310779. Email: rizaldiboer@gmail.com

Bogner Franz is Professor at the Bavarian Chair of Biology Didactics at the University of Bayreuth, Germany. He also is head of the Centre of Maths and Science Education (Z-MNU). His research interests concentrate on educational issues including environmental, health and experimental hands-on education.
Contact: Chair of Biology Education, University of Bayreuth, NW I, Universitaetsstrasse 30, 95447 Bayreuth, Germany.
Email: franz.bogner@uni-bayreuth.de

Burger Dieter Professor at the Institute of Geography and Geocecology (IfGG, KIT) Karlsruhe, Germany. Member of the Physical Geography Chair and Studies Advisor for Geography Diploma. Main interest research works are: soil sciences, applied climatology and physical geography. He has supervised several research works on soil physics in Brazil, Spain, Tunisia and Turkey.
Contact: Karlsruher Institut für Technologie (KIT) Campus Süd, Kaiserstrasse 12, 76128 Karlsruhe, Germany. Email: Dieter.Burger@kit.edu

Catenazzo Giuseppe is a Doctoral Assistant at HEC University of Geneva, Switzerland. He holds a Master of Science degree in Management from the University of Savoy, France, a Postgraduate Certificate in Applied Environmental Economics from the University of London, UK, and an undergraduate degree in Economics and Business Administration from the University of Aosta Valley, Italy. He was formerly a Research Assistant at Geneva School of Business Administration and co-headed several projects on business and environmental management. He is the co-author of a book dealing with services management, La gestion des services (Economica, Paris, 2008), and papers on risk, services and environmental management.
Contact: Haute École de Gestion de Genève Geneva School of Business Administration Campus Battelle—Bât. F7, Route de Drize, 1227 Carouge GE, Switzerland. Email: giuseppe.catenazzo@unige.ch

Clements Rebecca is a policy specialist at Practical Action Latin America where she is working on a number of climate change projects. She has a Master's degree in International Studies from the University of Sheffield in the UK. Her thematic areas of focus include governance and advocacy, rural livelihoods, small-scale bioenergy development, community-based adaptation to climate change, international climate change policy, and mitigation mechanisms (including clean development technologies and carbon markets). Country of origin: England.
Contact: Soluciones Practicas—ITDG Av. Jorge Chavez 275, Miraflores Lima 18, Perú. Telephone: +511-4475127. Email: rclement@solucionespracticas.org.pe

Colaiezzi Marina works at the Italian National Institute for Environmental Protection and Research (ISPRA). Her research focus is on monitoring and prevention of atmospheric impacts, emission scenarios, integrated models and indicators.
Email: marina.colaiezzi@isprambiente.it

Cóndor Rocío Dánica works at the Italian National Institute for Environmental Protection and Research. She is responsible for preparing, improving and reporting the National Agricultural Emission Inventory for the UNFCCC/Kyoto Protocol and the Convention on Long-Range Transboundary Air Pollution (CLRTAP). Member of UNFCCC and CLRTAP roster of experts in the area of GHG inventory. Her research focus is on assessing GHG/ammonia mitigation measures and policies, and the role of the Common Agricultural Policy on emission trends.
Contact: Istituto Superiore per la protezione e la ricerca ambientale (ISPRA), Dipartimento stato dell'ambiente e metrologia ambientale, Via Vitaliano Brancati 48, 00144 Rome, Italy. Email: rocio.condor@isprambiente.it

Corderi Novoa David is an economist working in the fields of climate change, water resources, energy and sustainable development. He has been working for international organizations such as the World Bank and the Asian Development Bank. David has previously worked as a researcher at Yale University and the

University of California. He has worked in countries across Latin America, Asia, Africa and Europe. David has a Master's degree in International and Development Economics from Yale University and is a Ph.D. candidate in Agricultural and Resource Economics at the University of California at Davis.
Email: david.corderi@aya.yale.edu

Cruickshank Abigail currently works as the executive director of the Alouette River Management Society, a community-based watershed management organization in Canada. She attained her Bachelor of Science in Physical Geography from Simon Fraser University and completed her Master's in Environmental Studies at York University. Abby has had previous experience working with different cultural groups on water security issues in South Africa and Kenya and completed a summer placement with Canada's International Development Research Centre's Rural Poverty and the Environment and Climate Change Adaptation in Africa programmes.
Email: abcruicks@gmail.com

D'Urso Jennifer is currently a teacher in Sierre, Switzerland. Holder of a Bachelor degree in Business Administration from the University of Applied Sciences of Wallis, Switzerland, she was formerly a Research Assistant at Geneva School of Business Administration. As a staff member of Geneva School of Business Administration Laboratory of Market Research (LEM–HEG), she co-headed several survey projects on water management and sustainable development.
Email: jenndurso@hotmail.com

Dasanto Dwi Bambang is a Senior Lecturer and Researcher at the Department of Geophysics and Meteorology, Bogor Agricultural University, Indonesia. He has a strong background in geographic information systems and remote sensing. He specializes in hydro-climatology and has been involved in many studies on climate risk assessments and climate change mitigation.
Contact: Department of Geophysics and Meteorology, Gedung FMIPA Level 4 Wing 19 Kampus IPB Darmaga Bogor 16680, Indonesia.
Email: bambangdwi@scientist.com

Danh Thao Nguyen received a Ph.D. degree from Yokohama National University, Yokohama, Japan in 2007. He is a lecturer at the Division of Port and Coastal Engineering and Department of Civil Engineering, HCMC University of Technology, Vietnam. His research interests include wave forces on coastal structures, interaction between waves and structures, port and coastal management, natural disasters: prediction and methods of prevention, damage reduction, and change of coastal areas due to human activities.
Contact: Faculty of Civil Engineering, HCMC University of Technology, 268 Ly Thuong Kiet Str., Dist. 10, Ho Chi Minh City, Vietnam. Telephone: +84-8-38652442, Fax: +84-8-38653823

De Castro Mariagrazia is a Ph.D. student in Analysis of Economic and Social Systems: Enterprise, Institutions, Territory at Sannio University. She is a learner in Ecology under the Chair of Prof. Eugenia Aloj.
Email: mgdeca@tin.it

Diaz Harry is the Director of the Canadian Plains Research Center and Professor of Sociology and Social Studies at the University of Regina, Canada. He has been involved in several projects and publications focused on social vulnerability and adaptation to climate change.
Email: harry.diaz@uregina.ca

Edwin Golda has obtained her Master of Science degree in Zoology and has over 10 years of work experience as an environmentalist, researcher and activist for sustainable development. She is currently pursuing her doctorate from Pondicherry University, India and presently conducting an extensive study to access the role of invertebrates in water treatment using constructed wetlands. The main objective of her research is to assess and provide factual solutions to the local municipalities to make them sustainable with special reference to water, energy and waste. She has also worked with local NGOs and has undertaken community development projects such as education, rainwater harvesting, health and hygiene workshops with guidance from the Center for Sustainable Development, Guatemala. Her honours include general proficiency for two consecutive years and she is also a gold medalist in the Master's programme.
Contact: No. 23, Glory Home, Kamaraj Salai, Thattanchavady, Pondicherry 605009, India. Email: golda_edwin@yahoo.in

El Karamany Mohamed Fraouk Professor of Agronomy, National Research Center. She has worked on many projects for increasing the productivity of field crops under Egyptian conditions.
Contact: El Bohooth Str., Dokki, Giza. Egypt, PO Box: 12311.
Email: karamanynrc@yahoo.com

Esteban Miguel is an Assistant Professor in the Department of Civil and Environmental Engineering at Waseda University, Japan. There, his research and teaching relates to the prevention and analysis of natural disasters, and how climate change might have an effect on them. He received his Ph.D. in Coastal Engineering from Yokohama National University in Japan in 2007, and then he continued his work with Post-Doctoral Fellowships at the United Nations University Institute of Advanced Studies (UNU-IAS) and at Kyoto University. He is currently also a Visiting Research Fellow at the UNU-IAS.
Contact: Department of Civil and Environmental Engineering, Waseda University, 3-4-1 Okubo, Shinjuku-ku, 51-goukan, 4-kai, 7-shitsu, Tokyo 169-8555, Japan.
Email: esteban.fagan@gmail.com

Fischer Daniel MA (Institute for Environmental & Sustainability Communication, Leuphana Universität Lüneburg) was educated and trained as a school teacher, holds a Master's degree in Education Management and School Development, and works as a Research Fellow at the Institute of Environmental and Sustainability Communication (INFU) in the Faculty of Sustainability Sciences of the Leuphana University of Lüneburg. He teaches and does research in the fields of education for sustainable development, sustainable consumption and sustainable school development and school/university culture.
Contact: Institute for Environmental and Sustainability Communication (INFU), UNESCO Chair Higher Education for Sustainable Development, Leuphana University of Lueneburg, Scharnhorststr. 1, 21335 Lueneburg, Germany. Telephone: +49-4131-6772927, Fax: +49-4131-6772819

Fragnière Emmanuel Certified Internal Auditor (CIA) is a Professor of Service Management at the Haute Ecole de Gestion of Geneva, Switzerland. He is also a Lecturer in Enterprise Risk Management at the Management School of the University of Bath, UK. Previously, he was a commodity risk analyst at Cargill (Ocean Transportation) and a senior internal auditor at Banque Cantonale Vaudoise, the fourth largest bank in Switzerland. His research is focused on the development of risk management models for decision-makers in the service sector. He has published several papers in academic journals such as *Annals of Operations Research, Environmental Modelling and Assessment, Interfaces and Management Science*. He is the author (with Sullivan) of the book entitled Risk Management: Safeguarding Company Assets, (Fifty-Minute Crisp Series, 2006).
Email: emmanuel.fragniere@hesge.ch

Freund Eva Dipl.-Environmental Science (Chair of Vocational and Economic Education/ Didactics of Business Studies, Leuphana Universität Lüneburg), studied at the Leuphana University of Lüneburg and holds a diploma degree in Environmental Sciences. She works as a Scientific Officer at the Chair of Vocational and Economic Education at the Leuphana University of Lüneburg in the project ZUKKER (Development of sustainable competences for future markets). The aim of the project is to provide additional qualifications for apprentices in retail business.

Garay-Flühmann Rosa social ecologist. She obtained her Ph.D. in Environmental Studies at Yale University. Her research focuses on water-stressed environments and social organization, socio-ecosystem-based marine conservation and management, and climate change and social institution adaptation.
Email: rogaflu@yahoo.com

Gaudioso Domenico works at the Italian National Institute for Environmental Protection and Research, where he is currently responsible for the Climate and Atmosphere Unit. He has published about 50 articles on air emissions, with special reference to methodologies for the estimation of greenhouse gas emissions.

In particular, he has contributed to the preparation of the UNECE-CORINAIR Atmospheric Emission Inventory Guidebook (Chapter 6, "Solvent and other product use"), the 1996 IPCC Revised Guidelines on Greenhouse Gas Inventories (Chapter 6, "Waste") and the 2006 IPCC Guidelines on Greenhouse Gas Inventories (Volume 3, "Industrial Processes and Product Use"). He is member of the roster of methodology experts of the UNFCCC, and has taken part in several reviews of GHG inventories and national communications.
Email: domenico.gaudioso@isprambiente.it

Geiger Sonja Maria is a cognitive psychologist who is currently working with Amigos de la Tierra, Argentina in Buenos Aires on programmes concerning water and climate justice. She is investigating the role of cognitive factors such as environmental knowledge and consciousness for sustainable behaviour of people and their consequent contribution to mitigation of climate change.
Contact: Donado 1076, 1427 Buenos Aires, Argentina.
Email: sonja.m.geiger@gmail.com

Gongwe Mhando David is a Senior Research Fellow and Head of the Socioeconomy Section at the Centre for Sustainable Rural Development, Sokoine University of Agriculture, Tanzania. He received his Masters and Ph.D. degrees in Rural Sociology from the University of Dar es Salaam, Tanzania and Graduate School of Asian and African Area Studies, Kyoto University, Japan, respectively. His research focus is in the areas of farmer groups' dynamics, livelihoods and rural economy, with special interest in rural development process. Country of origin: Tanzania.
Contact: Centre for Sustainable Rural Development, Sokoine University of Agriculture (SCSRD), P.O. Box 3035, Chuo Kikuu, Morogoro, Tanzania. Mobile: +255-787-074407; +255-773-074407, Telephone/Fax: +255-232-604279.
Email: David_mhando@yahoo.com

González González Leonel holds a Master's degree in Hydraulic Engineering and works as a consultant for the National Water Authorities of Mexico, Comisión Nacional del Agua (CONAGUA). His main activities include the planning and implementation of integrated water resources management in central Mexico, in the province of Mexico (Estado de México). Country of origin: Mexico.
Contact: Comisión Nacional del Agua (CONAGUA) Integrated Expert, Centre for International Migration and Development (CIM), Dirección Local, Estado de México, Conjunto SEDAGRO, Edifício C-1 Metepec, Estado de México, C.P. 52148, Mexico, Telephone: +52-722-2322660.
Email: leonel.gonzalez@conagua.gob.mx

Grover Velma manages the Freshwater Ecosystem Programme at United Nations University's International Network on Water, Environment, and Health and has had extensive experience in capacity building in developing countries, particularly

related to distance learning for integrated water resources management. She is an Adjunct Professor at the DOFASCO School of Engineering and Public Policy at McMaster University in Hamilton, Canada, and at the Faculty of Environmental Studies at York University in Toronto.
Email: grovervi@yorku.ca

Halliru Salisu Lawal is a teacher at the Federal College of Education Kano, Nigeria. He received his Bachelor of Science degree and Postgraduate Diploma in Environmental Management from Bayero University, Kano. As a young geographer he undertakes projects on climate change issues, sustainable development, poverty indices, presenting his research work at both national and international conferences, and he is a member of a research and development network. He is also an associate editor of *International Journal of Climate Change: Impacts and Responses*, Volume 2, Nos. 1 and 2, 2010. He was recently offered a place at the University of Sussex to read for the degree of Master of Science in Climate Change and Development. His research focus is in the areas of environmental management, sustainable development, population dynamic, water resources management, rural development and poverty, education sustainability, with special interest in climate change and food security.
Contact: Department of Geography, Staff Secondary School, Federal College of Education, Kano, P.M.B. 3045, Kano State, Nigeria.
Email: lhsalisu09@gmail.com

Hasan Rashedul is a lecturer at Chittagong University of Engineering and Technology (CUET). His major research areas are urban fringe, urban governance, land-use changes and hedonic pricing.
Email: udoy_ku@yahoo.com

Herrfahrdt-Pähle Elke is a researcher at the Department of Environmental Policy and Natural Resource Management at the German Development Institute/ Deutsches Institut für Entwicklungspolitik (DIE) in Bonn, Germany. She holds a Ph.D. in Applied System Sciences from the University of Osnabrück, Germany. Her fields of study include resilience, adaptability and transformability of water governance regimes, as well as integrated and adaptive water governance.
Contact: Tulpenfeld 6, 53113 Bonn, Germany. Email: elke.herrfahrdt@die-gdi.de

Hill Margot is a Ph.D. candidate at the University of Geneva, in the Research Group on Climate Change and Climate Change Impacts. She is researching the adaptive capacity of water governance systems to climate-related extreme events in two contrasting mountain watersheds (Aconcagua, Chile and Rhone, Switzerland).
Contact: Research Group on Climate Change and Climate Impacts, University of Geneva, Battelle, Building D, 7 Route de Drize, CH 1227 Carouge, Switzerland.
Email: margot.hill@unige.ch

Hossain Afjal works as Assistant Project Coordinator at Medecins Sans Frontieres, Belgium. He is also Director of the Safe Humanity Foundation. In the professional arena he has been involved in various development projects in nutrition, sustainable development, renewable energy and climate change in Bangladesh.
Contact: 382/1 North Kazipara, Mirpur, Kafrul, Dhaka-1216, Bangladesh. Telephone: +88-01715623485. Email: robinpln@gmail.com

Hossain Zakir is an Assistant Professor of Urban and Rural Planning at Khulna University, Khulna, Bangladesh. He has a Master's degree from the University Antwerp in Development Evaluation. His research areas deal with institutional aspects of climate change adaptation, social exclusion and poverty, capability approach and adaptation.
Email: zakir_urp9913@yahoo.com

Hugé Jean has a Master's in BioScience Engineering and a Master's in Development Studies from Ghent University, Belgium. He is a research fellow at the Vrije Universiteit Brussel Policy Research Centre for Sustainable Development and KLIMOS Research Platform for Climate Change and Development Cooperation. His research focuses on the theory and practice of sustainability assessment, applied in particular in the fields of regional sustainable development policy, development cooperation and energy and climate change.

Huq Nazmul is the Senior Research Officer at the Bangladesh Centre for Advanced Studies. He is currently involved with the project Action Research on Community Adaptation in Bangladesh (ARCAB). His major research focuses are community-based adaptation, social dimensions of climate change, and development cooperation and climate change adaptation. He has a Master's in Human Ecology from the Vrije Universiteit, Brussels, Belgium.

Hurlbert Margot Associate Professor, is jointly appointed to the Department of Justice Studies and Department of Sociology and Social Studies, University of Regina. Margot has authored numerous journal articles, book chapters and scholarly papers on a broad range of justice topics including Aboriginal justice, earth system governance and climate change adaptation. Prior to embarking on a full-time academic career, Margot practised law in private practice for 12 years and in corporate practice as the Assistant General Counsel for SaskPower for 7 years.
Contact: Department of Justice Studies and Department of Sociology and Social Studies, Canadian Plains Research Center, CL 235, University of Regina, 3737 Wascana Parkway, Regina, SK, S4S 0A2, Canada.
Email: Margot.Hurlbert@uregina.ca

Hutton Craig is a research scientist at the GeoData Institute based at the University of Southampton, an institution which is developing a considerable

portfolio of research and supporting activity in the field of risk, vulnerability and climate change. Craig has undertaken substantial research in Asia, Africa, the EU and the UK, with focus on vulnerability mapping of climate change. His research provides valuable context for decision support and policy formulation in water and land resource management, as well as food security.
Email: C.Hutton@soton.ac.uk

Justice Nindi Stephen is a Senior Research Fellow and heads the Resource Management Section at the Centre for Sustainable Rural Development, Sokoine University of Agriculture in Tanzania, where he teaches, consults, undertakes and supervises a number of projects on sustainable rural development, land use dynamics, climate change, GIS and remote sensing, livelihoods and rural economy across southern Africa. He is also Patron of the Young Tanzanian for Community Prosperity (YTCP). Country of origin: Tanzania.
Contact: Centre for Sustainable Rural Development, Sokoine University of Agriculture (SCSRD), P.O. Box 3035, Chuo Kikuu, Morogoro, Tanzania. Mobile: +255-756092344, Optional Telephone/Fax: +255-232-604279.
Email: nindistephen@yahoo.com, snindi@suanet.ac.tz

Kabiri Reza is a Research Assistant in a research project on database information systems and management funded by Asia Pacific Network for Global Change Research, Japan. He is pursuing his Ph.D. under the supervison of Ramani Bai and Andrew Chan at the University of Nottingham Malaysia Campus, Malaysia.

Kayes Imrul is a Project Coordinator of MSSUS Save the Children UK NSA Project. He received his Master's degrees in Forestry and Wood Technology from Khulna University, Bangladesh. His research focuses on the area climate change and its impacts on the livelihoods of vulnerable people in the southwestern coastal zone in Bangladesh.
Contact: 316, Khan Jahan Ali Hall, Khulna University, Khulna-9208, Bangladesh. Telephone: +88-1715605919. Email: kayes_fwt@yahoo.com

Kreutz Esther works as a Project Coordinator at the Union of the Baltic Cities—Environment and Sustainable Development Secretariat in Turku, Finland. She graduated from the University of Lüneburg, Germany as a Master of Environmental Sciences with specialization in Environmental Law and Policy and Sustainability Communications. For her Master's thesis about multi-level governance in the Baltic Sea Region, she also studied at Åbo Akademi University in Turku, Finland. Currently, Mrs. Kreutz coordinates the climate change-related project CHAMP.
Contact: Union of the Baltic Cities, Environment and Sustainable Development Secretariat, Vanha Suurtori 7, FIN-20500 Turku, Finland.
Email: esther.kreutz@ubc.net

Kumar Anil is a Professor of Soil and Water Conservation Engineering at the G.B. Pant University of Agriculture and Technology, Pantnagar, India. He has been involved with teaching and research in the fields of soil and water conservation engineering, hydrology, water resources development, and watershed management. His interests in field studies in hilly and mountainous areas of the Himalayas include water harvesting and management, rejuvenation of natural water springs, soil and water conservation measures, and adaptation mechanisms to face climate change.
Contact: Department of Soil and Water Conservation Engineering, College of Technology, G. B. Pant University of Agriculture and Technology, Pantnagar-263 145, Uttarakhand, India. Email: anilkumar_swce61@yahoo.co.in

Leal Filho Walter heads the Research and Transfer Centre "Applications of Life Sciences" at the Hamburg University of Applied Sciences in Germany, where he undertakes a number of projects on sustainable development, renewable energy and climate issues across the world. He is also the Chairman of the International Climate Change Information Programme (ICCIP).
Contact: Lohbruegger Kirchstrasse 65, Sector S4 / Room 0.38, 21033 Hamburg, Germany. Email: walter.leal@ls.haw-hamburg.de

Lonkila Kirsi-Marja is a Master of Social Sciences, majoring in political science. She graduated from the University of Turku, Finland in 2009. For her Master's thesis, she studied extended producer responsibility as an environment policy instrument in the EU. In 2010 she also graduated as a Bachelor of Administrative Sciences from the University of Tampere, Finland, majoring in Environmental Policy and conducted her Bachelor's thesis on climate policy in the city of Lahti, Finland. At the moment she is working at the Union of the Baltic Cities Environment and Sustainable Development Secretariat as Project Officer, for two climate-related projects called CHAMP and BaltCICA.
Email: kirsi-marja.lonkila@ubc.net

Maheu Audrey undertook the work presented in this book as a research intern in the Climate Change and Water programme at the International Development Research Center (IDRC). She now works as an Environmental Indicators Specialist at Environment Canada. She holds a Master's degree in Integrated Water Resources Management from McGill University and a Bachelor degree in Applied Geomatics from the Université de Sherbrooke.
Contact: 7 Des Frênes, Vaudreuil-sur-le-lac, QC, Canada, J7V 8P3.
Email: audrey.maheu@mail.mcgill.ca

Marth Reinhard is a Senior Scientific Officer at Project Management Jülich GmbH, Germany, where he manages scientific programmes in the field of environment technologies. He is a civil engineer and holds a Ph.D. in Hydraulic Engineering. He is especially interested in integrated water resources management (IWRM) and sustainable sanitation approaches. The present paper is a result of his

activities with the National Water Authorities of Mexico (CONAGUA), where he was working as an adviser for IWRM.
Contact: Ebersstr. 87, 10827 Berlin, Germany. Telephone: +49-30-2019930. Email: r.marth@fz-juelich.de

Marthinus Delon is an Executive Director of Carbon and Environmental Research, Indonesia. He has extensive experience in geographic information systems (GIS) and remote sensing. He has been involved in a lot of research related to climate risk assessment and climate change mitigation.
Contact: CER Indonesia, Jl. Aryawidura 8 No. 2 Perum Aryawidura Bogor 16152, West Java, Indonesia. Email: delonmarthinus@gmail.com

McDowell Julia is currently completing her Master of Public Health at the Rollins School of Public Health at Emory University. She served as the Principal Investigator of the Climate Change Unit at Agua Sustentable in Bolivia and, previously, as a Fulbright Scholar to Bolivia. She received her BA from Brown University in Latin American Studies and in Development Studies.
Email: julia.z.mcdowell@gmail.com

Miller Tony Tony is currently working at the University of British Columbia, and working as a researcher for the Vancouver Aquarium, on the Pacific Ocean Shelf Tracking Project (POST). He will be returning to Australia in 2011 to work on Geographical Information Science at the University of Queensland.
Email: amiller@live.com.au

Mirgartz Michael has been a Student Research Assistant at the Institute of Environmental Engineering since 2009. He has assisted in several projects concerning adaptation to climate change and wastewater treatment.

Mohan S. S. Mohan obtained his Ph.D. from the Indian Institute of Science, Bangalore, India. He is currently the Director of the National Institute of Technical Teachers Training and Research, Chennai. His research areas of interest include engineering education, environmental engineering, water resources management, groundwater assessment and management and evolutionary algorithms. Prof. Mohan has authored many publications, journals and papers while attaining multiple honours and awards from India and abroad.

Moniruzzaman Md. is an Assistant Professor in the Department of Geography and Environmental Studies, Rajshahi University, Bangladesh. He received his Master's degrees in Geospatial Technology from University Nova de Lisboa, Portugal, University Uaume I, Spain and the University of Münster, Germany, and in Geography and Environmental Studies from the University of Rajshahi, Bangladesh, respectively. He has seven articles and three international conference papers published in various journals and conference proceedings. His research

focus is in the areas of physical environment management with the application of GIS and remote sensing.
Contact: Dept. of Geography & Environmental Studies, Rajshahi University, Rajshahi, 6205, Bangladesh. Email: moniralvi@gmail.com

Montaña Elma is a CONICET (Science and Technology National Research Council) Researcher at the Human, Social and Environmental Sciences Institute and Full Professor at the National University of Cuyo, Mendoza, Argentina. She holds a Ph.D. and a DEA from the Sorbonne University, Paris in Geography and Planning. Her research focus is in the areas of space–society and nature–culture relations in drylands, political ecology of water and social dimensions of global environmental change.
Email: emontana@mendoza-conicet.gob.ar

Mühle Sarah is a member of the working group on urban drainage at the Institute of Environmental Engineering of RWTH Aachen University. She joined the group in 2010 and works as a student research assistant. Sarah will complete her course in Environmental Engineering in 2011.

Nursey-Bray Melissa is a Senior Lecturer in the Department of Geography, Environment and Population, University of Adelaide, Australia, where she undertakes a number of projects in climate change adaptation, training and resilience. She is Director of the Social Science for Climate Change Research Network, co-theme leader of the National Climate Change Adaptation Research Network for Marine Biodiversity and a Tyndall Fellow (2009).
Contact: University of Adelaide, North Terrace campus, North Terrace, Adelaide, South Australia, 5005, Australia. Email: Melissa.Nursey-Bray@adelaide.edu.au

Odekunle Theophilus is an Associate Professor of Geography at Obafemi Awolowo University, Ile Ife, Nigeria. He is a climatologist with special interest in climate change science. He has carried out studies on extended range weather forecasting as a strategy for enhancement of crop productivity, rainfall onset and retreat dates, climate variability and food security. His articles have appeared in *International Journal of Climatology and Weather*. He is also Head of the Department of Geography of the same institution.
Contact: Department of Geography, Obafemi Awolowo University, Ile Ife, Osun State, Nigeria. Email: odeyemitheophilus@yahoo.com

Omrani Nizar Researcher in water and soil management at the Institute of Arid Regions (IRA), Tunisia, and Ph.D. student in Geoecology at the Intistute of Geopgraphy and Geoecology (IfGGI, KIT), Karlsruhe, Germany. Main research interests are irrigation and drainage issues and water resources management under a climate change context. Involved in EU-funded research project MELIA and member of the Water History Association.

Contact: Institut des Régions Arides-(IRA) Direction Régionale de Kébili BP32, Route de Douz 4200, Kebili, Tunisia. Email: omrani_nizar@yahoo.fr

Oroud Ibrahim is a Professor of Physical Climatology, Hydroclimatology and Water Resources. He has been working on climate change impacts on water resources, irrigation water needs and green water fluxes in the eastern Mediterranean since 2003. He has authored a book, Climate Change in the Balance. He has done significant work on evaporation from fresh and hypersaline water bodies, and atmospheric radiation, with several scientific papers published in renowned journals in this regard. Prof. Oroud has been working with Glowa for the past nine years on climate change issues in the eastern Mediterranean. Many of the results were presented in scientific journals and international conferences. He served as a consultant on climate change issues, hydrological issues in arid and semi-arid areas, and also as an expert on modelling issues of evaporation from hypersaline water bodies.
Contact: Mu'tah University, P.O. Box 7, Karak, Jordan. Mobile: +962-795289874; +962-32372380. Email: ioroud@mutah.edu.jo

Patnaik Rasmi is a Doctoral Scholar at the Department of Ecology and Environmental Sciences, Pondicherry University, India. She received her Master's degree in Ecology and Environmental Sciences from Pondicherry University, India. Her research focus is in the areas of industrial ecology with special interest in eco-industrial parks, industrial symbiosis, by-product exchange, waste management and corporate environmental management.
Contact: Department of Ecology & Environmental Sciences, Pondicherry University, Kalapet, Puducherry-605014, India. Email: rasenvs@gmail.com

Perdinan is a lecturer and researcher in the Department of Geophysics and Meteorology, Bogor Agricultural University, Indonesia. He specializes in applied climatology and has been involved in many studies on assessing climate risks on economic sectors such as agriculture and water resources. Currently, he is pursuing a Ph.D. at the Department of Geography, Michigan State University, sponsored by the Fulbright Presidential Fellowship.
Contact: Department of Geography, 118 Geography Building, East Lansing, Michigan, 48824, USA. Email: perdinan@gmail.com

Pinnekamp Johannes has been the Director of the Institute for Environmental Engineering (ISA) of Aachen University (RWTH) since 2004. Under the direction of Prof. Pinnekamp, the Institute has been working successfully for many years in the fields of water supply, urban drainage, wastewater treatment and solid waste disposal.
Contact: Institute of Environmental Engineering, RWTH Aachen University, Mies-van-der-Rohe-Straße 1, 52074 Aachen, Germany.
Email: siekmann@isa.rwth-aachen.de

Poormohammadi Samaneh has an M.Sc. in Desert Management Engineering and graduated from the University of Yazd, Iran in 2009. She has experience in the fields of remote sensing and GIS, climate change and evapotranspiration studies. At present, she is Senior Expert at the National Cloud Seeding Research Center (NACSER), Iran and works on this topic.
Contact: National Cloud Seeding Research Center (NACSER), Daneshjoo Blvd., P.O. Box 89158-69445, Yazd, Iran. Telephone: +98-351-82505512, Fax: +98-351-08251403. Email: s.poormohammadi@gmail.com

Poyyamoli Gopalsamy is an Associate Professor at the Department of Ecology and Environmental Sciences, Pondicherry University, India, where he is focusing on eco-regional planning, ecotourism, solid waste management, agro-ecology and industrial ecology. He is also the Coordinator for the M.Sc. (Sustainable Development) programme. He is a member of CEM (IUCN), Global Experts Directory on Ecosystem Services and network member of the National Ecosystem Services Research Partnership, USEPA.
Contact: Department of Ecology & Environmental Sciences, Pondicherry University, Kalapet, Puducherry-605014, India. Email: gpoyya9@gmail.com

Prasad Devi is a Professor in Ecology and Environment Science and is interested in policy-related research. His other interests include behavioural ecology and modelling ecosystems. He is currently with the Department of Science and Technology, Government of India.
Contact: Department of Ecology & Environmental Sciences, Pondicherry University, Kalapet, Puducherry-605014, India.

Propastin Pavel born in Kazakhstan, received his MA degree in Geography and his Ph.D. in Remote Sensing and GIS from Georg-August University, Göttingen, Germany, in 2003 and 2007, respectively. He is currently a Postdoctoral Research Associate in the Department of Cartography, GIS and Remote Sensing at Georg-August University, Göttingen, Germany. His main research activities focus on the integration of remote sensing and GIS in studies on climate dynamics and their impacts on vegetation in Asia.
Contact: Department of GIS and Remote Sensing, Institute of Geography, Georg-August-University Göttingen, Goldschmidtstr. 5, 37077, Göttingen, Germany. Email: ppropas@uni-goettingen.de

Rahimian Hassan Mohammad is a Senior Expert in irrigation and drainage at the National Salinity Research Center (NSRC), Iran. He has six years of experience in the field of remote sensing and GIS. His main research interests are soil and water salinity, geo-statistics, vegetation cover monitoring, energy balance approaches for actual evapotranspiration estimation and climate change impact assessment.

Contact: Azadegan Blvd., Nahalestan Ave., National Salinity Research Center (NSRC), P.O. Box 89195-315, Yazd, Iran. Telephone: +98-351-7212311, Fax: +98-351-7212312; Mobile: +98-9133588494. Email: mhrahimian@gmail.com

Rahman Sania received her Master's in Urban Planning and Management from the International Institute for Geoinformation Science and Earth Observation (ITC). In the professional arena, she has been involved in various development projects in sustainable development, renewable energy and climate change in Bangladesh. Her research focus is vulnerability, with particular concern for climate change.
Contact: International Institute for Geoinformation Science and Earth Observation (ITC), UPM M.sc 09, P.O. Box 6, P.M. Box 197, 7500AA Enschede, The Netherlands. Email: sania.rahman@yahoo.com

Reyes Bernardo is President of Etica en los Bosques, an NGO specializing in the protection of Chilean native forest ecosystems. Since 1998 he has been the Director of the Ecological Economics Unit of the Institute for Political Ecology (IEP), Santiago, Chile.
Contact: bernardo@eticaenlosbosques.cl

Reza Imran works as Research Associate at Practical Action. He is also Executive Director of Safe Humanity Foundation. In the professional arena, he has been involved in various development projects in sustainable development, renewable energy, infrastructure development and climate change in Bangladesh.
Contact: Practical Action Bangladesh. Sajeda Mohol. First Floor, House 13, Road 1, Civil Station Road, DC More, Rangpur, Bangladesh.
Telephone: +88-018-18350177. Email: imrankuurp@gmail.com

Rüd Sören works as a Technical Advisor for the Technical Cooperation Programme Urban-Industrial Environmental Management in Mexico of the Deutsche Gesellschaft für Technische Zusammenarbeit (GTZ) GmbH, focusing on solid waste management and resource efficiency. He studied Environmental and Resource Management and has worked for GTZ since 2006, primarily in the sector project Ecosan–resource-oriented sanitation systems.
Contact: Deutsche Gesellschaft für Technische Zusammenarbeit (GTZ) GmbH, Edif. Semarnat, Av. San Jerónimo 458, Col. Jardines del Pedregal, 01900 Ciudad de México, Mexico. Email: soeren.rued@gtz.de

Salas Sonia is a Professor at the University of La Serena, Chile. She received her Master's degree from the Catholic University of Chile and her Ph.D. from the University of Tennessee, USA. She is currently a Research Associate at CEAZA (Advanced Research Center for Arid Zones). Her focus is on the areas of sustainable development, social vulnerability in rural areas and socio-clinical psychology.
Email: salassc@yahoo.com

Salvático Natalia is an activist for the NGO Amigos de la Tierra Argentina and the coordinator of the programme "Water and sustainability". She also works in the communication area of the organization. She studies Design at the University of Buenos Aires and investigates organic production of fabrics and non-polluting dyeing.

Sanni Maruf is a Research Officer at the National Centre for Technology Management, Nigeria. He attended Obafemi Awolowo University, Ile-Ife, Nigeria, where he obtained his M.Sc. degree in Geography. His area of focus is environmental change. He has participated in projects such as the Assessments of Impacts and Adaptations to Climate Change, Technology Transfer Requirements for Nigeria in relation to Climate Change, and Adaptation to Global Climate Change in Nigeria. His articles have appeared in *Int. J. Technol Manag*, Climate Change Management Series and World Association of Sustainable Development Publication.
Contact: Maruf Sanni, National Centre for Technology Management, PMB 012, Obafemi, Awolowo University, Ile-Ife, Osun State, Nigeria.
Email: marufsanni@yahoo.com

Sarkar A. N. is currently working as Senior Professor and Area Chairperson (International Business) in the Asia-Pacific Institute of Management, New Delhi, where he undertakes a number of research studies in the emerging fields of energy and environmental security, natural resources management, sustainable development, renewable energy, and climate change-related policy issues. He has authored 10 books and more than 100 research papers, published in reputed national and international journals.
Contact: Asia-Pacific Institute of Management, 3 & 4 Institutional Area, Jasola Vihar, New Delhi-110025, India. Telephone: +91-98-73824296.
Email: ansarkar1@gmail.com

Sellmann Daniela is a Ph.D. student at the Bavarian Chair of Biology Didactics at the University of Bayreuth, Germany. Her current research focuses on environmental education and its potential in areas of sustainability with special focus on global climate change.
Contact: Chair of Biology Education, University of Bayreuth, NW I, Universitaetsstrasse 30, 95447 Bayreuth, Germany.
Email: daniela.sellmann@uni-bayreuth.de

Shibayama Tomoya is a Professor at the Department of Civil and Environmental Engineering, Waseda University. His research interests and teaching subjects include coastal engineering, sediment transport mechanics, tsunami and storm surge simulation, coastal environment, hydraulics, international development engineering, sociology of construction, engineering ethics, and construction management. He obtained his Doctor of Engineering in Coastal Engineering and Hydraulics in Civil Engineering from the University of Tokyo in 1985. He is a

member of the Japan Society of International Development (JSID), fellow of Japan Society of Civil Engineers (JSCE), member of Japanese Association for Coastal Zone Management (JACZM), member of American Society of Civil Engineers (ASCE), member of International Association for Hydraulic Research (IAHR), and a member of Japan Sociological Society.
Contact: Department of Civil and Environmental Engineering, Waseda University, 3-4-1 Okubo, Shinjuku-ku, Tokyo 169-8555, Japan. Email: shibayama@waseda.jp

Siekmann Marko is the head of the working group on urban drainage at the Institute of Environmental Engineering of Aachen University. He completed his Civil Engineering studies 1993 in Hannover and worked at a consulting company for a period of 12 years. He moved to the Institute of Environmental Engineering in 2005. Since that time, numerous projects in the field of climate change, environmental engineering and urban planning have been part of his research activities.
Contact: Institute of Environmental Engineering, RWTH Aachen University, Mies-van-der-Rohe-Strasse 1, 52074 Aachen, Germany. Email: siekmann@isa.rwth-aachen.de

Steinel Anke works as a hydrogeologist at the Federal Institute for Geosciences and Natural Resources, Hannover, Germany. She conducted her Ph.D. at the Institut for Sustainable Water Resources (now Centre for Water Sensitive Cities) at the Civil Engineering Department, Monash University, Australia. Her research interests are in the area of sustainable groundwater management, water reuse and managed aquifer recharge.
Contact: Federal Institute for Geosciences and Natural Resources (BGR), Stilleweg 2, 30655 Hannover, Germany.
Email: anke.steinel@bgr.de

Sumarto Saroyo heads a conservation biology laboratory at the Faculty of Mathematics and Natural Sciences, Sam Ratulangi University, Manado, Indonesia. He has an interest in nature conservation. His current main research is conservation of Macaca nigra, an endemic fauna of North Sulawesi.
Contact: Department of Biology, FMIPA, University of Sam Ratulangi, Manado, 95115, Indonesia. Email: saroyos@yahoo.com

Takagi Hiroshi is in charge of various overseas projects, mainly in the field of disaster and climate change issues as an Associate Expert at Japan International Cooperation Agency (JICA). He received his Doctoral degree from Yokohama National University in 2008.
Contact: Disaster Management Division, Global Environment Department, Japan International Cooperation Agency (JICA), 5-25, Niban-cho, Chiyoda-ku, Tokyo 102-8012, Japan. Telephone: +81-3-52269589, Fax: +81-3-52266343.
Email: htakagi.jp@gmail.com

Tallei Trina heads a biotechnology service unit at the Faculty of Mathematics and Natural Sciences, Sam Ratulangi University, Manado, Indonesia. Her main interest is in genetic resources conservation. She has been involved in research about nature conservation and biodiversity.
Contact: Department of Biology, FMIPA, University of Sam Ratulangi, Manado, 95115, Indonesia. Email: trinatallei@yahoo.com; trina@daad-alumni.de

Tawfik Medhat Mekhail Associate Professor of Field Crop Physiology, National Research Center. He works on many projects to improve the livelihoods of the rural community through sustainable utilization of saline agriculture, climate change problems and their effect on the ecological footprint of field crops in Egypt and integrating drought management and water policy.
Contact: El Bohooth Str., Dokki, Giza, Egypt, P.O. Box 12311.
Email: medhatnrc@hotmail.com

Torres Juan is Practical Action Latin America's resident climate change expert with a Master's degree in Biology from the National Autonomous University in Mexico. His specialist areas include: community-based climate change adaptation, indigenous knowledge, environmental impact assessments, ecology, desertification and forests. Since 1991, Juan Torres has been working at the National Agrarian University of Peru as a Senior Lecturer at the Faculty of Science and since 1990 has worked as Director of the Centre for Research into Arid Zones at the same university. Over the past ten years, Juan Torres has held the position of Executive Director at the Andean Centre for Science and Technology. Country of origin: Peru.
Email: jtorres@solucionespracticas.org.pe

Totàro Michela is a primary school teacher. She is an expert in environmental education, environmental didactic approaches and integration of disabled people in environment.
Email: Ecolab@inwind.it

Tuberosa Jean is a UAS Professor of Accounting and Finance and Head of the Laboratory for Market Studies (LEM) at Geneva Haute Ecole de Gestion, Switzerland. He specifically focuses on consumer expectations and distributors' requirements. He provides support with regard to market expectations to Fair Trade products and traceability. He has previously obtained funding from RUIG-GIAN Geneva International Academic Network for a project entitled "Understanding the Private Demand for International Payments for Ecosystems Services", which was conducted in association with the United Nations Environment Programme, Economics and Trade Branch (UNEP-ETB) and the World Conservation Union (IUCN).
Contact: Jl. IKIP III No. 10, Kampus UNSRAT, Manado, 95115, Indonesia. Mobile: +62-8179255866. Email: jean.tuberosa@hesge.ch

Umar Da'u Abba was educated at Bayero University Kano and Federal University of Technology Yola, where he obtained his B.Sc. and M.Sc. degrees, respectively. He is a geographer, an environmental scientist, educationist and registered member of the Nigerian Environmental Society (NES). He is presently a research officer with Jigawa Research Institute, Kazaure and a part-time lecturer with the Geography Department at the Federal College of Education, Kano. He has four journal articles and he has presented a number of papers at various conferences, nationally and internationally. His research interests are soil geography, hydrology and water resources, climate change and environmental issues. He is married with children.

Varughese Susan is a Ph.D. scholar at the Department of Ecology and Environmental Sciences, Pondicherry University, India. She completed her Master's with a focus on membrane wastewater treatment technologies at Pondicherry University. Her research focus is in the area of agricultural water management and water transactions.

Vitullo Marina works at the Italian National Institute for Environmental Protection and Research. She is responsible for preparing, improving and reporting the LULUCF sector of National Emission Inventory for the UNFCCC and Kyoto Protocol. Member of UNFCCC roster of experts in the area of greenhouse gas inventory. Her research is focused on the development of methodologies to estimate removals and emissions of GHG in the LULUCF, and collecting, analysing and elaborating data concerning various processes affecting atmospheric emission at national and local level.
Email: marina.vitullo@isprambiente.it

Vomberg Nina graduated in Environmental Engineering at Aachen University in 2010. She continued working on her diploma thesis on multifunctional land use for stormwater retention purposes as a graduate assistant at the Institute of Environmental Engineering. Today she works at a consultant company for environmental technologies and civil works.
Contact: Dr.-Ing. Bernd Diering GmbH, Ingenieurgesellschaft für Umwelttechnologie, Jülicher Strasse 318-320, 52070 Aachen, Germany.
Email: nina.vomberg@rwth-aachen.de

Waagsaether Katinka Lund is a Master's student with the Climate Systems Analysis Group at the University of Cape Town, South Africa. In 2009 she did her honours in Atmospheric Science, also with the Climate Systems Analysis Group at the University of Cape Town, focusing on the communication of climate science to practitioners. Her current Master's research is concerned with small-scale farmers' vulnerability to climate change.
Contact: Climate System Analysis Group, Department of Environmental and Geographical Science University of Cape Town, Private Bag X3, Rondebosch, 7701, South Africa. Email: katinka@csag.uct.ac.za

Waas Tom is a Research Fellow at the Biology Department of the Vrije Universiteit Brussel, Belgium. Together with Jean Hugé, they form a multidisciplinary team focusing on the theory and practice of sustainability assessment. They develop and apply assessment processes in the fields of (sub-)national sustainable development policy, development co-operation, sustainble higher education and energy. They are part of the Flemish Policy Research Centre for Sustainable Development team, as well as of the KLIMOS Research Platform for Climate Change and Development Cooperation.
Contact: Vrije Universiteit Brussel, DBIO / APNA, Pleinlaan 2, 1050 Brussels, Belgium

Watmough Gary is a Ph.D. student in the School of Geography at the University of Southampton. He completed his Master's degree in Remote Sensing and Spatial Analysis at the University of Southampton and a Bachelor's degree in Geography at Lancaster University. Gary's research is focused on the use of remote sensing, geographic information systems (GIS) and spatial statistics for international development. His Ph.D. research is exploring the spatial links between census variables and remotely sensed environmental metrics in Assam, northeast India.
Email: G.Watmough@soton.ac.uk

Zenchanka Siarhei heads the Department of International Relations and Marketing at the Minsk Branch of Moscow State University for Economics, Statistics and Informatics. He has a Ph.D. in Applied Science. His research focus is in the areas of quality management and environmental management. He is a contact person for relations with the Baltic University Programme.
Contact: Minsk Branch of Moscow State University for Economics, Statistics and Informatics, 127, Building 2, Room 301, Mayakovskogo Str., Minsk, 220028, Belarus. Telephone/Fax: +375-17-2914497. Email: szenchenko@mfmesi.ru

Zidan Mohamed Soliman Professor of Agronomy, National Research Center. She has worked on many projects for increasing the productivity of field crops under Egyptian conditions.
Contact: El Bohooth str., Dokki, Giza, Egypt, P.O. Box 12311.
Email: Zeidan_ms@yahoo.com

Ziervogel Gina is a Lecturer at the University of Cape Town, South Africa. She completed her Ph.D. in Geography in 2002 at the University of Oxford, exploring issues around seasonal climate forecasts and use among smallholder farmers in Lesotho, as well as the national and district institutions supporting climate information dissemination. This work has since developed to focus on a broad range of issues related to vulnerability and adaptation to climate change in southern Africa.

Contact: Climate System Analysis Group, Department of Environmental and Geographical Science University of Cape Town, Private Bag X3, Rondebosch, 7701, South Africa. Email: gina@csag.uct.ac.za

Zollo Anna is Professor of Ecology at the L. Quaroni Faculty of Architecture at Sapienza University. She is Learner in Ecology under the Chair of Prof. Eugenia Aloj.
Email: annazollo2@virgilio.it